Lecture Notes in Computer Science 14451

Founding Editors

Gerhard Goos
Juris Hartmanis

Editorial Board Members

The series Lecture Notes in Computer Science (LNCS), including its subseries Lecture Notes in Artificial Intelligence (LNAI) and Lecture Notes in Bioinformatics (LNBI), has established itself as a medium for the publication of new developments in computer science and information technology research, teaching, and education.

LNCS enjoys close cooperation with the computer science R & D community, the series counts many renowned academics among its volume editors and paper authors, and collaborates with prestigious societies. Its mission is to serve this international community by providing an invaluable service, mainly focused on the publication of conference and workshop proceedings and postproceedings. LNCS commenced publication in 1973.

Biao Luo · Long Cheng · Zheng-Guang Wu ·
Hongyi Li · Chaojie Li
Editors

Neural
Information Processing

30th International Conference, ICONIP 2023
Changsha, China, November 20–23, 2023
Proceedings, Part V

 Springer

Editors
Biao Luo (iD)
Central South University
Changsha, China

Zheng-Guang Wu (iD)
Zhejiang University
Hangzhou, China

Chaojie Li (iD)
UNSW Sydney
Sydney, NSW, Australia

Long Cheng (iD)
Chinese Academy of Sciences
Beijing, China

Hongyi Li (iD)
Guangdong University of Technology
Guangzhou, China

ISSN 0302-9743 ISSN 1611-3349 (electronic)
Lecture Notes in Computer Science
ISBN 978-981-99-8072-7 ISBN 978-981-99-8073-4 (eBook)
https://doi.org/10.1007/978-981-99-8073-4

This Springer imprint is published by the registered company Springer Nature Singapore Pte Ltd.
The registered company address is: 152 Beach Road, #21-01/04 Gateway East, Singapore 189721, Singapore

Paper in this product is recyclable.

Preface

Welcome to the 30th International Conference on Neural Information Processing (ICONIP2023) of the Asia-Pacific Neural Network Society (APNNS), held in Changsha, China, November 20–23, 2023.

The mission of the Asia-Pacific Neural Network Society is to promote active interactions among researchers, scientists, and industry professionals who are working in neural networks and related fields in the Asia-Pacific region. APNNS has Governing Board Members from 13 countries/regions – Australia, China, Hong Kong, India, Japan, Malaysia, New Zealand, Singapore, South Korea, Qatar, Taiwan, Thailand, and Turkey. The society's flagship annual conference is the International Conference of Neural Information Processing (ICONIP). The ICONIP conference aims to provide a leading international forum for researchers, scientists, and industry professionals who are working in neuroscience, neural networks, deep learning, and related fields to share their new ideas, progress, and achievements.

ICONIP2023 received 1274 papers, of which 256 papers were accepted for publication in Lecture Notes in Computer Science (LNCS), representing an acceptance rate of 20.09% and reflecting the increasingly high quality of research in neural networks and related areas. The conference focused on four main areas, i.e., "Theory and Algorithms", "Cognitive Neurosciences", "Human-Centered Computing", and "Applications". All the submissions were rigorously reviewed by the conference Program Committee (PC), comprising 258 PC members, and they ensured that every paper had at least two high-quality single-blind reviews. In fact, 5270 reviews were provided by 2145 reviewers. On average, each paper received 4.14 reviews.

We would like to take this opportunity to thank all the authors for submitting their papers to our conference, and our great appreciation goes to the Program Committee members and the reviewers who devoted their time and effort to our rigorous peer-review process; their insightful reviews and timely feedback ensured the high quality of the papers accepted for publication. We hope you enjoyed the research program at the conference.

October 2023

Biao Luo
Long Cheng
Zheng-Guang Wu
Hongyi Li
Chaojie Li

Organization

Honorary Chair

Weihua Gui   Central South University, China

Advisory Chairs

Jonathan Chan   King Mongkut's University of Technology Thonburi, Thailand
Zeng-Guang Hou   Chinese Academy of Sciences, China
Nikola Kasabov   Auckland University of Technology, New Zealand
Derong Liu   Southern University of Science and Technology, China
Seiichi Ozawa   Kobe University, Japan
Kevin Wong   Murdoch University, Australia

General Chairs

Tingwen Huang   Texas A&M University at Qatar, Qatar
Chunhua Yang   Central South University, China

Program Chairs

Biao Luo   Central South University, China
Long Cheng   Chinese Academy of Sciences, China
Zheng-Guang Wu   Zhejiang University, China
Hongyi Li   Guangdong University of Technology, China
Chaojie Li   University of New South Wales, Australia

Technical Chairs

Xing He   Southwest University, China
Keke Huang   Central South University, China
Huaqing Li   Southwest University, China
Qi Zhou   Guangdong University of Technology, China

Local Arrangement Chairs

Wenfeng Hu	Central South University, China
Bei Sun	Central South University, China

Finance Chairs

Fanbiao Li	Central South University, China
Hayaru Shouno	University of Electro-Communications, Japan
Xiaojun Zhou	Central South University, China

Special Session Chairs

Hongjing Liang	University of Electronic Science and Technology, China
Paul S. Pang	Federation University, Australia
Qiankun Song	Chongqing Jiaotong University, China
Lin Xiao	Hunan Normal University, China

Tutorial Chairs

Min Liu	Hunan University, China
M. Tanveer	Indian Institute of Technology Indore, India
Guanghui Wen	Southeast University, China

Publicity Chairs

Sabri Arik	Istanbul University-Cerrahpaşa, Turkey
Sung-Bae Cho	Yonsei University, South Korea
Maryam Doborjeh	Auckland University of Technology, New Zealand
El-Sayed M. El-Alfy	King Fahd University of Petroleum and Minerals, Saudi Arabia
Ashish Ghosh	Indian Statistical Institute, India
Chuandong Li	Southwest University, China
Weng Kin Lai	Tunku Abdul Rahman University of Management & Technology, Malaysia
Chu Kiong Loo	University of Malaya, Malaysia

| Qinmin Yang | Zhejiang University, China |
| Zhigang Zeng | Huazhong University of Science and Technology, China |

Publication Chairs

Zhiwen Chen	Central South University, China
Andrew Chi-Sing Leung	City University of Hong Kong, China
Xin Wang	Southwest University, China
Xiaofeng Yuan	Central South University, China

Secretaries

| Yun Feng | Hunan University, China |
| Bingchuan Wang | Central South University, China |

Webmasters

| Tianmeng Hu | Central South University, China |
| Xianzhe Liu | Xiangtan University, China |

Program Committee

Rohit Agarwal	UiT The Arctic University of Norway, Norway
Hasin Ahmed	Gauhati University, India
Harith Al-Sahaf	Victoria University of Wellington, New Zealand
Brad Alexander	University of Adelaide, Australia
Mashaan Alshammari	Independent Researcher, Saudi Arabia
Sabri Arik	Istanbul University, Turkey
Ravneet Singh Arora	Block Inc., USA
Zeyar Aung	Khalifa University of Science and Technology, UAE
Monowar Bhuyan	Umeå University, Sweden
Jingguo Bi	Beijing University of Posts and Telecommunications, China
Xu Bin	Northwestern Polytechnical University, China
Marcin Blachnik	Silesian University of Technology, Poland
Paul Black	Federation University, Australia

Anoop C. S.	Govt. Engineering College, India
Ning Cai	Beijing University of Posts and Telecommunications, China
Siripinyo Chantamunee	Walailak University, Thailand
Hangjun Che	City University of Hong Kong, China
Wei-Wei Che	Qingdao University, China
Huabin Chen	Nanchang University, China
Jinpeng Chen	Beijing University of Posts & Telecommunications, China
Ke-Jia Chen	Nanjing University of Posts and Telecommunications, China
Lv Chen	Shandong Normal University, China
Qiuyuan Chen	Tencent Technology, China
Wei-Neng Chen	South China University of Technology, China
Yufei Chen	Tongji University, China
Long Cheng	Institute of Automation, China
Yongli Cheng	Fuzhou University, China
Sung-Bae Cho	Yonsei University, South Korea
Ruikai Cui	Australian National University, Australia
Jianhua Dai	Hunan Normal University, China
Tao Dai	Tsinghua University, China
Yuxin Ding	Harbin Institute of Technology, China
Bo Dong	Xi'an Jiaotong University, China
Shanling Dong	Zhejiang University, China
Sidong Feng	Monash University, Australia
Yuming Feng	Chongqing Three Gorges University, China
Yun Feng	Hunan University, China
Junjie Fu	Southeast University, China
Yanggeng Fu	Fuzhou University, China
Ninnart Fuengfusin	Kyushu Institute of Technology, Japan
Thippa Reddy Gadekallu	VIT University, India
Ruobin Gao	Nanyang Technological University, Singapore
Tom Gedeon	Curtin University, Australia
Kam Meng Goh	Tunku Abdul Rahman University of Management and Technology, Malaysia
Zbigniew Gomolka	University of Rzeszow, Poland
Shengrong Gong	Changshu Institute of Technology, China
Xiaodong Gu	Fudan University, China
Zhihao Gu	Shanghai Jiao Tong University, China
Changlu Guo	Budapest University of Technology and Economics, Hungary
Weixin Han	Northwestern Polytechnical University, China

Xing He	Southwest University, China
Akira Hirose	University of Tokyo, Japan
Yin Hongwei	Huzhou Normal University, China
Md Zakir Hossain	Curtin University, Australia
Zengguang Hou	Chinese Academy of Sciences, China
Lu Hu	Jiangsu University, China
Zeke Zexi Hu	University of Sydney, Australia
He Huang	Soochow University, China
Junjian Huang	Chongqing University of Education, China
Kaizhu Huang	Duke Kunshan University, China
David Iclanzan	Sapientia University, Romania
Radu Tudor Ionescu	University of Bucharest, Romania
Asim Iqbal	Cornell University, USA
Syed Islam	Edith Cowan University, Australia
Kazunori Iwata	Hiroshima City University, Japan
Junkai Ji	Shenzhen University, China
Yi Ji	Soochow University, China
Canghong Jin	Zhejiang University, China
Xiaoyang Kang	Fudan University, China
Mutsumi Kimura	Ryukoku University, Japan
Masahiro Kohjima	NTT, Japan
Damian Kordos	Rzeszow University of Technology, Poland
Marek Kraft	Poznań University of Technology, Poland
Lov Kumar	NIT Kurukshetra, India
Weng Kin Lai	Tunku Abdul Rahman University of Management & Technology, Malaysia
Xinyi Le	Shanghai Jiao Tong University, China
Bin Li	University of Science and Technology of China, China
Hongfei Li	Xinjiang University, China
Houcheng Li	Chinese Academy of Sciences, China
Huaqing Li	Southwest University, China
Jianfeng Li	Southwest University, China
Jun Li	Nanjing Normal University, China
Kan Li	Beijing Institute of Technology, China
Peifeng Li	Soochow University, China
Wenye Li	Chinese University of Hong Kong, China
Xiangyu Li	Beijing Jiaotong University, China
Yantao Li	Chongqing University, China
Yaoman Li	Chinese University of Hong Kong, China
Yinlin Li	Chinese Academy of Sciences, China
Yuan Li	Academy of Military Science, China

Yun Li	Nanjing University of Posts and Telecommunications, China
Zhidong Li	University of Technology Sydney, Australia
Zhixin Li	Guangxi Normal University, China
Zhongyi Li	Beihang University, China
Ziqiang Li	University of Tokyo, Japan
Xianghong Lin	Northwest Normal University, China
Yang Lin	University of Sydney, Australia
Huawen Liu	Zhejiang Normal University, China
Jian-Wei Liu	China University of Petroleum, China
Jun Liu	Chengdu University of Information Technology, China
Junxiu Liu	Guangxi Normal University, China
Tommy Liu	Australian National University, Australia
Wen Liu	Chinese University of Hong Kong, China
Yan Liu	Taikang Insurance Group, China
Yang Liu	Guangdong University of Technology, China
Yaozhong Liu	Australian National University, Australia
Yong Liu	Heilongjiang University, China
Yubao Liu	Sun Yat-sen University, China
Yunlong Liu	Xiamen University, China
Zhe Liu	Jiangsu University, China
Zhen Liu	Chinese Academy of Sciences, China
Zhi-Yong Liu	Chinese Academy of Sciences, China
Ma Lizhuang	Shanghai Jiao Tong University, China
Chu-Kiong Loo	University of Malaya, Malaysia
Vasco Lopes	Universidade da Beira Interior, Portugal
Hongtao Lu	Shanghai Jiao Tong University, China
Wenpeng Lu	Qilu University of Technology, China
Biao Luo	Central South University, China
Ye Luo	Tongji University, China
Jiancheng Lv	Sichuan University, China
Yuezu Lv	Beijing Institute of Technology, China
Huifang Ma	Northwest Normal University, China
Jinwen Ma	Peking University, China
Jyoti Maggu	Thapar Institute of Engineering and Technology Patiala, India
Adnan Mahmood	Macquarie University, Australia
Mufti Mahmud	University of Padova, Italy
Krishanu Maity	Indian Institute of Technology Patna, India
Srimanta Mandal	DA-IICT, India
Wang Manning	Fudan University, China

Piotr Milczarski	Lodz University of Technology, Poland
Malek Mouhoub	University of Regina, Canada
Nankun Mu	Chongqing University, China
Wenlong Ni	Jiangxi Normal University, China
Anupiya Nugaliyadde	Murdoch University, Australia
Toshiaki Omori	Kobe University, Japan
Babatunde Onasanya	University of Ibadan, Nigeria
Manisha Padala	Indian Institute of Science, India
Sarbani Palit	Indian Statistical Institute, India
Paul Pang	Federation University, Australia
Rasmita Panigrahi	Giet University, India
Kitsuchart Pasupa	King Mongkut's Institute of Technology Ladkrabang, Thailand
Dipanjyoti Paul	Ohio State University, USA
Hu Peng	Jiujiang University, China
Kebin Peng	University of Texas at San Antonio, USA
Dawid Połap	Silesian University of Technology, Poland
Zhong Qian	Soochow University, China
Sitian Qin	Harbin Institute of Technology at Weihai, China
Toshimichi Saito	Hosei University, Japan
Fumiaki Saitoh	Chiba Institute of Technology, Japan
Naoyuki Sato	Future University Hakodate, Japan
Chandni Saxena	Chinese University of Hong Kong, China
Jiaxing Shang	Chongqing University, China
Lin Shang	Nanjing University, China
Jie Shao	University of Science and Technology of China, China
Yin Sheng	Huazhong University of Science and Technology, China
Liu Sheng-Lan	Dalian University of Technology, China
Hayaru Shouno	University of Electro-Communications, Japan
Gautam Srivastava	Brandon University, Canada
Jianbo Su	Shanghai Jiao Tong University, China
Jianhua Su	Institute of Automation, China
Xiangdong Su	Inner Mongolia University, China
Daiki Suehiro	Kyushu University, Japan
Basem Suleiman	University of New South Wales, Australia
Ning Sun	Shandong Normal University, China
Shiliang Sun	East China Normal University, China
Chunyu Tan	Anhui University, China
Gouhei Tanaka	University of Tokyo, Japan
Maolin Tang	Queensland University of Technology, Australia

Shu Tian	University of Science and Technology Beijing, China
Shikui Tu	Shanghai Jiao Tong University, China
Nancy Victor	Vellore Institute of Technology, India
Petra Vidnerová	Institute of Computer Science, Czech Republic
Shanchuan Wan	University of Tokyo, Japan
Tao Wan	Beihang University, China
Ying Wan	Southeast University, China
Bangjun Wang	Soochow University, China
Hao Wang	Shanghai University, China
Huamin Wang	Southwest University, China
Hui Wang	Nanchang Institute of Technology, China
Huiwei Wang	Southwest University, China
Jianzong Wang	Ping An Technology, China
Lei Wang	National University of Defense Technology, China
Lin Wang	University of Jinan, China
Shi Lin Wang	Shanghai Jiao Tong University, China
Wei Wang	Shenzhen MSU-BIT University, China
Weiqun Wang	Chinese Academy of Sciences, China
Xiaoyu Wang	Tokyo Institute of Technology, Japan
Xin Wang	Southwest University, China
Xin Wang	Southwest University, China
Yan Wang	Chinese Academy of Sciences, China
Yan Wang	Sichuan University, China
Yonghua Wang	Guangdong University of Technology, China
Yongyu Wang	JD Logistics, China
Zhenhua Wang	Northwest A&F University, China
Zi-Peng Wang	Beijing University of Technology, China
Hongxi Wei	Inner Mongolia University, China
Guanghui Wen	Southeast University, China
Guoguang Wen	Beijing Jiaotong University, China
Ka-Chun Wong	City University of Hong Kong, China
Anna Wróblewska	Warsaw University of Technology, Poland
Fengge Wu	Institute of Software, Chinese Academy of Sciences, China
Ji Wu	Tsinghua University, China
Wei Wu	Inner Mongolia University, China
Yue Wu	Shanghai Jiao Tong University, China
Likun Xia	Capital Normal University, China
Lin Xiao	Hunan Normal University, China

Qiang Xiao	Huazhong University of Science and Technology, China
Hao Xiong	Macquarie University, Australia
Dongpo Xu	Northeast Normal University, China
Hua Xu	Tsinghua University, China
Jianhua Xu	Nanjing Normal University, China
Xinyue Xu	Hong Kong University of Science and Technology, China
Yong Xu	Beijing Institute of Technology, China
Ngo Xuan Bach	Posts and Telecommunications Institute of Technology, Vietnam
Hao Xue	University of New South Wales, Australia
Yang Xujun	Chongqing Jiaotong University, China
Haitian Yang	Chinese Academy of Sciences, China
Jie Yang	Shanghai Jiao Tong University, China
Minghao Yang	Chinese Academy of Sciences, China
Peipei Yang	Chinese Academy of Science, China
Zhiyuan Yang	City University of Hong Kong, China
Wangshu Yao	Soochow University, China
Ming Yin	Guangdong University of Technology, China
Qiang Yu	Tianjin University, China
Wenxin Yu	Southwest University of Science and Technology, China
Yun-Hao Yuan	Yangzhou University, China
Xiaodong Yue	Shanghai University, China
Paweł Zawistowski	Warsaw University of Technology, Poland
Hui Zeng	Southwest University of Science and Technology, China
Wang Zengyunwang	Hunan First Normal University, China
Daren Zha	Institute of Information Engineering, China
Zhi-Hui Zhan	South China University of Technology, China
Baojie Zhang	Chongqing Three Gorges University, China
Canlong Zhang	Guangxi Normal University, China
Guixuan Zhang	Chinese Academy of Science, China
Jianming Zhang	Changsha University of Science and Technology, China
Li Zhang	Soochow University, China
Wei Zhang	Southwest University, China
Wenbing Zhang	Yangzhou University, China
Xiang Zhang	National University of Defense Technology, China
Xiaofang Zhang	Soochow University, China
Xiaowang Zhang	Tianjin University, China

Contents – Part V

Applications

Text to Image Generation with Conformer-GAN

Zhiyu Deng[1], Wenxin Yu[1(✉)], Lu Che[1(✉)], Shiyu Chen[1], Zhiqiang Zhang[1], Jun Shang[1], Peng Chen[2], and Jun Gong[3]

[1] Southwest University of Science and Technology, Mianyang, China
yuwenxin@swust.edu.cn
[2] Chengdu Hongchengyun Technology Co., Ltd., Chengdu, China
[3] Southwest Automation Research Institute, Mianyang, China

Abstract. Text-to-image generation (T2I) has been a popular research field in recent years, and its goal is to generate corresponding photorealistic images through natural language text descriptions. Existing T2I models are mostly based on generative adversarial networks, but it is still very challenging to guarantee the semantic consistency between a given textual description and generated natural images. To address this problem, we propose a concise and practical novel framework, Conformer-GAN. Specifically, we propose the Conformer block, consisting of the Convolutional Neural Network (CNN) and Transformer branches. The CNN branch is used to generate images conditionally from noise. The Transformer branch continuously focuses on the relevant words in natural language descriptions and fuses the sentence and word information to guide the CNN branch for image generation. Our approach can better merge global and local representations to improve the semantic consistency between textual information and synthetic images. Importantly, our Conformer-GAN can generate natural and realistic 512 × 512 images. Extensive experiments on the challenging public benchmark datasets CUB bird and COCO demonstrate that our method outperforms recent state-of-the-art methods both in terms of generated image quality and text-image semantic consistency.

Keywords: Text-to-Image Synthesis · Computer Vision · Deep Learning · Generative Adversarial Networks

1 Introduction

In recent years, text-to-image generation (T2I) has received more attention in the field of computer vision [5,17,31], whose primary goal is to generate photorealistic high-quality images based on a given natural descriptive language.

This Research is Supported by National Key Research and Development Program from Ministry of Science and Technology of the PRC (No. 2021ZD0110600), Sichuan Science and Technology Program (No. 2022ZYD0116), Sichuan Provincial M. C. Integration Office Program, and IEDA Laboratory of SWUST.

This bird has wings that are grey and has a yellow belly.

The bird has a rather bland dark grey covering with a distinctive brown eye color.

This bird is mostly yellow with a light brown nape and dark brown wings with white wing stripes.

This small billed bird has a dark blue crown, black wings, and a white throat and belly.

A small plump bird with a silver majority of coloration.

This bird has a think pointed red bill, with a blue back.

Fig. 1. Examples of images generated by our Conformer-GAN sized with **512 × 512** on the test set of CUB dataset. (Color figure online)

Since Generative Adversarial Networks (GANs) [3] are introduced to solve the T2I task, existing T2I models have made significant progress [4,7,10].

StackGAN [29] generates photorealistic images by stacking several GANs and provides text information to the Generator by concatenating global text vectors along with input noise. AttnGAN [25] introduces a cross-modal attention mechanism to generate images with more detail by focusing on word-level information in a given text description. DF-GAN [20] proposes a single-stage structure to achieve high-resolution image generation through a generator and discriminator pair. However, the T2I task is still challenging because (1) it attempts to solve the cross-modal generation leads to great difficulty in the T2I task, and (2) it needs to synthesize high-quality images but also to ensure the semantic consistency of the text-image.

In the T2I task, the Generator in most previous works [2,11,15,20,25] has applied the Convolutional Neural Network (CNN) framework or the Transformer framework. Still, they didn't combine the best of both. CNN will pay more attention to local details, and Transformer will pay more attention to global information. A good representation should be global and local, but CNN and Transformer ignore one. Inspired by [14], we tried to combine the advantages of CNN and Transformer to propose Conformer-GAN.

The Conformer-GAN model can encode features from both global and local dimensions and use global information such as text descriptions to guide the generation results. The Conformer Block consists of a CNN and a Transformer branches, which comprises a comprehensive combination of local convolutional blocks, self-attention modules, and Multilayer Perceptron (MLP) units.

We design the linear layer as a bridge, extract the global features in Transformer, and combine the affine transformation to guide the image synthesis of the CNN branch. Since CNN and Transformer branches tend to capture different levels of features, the bridge is inserted into each Conformer block to narrow the semantic divergence between them. In addition, affine transformations also continuously act on image features in the same way. This fusion method can significantly enhance the global perception of local features of CNN branches.

We evaluate the Conformer-GAN model on the challenging benchmark Caltech-UCSD Birds 200 (CUB) dataset [23] and the Microsoft Common Objects in Context (COCO) dataset [13]. The quality of the generated images is measured using the CLIPSIM (CS) [24] and Fréchet Inception Distance (FID) [6]. Our experiments show that our Conformer-GAN outperforms the state-of-the-art T2I methods. Our model improves CS from 0.3125 to 0.3164 and reduces FID from 13.91 to 10.14 on the CUB dataset. For the coco dataset, we lowered the FID from 8.21 to 6.73. The contributions of this paper are as follows:

- We propose a novel framework, Conformer-GAN, which can encode features in both global and local dimensions, capable of generating images with higher quality and text-image semantic consistency.
- Extensive experiments on two challenging benchmark datasets demonstrate the superiority of our approach compared to the existing advanced methods. Importantly, our proposed Conformer-GAN can generate 512×512 text-image semantically consistent high-resolution images.

2   Related Work

2.1   GAN-Based Text-to-Image Synthesis

Reed et al. [18] proposed text-generated images based on GANs in 2016, an extension of Conditional GANs, capable of generating small images with a 64×64 resolution. Subsequently, GANs have become one of the most popular methods in text-to-image generation in recent years [20,25,26,31]. Regardless of the application of GANs in any form, the given text is first processed to obtain the text features, and then the text features are used to constrain the subsequent image generation process. For example, Zhang et al. [29] accomplished text-to-image generation in two stages. However, all their GANs are conditioned on the global sentence vector, which lacks word-level information for generating images. Xu et al. [25] generated finer images by exploiting the word-level cross-modal attention mechanism. Next, Zhu et al. [31] proposed a dynamic memory component, which can dynamically select important word information based on the initial image content. Then, Tao et al. [22] proposed a deep fusion method, and their generator consists of a series of UPBlocks with multiple affine layers in a block, which can generate high-quality images.

Similar to [22], our Conformer block also uses affine layers. Also, our framework adopts a single-stage structure to avoid the training difficulties in the stacked GANs structure.

2.2   Transformer-Based Text-to-Image Synthesis

Since Transformer is a sequential model, its performance on image generation tasks is relatively low. In recent years, with the increasing demand in the field of image generation, the development of Transformer in the field of image generation has gradually attracted attention [8,19,27].

Ramesh et al. [17] utilized Transformer to tokenize text and images into a single data stream for autoregressive modelling. Lin et al. [12] implemented unified pretraining on unimodal and multimodal data. Unlike the Transformer-based approach described above, we refer to [14] and propose Conformer-GAN, which can combine the structural advantages of CNN and Transformer and uses convolution operations and Self-attention mechanisms to enhance representation learning so that local features and global representations can be preserved to the greatest extent, resulting in better T2I performance.

3   Method

The proposed method shown in Fig. 2 mainly consists of a Generator and a Discriminator, where the Generator contains a feature encoding process and N Conformer blocks. At the same time, The Discriminator predicts the adversarial loss through the structure of the CNN branch and Transformer branch. Based on a text description and a noise vector z sampled from a normal distribution as input, Conformer-GAN outputs an RGB image of 256×256 or 512×512.

3.1   Model Overview

The model's input is a random vector $z \in \mathbb{R}^{100}$ that satisfies the Gaussian distribution and a text description used to guide the Generator to generate images. To meet the central part of the model, that is, the learning of global and

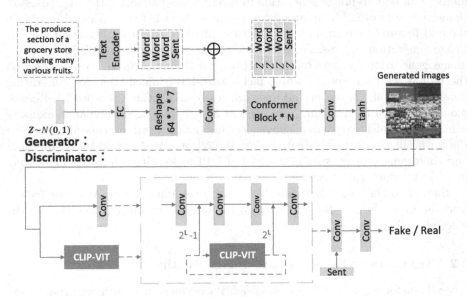

Fig. 2. The architecture of the proposed Conformer-GAN model for text-to-image generation. It has one generator-discriminator pair. The generator is mainly composed of N Conformer blocks.

local features by the Conformer block, we need to encode the input into features in different formats that can accommodate the CNN and Transformer branches using different processing.

First, we use the pre-trained text encoder provided by CLIP [16] to encode the text information and get the sentence embedding and word embeddings as the global semantic features to guide the image generation. We will refer to sentence and word embeddings as text embedding in the following contents.

To encode local information, we use a fully connected layer, then reshape the feature vector z into a feature map of $64 \times 7 \times 7$ and feed it into a 1×1 convolutional layer. At the same time, to satisfy the semantic consistency of the given text description and the generated image as much as possible and to make up for the weak global perception ability of the CNN network, we combine z with the text embedding as the global feature and input this combination into the Transformer branch.

The feature vectors and feature maps obtained in the two ways will be used together as the input of the Conformer block. Then the subsequent encoding/decoding of the feature vectors and maps will be performed from the global and local dimensions, respectively.

The Discriminator is also divided into two branches, the CNN and the Transformer. The Transformer branch uses a pre-trained ViT-based CLIP image encoder (CLIP-ViT), as shown in Fig. 2. Specifically, in the Transformer branch, we put the image into CLIP-ViT to extract the features and get the features of $2^L - 1$ layer and 2^L layer ($L = 2, 3$). In the CNN branch, we fuse the features provided by CLIP-ViT with the features extracted by CNN itself. Finally, we combine the output of the CNN branch concat with sentence embedding and get the final result after two layers of convolution authenticity score.

3.2 Conformer Block

The Conformer block is the core of Conformer-GAN. Figure 3 shows that each Conformer block consists of a CNN and a Transformer branch. Note that unlike the vanilla Conformer [14], they use convolution in multiple convolution blocks to reduce the resolution of the feature map. To complete the generation task, we must enlarge the feature map between Conformer blocks.

The CNN branch contains a variety of blocks, starting with an upsample block, and then two affine blocks and two convolutional layers are spaced in series to encode the feature of the CNN branch.

The affine transformation gives geometric functions such as zooming, rotating, translation, and offsetting the picture. If affine transformation is applied to image sub-regions or individual pixel blocks, they will impact the image's microscopic level. That's why the affine transformation can deepen the text-image fusion process. Primarily, we use an MLP to obtain the feature vector e that combines the global information of sentence embedding s and the local information of word embedding w:

$$e = MLP_0(s, w), \tag{1}$$

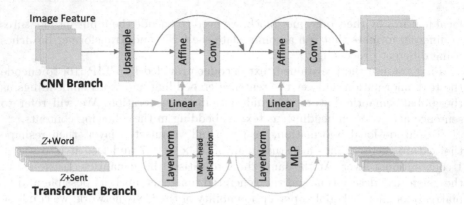

Fig. 3. A schematic of our framework Conformer Block. It contains two branches, the CNN branch and the Transformer branch. The Conformer block can improve the semantic consistency between text information and synthetic images by combining the global and local representations.

Next, we employ two MLPs to build the affine block. Among them, one MLP predicts the channel-wise scaling parameter γ under text conditions, and the other predicts its shifting parameter β:

$$\gamma = MLP_1(e), \qquad \beta = MLP_2(e). \tag{2}$$

both γ and β are learned from the encoded feature vector e, which will spatially process the image features and perform channel-wise scaling and shifting of the image features. The affine transformation can be formally expressed as follows:

$$\text{Affine}(x_i \mid e) = \gamma_i \cdot x_i + \beta_i, \tag{3}$$

where x_i is the ith channel of visual feature maps; γ_i and β_i are scaling parameter and shifting parameter for the ith channel of visual feature maps. We insert affine blocks in each Conformer block, which enables the CNN branch to get more text information, so text and visual features are fully integrated.

In the Transformer branch, the input is the z-encoded feature vectors, and the text embedding represents global semantic information. As shown in Fig. 3, before each layer and residual connections in both the self-attention layer and MLP block, we use LayerNorms [1]. The multi-head self-attention module establishes a global connection among all text feature vectors.

3.3   Loss Functions

Discriminator Loss. We use adversarial loss associated with MA-GP [20] loss to train the network; we also adopt their one-way discriminator. To assist the learning process of the discriminator, we adopt hinge loss [28]. The corresponding training objective of the discriminator is:

$$
\begin{aligned}
\mathcal{L}_D = & - \mathbb{E}_{x \sim P_r}[min(0, -1 + D(x,s))] \\
& - \frac{1}{2}\mathbb{E}_{G(z,s) \sim P_g}[min(0, -1 - D(G(z,s),s))] \\
& - \frac{1}{2}\mathbb{E}_{x \sim P_r}[min(0, -1 - D(x,\hat{s}))] \\
& + \lambda_{MA}\mathbb{E}_{x \sim P_r}[(\| \bigtriangledown_x D(x,s) \|_2 \\
& + \| \bigtriangledown_s D(x,s) \|_2)^p]
\end{aligned} \tag{4}
$$

where s is the given matching text description and $\hat{s}$ is the mismatched text description; x is the real image corresponding to s. $D(\cdot)$ is the result the Discriminator gives on whether the input image matches the input sentence. P_r represents the true data distribution and P_g represents the synthetic data distribution. The variables λ_{MA} and p are two hyperparameters of MA-GP loss capable of balancing gradient penalties.

Generator Loss. The adversarial loss and the cosine similarity between the encoded visual and text features of CLIP constitute the total loss of the Generator:

$$
\begin{aligned}
\mathcal{L}_G = & - \mathbb{E}_{G(z,s) \sim P_g}[D(G(z,s),s)] \\
& - \lambda \mathbb{E}_{G(z,s) \sim P_g} \frac{G(z,s) \times s}{\| G(z,s) \| \times \| s \|}
\end{aligned} \tag{5}
$$

where λ is the coefficients of the text-image similarity, the setup allows the network to generate realistic images at the sentence and word levels.

4   Experiments

Datasets. We evaluate the proposed model on two commonly used datasets: the CUB [23] and the COCO [13]. The CUB dataset contains 200 bird species with 10 text descriptions for each bird. We follow previous work [20,29,31] and divide the images into class disjoint training sets (150 species) and test sets (50 species). The COCO dataset has 82783 training images and 40504 validation images, and each image has 5 text descriptions.

Table 1. Training details related to **datasets and image size**.

Strategy	Dataset	Image Size	Batch Size	GPU	Epoch
B1	CUB	256 × 256	64	Tesla V100	1020
B2	CUB	512 × 512	32	Tesla V100	1180
C1	COCO	256 × 256	64	Tesla V100	750

Training and Evaluation Details. We use the Adam optimizer [9] to optimize the network, where $\beta_1 = 0.0$, $\beta_2 = 0.9$. The learning rate of the Generator is set to 0.0001, and that of the Discriminator is set to 0.0004. We choose the ViT-B/32 [16] model as the CLIP model in our Conformer-GAN. In Fig. 2, when the generated picture measures 256×256, the generator has 6 Conformer blocks ($N = 6$). For a picture size of 512×512, the Generator has 7 blocks ($N = 7$). Table 1 shows more training details about datasets and image sizes.

Following [20,21,24,31], we employ the widely used Fréchet Inception Distance (FID) [6] and CLIPSIM(CS) [24] to evaluate the performance of our network. The CS incorporates the CLIP [16] model to compute the semantic similarity between text and the generated images. Higher CS means higher semantic consistency of the text-image. Contrary to CS, a lower FID means a closer distance between the generated and real-world image distributions.

4.1   Quantitative Results

We quantitatively analyze the performance of the proposed method on the CUB and COCO datasets in Table 2. And we compare our method with several state-of-the-art methods in the field of text-to-image generation, including AttnGAN [25], DM-GAN [31], DF-GAN [20], SSA-GAN [11] LAFITE [30] and RAT-GAN [26].

On the CUB dataset, our model shows excellent superiority. Compared with the recently proposed method RAT-GAN [26], our Conformer-GAN reduces the FID metric from 13.91 to 10.14. For the CS metric, we improve the CS from 0.3125 to 0.3164 in LAFITE [30]. In addition, the FID score is 10.89, and the CS is 0.3076 for an image of size 512×512. Our model obtained the lowest FID score (10.14) on the COCO dataset compared to other state-of-the-art models.

Table 2. Results(256×256) on the CUB and COCO test set. The results are taken from the authors' papers. And $\triangle$ represents the performance improvement over state-of-the-art methods. All results are the average over 5 trials. The best results are in bold.

Methods	CUB		COCO	
	FID↓	CS↑	FID↓	CS↑
AttnGAN [25]	23.98	–	35.49	0.2772
DM-GAN [31]	16.09	–	32.64	0.2838
DF-GAN [20]	14.81	0.2920	21.42	0.2972
SSA-GAN [11]	15.61	–	19.37	–
LAFITE [30]	14.58	0.3125	8.21	**0.3335**
RAT-GAN [26]	13.91	–	14.60	–
Ours (256×256)	**10.14**	**0.3164**	**6.73**	0.3255
$\triangle$	**−3.77**	+0.0039	**−1.48**	−0.008

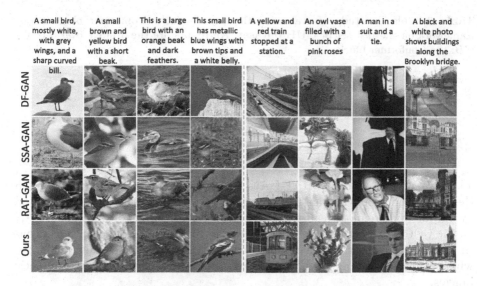

Fig. 4. Examples of images generated by DF-GAN [20], SSA-GAN [11], RAT-GAN [26] and our method conditioned on text descriptions from the test set of CUB and COCO datasets. (Color figure online)

Although our CS score is slightly lower than the one LAFITE gave (0.3255 *v.s.* 0.3335), it is still much higher than other recent methods. Many quantitative evaluation results demonstrate that the images generated by our Conformer-GAN possess better quality and higher semantic consistency of text-image.

4.2 Qualitative Results

We compare images generated by DF-GAN [20], SSA-GAN [11], RAT-GAN [26] and our method for qualitative evaluation.

For the CUB dataset, as shown in the 2nd and 4th columns of Fig. 4, except that our model can correctly synthesize images represented by text descriptions, the other three mentioned models cannot correctly synthesize "brown and yellow bird", "white belly" and "blue wings with brown tips". Furthermore, the RAT-GAN synthesized bird contains the wrong color, and the SSA-GAN and DF-GAN generated birds look unnatural (e.g., 1st and 2nd columns). However, the images generated by our Conformer-GAN have more vivid details and better object shapes. Notably, as shown in Fig. 1, our Conformer-GAN can generate text-image consistent fine images with high resolution (512 × 512).

For the COCO dataset, most notably the 8th column of Fig. 4, given the text "A black and white photo shows buildings along the Brooklyn bridge", our method generates images with all mentioned attributes. However, the images generated by the other three models mentioned do not fit the "black and white photo". As shown in the 6th and 7th columns, DF-GAN and SSA-GAN cannot synthesize the correct shapes of "pink roses" and "man", there is no "train" at

Table 3. Ablation study of evaluating the impact of the Conformer block in our framework on the test set of **CUB dataset**. All results are the average over 5 trials. CB: the Conformer block.

Case	CB	CS↑	FID↓
1	2	0.3162	19.66
2	4	0.3090	12.71
3	6	**0.3164**	**10.14**

all in 5th column. Although RAT-GAN can generate images that roughly match the text description, the quality is relatively low. Overall, the images generated by our method are more realistic. The qualitative advantage of Conformer-GAN on the COCO dataset is more evident because the Transformer branch can continuously provide text information to guide the CNN branch to generate images, which enables Conformer-GAN to achieve more complex control over synthetic images.

4.3   Ablation Studies

In this subsection, we verify the effectiveness of the Conformer block in our Conformer-GAN by conducting extensive ablation studies on the testing set of the CUB dataset [23]. We set different numbers of the Conformer blocks in Conformer-GAN for several comparison experiments. As shown in Table 3, when the number of the Conformer blocks is 2, the CS is 0.3162 and the FID is 19.66. It can be observed that the performance of our model is enhanced when the number of the Conformer blocks is gradually increased. When the framework uses 6 Conformer blocks, it has the highest CS (0.3164) and the lowest FID (10.14). Extensive results well prove the effectiveness of the Conformer block. The number of The Conformer block directly determines the fusion process of the text-image. When the number of the Conformer block increases, the CNN branch will get more text information and be able to generate finer pictures.

5   Conclusions

In this paper, we first try to combine the advantages of CNN and Transformer, apply them to the T2I field, and propose a novel architecture called Conformer-GAN. The core module of our model is the Conformer block, which contains the CNN and Transformer branches. To improve semantic consistency, we integrate both sentence and word information to guide the image generation of the CNN branch. The Transformer branch can continuously pay attention to the relationship between relevant words in the natural language description. Extensive experiments on the CUB dataset and the more challenging COCO dataset show that our model significantly improves the quality of generated images. The ablation experiments were done to demonstrate the effectiveness of our proposed method. Importantly, our Conformer-GAN can generate high-resolution 512×512 images with more photo-realistic details.

References

1. Ba, J.L., Kiros, J.R., Hinton, G.E.: Layer normalization. arXiv preprint arXiv:1607.06450 (2016)
2. Ding, M., et al.: CogView: mastering text-to-image generation via transformers. In: Advances in Neural Information Processing Systems, vol. 34, pp. 19822–19835 (2021)
3. Goodfellow, I.J., et al.: Generative adversarial nets. In: Proceedings of the 27th International Conference on Neural Information Processing Systems, NIPS 2014, vol. 2, pp. 2672–2680. MIT Press, Cambridge (2014)
4. Gou, Y., Wu, Q., Li, M., Gong, B., Han, M.: SegAttnGAN: text to image generation with segmentation attention. arXiv (2020)
5. Gu, S., et al.: Vector quantized diffusion model for text-to-image synthesis. arXiv e-prints (2021)
6. Heusel, M., Ramsauer, H., Unterthiner, T., Nessler, B., Hochreiter, S.: GANs trained by a two time-scale update rule converge to a local Nash equilibrium. In: Advances in Neural Information Processing Systems, vol. 30 (2017)
7. Hong, S., Yang, D., Choi, J., Lee, H.: Inferring semantic layout for hierarchical text-to-image synthesis. In: 2018 IEEE/CVF Conference on Computer Vision and Pattern Recognition (2018)
8. Huang, Y., Xue, H., Liu, B., Lu, Y.: Unifying multimodal transformer for bi-directional image and text generation. In: Proceedings of the 29th ACM International Conference on Multimedia, pp. 1138–1147 (2021)
9. Kingma, D.P., Ba, J.: Adam: a method for stochastic optimization. arXiv preprint arXiv:1412.6980 (2014)
10. Li, B., Qi, X., Lukasiewicz, T., Torr, P.: Controllable text-to-image generation. arXiv (2019)
11. Liao, W., Hu, K., Yang, M.Y., Rosenhahn, B.: Text to image generation with semantic-spatial aware GAN. In: Proceedings of the IEEE/CVF Conference on Computer Vision and Pattern Recognition, pp. 18187–18196 (2022)
12. Lin, J., et al.: M6: a Chinese multimodal pretrainer. arXiv preprint arXiv:2103.00823 (2021)
13. Lin, T.-Y., et al.: Microsoft COCO: common objects in context. In: Fleet, D., Pajdla, T., Schiele, B., Tuytelaars, T. (eds.) ECCV 2014. LNCS, vol. 8693, pp. 740–755. Springer, Cham (2014). https://doi.org/10.1007/978-3-319-10602-1_48
14. Peng, Z., et al.: Conformer: local features coupling global representations for visual recognition. In: Proceedings of the IEEE/CVF International Conference on Computer Vision, pp. 367–376 (2021)
15. Qiao, T., Zhang, J., Xu, D., Tao, D.: MirrorGAN: learning text-to-image generation by redescription. IEEE (2019)
16. Radford, A., et al.: Learning transferable visual models from natural language supervision. In: International Conference on Machine Learning, pp. 8748–8763. PMLR (2021)
17. Ramesh, A., et al.: Zero-shot text-to-image generation. In: International Conference on Machine Learning, pp. 8821–8831. PMLR (2021)

18. Reed, S., Akata, Z., Yan, X., Logeswaran, L., Schiele, B., Lee, H.: Generative adversarial text to image synthesis. In: International Conference on Machine Learning, pp. 1060–1069. PMLR (2016)
19. Sortino, R., Palazzo, S., Rundo, F., Spampinato, C.: Transformer-based image generation from scene graphs. Comput. Vis. Image Underst. **233**, 103721 (2023)
20. Tao, M., Tang, H., Wu, F., Jing, X.Y., Bao, B.K., Xu, C.: DF-GAN: a simple and effective baseline for text-to-image synthesis. arXiv e-prints (2020)
21. Tao, M., Bao, B.K., Tang, H., Xu, C.: GALIP: generative adversarial clips for text-to-image synthesis. arXiv preprint arXiv:2301.12959 (2023)
22. Tao, M., et al.: Deep fusion generative adversarial networks for text-to-image synthesis. arXiv preprint arXiv:2008.05865 (2020)
23. Wah, C., Branson, S., Welinder, P., Perona, P., Belongie, S.: The Caltech-UCSD Birds-200-2011 dataset (2011)
24. Wu, C., et al.: Nüwa: visual synthesis pre-training for neural visual world creation. In: Avidan, S., Brostow, G., Cissé, M., Farinella, G.M., Hassner, T. (eds.) ECCV 2022, Part XVI. LNCS, vol. 13676, pp. 720–736. Springer, Cham (2022). https://doi.org/10.1007/978-3-031-19787-1_41
25. Xu, T., et al.: AttnGAN: fine-grained text to image generation with attentional generative adversarial networks. In: Proceedings of the IEEE Conference on Computer Vision and Pattern Recognition, pp. 1316–1324 (2018)
26. Ye, S., Wang, H., Tan, M., Liu, F.: Recurrent affine transformation for text-to-image synthesis. IEEE Trans. Multimedia (2023)
27. Zhang, B., et al.: StyleSwin: transformer-based GAN for high-resolution image generation. In: Proceedings of the IEEE/CVF Conference on Computer Vision and Pattern Recognition, pp. 11304–11314 (2022)
28. Zhang, H., Goodfellow, I., Metaxas, D., Odena, A.: Self-attention generative adversarial networks. In: International Conference on Machine Learning, pp. 7354–7363. PMLR (2019)
29. Zhang, H., et al.: StackGAN: text to photo-realistic image synthesis with stacked generative adversarial networks. In: Proceedings of the IEEE International Conference on Computer Vision, pp. 5907–5915 (2017)
30. Zhou, Y., et al.: LAFITE: towards language-free training for text-to-image generation. arXiv e-prints (2021)
31. Zhu, M., Pan, P., Chen, W., Yang, Y.: DM-GAN: dynamic memory generative adversarial networks for text-to-image synthesis. In: 2019 IEEE/CVF Conference on Computer Vision and Pattern Recognition (CVPR) (2020)

MGFNet: A Multi-granularity Feature Fusion and Mining Network for Visible-Infrared Person Re-identification

BaiSheng Xu, HaoHui Ye, and Wei Wu[✉]

College of Computer and Information Science, Southwest University, Chongqing, China
{xu20161053,felixhui}@email.swu.edu.cn, ww@swu.edu.cn

Abstract. Visible-infrared person re-identification (VI-ReID) aims to match the same pedestrian in different forms captured by the visible and infrared cameras. Existing works on retrieving pedestrians focus on mining the shared feature representations by the deep convolutional neural networks. However, there are limitations of single-granularity for identifying target pedestrians in complex VI-ReID tasks. In this study, we propose a new Multi-Granularity Feature Fusion and Mining Network (MGFNet) to fuse and mine the feature map information of the network. The network includes a Local Residual Spatial Attention (LRSA) module and a Multi-Granularity Feature Fusion and Mining (MGFM) module to jointly extract discriminative features. The LRSA module aims to guide the network to learn fine-grained features that are useful for discriminating and generating more robust feature maps. Then, the MGFM module is employed to extract and fuse pedestrian features at both global and local levels. Specifically, a new local feature fusion strategy is designed for the MGFM module to identify subtle differences between various pedestrian images. Extensive experiments on two mainstream datasets, SYSU-MM01 and RegDB, show that the MGFNet outperforms the existing techniques.

Keywords: Visible-thermal Person re-identification · feature fusion · multi-granularity · cross-modality

1 Introduction

Person re-identification (ReID) aims to match the target pedestrians from multiple camera viewpoints and is widely used in the automatic tracking [1]. Most existing ReID methods [2,3] focus on retrieving RGB pedestrian images collected from visible cameras, which is also known as single-modal (RGB-RGB) matching tasks. To improve the applicability of ReID methods in real-world scenarios, scholars have proposed cross-modality visible-infrared person re-identification

B. Luo et al. (Eds.): ICONIP 2023, LNCS 14451, pp. 15–28, 2024.
https://doi.org/10.1007/978-981-99-8073-4_2

(VI-ReID) method to retrieve pedestrian images collected by infrared (IR) cameras in poor illumination conditions such as at night. VI-ReID gives a target person's RGB (or IR) image to query and then match it with the IR (or RGB) images in the gallery. However, VI-ReID has to deal with the inter-modal differences caused by camera views, pose variations, etc., and solve the impact of inter-modal variations caused by different device spectra and reflectance.

To handle the differences between two modalities simultaneously, existing solutions focus on learning and mining shared feature representations at a high level. This is achieved by mapping cross-modal images to a shared feature space using one- [4,5] or two- [6,7] stream networks and learning shared features that are both invariant and highly discriminable. Considering the complexity of pedestrian images, some works horizontally partition the feature output by flow networks to mine fine-grained local features. They believe [8,9] that the fine-grained local features can provide effective discriminative information in describing complex pedestrian images. Some scholars have also used Generative adversarial networks (GANs) [10,11] or designed more suitable loss functions for the task [12,13] to deal with modal differences. However, these methods have two drawbacks that prevent them from effective exploiting the learned pedestrian features. One drawback is that previous research usually updates the network by directly supervising the output of pedestrian features from the stream network. Single-granularity supervision does not fully utilize the information in the feature map, which may cause the network to ignore some discriminative information. The other drawback is that the performance of common fine-grained mining methods can be affected by part-level feature misalignment and the number of blocks. Directly aligning and supervising part-level features not only fails to obtain discriminative fine-grained features, but also affects the network performance and increases the difficulty of network convergence.

To overcome the above problems, we propose a new Multi-Granularity Feature Fusion and Mining net (MGFNet) to learn and mine robust multi-granularity pedestrian features. As shown in Fig. 1, a dual-stream network is used as the backbone network to extract pedestrian features. The local residual attention module (LRSA) is embedded inside the backbone network, and the multi-granularity feature fusion and mining (MGFM) module is placed at the end of the backbone network. LRSA enhances the feature representation with attention mechanism and thus the network can learn more local features. MGFM aims to fully mine the global and local granularity information of the feature map. At the global level, two different pooling methods are used to obtain coarse-grained and fine-grained global features. These features are further fused into more robust global features. Besides, a new local fine-grained feature fusion strategy is designed to obtain rich and discriminative local features.

The main contributions of this study can be summarized as follows: (1) LRSA is designed to obtain more discriminative feature maps with global-local discrimination. (2) MGFM is introduced to perform multi-granularity feature fusion and mining on feature maps from both global and local perspectives. (3) Experimental results show that our method achieves outstanding performance on two mainstream datasets (SYSU-MM01 [14] and RegDB [15]).

2 Related Work

Cross-modal pedestrian re-identification aims to match pedestrians of different modalities and has attracted the attention of scholars due to its high effectiveness in low-light environments. To overcome the huge differences between modalities, many VI-ReID methods have been proposed. Wu et al. [14] first defined the VI-ReID task, proposed a deep zero-padding network to learn modality-independent features, and open-sourced the first large-scale cross-modal pedestrian retrieval dataset SYSU-MM01. Ye et al. [1] proposed a hierarchical cross-modal matching model that uses a dual-stream network to map images of different modalities to a consistent feature space to eliminate the impact of modality differences. Wu et al. [5] proposed a joint Modality and Pattern Alignment Network (MPANet) based on a dual-stream network that significantly improved the accuracy of the VI-ReID task by designing a modality difference mitigation module and a modality alignment module. Liang et al. [16] proposed a cross-modal transformation network (CMTR) that makes the features generated by the network more discriminative by introducing learnable modality embedding (ME). Yang et al. [17] first considered the impact of mislabeling on cross-modal tasks based on the dual-stream network and proposed Dually Robust Training (DART) to correct the impact of noise by estimating confidence. However, these works mostly focus on eliminating the impact of modality differences by designing different network structures instead of mining complex pedestrian features.

Part-level information can improve the accuracy of re-identification tasks. For instance, Wei et al. [18] designed an Adaptive Body Part (ABP) model to distinguish discriminative part-level features. Zhang et al. [19] proposed a Hybrid Mutual Learning (HML) method to eliminate modality differences by learning the relationship between different modality parts and overall features. Considering the misalignment of part-level features caused by the variable posture of pedestrians, Wang et al. [6] used the shortest path algorithm to reduce the impact of misalignment errors on the task. However, most of these methods directly mine part-level features and ignore the relationship between part-level features. In addition, the attention mechanism has been used in the VI-ReID field [20, 21] due to its outstanding effects. Among them, spatial attention [22] is widely used. By generating a probability mask of feature maps in space, the network focuses on the important parts in space, which is very suitable for complex pedestrian images. For example, Wu et al. [23] proposed a Feature Aggregation Module (FAM) based on spatial attention to fully utilize global information and enhance information interaction between two modalities.

There are some scholars such as Zhu et al. [12] designed a heterogeneous center loss that supervises network learning by constraining the intra-class center distance between different modalities. Feng et al. [24] proposed a dual-constrained cross-modality (DCCM) loss to enhance the class and modal discrimination of features, which effectively reduces the difference between intra- and inter-modalities.

3   Methodology

In this section, we first introduce the backbone network structure and the details of the designed LRSA and MGFM modules. Finally, we explain the loss function used to supervise network learning.

3.1   Backbone Network

Figure 1 provides an overview of the proposed Multi-Granularity Feature Fusion and Mining Network (MGFNet), which utilizes a two-stream [1,5] pre-trained ResNet-50 [25] network as the backbone.

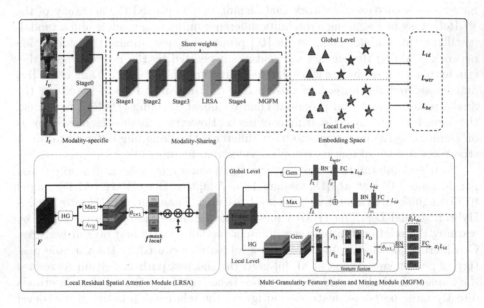

Fig. 1. The proposed Multi-Granularity Feature Fusion and Mining net (MGFNet) for VI-ReID. $\otimes$ and $\oplus$ indicate the element-wise multiplication and addition, respectively. HG stands for horizontal grouping, Gem and Max represent Gem pooling and Max pooling, respectively, to obtain features of different granularities. BN stands for batch normalization, θ represents 1x1 convolution, and FC stands for fully connected layer.

The feature extraction process of the network can be divided into two stages. First, in the specific modality feature learning stage, where two residual convolutional blocks (stage 0) with non-shared parameters are designed to learn information from two different modalities. Subsequently, the different modality feature maps learned by the specific modality are input into the modality sharing stage. The modality sharing stage is composed of four residual convolutional blocks (stage 1–4) and LRSA with shared parameters to learn multi-granularity information that is independent of modality. Finally, the MGFM module is used to perform multi-granularity feature fusion and mining on the feature maps.

3.2   Local Residual Spatial Attention

As shown in Fig. 1, the LRSA is placed behind the third residual block (Stage 3) of the modality sharing phase. Given the intermediate feature mapping of the input data $F \in R^{C \times H \times W}$, where C, H, and W represent the dimensions, height and width of the feature map, respectively.

Traditional spatial attention mechanisms [22] aggregate spatial information of feature maps through Average pooling and Max pooling to generate global spatial attention weights to guide network learning. However, for complex VI-ReID tasks, directly using the spatial attention mechanism to guide network learning is not stable in experiments and even shows a decline in performance compared to the baseline. This may be because the global spatial attention mechanism mistakenly focuses on common occlusions and variable scenes in pedestrian images, ignoring fine-grained information that is useful for discrimination.

To solve this problem, our proposed LRSA horizontally divides the feature map F according to body parts, generates local spatial attention weights f_i^{mask} to guide the network to focus on key areas at the part level, and then concatenates the part-level spatial attention weights to generate new attention weights f^{mask}, Eq. (1–2).

$$f_i^{mask} = \sigma \left(COV \left(Cat \left(AvgPool \left(F_i \right), MaxPool \left(F_i \right) \right) \right) \right) \tag{1}$$

$$f^{mask} = Cat_{i=1}^{p} f_i^{mask} \tag{2}$$

Here, σ represents the sigmoid function, COV represents a two-dimensional convolution kernel with a kernel size of 7, Cat represents the concatenation operation, and f_i represents the feature map at the component level.

In addition, to ensure the stability of training, the input feature F is directly added to the attention feature map generated by the LRSA module through residual connections. The attention feature map F^{new} with local residual can be represented as:

$$F^{new} = \tau \left(F \otimes f^{mask} \right) \oplus F \tag{3}$$

where τ is a local attention weight hyperparameter, which is set to 0.5 here.

3.3   Multi-granularity Feature Fusion and Mining Module

As shown in Fig. 1, MGFM aims to fuse and mine the final pedestrian feature map $F^* \in R^{C \times H \times W}$ extracted by the backbone network, mining F^* in both global and local branches.

At the global level, inspired by the work of [7], generalized mean (Gem) pooling [26] and Max pooling are used to obtain global features f_1 and f_2 of different granularities. For f_1, after batch normalization, a fine-grained feature f_g is obtained. For f_2, after element-wise addition with a fine-grained feature f_g, it is then passed through BN to obtain a global feature f_m that fuses coarse and fine-grained features, which is formulated as follows:

$$f_g = BN\left(f_1\right),$$
$$f_m = BN\left(f_g \oplus f_2\right) \tag{4}$$

In the local branch, a common practice [8,9] is to directly supervise and mine the body part local features. But this approach ignores the connection between part-level features. Considering the misalignment of part-level features caused by differences in camera angles, pedestrian posture, etc., and the correlation between pedestrian part-level features, we propose a new part-level feature fusion strategy (Fig. 1). Feature maps are first horizontally split into 6 blocks along the H dimension, and after Gem pooling, six body parts local feature maps $p_i \in R^{C \times \frac{H}{6} \times W}$ are obtained, where $i \in \{0, 1, ..., 5\}$. Then according to the upper and lower body partitions of the pedestrian, i.e., p_{l1} and p_{l2} are constructed by fusing the first and last three part-level features respectively. In addition, part-level features p_i are fused to form p_{l3} and p_{l4} to describe pedestrian body parts with respect to the complexity and correlation of pedestrian images. Finally, in order to explicitly express the correlation between new part features and local pedestrian description features, all part-level features p_i are fused to generate f_{lg}. It is calculated as follows,

$$\begin{cases} p_{l1} = Cat_{i=0}^2\left(p_i\right) \\ p_{l2} = Cat_{i=3}^5\left(p_i\right) \\ p_{l3} = Cat_{i=1}^3\left(p_i\right) \\ p_{l4} = Cat_{i=2}^4\left(p_i\right) \\ f_{lg} = Cat_{i=0}^5\left(p_i\right) \end{cases} \tag{5}$$

Here Cat_i^j represents concatenating part-level features i-j along the H dimension.

In fact, the new blocks after fusion avoid the occurrence of missing stripe boundaries in local features to a certain extent. Moreover, the block feature fusion strategy generates pedestrian part features with more information than fixed block partitioning. This makes it more adaptable to changes in pedestrian posture, camera angles, and other factors. Without introducing additional supervision information or discriminators, it minimizes the impact of misaligned block features on network learning and prediction. The final part-level features are obtained by COV and BN on the new part-level features using,

$$\begin{cases} p_{l1}^* = BN\left(COV\left(p_{l1}\right)\right) \\ p_{l2}^* = BN\left(COV\left(p_{l2}\right)\right) \\ p_{l3}^* = BN\left(COV\left(p_{l3}\right)\right) \\ p_{l4}^* = BN\left(COV\left(p_{l4}\right)\right) \\ f_{lg}^* = BN\left(COV\left(f_{lg}\right)\right) \end{cases} \tag{6}$$

Here BN and COV do not share parameters.

Based on MGFM, we obtain fine-grained features f_g and features f_m that integrate coarse and fine-grained features at the global level, and part-level features $\{p_{l1}^*, p_{l2}^*, p_{l3}^*, p_{l4}^*\}$ and part-level pedestrian description features f_{lg}^* at the local level.

3.4   Loss Function

In order to enhance the discriminative power of the network output features and maximize the multi-granularity mining ability of MGFM, we use ID loss (L_{id}), heterogeneous center loss triplet loss (L_{hc}), and weighted regularization triplet loss (L_{wtr}) to guide network learning. For basic discrimination learning, we treat the re-identification task as a multi-classification problem. For an input pedestrian image x, the calculation of ID loss is:

$$L_{id} = -\frac{1}{N}\sum_{i=1}^{N} log\left(P\left(y_i\middle| f\left(x_i\right);\theta^t\right)\right) \tag{7}$$

Here, N represents the total number of pedestrian images in each mini-batch, $f\left(x_i\right)$ generally refers to the pedestrian features output by the network, and θ^t represents the parameters of the identity classifier.

Weighted regularization triplet loss [1] aims to optimize the distance between all positive and negative samples from both intra-modal and inter-modal perspectives, which is expressed as:

$$L_{wtr} = -\frac{1}{N}\sum_{i=1}^{N} log\left(1 + exp\left(\sum_{ij} w_{ij}^p d_{ij}^p - \sum_{ik} w_{ik}^n d_{ik}^n\right)\right) \tag{8}$$

$$w_{ij}^p = \frac{exp\left(d_{ij}^p\right)}{\sum_{d_{ij}^p \in p_i} exp\left(d_{ij}^p\right)},\ w_{ik}^n = \frac{exp\left(-d_{ik}^n\right)}{\sum_{d_{ik}^n \in N_i} exp\left(-d_{ik}^n\right)}$$

Here, (i,j and k) respectively represent a triplet for anchor x_i in each training mini-batch. P_i represents the positive sample and N_i represents the negative sample. d_{ij}^p/d_{ik}^n is the pairwise distance of positive/negative sample pairs, where d_{ij} is the Euclidean distance between different sample pairs and calculated by $d_{ij} = \|f\left(x_i\right) - f\left(x_j\right)\|_2$.

Heterogeneous center triplet loss [12] focuses on optimizing the distance between the cross-modal positive sample center corresponding to the anchor point and the center of the hardest (intra- and inter-modality) negative sample. The calculation of heterogeneous center triplet loss is expressed as:

$$L_{hc} = \sum_{i=1}^{n} \left[m + \left\| f_{c_v}^i - f_{c_t}^i \right\|_2 - \min_{\substack{m \in \{v,t\} \\ j \neq i}} \left\| f_{c_v}^i - f_{c_m}^j \right\|_2 \right]_+$$

$$+ \sum_{i=1}^{n} \left[m + \left\| f_{c_t}^i - f_{c_v}^i \right\|_2 - \min_{\substack{m \in \{v,t\} \\ j \neq i}} \left\| f_{c_t}^i - f_{c_m}^j \right\|_2 \right]_+ \tag{9}$$

Here, m represents the margin, $f_{c_v}^i = \frac{1}{K}\sum_{j=1}^{K} f_{v,j}^i$, $f_{c_t}^i = \frac{1}{K}\sum_{j=1}^{K} f_{t,j}^i$ are the centers of the ith pedestrian identity under the visible or infrared modality. In addition, $f_{v,j}^i$ and $f_{t,j}^i$ represent the features of the j-th visible light image and infrared image of the i-th unit, respectively.

At the global level, considering the difference between f_g and f_m, L_{wtr} and L_{hc} are used to constrain f_g and f_m respectively,

$$L_{global} = L_{wtr}\left(f_g\right) + L_{id}\left(f_g\right) + L_{hc}\left(f_m\right) + L_{id}\left(f_m\right) \tag{10}$$

At the local level, each part-level feature is supervised by L_{hc} and L_{id}, where α and β are hyperparameters used to balance different losses.

$$L_{local} = L_{hc}\left(f_{lg}\right) + L_{id}\left(f_{lg}\right) + \sum_{i=1}^{p} \left(\alpha L_{hc}\left(p_{l\_i}^*\right) + \beta L_{id}\left(p_{l\_i}^*\right) \right) \tag{11}$$

The total loss of the network can be calculated as:

$$L = L_{global} + L_{local} \tag{12}$$

4   Experiments

4.1   Datasets and Settings

Datasets: SYSU-MM01 [14] and RegDB [15] datasets were used in our experiments. SYSU-MM01 is a large-scale and widely used VI-ReID dataset. The pedestrian images in SYSU-MM01 are captured by six cameras, including four RGB cameras and two infrared cameras, with scenes including indoor and outdoor. Following the evaluation criteria of SYSU-MM01, the training set includes 395 identities, 22258 RGB images, and 11,909 IR images. To test the model, we randomly select 3803 IR images of 96 identities as queries and 301 RGB images as gallery. This dataset includes two test modes: all search and indoor search. The former is more in line with real-world applications and is more challenging.

The RegDB dataset is captured by only one visible light and one infrared camera. It contains a total of 412 IDs, each with ten visible light and infrared images. According to the common VI-ReID setting, 206 identities are randomly selected for training, and the remaining 206 identities are used for testing. Training and testing are repeated ten times, and the average of the ten results is taken as the final outcome of the model.

Evaluation metrics: Like [27], we use rank-1 matching accuracy, mean average precision (map), and mean inverse negative penalty [27] (minp) as evaluation metrics. During testing, pedestrian features are processed using $L2$ normalization.

Settings: We use Pytorch to implement our method. All pedestrian images are scaled to $288 \times 144 \times 3$ and random erasure [28] and random grayscale transformation [27] are used as data augmentation strategies. In addition, the backbone network is pre-trained on the ImageNet dataset. During optimization, the stochastic gradient descent (SGD) optimizer is used for training, with an initial learning rate of 0.1 and a warm-up strategy [29] used in the first ten epochs. The learning rate decays to 0.1 and 0.01 after 20 and 50 epochs, respectively. The total number of training epochs is set to 100 rounds. In each batch, 8 pedestrians are randomly selected, each containing 4 visible light images and 4 infrared images.

4.2 Comparison with the State-of-the-Art Methods

In this section, we compared our proposed MGFNet with advanced VI-ReID methods released in recent years. The results on the SYSU-MM01 and RegDB datasets are listed in Tables 1 and 2, respectively. The results show that our proposed method outperforms existing solutions under various settings. In the All-search mode of the large-scale SYSU-MM01 dataset, R1, mAP, and mINP

Table 1. Comparison with the state-of-the-art methods on the SYSU-MM01 dataset.

Methods	Time	All search			Indoor search		
		rank-1	mAP	mINP	rank-1	mAP	mINP
AlignGAN [10]	2019	42.41	40.70	–	45.9	54.30	–
JSIA [30]	2020	38.10	36.90	–	43.80	52.90	–
DDAG [31]	2020	54.75	53.02	–	61.02	67.98	–
IIC [12]	2020	59.96	54.95	–	59.74	64.91	–
HAT [32]	2020	55.29	53.89	–	62.10	69.37	–
AGW [1]	2021	47.50	47.65	35.30	54.17	62.49	59.23
SFANet [33]	2021	65.74	60.83	–	71.60	80.05	–
MCLNet [34]	2021	65.40	61.98	47.39	72.56	76.58	72.10
CAJ [27]	2021	69.88	66.89	53.61	76.26	80.37	76.79
MPANet [5]	2021	70.58	68.24	–	76.74	80.95	–
MMN [35]	2021	70.60	66.90	–	76.20	79.60	–
CMTR [36]	2022	62.58	61.33	–	67.02	73.78	–
TSME [37]	2022	64.23	61.21	–	64.80	71.53	–
SPOT [38]	2022	65.34	62.52	48.86	69.42	74.63	70.48
DART [17]	2022	68.72	66.29	53.26	**78.17**	73.78	74.94
DCLNet [39]	2022	70.80	65.30	–	73.50	76.80	–
MAUM [40]	2022	71.68	68.79	–	71.97	81.94	–
TVTR [41]	2023	65.30	64.15	–	77.21	77.94	–
Ours	–	**72.63**	**69.64**	**56.79**	77.90	**82.28**	**79.33**

Table 2. Comparison with the state-of-the-art methods on the RegDB dataset.

Methods	Time	Visible-to-Infrared			Infrared-to-Visible		
		rank-1	mAP	mINP	rank-1	mAP	mINP
AlignGAN [10]	2019	57.90	53.60	–	56.30	53.40	–
JSIA [30]	2020	48.10	48.90	–	48.50	49.30	–
DDAG [31]	2020	69.34	63.46	–	68.06	61.80	–
HAT [32]	2020	71.83	67.56	–	70.02	66.30	–
AGW [1]	2021	70.05	66.37	50.19	–	–	–
SFANet [33]	2021	76.31	68.00	–	70.15	63.77	–
CMTR [36]	2021	80.62	74.42	–	81.06	73.75	–
VSD [42]	2021	73.20	71.60	–	71.80	70.10	–
MPANet [5]	2021	70.58	68.24	–	76.74	**80.95**	–
MCLNet [34]	2021	80.31	73.07	57.39	75.93	69.49	52.63
CAJ [27]	2021	85.03	79.14	65.33	84.75	77.82	61.56
TSME [37]	2022	87.35	76.94	–	84.41	75.70	–
SPOT [38]	2022	80.35	72.46	52.19	79.37	72.26	56.06
CMTR [36]	2022	80.62	74.42	–	81.06	73.75	–
DART [17]	2022	83.60	75.67	60.60	81.97	73.78	56.70
DCLNet [39]	2022	81.20	74.30	–	78.00	70.60	–
TVTR [41]	2023	84.10	79.50	–	83.70	78.00	–
Ours	–	**91.14**	**82.53**	**67.81**	**89.06**	80.50	**63.99**

of MGFNet increased by 2.74%, 2.75%, and 3.18%, respectively, compared to baseline (CAJ [27]).

Ablation: We conducted detailed ablation experiments to verify the effectiveness of our proposed module on the SYSU-MM01 and RegDB datasets. We first evaluated the effectiveness of each component, where B represents the baseline network, using the network output f_m as the pedestrian feature, and each pedestrian feature was l2 regularized during testing. Table 3 shows that the baseline network plus LRSA module and the baseline network plus MGFM module performed better than the baseline network. Moreover, our MGFNet that combined these two modules gave the best performance, since the local fine-grained features detected by LRSA can be captured by MGFM's multi-granularity mining method.

Table 3. Ablation experiment on SYSU-MM01 and RegDB datasets.

Method	SYSU-MM01			RegDB		
	R1	mAP	mINP	R1	mAP	mINP
B	69.88	66.89	53.61	85.03	79.14	65.33
B+**LRSA**	71.61	68.16	54.80	88.74	80.69	66.23
B+**MGFM**	72.30	68.86	55.68	90.12	81.07	66.59
Ours	**72.63**	**69.64**	**56.79**	**91.14**	**82.53**	**67.81**

Visualization: We used the Grad-CAM [43] method to locate the important areas of our network output features to further prove the effectiveness of the proposed method. As shown in Fig. 2, compared to the baseline, the areas of interest in the network output features tend to be more focused on the human body. In complex scenarios such as low light and occlusion, it can also focus well on the human body and is not easily disturbed.

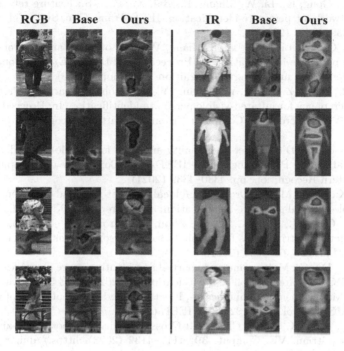

Fig. 2. The results of Grad-CAM visualization.

5 Conclusion

We propose a Multi-Granularity Feature Fusion and Mining Network (MGFNet) for VI-ReID. The innovation of MGFNet lies in: the LRSA guides the network to learn fine-grained features useful for discrimination through local attention mechanism, directly enhancing pedestrian feature representation; MGFM mines and fuses pedestrian features from local and global levels in a multi-granularity manner. At the global level, different pooling methods and simple fusion make pedestrian features more discriminative, while at the local level, the designed part-level feature fusion strategy enriches part-level features and reduces the impact of part misalignment with a small cost. Experimental results show that our method performs excellent. We hope that the proposed MGFNet can provide help in mining VI-ReID feature maps.

References

1. Ye, M., Shen, J., Lin, G., Xiang, T., Shao, L., Hoi, S.C.H.: Deep learning for person re-identification: a survey and outlook. IEEE Trans. Pattern Anal. Mach. Intell. **44**(6), 2872–2893 (2021)
2. Ning, X., Gong, K., Li, W., Zhang, L., Bai, X., Tian, S.: Feature refinement and filter network for person re-identification. IEEE Trans. Circ. Syst. Video Technol. **31**(9), 3391–3402 (2020)
3. Luo, H., Gu, Y., Liao, X., Lai, S., Jiang, W.: Bag of tricks and a strong baseline for deep person re-identification. In: Proceedings of the IEEE/CVF Conference on Computer Vision and Pattern Recognition Workshops (2019)
4. Wang, Z., Wang, Z., Zheng, Y., Chuang, Y.-Y., Satoh, S.: Learning to reduce dual-level discrepancy for infrared-visible person re-identification. In: Proceedings of the IEEE/CVF Conference on Computer Vision and Pattern Recognition, pp. 618–626 (2019)
5. Wu, Q., et al.: Discover cross-modality nuances for visible-infrared person re-identification. In: Proceedings of the IEEE/CVF Conference on Computer Vision and Pattern Recognition, pp. 4330–4339 (2021)
6. Wang, X., Li, C., Ma, X.: Cross-modal local shortest path and global enhancement for visible-thermal person re-identification. arXiv preprint arXiv:2206.04401 (2022)
7. Liu, H., Chai, Y., Tan, X., Li, D., Zhou, X.: Strong but simple baseline with dual-granularity triplet loss for visible-thermal person re-identification. IEEE Sig. Process. Lett. **28**, 653–657 (2021)
8. Yuan, Y., Du, G.: Multi-granularity partial and identity-aware global feature learning for RGB-infrared person re-identification. In: 2022 IEEE 5th Advanced Information Management, Communicates, Electronic and Automation Control Conference (IMCEC), vol. 5, pp. 323–329. IEEE (2022)
9. Wen, X., Feng, X., Li, P., Chen, W.: Cross-modality collaborative learning identified pedestrian. Vis. Comput. **39**, 4117–4132 (2023). https://doi.org/10.1007/s00371-022-02579-y
10. Wang, G., Zhang, T., Cheng, J., Liu, S., Yang, Y., Hou, Z.: RGB-infrared cross-modality person re-identification via joint pixel and feature alignment. In: Proceedings of the IEEE/CVF International Conference on Computer Vision, pp. 3623–3632 (2019)
11. Zhang, S., Yang, Y., Wang, P., Liang, G., Zhang, X., Zhang, Y.: Attend to the difference: cross-modality person re-identification via contrastive correlation. IEEE Trans. Image Process. **30**, 8861–8872 (2021)
12. Zhu, Y., Yang, Z., Wang, L., Zhao, S., Xiao, H., Tao, D.: Hetero-center loss for cross-modality person re-identification. Neurocomputing **386**, 97–109 (2020)
13. Yan, C., et al.: Beyond triplet loss: person re-identification with fine-grained difference-aware pairwise loss. IEEE Trans. Multimedia **24**, 1665–1677 (2021)
14. Wu, A., Zheng, W.-S., Yu, H.-X., Gong, S., Lai, J.: RGB-infrared cross-modality person re-identification. In: Proceedings of the IEEE International Conference on Computer Vision, pp. 5380–5389 (2017)
15. Nguyen, D.T., Hong, H.G., Kim, K.W., Park, K.R.: Person recognition system based on a combination of body images from visible light and thermal cameras. Sensors **17**(3), 605 (2017)
16. Liang, T., et al.: CMTR: cross-modality transformer for visible-infrared person re-identification. arXiv preprint arXiv:2110.08994 (2021)

17. Yang, M., Huang, Z., Hu, P., Li, T., Lv, J., Peng, X.: Learning with twin noisy labels for visible-infrared person re-identification. In: Proceedings of the IEEE/CVF Conference on Computer Vision and Pattern Recognition, pp. 14308–14317 (2022)
18. Wei, Z., Yang, X., Wang, N., Song, B., Gao, X.: ABP: adaptive body partition model for visible infrared person re-identification. In: 2020 IEEE International Conference on Multimedia and Expo (ICME), pp. 1–6. IEEE (2020)
19. Zhang, Z., Dong, Q., Wang, S., Liu, S., Xiao, B., Durrani, T.S.: Cross-modality person re-identification using hybrid mutual learning. IET Comput. Vis. **17**(1), 1–12 (2023)
20. Tan, L., et al.: Exploring invariant representation for visible-infrared person re-identification. arXiv preprint arXiv:2302.00884 (2023)
21. Li, W., Zhu, X., Gong, S.: Harmonious attention network for person re-identification. In: Proceedings of the IEEE Conference on Computer Vision and Pattern Recognition, pp. 2285–2294 (2018)
22. Woo, S., Park, J., Lee, J.-Y., Kweon, I.S.: CBAM: convolutional block attention module. In: Ferrari, V., Hebert, M., Sminchisescu, C., Weiss, Y. (eds.) ECCV 2018. LNCS, vol. 11211, pp. 3–19. Springer, Cham (2018). https://doi.org/10.1007/978-3-030-01234-2_1
23. Baotai, W., Feng, Y., Sun, Y., Ji, Y.: Feature aggregation via attention mechanism for visible-thermal person re-identification. IEEE Sig. Process. Lett. **30**, 140–144 (2023)
24. Feng, Y., Chen, Z.: Taking both the modality and class information for visible infrared person re-identification. In: 2022 34th Chinese Control and Decision Conference (CCDC), pp. 4338–4342. IEEE (2022)
25. He, K., Zhang, X., Ren, S., Sun, J.: Deep residual learning for image recognition. In: Proceedings of the IEEE Conference on Computer Vision and Pattern Recognition, pp. 770–778 (2016)
26. Radenović, F., Tolias, G., Chum, O.: Fine-tuning CNN image retrieval with no human annotation. IEEE Trans. Pattern Anal. Mach. Intell. **41**(7), 1655–1668 (2018)
27. Ye, M., Ruan, W., Du, B., Shou, M.Z.: Channel augmented joint learning for visible-infrared recognition. In: Proceedings of the IEEE/CVF International Conference on Computer Vision, pp. 13567–13576 (2021)
28. Zhong, Z., Zheng, L., Kang, G., Li, S., Yang, Y.: Random erasing data augmentation. In: Proceedings of the AAAI Conference on Artificial Intelligence, vol. 34, pp. 13001–13008 (2020)
29. Luo, H., et al.: A strong baseline and batch normalization neck for deep person re-identification. IEEE Trans. Multimedia **22**(10), 2597–2609 (2019)
30. Wang, G.-A., et al.: Cross-modality paired-images generation for RGB-infrared person re-identification. In: Proceedings of the AAAI Conference on Artificial Intelligence, vol. 34, pp. 12144–12151 (2020)
31. Ye, M., Shen, J., J. Crandall, D., Shao, L., Luo, J.: Dynamic dual-attentive aggregation learning for visible-infrared person re-identification. In: Vedaldi, A., Bischof, H., Brox, T., Frahm, J.-M. (eds.) ECCV 2020. LNCS, vol. 12362, pp. 229–247. Springer, Cham (2020). https://doi.org/10.1007/978-3-030-58520-4_14
32. Ye, M., Shen, J., Shao, L.: Visible-infrared person re-identification via homogeneous augmented tri-modal learning. IEEE Trans. Inf. Forensics Secur. **16**, 728–739 (2020)
33. Liu, H., Ma, S., Xia, D., Li, S.: SFANet: a spectrum-aware feature augmentation network for visible-infrared person reidentification. IEEE Trans. Neural Netw. Learn. Syst. **34**(4), 1958–1971 (2023)

34. Hao, X., Zhao, S., Ye, M., Shen, J.: Cross-modality person re-identification via modality confusion and center aggregation. In: Proceedings of the IEEE/CVF International Conference on Computer Vision, pp. 16403–16412 (2021)
35. Zhang, Y., Yan, Y., Lu, Y., Wang, H.: Towards a unified middle modality learning for visible-infrared person re-identification. In: Proceedings of the 29th ACM International Conference on Multimedia, pp. 788–796 (2021)
36. Jiang, K., Zhang, T., Liu, X., Qian, B., Zhang, Y., Wu, F.: Cross-modality transformer for visible-infrared person re-identification. In: Avidan, S., Brostow, G., Cissé, M., Farinella, G.M., Hassner, T. (eds.) ECCV 2022, Part XIV. LNCS, vol. 13674, pp. 480–496. Springer, Cham (2022). https://doi.org/10.1007/978-3-031-19781-9_28
37. Liu, J., Wang, J., Huang, N., Zhang, Q., Han, J.: Revisiting modality-specific feature compensation for visible-infrared person re-identification. IEEE Trans. Circ. Syst. Video Technol. **32**(10), 7226–7240 (2022)
38. Chen, C., Ye, M., Qi, M., Wu, J., Jiang, J., Lin, C.-W.: Structure-aware positional transformer for visible-infrared person re-identification. IEEE Trans. Image Process. **31**, 2352–2364 (2022)
39. Sun, H., et al.: Not all pixels are matched: dense contrastive learning for cross-modality person re-identification. In: Proceedings of the 30th ACM International Conference on Multimedia, pp. 5333–5341 (2022)
40. Liu, J., Sun, Y., Zhu, F., Pei, H., Yang, Y., Li, W.: Learning memory-augmented unidirectional metrics for cross-modality person re-identification. In: Proceedings of the IEEE/CVF Conference on Computer Vision and Pattern Recognition, pp. 19366–19375 (2022)
41. Yang, B., Chen, J., Ye, M.: Top-k visual tokens transformer: selecting tokens for visible-infrared person re-identification. In: ICASSP 2023–2023 IEEE International Conference on Acoustics, Speech and Signal Processing (ICASSP), pp. 1–5. IEEE (2023)
42. Tian, X., Zhang, Z., Lin, S., Qu, Y., Xie, Y., Ma, L.: Farewell to mutual information: variational distillation for cross-modal person re-identification. In: Proceedings of the IEEE/CVF Conference on Computer Vision and Pattern Recognition, pp. 1522–1531 (2021)
43. Selvaraju, R.R., Cogswell, M., Das, A., Vedantam, R., Parikh, D., Batra, D.: Grad-CAM: visual explanations from deep networks via gradient-based localization. In: Proceedings of the IEEE International Conference on Computer Vision, pp. 618–626 (2017)

Isomorphic Dual-Branch Network for Non-homogeneous Image Dehazing and Super-Resolution

Wenqing Kuang, Zhan Li[✉], Ruijin Guan, Weijun Yuan, Ruting Deng, and Yanquan Chen

Department of Computer Science, Jinan University, Guangzhou 510632, China
lizhan@jnu.edu.cn

Abstract. Removing non-homogeneous haze from real-world images is a challenging task. Meanwhile, the popularity of high-definition imaging systems and compute-limited smart mobile devices has resulted in new problems, such as the high computational load caused by haze removal for large-size images, or the severe information loss caused by the degradation of both the haze and image downsampling, when applying existing dehazing methods. To address these issues, we propose an isomorphic dual-branch dehazing and super-resolution network for non-homogeneous dehazing of a downsampled hazy image, which produces dehazed and enlarged images with sharp edges and high color fidelity. We quantitatively and qualitatively compare our network with several state-of-the-art dehazing methods under the condition of different downsampling scales. Extensive experimental results demonstrate that our method achieves superior performance in terms of both the quality of output images and the computational load.

Keywords: isomorphic dual-branch · image dehazing · super-resolution · non-homogeneous · loss attention

1 Introduction

As a common atmospheric phenomenon, haze reduces the transparency of the air and thus inevitably degrades the quality of images captured in the real world. Removing haze from natural images is a challenging task, particularly when the haze is non-homogeneously distributed and the image size is large with high resolution.

On the one hand, for real-world non-homogeneous hazy images that contain both densely and thinly hazed regions, uniform processing of dehazing has intrinsic limitations. For instance, edges may be blurred and vulnerable to the remaining haze for densely hazed regions, while over-dehazing of thinly hazed regions frequently leads to color distortions. On the other hand, the widespread use of high-definition imaging equipment and embedded or mobile devices results in new problems related to a contradiction between a large image size and limited computing resources. Removing haze from high-resolution (HR) images is

B. Luo et al. (Eds.): ICONIP 2023, LNCS 14451, pp. 29–40, 2024.
https://doi.org/10.1007/978-981-99-8073-4_3

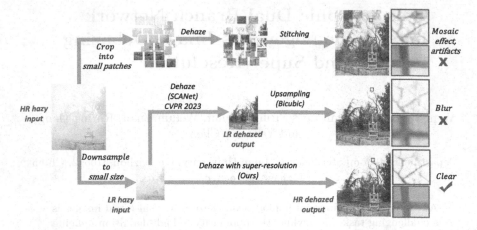

Fig. 1. For non-homogeneous dehazing of large-size images, we propose dehazing with SR, which produces clear edges and vibrant colors with the computational load significantly reduced.

a computationally intensive task, which requires both high-performance processors and a significant amount memory. Therefore, feasible solutions are either downsampling the input image to a small size or cropping it into small patches, as shown in Fig. 1. However, dehazing then stitching frequently results in mosaic effects and artifacts, because the information across patches cannot be transmitted and coordinated. To overcome these limitations, we can alternatively perform dehazing on a low-resolution (LR) hazy image resized from the original HR image. However, most off-the-shelf dehazing methods ignore the degradation and loss of details introduced by the downsampling, which can hardly be recovered by traditional interpolations and will lead to a blurry output.

To address these problems, we propose an end-to-end network for recovering details while dehazing. In our method, isomorphic dual-branch (IDB) is employed to separately learn "easy-" and "hard-" recovered pixels for regions of thin and dense haze in parallel. Moreover, a loss attention (LA) module is designed to differentiate the IDB module by calculating a changing loss mask. Further, we construct a feature fusion super-resolution (FFSR) module to restore abundant details in LR hazy images. Our method achieves state-of-the-art performance on dehazing with the computation cost significantly reduced, owing to the replacement of LR hazy inputs downsampled from original HR images.

Overall, our contributions can be summarized as follows:

- We design an IDB network for both image non-homogeneous dehazing and super-resolution reconstruction in an end-to-end manner.
- We propose an LA module to distinguish pixels "easy for learning" from those "hard" ones by constructing a dynamic mask, which differentiates functions of two branches in the training process.

– We construct an FFSR module for restoring details in hazy images of low resolutions. Therefore, large-size images can be downsampled and dehazed.

2 Related Works

2.1 Dual-Branch Network for Dehazing

A dual-branch structure is introduced to several dehazing networks [7,9,10,12, 16,23], owing to its strong capability of information aggregation. In addition to an encoder-decoder main branch, a branch based on a discrete wavelet transform is constructed in DWGAN [7] to extract signals of high and low frequency from images. A structure representation defog network (SRDefog) [12] incorporates a grayscale branch and a color branch to extract the background through consistency loss. A double-branch dehazing network based on a self-calibrated attentional convolution (DBDN-SCAC) [26] uses an auxiliary branch to assist the main branch to generate more identifiable information.

In addition to the aforementioned networks with two asymmetric branches, a knowledge transform dehazing network (KTDN) [22] uses two isomorphic networks, which separately accept clear and hazy images in training to learn their distribution, and only use one of them for dehazing during testing. Unlike these dual-branch models, our network employs two branches with identical structures to learn non-homogeneous distribution of haze regions only from the hazy images.

2.2 Super-Resolution in Image Dehazing

Super-resolution (SR) techniques reconstruct HR images from LR inputs, aiming at recovery of details [6,25]. It is essential to introduce SR to dehazing, because the details are frequently covered by haze and lost in images. SRKTDN [6] adds a detail refinement branch of a trident dehazing network (TDN) [14] to the KTDN to learn the high gradient in hazy images. A two-branch neural network (TNN) [23] incorporates a residual channel attention network (RCAN) [25] as an SR branch. Similarly, a transformer-based dehazing network (ITBdehaze) [16] uses RCAN as a second branch to provide additional information to the main branch.

By incorporating SR techniques, previous dehazing networks enhance details in the original input image size. Considering the degradation of both haze and downsampling, our method produces HR-dehazed and SR-reconstructed images for LR inputs, with preferable results at a low computation cost.

3 Proposed Method

3.1 Network Architecture

Figure 2 illustrates the architecture of the proposed network, which comprises IDB, LA, FFSR modules, and connections of picture or feature outputs.

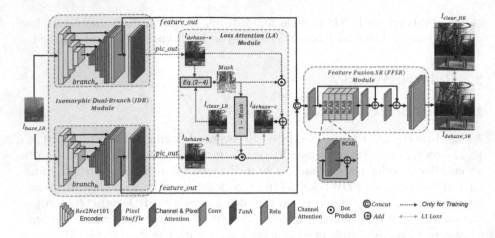

Fig. 2. Architecture of IDB dehazing and SR network.

Isomorphic Dual-Branch (IDB) Module: As shown in Fig. 2, we employed a structure of encoder-decoder as branches, which learns "easy-" and "hard-" training pixels separately. For the encoder, the first four stages of a Res2Net101 [8] pretrained on the ImageNet dataset [18] are used to introduce natural image priors for the dehazing network, and extract abundant and multiscale features. For the decoder, four pixel-shuffle [19] layers are applied for upsampling, corresponding to four times the downsampling in the encoder, each followed by a series connection of a channel attention and a pixel attention blocks. Moreover, skip connections are applied between the encoder and the decoder. Notably, the last convolutional layer and a Tanh activation are effective only in the training stage, and they are stacked behind the decoder to produce dehazed images used by the proposed LA module.

Loss Attention (LA) Module: Generally, a dual-branch network obtains better generalization ability by fusing complementary information from each other. Because the structures of the two branches in the IDB module are exactly the same, making the module learn a different knowledge is a key point. We designed an effective and efficient LA module to differentiate the dual-branch. Outputs of these two branches are defined as I_{dehaze_e} and I_{dehaze_h} as preliminary dehazed images. The average L1 loss of pixels in the LR haze-free image I_{clear_LR} (downsampled from the ground truth HR clear image I_{clear_HR}) and I_{dehaze_e} is calculated as follows:

$$LossL1 = \frac{1}{CWH} \sum_{c=1}^{C} \sum_{x=1}^{W} \sum_{y=1}^{H} |I_{clear\_LR}^{c}(x,y) - I_{dehaze\_e}^{c}(x,y)|, \qquad (1)$$

where c, x, and y denote the color channel, horizontal pixel coordinate, and vertical pixel coordinate, respectively. Further, distance (L1 Norm) between I_{dehaze_e}

and I_{clear_LR} for each pixel is computed as follows:

$$Distance(c, x, y) = |I^c_{clear\_LR}(x, y) - I^c_{dehaze\_e}(x, y)|, \tag{2}$$

where $Distance(c, x, y)$ denotes the distance between I_{clear_LR} and I_{dehaze_e} of the color channel c at pixel coordinate (x, y).

To indicate which points are easy to recover, a dynamic mask is estimated by Eq. (3). Essentially, it measures whether the pixel values are more approximate to their counterparts in the clear LR image than the average distance. Using this mask, a composite image I_{dehaze_LR} is obtained according to Eq. (4).

$$Mask = Mask(c, x, y) = \begin{cases} 1, & \text{if } Distance(c, x, y) < LossL1; \\ 0, & \text{otherwise.} \end{cases} \tag{3}$$

$$I_{dehaze\_LR} = I_{dehaze\_e} \times Mask + I_{dehaze\_h} \times (1 - Mask). \tag{4}$$

Figure 3(a) shows an example of $Mask$ and its inverse map $1 - Mask$. $Mask$ assumes a value of 1 in smooth or thinly hazed areas, and a value of 0 in edges or densely hazed areas. This is because regions with thin haze are generally easier to recover than those with dense haze. Meanwhile, edges are more vulnerable to the effect introduced by haze, making these areas commonly harder to recover than smooth regions.

I_{haze_LR} I_{clear_HR} $Mask$ $1 - Mask$ I_{haze_LR} I_{clear_HR} $Branch_e$ $Branch_h$

(a)　　　　　　　　　　　　　　　　(b)

Fig. 3. Differentiation of two branches. (a) $Mask$ distinguishing pixels that are easy to recover (white: smooth or thin haze regions) from hard ones (black: edges or dense haze regions) in training; (b) Attention heatmaps of two branches. Red indicates more attention. (Color figure online)

With the motivation of guiding two branches to learn "easy-" or "hard-" recovered regions during the training for images containing both dense and thin haze, instead of merely distinguishing edges from smooth regions, we use loss values to calculate a changing mask instead of any edge detection operators. Figure 3(b) displays pixel attention maps in the last attention layers of the decoders in two branches as heatmaps. It is observed that $branch_h$ focuses more on densely hazed regions such as bushes, grass, and parterre in the example images, while $branch_e$ focuses more on thinly hazed areas.

As shown in Fig. 3(a) and Fig. 3(b), $Mask$ has the function of indicating the learning ability of pixels at different locations in real-world images. Finally, the two branches in the IDB module is differentiated by using a constraint between I_{clear_LR} and I_{dehaze_LR} estimated by Eq. (4).

Feature Fusion SR (FFSR) Module: We designed an FFSR module to generate the final dehazed and SR reconstructed image. As shown in Fig. 2, feature output by two decoders in the IDB are concatenated and fed into a convolutional layer. Next, five residual channel attention blocks (RCABs) [25] are stacked to transfer LR features and restore image details. Finally, a pixel-shuffle layer and several convolutional layers are added to upsample LR images to HR outputs.

3.2   Loss Function

To train the proposed network, the LR dehazed image I_{dehaze_LR} produced by the LA module, and final output HR clear image I_{dehaze_SR}, are respectively constrained by L1 loss with I_{clear_LR} (downsampled haze-free image) and I_{clear_HR}, as shown in Fig. 2. The total loss is the sum of both LR and HR losses.

4   Experiment

4.1   Experimental Settings

Datasets: We conducted experiments on three most commonly used datasets NH-HAZE [1], NH-HAZE2 [2], and HD-NH-HAZE [3] for the task of non-homogeneous dehazing, which were released by the workshops of challenges of New Trends in Image Restoration and Enhancement (NTIRE) on image and video processing, in conjunction with the top conference Computer Vision and Pattern Recognition (CVPR), only for three years from 2020 to 2023, except 2022. All these datasets comprise real-world hazy and haze-free image pairs. Notably, the HD-NH-HAZE released in 2023 is a high-definition extremely large-size (6000 × 4000) image set, which can hardly be processed directly by most dehazing methods without any downsampling or cropping. In addition, because the validation and testing sets of haze-free images have not been released in 2021 and 2023, we have used the last five image pairs in the official training sets as the testing sets for a full-reference evaluation, as in most papers [6, 7, 10, 16, 23].

Training Details: To train our network, images are resized by a bicubic downsampling of two or four times in both width and height, and cropped into patches randomly with a size of 256 × 256. Random horizontal, vertical flipping and rotation is applied as the data augmentation. The batch size is set to 4. The Adam optimizer ($\beta_1 = 0.9$, $\beta_2 = 0.999$) is implemented. The learning rate was initially set to $1e^{-4}$, and reduced linearly after 1500 epochs, with a total of 6000 epochs. The NVIDIA 3090 GPU was utilized for network training.

4.2   Results and Discussion

We adopted the peak signal to noise ratio (PSNR) [21] and structural similarity (SSIM) [20] to evaluate the accuracy of dehazed outputs related to haze-free

Table 1. Quantitative comparison on the NTIRE datasets. The 1st and 2nd winners of each metric are displayed in **bold** and <u>underlined</u>, respectively. "↓" indicates the smaller the better, and "↑" indicates the larger the better.

Method		NH-HAZE				NH-HAZE2				HD-NH-HAZE				AVERAGE			
		PSNR ↑	SSIM ↑	LPIPS ↓	PI ↓	PSNR ↑	SSIM ↑	LPIPS ↓	PI ↓	PSNR ↑	SSIM ↑	LPIPS ↓	PI ↓	PSNR ↑	SSIM ↑	LPIPS ↓	PI ↓
2× downsampling	DCP '10	13.41	0.429	0.642	<u>5.002</u>	11.80	0.563	0.583	<u>4.662</u>	11.10	0.496	0.679	5.893	12.10	0.496	0.635	<u>5.186</u>
	AOD '17	15.88	0.465	0.633	5.749	15.83	0.597	0.551	5.029	14.30	0.584	0.577	6.101	15.34	0.549	0.587	5.626
	FFA '19	18.28	0.586	0.554	5.797	19.73	0.690	0.490	5.655	18.56	0.616	0.600	6.820	18.86	0.631	0.548	6.091
	TDN '20	20.39	0.596	0.476	5.407	18.29	0.706	0.473	4.920	14.28	0.562	0.640	<u>5.699</u>	17.65	0.621	0.530	5.342
	KTDN '20	19.83	0.568	0.507	5.769	19.01	0.520	0.624	6.973	18.18	0.587	0.632	7.787	19.01	0.558	0.588	6.843
	4KDehazing '21	17.48	0.471	0.681	5.982	18.00	0.599	0.616	5.421	14.32	0.531	0.667	7.132	16.60	0.534	0.655	6.178
	TNN '21	20.40	0.609	<u>0.461</u>	5.069	20.62	0.724	<u>0.403</u>	4.992	19.54	0.646	<u>0.459</u>	6.232	<u>20.19</u>	0.660	<u>0.441</u>	5.431
	DWGAN '21	19.82	<u>0.620</u>	0.498	5.259	<u>20.78</u>	<u>0.745</u>	0.405	5.038	18.75	0.638	0.494	6.142	19.78	<u>0.668</u>	0.466	5.480
	SRDefog '22	<u>20.49</u>	0.452	0.602	6.443	16.92	0.482	0.724	7.005	15.62	0.499	0.764	9.564	17.68	0.478	0.697	7.671
	SCANet '23	18.81	0.501	0.601	6.822	20.19	0.555	0.571	7.153	<u>20.24</u>	<u>0.643</u>	0.494	6.173	19.75	0.566	0.555	6.716
	Ours	**21.36**	**0.679**	**0.371**	**3.802**	**21.82**	**0.786**	**0.323**	**4.019**	**21.43**	**0.728**	**0.424**	**5.540**	**21.54**	**0.731**	**0.373**	**4.454**
4× downsampling	DCP '10	13.21	0.337	0.794	<u>7.060</u>	11.56	0.417	0.761	<u>6.999</u>	11.14	0.470	0.755	6.958	11.97	0.408	0.770	<u>7.006</u>
	AOD '17	15.60	0.362	0.787	7.583	15.36	0.441	0.756	7.331	14.15	0.543	0.675	7.837	15.04	0.449	0.739	7.584
	FFA '19	16.95	0.431	0.766	7.591	18.52	0.493	0.747	7.585	18.02	0.558	0.712	8.407	17.83	0.494	0.742	7.861
	TDN '20	18.37	0.431	0.714	7.151	17.89	0.497	0.731	7.025	16.95	0.559	0.665	<u>6.923</u>	17.74	0.496	0.703	7.033
	KTDN '20	17.73	0.392	0.755	7.472	17.55	0.361	0.908	8.694	17.33	0.529	0.726	9.195	17.54	0.427	0.796	5.646
	4KDehazing '21	17.20	0.406	0.811	7.633	17.43	0.470	0.791	7.531	14.24	0.506	0.746	8.665	16.29	0.461	0.783	7.943
	TNN '21	18.24	0.437	0.706	7.189	18.45	0.508	0.698	7.351	18.02	<u>0.568</u>	<u>0.613</u>	7.405	18.24	0.504	0.672	7.315
	DWGAN '21	18.50	<u>0.440</u>	0.718	7.402	<u>19.00</u>	<u>0.520</u>	<u>0.678</u>	7.521	17.61	0.565	0.670	8.324	<u>18.37</u>	<u>0.508</u>	<u>0.689</u>	7.749
	SRDefog '22	**20.21**	0.420	<u>0.684</u>	7.175	16.72	0.439	0.792	7.435	15.62	0.501	0.766	9.329	17.52	0.453	0.747	7.980
	SCANet '23	16.97	0.340	0.833	8.369	18.27	0.373	0.868	8.551	<u>18.77</u>	0.565	0.650	7.712	18.00	0.426	0.784	8.211
	Ours	<u>20.03</u>	**0.532**	**0.565**	**6.324**	**19.72**	**0.579**	**0.572**	**6.583**	**21.01**	**0.686**	**0.438**	**5.895**	**20.25**	**0.599**	**0.525**	**6.267**

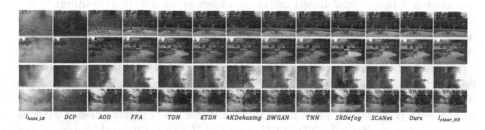

I_{haze_LR} DCP   AOD   FFA   TDN   KTDN   4KDehazing   DWGAN   TNN   SRDefog   SCANet   Ours I_{clear_HR}

Fig. 4. Dehazed images of various methods when the input hazy image is downsampled by a factor of 2 (the first two rows) and 4 (the 3rd to 4th rows).

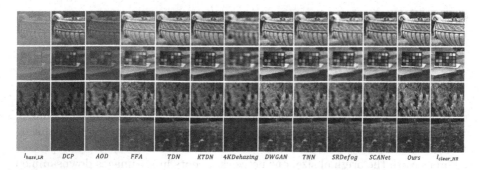

I_{haze_LR} DCP   AOD   FFA   TDN   KTDN   4KDehazing   DWGAN   TNN   SRDefog   SCANet   Ours I_{clear_HR}

Fig. 5. Local regions of dehazed images with a 2× downsampling of input images.

images, as well as the perceptual index (PI) [4] and learned perceptual image patch similarity (LPIPS) [24] as indicators of the perceptual quality.

Our method is compared with other dehazing methods, including [7, 10–14, 17, 22, 23, 27]. Table 1 lists metrics on all datasets. To evaluate the performance of dehazing LR hazy images, bicubic downsampling of 2× or 4× is applied to input hazy images. Meanwhile, for all the other methods except ours, the output dehazed images are upsampled to the original size using bicubic interpolation.

As summarized in Table 1, our method outperforms all other methods in terms of almost all metrics when the input image contains both the non-homogeneous haze and the downsampling degradation. Specifically, with 2× downsampling, our PSNR values are higher than those of the second-best ones by 0.87 dB on NH-HAZE, 1.04 dB on NH-HAZE2, and 1.19 dB on HD-NH-HAZE, respectively. Significant improvement is also observed for other indices of SSIM, LPIPS, and PI. When processing the worse degradation of 4× downsampling, our method still achieves superior performance, indicated by the best metrics on most datasets and the best average values.

Figure 4 shows a qualitative comparison of our method and others at a downsampling factor of 2 and 4 on the NTIRE datasets. To further demonstrate the advantages of our method and its capability to recover details, some local regions of dehazed images are enlarged in Fig. 5. Compared to all the other methods, our approach produces dehazed images with better quality of visual perception and higher color fidelity, which contains sharper edges, finer textures, and less noises, as shown by the sky and the pavement regions in Fig. 4, as well as the color palette and the purple streetlight in Fig. 5.

To evaluate performance on dehazing images of the original size without the degradation introduced by the downsampling, we removed the pixel shuffle layer used for upsampling in the FFSR module of our network, which outputs dehazed images with an identical size of input images. Table 2 lists metrics tested without downsampling on datasets NH-HAZE and NH-HAZE2. On both datasets, our method produces competitive results.

Table 2. Quantitative comparison on original size (without downsampling).

Dataset	Metrics	DCP '10	AOD '17	FFA '20	TDN '20	KTDN '20	4KDehazing '21	DWGAN '21	TNN '21	SRDefog '22	DeHamer '22	SCANet '23	ITBdehaze '23	Ours
NH-HAZE	PSNR ↑	12.35	14.04	18.82	21.43	20.84	17.37	**21.51**	21.44	20.99	20.66	19.52	21.44	<u>21.50</u>
	SSIM ↑	0.448	0.445	0.645	0.709	0.695	0.468	**0.711**	0.704	0.488	0.684	0.649	<u>0.710</u>	0.708
NH-HAZE2	PSNR ↑	10.57	14.52	20.90	21.24	21.32	18.28	<u>21.99</u>	21.92	16.95	19.18	21.14	21.67	**22.39**
	SSIM ↑	0.603	0.674	0.800	0.788	0.733	0.651	<u>0.856</u>	0.848	0.488	0.794	0.769	0.838	**0.857**

To validate the superiority of our model for haze removal on large-size images, we tested the HD-NH-HAZE dataset, which contains images with a resolution of 6000 × 4000. Unlike other methods [5, 9, 10, 13, 15–17, 23] that directly process the image with the original size, our network accepts input images downsampled 2× (1/4 size) and 4× (1/16 size), and performs dehazing and SR reconstruction. Notably, for testing images with such a large size, most dehazing networks require high-performance computing resources, which are unaffordable to common users, for instance, the self-paced semi-curricular attention network (SCANet) [10] testing this dataset using Nvidia A100 80G graphics card. Therefore, metrics except ours are copied from another paper [10,16] in Table 3.

As summarized in Table 3, and Fig. 6, while saving a significant number of computational resources, our method still outperforms other dehazing methods in terms of PSNR and SSIM metrics, even though other models are fed 4 (2× downsampling) or 16 (4× downsampling) times as much image information as ours. Notably, the two most advanced methods, SCANet [10] and ITBdehaze [16] (the first place of the NTRIE 2023 High Definition Non-Homogeneous Dehazing Challenge), have lower PSNR values than ours by 0.99 and 0.87dB, respectively. Moreover, although both TNN [23] and ITBdehaze use the RCAN branch to extract fine details, our method performs much better than them by using only 1/10 numbers of their RCABs.

Table 3. Quantitative comparison on the HD-NH-HAZE dataset.

Metrics	DCP '10	AOD '17	GCANet '19	GDN '19	FFA '20	TNN '21	DeHamer '22	SCANet '23	ITBdehaze '23	Ours 1/16 size	Ours 1/4 size
PSNR↑	10.98	13.75	16.36	16.85	17.85	18.19	17.61	20.44	20.56	21.01	**21.43**
SSIM↑	0.478	0.562	0.512	0.607	0.649	0.642	0.605	0.662	0.636	0.687	**0.728**

Further, we compared the computational cost, inference time, and memory consumption by testing a real-world color image with the size of $3 \times 1600 \times 1200$. As shown in Fig. 6, both of our models are located in the top left corner, indicating that compared to others, our method achieves the best performance with low computational cost and fast speed. As to GPU memory consumption, by feeding an input image with a size of 1/16 (4× downsampling), the memory is significantly saved up to 45.54% compared to the original input size (from 8589 to 4677 MB).

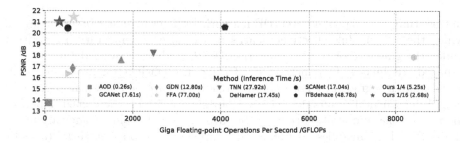

Fig. 6. Comparison of the computational efficiency and PSNR.

4.3 Ablation Study

In this section, we conducted a series of ablation experiments to demonstrate the effectiveness of each component of the network structure.

To verify the effectiveness of IDB, LA, and FFSR modules, we designed seven network models: M1: a single branch with picture output ("picture_out" in Fig. 2) directly; M2: IDB with a concatenation of "picture_out" and a convolutional layer; M3: M2 with LA module; M4: IDB and LA modules with a concatenation of feature outputs ("feature_out" in Fig. 2) and a convolutional layer; M5: M1 with feature outputs and FFSR comprising five RCABs. M6: M4 with FFSR comprising five RCABs (the proposed method); M7: M4 with FFSR comprising 15 RCABs. Table 4 lists the testing results of these models.

Table 4. Ablation experiments of each module.

Model	M1	M2	M3	M4	M5	M6	M7
PSNR ↑	20.77	20.85	20.97	21.12	21.22	21.36	21.38
SSIM ↑	0.6420	0.6419	0.6421	0.6511	0.6736	0.6789	0.6801

We further compared the dehazed outputs of models with (w/) or without (w/o) FFSR in Fig. 7, indicating the FFSR module recovers sharp edges and vibrant colors.

w/o FFSR   w/FFSR   w/o FFSR   w/FFSR   w/o FFSR   w/FFSR   w/o FFSR   w/FFSR

Fig. 7. Output images with (w/) or without (w/o) attaching FFSR module.

5 Conclusion

In this study, we propose a novel IDB network for dehazing and SR enhancing, in which the IDB module effectively learns beneficial complementary information under the guidance of the LA, and the FFSR module recovers details for the dehazed HR images. Our method is effective and efficient for processing real-world images degraded by both non-homogeneous haze and downsampling with blur. Experimental results demonstrate our superiority on the dehazing performance and the computation cost. In the future, we will explore some variants of the IDB and LA modules, as well as their applications to other visual tasks by replacing the branch backbones.

Acknowledgements. This work was financially supported by the National Natural Science Foundation of China (No. 62071201, No. U2031104), and Guangdong Basic and Applied Basic Research Foundation (No. 2022A1515010119).

References

1. Ancuti, C.O., Ancuti, C., Vasluianu, F.A., Timofte, R.: NTIRE 2020 challenge on nonhomogeneous dehazing. In: Proceedings of the IEEE Conference on Computer Vision and Pattern Recognition Workshops, NJ, pp. 490–491. IEEE (2020)
2. Ancuti, C.O., Ancuti, C., Vasluianu, F.A., Timofte, R.: NTIRE 2021 nonhomogeneous dehazing challenge report. In: Proceedings of the IEEE Conference on Computer Vision and Pattern Recognition Workshops, NJ, pp. 627–646. IEEE (2021)
3. Ancuti, C.O., et al.: NTIRE 2023 HR nonhomogeneous dehazing challenge report. In: Proceedings of the IEEE Conference on Computer Vision and Pattern Recognition Workshops, NJ, pp. 1808–1824. IEEE (2023)
4. Blau, Y., Mechrez, R., Timofte, R., Michaeli, T., Zelnik-Manor, L.: The 2018 PIRM challenge on perceptual image super-resolution. In: Leal-Taixé, L., Roth, S. (eds.) ECCV 2018. LNCS, vol. 11133, pp. 334–355. Springer, Cham (2019). https://doi.org/10.1007/978-3-030-11021-5_21
5. Chen, D., et al.: Gated context aggregation network for image dehazing and deraining. In: 2019 IEEE Winter Conference on Applications of Computer Vision, NJ, pp. 1375–1383. IEEE (2019)
6. Chen, T., Fu, J., Jiang, W., Gao, C., Liu, S.: SRKTDN: applying super resolution method to dehazing task. In: Proceedings of the IEEE Conference on Computer Vision and Pattern Recognition Workshops, NJ, pp. 487–496. IEEE (2021)
7. Fu, M., Liu, H., Yu, Y., Chen, J., Wang, K.: DW-GAN: a discrete wavelet transform GAN for nonhomogeneous dehazing. In: Proceedings of the IEEE Conference on Computer Vision and Pattern Recognition Workshops, NJ, pp. 203–212. IEEE (2021)
8. Gao, S.H., Cheng, M.M., Zhao, K., Zhang, X.Y., Yang, M.H., Torr, P.: Res2Net: a new multi-scale backbone architecture. IEEE Trans. Pattern Anal. Mach. Intell. **43**(2), 652–662 (2019)
9. Guo, C.L., Yan, Q., Anwar, S., Cong, R., Ren, W., Li, C.: Image dehazing transformer with transmission-aware 3D position embedding. In: Proceedings of the IEEE Conference on Computer Vision and Pattern Recognition, NJ, pp. 5812–5820. IEEE (2022)
10. Guo, Y., et al.: SCANet: self-paced semi-curricular attention network for nonhomogeneous image dehazing. In: Proceedings of the IEEE Conference on Computer Vision and Pattern Recognition Workshops, NJ, pp. 1884–1893. IEEE (2023)
11. He, K., Sun, J., Tang, X.: Single image haze removal using dark channel prior. IEEE Trans. Pattern Anal. Mach. Intell. **33**(12), 2341–2353 (2010)
12. Jin, Y., Yan, W., Yang, W., Tan, R.T.: Structure representation network and uncertainty feedback learning for dense non-uniform fog removal. In: Wang, L., Gall, J., Chin, T.J., Sato, I., Chellappa, R. (eds.) ACCV 2022, Part III. LNCS, vol. 13843, pp. 155–172. Springer, Cham (2023). https://doi.org/10.1007/978-3-031-26313-2_10
13. Li, B., Peng, X., Wang, Z., Xu, J., Feng, D.: AOD-Net: all-in-one dehazing network. In: Proceedings of the IEEE International Conference on Computer Vision, NJ, pp. 4770–4778. IEEE (2017)
14. Liu, J., Wu, H., Xie, Y., Qu, Y., Ma, L.: Trident dehazing network. In: Proceedings of the IEEE Conference on Computer Vision and Pattern Recognition Workshops, NJ, pp. 430–431. IEEE (2020)

15. Liu, X., Ma, Y., Shi, Z., Chen, J.: GridDehazeNet: attention-based multi-scale network for image dehazing. In: Proceedings of the IEEE International Conference on Computer Vision, NJ, pp. 7314–7323. IEEE (2019)

16. Liu, Y., Liu, H., Li, L., Wu, Z., Chen, J.: A data-centric solution to nonhomogeneous dehazing via vision transformer. In: Proceedings of the IEEE Conference on Computer Vision and Pattern Recognition Workshops, NJ, pp. 1406–1415. IEEE (2023)

17. Qin, X., Wang, Z., Bai, Y., Xie, X., Jia, H.: FFA-Net: feature fusion attention network for single image dehazing. In: Proceedings of the AAAI Conference on Artificial Intelligence, Menlo Park, vol. 34, pp. 11908–11915. AAAI (2020)

18. Russakovsky, O., et al.: ImageNet large scale visual recognition challenge. Int. J. Comput. Vis. **115**, 211–252 (2015). https://doi.org/10.1007/s11263-015-0816-y

19. Shi, W., et al.: Real-time single image and video super-resolution using an efficient sub-pixel convolutional neural network. In: Proceedings of the IEEE Conference on Computer Vision and Pattern Recognition, NJ, pp. 1874–1883. IEEE (2016)

20. Wang, Z., Bovik, A.C., Sheikh, H.R., Simoncelli, E.P.: Image quality assessment: from error visibility to structural similarity. IEEE Trans. Image Process. **13**(4), 600–612 (2004)

21. Wang, Z., Li, Q.: Information content weighting for perceptual image quality assessment. IEEE Trans. Image Process. **20**(5), 1185–1198 (2010)

22. Wu, H., Liu, J., Xie, Y., Qu, Y., Ma, L.: Knowledge transfer dehazing network for nonhomogeneous dehazing. In: Proceedings of the IEEE Conference on Computer Vision and Pattern Recognition Workshops, NJ, pp. 478–479. IEEE (2020)

23. Yu, Y., Liu, H., Fu, M., Chen, J., Wang, X., Wang, K.: A two-branch neural network for non-homogeneous dehazing via ensemble learning. In: Proceedings of the IEEE Conference on Computer Vision and Pattern Recognition Workshops, NJ, pp. 193–202. IEEE (2021)

24. Zhang, R., Isola, P., Efros, A.A., Shechtman, E., Wang, O.: The unreasonable effectiveness of deep features as a perceptual metric. In: Proceedings of the IEEE Conference on Computer Vision and Pattern Recognition, NJ, pp. 586–595. IEEE (2018)

25. Zhang, Y., Li, K., Li, K., Wang, L., Zhong, B., Fu, Y.: Image super-resolution using very deep residual channel attention networks. In: Ferrari, V., Hebert, M., Sminchisescu, C., Weiss, Y. (eds.) ECCV 2018. LNCS, vol. 11211, pp. 294–310. Springer, Cham (2018). https://doi.org/10.1007/978-3-030-01234-2_18

26. Zheng, C., Zhang, J., Hwang, J.N., Huang, B.: Double-branch dehazing network based on self-calibrated attentional convolution. Knowl.-Based Syst. **240**, 108148 (2022)

27. Zheng, Z., et al.: Ultra-high-definition image dehazing via multi-guided bilateral learning. In: 2021 IEEE Conference on Computer Vision and Pattern Recognition, NJ, pp. 16180–16189. IEEE (2021)

Hi-Stega: A Hierarchical Linguistic Steganography Framework Combining Retrieval and Generation

Huili Wang[1(✉)], Zhongliang Yang[2], Jinshuai Yang[3], Yue Gao[3], and Yongfeng Huang[3,4]

[1] Institute for Network Sciences and Cyberspace, Tsinghua University, Beijing 100084, China
whl21@mails.tsinghua.edu.cn
[2] School of Cyberspace Security, Beijing University of Posts and Telecommunications, Beijing 100876, China
yangzl@bupt.edu.cn
[3] Department of Electronic Engineering, Tsinghua University, Beijing 100084, China
[4] Zhongguancun Laboratory, Beijing 100094, China

Abstract. Due to the widespread use of social media, linguistic steganography which embeds secret message into normal text to protect the security and privacy of secret message, has been widely studied and applied. However, existing linguistic steganography methods ignore the correlation between social network texts, resulting in steganographic texts that are isolated units and prone to breakdowns in cognitive-imperceptibility. Moreover, the embedding capacity of text is also limited due to the fragmented nature of social network text. In this paper, in order to make the practical application of linguistic steganography in social network environment, we design a hierarchical linguistic steganography (**Hi-Stega**) framework. Combining the benefits of retrieval and generation steganography method, we divide the secret message into data information and control information by taking advantage of the fact that social network contexts are associative. The data information is obtained by retrieving the secret message in normal network text corpus and the control information is embedded in the process of comment or reply text generation. The experimental results demonstrate that the proposed approach achieves higher embedding payload while the imperceptibility and security can also be guaranteed. (All datasets and codes used in this paper are released at https://github.com/wanghl21/Hi-Stega.)

Keywords: Linguistic steganography · Text Generation · Information Security

1 Introduction

With the advancement of technology, social network has gradually become an integral part of people's lives and people pay increasing attention on their privacy. The technology with a long history named steganography [2] is designed

B. Luo et al. (Eds.): ICONIP 2023, LNCS 14451, pp. 41–54, 2024.
https://doi.org/10.1007/978-981-99-8073-4_4

to safeguard the privacy and security of secret message by embedding the message into normal carriers thereby making it difficult for monitors to detect the transmission of secret message.

According to the International Data Corporation (IDC) [18], the global data circle will expand to 163 ZB (1 ZB equals 1 trillion GB) by 2025, which is approximately ten times the data generated in 2016. It is worth mentioning that steganography can be employed with various forms of carriers including image [8], text [11], audio [1], video [12] and so on. The vast volume of data available on social media platforms provides numerous carrier options and scenarios for transmitting secret messages using steganography. However, affected by the transmission channel, the carriers of some media forms like image, audio, and video may be damaged during the transmission process, resulting in a decrease in the robustness of secret information. Among the various media carriers, text steganography is less affected by transmission channels, exhibiting higher levels of robustness and practicality, thus rendering it particularly suitable for social network scenarios.

The linguistic steganography technique based on text context can be categorized into three primary methods: modification-based method, retrieval-based method, and generation-based method. Modification-based methods [3,19,20] embed information by lexically modifying the text content. These methods generally have a lower embedding rate and tend to alter word frequencies, which can make them more susceptible to steganalysis techniques. Retrieval-based methods [4,21,22,30] first encode the samples in the text corpus, and then select the corresponding sentence for transmission by mapping sample with secret message. These methods need to encode and construct the text corpus, and their capacity is relatively low. However, since the text carrier itself has not changed, it is difficult for the retrieval-based method to be detected by the existing steganalysis methods. In recent years, with the development of Natural Language Processing (NLP) technology, generation-based methods [9,10,24,26,27,33] mainly embed secret messages by encoding the conditional probability distribution of each word reasonably in the process of text generation using neural networks. These methods can improve the embedding payload to some extent, but the generated text is semantically random resulting in ineffective resistance to some steganalysis methods [23,28]. The flexibility of generation-based methods, including the ability to adapt to any context and any scene, has not been fully explored and utilized.

There are two significant challenges associated with the application of linguistic steganography methods in social network scenarios. Firstly, in the social network environment, information carriers are no longer isolated semantic expression units. Instead, they are interconnected through social and cognitive relationships. For example, two parties communicate in social network as shown in Fig. 1. Alice says "Hi, long time no see!". Bob will probably reply that "Yes, how's it going?". It will not be that "Eve's experiment isn't done.". Previous text steganography algorithms paid less attention on the relationship between texts in social networks, making them inconsistent in contextual semantics, which may

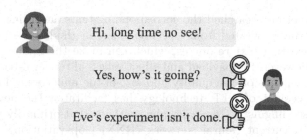

Fig. 1. Two parties communicate in social network.

bring potential security risks. Secondly, the fragmentation of social networks is serious, and we need steganography with higher embedding payload in order to embed certain secret message in short sentences and avoid the security risks associated with continuous communication.

According to the contextual relevance of social network text and the demand for high embedding payload, we propose a new **hierarchical linguistic steganography (Hi-Stega)** framework that combines the strengths of strong imperceptibility of retrieval method and large capacity of generation method. The secret message embedding process is divided into data information process layer and control information embedding layer according to different processing phases. The **data information** is the context data carrier obtained by retrieving the normal social network text (e.g. news) according to the secret message, while the **control information** is the identification signal of the secret message in the context data carrier and is embedded in the process of comment text generation. The proposed framework has the following contributions to application of linguistic steganography in social network. Firstly, it can strengthen the semantic coherence between steganographic text (stegotext) and normal text to improve cognitive imperceptibility [29]. Secondly, the data layer and the control layer are separated, and secret message cannot be decoded only by obtaining a single layer of information, which improves the security of secret message. Thirdly, experimental results demonstrate that by properly designing the language model (LM), the proposed framework has a large embedding payload while the imperceptibility and security can also be guaranteed.

2   Related Works

Steganography is the practice of hiding secret or sensitive information within an ordinary-looking file or message without arousing suspicion. Retrieval-based methodologies initially encode the samples within the textual corpus, subsequently selecting the suitable sentence for transmission by associating the samples with the concealed message through mapping [4,21,22,30]. For example, Chen et al. [4] proposed an efficient method of coverless text steganography based on the Chinese mathematical expression. By exploiting the space mapping concept, Wang et al. [21] initialized a binary search tree based on the texts from the Internet and searched the corresponding texts according to the secret binary digit string generated from a secret. These methods have the advantage of leveraging existing text resources, which can make the hidden message less conspicuous. However, the payload and effectiveness of retrieval-based steganography heavily depends on the quality and size of the underlying dataset.

The advancement of NLP technology has facilitated the development of generation-based linguistic steganography methods that primarily utilize neural networks like Recurrent Neural Network (RNN) [26], variational autoencoder (VAE) [27] and transformers [9,10,24,33] to learn the statistical language model and then employ the encoding algorithm to encode the conditional probability distribution of each word in the generation process to embed secret information. These approaches offer several advantages, including improved quality of steganographic texts and higher embedding payload. Besides, some linguistic steganography algorithms are concerned with the semantics of generating text [9,24,31]. For instance, Zhang et al. [31] introduced a generation-based linguistic steganography method that operates in the latent space, encoding secret messages into implicit attributes (semantemes) present in natural language. Yang et al. [24] involved pivoting the text between two distinct languages and embedded secret data using a strategy that incorporated semantic awareness into the information encoding process. Nevertheless, these approaches overlook the semantic relationship between sentence contexts, rendering them vulnerable to some steganalysis methods [23,28]. Consequently, this paper aims to address these limitations and proposes an improved method combined advantages both of retrieval-based and generation-based methods.

3   The Proposed Framework

As analyzed above, linguistic steganography in the social network environment should not only satisfy perceptual and statistical imperceptibility, but also make stegotext and context semantics coherent, and expand the steganographic embedding payload as much as possible. The proposed framework combines retrieval and generation of text and focuses on addressing these challenges to make linguistic steganography practical and applicable to social network scenarios.

The core idea is to retrieve the appropriate context data carrier c in the network social text corpus C according to the secret message m and then we can perform information embedding when we make comments or replies on it. Therefore, the linguistic steganography adapted to the social network can be expressed as:

$$Embed(m, c) \rightarrow s,$$

$$\text{s.t.} \begin{cases} \min D_{SD}(P_S \| P_C) \\ | \Delta D_{SC}(\cdot) | = | D_{SC}(s, c) - D_{SC}(c, c) | \leq \varepsilon (\varepsilon \rightarrow 0), \end{cases} \quad (1)$$

where $Embed(\cdot)$ is the overall description of the Hi-Stega, and s is the stegotext and also the comment to the context c. $D_{SD}(\cdot)$ is implemented to measure

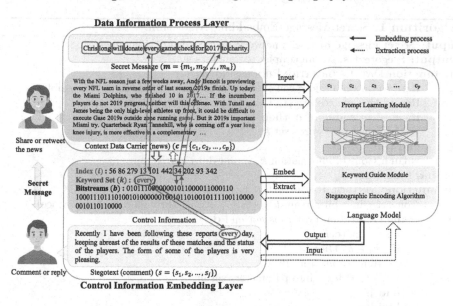

Fig. 2. The proposed Hi-Stega framework.

differences between the statistical distribution of the stegotext P_S and normal text P_C. $D_{SC}(\cdot)$ is the function that measures the degree of contextual semantic coherence. Based on an example of the steganographic process, the overall design of the proposed framework is shown in Fig. 2 and the pseudocode of the embedding process of the proposed framework is demonstrated in Algorithm 1. We divide the secret message embedding process into data information process layer and control information embedding layer. It is reasonable to treat social networks as a huge real-time updated corpus C. The main task for the upper layer is to retrieve secret message m in the social network environment to find the appropriate context data carrier m (ensure that as many words in the secret message as possible appear in the context data carrier). The secret message and the context data carrier are processed to obtain control information which includes the indexes i (the positions of the words in the secret message appear in the context data carrier) and the words k (named keyword set) that do not appear. In the lower layer, the information returned from the upper layer needs to be processed and transformed into bit stream b through a format protocol. Bit stream embedding is performed during the process of generating text with the context data carrier as model input. It is worth noting that the process of generation is guided using k to ensure that some of the words in the secret message which may not be covered by the context data carrier will appear in the generated stegotext s. At the same time, the semantics of the generated text is directed in the relevant direction and semantic coherence is enhanced.

Algorithm 1. Secret Message Embedding Algorithm.

Input: Secret message m, social network text corpus C, Language Model $LM(\cdot)$.
Output: Stegotext s, steganography parameters Φ.

 1: **function** DATAINFORMATIONPROCESSING(m, C)
 2:     Retrieve in C to find c can that cover as many words in m as possible;
 3:     Record the indexes $i = \{i_1, i_2, \ldots, i_p\}$ of these words in c;
 4:     **if** $m_i \in m$ and $m_i \notin c$ **then**
 5:         Add m_i to keywords set k;
 6:     **end if**
 7:     **return** Control information k, i, and data carrier c;
 8: **end function**
 9: **function** CONTROL INFORMATION EMBEDDING(k, i, c)
10:     Convert i into bit stream $b = \{b_1, b_2, \ldots, b_l\}$ through a format protocol;
11:     **while** Generation process is not end **do**
12:         $w_j \leftarrow LM(c, k, b)$;
13:         **if** $w_j \in k$ **then**
14:             Remove w_j from k;
15:             Add index j into parameters set Φ;
16:         **end if**
17:     **end while**
18:     **return** Stegotext $s = \{w_1, w_2, \ldots, w_s\}$ and Φ.
19: **end function**
20: $(k, i, c) \leftarrow$ DataInformationProcessing(m, C);
21: $s, \Phi \leftarrow$ ControlInformationEmbedding(k, i, c);
22: Negotiate parameters with the receiver, such as comment time means Φ, etc.

The extraction algorithm is the reverse process of the embedding algorithm, where the control information is obtained and then mapped to the context data carrier to recover the secret message. The pseudocode is demonstrated in Algorithm 2.

4   Methodology of Our Implementation

To verify the effectiveness of the proposed Hi-Stega framework, we construct a concrete model for further assessment.

4.1   Steganographic Text Generation

As shown in the right half of Fig. 2, our steganographic text is finally obtained by using a generative steganography algorithm, so in the first step we describe its overall process. Text can be regarded as a token sequence, composed of specific words according to the semantic association and syntactic rules. In Hi-Stega, the context data carrier c retrieved according to the secret message m can be regarded as the input of the language model and the chain rule is used to describe

---

Algorithm 2. Secret Message Extraction Algorithm.

Input: Context c, steganography parameters Φ, stegotext s, Language Model $LM(\cdot)$.
Output: Secret message $m = \{m_1, m_2, m_3, \ldots, m_n\}$.
  1: Input c and prompt phase to $LM(\cdot)$.
  2: **if** $\Phi == \phi$ **then**
  3:    Decode algorithm outputs embedded bits b;
  4: **else**
  5:    Recover keywords set k from c and Φ;
  6:    Decode algorithm outputs embedded bits b with k;
  7: **end if**
  8: $m \leftarrow (i) \leftarrow b$;

---

the probability distribution of word sequences:

$$p_\theta(y) = \prod_{t=1}^{|y|} p_\theta(y_t \mid (c, y_{<t})), \tag{2}$$

where the distributional probability p_θ is typically parameterized by a neural network with parameters θ, e.g., a transformer [5,14,17]. Generative steganography algorithm mainly embeds secret information by adjusting and encoding the probability distributions p_θ of candidate words.

4.2 Prompt Learning Module (PL Module)

After inputting context data carrier c into the language model, how to make the generated stegotext s semantically coherent like the normal social text is the problem that proposed Hi-Stega needs to pay attention to. We adopt the prompt learning [7] by adding some specific words as prompts after the input, which can guide LM to generate better sentence vectors and improve the consistency of contextual semantics. According to the different social scenarios, different prompt templates can be used. For example, here we exploit the correlation between news and comments for Hi-Stega, the template can be "[c] comment is [mask]". Given the data context carrier c, we can map c to c_{prompt} with the template, then generate better sentence representation.

4.3 Keyword Guide Module (KG Module)

As introduced in Algorithm 1, sometimes the secret message m cannot be completely covered in the process of retrieving the context data carrier c. At this time, it is necessary to include the uncovered secret words in the stegotext generation process. The keyword guide module is used to adjust the p_θ in Eq. 2 to make the semantics of the generated text closer to the keywords set $k = \{k_1, k_2, \ldots, k_n\}$. Here, inspired by [15], we make a simple modification to the distributional probability $p_\theta(\cdot)$ to guide generation towards keyword $k_i \in k$:

$$p_\theta'(y_t, k_i \mid y_{<t}) = p_\theta(y_t \mid y_{<t}) + \alpha \cdot \frac{1 + D_{SC}(y_t, k_i)}{2}. \tag{3}$$

Here we use the $D_{SC}(y_t, k_i) = cosine_{similarity}(glove(y_t), glove(k_i))$[1]. As explained in [15], we can control the parameters α in Eq. 3 to guarantee the appearance of specified keyword k_i in the generated sentences. For instance, we can increase α on an exponential schedule which means that as the generated sequence increases in length, so does the strength of the semantic shift, until we deterministically choose the guide keyword k_i. The parameters α at time t can be expressed:

$$\alpha_t = \begin{cases} \alpha_0 exp\{\frac{t - t_j}{T - |k| - t_j}\}, & \text{if } t < T - |k| \\ \infty, & \text{otherwise} \end{cases} \tag{4}$$

where T is the maximum length of the sentence, and previous keyword appeared at t_j.

5   Experiments and Analysis

5.1   Dataset

To simulate the real social network environment, we adopted a natural corpus including news and its comments from Yahoo! News [25]. Here we preprocessed the raw data and calculated the basic statistical information. On average news titles, bodies, and comments contain 12, 498, and 32 words respectively. The titles of news generally introduce the main information such as time, place, people, and events clearly and concisely, which have similarity with the intelligence message delivered. Therefore, we treated news titles as secret messages to be delivered in our experiments. In addition to the comments themselves, the dataset includes the number of upvotes and downvotes. It can be speculated that the comments with the maximum sum of upvotes and downvotes which can stimulate people's discussion are more in line with the comment relationship of news in the social network environment. We calculated the semantic coherence metric between the two as the baseline.

5.2   Experimental Setting

We employed the GPT-2 [17] as our pre-training language model. Based on this, comparison experiments are conducted by different fine-tuning methods as well as information embedding methods.

- *GPT*-2: Without fine-tuning GPT-2, random bit streams are embedded during the generation.
- $model_{base}^*$: GPT-2 is fine-tuned with above dataset and random bit streams are embedded during the generation.

---

[1] Glove for Word Representation in https://github.com/stanfordnlp/GloVe.

- $model_{base}$: GPT-2 is fine-tuned with above dataset with PL and random bit streams are embedded during the generation.
- Hi^*_{Stega}: GPT-2 is fine-tuned with above dataset and the secret message is embedded according to the process of Algorithm 1.
- Hi_{Stega}: GPT-2 is fine-tuned with above dataset with PL and the secret message is embedded according to the process of Algorithm 1.

Specifically, when fine-tuning the pre-trained language model, we used AdamW optimizer [13] with an initial learning rate which was 8e−4 and a linear scheduler type. The parameters α_0 was set to 5.0. The number of training epochs was set to 100.

To fully verify the effectiveness of the proposed framework, we employed three commonly used steganographic encoding algorithms including Huffman Coding (HC) [26], Arithmetic Coding (AC) [33], and Adaptive Dynamic Grouping (ADG) [32]. We tried to retrieve the news titles (2,400 pieces) in the news dataset (4,500 pieces) (news corresponding to the above titles is not included) and generated stegotext according to Algorithm 1. For statistical and perceptual imperceptibility evaluation, we calculated the average $perplexity(ppl)$, $distinct\text{-}n$ and $mauve$ [16]. For cognitive imperceptibility evaluation, we calculated two standard metrics including $\Delta(cosine)^2$ and $\Delta(simcse)^3$ [6], both of which measure the gap of the semantic coherence between the stegotext and the context data carrier and the normal news-comment semantic coherence. At the same time, we defined two types of embedding payload: practice embedding payload ER_1 which means how many bits per word are embedded on average in the text generation process and effective embedding payload ER_2 which computes the average equivalent information bits embedded per word contained in the stegotext under the Hi-Stega framework:

$$\begin{cases} ER_1 = \frac{|b|}{|s|} & \text{if } s \leftarrow Embed(b), \\ ER_2 = \frac{8 \cdot \sum_0^n |m_i|}{|s|} & \text{if } s \leftarrow Embed(m, c) \end{cases} \tag{5}$$

where a letter needs 8 bits to be represented and b is the bit stream embedded in the language model generation process.

5.3 Imperceptibility Analysis

Metrics including text fluency and diversity are recorded in Table 1, and these values reflect the perceptual and cognitive imperceptibility of the stegotext generated by different methods. Firstly, the ppl values of Hi_{Stega} is much smaller than other models, indicating that the proposed Hi_{Stega}, which only embeds control information, can generate higher quality text than the directly embedding secret information bit stream methods. Secondly, the $mauve$ value of our method is much higher than that of the comparison methods, even about 10

---

[2] It is the cosine distance of the sentence embedding using https://huggingface.co/sentence-transformers.

[3] https://github.com/princeton-nlp/SimCSE.

Table 1. Comparison of fluency and diversity coherence of steganographic texts generated by different methods.

Algorithm	Model	Fluency		Diversity		
		ppl (↓)	$mauve$(↑)	$distinct_2$	$distinct_3$	$distinct_4$
HC [26]	GPT-2	213.94	0.0135	0.925	0.976	0.984
	$model^*_{base}$	152.64	0.0712	**0.961**	**0.991**	0.996
	$model_{base}$	211.50	0.0403	0.959	0.990	**0.998**
	Hi^*_{Stega}	112.34	0.1068	0.856	0.905	0.925
	Hi_{Stega}	**109.60**	**0.1341**	0.869	0.920	0.938
AC [33]	GPT-2	251.53	0.0184	0.960	0.979	0.982
	$model^*_{base}$	279.90	0.0668	**0.967**	**0.991**	**0.998**
	$model_{base}$	290.34	0.0425	0.964	0.987	0.994
	Hi^*_{Stega}	182.44	0.1499	0.853	0.895	0.914
	Hi_{Stega}	**174.44**	**0.2051**	0.870	0.909	0.926
ADG [32]	GPT-2	395.11	0.0175	0.946	0.971	0.975
	$model^*_{base}$	287.04	0.1253	**0.966**	**0.990**	**0.995**
	$model_{base}$	259.88	0.0534	0.959	0.985	0.992
	Hi^*_{Stega}	147.70	0.0937	0.860	0.905	0.924
	Hi_{Stega}	**143.77**	**0.1413**	0.864	0.911	0.930

times that of the comparison method GPT-2, indicating that there is little difference between the distribution of stegotext and the distribution of normal text. Thirdly, the diversity of the text generated by the proposed method is slightly lower due to the KG module, if we can expand the scope of text retrieval and find a more suitable context data carrier, we can further reduce the semantic deviation and statistical bias brought by the KG module. In addition, it can be seen that the results are further improved by adding the PL module to the model, indicating that the PL module is of enhanced significance for the generation of contextual text.

The results of embedding payload and contextual semantic coherence are recorded in Table 2. While the embedding payload ER_1 of the comparison methods including GPT-2, $model^*_{base}$ and $model_{base}$ is larger, such large embedding capacity leads to semantic consistency deviation of context. And the more bits embedded in the text generation process, the worse the quality of the generated text. The embedding payload ER_1 of our method is small, but equivalent effective information embedding payload ER_2 is twice or more than that of the comparison methods. In addition, the semantic coherent metrics of Hi_{Stega} are the closest to those of the normal yahoo news dataset and the magnitude of the difference is around $10-e2$ which can be ignored almost. This indicates that although we embed secret information in our generated comment, the comment itself does not semantically draw the attention of third parties. It remains consistent with the context data carrier, ensuring strong cognitive imperceptibility.

Table 2. Comparison of embedding payload and semantic coherence of steganographic texts generated by different methods.

Algorithm	Model	Payload		Semantic Coherent	
		ER_1	ER_2	$\mid \Delta(cosine) \mid (\downarrow)$	$\mid \Delta(simcse) \mid (\downarrow)$
HC [26]	GPT-2	3.52	–	0.2266	0.3114
	$model^*_{base}$	**3.56**	–	0.2209	0.3755
	$model_{base}$	3.50	–	0.2137	0.3331
	Hi^*_{Stega}	2.15	9.16	0.0680	0.1032
	Hi_{Stega}	2.21	**9.34**	**0.0282**	**0.0321**
AC [33]	GPT-2	**6.26**	–	0.2311	0.3226
	$model^*_{base}$	5.20		0.2187	0.2699
	$model_{base}$	4.72	–	0.2100	0.3632
	Hi^*_{Stega}	2.32	9.89	0.0542	0.0889
	Hi_{Stega}	2.46	**10.42**	**0.0088**	**0.0191**
ADG [32]	GPT-2	**4.91**	–	0.2306	0.3066
	$model^*_{base}$	3.88	–	0.2213	0.3700
	$model_{base}$	3.58	–	0.2121	0.3583
	Hi^*_{Stega}	2.12	9.25	0.0593	0.0933
	Hi_{Stega}	2.18	**9.40**	**0.0275**	**0.0376**

5.4 Anti-steganalysis Ability

We utilized two of most commonly used steganalysis methods [23,28] to distinguish stegotext from normal news comment. Figure 3 records the detection accuracy and F1 score of each steganalysis model for different steganographic methods and encoding algorithms. We can find that the stegotext generated by Hi^*_{Stega} and Hi_{Stega} has a stronger ability to resist various steganalysis models when using different encoding algorithms. This shows that the stegotext generated by the proposed method is more consistent with the distribution of normal comments in the social network environment and more secure than the previous methods. In addition, we only detect the stegotext here. Actually, the secret message is scattered in the context data carrier and the generated stegotext, and it is not possible to extract the complete secret message by obtaining only the stegotext itself. Besides, it is more difficult to detect when the two are considered as a whole according to the previous detection methods.

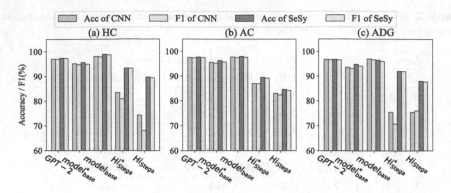

Fig. 3. Performance of different methods when using different steganalysis methods.

6   Conclusion

In order to make linguistic steganography applied in the real social network environment, we design a hierarchical steganography framework that combines retrieval and generation by exploiting the relevance of the social network context. Specific implementation results demonstrate that the proposed framework synthesizes the advantages of both steganography paradigms and enhances the contextual semantic coherence, while also extending the steganographic capacity and improving the resistance to steganalysis. We believe it can be well applied in the social network environment.

References

1. AlSabhany, A.A., Ali, A.H., Ridzuan, F., Azni, A., Mokhtar, M.R.: Digital audio steganography: systematic review, classification, and analysis of the current state of the art. Comput. Sci. Rev. **38**, 100316 (2020)
2. Anderson, R.J., Petitcolas, F.A.: On the limits of steganography. IEEE J. Sel. Areas Commun. **16**(4), 474–481 (1998)
3. Chang, C.Y., Clark, S.: Practical linguistic steganography using contextual synonym substitution and a novel vertex coding method. Comput. Linguist. **40**(2), 403–448 (2014)
4. Chen, X., Sun, H., Tobe, Y., Zhou, Z., Sun, X.: Coverless information hiding method based on the Chinese mathematical expression. In: Huang, Z., Sun, X., Luo, J., Wang, J. (eds.) ICCCS 2015. LNCS, vol. 9483, pp. 133–143. Springer, Cham (2015). https://doi.org/10.1007/978-3-319-27051-7_12
5. Devlin, J., Chang, M.W., Lee, K., Toutanova, K.: BERT: pre-training of deep bidirectional transformers for language understanding. arXiv preprint arXiv:1810.04805 (2018)
6. Gao, T., Yao, X., Chen, D.: SimCSE: simple contrastive learning of sentence embeddings. In: Proceedings of the 2021 Conference on Empirical Methods in Natural Language Processing, pp. 6894–6910 (2021)
7. Jiang, T., et al.: PromptBERT: improving BERT sentence embeddings with prompts. arXiv preprint arXiv:2201.04337 (2022)

8. Kadhim, I.J., Premaratne, P., Vial, P.J., Halloran, B.: Comprehensive survey of image steganography: techniques, evaluations, and trends in future research. Neurocomputing **335**, 299–326 (2019)
9. Kang, H., Wu, H., Zhang, X.: Generative text steganography based on LSTM network and attention mechanism with keywords. Electron. Imaging **2020**(4), 291-1 (2020)
10. Kaptchuk, G., Jois, T.M., Green, M., Rubin, A.D.: Meteor: cryptographically secure steganography for realistic distributions. In: Proceedings of the 2021 ACM SIGSAC Conference on Computer and Communications Security, pp. 1529–1548 (2021)
11. Krishnan, R.B., Thandra, P.K., Baba, M.S.: An overview of text steganography. In: 2017 Fourth International Conference on Signal Processing, Communication and Networking (ICSCN), pp. 1–6. IEEE (2017)
12. Liu, Y., Liu, S., Wang, Y., Zhao, H., Liu, S.: Video steganography: a review. Neurocomputing **335**, 238–250 (2019)
13. Loshchilov, I., Hutter, F.: Decoupled weight decay regularization. arXiv preprint arXiv:1711.05101 (2017)
14. Mikolov, T., Zweig, G.: Context dependent recurrent neural network language model. In: 2012 IEEE Spoken Language Technology Workshop (SLT), pp. 234–239. IEEE (2012)
15. Pascual, D., Egressy, B., Meister, C., Cotterell, R., Wattenhofer, R.: A plug-and-play method for controlled text generation. In: Findings of the Association for Computational Linguistics: EMNLP 2021, pp. 3973–3997 (2021)
16. Pillutla, K., et al.: MAUVE: measuring the gap between neural text and human text using divergence frontiers. In: Advances in Neural Information Processing Systems, vol. 34, pp. 4816–4828 (2021)
17. Radford, A., et al.: Language models are unsupervised multitask learners. OpenAI Blog **1**(8), 9 (2019)
18. Reinsel, D., Gantz, J., Rydning, J.: Data age 2025: the evolution of data to life-critical. Don't Focus on Big Data 2 (2017)
19. Wai, E.N.C., Khine, M.A.: Modified linguistic steganography approach by using syntax bank and digital signature. Int. J. Inf. Educ. Technol. **1**(5), 410 (2011)
20. Wang, F., Huang, L., Chen, Z., Yang, W., Miao, H.: A novel text steganography by context-based equivalent substitution. In: 2013 IEEE International Conference on Signal Processing, Communication and Computing (ICSPCC 2013), pp. 1–6. IEEE (2013)
21. Wang, K., Gao, Q.: A coverless plain text steganography based on character features. IEEE Access **7**, 95665–95676 (2019)
22. Xiang, L., Wu, W., Li, X., Yang, C.: A linguistic steganography based on word indexing compression and candidate selection. Multimedia Tools Appl. **77**(21), 28969–28989 (2018). https://doi.org/10.1007/s11042-018-6072-8
23. Yang, J., Yang, Z., Zhang, S., Tu, H., Huang, Y.: SeSy: linguistic steganalysis framework integrating semantic and syntactic features. IEEE Sig. Process. Lett. **29**, 31–35 (2021)
24. Yang, T., Wu, H., Yi, B., Feng, G., Zhang, X.: Semantic-preserving linguistic steganography by pivot translation and semantic-aware bins coding. IEEE Trans. Dependable Secure Comput. (2023)
25. Yang, Z., Xu, C., Wu, W., Li, Z.: Read, attend and comment: a deep architecture for automatic news comment generation. In: Proceedings of the 2019 Conference on Empirical Methods in Natural Language Processing and the 9th International Joint Conference on Natural Language Processing (EMNLP-IJCNLP), pp. 5077–5089 (2019)

26. Yang, Z.L., Guo, X.Q., Chen, Z.M., Huang, Y.F., Zhang, Y.J.: RNN-Stega: linguistic steganography based on recurrent neural networks. IEEE Trans. Inf. Forensics Secur. **14**(5), 1280–1295 (2018)
27. Yang, Z.L., Zhang, S.Y., Hu, Y.T., Hu, Z.W., Huang, Y.F.: VAE-Stega: linguistic steganography based on variational auto-encoder. IEEE Trans. Inf. Forensics Secur. **16**, 880–895 (2020)
28. Yang, Z., Wei, N., Sheng, J., Huang, Y., Zhang, Y.J.: TS-CNN: text steganalysis from semantic space based on convolutional neural network. arXiv preprint arXiv:1810.08136 (2018)
29. Yang, Z., Xiang, L., Zhang, S., Sun, X., Huang, Y.: Linguistic generative steganography with enhanced cognitive-imperceptibility. IEEE Sig. Process. Lett. **28**, 409–413 (2021)
30. Zhang, J., Xie, Y., Wang, L., Lin, H.: Coverless text information hiding method using the frequent words distance. In: Sun, X., Chao, H.-C., You, X., Bertino, E. (eds.) ICCCS 2017. LNCS, vol. 10602, pp. 121–132. Springer, Cham (2017). https://doi.org/10.1007/978-3-319-68505-2_11
31. Zhang, S., Yang, Z., Yang, J., Huang, Y.: Linguistic steganography: from symbolic space to semantic space. IEEE Sig. Process. Lett. **28**, 11–15 (2020)
32. Zhang, S., Yang, Z., Yang, J., Huang, Y.: Provably secure generative linguistic steganography. In: Findings of the Association for Computational Linguistics: ACL-IJCNLP 2021, pp. 3046–3055 (2021)
33. Ziegler, Z., Deng, Y., Rush, A.M.: Neural linguistic steganography. In: Proceedings of the 2019 Conference on Empirical Methods in Natural Language Processing and the 9th International Joint Conference on Natural Language Processing (EMNLP-IJCNLP), pp. 1210–1215 (2019)

Effi-Seg: Rethinking EfficientNet Architecture for Real-Time Semantic Segmentation

Tanmay Singha[✉], Duc-Son Pham, and Aneesh Krishna

School of EECMS, Curtin University, Bentley, WA 6102, Australia
tanmay.singha@postgrad.curtin.edu.au

Abstract. A popular strategy for designing a semantic segmentation model is to utilize a well-established pre-trained Deep Convolutional Neural Network (DCNN) as a feature extractor and replace the classification head with a decoder to generate segmented outputs. The advantage of this strategy is the ability to obtain a ready-made backbone with additional knowledge. However, there are several disadvantages, such as a lack of architectural knowledge, a significant semantic gap among the deep feature maps, and a lack of control over architectural changes to reduce memory overhead. To overcome these issues, we first study the complete architecture of EfficientNetV1 and EfficientNetV2, analyzing the architectural and performance gaps. Based on this analysis, we develop an efficient segmentation model called Effi-Seg by implementing several architectural changes to the backbone. This approach leads to better semantic segmentation results with improved efficiency. To enhance contextualization and achieve accurate object localization in the scene, we introduce a feature refinement module (FRM) and a semantic aggregation module (SAM) in the decoder. The complete segmentation network comprises only 1.49 million parameters and 8.4 GFLOPs. We evaluate the performance of the proposed model using three popular benchmarks, and it demonstrates highly competitive results on all three datasets while maintaining excellent efficiency.

Keywords: deep learning · deep convolutional neural networks · semantic segmentation · encoder-decoder and feature aggregation · real-time applications

1 Introduction

Scene understanding is one of the most challenging tasks in the field of computer vision. Over the past decade, several deep convolutional neural network (DCNN) models [10, 24, 25, 27] have been proposed for classifying an image that specifically contains an object in the scene, while ignoring other objects or the background. Although these models can identify the object, they lack information about the object's position and the surrounding details. Subsequently, several object

B. Luo et al. (Eds.): ICONIP 2023, LNCS 14451, pp. 55–68, 2024.
https://doi.org/10.1007/978-981-99-8073-4_5

detection models [7,8,26] have been introduced to identify the position of each object in the scene. However, they still do not capture the background details. To obtain semantic details of each object and the background of the scene, semantic segmentation tasks are needed. Semantic segmentation assigns a class to each pixel of the scene and groups them together based on the same class label. Therefore, for a better understanding of the scene, semantic segmentation is necessary.

Over the past decade, several semantic segmentation models [4,11,33] have been proposed. These models typically follow an encoder-decoder architecture [16], where an encoder is used to extract semantic feature maps from different stages, and a decoder is employed to incorporate contextual details from the extracted feature maps and assign a label to each pixel based on their softmax score. Most existing semantic segmentation models utilize a popular pre-trained classification DCNN model as the encoder for feature extraction. This approach allows the pre-trained knowledge to guide the model in the subsequent training process, enabling researchers to focus more on the design of the decoder. However, this approach has drawbacks, such as a lack of control over the architectural design of the encoder. Furthermore, all these existing classification DCNNs are large, resulting in a significant memory footprint during training and inference. Additionally, while classification models focus on contextual details for identifying objects in a scene, local information, such as boundary and texture details, is also necessary for object localization. Deploying a pre-trained backbone as the encoder in a semantic segmentation model may lead to the omission of such local information.

In recent times, efficient semantic segmentation for improved scene understanding has gained popularity in various fields, including the autonomous car industry [21,22], medical fields [3,15], civil engineering [19], and online video surveillance [9]. To achieve real-time performance, significant architectural changes are required in the overall network design. Therefore, we present an efficient real-time semantic segmentation model that maintains a balance between accuracy and efficiency, allowing it to handle high-resolution input images with a high frames per second (FPS) rate. The major contributions of this study are:

– After analyzing the architectural and performance differences between EfficientNetV1 [24] and EfficientNetV2 [25] in semantic segmentation, we introduce an efficient deep branch with reduced memory overhead.
– We also introduce a shallow branch, parallel to deep branch, to guide the deep branch by providing local information.
– We introduce a Feature Refinement Module (FRM) in the decoder for coarse-to- fine refinement.
– We also introduce a Semantic Aggregation Module (SAM) in the decoder for accurate object localization.

2   Related Work

In semantic segmentation, a revolutionary approach was introduced by FCN-8S [11], where the classification head of VGG-16 [17] was replaced by a series of convolutional layers to obtain the spatial feature map. Following this architectural revolution, several semantic segmentation models [12,16] were introduced to improve semantic representation. To enhance the receptive field of the neural network, later models such as PSPNet [33] and DeepLab [4] proposed techniques like image pooling and dilated convolutions with higher dilation rates, respectively. These models utilize large DCNN models such as ResNet-101 [27] as the network's encoder. By using large backbone and multi-scale feature scaling technique, these models achieve outstanding results on different datasets [1,6]. Inspired by their performance, several other models [5,34] employ their pre-trained backbones for feature extraction and introduce attention mechanisms to guide the global feature maps and improve semantic performance.

In recent times, transformer-based approaches have gained popularity in the field of semantic segmentation. Several transformer models, such as Segmenter [23] and TopFormer [31], have been introduced. Most of these transformer-based models are larger than the off-line DCNN models. Additionally, they are data-driven and trained on extensive datasets. Although these models achieve higher accuracy, their large network architectures make them unsuitable for real-time performance.

To achieve real-time semantic performance, the multi-branch encoder design was introduced by ICNet [32]. In this design, a dedicated deep branch is created to extract deep feature maps from a low-resolution input image, while multiple shallow branches are deployed to extract local features from the high-resolution input image. This allows for controlling the memory overhead of the model. Following this approach, several models [18,29] were introduced. Building upon the advantages of the multi-branch approach, we introduce a shallow branch parallel to the deep branch at the intermediate stage. This shallow branch shares the boundary and texture details with the deep feature maps, enhancing the model's ability to capture fine-grained information.

Additionally, there are several single-branch real-time semantic segmentation models [2,14,20,22,28] that also achieve competitive results in real-time environments. These models have relatively less deep encoders and utilize feature scaling and feature fusion modules to improve performance. Similar to this approach, we also incorporate coarse-to-fine refinement module and semantic aggregation module in our network design.

3   Proposed Method

In this section, we illustrate the end-to-end design of the proposed model. It comprises two main parts: encoder and decoder design. We focus on shallow and deep backbone at the encoder side and at the decoder side, we demonstrate feature refinement module (FRM) and semantic aggregation module (SAM). The complete network architecture can be seen in Fig. 1.

3.1 Encoder Design

The design of the proposed deep backbone is inspired by the architecture of EfficientNetV1 [24] and EfficientNetV2 [25]. The layer architectures of EfficientNet-B0 and EfficientNet-V2-S are displayed in Table 1. The "Operator" column in Table 1 defines the type of operation, kernel size, and the number of output channels. The "Layers (n)" column illustrates the repetition of each operator at a particular stage. In comparison to EfficientNet-B0, the major changes incorporated in EfficientNetV2-S are as follows: (1) the introduction of Fused-MBConv (F-MBConv) blocks at early stages to reduce training time, (2) the use of a smaller expansion ratio to reduce memory footprint, (3) the use of smaller 3×3 filter size across all layers, and (4) the removal of the last stage (8th) of the EfficientNet-B0 backbone. The reason for the last two distinctions is to reduce the number of parameters and FLOPs, which subsequently reduces the memory overhead. However, to improve the model's performance, EfficientNetV2 gradually increases the number of layers in all stages except the first one, contributing more than 18 million parameters compared to EfficientNet-B0. Moreover, both classification models incorporate MBConv blocks from MobileNetV3 [10], which have a squeeze and excitation (SE) attention mechanism in every block. This contributes a large number of parameters and FLOPs to the overall backbone design.

Considering the merits and demerits of both networks, we propose an efficient deep backbone, as shown in Table 1. The major distinctions of our proposed

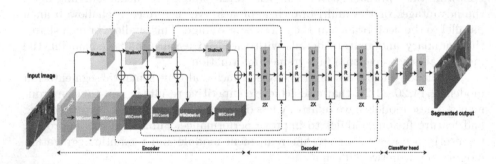

Fig. 1. Complete architecture of the proposed model

Table 1. Layered architecture of encoder

St.	EfficienNet-B0			EfficientNetV2-S			Effi-Segbackbone		
	Operators	Stride	Layers (n)	Operators	Stride	Layers (n)	Operators	Stride	Layers (n)
1	Conv, k3 × 3, 32	1	1	Conv, k3 × 3, 24	2	1	Conv, k3 × 3, 32	2	1
2	MBConv1, k3 × 3, 16	1	1	F-MBConv1, k3 × 3, 24	1	2	MBConv1, k3 × 3, 16	1	1
3	MBConv6, k3 × 3, 24	2	2	F-MBConv4, k3 × 3, 48	2	4	MBConv4, k3 × 3, 24	2	2
4	MBConv6, k5 × 5, 40	2	2	F-MBConv4, k3 × 3, 64	2	4	MBConv4, k3 × 3, 40	2	2
5	MBConv6, k3 × 3, 80	2	3	MBConv4, k3 × 3, 128	2	6	MBConv6, k3 × 3, 80	2	3
6	MBConv6, k5 × 5, 112	1	3	MBConv6, k3 × 3, 160	1	9	MBConv6, k3 × 3, 96	2	3
7	MBConv6, k5 × 5, 192	2	4	MBConv6, k3 × 3, 272	2	15	MBConv6, k3 × 3, 128	1	4
8	MBConv6, k3 × 3, 320	2	1	-	-	-	-	-	-

design are as follows: (1) like EfficientNetV2-S, we exclude the last stage of EfficientNet-B0 and use the same 3×3 kernel across all MBConv blocks, (2) we also use smaller expansion ratios (1, 4, 6) at different stages, (3) we remove the SE attention block from each MBConv block, and (4) we keep the same number of output channels as EfficientNet-B0, except for the last two stages. All these distinctions are made to develop an efficient backbone with reduced memory overhead. The impact of adding F-MBConv, higher expansion ratio, SE block, and a large number of channels at the later part of the backbone of EfficientNet is described in our ablation study.

Shallow Branch. The reason for introducing the shallow branch is to extract boundary and texture details and guide the deep branch feature maps. In a classification task, localizing an object in a scene is not required; rather, identifying the object based on contextual details is more important. Hence, classification models like EfficientNet do not need local features for object identification. However, in semantic segmentation, accurate object localization in the scene is important. Many existing semantic segmentation models lack this capability. To address this issue, we introduce a parallel shallow branch alongside our deep branch, as shown in Fig. 1. As the name suggests, the design of this branch

Table 2. Layered architecture of FRM

Layer	Input	Operator	Filter/P.size	Dilation rate	Output
ConvX1	$1 \times 1 \times c$	Conv1, BN, ReLU	1×1	-	$1 \times 1 \times c$
Pool	$h \times w \times c$	GlobalAveragePool	Pool Size (h, w)	-	$1 \times 1 \times c$
ConvX2	$1 \times 1 \times c$	Conv1, BN, ReLU	1×1	-	$1 \times 1 \times c$
Multiply ConvX2 output with the actual input					
DSConvX1	$h \times w \times c$	DSConv, BN, ReLU	3×3	6	$h \times w \times c$
Add DSConvX1 output with ConvX2 output					
DSConvX2	$h \times w \times c$	DSConv, BN, ReLU	3×3	12	$h \times w \times c$
Add DSConvX2 output with DSConvX1 output					
DSConvX3	$h \times w \times c$	DSConv, BN, ReLU	3×3	18	$h \times w \times c$
Add DSConvX3 output with DSConvX2 output and with the actual input					

Table 3. Layered architecture of SAM

Layer	Input	Operator	Filter/Pool size	Output
Concat	$h \times w \times c$ $h \times w \times c'$ $h \times w \times c'$	Concatenate	-	$h \times w \times (c + 2 \times c')$
DSConvX	$h \times w \times (c+2 \times c')$	DSConv, BN, ReLU	1×1	$h \times w \times c'$
Pool	$h \times w \times c'$	AveragePool	Pool Size (1, 1)	$h \times w \times c'$
Conv1	$h \times w \times c'$	Conv, Sig.	1×1	$h \times w \times c'$
Add Conv1 output with the output of DSConvX				

is shallow. It consists of four shallowX blocks, each down-sampling the input tensor by a factor of two. The sequential layered architecture of each shallowX block includes a ConvX block, a global average pool layer, a Conv layer, and a multiplication operation. The ConvX block filters the input using a 3×3 kernel and downsamples the input by a factor of two. The global average pool layer transforms the output of the ConvX block into a 1×1 tensor, and the point-wise Conv layer filters each pixel using a 1×1 kernel. Finally, the output of the 1×1 Conv layer is multiplied with the output of the ConvX block to restore the same spatial dimensions. Thus, the Layered architecture of each shallowX block is designed. The proposed shallow branch consists of four shallowX blocks in total (refer Fig. 1).

3.2 Decoder Design

The decoder consists of three main parts: FRM, SAM, and a classifier. However, a series of upsample and Conv layers are also used to provide the compatibility among the spatial and channel dimensions between the feature maps at different stages.

Feature Refinement Module. The motivation behind introducing this module is to refine the global feature maps at different stages. In urban street scenes, there are numerous geometric objects. Often, the presence of a large object can overshadow smaller or distant objects in the scene, affecting the model's performance. To address this issue, we propose FRM, which provides multiple fields-of-view at different scales to capture objects of different sizes. The sequential layered architecture of FRM is displayed in Table 2. It consists of two point-wise ConvX layers, one global pooling layer, and three dilated DSConvX blocks. Similar to ASPP [4], we use dilation rates of (6, 12, 18) in the dilated DSConvX blocks. Instead of using parallel branches, we utilize sequential connections, where the output of each block is fed into the input of the next block. Additionally, a feed-forward connection is introduced between the input and output of each block. This allows the input features to be refined through multiple blocks with varying receptive fields. We deploy three FRMs at the decoder side to refine the global feature maps at different stages (refer Fig. 1).

Semantic Aggregation Module. The main motivation behind introducing this module is to achieve accurate object localization in semantic segmentation. By refining the global feature maps using FRM, objects can be identified based on fine contextual details. However, for precise object localization, local feature maps are also required. Therefore, it can be observed that each SAM takes three inputs, including the local features from the shallow blocks as shown in Fig. 1. The layered architecture of SAM is shown in Table 3. Firstly, it takes three feature maps and concatenates them along the channel dimension. Then, the concatenated output is passed through a DSConvX block, which refines it using a 1×1 filter. Subsequently, an average pooling layer, a point-wise Conv layer, and an addition layer are added to enhance the feature representation. Similar to FRM, we also deploy three SAMs at different levels of the decoder network.

Classifier. The design of the classifier head is kept simple. It comprises of one ConvX block, one softmax layer, and one upsample layer. The ConvX block is utilized to filter the final feature map and assign the number of channels equal to the total number of classes in the dataset. Next, the softmax layer assigns a class to each pixel based on their scores. Furthermore, the upsample layer scales up the output to the original size of the input. To address model overfitting, a dropout layer with a dropout rate of 0.3 is also incorporated in the classifier.

4 Experiment

4.1 Datasets, Implementation Details, and Performance Metrics

To evaluate the performance of the proposed model, we utilize three public datasets: Cityscapes [6], BDD100K [30], and KITTI [1]. Cityscapes provides 5,000 finely annotated and 20,000 coarsely annotated images at a resolution of 1204×2048. BDD100K consists of 7,000 training, 1,000 validation, and 2,000 test images at a resolution of 720×1280. Compared to these two datasets, KITTI is a smaller dataset, consisting of 200 training and 200 test images at a resolution of 375×1242. Test set annotations are not available for all three datasets. Therefore, for evaluating model's performance on test set, we submitted our predictions to the respective evaluation server. All datasets follow the same class configuration and use 19 classes for evaluation.

A computer equipped with three Nvidia TITAN RTX5000 GPUs, each having 24 GB of VRAM, was used for conducting all the experiments. For parallel computing, we used `cuda` 10.2 and for effective resource utilization, `horovod` 19.5.0 is installed in the system environment. We used `tensorflow` 2.1.0 for designing and testing the proposed model. We selected a batch size of 4 and 8 in each GPU for low and high resolution input images, respectively. Following the literature [20,22], a learning rate strategy with a base learning rate 0.045 and a power of 0.9 is deployed which eventually finds the best optimized learning rate at each epoch. To address class imbalance issue, the categorical-cross entropy loss function is used and for optimizing the model loss, the stochastic gradient decent (SGD) optimizer is deployed. To address the limited size of the dataset, we used several data augmentation techniques on the fly such as random horizontal and vertical flips, adjusting brightness and contrast, random cropping, random resizing, and clipping by value. We also utilized ℓ_2 and Dropout regularization methods to address model over-fitting issue. For inference measurement, we convert our `tensorflow` model into a TRT-based model using a `TensorRT` 8.5.0 engine.

Various performance metrics, including mean Intersection over Union (mIoU), Instance IoU (iIoU) based on class and category, are used for accuracy measurement. Additionally, metrics such as parameters, FLOPs, and frames per second (FPS) are used to measure efficiency.

4.2 Ablation Study

In the first stage of our ablation study, we present an architectural and performance comparison between the proposed backbone and different versions of EfficientNet. Table 4 displays model's parameters, FLOPs, training time (t) per epoch, mIoU, and FPS at a resolution of 512 × 1024. We initially developed semantic segmentation models using the backbones of EfficientNetV1-B0 [24] and EfficientNetV2-S [25]. The corresponding results of these two models are shown in the first two rows of Table 4. The last three rows present the performance of the proposed model with a deep branch with squeeze and excitation (SE) units, with a deep branch without SE units, and with both deep and shallow branches combined without SE unit.

Table 4. Comparison of parameters, FLOPs, training time per epoch, mean Intersection over Union (mIoU), and FPS between the proposed model and different versions of EfficientNet

Model	backbone	param. (M)	FLOPs	Time (t)	val. mIoU	FPS
EfficientSegV1	E.NetV1-B0	3.5	25.2G	318 s	59.7	92
EfficientSegV2	E.NetV2-S	23.4	152G	322 s	58.2	44
Base Effi-Seg	Deep branch with SE	1.76	17.7G	152 s	58.2	155
Base Effi-Seg	Deep branch	1.36	5.57G	124 s	58.3	258
Base Effi-Seg	Deep+Shallow branch	1.41	5.69G	127 s	59.6	250

Table 5. Ablation results of the proposed model

D.branch	S.branch	FRM	SAM	Param. (M)	FLOPs	mIoU (%)	FPS
✓	✓	-	-	1.41M	5.69G	59.6	250
✓	✓	✓	-	1.46M	5.4G	63.5	242
✓	✓	✓	✓	1.49M	8.4G	65.1	188

In this study, a series of convolution and upsample layers were used to design a simple decoder. All models were trained for 500 epochs with a batch size of 4 at a 512 × 1024 input resolution using the Cityscapes [6] dataset. The results show that the model with the EfficientNetV1-B0 backbone achieves better validation mIoU (59.7%) compared to the model with the EfficientNetV2-S backbone. Due to the larger network architecture, the model with the EfficientNetV2-S backbone may require more epochs to achieve the best result. The proposed model with both deep and shallow branches produces similar results to the EfficientNetV1-B0 backbone model. We excluded the SE unit from our backbone design as the results indicate that adding the SE unit decreases model efficiency without any performance gain. The proposed backbone can process around 250 frames per second, which is 2.5 to 6 times higher than the different variants of EfficientNet. Thus, we conclude that deploying the proposed backbone in our semantic segmentation model is preferable.

In the second stage, we demonstrate the performance improvement of the proposed model by progressively adding FRM and SAM. Table 5 clearly shows that the model's performance improves with the addition of these modules, reaching a validation mIoU of 65.1% on Cityscapes after 500 epochs. Thus, we complete the design of the entire pipeline. Once the performance of the entire pipeline is examined, we fine-tuned the model and trained it for a large number of epochs to achieve better results. The final outcome is presented in the model evaluation section.

Fig. 2. Demonstration of coarse-to-fine feature refinement using heat map. Intermediate weights of the model are used to generate these feature maps. The sequential arrangement is: (a) RGB input, (b) corresponding feature map after the encoder, (c) After FRM, (d) After SAM.

To visually demonstrate the improvement in model performance, we display the feature maps generated from the intermediate stages of the model in Fig. 2. Brighter and hotter pixels in the heat map represent pixels with similar characteristics. From Fig. 2, it is evident that the input coarse feature map undergoes better refinement with the addition of each module. After the ablation study, we fine-tune our model and train it for a greater number of epochs to achieve the best accuracy. The final results of the proposed model are discussed next.

4.3 Model Evaluation

Model Performance on Cityscapes. The model's performance on the Cityscapes test and validation sets is displayed in Table 6. We only use Cityscapes fine-tune training set to train the proposed model. However, most of the existing models are pre-trained by other datasets. Among the offline models, DeepLabV3+ [4] achieves state-of-the-art performance with an mIoU of 82.1% on the Cityscapes test set. The transformer model, Segmenter [23], reports only validation performance of 81.3% mIoU, which is 1.8% higher than DeepLabV3+, despite having seven times more parameters.

In the real-time category, we have listed models with fewer than 8 million parameters. Among them, ICNet [32] achieves the highest test mIoU (69.5%). However, the model's efficiency is low, with a frame rate of 30.5 FPS, primarily due to its 6.7 million parameters and 28.3 GFLOPs. Another transformer-based

Table 6. Model performance on Cityscapes validation and test sets

Type	Model	Parameters (Million)	FLOPs	Val. Class mIoU (%)	Val. Category mIoU (%)	Test Class mIoU (%)	Test Class iIoU (%)	Test Category mIoU (%)	Test Category iIoU (%)	FPS
Offline	PSPNet [33]	250.8	412.2G	-	-	81.2	59.6	91.2	79.2	0.78
	HANet [5]	65.4	2138.0G	80.3	-	80.9	58.6	91.2	79.5	-
	DeepLabV3+ [4]	43.0	1550.0G	79.6	-	82.1	62.4	92.0	81.9	0.25
	Segmenter [23]	>86.0	-	81.3	-	-	-	-	-	-
Real-time	ESPNet [22]	7.6	6.5G	60.8	-	-	-	-	-	-
	ICNet [32]	6.7	28.3G	-	-	69.5	-	-	-	30.5
	TopFormer-B [31]	5.9	53.6G	65.5	-	64.9	35.9	83.2	63.4	-
	BiseNet [29]	5.8	-	69.0	-	68.4	-	-	-	105.8
	ThunderNet [28]	4.7	-	64.77	-	64.0	40.4	84.1	69.3	96.2
	SCMNet [18]	1.2	38.3G	66.5	84.2	67.9	37.1	86.8	68.0	117
	FANet [20]	1.1	11.4G	65.9	83.6	64.1	33.2	83.1	61.1	253
	ENet [14]	0.4	3.8G	-	-	58.3	34.4	80.4	64	21
	QNet [2]	0.2	19.8G	-	-	49.2	-	70.1	-	18.2
	Effi-Seg	**1.49**	**8.4G**	**68.2**	**83.8**	**69.3**	**40.5**	**86.9**	**69.6**	**188**

The '-' sign means the absence of results either in the literature or on the public leaderboard.

Table 7. Performance evaluation on the BDD100K validation set

Type	Model	Input size	Parameters (Million)	GFLOPs	Class mIoU (%)	FPS
Off-line	HANet (MobileNetV2) [5]	608 × 608	14.8	142.7	58.9	-
Real-time	FANet [20]	768 × 1280	1.1	11.4	50.0	-
	SCMNet [18]	768 × 1280	1.2	38.9	51.2	-
	Effi-Seg	**768 × 1280**	**1.49**	**15.7**	**54.9**	103

The '-' sign means the absence of results in the literature.

model called TopFormer [31] achieved a test accuracy of 64.9% while having 5.9 million parameters. In comparison, our proposed model, Effi-Seg, achieves validation and test mIoU of 68.2% and 69.3% respectively on Cityscapes, with only 1.49 million parameters. Furthermore, it can process 188 frames per second at an input resolution of 512 × 1024, while requiring significantly fewer FLOPs (8.4G). Accordingly, our proposed model strikes a competitive trade-off between accuracy and efficiency, especially when handling large images.

Model Performance on BDD100K. The proposed model's performance on the BDD100K [30] dataset is shown in Table 7. Since a BDD100K test server was unavailable, we evaluated the model on the validation set only. Few models, including one offline model, have been evaluated on BDD100K, likely due to the dataset's challenging nature. The HANet [5] model achieves state-of-the-art performance (58.9%), but its MobileNet version has much more parameters (14.8 million) and higher GFLOPs (142.7) compared to our proposed model. In comparison to real-time models, our proposed model achieves the best validation mIoU of 54.9% and operates at 103 FPS at the resolution 768 × 1280.

Table 8. Model performance on KITTI test set

Type	Model	Dataset used for pre-training	Class mIoU (%)	ClassiIoU (%)	Cat. mIoU (%)	Cat. iIoU (%)	FPS
Offline	VideoProp. [34]	Mapilary, Ciy., synthetic dataset	72.8	48.7	88.9	75.3	-
	SGDepth (Seg.)	Cityscapes	53.0	24.4	78.7	55.9	-
	SDNet [13]	Cityscapes	51.1	17.7	79.6	50.5	-
Real-time	**Effi-Seg**	Cityscapes	**56.2**	**22.5**	**79.9**	49.9	201

The '-' sign means the absence of results in the literature.

Model Performance on KITTI. The KITTI [1] test result is obtained from the online evaluation server. To the best of our knowledge, most of the models on the KITTI leaderboard are offline models. In Table 8, we present three offline models, among which VideoProp [34] achieved state-of-the-art performance (72.8%) by employing a joint propagation strategy, a large DeepLab3+ [4] semantic architecture, and utilizing a large synthetic dataset as well as other datasets. Our proposed model, Effi-Seg, achieves a test mIoU of 56.2%, surpassing the performance of many offline models. It is capable of processing 201 KITTI frames per second at a resolution of 384 × 1280. As the input resolution varies across the three datasets, the model's frames per second (FPS) also change accordingly.

Qualitative Performance Analysis. For visualization, we also present the model's predictions using the samples from different datasets. Figure 3 and Fig. 4 display the model's predictions using Cityscapes [6] and BDD100K [30] samples. It can be observed that the model accurately identifies and localizes small objects such as poles, traffic lights, and motorcycles in the scene. Figure 5 shows the model's prediction using a KITTI [1] test sample. The error image in Fig. 5(c) is generated by the KITTI server to display the correct and incorrect pixel classifications made by the proposed model. In the error image, red pixels indicate incorrect classifications, while green pixels represent correct labeling by the proposed model. These three figures clearly demonstrate the model's qualitative performance in the field of semantic segmentation.

Fig. 3. Prediction using Cityscapes (a) validation sample, (b) test sample (Color figure online)

Fig. 4. Prediction using BDD100K (a) validation sample, (b) test sample (Color figure online)

Fig. 5. Prediction using KITTI test sample. (a) RGB image, (b) model prediction, (c) error image (Color figure online)

5 Conclusion

Based on an efficient backbone, this study introduces an end-to-end real-time semantic segmentation model for resource-constrained embedded devices. The proposed model, Effi-Seg, achieves competitive results on the Cityscapes dataset and demonstrates state-of-the-art performance on the BDD100K and KITTI datasets. It is worth noting that the model's predictions exhibit slightly squarish boundaries for each object in the scene. This phenomenon may be attributed to the four times upsampling at the end and the lack of detailed boundary analysis. Therefore, in future research, we plan to investigate the boundary details of each object in the scene and incorporate this additional knowledge into the training process of the model. For reproducibility, our official implementation is available at https://github.com/tanmaysingha/Effi-Seg.

References

1. Abu Alhaija, H., Mustikovela, S.K., Mescheder, L., Geiger, A., Rother, C.: Augmented reality meets computer vision: efficient data generation for urban driving scenes. Int. J. Comput. Vis. **126**(9), 961–972 (2018). https://doi.org/10.1007/s11263-018-1070-x
2. Cai, J., Liu, Y., Qin, P.: Attention based quick network with optical flow estimation for semantic segmentation. IEEE Access **11**, 12402–12413 (2023)
3. Cai, W., Wang, B.: DSE-Net: deep semantic enhanced network for mobile tongue image segmentation. In: Tanveer, M., Agarwal, S., Ozawa, S., Ekbal, A., Jatowt, A. (eds.) ICONIP 2022. CCIS, vol. 1794, pp. 138–150. Springer, Singapore (2023). https://doi.org/10.1007/978-981-99-1648-1_12

4. Chen, L.-C., Zhu, Y., Papandreou, G., Schroff, F., Adam, H.: Encoder-decoder with atrous separable convolution for semantic image segmentation. In: Ferrari, V., Hebert, M., Sminchisescu, C., Weiss, Y. (eds.) ECCV 2018. LNCS, vol. 11211, pp. 833–851. Springer, Cham (2018). https://doi.org/10.1007/978-3-030-01234-2_49

5. Choi, S., Kim, J.T., Choo, J.: Cars can't fly up in the sky: improving urban-scene segmentation via height-driven attention networks. In: Proceedings of the CVPR, pp. 9373–9383 (2020)

6. Cordts, M., et al.: The cityscapes dataset for semantic urban scene understanding. In: Proceedings of the CVPR (2016)

7. Du, J.: Understanding of object detection based on CNN family and YOLO. In: Journal of Physics: Conference Series, vol. 1004, p. 012029. IOP Publishing (2018)

8. Girshick, R., Donahue, J., Darrell, T., Malik, J.: Rich feature hierarchies for accurate object detection and semantic segmentation. In: Proceedings of the CVPR, pp. 580–587 (2014)

9. Gruosso, M., Capece, N., Erra, U.: Human segmentation in surveillance video with deep learning. Multimedia Tools Appl. **80**, 1175–1199 (2021). https://doi.org/10.1007/s11042-020-09425-0

10. Howard, A., et al.: Searching for MobileNetV3. In: Proceedings of the ICCV, pp. 1314–1324 (2019)

11. Long, J., Shelhamer, E., Darrell, T.: Fully convolutional networks for semantic segmentation. In: Proceedings of the CVPR, pp. 3431–3440 (2015)

12. Noh, H., Hong, S., Han, B.: Learning deconvolution network for semantic segmentation. In: Proceedings of the ICCV, pp. 1520–1528 (2015)

13. Ochs, M., Kretz, A., Mester, R.: SDNet: semantically guided depth estimation network. In: Fink, G.A., Frintrop, S., Jiang, X. (eds.) DAGM GCPR 2019. LNCS, vol. 11824, pp. 288–302. Springer, Cham (2019). https://doi.org/10.1007/978-3-030-33676-9_20

14. Paszke, A., Chaurasia, A., Kim, S., Culurciello, E.: ENet: a deep neural network architecture for real-time semantic segmentation. arXiv preprint arXiv:1606.02147 (2016)

15. Progga, P.H., Shatabda, S.: iResSENet: an accurate convolutional neural network for retinal blood vessel segmentation. In: Tanveer, M., Agarwal, S., Ozawa, S., Ekbal, A., Jatowt, A. (eds.) ICONIP 2022. LNCS, vol. 13625, pp. 567–578. Springer, Cham (2023). https://doi.org/10.1007/978-3-031-30111-7_48

16. Ronneberger, O., Fischer, P., Brox, T.: U-Net: convolutional networks for biomedical image segmentation. In: Navab, N., Hornegger, J., Wells, W.M., Frangi, A.F. (eds.) MICCAI 2015. LNCS, vol. 9351, pp. 234–241. Springer, Cham (2015). https://doi.org/10.1007/978-3-319-24574-4_28

17. Simonyan, K., Zisserman, A.: Very deep convolutional networks for large-scale image recognition. arXiv preprint arXiv:1409.1556 (2014)

18. Singha, T., Bergemann, M., Pham, D.S., Krishna, A.: SCMNet: shared context mining network for real-time semantic segmentation. In: Proceedings of the DICTA, pp. 1–8. IEEE (2021)

19. Singha, T., Bergemann, M., Pham, D.S., Krishna, A.: SC-CrackSeg: a real-time shared feature pyramid network for crack detection and segmentation. In: Proceedings of the DICTA, pp. 1–8 (2022)

20. Singha, T., Pham, D.S., Krishna, A.: FANet: feature aggregation network for semantic segmentation. In: Proceedings of the DICTA, pp. 1–8. IEEE (2020)

21. Singha, T., Pham, D.S., Krishna, A.: A real-time semantic segmentation model using iteratively shared features in multiple sub-encoders. Pattern Recogn. **140**, 109557 (2023)

22. Singha, T., Pham, D.-S., Krishna, A., Dunstan, J.: Efficient segmentation pyramid network. In: Yang, H., Pasupa, K., Leung, A.C.-S., Kwok, J.T., Chan, J.H., King, I. (eds.) ICONIP 2020. CCIS, vol. 1332, pp. 386–393. Springer, Cham (2020). https://doi.org/10.1007/978-3-030-63820-7_44
23. Strudel, R., Garcia, R., Laptev, I., Schmid, C.: Segmenter: transformer for semantic segmentation. In: Proceedings of the CVPR, pp. 7262–7272 (2021)
24. Tan, M., Le, Q.: EfficientNet: Rethinking model scaling for convolutional neural networks. In: Proceedings of the ICML, pp. 6105–6114. PMLR (2019)
25. Tan, M., Le, Q.: EfficientNetV2: smaller models and faster training. In: Proceedings of the ICML, pp. 10096–10106. PMLR (2021)
26. Tan, M., Pang, R., Le, Q.V.: EfficientDet: scalable and efficient object detection. In: Proceedings of the CVPR, pp. 10781–10790 (2020)
27. Targ, S., Almeida, D., Lyman, K.: ResNet in ResNet: generalizing residual architectures. arXiv preprint arXiv:1603.08029 (2016)
28. Xiang, W., Mao, H., Athitsos, V.: ThunderNet: a turbo unified network for real-time semantic segmentation. In: Proceedings of the WACV, pp. 1789–1796. IEEE (2019)
29. Yu, C., Wang, J., Peng, C., Gao, C., Yu, G., Sang, N.: BiSeNet: bilateral segmentation network for real-time semantic segmentation. In: Ferrari, V., Hebert, M., Sminchisescu, C., Weiss, Y. (eds.) ECCV 2018. LNCS, vol. 11217, pp. 334–349. Springer, Cham (2018). https://doi.org/10.1007/978-3-030-01261-8_20
30. Yu, F., et al.: BDD100K: a diverse driving dataset for heterogeneous multitask learning. In: Proceedings of the CVPR, pp. 2636–2645 (2020)
31. Zhang, W., et al.: TopFormer: token pyramid transformer for mobile semantic segmentation. In: Proceedings of the CVPR, pp. 12083–12093 (2022)
32. Zhao, H., Qi, X., Shen, X., Shi, J., Jia, J.: ICNet for real-time semantic segmentation on high-resolution images. In: Ferrari, V., Hebert, M., Sminchisescu, C., Weiss, Y. (eds.) ECCV 2018. LNCS, vol. 11207, pp. 418–434. Springer, Cham (2018). https://doi.org/10.1007/978-3-030-01219-9_25
33. Zhao, H., Shi, J., Qi, X., Wang, X., Jia, J.: Pyramid scene parsing network. In: Proceedings of the CVPR, pp. 2881–2890 (2017)
34. Zhu, Y., et al.: Improving semantic segmentation via video propagation and label relaxation. In: Proceedings of the CVPR, pp. 8856–8865 (2019)

Quantum Autoencoder Frameworks
for Network Anomaly Detection

Moe Hdaib[✉], Sutharshan Rajasegarar, and Lei Pan

School of IT, Deakin University, Geelong, Australia
{mhdaib,srajas,l.pan}@deakin.edu.au

Abstract. Detecting anomalous activities in network traffic is important for the timely identification of emerging cyber-attacks. Accurate
analysis of the emerging patterns in the network traffic is critical to
identify suspicious behaviors. This paper proposes novel quantum deep
autoencoder-based anomaly detection frameworks for accurately detecting the security attacks that emerge in the network. In particular, we
propose three frameworks, one by constructing several reconstruction
error thresholds-based methods; second, a union of a quantum autoencoder and a one-class support vector machine-based method; and third
a union of a quantum autoencoder and quantum random forest-based
method. The quantum frameworks' effectiveness in accurately detecting
the attacks is evaluated using a publicly available benchmark dataset.
Our empirical evaluations demonstrate the improvements in accuracy
and F1-score for the three frameworks.

Keywords: Quantum Machine Learning · Quantum Autoencoder ·
Quantum Anomaly Detection · Network Anomaly Detection

1 Introduction

Quantum computing plays a significant role in machine learning (ML) in two ways.
The first is to enhance frequently utilized ML techniques; the other utilizes quantum computing techniques when the classical model cannot sufficiently capture the
correlation between the features. Quantum machine learning is based on elements
of quantum physics and a fast-developing field of research and technology.

Due to the rapid improvements in the field, considerable research has been
done in Quantum Machine Learning (QML) over the years. QML uses quantum devices to boost the processing speed of traditional machine learning algorithms. Applications using QML might consequently perform substantially better. According to Trugenberger [19], QML is especially useful for pattern recognition tasks involving data fitting and classification, where quantum techniques
may reduce the data dimensionality and data volume required to train the system. Additionally, it describes potential changes to the current classification and
grouping methods. Quantum state data may be preprocessed in QML by employing amplitude encoding. Using this encoding approach, we can convert the data

© The Author(s), under exclusive license to Springer Nature Singapore Pte Ltd. 2024
B. Luo et al. (Eds.): ICONIP 2023, LNCS 14451, pp. 69–82, 2024.
https://doi.org/10.1007/978-981-99-8073-4_6

in classical machine learning to a form suitable for quantum devices. Moreover, QML can identify new complex correlations between the data overlooked by its classical counterparts.

In QML anomaly detection, quantum algorithms outperform traditional computer approaches, according to Kyriienko and Magnusson [7]. If there is a vast majority of input data collected, some of which originates from questionable or ambiguous sources, it is important to automatically identify the outliers. This is particularly crucial when there are insufficient examples of outliers to base an expectation on. These anomalies might signal the emergence of an unanticipated phenomenon in the network, such as a broken system or a hostile assault. Anomaly detection aims to learn how to precisely identify outliers from training data.

To tackle the anomaly detection problem, we explore the potential and viability of merging a parameterized quantum circuit with a deep neural network. We propose hybrid deep learning- and quantum-based models to achieve effective anomaly detection performance. This paper's contributions are listed as follows:

- We investigate current implementation of quantum machine learning for cyber-security anomaly detection using network traffic information.
- We develop three novel approaches for anomaly detection using quantum deep learning with autoencoders.
- We conduct a systematic study of the findings and evaluations on a benchmark dataset.

The rest of this paper is organized as follows: Sect. 2 presents the related work. Section 3 introduces the fundamentals of autoencoders and quantum autoencoders. Section 4 proposes three frameworks. Section 5 details the evaluation, and the paper is concluded in Sect. 6.

2   Related Work

Anomaly detection helps identify unexpected behavior that differs from customary, anticipated behavior. The two underlying presumptions are that the anomalous points are rare and that they differ significantly from the typical (normal) points. Deep learning algorithms become a popular trend in anomaly detection. Quantum Deep-learning methods help handle higher dimensional data and detect sophisticated anomalies. Current research on quantum deep learning is relatively low. However, some existing works have important implications even though they do not address anomaly detection particularly; their work forms a significant building block in the quantum field.

A quantum machine learning strategy was proposed in [10] to address a genuine industrial issue. A quantum neural network (QNN) architecture consists of a classifier with an autoencoder used for achieving quality control. The QNN classifier is used to estimate the quality of the industrial samples, the QNN autoencoder reduces dimensionality and extracts features. An appropriate encoding strategy is used to translate classical data onto quantum states, and then a parameterized quantum circuit is made to learn a lower-dimensional representation of the input data, which is subsequently used for feature extraction.

A novel QNN technique for data reconstruction was proposed in [21]. A QNN-based technique reconstructs missing or erroneous data with three components, 1) data encoding that translates the input data onto a quantum state, 2) a quantum neural network that describes the link between the available data and the missing or corrupted values, and 3) a parameterized quantum circuit.

Quantum autoencoders were used to model Hamiltonian dynamics in [12]. It uses the physical system's underlying structure to decrease the computer resources needed for the operation. Hamiltonian simulation is a key challenge in quantum computation because the exponential development of the system's Hilbert space makes it difficult to computationally simulate the temporal evolution of quantum systems regulated by Hamiltonians. Data compression and dimensionality reduction are the goals of a category of quantum machine learning algorithms known as quantum autoencoders [16]. Quantum autoencoders compress the quantum state before a Hamiltonian simulator decompresses the state using the quantum encoders' inverse.

There are little to no implementations of deep quantum learning in anomaly detection. However, some potential classical deep learning for anomaly detection methods includes Variational AutoEncoders (VAE) [1,14]. Based on Support Vector Data Description (SVDD), which encircles normal points on a hypersphere in a replicating Hilbert space, the deep support vector data description [17] describes the data in detail. Last but not least, the LAKE approach in [9] is invented by combining VAE and kernel density estimate.

Based on the survey above, we observed that no works have been observed on quantum circuit based anomaly detection on traffic data. In this paper, we propose three frameworks based on quantum circuits to handle the challenges of detecting anomalies in network traffic.

3   Autoencoders and Quantum Autoencoders

3.1   Autoencoders

Autoencoders are neural networks taught to reconstruct their input [11]. Classical autoencoders are neural networks for efficiently training low-dimensional representations of data in higher-dimensional space. An autoencoder takes an input x and map it to a lower-dimensional point y so that x may probably be recovered from y. By changing the layout of the underlying autoencoder network to represent the data in a lower dimension, the input can be effectively compressed which is represented in Fig. 1a.

An autoencoder comprises an Encoder and a Decoder. Encoder and Decoder are neural networks in the most common type of autoencoders [15]. A linear autoencoder [2] has linear Encoding and Decoding components. The autoencoder could accomplish the same latent representation as Principal Component Analysis (PCA) in the case of a linear autoencoder when the non-linear operations are eliminated [13].

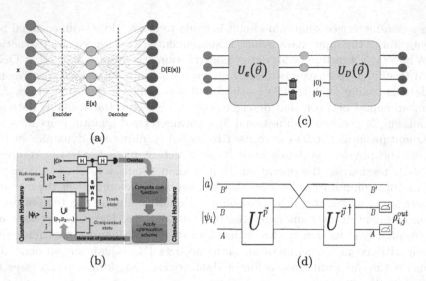

Fig. 1. (a) Autoencoder (b) Quantum autoencoder (QAE) [16] (c) QAE circuit (d) QAE architecture [16]

The Variational Autoencoders (VAE) model has significantly enhanced the autoencoders' representational capabilities [5]. VAE are generative models that seek to characterize data production using a probabilistic distribution, following Variational Bayes (VB) Inference [3]. VAEs are defined as autoencoders whose training is controlled to avoid overfitting. VAE's latent space possesses beneficial properties for generative processes. A decoder and an encoder are combined in their architecture, and their training aims to decrease the reconstruction error between the initial data and the encoded-decoded data. We encode an input as a distribution over the latent space rather than a single point to modify the encoding-decoding process somewhat and add some regularisation of the latent space.

3.2   Quantum Autoencoders

A quantum autoencoder (QAE) is a paradigm that allows quantum systems to do machine learning tasks without exponentially costly conventional memory. QAE compresses a dataset of quantum states when employing a conventional compression method is not feasible. The QAE's parameters are learned using traditional optimization methods.

The QAE is implemented analogous to classical autoencoders, but the data and operations are quantum mechanical. The basic idea is to encode an input state $|\psi\rangle$ to a lower-dimensional quantum state $|\phi\rangle$, and then decode $|\phi\rangle$ back to $|\psi\rangle$. This process has the potential to compress data for our anomaly detection application significantly. Our QAE uses a quantum circuit parameterized by a set of parameters where the aim is to find the optimal set of parameters that minimizes the difference between the input state and the reconstructed state. The encoding process is described as $U(\theta)\,|\psi\rangle = |\phi\rangle$ where $|\psi\rangle$ is the input state, $U(\theta)$ is the unitary transformation representing the quantum circuit parameterized by θ, and $|\phi\rangle$ is the encoded (compressed) state. The decoding process is described as $U'(\theta)\,|\phi\rangle = |\psi'\rangle$ where $|\phi\rangle$ is the encoded state, $U'(\theta)$ is the decoding unitary transformation, and $|\psi'\rangle$ is the reconstructed state. The input might be efficiently compressed by altering the topology of the underlying autoencoder network to represent the data on a smaller dimension [16].

However, such a scenario has no data compression since unitary transformations are probability-conserving and act on spaces with similar dimensions. Some qubits at the initial encoding stage are rejected and replaced by freshly made reference states to achieve data compression. Such a configuration is suitable for a two-dimensional latent space and a four-feature input (Fig. 1c). The unitary operators produce the same amount of qubits, but during the encoding process, two of their outputs are swapped out for newly created reference states that do not know the input states. The description of QAE for data compression from [16] describes the major source for the following explanation of the fundamentals of quantum autoencoding. The quantum anomaly detection of simulated quantum states was studied in [6].

This paper differs significantly from previous research in that our input states are fundamentally quantum mechanical. However, in our case, the input embedding of the classical data determines the type of quantum state. The qubits that make up the quantum states may be divided during angular encoding. Therefore, to entangle the different qubits, we use CNOT gates throughout the unitary evolution.

A well-known unsupervised challenge is the identification of anomalies. More specifically, it aims to train a normal profile only using the normal examples before recognizing the samples that do not fit the normal profile as anomalies. In this paper, we hypothesize that a trained autoencoder will learn the latent subspace of normal samples. We expect that the trained autoencoder will produce low reconstruction error for normal data and high reconstruction error for anomalies. It represents the design principle of our first quantum anomaly detection framework.

Note that the QAE is a combination of classical and quantum methods. After the input state has been prepared, the parameterized unitary is used to compress the state. A SWAP test is used to determine how much of the reference state and the trashed state by compression overlap. We use a classical optimization approach to construct the cost function.

The optimal compressed state is determined via a SWAP gate between the trashed state and the reference state. These states usually have fewer qubits, resulting in a straightforward comparison of fewer gates. Maximizing the fidelity between the trash state and the reference state is necessary to determine the optimal compression for our input circuit. We adjust the parameters of our encoder before conducting a SWAP test to evaluate the accuracy between these trash and reference states during training. It requires the addition of an additional qubit, which our auxiliary qubit will be used to measure the overall fidelity of the trash and reference states during the SWAP test.

The model is formulated using a variational quantum circuit technique. Hence, the optimal settings of the quantum gates can be derived gradually through learning. Such gates can only be achieved by unitaries with $2^n \times 2^n$ dimensions, resulting in an n-qubit unitary gate. However, it results in exponentially more parameters relative to the number of qubits, rendering an unsolvable optimization process. Therefore, we leverage the programmable circuit strategy [16] to decode the big unitary into single qubit rotation gates and CNOTs.

Once the programmable circuit is chosen, the architecture's structural foundation is established, and the defined model will function as an encoder. The decoder of a quantum autoencoder can be the inverse of the encoder, as opposed to the traditional autoencoders whose decoder is learned from scratch. It is feasible for quantum autoencoders because unitary matrices can be effectively inverted. Because the entire encoder will produce a unitary, if we describe the encoder network as $U^{\vec{p}}$ where $\vec{p}$ denotes the optimum network parameters, the decoder network will be defined as $(U^{\vec{p}})'$.

The method aims to split a quantum system into two subsystems, denoted by A and B, after applying the encoder to the system. The encoder creates a "latent code" in subsystem A, which may subsequently be used to reconstruct the input. As opposed to subsystem B, where the objective is to create a "reference state" that is optimal for all conceivable data occurrences. The same reference qubits may be added to the latent space to recreate the output if the encoding is completed with a considerable degree of precision since the model can design one subsystem that produces identical outcomes for each given data instance. For the sake of simplicity, the reference state in our design is chosen to be $|0\rangle$, whose qubit count may vary depending on how large the latent space is. As a result, the required functionality following the encoder is for subsystem A to consist of the latent code and subsystem B to produce the state $|0\rangle$ for any provided input values.

Applying a series of SWAP gates between the subsystems B and B', which are made up of the reference state ansatz, is one technique to achieve this result. It indicates that the network can produce the input using the fixed reference state switched into subsystem B if it can establish a latent space. Figure 1c depicts the model's overall view. The model's last component is to develop a loss function for the variational circuit's training. The $L2$ norm of the input and the output, which is the conventional loss function for autoencoders, can be converted into QAE as follows:

$$C_1\left(\vec{P}\right) = \sum_i p_i.F\left(|\psi_i\rangle, \rho_{i,\vec{p}}^{out}\right) \tag{1}$$

$$F\left(|\psi_i\rangle_{AB} \otimes |a\rangle_{B'}, U'_{AB}V_{BB'}U_{AB}|\psi_i\rangle_{AB} \otimes |a\rangle_{B'}\right), \tag{2}$$

where $\rho_{i,p\rightarrow}^{out}$ describes the density matrix of the output at the subsystems A and B, considering the parameterized unitaries, $|a\rangle$ is the reference state, and V is for the unitary of the SWAP gate. The cost function is defined in Eq. (1) as the similarity of the output, which is a reconstruction of the input, to the original input, $|\psi\rangle$. Subsystem B' is traced out at this level, and only subsystems A and B need to be monitored. A metric to gauge how comparable such states are is called the *fidelity* of quantum systems. Therefore, we define a successful autoencoding as one that $F(\psi_i, \rho_i^{out}) \approx 1$ for all input states. By considering the inner product of the states, the fidelity may be determined whether they are pure. Equation (3) illustrates this.

$$F(\rho, \sigma) = |\langle \psi_A | \psi_B \rangle|^2,\tag{3}$$

where $\rho = |\psi_A\rangle\langle\psi_A|$ and $\sigma = |\psi_B\rangle\langle\psi_B|$. The following results from an additional simplification of the cost function in Eq. (1):

$$C_2(\overrightarrow{p}) = \sum_i p_i.F\left(Tr_A\left[\overrightarrow{U}\,|\psi_i\rangle\,\langle\psi_i|_{AB}\left(\overrightarrow{U}\right)'\right], |a\rangle_B\right).\tag{4}$$

The initial cost function and the reduced cost version provide the same outcome, suggesting that it is possible to determine the fidelity between the reference state and the predicted value of the subsystem B after the encoder. By not measuring the subsystem A's qubits, the subsystem A may be traced out. The measured component is sometimes known as the "*trash state*" since it must match the set reference state exactly for every conceivable input state. The "compressed state" or "latent space," produced by the remaining qubits that are traced out, can be saved or used for some future inference or learning tasks. Instead of monitoring subsystem A during testing, we can measure subsystem B and obtain the latent space state. To avoid irrevocably changing the state, it would be more relevant to conduct the desired process sequentially following the encoding instead of measuring the subsystem because the compressed state might be entangled.

4 Our Novel Quantum Autoencoder Frameworks

We propose three novel frameworks for using quantum autoencoder for anomaly detection. The first framework (framework 1) is the reconstruction error threshold-based method, the second is a union of a quantum autoencoder and one-class SVM, and the third is a union of a quantum autoencoder and quantum random forest. In the context of anomaly detection, an autoencoder is trained on a dataset of "normal" data samples and then used to reconstruct test samples. The reconstruction error, indicating the difference between the original sample and its reconstruction, can be used as a measure of how "normal" or "anomalous" the sample is.

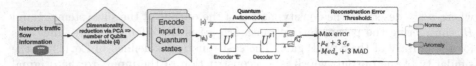

(a) Framework 1: QAE and reconstruction error-based approach.

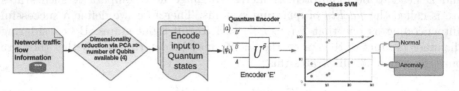

(b) Framework 2: Union of QAE and one-class SVM.

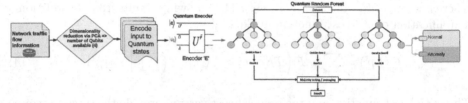

(c) Framework 3: Union of QAE and quantum random forest.

Fig. 2. Three novel quantum autoencoder frameworks for anomaly detection.

The reconstruction error is used as a measure of how well the autoencoder can reconstruct a given sample. If the autoencoder is trained on normal data, it should be able to reconstruct normal samples with low error. Anomalies, on the other hand, may be difficult for the autoencoder to reconstruct accurately, leading to higher reconstruction errors. By setting a threshold on the reconstruction error, it is possible to flag samples with higher errors as potentially anomalous. To define the threshold, we propose to experiment with three methods, namely (1) Max reconstruction error (Max_e), (2) Mean reconstruction error + 3 standard deviation of reconstruction error ($\mu_e + 3\sigma_e$), and (3) Median reconstruction error + 3 median absolute deviation reconstruction error ($Med_e + 3MAD$).

Figure 2a illustrates framework 1 with the steps of using the reconstruction error threshold-based approach. First, we reduce the dimensionality of the network traffic data using PCA where the number of features equals the number of accessible qubits, then the data input is encoded to quantum states before the data is processed in the quantum autoencoder (QAE) quantum circuit. After the data is successfully encoded and decoded with QAE, the reconstruction error is calculated for each input using the classical cost function. We define the reconstruction error threshold using the later mentioned three methods any data above the threshold is labeled as an anomaly and the remaining data are labeled as normal.

Figure 2b demonstrates the framework 2. The steps involve using the quantum autoencoder as a dimensionality reduction technique and then the data is taken as an input for a one-class SVM, which separates normal data from anomalies with a hyperplane, after mapping the data via a kernel. We have experimented with two methods of one-class SVM, one with *classical kernel* and the other with *quantum kernel*. The first steps are the same from framework 1 (Fig. 2a) until the quantum data is fed to the QAE circuit. In this approach we separate the encoder from the decoder part in the QAE, thus the dimensionality is reduced. Then, the data is taken as input for a one-class SVM algorithm to eventually detect anomalies.

Figure 2c illustrates the third framework. The quantum autoencoder is used as a dimensionality reduction technique before the data is taken as input for a quantum random forest algorithm. It is a machine learning ensemble approach that integrates many Variational Quantum Classifier (VQC) models to create predictions. After training a predetermined number of VQC models (trees), the function employs majority voting to provide predictions for the input test data. In this approach, we separate the encoder from the decoder part in the QAE, so that dimensionality is reduced. Then, the data is taken as input for a quantum RF algorithm to eventually detect anomalies.

4.1 Experimental Setup

We implemented the quantum anomaly detection methods, and used the IBM Qiskit Python library version 0.37 and PennyLane with Python 3.9 carried out in Jupyter notebooks to specify and simulate our quantum circuits. Scikitlearn, PyTorch, and Matplotlib are used for evaluation, cost function optimization, and visualization, respectively. One-class support vector machines, cross-validation, machine learning evaluation metrics, and Principle component analysis (PCA) for dimensional reduction were all implemented using the Scikit-Learn library. PyTorch was utilized in this instance for the traditional optimization of the quantum gate's parameters. The hardware we used for implementation is an Apple M1 Pro ARM-based system with 16 core GPU and 10 core CPU for processing.

4.2 Implementation

We implemented the quantum autoencoder in Python, based on the approach presented in [16]. It is trained to compress a network flow information dataset of quantum states when using a traditional compression technique is not possible. Utilizing conventional optimization approaches, the quantum autoencoder's parameters are trained.

The quantum autoencoder as a combination of the classical and quantum methods is realized as follows. After the input state has been prepared, the parameterized unitary is used to compress the state. A SWAP test is used to determine how much of the reference state and the trash state by compression overlap. A classical optimization approach is used to construct the cost function that is minimized from the outcomes for all the states in the train data.

The AE quantum circuit includes controlled rotational gates and Hadamard gates. It is implemented using the quantum Python library PennyLane. The quantum process includes angle embedding and amplitude embedding. The circuit shown in Fig. 3a depicts the QAE circuit SWAP test consisting of 4 total qubits with 1 trash qubit and 1 auxiliary qubit.

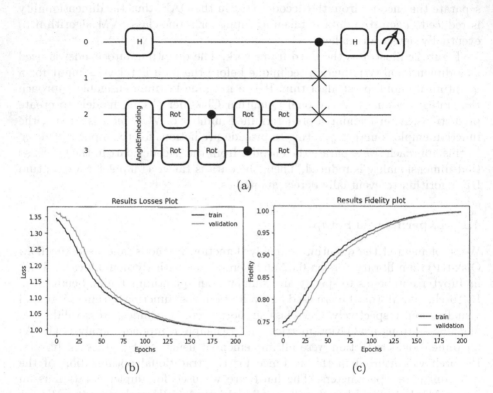

Fig. 3. (a) QAE circuit (b) QAE reconstruction loss (c) QAE Fidelity

The compressed qubits (2 and 3) first go through an angle embedding function then the first layer of rotational gates, after that a controlled rotational gate between each qubit, then the second layer of rotational gates. Moreover, it demonstrates the SWAP test where the qubit 2 is swapped with qubit 0 which went through 2 Hadamard gates then it is measured to calculate the fidelity of the 2 quantum states that pass through a classical cost function to apply optimization. Finally, qubit 1 illustrates the auxiliary qubit.

5 Experimental Results and Discussion

In this section, we will discuss the results obtained from our different experiments. The experiments were conducted using actual NISQ quantum devices and a quantum simulator. In our trials, we employed the KDD99 dataset.

The quantum autoencoder circuit consists of a networked collection of nodes, each representing a qubit. The network's first layer represents the input register, while the output register is represented by the network's last layer. A unitary change from one layer to the next is represented by the edges separating adjacent layers. The distance between the first and second layers is reduced via autoencoders. After the initial "encoding" E, the data present in some of the input nodes must be removed for a quantum circuit to represent an autoencoder network. The final "decoding" evolution D is then implemented using brand-new qubits that are created to some reference state, and the result is contrasted with the initial state.

Figures 3b and 3c illustrate the reconstruction loss and fidelity of the quantum autoencoder. This gives us a detailed idea of how the autoencoder model performs on the KDD99 dataset. The model was trained on normal data, expecting each input to be recovered as close as feasible to the learned normal patterns by an autoencoder-based anomaly detection system. Therefore, if the reconstruction error of an input instance exceeds a certain threshold, we may categorize it as an attack; otherwise, we can classify it as normal. Out of the three threshold options we experimented with, the threshold calculated using mean + 3 standard deviation gave better results.

The efficacy of anomaly detection models is assessed using the metrics of accuracy, F1-score, precision, recall, and ROC curve. However, some evaluation metrics have more weight in evaluating our model. Our dataset is unbalanced since there are not many data samples from the minority class, and the accuracy measure does not accurately reflect the model's performance because a subpar model will perform well for the majority class. On the other hand, F1-score is the metric that accurately represents model performance when dealing with unbalanced datasets, as it is a natural property of anomaly detection datasets because the F1 score considers the distribution of the data (Table 1).

Table 1. Comparison of QAE and state-of-the-art methods.

	QNN [21]	QAE [21]	SRCNN [21]	QAE [10]	QAE [12]	Our QAE
MSE	0.002	0.002	**0.001**	0.006	0.004	0.004
Fidelity	0.953	0.903	0.959	0.984	0.961	**0.996**

Our implementations of the quantum autoencoder recorded the highest fidelity value of 0.9963 and an average reconstruction error value of 0.0037 compared to state-of-the-art quantum and classical methods. The primary distinction between previous research and ours is that other implementations' input states are fundamentally quantum mechanical. In contrast, in our work, the kind of quantum state depends on the input embedding of the classical data. In angular encoding, the qubits that make up the quantum states may be separated. However, we utilize CNOT gates in the unitary evolution to entangle the various qubits. Finally, compared to previous datasets utilized in prior research, our version of the quantum autoencoder had a greater likelihood of compressing the input data into a latent-space representation and discovering novel correlations of the network traffic flow information.

Table 2. Comparison among the proposed quantum frameworks

	Accuracy	Precision	Recall	F1-score
Framework 1	83.23%	**100.00%**	81.12%	77.60%
Framework 2 with classical kernel	**97.48%**	95.06%	**99.43%**	**97.19%**
Framework 2 with quantum kernel	60.84%	98.10%	63.01%	73.57%
Framework 3	92.39%	89.63%	97.16%	93.41%

Table 2 compares the performance of our proposed novel frameworks—framework 1 (Reconstruction error threshold), framework 2 (union of QAE and OC-SVM), and framework 3 (union of QAE and quantum random forest). Accuracy measures how accurately an approach predicts anomalies, while F1-score is the harmonic mean of precision and recall and is a balanced indicator of the two. The best-performing framework is the combination of QAE and traditional OC-SVM, which has 97% accuracy and 97% F1-score. Combining QAE plus quantum random forest yields 92% accuracy and 93% F1-score, making it the second-best strategy. Although it has 61% accuracy and 73% F1-score, the merger of QAE with quantum OC-SVM is the least effective solution. However, the F1-score was chosen as the main evaluation metric as it considers both precision and recall, it provides a more balanced measure than accuracy, which can be misleading on imbalanced datasets which is a natural property of anomaly detection datasets (Table 3).

Table 3. Comparison with other state-of-the-art works

	OC-SVM [18]	QOC-SVM [8]	VAE-Bayes [1]	QOCC [4]	Deep-SVDD [17]	DMKDE [20]	Framework 2
F1-score	74.4%	92.3%	80.3%	87.3%	75.5%	77.1%	**97.1%**

Our best framework outperformed the state-of-the-art approaches as it yields 97.1% F1-score while being superior to all traditional algorithms and being on par with some quantum deep learning-based techniques.

Table 4 demonstrates the time taken for training the QAE with a different number of qubits and we can see the time taken increase exponentially with the number of qubits used while having a lower fidelity score and based on our quantum autoencoder approach we used 4 total qubits and latent qubits is 2 after the SWAP test is done which in turn transform the data into linear one. Quantum kernels struggle with separating normal and anomaly data compared with a classical linear kernel. Table 2 compares the three novel frameworks we

Table 4. Time comparison with numbers of Qubits used.

Total Qubits	Latent Qubits	Time Taken	Fidelity
4	2	1.64 h	0.9963
4	3	1.49 h	0.9812
6	4	7.85 h	0.8610

implemented. Each method was evaluated using the KDD99 dataset. We observe that the best-performed method is the union of QAE and classical OC-SVM that achieved 97% accuracy and 97% F1-score. The second best approach is the union of QAE and quantum random forest with 92% accuracy and 93% F1-score. Nevertheless, the worst-performed approach is the union of QAE and quantum OC-SVM with 61% accuracy and 73% F1-score.

6 Conclusion

The primary aim of this research paper was to explore the potential of leveraging quantum machine learning to detect anomalies in network traffic. To achieve this objective, we have developed three frameworks that incorporate quantum autoencoder anomaly detection. The first framework is grounded on error reconstruction and identifies any attacks that exceed a particular threshold. This approach is instrumental in recognizing unknown attack types that depart from the established regular patterns in a quantum autoencoder-based anomaly detection system. The second and third frameworks amalgamate quantum autoencoder with quantum one-class support vector machine and quantum random forest, respectively. We have evaluated the efficacy of these methods in accurately identifying attacks, and framework 2 has been found to have the highest accuracy and F1 score compared to other frameworks and state-of-the-art quantum and classical anomaly detection methods.

References

1. An, J., Cho, S.: Variational autoencoder based anomaly detection using reconstruction probability. Spec. Lect. IE **2**(1), 1–18 (2015)
2. Baldi, P., Hornik, K.: Neural networks and principal component analysis: learning from examples without local minima. Neural Netw. **2**(1), 53–58 (1989)
3. Bishop, C.M., Nasrabadi, N.M.: Pattern Recognition and Machine Learning. Springer, New York (2006)
4. De Oliveira, N.M., Lucas, P., De Oliveira, W.R., Ludermir, T.B., Da Silva, A.J.: Quantum one-class classification with a distance-based classifier. In: Proceedings of the International Joint Conference on Neural Networks (IJCNN), pp. 1–7. IEEE (2021)
5. Kingma, D.P., Welling, M.: Auto-encoding variational bayes. arXiv preprint arXiv:1312.6114 (2013)

6. Kottmann, K., Metz, F., Fraxanet, J., Baldelli, N.: Variational quantum anomaly detection: unsupervised mapping of phase diagrams on a physical quantum computer. Phys. Rev. Res. **3**(4), 043184 (2021)
7. Kyriienko, O., Magnusson, E.B.: Unsupervised quantum machine learning for fraud detection (2022)
8. Liang, J.M., Shen, S.Q., Li, M., Li, L.: Quantum anomaly detection with density estimation and multivariate Gaussian distribution. Phys. Rev. A **99**(5), 052310 (2019)
9. Lv, P., Yu, Y., Fan, Y., Tang, X., Tong, X.: Layer-constrained variational autoencoding kernel density estimation model for anomaly detection. Knowl.-Based Syst. **196**, 105753 (2020)
10. Mangini, S., Marruzzo, A., Piantanida, M., Gerace, D., Bajoni, D., Macchiavello, C.: Quantum neural network autoencoder and classifier applied to an industrial case study. Quantum Mach. Intell. **4**(2), 13 (2022). https://doi.org/10.1007/s42484-022-00070-4
11. McClelland, J.L., Rumelhart, D.E., Group, P.R., et al.: Parallel Distributed Processing, Volume 2: Explorations in the Microstructure of Cognition: Psychological and Biological Models, vol. 2. MIT Press, Cambridge (1987)
12. Mete, B., Gutierrez, I.L., Mendl, C.: Hamiltonian simulation using quantum autoencoders (2021)
13. Plaut, E.: From principal subspaces to principal components with linear autoencoders. arXiv preprint arXiv:1804.10253 (2018)
14. Pol, A.A., Berger, V., Germain, C., Cerminara, G., Pierini, M.: Anomaly detection with conditional variational autoencoders. In: Proceedings of the 2019 18th IEEE International Conference on Machine Learning and Applications (ICMLA), pp. 1651–1657. IEEE (2019)
15. Ranzato, M., Huang, F.J., Boureau, Y.L., LeCun, Y.: Unsupervised learning of invariant feature hierarchies with applications to object recognition. In: IEEE Conference on Computer Vision and Pattern Recognition, pp. 1–8. IEEE (2007)
16. Romero, J., Olson, J.P., Aspuru-Guzik, A.: Quantum autoencoders for efficient compression of quantum data. Quantum Sci. Technol. **2**(4), 045001 (2017)
17. Ruff, L., et al.: Deep one-class classification. In: Proceedings of the 2018 International Conference on Machine Learning, pp. 4393–4402. PMLR (2018)
18. Schölkopf, B., Platt, J.C., Shawe-Taylor, J., Smola, A.J., Williamson, R.C.: Estimating the support of a high-dimensional distribution. Neural Comput. **13**(7), 1443–1471 (2001)
19. Trugenberger, C.A.: Quantum pattern recognition. Quantum Inf. Process. **1**, 471–493 (2002). https://doi.org/10.1023/A:1024022632303
20. Useche, D.H., Bustos-Brinez, O.A., Gallego, J.A., González, F.A.: Computing expectation values of adaptive Fourier density matrices for quantum anomaly detection in NISQ devices. arXiv: 2201.10006 (2022)
21. Wang, M.M., Jiang, Y.D.: Data reconstruction based on quantum neural networks. arXiv preprint arXiv:2209.05711 (2022)

Spatially-Aware Human-Object Interaction Detection with Cross-Modal Enhancement

Gaowen Liu[1], Huan Liu[1($\boxtimes$)], Caixia Yan[1], Yuyang Guo[1], Rui Li[2],
and Sizhe Dang[1]

[1] School of Computer Science and Technology, Xi'an Jiaotong University,
Xi'an, China
{huanliu,yancaixia}@xjtu.edu.cn, darknight1118@stu.xjtu.edu.cn
[2] School of Continuing Education, Xi'an Jiaotong University, Xi'an, China

Abstract. We propose a novel two-stage HOI detection model that incorporates cross-modal spatial information awareness. Human-object relative spatial relationships are highly relevant for specific HOI species, but current approaches fail to model such crucial cues explicitly. We observed that relative spatial relationships possess properties that can be described in natural language easily and intuitively. Building on this observation and inspired by recent advancements in prompt-tuning, we design a Prompt-Enhanced Spatial Modeling (PESM) module that generates linguistic descriptions of spatial relations between humans and objects. PESM is capable of merging the explicit spatial information obtained by the aforementioned text descriptions with the implicit spatial information of the visual modality. Moreover, we devise a two-stage model architecture that effectively incorporates auxiliary cues to exploit the enhanced cross-modal spatial information. Extensive experiments conducted on the HICO-DET benchmark demonstrate that the proposed model outperforms state-of-the-art methods, indicating its effectiveness and superiority. The source code is available at https://github.com/liugaowen043/tsce.

Keywords: HOI Detection · Transformer · Spatial Feature · Cross-Modal Enhancement

1 Introduction

Human-Object Interaction (HOI) detection has attracted significant research interest in recent years, as it serves as the foundation for high-level vision tasks and has a wide range of applications. Specifically, the HOI detection task aims to detect both the interacting human-object pairs and the type of interaction, which can be generally represented as sets of triples, *i.e.*, <human, verb, object> [2]. In this context, both humans and objects can be represented as bounding boxes with corresponding labels.

Recent studies [11,25] have accomplished great advances in HOI detection, depending on the powerful ability of DETR [1] for object detection. With the

© The Author(s), under exclusive license to Springer Nature Singapore Pte Ltd. 2024
B. Luo et al. (Eds.): ICONIP 2023, LNCS 14451, pp. 83–96, 2024.
https://doi.org/10.1007/978-981-99-8073-4_7

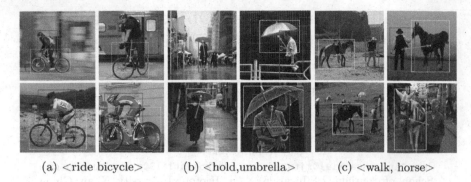

(a) <ride bicycle>   (b) <hold,umbrella>   (c) <walk, horse>

Fig. 1. Samples of HOI categories with significant relative spatial relationships. As shown in (a), the bicycle is always underneath the person. While in (b), the umbrella is always above the person. Similarly, in (c), the person and horse are often aligned on the same horizontal line.

global attention mechanism in Transformer [19], these DETR-based methods effectively capture the global image features or image-wide contextual information that is beneficial for interaction classification. Although these methods are able to find as much task-relevant information as possible, some crucial aspects may still be difficult to reveal due to the complexity of the HOI detection task. Specifically, we notice that the interactions between humans and objects exhibit significant characteristics in terms of their relative spatial relationships. For instance, as illustrated in Fig. 1, in the HOI <human, ride, bicycle>, the bicycle is always situated below the person, while in <human, hold, umbrella>, the umbrella is always positioned above the person. Similarly, in <human, walk, horse>, the person and horse are often aligned on the same horizontal line.

Given this observation, we argue that explicitly modeling the spatial relations between humans and objects within an image could be advantageous for accurately classifying their interactions. Previous research [2,5,23] has conducted a preliminary exploration of the spatial relationships between human-object pairs. Generally, by using detection boxes, they can preserve the visual features of humans and objects within an image, and then extract the spatial features of the human-object pairs. Regrettably, these methods learn spatial relationship features implicitly and indirectly, resulting in limited performance improvement. In contrast, it is easy and intuitive to describe the relative spatial relationship between a human and an object in natural language. For example, the person and the bicycle in Fig. 1a can be explicitly described as "the bicycle is underneath the person". Therefore, combining both explicit and implicit spatial relationship information would be beneficial for human-object interaction classification.

Based on the above motivation, we propose a novel two-stage HOI detection model with cross-modal spatial information awareness. Inspired by the latest advances in prompt-tuning [16], we propose a Prompt-Enhanced Spatial Modeling (PESM) module to generate linguistic description of the spatial relationship between humans and objects. The constructed explicit spatial information of the text modality is then merged with the implicit spatial information of the visual modality to achieve information enhancement. Moreover, in order to enable the model to better utilize the enhanced spatial information, we devise a two-stage model architecture, which can effectively incorporate auxiliary cues compared to a one-stage one [12,25]. In addition, we further employ a Transformer encoder to extract appearance features and perform final feature fusion for spatially-aware interaction recognition. Our main contributions are summarized as follows:

1) We explore both explicit and implicit spatial relationship information for HOI detection, and accordingly propose a novel Transformer-based Spatially-aware with Cross-modal Enhancement (TSCE) two-stage HOI Detector.
2) We propose the PESM module for generating linguistic descriptions of spatial relationships between humans and objects and constructing enhanced spatial features. Additionally, an effective two-stage architecture is devised to introduce enhanced spatial information.
3) We conduct extensive experiments on two popular HOI datasets for evaluation. The results show that the proposed model outperforms the state-of-the-arts, demonstrating its effectiveness and superiority.

2 Related Work

2.1 Two-Stage Methods

The two-stage methods perform object detection and interaction recognition sequentially. Specifically, this approach first uses a pre-trained object detector in the first stage to obtain human and object bounding boxes and class labels. In the second stage, a multi-branch interaction classification network is used, where each branch extracts a specific kind of feature, to obtain the predicted interactions for human-object pairs using fused features. The most fundamental two-stage model [2] utilizes human and object appearance features and relative spatial features. To improve the performance of interaction recognition, subsequent studies have incorporated attention mechanisms [5,6], human pose information [14,20] and graphical neural networks [4,18] to assist with reasoning. It is worth mentioning that studies, such as the SCG [23] model, have verified the effectiveness of spatial features for interaction recognition. Two-stage models have inherent drawbacks of high complexity and low efficiency due to their serial and decoupled structure. However, the advantages of two-stage models are also obvious, as they are easier to incorporate multiple auxiliary cues, use multi-scale features, and optimize the results of each stage for the current stage.

2.2 One-Stage and DETR-Based Methods

One-stage methods often use a two-branch architecture to detect instances and interactions in parallel, and the results from both branches are combined using a hand-designed matching strategy. One-stage methods use different forms to represent instances and interactions, including PPDM [12] and IP-Net [21], which

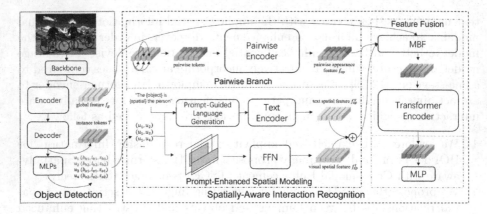

Fig. 2. The framework of our TSCE includes three stages: 1) TSCE first utilizes DETR for object detection, and outputs classification and instance results. 2) The classification heads and instance tokens are fed into the PESM module and pairwise branch to extract multi-modal spatial features and pairwise appearance features, respectively. 3) These two kinds of features are fused with the MBF module and then represented with a Transformer encoder. Finally, the interaction results are obtained by inputting multi-scale features into an MLP head.

use key points to represent instances and vectors to represent interactions, and UnionDet [9], which use joint regions of human-object pairs to represent interactions. The DETR-based approach is based on the encoder-decoder structure of the transformer for HOI detection in an end-to-end manner. Basic DETR-based models [25] are made applicable to HOI detection problems by modifying the classification head of DETR, while some models [10] decouple the instance decoder and the interaction decoder. One-stage and DETR-based approaches have led to significant improvements in the efficiency and accuracy of HOI detection. However, these two approaches still suffer from the disadvantages of using only global image features and discarding auxiliary cues, which have been proven to be beneficial for interaction classification in the past.

3   Method

In this section, we provide a detailed introduction to our model. In Sect. 3.1, we present the overall framework of our model and provide a brief description of its pipeline. In Sect. 3.2, we describe in detail how our PESM module models spatial information and utilizes linguistic augmentation. Then, in Sect. 3.3, we introduce the module for interactive recognition. Finally, in Sect. 3.4, the training and inference are discussed.

3.1   Overview

The overall architecture of the proposed TSCE model is illustrated in Fig. 2. As a two-stage approach, TSCE achieves HOI detection by performing object detection and interaction recognition sequentially. More specifically, given an image, we first perform object detection with DETR [1]. The image global feature f_g from backbone, the instance outputs $\{U|u = (b, l, s)\}$ from classification head, where b, l, s respectively represent the instance bounding box, category label and confidence score, and instance tokens T from DETR decoder are then input into Interaction Recognition Module.

In the interaction recognition phase, the multi-modal spatial features of human-object pairs are generated by the Prompt-Enhanced Spatial Modeling (PESM) Branch, while the pairwise appearance features are extracted by using the Pairwise Encoding Branch. Finally, these features are then fed into the spatially-aware interaction classification module to obtain the corresponding categories for each human-object pair. A detailed description of each key component is presented below.

3.2 Prompt-Enhanced Spatial Modeling

To achieve cross-modal enhanced spatial modelling, the PESM module employs a two-stream structure to extract multi-modal spatial features of human-object pairs, including both visual and textual modalities. As illustrated in Fig. 2, we first pair the results of the DETR classification heads $u = (b, l, s)$ to create human-object pairs (u_i, u_j), with special emphasis on ensuring that the label of u_j must be human. Next, we utilize a two-stream structure to extract spatial features of human-object pairs separately. The visual spatial branch is responsible for extracting spatial features of the visual modality, while the text spatial branch extracts spatial features of the text modality. Finally, we fuse the spatial features of the two branches to obtain multi-modal spatial features. The fusion process strengthens our spatial information for interaction classification with text modality information.

Visual Spatial Branch. For the spatial features of the visual modality, we use a two-channel image to encode the spatial relationship between a human-object pair. One channel represents the human bounding box, while the other channel represents the object bounding box. The inside of the bounding box is set to 1, and the outside is set to 0, thereby representing the location of the bounding box. Finally, a feed-forward neural network (FFN) is utilized to extract the visual spatial features. We denote the visual spatial features of the human-object pair (u_h, u_o) as $f^v_{h,o,sp}$.

Text Spatial Branch. When describing spatial relationships using natural language, for example, when describing the position of a bicycle relative to a person in the sample image shown in Fig. 2, we might use the phrase "the bicycle is underneath the person". This type of textual description is essentially an encoding of the relative spatial position from a linguistic perspective. Therefore, in this branch, we are essentially attempting to encode spatial information using natural language. In brief, as illustrated in Fig. 2, given a human-object pair (u_h, u_o), we first use the Prompt-Guided Language Generation module to generate a textual description of the pair's relative spatial position. Next, we use a

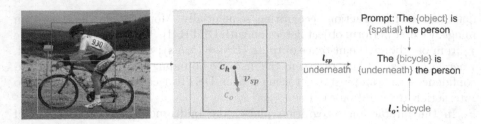

Fig. 3. Illustration of the Prompt-Guided Language Generation module. Based on the centroids of the human and object boxes, we get spatial vector v_{sp} containing spatial information. And according to the length and perspective of v_{sp}, we get the spatial label l_{sp}. Finally, by embedding l_{sp} and the object label l_o into the prompt, we can get the textual description of the spatial relationship.

Table 1. Correspondence between the spatial vector v_{sp} and the spatial relation category l_{sp}.

v_{sp}	l_{sp}	v_{sp}	l_{sp}	v_{sp}	l_{sp}
$\lvert v_{sp} \rvert < \lvert v_h \rvert$	center	(345, 15]	left	(15, 75]	lower left
(75, 105]	underneath	(105, 165]	lower right	(165, 195]	right
(195, 255]	top right	(255, 285]	above	(285, 345]	top left

text encoder to obtain the text spatial feature. Subsequently, we will provide a detailed explanation of both the Prompt-Guided Language Generation module and the selection of the text encoder.

As for Prompt-Guided Language Generation module, to facilitate the generation of textual descriptions that reflect spatial relationships, we first categorized the relative spatial relationships between humans and objects into nine categories based on human language habits, as shown in the l_{sp} column of Table 1. We then designed a prompt, "the {object label} is {spatial label} the person", to describe the relative spatial position between human and objects. Given a human-object pair (u_h, u_o), where u_o already contains the object label, we only need to distinguish which of the nine categories the relative spatial relationship between the pair belongs to. We achieve this through vector operations. Specifically, as illustrated in Fig. 3, we first calculate the center point of the human bounding box c_h and the center point of the object bounding box c_o. We then obtain a spatial vector v_{sp} containing relative spatial information from the human center point to the object center point. Based on the length and angle of the spatial vector, we can determine the spatial relationship label for the human-object pair, as shown in Table 1, where $\lvert \cdot \rvert$ represents the length of a vector and $(\cdot, \cdot]$ represents the range of the angle between v_{sp} and the x-axis. Finally, by incorporating the object label l_o of t_o and the calculated spatial label l_{sp} into the prompt, we can obtain a textual description of the spatial relationship. This process can be viewed as an encoding of relative spatial positions using language.

The text encoder is utilized to convert the textual descriptions of spatial information into feature vectors. In this regard, we chose to use the text encoder of CLIP [16] for textual feature encoding. Compared to the normal text encoder, the text encoder of CLIP is co-trained with the image encoder, thus the extracted features are more suitable for the visual task of HOI detection. We denote the text spatial features of the human-object pair (u_h, u_o) as $f^t_{h,o,sp}$.

Spatial Feature Fusion. At this stage, we aim to enhance the spatial information corresponding to human-object pairs through spatial feature fusion, where both the visual and text spatial features are fused to obtain multi-modal spatial features. For each human-object pair, the information contained in both features is similar at a deep semantic-level. Thus, we opt to fuse them by directly summing $f^t_{h,o,sp}$ and $f^v_{h,o,sp}$:

$$f_{h,o,sp} = f^t_{h,o,sp} + f^v_{h,o,sp}, \tag{1}$$

where $f_{h,o,sp}$ refers to the fused multi-modal spatial feature.

3.3 Spatially-Aware Interaction Recognition

This section provides a detailed description of the Spatially-Aware Interaction Recognition module developed for identifying human-object pair interactions. As illustrated in Fig. 2, this module is composed of three key components: Prompt-Enhanced Spatial Modeling, Pairwise Feature Encoding, and Interaction Classification. While Prompt-Enhanced Spatial Modeling has been elaborated in Sect. 3.2, the remaining two primary components will be elaborated below.

Pairwise Feature Encoding. This branch is used to extract pairwise appearance features of human-object pairs. Specifically, we connect the instance tokens from the DETR decoder corresponding to the human-object pair (t_h, t_o) mentioned above to form pairwise tokens. Based on these instance tokens, we can obtain the instance-level appearance features of the human-object pair (t_h, t_o), denoted as $f_{h,ap}, f_{o,ap}$. The pairwise tokens are then fed to the pairwise encoder encoder$_{pair}(\cdot)$, i.e., the modified transformer encoder in UPT [24], to extract the pairwise appearance features $f_{h,o,ap}$:

$$f_{h,o,ap} = \text{encoder}_{pair}(f_{h,ap} \oplus f_{o,ap}), \tag{2}$$

where $\oplus$ refers to vector concatenation.

Interaction Classification. The interaction classification module is utilized to determine the interaction categories of human-object pairs based on the pairwise appearance feature, multi-modal spatial feature, and image global features from backbone. The module consists of the Feature Fusion Module that fuses the above-mentioned features, and the final classification head that performs the interaction classification.

The Feature Fusion Module first fuses the pairwise appearance feature and the multi-modal spatial feature for each human-object pair. By fusing these two kind of features, we can obtain the local pairwise features that contain both appearance and spatial information. However, since this information is limited to human-object pairs, contextual information is also essential in inferring the type of interaction between human-object pairs. This contextual information, such as the background and scene, as well as other people and objects, is generally included in the global image features. To incorporate such contextual information, we joint the local pairwise features with the global image features and then fuse them using a standard transformer encoder to obtain multi-scale pairwise features. Finally, we forward the multi-scale features to the classification head, which is implemented as a simple MLP network, to obtain the corresponding human-object pair interaction categories:

$$f_{h,o} = \text{encoder}(\text{MBF}(f_{h,o,ap}, f_{h,o,sp}) \oplus f_{global}), \qquad (3)$$

$$s_{hoi} = \text{MLP}(f_{h,o}), \qquad (4)$$

where encoder($\cdot$) denotes the transformer encoder for multi-scale pairwise feature encoding; s_{hoi} refers to the interaction category corresponding to human-object pair (t_h, t_o); MBF($\cdot$) denotes the MBF module in SCG [23] specifically designed for fusing spatial and appearance features.

3.4   Training and Inference

During the training phase, we freeze the parameters of DETR and train the interaction recognition module using focal loss with the same way as UPT [24]. During the inference phase, we follow the traditional two-stage approach [5] to compute the final interaction score $s_{h,o}^a$ for a given human-object pair (t_h, t_o):

$$s_{h,o}^a = (s_h)^\lambda (s_o)^\lambda s_{hoi}^a, \qquad (5)$$

where s_h and s_o refer to the human confidence and object confidence score respectively; s_{hoi}^a is the output of the interaction classification head; λ is a hyper parameter greater than 1 to mitigate the impact of target detection results with excessively high confidence levels on the final result.

4   Experiments

In this section, we present a comprehensive set of experiments to demonstrate the effectiveness of our model and the PESM module. In Sect. 4.1, we provide a brief description of the experimental setup, including the dataset used, evaluation metrics, and implementation details. In Sect. 4.2, we compare our approach with existing methods. The ablation study will be presented in Sects. 4.3. We also investigate the generalization capability of the PESM module in Sect. 4.4. Finally, we perform some quantitative analysis in Sect. 4.5.

4.1 Experiment Setting

Datasets. We conducted experiments on two most commonly used datasets for HOI detection, i.e., HICO-DET [2] and V-COCO [7]. Specifically, HICO-DET comprises 47,776 images and over 150,000 human-object pairs (38,118 images in the training set and 9,658 images in the test set). It contains 600 HOI categories, with 117 interactions and 80 objects. Additionally, the 600 HOI categories are categorized into 138 rare and 462 non-rare categories based on the number of training instances. V-COCO is a subset of the MS-COCO [13] dataset, much smaller than HICO-DET, and includes 2,533 training images, 2,867 validation images, and 4,946 test images. It is labeled with 80 objects and 29 actions.

Evaluation Metrics. Following standard evaluation practices, we use mAP to measure the model's performance. Furthermore, we consider a detection result as true positive when the HOI detector accurately localizes humans and objects (i.e., the interaction association (IOU) ratio between the predicted box and the ground truth is greater than 0.5) and correctly predicts the interaction.

Implementation Details. We adopt the same data augmentation approach as HOI-Transformer [25]. During the object detection phase, we use DETR pre-trained on MS COCO. Parameters of both the object detector DETR and CLIP [16] text encoder are frozen. In the interaction recognition phase, we filter out human boxes with a confidence less than 0.7 and object boxes with a confidence less than 0.3. The dimension of the hidden layer is 256, the same as in DETR. Similar to SCG [23], we set λ to 1.0 for training and 2.8 for inference, and the hyper parameters of the focal loss are set to $\beta = 0.5, \gamma = 0.2$. We use AdamW [15] as the optimizer with the learning rate initially set to 10^{-4} and decrease the learning rate by a factor of 10 every 10 epochs. We train the model on four NVIDIA GeForce RTX 3090 s with a batch size of four on each GPU for 15 epochs to obtain the final results.

4.2 Comparison to State-of-the-Art

Tables 2 present the experimental results of our model TSEC on HICO-DET and V-COCO. As shown in Table 2, our model achieves 32% mAP for the "full" category, 27.10% mAP for the "rare" category, and 33.46% mAP for the "non-rare" category under the "default" setting of HICO-DET, surpassing the existing models in both settings of HICO-DET. In the default mode, our model improves the "full" category by more than 1% compared to the state-of-art two-stage model UPT [24]. This improvement is mainly due to the significant increase in the "rare" category, from 25.94% mAP to 27.10% mAP, which represents an improvement of nearly 4.5% compared to UPT, indicating the higher generalization ability of our model. In contrast, compared to the one-stage model MSTR [11], the improvement of our model is more significant.

Our model also performs well on the V-COCO dataset, surpassing all existing two-stage models and achieving performance comparable to MSTR on AP^2_{role}. However, our model's performance on AP^1_{role} differs significantly from that of

Table 2. Performance comparison (mAP) on the HICO-DET and V-COCO.

Method	Backbone	HICO-DET						V-COCO	
		Default			Known Object			AP_{role}^{S1}	AP_{role}^{S2}
		Full	Rare	Non-Rare	Full	Rare	Non-Rare		
Two-stage									
InteractNet [6]	ResNet-50-FPN	9.94	7.16	10.77	–	–	–	40.0	–
iCAN [5]	ResNet-50	14.84	10.45	16.15	16.26	11.33	17.73	45.3	52.4
PMFNet [20]	ResNet-50-FPN	17.46	15.65	18.00	20.34	17.47	21.20	52.0	–
DRG [4]	ResNet-50-FPN	19.26	17.74	19.71	23.40	21.75	23.89	51.0	–
VSGNet [18]	ResNet-152	19.80	16.05	20.91	–	–	–	51.8	57.0
FCMNet [14]	ResNet-50	20.41	17.34	21.56	22.04	18.97	23.12	53.1	–
SCG [23]	ResNet-50-FPN	29.26	24.61	30.65	32.87	27.89	34.35	54.2	60.9
UPT [24]	ResNet-50	31.66	25.94	33.36	35.05	29.27	36.77	59.0	64.5
One-stage									
UnionDet [9]	ResNet50	17.58	11.72	19.33	19.76	14.68	21.27	47.5	56.2
IP-Net [21]	Hourglass-104	19.56	12.79	21.58	22.05	15.77	23.92	52.9	–
PPDM [12]	Hourglass-104	21.94	13.97	24.32	24.81	17.09	27.12	–	–
HOI-Trans [25]	ResNet-50	23.46	16.91	25.41	23.15	19.24	28.22	52.9	–
ATL [8]	ResNet-50	23.81	17.43	25.72	27.38	22.09	28.96	–	–
AS-Net [3]	ResNet-50	28.87	24.25	30.25	31.74	27.07	33.14	53.9	–
QPIC [17]	ResNet-50	29.07	21.85	31.23	31.68	24.14	33.93	58.8	61.0
MSTR [11]	ResNet-50	31.17	25.31	32.92	34.02	28.83	35.57	**62.0**	**65.2**
Ours	ResNet-50	**32.00**	**27.10**	**33.46**	**35.17**	**29.67**	**36.82**	59.4	65.0

Table 3. Ablation study of the spatial feature on the HICO-DET.

Method	Full	Rare	Non-Rare
$TSCE_a$	31.27	26.07	32.83
$TSCE_{sp}^v$	31.49	26.03	33.12
$TSCE_{sp}^t$	31.46	26.77	32.87
TSCE	32.00	27.10	33.46

Table 4. Ablation study of spatial feature fusion on the HICO-DET.

Method	Full	Rare	Non-Rare
$TSCE_{L1}$	30.92	24.91	32.71
$TSCE_{L3}$	31.04	25.42	32.72
TSCE	32.00	27.10	33.46

MSTR, possibly because our model requires human and object to be paired before interaction inference, leading to the object detection box of the model results not being empty for "no object" actions, which reduces our model's performance on AP_{role}^1. This is a common drawback of two-stage models.

4.3 Ablation Study

The experimental results in the previous section have confirmed the validity of the model proposed in this paper. To further verify the validity of PESM module and the method of multi-modal spatial feature fusion, we conducts an ablation study on these two aspects.

Multi-modal Spatial Features. In this subsection, we conduct an ablation study of each spatial feature in the PESM module in three aspects: no spatial feature ($TSCE_a$), only visual spatial feature ($TSCE_{sp}^v$), and only textual spatial

Table 5. Result of PESM module generalizability study.

Method	Full	Rare	Non-Rare
CDN	31.29	26.57	32.71
CDN$_P$	31.87	27.01	33.33

feature ($TSCE_{sp}^t$). We compare the experimental results of these three models with the results of the original model TSCE in Table 3. From the results, we observe that our model TSCE using fused multi-modal spatial features outperforms the other three models in all three categories, which confirms the effectiveness of the PESM module. Additionally, the model $TSCE_{sp}^t$ performs better than the other two models on "rare" category, which shows that incorporating textual modality effectively improves the model's generalization ability and verifies the feasibility and effectiveness of organizing auxiliary cues into text.

Multi-modal Spatial Feature Fusion Method. Table 4 shows the experimental results of the three spatial feature fusion methods: $TSCE_{L1}$ fuses the spatial features through a single linear layer after concatenation, $TSCE_{L3}$ utilizes a three-layers feed-forward neural network, and TSCE fuses the spatial features through vector summation.

From the results in Table 3 and Table 4, we observe that fusing multi-modal spatial vectors through summation yields the best performance. This may be due to the semantic consistency of the information contained in the spatial features of the two modalities. It is worth noting that the performance of $TSCE_{L1}$ and $TSCE_{L3}$ after fusion, compared to the models $TSCE_{sp}^v$ and $TSCE_{sp}^t$ without fusion, not only does not improve but also shows a decrease. This may be due to the loss of some spatial information when using feed forward neural networks for dimension reduction transformations, while fusing features through vector summation can ensure the integrity of spatial information.

4.4 Generalizability of PESM Modules

Our proposed PESM input requires only three types of information: the human bounding box, the object bounding box, and the object label. Therefore, the PESM module can theoretically be applied to any two-stage model with compatible and generalizable features. In this section, we have combined the PESM module with CDN [22] to demonstrate this capability. Specifically, we put PESM module behind the HO-PD of CDN to model the relative spatial relationship between humans and objects, and the modified model is denoted as CDN$_P$. Table 5 shows our reproduced CDN result and CDN$_P$ result. After incorporating the PESM module, the model is significantly improved in all three categories, where CDN$_P$ improves more than 1.8% in the "full" category, nearly 1.7% in the "rare" category, and nearly 1.9% in the "non-rare" class. The experimental results demonstrate the compatibility and generalization of the PESM module and further prove the effectiveness of the module in improving HOI detection.

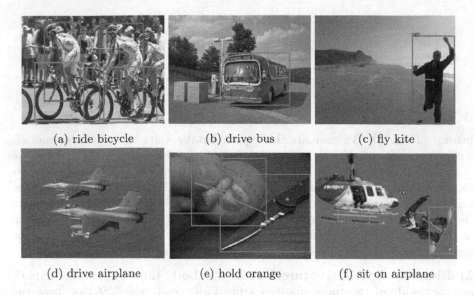

(a) ride bicycle   (b) drive bus   (c) fly kite

(d) drive airplane   (e) hold orange   (f) sit on airplane

Fig. 4. Some sample images of the test results. (a–c) show examples of correct detection while (d–f) show incorrect.

4.5   Qualitative Analysis

Figure 4a–c displays some correct prediction results of our model, while Fig. 4d–f illustrates some classical cases of incorrect predictions, highlighting potential areas for improvement in future work. As shown in Fig. 4d, due to the poor performance of DETR in detecting small objects, it failed to detect the human in the aircraft, which leads to the failure of the model to detect the interaction of "human flying the aircraft", which is also due to the dependence of the two-stage model on the target detector, thus affecting the classical performance of HOI detection. In Fig. 4e, the model incorrectly matches the person with the object "knife", which does not interact with the aircraft, mainly because of the close distance between the person and the knife, which leads to the incorrect matching of the person and the object. In Fig. 4f, the model only detects the person skateboarding in the image and fails to detect the other person sitting on the plane, which indicates the difficulty of our model handling low-pixel images.

5   Conclusion

In this paper, we leverage the property of spatial information that can be easily expressed in natural language to propose a Prompt-Enhanced Spatial Modeling module, to generate textual descriptions and fuse explicit and implicit spatial features. In order to effectively utilize the enhanced spatial feature, a Transformer-based two-stage model TSCE, is designed. Extensive experiments validate the effectiveness of our model and module. In the future, we are going to try to design end-to-end models that can utilize the PESM module to address the inefficiency problem of the current model.

Acknowledgements. This work was supported by National Key Research and Development Program of China (2020AAA0108800), National Natural Science Foundation of China (62202367, 62192781), Innovative Research Group of the National Natural Science Foundation of China(61721002), Innovation Research Team of Ministry of Education (IRT_17R86), National Education Examinations Authority (GJK2021009), Project of China Knowledge Centre for Engineering Science and Technology, Project of Chinese academy of engineering "The Online and Offline Mixed Educational Service System for 'The Belt and Road' Training in MOOC China".

References

1. Carion, N., Massa, F., Synnaeve, G., Usunier, N., Kirillov, A., Zagoruyko, S.: End-to-end object detection with transformers. In: Vedaldi, A., Bischof, H., Brox, T., Frahm, J.-M. (eds.) ECCV 2020. LNCS, vol. 12346, pp. 213–229. Springer, Cham (2020). https://doi.org/10.1007/978-3-030-58452-8_13

2. Chao, Y.W., Liu, Y., Liu, X., Zeng, H., Jia, D.: Learning to detect human-object interactions. In: Workshop on Applications of Computer Vision (2018)

3. Chen, M., Liao, Y., Liu, S., Chen, Z., Wang, F., Qian, C.: Reformulating HOI detection as adaptive set prediction. In: Proceedings of the IEEE/CVF Conference on Computer Vision and Pattern Recognition, pp. 9004–9013 (2021)

4. Gao, C., Xu, J., Zou, Y., Huang, J.-B.: DRG: dual relation graph for human-object interaction detection. In: Vedaldi, A., Bischof, H., Brox, T., Frahm, J.-M. (eds.) ECCV 2020. LNCS, vol. 12357, pp. 696–712. Springer, Cham (2020). https://doi.org/10.1007/978-3-030-58610-2_41

5. Gao, C., Zou, Y., Huang, J.B.: iCAN: instance-centric attention network for human-object interaction detection. arXiv preprint arXiv:1808.10437 (2018)

6. Gkioxari, G., Girshick, R., Dollár, P., He, K.: Detecting and recognizing human-object interactions. In: Proceedings of the IEEE Conference on Computer Vision and Pattern Recognition, pp. 8359–8367 (2018)

7. Gupta, S., Malik, J.: Visual semantic role labeling. arXiv preprint arXiv:1505.04474 (2015)

8. Hou, Z., Yu, B., Qiao, Y., Peng, X., Tao, D.: Affordance transfer learning for human-object interaction detection. In: Proceedings of the IEEE/CVF Conference on Computer Vision and Pattern Recognition, pp. 495–504 (2021)

9. Kim, B., Choi, T., Kang, J., Kim, H.J.: UnionDet: union-level detector towards real-time human-object interaction detection. In: Vedaldi, A., Bischof, H., Brox, T., Frahm, J.-M. (eds.) ECCV 2020. LNCS, vol. 12360, pp. 498–514. Springer, Cham (2020). https://doi.org/10.1007/978-3-030-58555-6_30

10. Kim, B., Lee, J., Kang, J., Kim, E.S., Kim, H.J.: HOTR: end-to-end human-object interaction detection with transformers. In: Proceedings of the IEEE/CVF Conference on Computer Vision and Pattern Recognition, pp. 74–83 (2021)

11. Kim, B., Mun, J., On, K.W., Shin, M., Lee, J., Kim, E.S.: MSTR: multi-scale transformer for end-to-end human-object interaction detection. In: Proceedings of the IEEE/CVF Conference on Computer Vision and Pattern Recognition, pp. 19578–19587 (2022)

12. Liao, Y., Liu, S., Wang, F., Chen, Y., Qian, C., Feng, J.: PPDM: parallel point detection and matching for real-time human-object interaction detection. In: Proceedings of the IEEE/CVF Conference on Computer Vision and Pattern Recognition, pp. 482–490 (2020)

13. Lin, T.-Y., et al.: Microsoft COCO: common objects in context. In: Fleet, D., Pajdla, T., Schiele, B., Tuytelaars, T. (eds.) ECCV 2014. LNCS, vol. 8693, pp. 740–755. Springer, Cham (2014). https://doi.org/10.1007/978-3-319-10602-1_48

14. Liu, Y., Chen, Q., Zisserman, A.: Amplifying key cues for human-object-interaction detection. In: Vedaldi, A., Bischof, H., Brox, T., Frahm, J.-M. (eds.) ECCV 2020. LNCS, vol. 12359, pp. 248–265. Springer, Cham (2020). https://doi.org/10.1007/978-3-030-58568-6_15

15. Loshchilov, I., Hutter, F.: Decoupled weight decay regularization. arXiv preprint arXiv:1711.05101 (2017)

16. Radford, A., et al.: Learning transferable visual models from natural language supervision. In: International Conference on Machine Learning, pp. 8748–8763. PMLR (2021)

17. Tamura, M., Ohashi, H., Yoshinaga, T.: QPIC: query-based pairwise human-object interaction detection with image-wide contextual information. In: Proceedings of the IEEE/CVF Conference on Computer Vision and Pattern Recognition, pp. 10410–10419 (2021)

18. Ulutan, O., Iftekhar, A., Manjunath, B.S.: VSGNet: spatial attention network for detecting human object interactions using graph convolutions. In: Proceedings of the IEEE/CVF Conference on Computer Vision and Pattern Recognition, pp. 13617–13626 (2020)

19. Vaswani, A., et al.: Attention is all you need. In: Advances in Neural Information Processing Systems, vol. 30 (2017)

20. Wan, B., Zhou, D., Liu, Y., Li, R., He, X.: Pose-aware multi-level feature network for human object interaction detection. In: Proceedings of the IEEE/CVF International Conference on Computer Vision, pp. 9469–9478 (2019)

21. Wang, T., Yang, T., Danelljan, M., Khan, F.S., Zhang, X., Sun, J.: Learning human-object interaction detection using interaction points. In: Proceedings of the IEEE/CVF Conference on Computer Vision and Pattern Recognition, pp. 4116–4125 (2020)

22. Zhang, A., et al.: Mining the benefits of two-stage and one-stage HOI detection, vol. 34, pp. 17209–17220 (2021)

23. Zhang, F.Z., Campbell, D., Gould, S.: Spatially conditioned graphs for detecting human-object interactions. In: Proceedings of the IEEE/CVF International Conference on Computer Vision, pp. 13319–13327 (2021)

24. Zhang, F.Z., Campbell, D., Gould, S.: Efficient two-stage detection of human-object interactions with a novel unary-pairwise transformer. In: Proceedings of the IEEE/CVF Conference on Computer Vision and Pattern Recognition, pp. 20104–20112 (2022)

25. Zou, C., et al.: End-to-end human object interaction detection with HOI transformer. In: Proceedings of the IEEE/CVF Conference on Computer Vision and Pattern Recognition, pp. 11825–11834 (2021)

Intelligent Trajectory Tracking Control of Unmanned Parafoil System Based on SAC Optimized LADRC

Yuemin Zheng[1], Jin Tao[2(✉)], Qinglin Sun[1(✉)], Jinshan Yang[1], Hao Sun[1], Mingwei Sun[1], and Zengqiang Chen[1]

[1] Nankai University, Tianjin 300350, China
{sunql,sunh,chenzq}@nankai.edu.cn, yangjs@mail.nankai.edu.cn
[2] Silo AI, 00100 Helsinki, Finland
taoj@nankai.edu.cn

Abstract. The unmanned parafoil system has become increasingly popular in a variety of military and civilian applications due to its remarkable carrying capacity, as well as its capacity to modify its flight path by adjusting the left or right paracord. To ensure the unmanned system's safe completion of its flight mission, precise trajectory tracking control of the parafoil is essential. This paper presents an intelligent trajectory tracking approach that employs a soft actor-critic (SAC) algorithm optimized linear active disturbance rejection control (LADRC). Using the eight-degree-of-freedom (DOF) parafoil model as a basis, we have developed a trajectory tracking guidance law to address the underactuated problem. To ensure that the system's yaw angle accurately tracks the guided yaw angle, we have designed a second-order LADRC. Additionally, SAC algorithm is used to obtain adaptive parameters for the controller, ultimately enhancing tracking performance. Simulation results show that the proposed method can overcome the wind disturbance and achieve the convergence of tracking errors.

Keywords: Trajectory tracking · Parafoil system · SAC · LADRC

1 Introduction

The powered parafoil is an exceptional unmanned aerial system consisting of a parafoil canopy, connecting rope and payload. It possesses the remarkable capability to carry heavy objects and remain airborne for extended periods, making it a preferred choice for fixed-point airdrops or aircraft recovery operations [1,2]. While parafoils offer great potential, they present significant challenges. On the one hand, they are complex nonlinear systems with coupling, and on the other hand, they are highly influenced by the environment during flight. Therefore,

---

Supported by College of Artificial Intelligence, Nankai University.

B. Luo et al. (Eds.): ICONIP 2023, LNCS 14451, pp. 97–108, 2024.
https://doi.org/10.1007/978-981-99-8073-4_8

it is essential to overcome these challenges for powered parafoils to effectively carry out their missions in practical applications.

The high cost of conducting actual flight experiments necessitates the use of theoretical simulation to verify control methods before flight. This crucial process relies heavily on accurate modeling of the parafoil system. Currently, research on parafoil models mainly involves six-degree-of-freedom (DOF) models [3], eight-DOF models [4,5], and nine-DOF models [6]. The selection of varying degrees of freedom depends on diverse application scenarios and specific problem requirements.

Scholars have conducted a series of research on motion tracking control, particularly in the area of path following control, utilizing these models. Path following control ensures that the controlled object converges to a desired path without temporal constraints [7]. For example, Sun et al. [8] developed a sliding mode surface based on the heading angle deviation and lateral distance deviation, and designed a sliding mode controller (SMC) based on the linear extended state observer (LESO). The hardware-in-the-loop simulation verified the effectiveness of the proposed trajectory tracking control method. Luo et al. [9] proposed a decoupling control approach using linear active disturbance rejection control (LADRC)-based feedforward coupling compensation for trajectory tracking control based on the 8-DOF powered parafoil. However, these methods are still path following control in nature, because the convergence of the longitudinal error is not achieved. Based on the six-DOF model, Li et al. [10] designed the lateral, longitudinal and velocity PID controllers for powered parafoil, and optimized the controller parameters using the ecosystem particle swarm optimization (ESPSO) algorithm. However, powered parafoils are usually not configured with forward thrust, making this approach difficult to implement. In general, research on parafoil trajectory tracking is relatively limited compared to path following control, and the majority of existing studies rely on traditional control methods.

There are currently limited intelligent control methods available for parafoil tracking control. However, with the emergence of the artificial intelligence era, intelligent controllers are becoming an inevitable trend in this field. In our previous work, we developed the path following control strategy for parafoils using deep reinforcement learning (DRL) algorithms [11,12]. The DRL algorithm combines the independent decision-making concept of reinforcement learning with the computing capabilities of deep learning, resulting in excellent performance in intelligent control of complex systems [13–15]. This paper applies the soft actor-critic (SAC) algorithm [16,17] in DRL to trajectory tracking control of parafoil system, which is capable of handling systems with continuous state and action spaces.

The main contributions of this paper are summarized as follows:

- The trajectory tracking control of parafoil suffers from an underactuation problem. To address this challenge, we propose a trajectory tracking guidance law.

- LADRC is utilized to track the plane trajectory. The LADRC exhibits model-free characteristic, making it an ideal choice for this type of control.
- By utilizing the SAC algorithm in DRL to optimize controller parameters, the proposed control system is able to more effectively counteract wind disturbances, while also advancing the development of intelligent controllers.

2   A Brief Introduction to a 8-DOF Parafoil Model

A parafoil is mainly composed of a canopy and a payload, as shown in Fig. 1. Parafoil canopy and payload are connected through parafoil ropes at points c_1 and c_2, then by pulling the left or right ropes, the shape and angle of the canopy can be altered, changing the aerodynamic forces and enabling control of the flight direction. During the modeling process of the eight-DOF parafoil, the forces acting on the canopy include aerodynamic force, gravity, and tension from the connecting rope. Meanwhile, the forces acting on the load consist of aerodynamic force, gravity and pulling force from the connecting rope. This paper will not delve into the specifics of the modeling process, but interested readers can refer to Ref. [4] for further details.

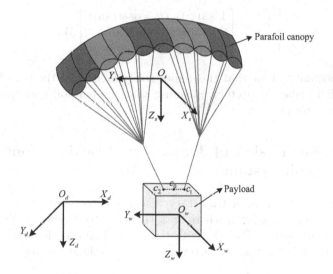

Fig. 1. Schematic diagram of a powered parafoil coordinate system

In Fig. 1, $O_d X_d Y_d Z_d$, $O_s X_s Y_d Z_s$ and $O_w X_w Y_w Z_w$ are earth coordinate system, canopy coordinate system and payload coordinate system respectively. Define system state variables as $x_s = \left[V_w^T, W_w^T, V_s^T, W_s^T, \dot{\psi}_r, \dot{\theta}_r\right]^T$, where $V_w = [u_w, v_w, w_w]^T$ and $W_w = [p_w, q_w, r_w]^T$ are velocity vector and angular velocity vector in payload coordinate system, respectively; $V_s = [u_s, v_s, w_s]^T$ and

$W_s = [p_s, q_s, r_s]^T$ represent the velocity vector and the angular velocity vector in the canopy coordinate system, respectively; θ_r and ψ_r denote the relative pitch angle and relative yaw angle between the parafoil canopy and payload, respectively. And the dynamic equation of powered parafoil can be abbreviated as

$$\dot{x}_s = f(x_s, u) \tag{1}$$

where f represents the function between the state variables and input variables; $u(cm)$ represents the flap deflection, and $u > 0$ means the parafoil turns left, otherwise $u < 0$ means the parafoil turns right.

Assuming (x, y, z) represents the position coordinates in the earth coordinate system, then there is

$$
\begin{bmatrix} \dot{x} \\ \dot{y} \\ \dot{z} \end{bmatrix} = \begin{bmatrix} \cos\theta\cos\psi & \cos\theta\sin\psi & -\sin\theta \\ \sin\phi\sin\theta\cos\psi - \cos\phi\sin\psi & \sin\phi\sin\theta\sin\psi + \cos\phi\cos\psi & \sin\phi\cos\theta \\ \cos\phi\sin\theta\cos\psi + \sin\phi\sin\psi & \cos\phi\sin\theta\sin\psi - \sin\phi\cos\psi & \cos\phi\cos\theta \end{bmatrix}^T V_s \tag{2}
$$

where ϕ, θ, and ψ stand for the roll, pitch, and yaw angles, respectively. And these Euler angles can be obtained by the following expression:

$$
\begin{bmatrix} \dot{\phi} \\ \dot{\theta} \\ \dot{\psi} \end{bmatrix} = \begin{bmatrix} 1 & \sin\phi\tan\theta & \cos\phi\tan\theta \\ 0 & \cos\phi & -\sin\phi \\ 0 & \sin\phi/\cos\theta & \cos\phi/\cos\theta \end{bmatrix} W_s \tag{3}
$$

It is important to note that this paper solely focuses on the tracking control of the parafoil plane's trajectory, meaning that the parafoil undergoes free fall during the control process.

3   Controller Design of Trajectory Tracking Control for Parafoil System

In this paper, the goal of trajectory tracking is to make the parafoil track a target point $(x_d(\varpi), y_d(\varpi))$, where ϖ is the trajectory variable. Establish the target coordinate system $O_v X_v Y_v$ with the target point as the origin, shown in Fig. 2, then there is the tracking error in target coordinate system as

$$
\begin{bmatrix} x_e \\ y_e \end{bmatrix} = \begin{bmatrix} \cos(\psi_d) & \sin(\psi_d) \\ -\sin(\psi_d) & \cos(\psi_d) \end{bmatrix} \begin{bmatrix} x'_e \\ y'_e \end{bmatrix} \tag{4}
$$

with

$$
\begin{cases} x'_e = x - x_d \\ y'_e = y - y_d \end{cases} \tag{5}
$$

where $\psi_d(\varpi) = \arctan(\dot{y}_d(\varpi)/\dot{x}_d(\varpi))$.

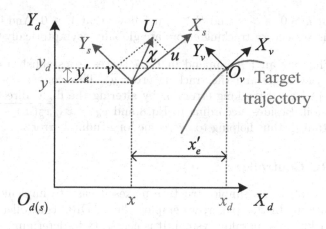

Fig. 2. Schematic diagram of plane trajectory tracking

3.1 Trajectory Tracking Guidance Law

According to Eq. 4, the tracking errors can be further derived as

$$\begin{cases} \dot{x}_e = U\cos\left(\psi + \chi - \psi_d\right) - u_d + \dot{\psi}_d y_e \\ \dot{y}_e = U\sin\left(\psi + \chi - \psi_d\right) - \dot{\psi}_d x_e \end{cases} \tag{6}$$

where $U = \sqrt{u^2 + v^2}$, $\chi = \arctan(v/u)$, and $u_d = \dot{\varpi}\sqrt{\dot{x}_d^2(\varpi) + \dot{y}_d^2(\varpi)}$.

Given the requirement of using a control variable u to achieve convergence of both yaw angle and tracking errors, it is evident that the trajectory tracking is an underactuation problem. Thus a trajectory tracking guidance law is designed

$$\psi_G = \psi_d + \arctan\left(-\alpha y_e\right) - \chi \tag{7}$$

where $\alpha > 0$. Besides, take u_d as

$$u_d = U\cos\left(\psi + \chi - \psi_d\right) + k x_e \tag{8}$$

with $k > 0$.

Theorem 1. *The convergence of plane trajectory tracking errors x_e and y_e can be guaranteed if the yaw angle (ψ) is able to accurately track the guidance yaw angle (ψ_G).*

Proof. Construct a Lyapunov function as $V = \frac{1}{2}x_e^2 + \frac{1}{2}y_e^2$, then there is

$$\dot{V} = x_e\dot{x}_e + y_e\dot{y}_e \tag{9}$$

By utilizing $\psi = \psi_G$ and incorporating Eqs. 6, 7, and 8, Eq. 9 can be expressed as

$$\dot{V} = -kx_e^2 - \frac{\alpha U y_e^2}{\sqrt{1 + \alpha^2 y_e^2}} \tag{10}$$

Given that $k > 0$, $\alpha > 0$, and $U \geq 0$, it follows that $\dot{V} \leq 0$. And it indicates that the plane trajectory tracking errors are globally asymptotically stable.

Remark 1. The yaw angle command in Eq. 7 serves to establish a connection between the yaw angle and the tracking error y_e. This connection allows for the reducing of lateral tracking errors y_e by altering the flight direction of the powered parafoil. Besides, according to Eq. 8 and $u_d = \dot{\varpi}\sqrt{\dot{x}_d^2(\varpi) + \dot{y}_d^2(\varpi)}$, the $\dot{\varpi}$ can be obtained, thus helping to overcome longitudinal error x_e.

3.2   LADRC Controller

The LADRC controller's simple structure makes it easy to implement in real-world applications. In this paper, we employ the LADRC controller to achieve trajectory tracking of a parafoil system. It is necessary to determine the order of the controlled system, specifically the relationship between the output variable ψ and the control variable u.

According to Eq. 3, there is

$$
\begin{aligned}
\ddot{\psi} = {} & \frac{\sin\phi}{\cos\theta}\dot{q} + \frac{\cos\phi}{\cos\theta}\dot{r} + \frac{\sin\theta\sin 2\phi}{\cos^2\theta}q^2 - \frac{\sin\theta\sin 2\phi}{\cos^2\theta}r^2 \\
& + \frac{\cos\phi}{\cos\theta}pq - \frac{\sin\phi}{\cos\theta}pr + \frac{2\sin\theta\cos 2\phi}{\cos^2\theta}qr
\end{aligned}
\tag{11}
$$

Then a second-order system can be expressed

$$
\ddot{\psi} = f_1(\cdot) + f_2(u)
\tag{12}
$$

where $f_1(\cdot)$ represents the disturbance function, including the internal state and external disturbance information, and $f_2(u)$ is an function about the control variable. By defining $y = \psi - \psi_G$, it becomes apparent that the desired value of y is 0. Additionally, y can be seen as a second-order system with respect to the control variable u.

Hence, we can obtain the following expression

$$
\ddot{y} = f + b_0 u
\tag{13}
$$

where f denotes the total unknown disturbances of the system, and b_0 is an adjustable parameter. Then define the states as $x_1 = y$, $x_2 = \dot{y}$, and $x_3 = f$, and a LESO can be designed to estimate the disturbance,

$$
\begin{cases}
\dot{\hat{x}} = A\hat{x} + Bu + L(y - \hat{y}) \\
\hat{y} = C\hat{x}
\end{cases}
\tag{14}
$$

where $A = \begin{bmatrix} 0 & 1 & 0 \\ 0 & 0 & 1 \\ 0 & 0 & 0 \end{bmatrix}$, $B = \begin{bmatrix} 0 \\ b_0 \\ 0 \end{bmatrix}$, $L = \begin{bmatrix} l_1 \\ l_2 \\ l_3 \end{bmatrix}$, $C = \begin{bmatrix} 1 \\ 0 \\ 0 \end{bmatrix}^T$, and $\hat{x} = \begin{bmatrix} \hat{x}_1 \\ \hat{x}_2 \\ \hat{x}_3 \end{bmatrix}$. The vector L represents the gain of the observer, while $\hat{x}$ denotes the observed states of x. By adjusting the observer gains, it is possible to get $\hat{x} \approx x$. After that, the following PD control law is designed to eliminate the estimated disturbance,

$$u = \frac{-k_p \hat{x}_1 - k_d \hat{x}_2 - \hat{x}_3}{b_0} \tag{15}$$

In Eq. 15, k_p and k_d are controller parameters. And by substituting Eq. 15 into Eq. 13, it is apparent that the disturbance f is canceled out. In addition, during the controller design process, there are many parameters that need to be adjusted, i.e., b_0, l_1, l_2, l_3, k_p and k_d. With the help of the pole configuration method in Ref. [18], it can be obtained that $l_1 = 3\omega_o$, $l_2 = 3\omega_o{}^2$, $l_3 = \omega_o{}^3$, where ω_o is positive.

4 SAC Optimized LADRC

The DRL algorithm exhibits robust autonomous decision-making capabilities. Therefore, this paper employs a SAC algorithm to derive the k_p and k_d in real-time for the controller, as shown in Fig. 3. This approach facilitates adaptation of the LADRC controller to changing environmental conditions and enables adaptive trajectory tracking.

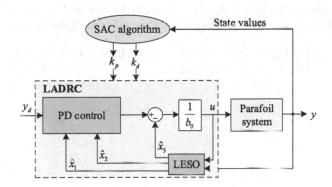

Fig. 3. Schematic diagram of SAC-optimized LADRC

The SAC algorithm comprises of two critic networks $Q_{\theta_1}(s, a)$, $Q_{\theta_2}(s, a)$, two target networks $Q_{\theta_1'}(s, a)$ and $Q_{\theta_2'}(s, a)$, and an actor network $\pi_\phi(s)$, as shown in Algorithm 1. The difference between the Q network in the SAC algorithm and other DRL algorithms lies in the addition of entropy $H(\pi_\phi(\cdot|s)) = -\log(\pi_\phi(\cdot|s))$, which improves the exploration performance of SAC.

In this paper, our objective is to utilize the SAC algorithm to train an agent so as to acquire the adaptive controller parameters based on the observed states. The environment is composed of the established parafoil model and LADRC controller, while the agent primarily refers to the actor network in SAC. To

---

Algorithm 1. SAC Algorithm

Initialize critic network weights $Q_{\theta_1}(s,a)$, $Q_{\theta_2}(s,a)$ and actor network $\pi_\phi(s)$ with random parameters θ_1, θ_2 and ϕ.

Initialize target networks $Q_{\theta_1'}(s,a)$ and $Q_{\theta_2'}(s,a)$ with weights $\theta_1' \leftarrow \theta_1$ and $\theta_2' \leftarrow \theta_2$.

Initialize replay buffer $\mathcal{D}$.

if $t \leq T$ **then**

    Sample mini-batch of m transitions $\mathcal{B} = \{(s,a,r,s')\}$ from $\mathcal{D}$.

    Compute targets for the Q functions:

$$y(s',r) = r + \gamma \left(\min_{i=1,2} Q_{\theta'_i}(s',a') - \alpha \log \pi_\phi(a'|s') \right).$$

    Update critics by gradient decent $\nabla_{\theta_i} \frac{1}{m} \sum_{(s,a,r,s')\in\mathcal{B}} (Q_{\theta_i}(s,a) - y(s',r))^2$.

    Update policy by gradient ascent $\nabla_\phi \frac{1}{m} \sum_{(s,a,r,s')\in\mathcal{B}} \left(\min_{i=1,2} Q_{\theta_i}(s,\cdot) - \alpha \log \pi_\phi(\cdot|s) \right)$

    Update target networks by moving average method: $\theta_i' \leftarrow \tau\theta_i + (1-\tau)\theta_i'$.

end if

---

ensure proper interaction between the environment and the agent, it is necessary to first define the state space, action space, and reward value.

Since the parameters to be optimized are k_p and k_d, the action space is expressed as $\{a_1, a_2 \in \mathcal{A} \,|\, a_1 = k_p, a_2 = k_d\}$. As for the state space, it is defined as $\{s_1, s_2 \in \mathcal{S} \,|\, s_1 = y, s_2 = \dot{y}\}$. Then the following reward function is designed

$$r = \begin{cases} -(|y| + |\dot{y}|) + 10\tanh(1/|y|), & \text{if } |y| \leq 0.1 \,\&\, |\dot{y}| \leq 1 \\ -(|y| + |\dot{y}|), & \text{otherwise} \end{cases} \tag{16}$$

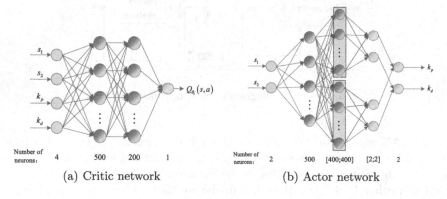

(a) Critic network   (b) Actor network

Fig. 4. Structure of critic and actor network

The objective behind formulating Formula 16 is to establish a direct relationship between the state value y, $\dot{y}$ and the magnitude of the reward value, whereby the closer the state value approaches zero, the higher the corresponding reward value will be. Also, to encourage the power parafoil to stay in a steady state ($|y| \leq 0.1 \& |\dot{y}| \leq 1$), a positive reward $10 \tanh(1/|y|)$ is incorporated.

5   Simulation Results

In this section, the effectiveness of the proposed SAC optimized LADRC method is verified by simulations. The main dimensions of the parafoil system studied are shown in Table 1. Moreover, the SAC algorithm employs the following parameters $\gamma = 0.9$ and $\tau = 0.001$. Figure 4 displays the structure diagram of the critic and actor networks.

Table 1. Main dimensions of the powered parafoil system

Parameter description	Value
Wing span	4.5 m
Chord length	1.3 m
Mass of parafoil	1.7 kg
Mass of payload	20 kg
Wing area	6.5 m^2
Aspect ratio	3.46
Rope length	3 m
Installation angle	10°

For the control system, the simulation sample step is 0.01 s. The parameters are set as $b_0 = 0.5$, $\omega_o = 6$, $k = 0.1$ and $\alpha = 0.2$, and the action space for SAC is limited as

$$\begin{cases} k_p \in [0.001, 0.02] \\ k_d \in [0.02, 0.5] \end{cases} \tag{17}$$

where this range is determined by trial and error on conventional controllers.

The parafoil system was initially positioned at $(0, 20)$ m with an initial velocity of $V_s = V_w = [14.9, 0, 2.1]^T$ m/s and $W_s = W_w = [0, 0, 0]^T$ rad/s. And the target trajectory is described as

$$\begin{cases} x_d(\varpi) = 8\varpi \\ y_d(\varpi) = 8\varpi \end{cases} \tag{18}$$

with the update rule $\dot{\varpi} = \dfrac{u_d}{\sqrt{\dot{x}_d^2(\varpi) + \dot{y}_d^2(\varpi)}}$. Besides, the following two cases are considered:

Case A: The influence of wind on the system is not considered;

Case B: The wind of 3 m/s in the X-direction, 2 m/s in the Y-direction is added at 60 s, and lasts until the end of the simulation.

The simulation results are shown in Figs. 5, 6 and 7. Figure 5(a) visually depicts the flight trajectory of the parafoil in the plane, with 'LADRC with fixed parameter' representing the trajectory obtained using the boundary value $k_p = 0.001$ and $k_d = 0.02$ in windless conditions. Figure 5(b) shows the change process of the control variable u. By observing Fig. 6, it is evident that the proposed method is a highly effective approach for trajectory tracking control. And even under the influence of wind, the tracking error can reach a steady state. Figure 7 shows the controller parameters obtained by the SAC algorithm without wind disturbance. It should be pointed out that the output of the SAC algorithm is that the action is obtained after passing through the Gaussian distribution, so the parameter changes in Fig. 7 are relatively drastic.

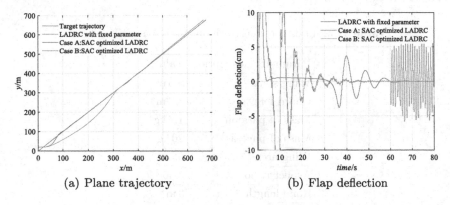

(a) Plane trajectory   (b) Flap deflection

Fig. 5. Trajectory tracking and corresponding control variable

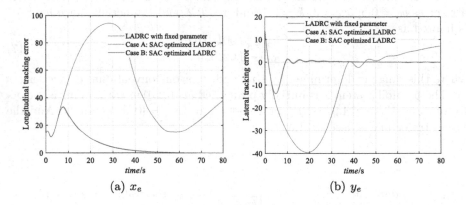

(a) x_e (b) y_e

Fig. 6. Trajectory tracking errors

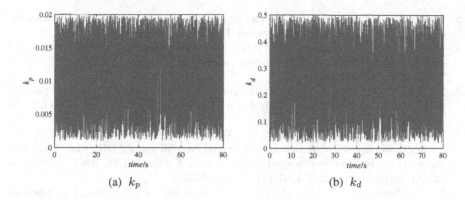

(a) k_p (b) k_d

Fig. 7. Adaptive parameters

6 Conclusion

This paper presents an intelligent trajectory tracking approach for unmanned parafoil systems, based on a SAC optimized LADRC strategy. The parafoil system's plane trajectory tracking requires the pulling of either the left or right paracord, resulting in fewer control variables than controlled variables. To address this underactuated problem, a guidance law is proposed. Next, a second-order LADRC is designed, and the SAC algorithm is employed to obtain the adaptive parameters of the LADRC. At last, we utilized a straight-line trajectory for conducting the simulation test and observed that the proposed method effectively achieves trajectory tracking control of the parafoil.

Acknowledgements. This work was supported by the National Natural Science Foundation of China (Grant Nos. 61973172, 61973175, 62003175, 62003177 and 62073177) and Key Technologies Research and Development Program of Tianjin (Grant No. 19JCZDJC32800).

References

1. Tao, J., Sun, Q., Tan, P., Chen, Z., He, Y.: Active disturbance rejection control (ADRC)-based autonomous homing control of powered parafoils. Nonlinear Dyn. **86**(3), 1461–1476 (2016). https://doi.org/10.1007/s11071-016-2972-1
2. Sun, H., Sun, Q., Sun, M., et al.: Accurate modeling and homing control for parafoil delivery system based on wind disturbance rejection. IEEE Trans. Aerosp. Electron. Syst. **58**(4), 2916–2934 (2022)
3. Wachlin, J., Ward, M., Costello, M.: In-canopy sensors for state estimation of precision guided airdrop systems. Aerosp. Sci. Technol. **90**, 357–367 (2019)
4. Zhu, E., Sun, Q., Tan, P., et al.: Modeling of powered parafoil based on Kirchhoff motion equation. Nonlinear Dyn. **79**, 617–629 (2015)
5. Tan, P., Sun, M., Sun, Q., et al.: Dynamic modeling and experimental verification of powered parafoil with two suspending points. IEEE Access **8**, 12955–12966 (2020)

6. Guo, Y., Yan, J., Wu, C., et al.: Modeling and practical fixed-time attitude tracking control of a paraglider recovery system. ISA Trans. **128**, 391–401 (2022)
7. Lakhekar, G.V., Waghmare, L.M.: Robust self-organising fuzzy sliding mode-based path-following control for autonomous underwater vehicles. J. Mar. Eng. Technol. **22**(3), 131–152 (2023)
8. Sun, Q., Yu, L., Zheng, Y., et al.: Trajectory tracking control of powered parafoil system based on sliding mode control in a complex environment. Aerosp. Sci. Technol. **122**, 107406 (2022)
9. Luo, S., Sun, Q., Sun, M., Tan, P., Wu, W., Sun, H., Chen, Z.: On decoupling trajectory tracking control of unmanned powered parafoil using ADRC-based coupling analysis and dynamic feedforward compensation. Nonlinear Dyn. **92**(4), 1619–1635 (2018). https://doi.org/10.1007/s11071-018-4150-0
10. Li, Y., Zhao, M., Yao, M., et al.: 6-DOF modeling and 3D trajectory tracking control of a powered parafoil system. IEEE Access **8**, 151087–151105 (2020)
11. Zheng, Y., Tao, J., Sun, Q., Zeng, X., Sun, H., Sun, M., Chen, Z.: DDPG-based active disturbance rejection 3D path-following control for powered parafoil under wind disturbances. Nonlinear Dyn. **111**, 11205–11221 (2023)
12. Zheng, Y., Tao, J., Sun, Q., et al.: Deep-reinforcement-learning-based active disturbance rejection control for lateral path following of parafoil system. Sustainability **15**(1), 435 (2022)
13. Gaudet, B., Linares, R., Furfaro, R.: Deep reinforcement learning for six degree-of-freedom planetary powered descent and landing. arXiv preprint arXiv:1810.08719. (2018) https://doi.org/10.48550/arXiv.1810.08719
14. Challita, U., Saad, W., Bettstetter, C.: Interference management for cellular-connected UAVs: a deep reinforcement learning approach. IEEE Trans. Wirel. Commun. **18**(4), 2125–2140 (2019)
15. Wang, C., Wang, J., Shen, Y., et al.: Autonomous navigation of UAVs in large-scale complex environments: a deep reinforcement learning approach. IEEE Trans. Veh. Technol. **68**(3), 2124–2136 (2019)
16. Haarnoja, T., Zhou, A., Hartikainen, K., et al.: Soft actor-critic algorithms and applications. arXiv preprint arXiv:1812.05905. (2018) https://doi.org/10.48550/arXiv.1812.05905
17. Haarnoja, T., Zhou, A., Abbeel, P., et al.: Soft actor-critic: off-policy maximum entropy deep reinforcement learning with a stochastic actor. In: Proceedings of the 35th International Conference on Machine Learning, PMLR 2018, vol. 80, pp. 1861–1870 (2018)
18. Gao, Z.: Scaling and bandwidth-parameterization based controller tuning. In: Proceedings of the 2003 American Control Conference, pp. 4989–4996. IEEE (2003) https://doi.org/10.1109/ACC.2003.1242516

CATS: Connection-Aware and Interaction-Based Text Steganalysis in Social Networks

Kaiyi Pang[1], Jinshuai Yang[1], Yue Gao[1(✉)], Minhao Bai[1], Zhongliang Yang[3], Minghu Jiang[1], and Yongfeng Huang[1,2]

[1] Tsinghua University, Beijing 100084, China
{pky22,yjs20,bmh20}@mails.tsinghua.edu.cn,
gaoyue2022@mail.tsinghua.edu.cn, {jiang.mh,yfhuang}@tsinghua.edu.cn
[2] Zhongguancun Laboratory, Beijing 100094, China
[3] School of Cyberspace Security, Beijing University of Posts and Telecommunications, Beijing 100876, China
yangzl@bupt.edu.cn

Abstract. The generative linguistic steganography in social networks have potential huge abuse and regulatory risks, with serious implications for information security, especially in the era of large language models. Many works have explored detecting steganographic texts with progressively enhanced imperceptibility, but they can only achieve poor performance in real social network scenarios. One key reason is that these methods primarily focus on linguistic features, which are extremely insufficient owing to the fragmentation of social texts. In this paper, we propose a novel method called CATS (**C**onnection-aware and inter**A**ction-based **T**ext **S**teganalysis) to effectively detected the potentially malicious steganographic texts. CATS captures social networks connection information by graph representation learning, enhances linguistic features by contrastive learning and fully integrates features above via a novel features interaction module. Our experimental results demonstrate that CATS outperforms existing methods by exploiting social network graph structure features and interactions in social network environments.

Keywords: linguistic steganalysis · social networks · graph structure

1 Introduction

Steganography is a significant technology to embed secret messages such as malicious programs or criminal schemes in seemingly innocuous carriers, making these messages imperceptible by surveillance in public channels. It is intuitive that steganography can be vulnerable to exploitation by unscrupulous individuals [1]. To avoid these potential malicious abuses, research on detecting steganographic carriers (termed "steganalysis") has emerged in recent years. However,

B. Luo et al. (Eds.): ICONIP 2023, LNCS 14451, pp. 109–121, 2024.
https://doi.org/10.1007/978-981-99-8073-4_9

to escape detection, recent advanced steganography methods try to minimize the statistical distribution difference of steganographic carriers (referred to as "stegos") from the statistical distribution of normal carriers (referred to as "covers") when embedding secret messages. The evolution of steganography methods calls steganalysis to keep designing more powerful feature extractors f that can keenly perceive the statistical difference of stegos and covers.

Considerable kinds of information carriers can be used by steganography to transfer messages, where some are tremendously harmful. Especially, in this era social platforms can be easily accessible to exchange information. Social networks have become one of the best platforms for information hiding due to its wide user base, convenient information dissemination channels and natural anonymity. Social medias such as image [2], audio [3], text [4] can be employed for steganography. Among these carriers, the popularity, accessibility and robustness make the text one of the most widely-used media steganographic carriers in social networks, leading to a surge of research on text steganography (also called linguistic steganography) [5,6], especially on generative linguistic steganography [4,7–10], due to its high payload and flexibility in recent years. The state-of-the-art generative linguistic steganography methods mainly achieve the embedding of secret information by encoding the conditional probabilities of the language model. As more powerful language models [13,14,36] and encoding algorithms [4,7,9,10,15] are developed, the statistical divergence between stegos and covers diminishes, posing a serious challenge for steganalysis. Especially in this stage recent large language models (such as ChatGPT [11], LLaMA [12]) have shown impressive generation ability, which will further expand the possibilities of generative linguistic steganography and will also further challenges linguistic steganalysis.

Notwithstanding the challenges, linguistic steganalysis researchers have been working towards identifying statistical differences between stegos and covers. Earlier researchers applied classical models in text classification to steganalysis, but these traditional methods could only capture shallow statistical information of texts by constructing manual features such as word frequencies or transition probability between words [16–20]. Recently, natural language processing enables deeper semantic information mining for steganalysis without manual features. Researchers have fully explored linguistic features from multiple aspects, ranging from single feature [14,21,22] to multi-specific scale features fusion [23–26], from features between adjacent words [21] to graph features [27,28], from sentence semantic features [29] to combined syntactic structure information [30].

However, these methods suffer a fatal setback in the face of real-world social network scenarios, as pointed out by *Yang et al.* [31]. One important reason is that texts on social networks are extremely fragmented. On the one hand, texts on social networks are usually casual short messages with few linguistic feature patterns. On the other hand, these texts usually can not express completed semantics without considering their social context. This extreme fragmentation of social texts makes existing methods that highly rely on linguistic features can only get insufficient features to make an unwise decision.

It is easy to notice that social texts are not isolated but are usually interconnected with comments and replies. There is an intuitive idea that existing linguistic features-based steganalysis methods can be enhanced by exploiting connection information. Some group [31] has already recognized the need for designing steganalysis methods that are suitable for social network scenarios, and they propose a naive framework. To our best knowledge, the naive framework is the only work for this problem. However, this work equally treats linguistic features and connection features that are derived from different spaces, and only simply concatenates the features from different aspects without proper fusion, which will cause a misalignment problem and lead to limited performance in social network scenarios.

To effectively detect stegos in social networks, in this paper, we propose CATS, a new method for steganalysis that makes full use of the interaction information between text and their associated texts. Specifically, our model is enhanced from both external and internal aspects of the texts. We use Graph Neural Networks(GNNs) and graph representation learning to capture the social network context information of each text and bring in a contrast learning module to further enhanced the linguistic information of the text itself. After that, we introduce a module that facilitates the full interaction between graph features and linguistic features to get fully-integrated fusion features. Finally, CATS will incorporate linguistic features, context features, and fusion features to make decisions. Extensive experiments demonstrate that the interaction between texts and their associated texts can significantly enhance the detection in social network scenarios.

2 Proposed Method

2.1 Task Modeling

The essential task of steganalysis is to construct a classifier to distinguish stegos from covers. In general, for a text x_i, x_i will be judged by the classifier as steganographic text if

$$p(S|f(x_i)) > p(C|f(x_i)), \tag{1}$$

where $x_i = \{w_1, w_2, ..., w_n\}$ denotes that the i-th text with n words, S represents the category of stegos, C represents the category of covers and $f(\cdot)$ is the feature extractor learned by the steganalysis method. Existing methods, such as [21,23,27,32], have focused on designing better linguistic feature extractors. These methods are limited to social network scenarios where they can only get insufficient information from a single fragmented text. Besides, advanced steganographic algorithms [9] increasingly aim to reduce the differences in statistical distributions between stegos and covers, leading to more difficult detection.

In social network scenarios, texts and their interconnections form a graph $G(V, E)$ where V represents the set of nodes consisting of social network texts and E represents the set of edges corresponding to inter-textual relations such as comments and replies. When we focus on a particular social text x_i (equivalently, node v_i) and its connected relations, we are actually focusing on a subgraph $G_i(V_i, E_i)$, where $V_i \subseteq V$ and $E_i = \{e_{jk} = (v_j, v_k)|v_j, v_k \in V_i\}$. To enhance the analysis of stegos in social networks that often exhibit fragmentation and sparsity, we propose utilizing the graph structure $G_i(V_i, E_i)$ to its fullest extent. In addition to designing an improved textual linguistic feature extractor $f_{lin}(\cdot)$, we propose a graph structure feature extractor $f_g(\cdot)$ and an appropriate function $I_{Integrated}(\cdot)$ that effectively aligns linguistic features and graph structure features within a unified space to facilitate seamless interaction and fusion of the two feature types.

Therefore, we propose a new steganalysis model that will discriminate a text as a stego if:

$$p(S|I(f_{lin}(x_i), f_g(G_i))) > p(C|I(f_{lin}(x_i), f_g(G_i))), \tag{2}$$

where $f_{lin}(x_i)$ and $f_g(G_i)$ are the linguistic features and graph features of text x_i, respectively. And the $I(f_{lin}(x_i), f_g(G_i))$ is the integrated feature fed to the discriminator. Ultimately, we design a discriminator to decide whether a text is a steganographic text based on probability $p(\cdot|I(f_{lin}(x_i), f_g(G_i)))$. In other words, in order to capture more information of social network texts, we enriched and refined the features from external, internal and internal-external fusion respectively. The overall architecture of our method can be found in Fig. 1.

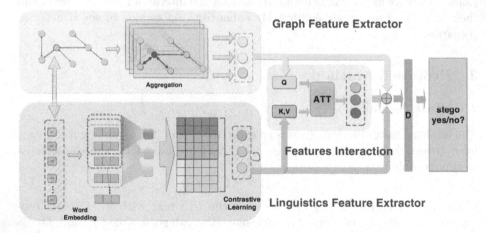

Fig. 1. The overall architecture of the proposed CATS. Our method includes a linguistic feature extractor $f_{lin}(\cdot)$, a social networks graph structure feature extractor $f_g(\cdot)$, a feature interaction module $I(f_{lin}, f_g)$ and a Discriminator.

2.2   Internal Feature Augmentation: Extracting Linguistic Features

In the realm of combating increasingly imperceptible steganography algorithms, researchers have proposed a sequence of potent linguistic feature extractors tailored to individual texts. In CATS, the backbone of the linguistic feature extraction module actually can be any existing method [22,23,27,30]. Specifically, taking into account the efficiency requirements for detecting massive amounts of text in social networks, we deployed an effective TS-CSW [23] method as the backbone of f_{lin}. Furthermore, we augmented our model with a contrast learning module, namely unsupervised SimCSE [33], to improve sentence representation in a self-supervised learning manner where dropout acts as minimal data augmentation. We produced two embeddings of the same input sentence x_i as "positive pairs" with different dropout masks and the other sentences in the same mini-batch are regarded as "negatives". Then the module predicts the positive one among the negatives. The final training target is incorporated into the contrastive loss function. which, for a batch of N sentences, is defined as follows:

$$loss_{CL} = -log \frac{e^{sim(h_i, h_i')}}{\sum_{j=1}^{N} e^{sim(h_i, h_j')}}, \tag{3}$$

where h_i and , h_i' are the embeddings of the sentence x_i produced using different dropout masks, $sim(\cdot)$ is a similarity measure function, which can be the cosine similarity.

2.3   External Feature Augmentation: Extracting Social Networks Graph Features

The limited applicability of existing methods to social networks is primarily attributed to their neglect of the potential graph structure. The problem of classifying steganographic texts in social networks can be cast as a node classification problem on the graph. To effectively leverage the graph structure of social network texts, we propose to use Gated Attention Networks (GAaN) [34] as the aggregation function $f_g(\cdot)$. GAaN is a multi-head attention-based aggregator with additional gates on the attention heads. The graph attention networks can capture the representation of G_i by aggregating the sub-graph information from neighbor nodes. We employ L-layer GNNs. For each GNN layer l, we use function M_l to aggregate the neighborhood information and function U_l to update the state information of node i :

$$z_i^l = \sum_{j \in \mathcal{N}(i)} M_l(h_i^{l-1}, h_j^{l-1}, e_{ij}), \tag{4}$$

$$h_i^l = U_l(h^{l-1}, z_i^l), \tag{5}$$

where z_i^l denotes for the central node i, the aggregated information obtained from the aggregation of neighbor nodes within l hops. $\mathcal{N}(i)$ is the set of neighbors of node i. e_{ij} is the edge connecting node i to node j, The initialization value of state information h_i^0 is the result of the corresponding linguistic feature $f_{lin}(x_i)$ of node i. The attention mechanism is used to assign more weight to the more critical information about different neighbors. For k-th attention head, the attention weight between node i and node j is calculated by:

$$\alpha_{ji} = \frac{e^{(d(Wh_i^{l-1}, Wh_j^{l-1}))}}{\sum_{u \in N(i)} e^{(d(Wh_i^{l-1}, Wh_u^{l-1}))}}, \tag{6}$$

where W is the shared weight matrix and $d(Wh_i^{l-1}, Wh_j^{l-1})$ is calculated by:

$$d(Wh_i^{l-1}, Wh_j^{l-1}) = LeakyReLU(\beta[Wh_i^{l-1}||Wh_j^{l-1}]), \quad (7)$$

where β is a vector parameters and $[\cdot||\cdot]$ denotes concatenation. Then the state information of center node i is calculated by:

$$h_i^{lk} = \sigma(\sum_{j \in N(i)} \alpha_{ji} Wh_i^{l-1}), \quad (8)$$

where σ is non-linear activation function, h_i^{lk} is the aggregated information of node i from k-th head. We also compute a soft gate between 0 and 1 to assign different importance to each head by employing a convolutional network ψ_g. ψ_g takes the node i features and neighboring node features to generate $\mathbf{g}_i$,

$$\mathbf{g}_i = [g_i^{(1)}, ..., g_i^{(K)}] = \psi_g(h_i^{l-1}, z_i^l), \quad (9)$$

where g_i^k is the gate value of the k-th head at node i. we concatenate h_i^{lk} from all attention heads with weights to obtain the final aggregated information:

$$h_i^l = \|_{k=1}^{K} g_i^{(k)} h_i^{lk}. \quad (10)$$

Finally, after passing through the L-layer GNN network, we obtain the graph structure G_i feature vectors of L-hop neighbors in an iterative manner.

2.4   Internal and External Hybrid Augmentation: Feature Interaction and Comprehensive Discriminator

After extracting linguistic and graph structure features, we obtained $f_{lin}(x_i)$ and $f_g(G_i)$, respectively. However, these features come from different modalities. So we need a suitable fusion module $I(f_{lin}(x_i), f_g(G_i))$ to fully interact with linguistic features and graph structure project them to the same space. and in this paper, we adopt an attention module to implement the fine-grained fusion, where the linguistic feature $f_{lin}(x_i)$ serves as the key and value, and the graph structure feature $f_g(G_i)$ serves as the query:

$$f_{fusion}(x_i, G_i) = att(flin(x_i), f_g(G_i)), \quad (11)$$

where the $f_{fusion}(x_i, G_i)$ is the fusion features of linguistic features and graph structure features. $att(\cdot)$ can be the existing attention function and we use the multi-head attention function here. To facilitate classification, we concatenate the linguistic features, graph structure features, and integrated features together as a basis for classification, denoted as :

$$f_{classify}(x_i, G_i) = flin(x_i)\|f_g(G_i))\|f_{fusion}(x_i, G_i). \quad (12)$$

To discriminate whether a text x_i (node i) is stego, we design a discriminator that makes a decision based on $f_{classify}(x_i, G_i)$. We propose a simple weight matrix W_C to evaluate each feature's power for detecting stegos, taking into account the different weights of features in the final judgment:

$$p(\cdot|I(f_{lin}(x_i), f_g(G_i))) = W_C f_{classify}(x_i, G_i) + Bias. \quad (13)$$

3 Experiments and Analysis

3.1 Experiment Settings

In this paper, we evaluated our proposed method using the Stego-Sandbox [31] which contains texts and connections information collected from various mainstream social networks, including Reddit, Twitter and Sina Weibo. Our experiments simulate different social network scenarios by employing various linguistic steganography algorithms, embedding payloads, and sparsity ratios of stegos. We employed generative linguistic steganography methods (Huffman Coding (HC) [14], Arithmetic Coding (AC) [7] and Adaptive Dynamic Grouping (ADG) [9]) to obtain steganographic texts.

Additionally, we consider the sparsity characteristics of stegos in real social networks and set the sparsity ratios (SRS) from 10% to 50%. SRS = 10% and SRS = 50% represent 2 typical scenarios: the sparse scenario and the general balanced scenario targeted by other baseline methods, respectively. We evaluated the proposed method compared with six representative linguistic steganalysis methods as baselines, namely TS-FCN [21], TS-RNN [32], TS-CSW [23], TS-ATT [22], TS-GNN [27] and TS-FCN+GRAPH [31]. These methods are briefly introduced in the Introduction. TS-FCN [21] is a deep learning method that uses single-layer fully connected networks to capture word-level correlation information. TS-RNN [32] employs a two-layer bidirectional RNN network to decide on the temporal connection of words. TS-CSW [23] uses multi-kernel CNNs to capture multi-gram statistical features. TS-ATT [22] uses LSTM to obtain sequential features and an attention mechanism to identify local features that are discordant in texts. TS-GNN [27] employs GNNs to capture the global graph information between words, yet still limits to single isolated text. TS-ATT+GRAPH [31] is a typical linguistic steganalysis method that initially considers social network connections information.

In practice, we chose subgraphs with 1000 random walk roots and a walk length of 2 for sampling purposes. Following this, we predicted the label for each node in the chosen subgraphs. We evaluated the validation set for each epoch and selected the best model based on the F_1 score, which balances precision and recall, as accuracy is not a reliable indicator in unbalanced cases. To obtain the average F_1 score, we repeated the process 10 times. Besides, we employed AdamW [35] optimizer with a learning rate of 0.01 and the Cross-Entropy loss function. The embedding dimension of the linguistic feature was 128, while for the graph feature extractor, we used 2 GNN layers. This choice was based on our observations that too small hops were insufficient for capturing sufficient graph information, and too large hops resulted in too much noise. A dropout rate of 0.2 was also utilized.

3.2 Evaluation Results and Discussion

We extensively tested nearly every scenario in Stego-Sandbox [31]. It is clear from Table 1, Table 2 and Table 3 that our approach outperforms existing typical linguistic steganalysis methods [21–23,27,31,32] in virtually all circumstances, particularly at high embedding rates where our model consistently achieves more significant gains. Specifically, we get three findings: Firstly, introducing and interacting with graph structures in social networks can effectively improve steganalysis detection in social networks. The detection power of the detectors decreases tremendously in the case of high sparsity (SRS = 10%), which reflects the fragmentation and sparsity challenges in social networks. All previous classifiers focusing on linguistic features failed to perform satisfactorily in real, sparse, and fragmented social network scenarios, while CATS can relieve the decreases and bring significant F_1 gains. For instance, when dealing with the so-called provably secure steganography algorithm ADG in sparse scenarios, previous models had a meager average F_1 score of under 3.5 on the Sina dataset. However, our model can achieve an average F_1 gain of 19.21 on the same dataset when facing the ADG challenge by introducing and interacting with graph structures, while still using the same linguistic feature extraction as TS-CSW.

Secondly, CATS can obtain a significant improvement over the generally poorly performing baseline methods in scenarios with high embedding rates or in Chinese language environments. As the embedding rate increases, the performance of steganalysis methods decreases, since the statistical difference between stegos and covers decreases. This phenomenon has been previously analyzed as a Psic effect [36]. But our method can get more gains in higher embedding capacity. We attribute this improvement to the critical role that social network graph features play in situations where linguistic statistical differences in a single text are minimal. Our results suggest that the classifier can learn from these graph features to compensate for the lack of linguistic features in the text. Also, baselines performed poorly in the Chinese-based social network Sina compared to

Table 1. F1 Result On Sina

Encoding Algorithm		HC					AC						ADG
BPW		1.00	1.86	2.59	3.23	3.82	0.24	1.17	2.00	4.52	4.88	5.09	4.13
SRS = 10%	TS-FCN	48.86	37.72	4.60	3.36	0.00	49.21	38.29	11.48	0.00	0.00	0.00	0.00
	TS-RNN	63.93	57.88	20.87	17.48	7.02	66.58	52.31	35.89	3.75	1.32	2.41	1.95
	TS-CSW	63.43	55.16	23.21	10.64	9.12	65.12	52.25	36.00	5.12	5.26	5.97	7.46
	TS-ATT	65.78	58.78	21.39	24.13	7.60	64.87	53.77	36.46	4.05	3.11	1.14	1.31
	TS-GNN	56.72	44.74	10.44	15.59	5.54	61.86	46.09	28.40	1.07	1.16	0.98	1.61
	TS-CSW+ GRAPH	71.15	65.32	32.93	34.22	19.25	78.42	61.91	51.19	14.33	11.90	7.79	8.22
	ours	70.35	64.09	**37.75**	**40.03**	**27.22**	74.49	**62.00**	47.22	**20.07**	**20.01**	**20.78**	**22.63**
SRS = 50%	TS-FCN	74.23	61.84	50.20	35.77	29.37	82.28	66.34	52.07	2.44	3.42	0.80	2.32
	TS-RNN	85.23	74.73	64.81	52.70	46.52	89.32	80.84	68.86	23.20	19.64	16.92	13.26
	TS-CSW	84.33	74.39	62.55	48.71	41.06	89.23	80.14	69.45	15.12	19.39	13.16	9.85
	TS-ATT	84.76	75.65	63.97	53.95	46.89	89.17	80.38	69.97	21.17	16.93	17.47	12.45
	TS-GNN	81.25	68.12	56.23	45.49	41.25	88.00	76.30	60.34	27.16	37.52	23.42	26.88
	TS-CSW+ GRAPH	91.44	86.08	79.53	74.74	70.25	93.05	87.56	83.06	55.24	52.97	47.31	39.46
	ours	91.38	**87.52**	**84.39**	**79.47**	**77.58**	**93.21**	**88.94**	**85.17**	**68.59**	**68.09**	**68.53**	**68.48**

Table 2. F1 Result On Tweet

Encoding Algorithm		HC					AC						ADG
BPW		1.00	1.90	2.72	3.57	4.35	0.37	1.39	2.40	6.33	6.94	7.54	8.91
SRS = 10%	TS-FCN	60.59	60.81	60.15	62.97	60.98	58.34	59.77	61.49	4.98	2.21	0.83	0.21
	TS-RNN	69.27	68.53	71.20	69.86	70.40	62.71	66.08	64.68	53.63	47.29	32.71	9.62
	TS-CSW	70.07	67.94	70.97	68.36	70.77	63.42	65.55	64.19	54.18	51.19	33.05	13.30
	TS-ATT	68.41	67.11	69.8	69.14	70.89	62.98	64.18	62.44	52.56	44.02	29.73	7.85
	TS-GNN	63.69	62.80	65.03	64.55	66.91	60.43	61.99	62.11	44.48	39.69	26.09	9.88
	TS-CSW+ GRAPH	71.44	69.96	72.18	69.25	71.17	70.50	70.71	69.98	54.87	53.39	34.64	14.07
	ours	**73.62**	**71.91**	71.56	**69.61**	**71.17**	67.65	69.67	67.25	**55.47**	**55.81**	**38.46**	**19.25**
SRS = 50%	TS-FCN	88.02	86.89	85.72	80.05	78.47	89.83	88.57	86.07	63.52	46.30	36.14	43.31
	TS-RNN	92.83	92.07	91.76	90.76	89.41	93.74	92.90	91.41	85.18	80.29	73.27	56.99
	TS-CSW	94.20	92.94	92.14	91.08	89.95	94.72	94.33	92.91	86.19	80.40	75.37	64.22
	TS-ATT	93.01	91.94	91.69	90.54	89.43	93.88	92.73	91.44	85.34	81.04	74.11	50.85
	TS-GNN	91.72	90.65	90.13	89.13	88.30	92.63	91.98	90.19	82.90	76.03	72.22	62.41
	TS-CSW+ GRAPH	93.99	92.61	92.08	90.36	89.24	94.88	94.04	92.85	86.12	82.43	77.81	52.63
	ours	**94.27**	**93.18**	**92.56**	**91.73**	**90.61**	**95.02**	**94.35**	**93.19**	**87.36**	**84.31**	**80.18**	**68.22**

Table 3. F1 Result On Reddit

Encoding Algorithm		HC					AC						ADG
BPW		1.00	1.82	2.63	3.42	4.17	0.45	1.46	2.38	5.82	6.28	6.71	6.93
SRS = 10%	TS-FCN	88.04	82.58	81.06	75.88	67.73	86.21	84.93	79.33	12.97	3.36	0.00	0.09
	TS-RNN	90.34	85.70	85.79	81.24	77.25	89.21	89.00	84.19	54.77	35.02	24.23	5.08
	TS-CSW	90.77	85.94	86.66	80.32	75.99	88.88	89.51	84.86	55.01	38.87	27.33	15.13
	TS-ATT	90.25	86.00	84.79	80.93	76.74	88.88	88.92	84.24	52.35	34.53	26.11	4.70
	TS-GNN	88.58	83.50	83.53	76.76	71.28	87.72	86.86	81.87	45.73	26.82	23.84	7.00
	TS-CSW+ GRAPH	90.90	86.57	85.60	80.05	78.60	90.43	89.07	84.90	50.50	40.04	39.00	14.94
	ours	90.03	86.10	**85.62**	**81.16**	**78.91**	90.25	**89.08**	**84.93**	61.22	50.23	39.46	20.13
SRS = 50%	TS-FCN	92.49	93.35	90.04	89.13	84.24	90.29	90.55	88.70	70.56	58.70	36.35	33.10
	TS-RNN	95.10	95.16	93.13	91.70	88.04	95.33	94.70	93.12	81.29	74.03	65.55	41.27
	TS-CSW	95.37	95.21	93.29	91.65	87.93	95.50	95.12	93.63	83.19	77.62	72.30	49.36
	TS-ATT	94.97	94.99	92.64	92.07	87.45	95.31	94.62	93.28	80.77	74.75	64.16	41.01
	TS-GNN	94.38	94.17	91.57	90.26	85.68	95.01	93.84	92.12	79.10	74.61	68.81	45.43
	TS-CSW+ GRAPH	95.53	95.46	94.07	92.44	88.73	95.59	95.23	93.88	83.24	77.86	73.13	54.52
	ours	**95.86**	**95.70**	**94.41**	**92.61**	**90.86**	**96.19**	**95.45**	**94.68**	**86.89**	**83.15**	**78.45**	**69.54**

English-based social networks like Twitter and Reddit, while CATS can improve the detection even further in Sina, with an average F1 gain of 35.16 against the ADG algorithm, demonstrating its effectiveness.

Thirdly, we did ablation experiments to explore the effect of each part of the proposed model on the detection performance, as presented in Fig. 2. We tested the performance of each model part of the model under 10% SRS, on maximum payloads of different steganography algorithms, representing a typically challenging detection scenario in social networks. Four different model configurations were evaluated, including "Linguistic", which solely used the linguistic extractor, "Graph", which solely used the graph structure information, "Linguistic+Graph", which combined both linguistic and graph features through simple concatenation, and "Ours", which extracted the interaction between linguistic text and graph structure more deeply, based on the mutual use of both linguistic features and graph structure information. Our results indicate that the "Ours" configuration outperformed the other three configurations, while simple concatenation or using only some of the features led to performance degradation. This finding highlights the importance of mining features and their interrelationships in-depth. Interestingly, the performance difference between using only graph features and using only linguistic features is small, and sometimes the performance of "graph" can even exceed that of using only linguistic features, which indicates that the associative nature of the social networks enables text to learn its own linguistic features to some extent when learning the graph structure.

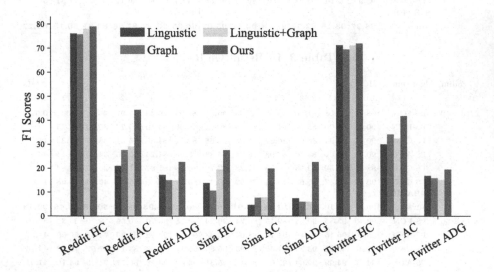

Fig. 2. The ablation study results. F_1 scores in high sparsity (SRS = 10%), high embedding rate scenarios. For HC and AC algorithms, we adopted the maximum payloads. TS-CSW [23] is employed as a linguistic extractor.

4   Conclusion

In this paper, we propose a novel linguistic steganalysis model specifically designed for social network scenarios. Fragmentation and sparsity pose a significant challenge for steganalysis in social networks. Our study shows that taking into account the connections between texts and social networks can improve the detection of powerful generative steganography methods. Therefore, we propose a method that focuses on extracting text interaction features with social networks from a global perspective. Experimental results show that the proposed method outperforms general steganalysis methods. In the future, we plan to explore more in-depth steganalysis approaches tailored to real social network scenarios. We hope that our research will motivate more researchers to explore steganalysis toward real social network scenarios.

Acknowledgements. This work was supported by the National Natural Science Foundation of China under Grant U1936216 and Grant 61862002.

References

1. Bak, P., Bieniasz, J., Krzeminski, M., Szczypiorski, K.: Application of perfectly undetectable network steganography method for malware hidden communication. In: 2018 4th International Conference on Frontiers of Signal Processing (ICFSP), pp. 34–38. IEEE (2018)
2. Li, F., Yu, Z., Qin, C.: GAN-based spatial image steganography with cross feedback mechanism. Signal Process. **190**, 108341 (2022)
3. Huang, Y.F., Tang, S., Yuan, J.: Steganography in inactive frames of VoIP streams encoded by source codec. IEEE Trans. Inf. Forensics Secur. **6**(2), 296–306 (2011). https://doi.org/10.1109/TIFS.2011.2108649
4. Kaptchuk, G., Jois, T.M., Green, M., Rubin, A.D.: Meteor: cryptographically secure steganography for realistic distributions. In: Proceedings of the 2021 ACM SIGSAC Conference on Computer and Communications Security, ser. CCS 2021, pp. 1529–1548. Association for Computing Machinery, New York (2021). https://doi.org/10.1145/3460120.3484550
5. Wayner, P.C.: Mimic functions. Cryptologia **16**, 193–214 (1992)
6. Shirali-Shahreza, M.: Text steganography by changing words spelling. In: 2008 10th International Conference on Advanced Communication Technology, Gangwon, Korea (South), 2008, pp. 1912–1913 (2008). https://doi.org/10.1109/ICACT.2008.4494159
7. Ziegler, Z.M., Deng, Y., Rush, A.M.: Neural linguistic steganography. arXiv preprint arXiv:1909.01496 (2019)
8. Zhou, X., Peng, W., Yang, B., Wen, J., Xue, Y., Zhong, P.: Linguistic steganography based on adaptive probability distribution. IEEE Trans. Depend. Secure Comput. **19**(5), 2982–2997 (2021)
9. Zhang, S., Yang, Z., Yang, J., Huang, Y.: Provably secure generative linguistic steganography, pp. 3046–3055 (2021). https://aclanthology.org/2021.findings-acl.268
10. de Witt, C.S., Sokota, S., Kolter, J.Z., Foerster, J., Strohmeier, M.: Perfectly secure steganography using minimum entropy coupling. arXiv preprint arXiv:2210.14889 (2022)
11. Ouyang, L., et al.: Training language models to follow instructions with human feedback. arXiv preprint arXiv:2203.02155 (2022)
12. Touvron, H., et al.: Llama: open and efficient foundation language models (2023)
13. Dai, W., Yu, Y., Dai, Y., Deng, B.: Text steganography system using markov chain source model and des algorithm. JSW **5**(7), 785–792 (2010)

14. Yang, Z.L., Guo, X.Q., Chen, Z.M., Huang, Y.F., Zhang, Y.J.: RNN-stega: linguistic steganography based on recurrent neural networks. IEEE Trans. Inf. Forensics Secur. **14**(5), 1280–1295 (2018)

15. Dai, F., Cai, Z.: Towards near-imperceptible steganographic text. In: Proceedings of the 57th Annual Meeting of the Association for Computational Linguistics, pp. 4303–4308. Association for Computational Linguistics, Florence (2019). https:// aclanthology.org/ P19-1422 (2019)

16. Yang, H., Cao, X.: Linguistic steganalysis based on meta features and immune mechanism. Chin. J. Electron. **19**(4), 661–666 (2010)

17. Chen, Z., Huang, L., Miao, H., Yang, W., Meng, P.: Steganalysis against substitution-based linguistic steganography based on context clusters. Comput. Electr. Eng. **37**(6), 1071–1081 (2011)

18. Xiang, L., Sun, X., Luo, G., Xia, B.: Linguistic steganalysis using the features derived from synonym frequency. Multimedia Tools Appl. **71**(3), 1893–1911 (2014)

19. Meng, P., Hang, L., Chen, Z., Hu, Y., Yang, W.: STBS: a statistical algorithm for steganalysis of translation-based steganography. In: Böhme, R., Fong, P.W.L., Safavi-Naini, R. (eds.) IH 2010. LNCS, vol. 6387, pp. 208–220. Springer, Heidelberg (2010). https://doi.org/10.1007/978-3-642-16435-4_16

20. Meng, P., Hang, L., Yang, W., Chen, Z., Zheng, H.: Linguistic steganography detection algorithm using statistical language model. In: 2009 International Conference on Information Technology and Computer Science, vol. 2, pp. 540–543. IEEE (2009)

21. Yang, Z., Huang, Y., Zhang, Y.J.: A fast and efficient text steganalysis method. IEEE Signal Process. Lett. **26**(4), 627–631 (2019)

22. Zou, J., Yang, Z., Zhang, S., Rehman, S., Huang, Y.: High-performance linguistic steganalysis, capacity estimation and steganographic positioning. In: Zhao, X., Shi, Y.-Q., Piva, A., Kim, H.J. (eds.) IWDW 2020. LNCS, vol. 12617, pp. 80–93. Springer, Cham (2021). https://doi.org/10.1007/978-3-030-69449-4_7

23. Yang, Z., Huang, Y., Zhang, Y.J.: TS-CSW: text steganalysis and hidden capacity estimation based on convolutional sliding windows. Multimedia Tools Appl. **79**, 18293–18316 (2020)

24. Niu, Y., Wen, J., Zhong, P., Xue, Y.: A hybrid r-bilstm-c neural network based text steganalysis. IEEE Signal Process. Lett. **26**(12), 1907–1911 (2019)

25. Xu, Q., Zhang, R., Liu, J.: Linguistic steganalysis by enhancing and integrating local and global features. IEEE Signal Process. Lett. **30**, 16–20 (2023)

26. Wen, J., Deng, Y., Peng, W., Xue, Y.: Linguistic steganalysis via fusing multi-granularity attentional text features. Chin. J. Electron. **32**(1), 76–84 (2023)

27. Wu, H., Yi, B., Ding, F., Feng, G., Zhang, X.: Linguistic steganalysis with graph neural networks. IEEE Signal Process. Lett. **28**, 558–562 (2021)

28. Fu, Z., Yu, Q., Wang, F., Ding, C.: HGA: hierarchical feature extraction with graph and attention mechanism for linguistic steganalysis. IEEE Signal Process. Lett. **29**, 1734–1738 (2022)

29. Peng, W., Zhang, J., Xue, Y., Yang, Z.: Real-time text steganalysis based on multi-stage transfer learning. IEEE Signal Process. Lett. **28**, 1510–1514 (2021)

30. Yang, J., Yang, Z., Zhang, S., Tu, H., Huang, Y.: SeSy: linguistic steganalysis framework integrating semantic and syntactic features. IEEE Signal Process. Lett. **29**, 31–35 (2021)

31. Yang, J., Yang, Z., Zou, J., Tu, H., Huang, Y.: Linguistic steganalysis toward social network. IEEE Trans. Inf. Forensics Secur. **18**, 859–871 (2023)

32. Yang, Z., Wang, K., Li, J., Huang, Y., Zhang, Y.-J.: TS-RNN: text steganalysis based on recurrent neural networks. IEEE Signal Process. Lett. **26**(12), 1743–1747 (2019)

33. Gao, T., Yao, X., Chen, D.: Simcse: simple contrastive learning of sentence embeddings. arXiv preprint arXiv:2104.08821 (2021)

34. Zhang, J., Shi, X., Xie, J., Ma, H., King, I., Yeung, D.Y.: Gaan: gated attention networks for learning on large and spatiotemporal graphs. In: 34th Conference on Uncertainty in Artificial Intelligence 2018, UAI 2018 (2018)

35. Loshchilov, I., Hutter, F.: Decoupled weight decay regularization. arXiv preprint arXiv:1711.05101 (2017)

36. Yang, Z.L., Zhang, S.Y., Hu, Y.T., Hu, Z.W., Huang, Y.F.: VAE-Stega: linguistic steganography based on variational auto-encoder. IEEE Trans. Inf. Forensics Secur. **16**, 880–895 (2020)

Syntax Tree Constrained Graph Network for Visual Question Answering

Xiangrui Su[1], Qi Zhang[2,3], Chongyang Shi[1(✉)], Jiachang Liu[1], and Liang Hu[2,3]

[1] Beijing Institute of Technology, Beijing, China
cy_shi@bit.edu.cn
[2] Tongji University, Shanghai, China
[3] DeepBlue Academy of Sciences, Shanghai, China

Abstract. Visual Question Answering (VQA) aims to automatically answer natural language questions related to given image content. Existing VQA methods integrate vision modeling and language understanding to explore the deep semantics of the question. However, these methods ignore the significant syntax information of the question, which plays a vital role in understanding the essential semantics of the question and guiding the visual feature refinement. To fill the gap, we suggested a novel Syntax Tree Constrained Graph Network (STCGN) for VQA based on entity message passing and syntax tree. This model is able to extract a syntax tree from questions and obtain more precise syntax information. Specifically, we parse questions and obtain the question syntax tree using the Stanford syntax parsing tool. From the word level and phrase level, syntactic phrase features and question features are extracted using a hierarchical tree convolutional network. We then design a message-passing mechanism for phrase-aware visual entities and capture entity features according to a given visual context. Extensive experiments on VQA2.0 datasets demonstrate the superiority of our proposed model.

Keywords: Visual question answering · Syntax tree · Message passing · Tree convolution · Graph neural network

1 Introduction

Visual Question answering (VQA) aims to automatically answer natural language questions related to a given image content. It requires both computer vision technology to understand the visual content of images and natural language processing technology to understand the deep semantics of questions. VQA has various potential applications, including image retrieval, image captioning, and visual dialogue systems, therefore becoming an important research area.

Recently, various VQA methods [4,7,9,12] have been proposed to capture significant question semantics and visual features by mining explicit or implicit entity relationships. For example, BAN [9] considers the bilinear interaction

---

X. Su and Q. Zhang—The first two authors contribute equally to this work.

© The Author(s), under exclusive license to Springer Nature Singapore Pte Ltd. 2024
B. Luo et al. (Eds.): ICONIP 2023, LNCS 14451, pp. 122–136, 2024.
https://doi.org/10.1007/978-981-99-8073-4_10

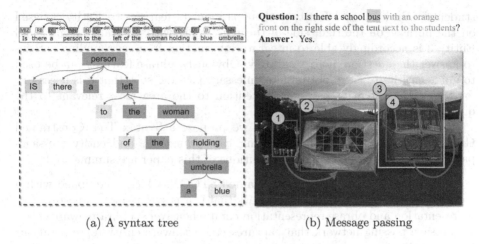

(a) A syntax tree   (b) Message passing

Fig. 1. Examples of a syntax tree and message passing respectively derived from the VQA2.0 dataset.

between two sets of input channels of images and questions by calculating the bilinear distribution of attention to fuse visual and textual information. Murel et al. [4] design an atomic inference unit to enrich the interaction between the problem region and the image region and optimize the visual and problem interaction by using a unit sequence composed of multiple atomic units. LCGN [7] designs a question-aware messaging mechanism, uses question features to guide the refinement of entity features based on the entity complete graph, and realizes the integration of entity features and context information. However, these methods typically capture explicit or implicit entity relationships in images while ignoring the syntax relation between words, which contributes to capturing the deep semantics of the question.

Intuitively, using a syntax tree in VQA tasks has two major benefits. First, questions are usually short in length, and adding more syntactic information is necessary to understand the semantics of the questions. Secondly, the syntax tree hierarchically organizes keywords and context words through a tree structure, which is effective for capturing key information in the questions. As shown in the illustration in Fig. 1(a), the words "person", "left", and "woman" which are far apart in the original question are adjacent in the syntax tree. These three words are the core information of this question. Therefore, the syntax tree can better capture the key information in terms of feature extraction.

Besides, in the field of VQA, images are the key information source to infer answers. It is also one core objective of the VQA model to understand the information in images. Since images are composed of many visual entities, there are many implicit relationships between these entities. Intuitively, it is necessary to perceive these implicit relationships and achieve effective message passing between entities for obtaining the entity features of scene context awareness.

As shown in Fig. 1(b), in the process of answering the question, Entity 1 firstly transmits its own features to Entity 2 based on the phrase "next to the students"; Entity 2 and Entity 4 then pass on its features to Entity 3 based on the phrase "on the right side of the tent" and "with an orange front", and Entity 3 is accordingly able to integrate information from surrounding entities to answer the question more accurately. Obviously, phrase features can be used to guide entities to carry out targeted message passing, such that it is necessary for the VQA system to pay more attention to the area most relevant to the question.

In light of the above observation, we propose a Syntax Tree Constrained Graph Network (STCGN) by modeling the syntax tree and entity message-passing mechanism. The main contributions of this paper are summarized:

- We propose a novel VQA model, named STCGN, which is equipped with a tree hierarchical convolutional network that learns syntax-aware question representation and phrase representation comprehensively, a phrase-aware entity message passing network that captures context-aware entity representations of visual scenes.
- Extensive experimental results on VQA2.0 datasets demonstrate the significant superiority of our STCGN and prove the validity of phrase-aware messaging network through visualization experiments.

2   Related Work

Recently, various methods have been proposed to improve the accuracy of visual question answering. The image is typically represented by a fixed-size grid feature extracted from a pre-trained model such as VGG [17] and AlexNet [22], and the question is typically embedded with LSTM [6]. Both of these features are combined by addition or element-wise multiplication [2,21]. But using the base fusion method to achieve feature fusion is too simple to capture a key part of the image. Therefore some neural Network based method [16] is proposed to fuse both the visual and question features. For example, Ren [16] propose an LSTM-based fusion strategy by projecting extracted visual feature into the same word embedding space. However, not all of the pictures are strongly related to the question, thus some of the non-relevant grids in the picture should be filtered.

The attention mechanism [1,8,14,19] is used to calculate the importance of each grid area. For example, Bottom-Up and Top-Down attention (BUTD) [1] uses a bottom-up and top-down approach to capture attention. In the bottom-up process, they used Faster-Rcnn to extract the features of all objects and their significant regions. In a top-down process, each attention candidate is weighted using task-specific context to obtain a final visual representation. Besides, the bilinear models [3,5,14,20] show a strong ability of cross-modal feature interaction. For example, MFH [20] fully mines the correlation of multi-mode features through factorization and high-order pooling, effectively reducing the irrelevant features, and obtaining more effective multi-mode fusion features. BGN [5] on the basis of the BAN [9] model designs a bilinear graph structure to model the context relationship between text words and visual entities and fully excavates the implicit relationship between text words and visual entities. However, these methods do not pay enough attention to the entity relationship in the pictures and the relationship between the words in the questions. Extracting entity features can also be optimized through implicit or explicit relational modeling.

To address the above problem, some relation-based VQA methods, e.g., Murel [4], LCGN [7], and ReGAT [12], and more effective attention mechanisms, e.g., UFSCAN [23] and MMMAN [11] have been proposed. Murel, LCGN, and ReGAT refine visual representations by explicitly or implicitly modeling relationships and enhance feature interactions between modalities. UFSCAN and MMMAN improve attention scores in visual areas through more efficient attention mechanisms. Compared to these prior works, Our model proposes a syntax-aware tree hierarchical convolutional network to extract syntax relation-aware question representation and phrase representation from questions. It further proposes a phrase-aware message passing network to capture the entity features of visual scene context-awareness based on implicit entity relations.

3   The STCGN Model

Given a question q and a corresponding image I, the goal of VQA is to predict an answer $\hat{a} \in \mathcal{A}$ that best matched the ground-truth answer a^*. Following previous VQA models, it can be defined as a classification task:

$$\hat{a} = arg \max_{a \in \mathcal{A}} F_\theta(a|I, q) \tag{1}$$

where F is the trained model, and θ is the model parameter.

Denote the given picture as $\mathcal{I}$, K visual entities are extracted from it using Faster R-CNN, and the feature representation of the i-th entity is denoted as $\mathbf{v}_i$. Meanwhile, we have a picture-related question consisting of N words $\mathcal{Q} = (q_1, q_2, ..., q_N)$. The features of each word are initialized using the Glove [15] word embedding model and the word sequence feature is $\mathbf{X} = (\mathbf{x}_1, \mathbf{x}_2, ..., \mathbf{x}_N)$. The task of the visual question answering is to use image and text information, extract relevant features and fuse them to generate an answer probability distribution vector $\hat{\mathbf{p}} = (\hat{p}_1, \hat{p}_2, ..., \hat{p}_{N_{ans}})$, where N_{ans} denotes the number of categories of answers and $\hat{p}_i$ denotes the probability that the i-th answer is the final answer. We take the answer with the highest probability in $\hat{\mathbf{p}}$ as the final predicted answer of the visual question answering system.

3.1   Network Architecture

The architecture of STCGN is shown in Fig. 2, which consists of three main modules: (1) **Syntax-aware Tree Convolution** module utilizes the syntax tree of the question and uses the tree hierarchical convolution model to extract the syntax-aware phrase features and question features. (2) **Phrase-aware Entity Message Passing** module discovers the implicit connections between visual entities and relevant message features based on the syntax-aware phrase features, question features, and entity complete graphs, and then builds the scene context-aware visual entity features. (3) **Top-down Attention-based Answer Prediction** module fuses the question features and visual entity features by using Top-down attention and performs the final answer prediction.

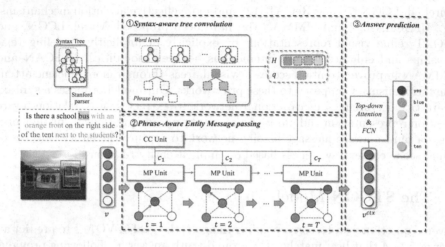

Fig. 2. The network architecture of our STCGN model.

3.2 Syntax-Aware Tree Convolution

In VQA tasks, understanding the essence of the question correctly is the first priority to make a good answer. Syntax information plays a very important role in the question text because it can offer important help in understanding the question. The syntax information contains the dependencies between words and the part of speech (POS) of these words. Therefore, we can construct a syntax tree $\mathcal{T} = (\mathcal{Q}, \mathcal{E})$ based on the dependencies between words. By observing Fig. 1(b), we can see that the phrases in question play an important role in instructing visual entities for message passing. Therefore, we propose a tree-based hierarchical convolutional network to model the syntax-aware phrase features. The network consists of two layers: word-level convolution and phrase-level convolution.

For word-level convolution, we first construct the syntax subtree with each non-leaf word node and its direct children in the syntax tree. In this way, we can decompose the syntax tree into a set of syntax subtree $F = (f_1, f_2, ..., f_s)$, where each subtree $f_i = (q_i, q_{c1}, q_{c2}, ..., q_{cn})$ as the convolution unit, cn denotes the number of children of i word nodes. Furthermore, we propose a convolution method based on text convolution [24]. First, we use the Glove word embedding model to map f_i to high-dimensional dense word features. Then, we obtain the learnable POS feature vector by using a POS feature dictionary of length 42 for random initialization. As a result, we can obtain the sequence of POS features in the question $\mathbf{T}_i = (\mathbf{t}_i, \mathbf{t}_{c1}, ..., \mathbf{t}_{cn})$. Finally, we concatenate word features and POS features to obtain word-level convolutional input features $\mathbf{X}_i^{cat}$ of words:

$$\mathbf{X}_i = \text{Glove}(f_i), \quad \mathbf{T}_i = \text{RandomInit}(t_i, t_{c1}, ..., t_{cn}), \quad \mathbf{X}_i^{cat} = [\mathbf{X}_i \oplus \mathbf{T}_i] \quad (2)$$

We define a text convolution kernel $\mathbf{G}$ for each syntax subtree. First, we concatenate the input features $\mathbf{X}_i^{cat}$ to extract the key information in the text. Then, we use the maximum pooling for further feature filtering.

$$\mathbf{g}_i = \max(\hat{\mathbf{g}}_i), \quad s.t., \quad \hat{\mathbf{g}}_i = \text{ReLU}(\mathbf{G} * [\mathbf{X}_i^{cat} \oplus \mathbf{X}_{c1}^{cat} \oplus ... \oplus \mathbf{X}_{cn}^{cat}] + \mathbf{b}_i) \quad (3)$$

$*$ indicates the text convolution, and $\mathbf{b}_i$ is the offset term.

For the phrase-level convolution, we design the syntax relation-aware graph attention network for capturing syntax phrase features $\mathbf{h}_i$ from multiple aspects based on the convolution features g_i of each subtree and the syntax tree $\mathcal{T}$:

$$\mathbf{h}_i^* = ||_{m=1}^M \sigma \left(\sum_{j \in \mathcal{N}_i} \alpha_{ij} \cdot \mathbf{W}_{dir(i,j)} \mathbf{g}_j + \mathbf{b}_{dep(i,j)} \right) \quad (4)$$

$$\alpha_{ij} = \frac{exp((\mathbf{U}\mathbf{x}_i')^\top \cdot \mathbf{V}_{dir(i,j)} \mathbf{g}_j + \mathbf{b}'_{dep(i,j)})}{\sum_{j \in \mathcal{N}_i} exp((\mathbf{U}\mathbf{x}_i')^\top \cdot \mathbf{V}_{dir(i,j)} \mathbf{g}_j + \mathbf{b}'_{dep(i,j)})} \quad (5)$$

where $\mathbf{W}_{\{\cdot\}} \in \mathbb{R}^{(d_h/M) \times (d_x+d_t)}, \mathbf{V}_{\{\cdot\}} \in \mathbb{R}^{(d_h/M) \times (d_x+d_t)}$, and $\mathbf{U} \in \mathbb{R}^{(d_h/M) \times (d_x+d_t)}$ are parameters matrices, $\mathbf{b}_{\{\cdot\}}$, $\mathbf{b}'_{\{\cdot\}}$ are offset terms, $dir(i,j)$ denotes the direction of each relation, and $dep(i,j)$ denotes the different kinds of dependencies.

Then, to capture the sequence correlation between phrase features and to obtain the features of the whole sentence, we performed phrase sequence feature extraction using a bidirectional GRU (biGRU) network:

$$h_l = \text{biGRU}([h_{l-1}, \mathbf{h}_i^*]) \quad (6)$$

where $\mathbf{W}_z \in \mathbb{R}^{d_h \times (2*d_h)}$, $\mathbf{W}_r \in \mathbb{R}^{d_h \times (2*d_h)}$, $\mathbf{W}_h \in \mathbb{R}^{d_h \times (2*d_h)}$ are the parameter matrices shared by the network, and σ denotes the $sigmoid$ function.

The final output is $\mathbf{H} = (\mathbf{h}_1, \mathbf{h}_2, ..., \mathbf{h}_s) \in \mathbb{R}^{s \times (2*d_h)}$ with $\mathbf{q} \in \mathbb{R}^{2*d_h}$, where $\mathbf{h}_i = [\overrightarrow{h}_i \oplus \overleftarrow{h}_i]$, $\mathbf{H}$ represents the syntax-aware phrase feature sequence, $\mathbf{q} = \mathbf{h}_s$, and $\mathbf{q}$ is the output of the last hidden layer, which incorporates the information of all iteration steps and can represent the syntax-aware question features.

3.3   Phrase-Aware Entity Message Passing

The inputs to the phrase-aware entity message passing module include the syntax-aware question feature $\mathbf{q}$, the syntax-aware phrase feature $\mathbf{H} = (\mathbf{h}_1, \mathbf{h}_2, ..., \mathbf{h}_s)$ and the original visual entity features $\mathbf{V} = (\mathbf{v}_1, \mathbf{v}_2, ..., \mathbf{v}_K)$. We propose a phrase-aware multi-step instruction calculation method, which sends each word separately with question features into the instruction calculation network at each time step to calculate the contribution of each word to the question.

Then, we weight the words according to their contribution levels to obtain the instruction vector $\mathbf{c}_t \in \mathcal{R}^{d_c}$ that guides the visual entities for message passing:

$$\mathbf{c}_t = \sum_{i=1}^{N} \alpha_{t,i} \cdot \mathbf{h}_i, \quad \alpha_{t,i} = softmax_i \left(\mathbf{W}_1 \left(\mathbf{h}_i \odot \left(\mathbf{W}_2^{(t)} \text{ReLU} \left(\mathbf{W}_3 \mathbf{q} \right) \right) \right) \right) \quad (7)$$

where $\mathbf{W}_1, \mathbf{W}_3$ are parameter matrices shared by all iteration steps, while $\mathbf{W}_2^{(t)}$ specifies to time step t to generate different messages at different iteration steps.

We use $w_{j,i}^{(t)}$ to denote the weight of message delivered from j entities to i entities at time step t and $m_{j,i}^{(t)}$ to denote the sum of messages received by the ith entity from other entities at time step t. At time step t, taking the i-th entity as an example, we compute the features of messages delivered by adjacent entities to the i-th vector based on the instruction vector $\mathbf{c}_t$:

$$\tilde{\mathbf{v}}_{i,t} = \left[\mathbf{v}_i \oplus \mathbf{v}_{i,t-1}^{ctx} \oplus \left((\mathbf{W}_4 \mathbf{v}_i) \odot (\mathbf{W}_5 \mathbf{v}_{i,t-1}^{ctx}) \right) \right] \quad (8)$$

$$w_{j,i}^{(t)} = softmax \left((\mathbf{W}_6 \tilde{\mathbf{v}}_{i,t})^{\top} (\mathbf{W}_7 \tilde{\mathbf{v}}_{j,t}) \odot (\mathbf{W}_8 \mathbf{c}_t) \right) \quad (9)$$

$$m_{j,i}^{(t)} = w_{j,i}^{(t)} \cdot \left((\mathbf{W}_9 \tilde{\mathbf{v}}_{j,t}) \odot (\mathbf{W}_{10} \mathbf{c}_t) \right) \quad (10)$$

And then we use the residual network for message aggregation to achieve the recognition of the scene context by the current visual entity $\mathbf{v}_{i,t}^{ctx}$. Finally, we concatenate the original entity features $\mathbf{v}_i$ with the scene context features $\mathbf{v}_{i,T}^{ctx}$ and obtain the final scene context-aware entity features $\mathbf{v}_i^{out}$:

$$\mathbf{v}_i^{out} = \mathbf{W}_{12} \left[\mathbf{v}_i \oplus \mathbf{v}_{i,T}^{ctx} \right], \quad \mathbf{v}_{i,t}^{ctx} = \mathbf{W}_{11} \left[\mathbf{v}_{i,t-1}^{ctx} \oplus \sum_{j=1}^{K} m_{j,i}^{(t)} \right] \quad (11)$$

where $\mathbf{W}_{\{.\}}$ denotes the parameter matrix shared by all iteration steps.

3.4 Top-Down Attention-Based Answer Prediction

Given a picture $\mathcal{I}$, we combine the set of visual entity features $\{\mathbf{v}_i\}_{i=1}^{K}$ with syntax-aware question features $\mathbf{q}$ and use a top-down attention mechanism for feature fusion. This mechanism first calculates the attention scores between each visual entity feature and the question feature as follows:

$$\beta_i = softmax_i \left(\mathbf{W}_{13} tanh(\mathbf{W}_{14} \mathbf{v}_i^{out} + \mathbf{W}_{15} \mathbf{q})) \right) \quad (12)$$

where $\mathbf{W}_{13}, \mathbf{W}_{14} \mathbf{W}_{15}$ denote the weight matrices. Then, we weigh each visual entity feature using a top-down attention mechanism and perform a nonlinear transformation of the joint features by a two-layer perceptron model. Then, we calculate the probability score for each answer using the $softmax$ function:

$$\mathbf{p} = \mathbf{W}_{16} ReLU \left(\mathbf{W}_{17} \left[\sum_{i=1}^{N} \beta_i \mathbf{v}_i^{out} \oplus \mathbf{q} \right] \right) \quad (13)$$

where $\mathbf{W}_{16}, \mathbf{W}_{17}$ are the weight matrices, and $ReLU$ is the activation function.

3.5   Loss Function

Finally, we train our model by minimizing the cross-entropy loss function:

$$\mathcal{L} = -\frac{1}{N_{ans}} \sum_{i=1}^{N_{ans}} \left(y_i \cdot log(\hat{p}(y_i)) + (1 - y_i) \cdot log((1 - \hat{p}(y_i))) \right) \tag{14}$$

where N_{ans} denotes the number of answer categories, and y_i is defined as $y_i = min(\frac{\#humans\ provided\ ans}{3}, 1)$ where $\#humans\ provided\ ans$ denotes the total times by which the answer was selected during the data annotation period. $\hat{p}(y_i)$ denotes the probability that the output belongs to the i-th class of answers.

4   Experiments

4.1   Experimental Settings

Datasets: We adopt two datasets: 1) MSCOCO [13] with more than 80k images and 444k questions (training), 40k images and 214k questions (test), 80k images, and 448k questions (validation), respectively. **Baselines:** We select the following 11 state-of-the-art methods as baselines to evaluate the performance of our STCGN: LSTM-VGG [2], SAN [18], DMN [10], MUTAN [3], BUTD [1], BAN [9], LCGN [7], Murel [4], ReGAT [12], UFSCAN [23], MMMAN [11]. Refer to Appendix A for more details about datasets, baselines, and settings.

4.2   Performance Comparison

Table 1 shows the overall performance of all comparative methods, with the best results highlighted in boldface, where we draw the following conclusions:

1) LSTM+VGG adopts the classic VQA framework with a single model module and only uses a simple vector outer product to achieve feature fusion. The modal information interaction is too simple, resulting in the performance of the final model being inferior to other models.
2) SAN and DMN adopt the typical NMN architecture to decompose the VQA task into a sequence of independent neural networks executing subtasks. The SAN achieves multi-step queries by stacking multiple attention modules, which significantly outperforms the LSTM+VGG network in terms of performance. DMN structure is more modular, using a dynamic memory network module to achieve episodic memory ability, all modules cooperate to complete the question-answering task, in the ability to answer three types of questions is better than the previous two models.

Table 1. Performance comparison on the VQA2.0 dataset.

method	Test-std				Test-dev			
	Overall	Y/N	Num	Other	Overall	Y/N	Num	Other
LSTM+VGG	58.2	80.41	37.12	44.21	57.13	80.52	36.74	43.18
SAN	58.9	80.56	36.12	46.73	58.70	79.3	36.6	46.10
DMN	61.28	81.64	37.81	48.94	60.30	80.50	36.8	48.3
MUTAN	63.58	80.35	40.68	53.71	62.70	79.74	40.86	53.13
BUTD	65.42	82.13	44.25	56.96	64.37	81.12	43.29	56.87
BAN	70.31	86.21	50.57	60.51	69.11	85.24	50.03	60.32
BAN + Counter	70.89	86.44	55.12	60.67	70.02	85.44	**54.21**	60.25
MuRel	68.41	84.22	50.31	58.49	68.03	84.77	49.84	57.75
LCGN	69.44	85.23	49.78	59.37	68.49	84.81	49.02	58.47
ReGAT	70.54	86.34	54.14	60.76	70.12	86.02	53.12	60.31
UFSCAN	70.09	85.51	**61.22**	51.46	69.83	85.21	50.98	60.14
MMMAN	70.28	**86.69**	51.83	60.22	70.03	**86.32**	52.21	60.16
STCGN	**70.99**	85.77	50.46	**60.79**	**70.21**	85.49	50.11	**60.40**

3) MUTAN, BUTD, BAN, and BAN+Counter are typical VQA models based on bilinear pooling and attentional mechanisms. Compared with classical frameworks and NMN architectures, these four models have more fine-grained feature extraction and feature fusion, significantly improving performance. In particular, the bilinear attention mechanism of BAN fully exploits the implicit interaction information between the question and the picture. BAN+Counter further integrates Counter's counting module on the basis of BAN, effectively improving the performance of counting tasks and the overall performance.

4) MuRel, LCGN, and ReGAT outperform other comparison models, focusing on extracting information about entity relationships in images and playing a key role in joint feature learning. LCGN constructs a complete graph of the entity and implements implicit message passing between entities through problem guidance, while MuRel models the implicit relationship between detailed image regions through cross-modal attention, gradually refining the interaction between the picture and the problem. ReGAT explicitly extracts entity relationships in images and uses relational aware graph attention to realize accurate learning of joint features, so its performance is better than the other three models.

5) UFSCAN and MMMAN used more effective attention mechanisms. UFSCAN adopted a feature-based attention mechanism and obtained better results on counting questions by suppressing irrelevant features and emphasizing informative features. MMMAN proposed a multi-level mesh mutual attention, utilizing mutual attention to fully explore the information interaction between visual and language modalities and improve the model accuracy on Y/N questions.

6) Finally, these methods are inferior to our proposed method in accuracy. It is attributed to two points: (1) STCGN uses a syntax tree to model the question features, introduces syntactic structure features, and designs a tree hierarchical convolutional network to convolute the syntax tree structure, to fully extract the grammatical information of the question and improve its performance. (2) STCGN introduces a phrasing-aware message-passing module, which uses different phrase information in the question representation to guide the message passing between entities in multiple steps and extracts the context-aware entity features of the scene, which further promotes the performance of the model.

4.3 Ablation Study

Figure 3 shows the performance of STCGN Variants. The results show that when any module is removed, the performance of STCGN on the Test-dev and Test-std subsets of VQA2.0 will decrease significantly. When the SHA module is lost, it has the greatest impact on the model, indicating that feature fusion is the module that has the greatest impact on the accuracy of visual question answering. Secondly, the influence of the TCN module on model performance is greater than that of the MPN module, which may be due to the fact that the tree convolution module is based on a syntax tree and plays an important role in extracting question features and guiding entities for message passing.

4.4 Parameter Sensitivity

In this section, we analyze the effect of different iteration steps T. We fixed other parameters as the optimal parameters, gradually increased T, and obtained the curve of answering accuracy of different types of questions with the change of T, as shown in Fig. 4. As T increases, the rotation of messages between entities increases to incorporate more scene context information. The performance of

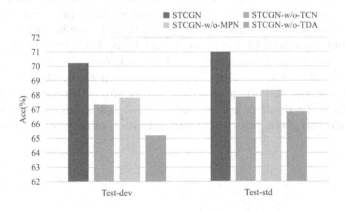

Fig. 3. The overall accuracy of STCGN Variants.

all questions is progressively improved and optimally reached at $T = 4$. As T continues to increase, the performance of the model gradually decreases, since message passing leads to receive redundant information of the entity and reduce the accuracy of entity representation. The reason why the outliers appear is that the binary question has lower requirements for understanding questions and pictures than other questions, leading to better performance of the model in the binary questions but the inferior overall performance. Therefore, we chose $T = 4$ as the final total number of messaging iteration steps.

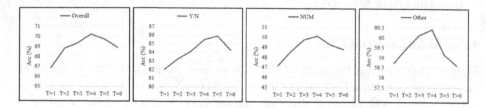

Fig. 4. Parameter sensitivity of both iteration steps.

4.5   Attention Visualization

To better illustrate the effectiveness of a phrase-based messaging mechanism, we experiment with visualizations in this section. We visualized the attention score between different entities in multiple iteration steps and different words of the question, as shown in Fig. 5. In the attention diagram, we can see: (1) Entity 2, Entity 4, and Entity 10 have significantly higher attention weights related to multiple phrases "the man", "in orange shoes" and "the other players" than other entity-word attention blocks. This suggests that the degree to which an entity is important in answering a question is closely related to multiple phrases. Syntax-aware phrase features in the messaging module provide guidance so that the VQA system can gradually understand entities that contribute more to the task. (2) The initial attention map can only initially locate important entities 2, 4, and 10, but their attention weights are not high. The initial visual attention map is very messy, and the contribution of each entity to the answer is not different. As the message passing iteration steps increase, the attention map becomes clearer and the attention weight of the key entity increases from 0.03 to 0.35. This is the result of non-critical entities passing information to critical entities, while also getting the entity representation of scene context awareness.

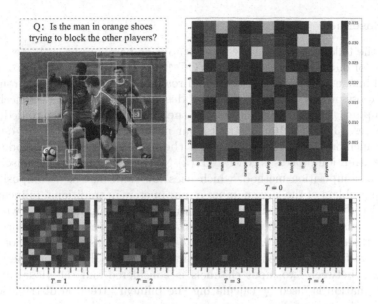

Fig. 5. Visualization of attention score.

5 Conclusion

In this work, we propose a Syntax Tree Constrained Graph Network. We design a hierarchical tree convolutional network and extract phrase representation and question representation of syntactic structure perception from syntactic tree structure by combining text convolution with graph attention. At the same time, we also suggest a phrase-aware entity message-passing mechanism based on the observation of the data set. In multiple iteration steps, different instruction vectors are calculated using phrase features and question features to capture the scene context-aware entity features.

A Experimental Settings

A.1 Datasets

The dataset consists of real images from MSCOCO [13] with the same training, test, and validation set separation, with more than 80k images and 444k questions, 40k images and 214k questions, 80k images, and 448k questions. For each image, there are an average of at least three questions. The questions fall into three categories based on the type of response: yes/no, number, and others. Each pair collected 10 answers from human annotators, choosing the most frequent answer as the correct answer. This data set contains both open-ended and multiple-choice questions. In this paper, we focus on open-ended questions, taking answers that appear more than 9 times in the training set as candidate answers, and generating 3129 candidate answers.

A.2 Baselines

We use the following state-of-the-art methods as baselines to evaluate the performance of our STCGN: (1) LSTM-VGG [2] uses LSTM and VGG to extract text and image features respectively, and realizes the feature fusion through the element cross product form. (2) SAN [18] designs a stacked attention mechanism that computes the most relevant areas of the question step by step to deduce the final answer. (3) DMN [10] constructs four network modules of input, question, episodic memory, and answer. Through the cyclic work of these modules, an iterative attention process is generated. (4) MUTAN [3] proposed a tensor-based Tracker decomposition method, which uses low-rank matrix decomposition to solve the problem of the large number of parameters in traditional bilinear models. (5) BUTD [1] proposed a combination of bottom-up and top-down attention mechanisms. Through the top-down attention mechanism, the correlation between the question and each region is calculated, so as to obtain more accurate picture features related to the question. (6) BAN [9] considers the bilinear interaction between two input channels is considered and the visual and textual information is fused by calculating the bilinear attention distribution. (7) LCGN [7] designs a question-aware messaging mechanism and uses question features to guide the refinement of entity features through multiple iteration steps, and realizes the integration of entity features and context information. (8) Murel [4] enriches the interaction between the problem region and the image region by vector representation and a sequence of units composed of multiple MuRel units. (9) ReGAT [12] extracted explicit relation and implicit relation in each image and constructed explicit relation graph and implicit relation graph. At the same time, the graph attention network of question awareness is used to integrate the information of different relation spaces. (10)UFSCAN [23] proposed a multimodal feature-wise attention mechanism and modeled feature-wise co-attention and spatial co-attention between image and question modalities simultaneously. (11)MMMAN [11] proposes a multi-level mesh mutual attention model to utilize low-dimensional and high-dimensional question information at different levels, providing more feature information for modal interactions.

A.3 Settings

In the experiments, Adamx was selected as the optimizer, and the initial learning rate was set at 0.001. With the increase in learning rounds, the learning rate gradually increased from 0.001 to 0.004. When the model accuracy reached its peak, the learning rate began to decay. During training, the decay rate of the learning rate was set to 0.5, and the model training rounds were 30 rounds. Regarding text, the word representation dimension of the question was 300, and the part of the speech embedding dimension was 128. In the tree hierarchical convolution module, word-level convolution uses a convolution kernel with the size of 3*428 for text convolution operation and the maximum syntax subtree length set to 4, and phrase-level convolution uses multi-head attention mechanism with the number of 8. To extract question features, we use a bidirectional

GRU network with a hidden layer dimension of 1024. For picture features, we set the dimension as 2048, each picture contains 10–100 visual entity areas, the total time step of message passing T is set as 4, and the dimension of scene context-aware entity feature is set as 1024. In our experiment, the VQA score provided by the official VQA competition is used as the evaluation metrics, shown as follows:

$$acc(\textbf{ans}) = min(\frac{\#human\ provides\ \textbf{ans}}{3}, 1) \tag{15}$$

References

1. Anderson, P., et al.: Bottom-up and top-down attention for image captioning and visual question answering. In: CVPR, pp. 6077–6086 (2018)
2. Antol, S., et al.: VQA: visual question answering. In: ICCV, pp. 2425–2433 (2015)
3. Ben-Younes, H., Cadène, R., Cord, M., Thome, N.: MUTAN: multimodal tucker fusion for visual question answering. In: ICCV, pp. 2631–2639 (2017)
4. Cadène, R., Ben-Younes, H., Cord, M., Thome, N.: MUREL: multimodal relational reasoning for visual question answering. In: CVPR, pp. 1989–1998 (2019)
5. Guo, D., Xu, C., Tao, D.: Bilinear graph networks for visual question answering. IEEE Trans. Neural Netw. Learn. Syst. **34**, 1023–1034 (2021)
6. Hochreiter, S., Schmidhuber, J.: Long short-term memory. Neural Comput. **9**(8), 1735–1780 (1997)
7. Hu, R., Rohrbach, A., Darrell, T., Saenko, K.: Language-conditioned graph networks for relational reasoning. In: ICCV, pp. 10293–10302 (2019)
8. Ilievski, I., Feng, J.: Generative attention model with adversarial self-learning for visual question answering. In: Thematic Workshops of ACM Multimedia 2017, pp. 415–423 (2017)
9. Kim, J., Jun, J., Zhang, B.: Bilinear attention networks. In: NeurIPS, pp. 1571–1581 (2018)
10. Kumar, A., et al.: Ask me anything: dynamic memory networks for natural language processing. In: ICML, vol. 48, pp. 1378–1387 (2016)
11. Lei, Z., Zhang, G., Wu, L., Zhang, K., Liang, R.: A multi-level mesh mutual attention model for visual question answering. Data Sci. Eng. **7**(4), 339–353 (2022)
12. Li, L., Gan, Z., Cheng, Y., Liu, J.: Relation-aware graph attention network for visual question answering. In: ICCV, pp. 10312–10321 (2019)
13. Lin, T.-Y., et al.: Microsoft COCO: common objects in context. In: Fleet, D., Pajdla, T., Schiele, B., Tuytelaars, T. (eds.) ECCV 2014. LNCS, vol. 8693, pp. 740–755. Springer, Cham (2014). https://doi.org/10.1007/978-3-319-10602-1_48
14. Nguyen, D.K., Okatani, T.: Improved fusion of visual and language representations by dense symmetric co-attention for visual question answering. In: CVPR, pp. 6087–6096 (2018)
15. Pennington, J., Socher, R., Manning, C.D.: Glove: global vectors for word representation. In: EMNLP, pp. 1532–1543 (2014)
16. Ren, M., Kiros, R., Zemel, R.: Exploring models and data for image question answering. In: NIPS, vol. 28 (2015)
17. Simonyan, K., Zisserman, A.: Very deep convolutional networks for large-scale image recognition. In: ICLR (2015)
18. Yang, Z., He, X., Gao, J., Deng, L., Smola, A.J.: Stacked attention networks for image question answering. In: CVPR, pp. 21–29 (2016)

19. Ye, T., Hu, L., Zhang, Q., Lai, Z.Y., Naseem, U., Liu, D.D.: Show me the best outfit for a certain scene: a scene-aware fashion recommender system. In: WWW, pp. 1172–1180 (2023)

20. Yu, Z., Yu, J., Xiang, C., Fan, J., Tao, D.: Beyond bilinear: generalized multimodal factorized high-order pooling for visual question answering. IEEE Trans. Neural Netw. Learn. Syst. **29**(12), 5947–5959 (2018)

21. Zhang, Q., Cao, L., Shi, C., Niu, Z.: Neural time-aware sequential recommendation by jointly modeling preference dynamics and explicit feature couplings. IEEE Trans. Neural Netw. Learn. Syst. **33**(10), 5125–5137 (2022)

22. Zhang, Q., Hu, L., Cao, L., Shi, C., Wang, S., Liu, D.D.: A probabilistic code balance constraint with compactness and informativeness enhancement for deep supervised hashing. In: IJCAI, pp. 1651–1657 (2022)

23. Zhang, S., Chen, M., Chen, J., Zou, F., Li, Y., Lu, P.: Multimodal feature-wise co-attention method for visual question answering. Inf. Fusion **73**, 1–10 (2021)

24. Zhang, Y., Wallace, B.: a sensitivity analysis of (and practitioners' guide to) convolutional neural networks for sentence classification. arXiv preprint arXiv:1510.03820 (2015)

CKR-Calibrator: Convolution Kernel Robustness Evaluation and Calibration

Yijun Bei[1], Jinsong Geng[1], Erteng Liu[1], Kewei Gao[1], Wenqi Huang[2], and Zunlei Feng[1,3(✉)]

[1] Zhejiang University, Hangzhou, China
{beiyj,22251346,let,gaokw,zunleifeng}@zju.edu.cn
[2] Digital Grid Research Institute, China Southern Power Grid,
Guangzhou 510670, China
huangwq@csg.cn
[3] Zhejiang University - China Southern Power Grid Joint Research Centre on AI,
Hangzhou 310058, China

Abstract. Recently, Convolution Neural Networks (CNN) have achieved excellent performance in some areas of computer vision, including face recognition, character recognition, and autonomous driving. However, there are still many CNN-based models that cannot be deployed in real-world scenarios due to poor robustness. In this paper, focusing on the classification task, we attempt to evaluate and optimize the robustness of CNN-based models from a new perspective: the convolution kernel. Inspired by the discovery that the root cause of the model decision error lies in the wrong response of the convolution kernel, we propose a convolution kernel robustness evaluation metric based on the distribution of convolution kernel responses. Then, we devise the Convolution Kernel Robustness Calibrator, termed as CKR-Calibrator, to optimize key but not robust convolution kernels. Extensive experiments demonstrate that CKR-Calibrator improves the accuracy of existing CNN classifiers by 1%–4% in clean datasets and 1%–5% in corrupt datasets, and improves the accuracy by about 2% over SOTA methods. The evaluation and calibration source code is open-sourced at https://github.com/cym-heu/CKR-Calibrator.

Keywords: Robustness · Convolution kernel · Evaluation · Calibration

1 Introduction

With the rise of computing power, Convolution Neural Networks (CNN) have become increasingly effective in various computer vision domains, such as facial recognition [19], autonomous driving [13], healthcare [23], etc. Image classification is a commonly researched topic in computer vision. CNN classifiers have made significant strides in improving classification results.

© The Author(s), under exclusive license to Springer Nature Singapore Pte Ltd. 2024
B. Luo et al. (Eds.): ICONIP 2023, LNCS 14451, pp. 137–148, 2024.
https://doi.org/10.1007/978-981-99-8073-4_11

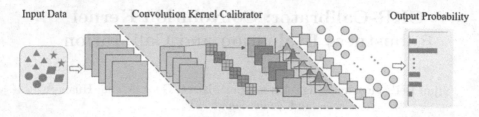

Fig. 1. The comparison between existing methods and the proposed method. The former focuses on evaluating and optimizing the robustness of the model through the input data and output probability aspects. While the proposed CKR-Calibrator evaluates and calibrates the convolutional kernel's robustness from a new perspective.

However, numerous CNN models still have insufficient robustness, making their application in real-world scenarios challenging. Model robustness refers to the ability of a machine learning model to maintain its performance and make reliable predictions when faced with various challenges or perturbations in the input data. Usually, the performance of deep learning-based models highly depends on the quality of the training dataset. Insufficient data, unbalanced categories, or domain differences often result in poor robustness on specific tasks.

Some works are developed to evaluate the model's robustness, particularly focusing on adversarial attacks and Out-Of-Distribution (OOD) transform. For the adversarial attack [6,22,30], the model robustness metric is based on their performance on adversarial samples, which are generated in a single or iterative manner during the training process. For the OOD transform [8,24], the performance of the model on the corrupted datasets is used as the model robustness metric. Another kind of works [4,34] focuses on measuring the model decision-making.

In this paper, unlike the above methods, we evaluate and optimize the model robustness from the perspective of the most basic component of the model: the convolution kernel. And the basic reason for the model's decision failure is the wrong response of some of the convolution kernels [5,11]. As shown in the Fig. 1, compared to other methods, our approach directly measures the robustness of each convolution kernel and performs optimization, which can fundamentally improve the robustness of the model.

Our inspiration comes from a simple observation that convolution kernels exhibit similar responses for samples with correct predictions within the same category, but significant differences for samples with incorrect predictions. Therefore, in the stage of robustness evaluation, convolution kernels are classified as robust and non-robust based on the variance of the convolution kernel response. In the stage of robustness optimization, a gradient aggregation technique [5] is adopted to find convolution kernels, which play a key role in decision making. To improve the robustness of the convolution kernel that is important but not robust, the triplet loss [26] is used to approximate the output response of samples in the same category.

In summary, the main contributions are given as follows:

- A new viewpoint on CNN model evaluation and optimization is explored from the perspective of convolution kernel response.
- A pluggable CNN model convolution kernel robustness evaluation and calibration tool called CKR-calibration is proposed.
- Experimental results show that the proposed method can improve the accuracy of mainstream CNN classifiers by 1%-4% on clean datasets and 1%-5% on corrupted datasets, while the proposed method can be combined with SOTA method and improves the accuracy by about 2% over the SOTA method when combined with data enhancement methods.

2   Related Work

In this section, we briefly review the mostly related techniques model of robustness evaluation, robustness optimization, and Feature Attribution.

2.1   Robustness of Deep Learning Based Model

Most existing methods evaluate model robustness based on its performance on imperfect datasets, which can be divided into adversarial sample datasets and the OOD datasets. The concept of adversarial samples is first proposed in [30]. The paper demonstrates that the model can be misclassified by adding small perturbations to the original data. FGSM [6] generates adversarial samples by adding perturbations along the direction of the gradient during model training. Madry et al. [22] employs multi-step gradient update (PGD attack) but suffers from the drawback of long training time. OOD datasets like Imagenet-C [8], Cifra10-C [8], Cifar100-C [8] and MNIST-C [24] are proposed to measure assess the model's ability to deal with natural disruptions. For the OOD transform task, methods like Coutout [3], Mixup [38], CutMix [35], AutoAug [32] and AugMix [9] augment the training data by cropping and blending.

In addition, some researchers measure model robustness from the perspective of decision making. Ding et al. [4] use margin to measure model robustness, where margin is defined as the distance from the input to the decision boundary of the classifier. Yang et al. [34] introduce the notion of classifier boundary thickness, demonstrate that robust models possess thick decision boundaries, and enhances model boundary thickness by noise-augmentation mixup training.

Unlike the above methods, our approach performs robustness evaluation and optimization on the basic component of the model, the convolution kernel.

2.2   Deep Feature Attribution

Existing deep feature attribution techniques can be categorized into forward propagation-based methods, back propagation-based methods, and activation-based methods. Forward propagation based methods [21,37,40] interfere with the inputs or feature maps and then put them into the original model for forward propagation to obtain new output results and compare the difference with the actual output to find the model sensitive region to the sample. Back propagation-based methods [1,17,27] back propagate the significant signal from the output neuron to the input neuron through each layer, and thus calculate the contribution of each neuron to the output neuron. The activation-based approach [2,25,31,39] calculates a set of weights on the activation graph generated by the middle layer and sums them to visualize the important features. This class of methods can be classified as grad-based or grad-free according to whether the weights are calculated using gradients or not.

3   CKR-Calibrator

In this section, we analyze the convolutional kernel responses for samples of the same category with different confidence levels. Inspired by the discovery that the root cause of the model decision error lies in the wrong response of the convolution kernel, a pluggable convolution kernel robustness evaluation and a calibrator tool (CKR-Calibrator) are proposed. The details about Preliminary Analysis and CKR-Calibrator are elaborated as follows.

3.1   Preliminary Analysis

For the input image I, the feature maps of the $(r\text{-}1)$-th layer are denoted as $\{m_{r-1}^1, m_{r-1}^2, m_{r-1}^3, ..., m_{r-1}^P\}$ and the convolution kernels of the r-th layer are denoted as $\{c_r^1, c_r^2, c_r^3, ..., c_r^K\}$, features maps of the r-th layer are calculated as follows:

$$m_r^k = f(c_r^k \otimes \{m_{r-1}^1, m_{r-1}^2, m_{r-1}^3, ..., m_{r-1}^P\}), \tag{1}$$

where $\otimes$ denotes convolution operation, f denotes the following operation, such as batch normalization, pooling operation and activation function. The convolution kernel output response is defined as the sum of the values of the feature map corresponding to the convolution kernel after filtering out the negative values. The convolution kernel output responses a_r^k of the r-th layer are calculated as follows:

$$a_r^k = \uplus ReLU(m_r^k), \tag{2}$$

where $\uplus$ denotes the summing values of the matrix.

As shown in the Fig. 2(a), we count the convolution kernel responses of 10 same-class samples with the confidence level higher than 0.9 in the last convolution layer of Squeezenet [14] on the Cifar10 dataset. We can find that high-confidence samples of same category have similar convolution kernel responses in the same layer. As shown in the Fig. 2(b), in contrast, we count the convolution kernel responses for samples with confidence levels lower than 0.1, which means they were incorrectly judged by the model as other classes, other settings are the same as the above. We can find that convolution kernel responses for low-confidence samples of same category are different.

In conclusion, samples of the same category that are correctly classified by the model have similar convolution kernel responses, while samples that are misclassified have different convolution kernel responses from those that are correctly classified, it can be inferred that the root cause of the model decision error lies in the wrong response of the convolution kernel.

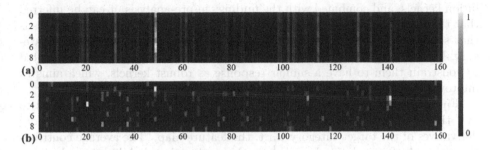

Fig. 2. The convolution kernel responses of 10 high-confidence samples(a) and 10 low-confidence samples(b) with the same category in the last convolution layer.

3.2   Convolution Kernel Robustness Evaluation

Inspired by the above discovery, we attempt to evaluate the robustness of CNN-based model through measuring the response of the convolution kernel. Specifically, for a particular convolution kernel c_r^k in the r-th layer, a response set $\{a_{r,1}^k, a_{r,2}^k, ..., a_{r,n}^k, ..., a_{r,N}^k\}$ corresponding to it for all samples of a specific class is calculated according to Eq.(2). The robustness score s_r^k for the convolution kernel c_r^k of a specific class is calculated as follows:

$$s_r^k = \sum_{n=1}^{N}(a_{r,n}^k - \bar{a}_r^k)^2, \bar{a}_r^k = \frac{1}{N}\sum_{n=1}^{N} a_{r,n}^k, \tag{3}$$

where $\bar{a}_r^k$ denotes the mean of the response set.

In addition, the normalization operation is adopted in the r-th layer, the normalized robustness score h_r^k is calculated as follows:

$$h_r^k = \frac{s_r^k - min(S_r)}{max(S_r) - min(S_r) + \varepsilon}, S_r = \{s_r^k\}_{k=1}^{K}, \tag{4}$$

where ε denotes a minimum to avoid the situation of divide-by-zero, S_r denotes the robustness sore set in the r-th layer.

For all convolution kernels in the r-th layer, a threshold α is adopted to distinguish robust and non-robust convolution kernel. Then, the indicator b_r^k for convolution kernel c_r^k is calculated as follows:

$$b_r^k = \begin{cases} 0 & h_r^k \geq \alpha, \\ 1 & h_r^k < \alpha \end{cases}, \tag{5}$$

where $\alpha > 0$, $b_r^k = 1$ denotes convolution kernel c_r^k is robust, $b_r^k = 0$ denotes convolution kernel c_r^k is not robust.

3.3 Convolution Kernel Robustness Calibration

Some studies have shown that model decisions depend on some of the key convolution kernels, and combined with the findings in pre-analysis, it can be inferred that the reason for the poor robustness of the deep model in some categories is that some of the key convolution kernels are not robust enough for that category. For the important but not robust kernels, the triplet loss [26] is adopted to constrain them to have a similar response as robust kernels, which aims to improve their robustness.

Firstly, a gradient aggregation-based contribution method [5] is adopted to find the key convolution kernels, which computes the aggregated first-order derivatives of the target category w.r.t. the feature map. The average contribution value g_r^k of convolution kernel c_r^k for the predicted probability $\hat{y}$ is calculated as follows:

$$g_r^k = \uplus |\frac{\partial \hat{y}}{\partial m_r^k}|, \tag{6}$$

where $\uplus$ denotes the summing values of the first-order derivatives matrix. For T samples with confidence levels higher than 0.9 of a specific category, a contribution set $\{g_{r,1}^k, g_{r,2}^k, g_{r,3}^k, ...g_{r,T}^k\}$ of convolution kernel c_r^k can be obtained according to Eq. (6). Then, the normalization operation is adopted in the r-th layer, the normalized contribution score u_r^k is calculated as follows:

$$u_r^k = \frac{g_r^k - min(G_r)}{max(G_r) - min(G_r) + \varepsilon}, G_r = \{g_r^k\}_{k=1}^K, g_r^k = \sum_{t=1}^{T} g_{r,t}^k, \tag{7}$$

where ε denotes a minimum to avoid the situation of divide-by-zero, G_r denotes the contribution set in the r-th layer.

There is also a threshold β to distinguish between important and unimportant convolution kernels, the indicator d_r^k is calculated as follows.

$$d_r^k = \begin{cases} 0 & u_r^k \geq \beta \\ 1 & u_r^k < \beta \end{cases}, \tag{8}$$

where $\beta > 0$, $d_r^k = 1$ denotes convolution kernel c_r^k is important, $d_r^k = 0$ denotes convolution kernel c_r^k is not unimportant.

Secondly, the critical but not robust convolution kernels are filtered out according to Eq. (5) and Eq. (8), the indicator e_r^k is calculated as follows:

$$e_r^k = \begin{cases} 0 & b_r^k = 1 \, or \, d_r^k = 0 \\ 1 & b_r^k = 0 \, and \, d_r^k = 1 \end{cases}, \tag{9}$$

where $e_r^k = 1$ denotes convolution kernel c_r^k is important but not robust, $e_r^k = 0$ denotes the other case.

Thirdly, for the important but not robust kernels, the triplet loss [26] is adopted to improve their robustness by constraining its response is similar to robust kernels'. When a batch of samples is fed to the model, for one of them x_o, its list of important but not robust convolution kernel responses $\widetilde{A}_r^o$ in the r-th layer is calculated as follows:

$$\widetilde{A}_r^o = f_C(A_r^o), A_r^o = [a_{r,o}^k]_{k=1}^K, C_r^o = [c_{r,o}^k]_{k=1}^K, \tag{10}$$

where A_r^o denotes all convolution kernel responses in the r-th layer, f_C denotes filter operation with C_r^o as condition.

For the anchor sample x_o, in the space of the convolution kernel response, the sample of the same class with the farthest distance and the sample of the different class with the closest distance are donated as x_p and x_n respectively, and the list of important but not robust convolution kernels are denotes as $\widetilde{A}_r^p$ and $\widetilde{A}_r^n$ respectively. Thus, the triplet loss $\mathcal{L}_{trip}^{r,o}$ for convolution kernels in the r-th layer is calculated as follows:

$$\mathcal{L}_{trip}^{r,o} = ||\widetilde{A}_r^p - \widetilde{A}_r^o||^2 - ||\widetilde{A}_r^n - \widetilde{A}_r^o||^2 + \gamma, \tag{11}$$

where γ is a margin that is enforced between positive and negative pairs.

The triplet loss for all samples of the current batch in all layers is summed and added to the original loss $\mathcal{L}_{ori}$. The total loss $\mathcal{L}_{all}$ of the current batch is calculated as follows:

$$\mathcal{L}_{all} = \mathcal{L}_{ori} + \sum_{r \in R} \sum_{o=1}^{w} \mathcal{L}_{trip}^{r,o}, \tag{12}$$

where w denotes batch size, R denotes layer sets.

4   Experiments

In the experiments, we verify the proposed CKR-Calibrator with nine mainstream CNN classifers (ResNet50 [7], GoogleNet [29], WideResNet [36], Vgg16 [28], AlexNet [16], Squeezenet [14], DenseNet [12], ResNeXt50 [33], MobileNet [10]) on four clean benchmark datasets (Cifar10 [15], Cifar100 [15], MNIST [20], Tiny-Imagenet [18]) and four corrupted benchmark dataset (MNIST-C [24], Cifar10-C [8], Cifar100-C [8], Tiny-Imagenet-C [8]). The corrupted datasets are the original test data poisoned by everyday image corruptions. Each noise has five intensity levels when injected into images. Unless specified, the proposed CKR-Calibrator is inserted into the last convolution layer during the optimization phase.

Parameter Setting. Unless stated otherwise, the default experiment settings are given as follows: thresholds $\alpha = 0.7$, $\beta = 0.5$, margin $\gamma = 100$. In the training stage, the default optimizer is SGD, the learning rate is 0.01, and the epoch number is 200.

Metric. For clean datasets, we adapt the top-1 accuracy as a metric. For corrupt datasets, we adapt the mean top-1 accuracy over the different corruption types as a metric.

4.1   The Effect of CKR-Calibrator for SOTA CNN Classifiers

In this section, we verify the effectiveness of CKR-Calibrator on four image classification datasets and the corresponding corrupted datasets. All results are reported as an average of three runs to ensure fairness. From Table 1, we can see that CKR-Calibrator helps the CNN classifier improve the accuracy by 1%–4% on the clean datasets and 1%–5% on the corrupted datasets, which verifies that the proposed method can improve the robustness of the model effectively. The accuracy improvement on the corrupted datasets is higher than that on the clean datasets, the possible reason is that the optimized convolution kernel can filter out perturbations and extract the key features. However, CKR-Calibrator has little effect for models trained on the MNIST dataset, the latent reason is that the convolution kernels of the original model are robust enough for handling handwritten digits.

Table 1. The calibration performance of CKR-Calibrator for nine mainstream classifiers on four benchmark datasets with clean and corrupted samples. 'score 1/+score 2' denotes the base accuracy and incremental accuracy with the proposed CKR-Calibrator, respectively. (All score are in %)

	models	MNIST	Cifar10	Cifar100	Tiny_Imagenet
clean samples	Vgg16	99.76/+0.05	93.39/+0.85	72.14/+2.30	63.82/+3.41
	GoogLeNet	99.69/+0.02	93.47/+1.07	70.38/+1.85	64.25/+3.72
	ResNet50	99.68/-0.04	95.00/+0.32	77.25/+3.03	65.36/+2.95
	WideResNet	99.67/+0.01	90.15/+1.78	66.29/+2.19	74.83/+3.87
	SqueezeNet	99.72/-0.02	90.95/+1.98	67.98/+1.49	67.12/+2.91
	MobileNet	99.54/-0.09	90.83/+2.06	68.16/+2.23	66.92/+3.25
	AlexNet	99.59/+0.03	85.69/+2.46	56.87/+3.77	52.14/+3.82
	ResNeXt50	99.74/-0.07	94.44/+0.76	75.27/+1.07	71.67/+3.12
	DenseNet	99.87/+0.06	95.04/+1.08	77.96/+1.74	72.19/+2.67
	models	MNIST	Cifar10	Cifar100	Tiny_Imagenet
corrupted samples	Vgg16	78.66/+1.04	73.12/+2.04	47.77/+2.76	35.91/+4.22
	GoogLeNet	79.05/+0.53	73.47/+2.96	48.52/+4.27	36.72/+3.76
	ResNet50	70.28/+0.63	75.38/+1.47	52.62/+1.96	37.86/+4.75
	WideResNet	69.70/+1.16	66.93/+3.41	41.20/+4.73	41.21/+3.58
	SqueezeNet	77.50/+0.33	71.55/+4.02	45.11/+1.82	35.73/+5.24
	MobileNet	76.29/+1.24	70.84/+2.58	47.81/+2.27	36.96/+4.30
	AlexNet	86.05/+0.23	68.93/+3.97	38.27/+4.83	21.17/+5.58
	ResNeXt50	67.64/+2.04	72.27/+1.70	49.43/+2.74	39.72/+4.95
	DenseNet	75.04/+1.89	74.76/+2.57	51.41/+2.87	40.16/+5.03

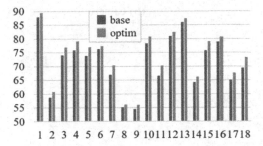

1: brightness   10: jpeg_compression
2: contrast   11: motion_blur
3: defocus_blur   12: pixelate
4: elastic_transform   13: saturate
5: fog   14: shot_noise
6: frost   15: snow
7: gaussian_blur   16: spatter
8: gaussian_noise   17: speckle_noise
9: impulse_noise   18: zoom_blur

Fig. 3. The base (blue) and improved (orange) accuracy of SqueezeNet on Cifar10 under different types of perturbation. (Color figure online)

Table 2. The base and improved accuracy of CKR-Calibrator combined with different robustness optimization methods on four mainstream CNN-based Classifiers.

	Coutout [3]	Mixup [38]	CutMix [35]	AutoAug [32]	AugMix [9]
AlexNet	62.23/+1.22	64.37/+1.52	62.45/+1.19	67.87/+2.12	69.67/+2.78
Vgg16	78.86/+2.07	83.52/+0.45	79.51/+0.92	85.92/+1.44	87.83/+1.62
SqueezeNet	74.61/+0.97	77.53/+1.43	73.79/+0.83	79.55/+1.93	84.46/+2.49
ResNet50	81.26/+0.45	85.78/+1.13	83.26/+0.92	87.15/+1.02	89.14/+1.24

Figure 3 shows the effect of the CKR-Calibrator under different types of perturbations. We can observe that the proposed CKR-Calibrator achieves improvement on various types of perturbations, which verifies that the CKR-Calibrator can comprehensively enhance the ability of the model to cope with disturbances. Moreover, the proposed CKR-Calibrator achieves the highest increase on blur types of perturbations, the possible reason is that the blur samples retain more features, and the convolution kernel optimized by the CKR-Calibrator can effectively extract these features.

4.2 The Effect of CKR-Calibrator Combined with SOTA Method

Due to the proposed CKR-Calibrator optimizing the robustness of the mode from a new perspective, the proposed method can be combined with all existing model robustness optimization methods. In this section, we verify the effectiveness of the CKR-Calibrator combined with five mainstream model robustness optimization strategies for four mainstream CNN classifiers on Cifar10-C. From Table 2, we can see that the CKR-Calibrator can improve the accuracy by about 2% and achieve the best results when combined with AugMix.

4.3 The Ablation Study

In this section, we conduct the ablation study on different layer depths where the CKR-Calibrator is plugged. From Fig. 4, we can see that the model with

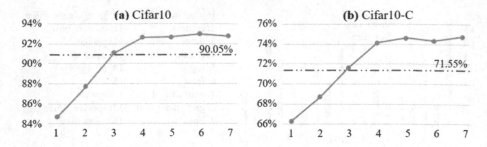

Fig. 4. The ablation study on different layer depths on clean samples (a) and corrupted samples (b). Red dash line denotes the original accuracy. (Color figure online)

the CKR-Calibrator achieves better performance along the layer goes deeper. It reveals that the CKR-Calibrator is more suitable for deep layers. The reason behind this observation can be attributed to the nature of the convolution kernels in shallow layers, which primarily capture low-dimensional features of the input samples. In these shallow layers, the low-dimensional features of samples belonging to the same class may not exhibit sufficient similarity. As a result, the presence of dissimilar features can interfere with the CKR-Calibrator's ability to identify and select robust convolution kernels. In contrast, as the network's depth increases, the convolution kernels in deeper layers can extract more abstract and high-dimensional features. These higher-level features tend to exhibit greater similarity among samples of the same class. Consequently, the CKR-Calibrator becomes more effective at identifying and screening robust convolution kernels, leading to enhanced performance in deeper layers.

5   Conclusion

In this paper, we put forward a universal and pluggable CKR-Calibrator for evaluating and optimizing the robustness of convolution kernels. Unlike other model robustness measures and optimization methods, we focus on the basic component of the model: the convolution kernel. Extensive experiments demonstrate that the proposed robustness evaluation method can effectively filter out robust convolution kernels, and the proposed CKR-Calibrator can improve the performance of mainstream CNN-based classifiers especially on corrupted datasets. In the future, we will focus on exploring the evaluation and calibration of convolution kernels' robustness for other vision tasks including image segmentation and object detection.

Acknowledgements. This work is funded by Public Welfare Technology Applied Research Projects of Zhejiang Province, China (LGG21F020004), Basic Public Welfare Research Project of Zhejiang Province (LGF21F020020), Ningbo Natural Science Foundation (2022J182), and the Fundamental Research Funds for the Central Universities (2021FZZX001-23, 226-2023-00048).

References

1. Bach, S., Binder, A., Montavon, G., Klauschen, F., Müller, K.R., Samek, W.: On pixel-wise explanations for non-linear classifier decisions by layer-wise relevance propagation. PLoS ONE **10**(7), e0130140 (2015)
2. Chattopadhay, A., Sarkar, A., Howlader, P., Balasubramanian, V.N.: Grad-cam++: generalized gradient-based visual explanations for deep convolutional networks. In: 2018 IEEE Winter Conference on Applications of Computer Vision (WACV), pp. 839–847. IEEE (2018)
3. DeVries, T., Taylor, G.W.: Improved regularization of convolutional neural networks with cutout. arXiv (2017)
4. Ding, G.W., Sharma, Y., Lui, K.Y.C., Huang, R.: Mma training: direct input space margin maximization through adversarial training. arXiv (2018)
5. Feng, Z., Hu, J., Wu, S., Yu, X., Song, J., Song, M.: Model doctor: a simple gradient aggregation strategy for diagnosing and treating CNN classifiers. In: AAAI, vol. 36, pp. 616–624 (2022)
6. Goodfellow, I.J., Shlens, J., Szegedy, C.: Explaining and harnessing adversarial examples. arXiv (2014)
7. He, K., Zhang, X., Ren, S., Sun, J.: Deep residual learning for image recognition. In: Proceedings of the IEEE Conference on Computer Vision and Pattern Recognition, pp. 770–778 (2016)
8. Hendrycks, D., Dietterich, T.: Benchmarking neural network robustness to common corruptions and perturbations. arXiv preprint arXiv:1903.12261 (2019)
9. Hendrycks, D., Mu, N., Cubuk, E.D., Zoph, B., Gilmer, J., Lakshminarayanan, B.: Augmix: a simple data processing method to improve robustness and uncertainty. arXiv (2019)
10. Howard, A.G., et al.: Mobilenets: efficient convolutional neural networks for mobile vision applications. arXiv (2017)
11. Hu, J., et al.: CNN LEGO: disassembling and assembling convolutional neural network. arXiv (2022)
12. Huang, G., Liu, Z., Laurens, V.D.M., Weinberger, K.Q.: Densely connected convolutional networks. IEEE Computer Society (2016)
13. Huang, Y., Chen, Y.: Autonomous driving with deep learning: a survey of state-of-art technologies. arXiv (2020)
14. Iandola, F.N., Han, S., Moskewicz, M.W., Ashraf, K., Dally, W.J., Keutzer, K.: Squeezenet: alexnet-level accuracy with 50× fewer parameters and¡ 0.5 mb model size. arXiv preprint arXiv:1602.07360 (2016)
15. Krizhevsky, A., Hinton, G., et al.: Learning multiple layers of features from tiny images (2009)
16. Krizhevsky, A., Sutskever, I., Hinton, G.: Imagenet classification with deep convolutional neural networks. In: NeurIPS, vol. 25, no. 2 (2012)
17. Lapuschkin, S., Wäldchen, S., Binder, A., Montavon, G., Samek, W., Müller, K.R.: Unmasking clever hans predictors and assessing what machines really learn. Nat. Commun. **10**(1), 1096 (2019)
18. Le, Y., Yang, X.: Tiny imagenet visual recognition challenge. CS 231N **7**(7), 3 (2015)
19. LeCun, Y., Bengio, Y., Hinton, G.: Deep learning. Nature **521**(7553), 436–444 (2015)
20. LeCun, Y., Bottou, L., Bengio, Y., Haffner, P.: Gradient-based learning applied to document recognition. Proc. IEEE **86**(11), 2278–2324 (1998)

21. Lengerich, B.J., Konam, S., Xing, E.P., Rosenthal, S., Veloso, M.: Towards visual explanations for convolutional neural networks via input resampling. arXiv (2017)
22. Madry, A., Makelov, A., Schmidt, L., Tsipras, D., Vladu, A.: Towards deep learning models resistant to adversarial attacks. arXiv (2017)
23. Miotto, R., Wang, F., Wang, S., Jiang, X., Dudley, J.T.: Deep learning for healthcare: review, opportunities and challenges. Brief. Bioinform. **19**(6), 1236–1246 (2018)
24. Mu, N., Gilmer, J.: Mnist-c: a robustness benchmark for computer vision (2019)
25. Omeiza, D., Speakman, S., Cintas, C., Weldermariam, K.: Smooth grad-cam++: an enhanced inference level visualization technique for deep convolutional neural network models. arXiv (2019)
26. Schroff, F., Kalenichenko, D., Philbin, J.: Facenet: a unified embedding for face recognition and clustering. In: Proceedings of the IEEE Conference on Computer Vision and Pattern Recognition, pp. 815–823 (2015)
27. Simonyan, K., Vedaldi, A., Zisserman, A.: Deep inside convolutional networks: visualising image classification models and saliency maps. arXiv (2013)
28. Simonyan, K., Zisserman, A.: Very deep convolutional networks for large-scale image recognition. arXiv (2014)
29. Szegedy, C., et al.: Going deeper with convolutions. In: Proceedings of the IEEE Conference on Computer Vision and Pattern Recognition, pp. 1–9 (2015)
30. Szegedy, C., et al.: Intriguing properties of neural networks. arXiv (2013)
31. Wang, H., et al.: Score-cam: score-weighted visual explanations for convolutional neural networks. In: Proceedings of the IEEE/CVF Conference on Computer Vision and Pattern Recognition Workshops, pp. 24–25 (2020)
32. Wei, W., Zhou, J., Wu, Y.: Beyond empirical risk minimization: local structure preserving regularization for improving adversarial robustness. arXiv (2023)
33. Xie, S., Girshick, R., Dollár, P., Tu, Z., He, K.: Aggregated residual transformations for deep neural networks. IEEE (2016)
34. Yang, Y., et al.: Boundary thickness and robustness in learning models. NeurIPS **33**, 6223–6234 (2020)
35. Yun, S., Han, D., Oh, S.J., Chun, S., Choe, J., Yoo, Y.: Cutmix: regularization strategy to train strong classifiers with localizable features. In: ICCV, pp. 6023–6032 (2019)
36. Zagoruyko, S., Komodakis, N.: Wide residual networks. arXiv (2016)
37. Zeiler, M.D., Fergus, R.: Visualizing and understanding convolutional networks. In: Fleet, D., Pajdla, T., Schiele, B., Tuytelaars, T. (eds.) ECCV 2014. LNCS, vol. 8689, pp. 818–833. Springer, Cham (2014). https://doi.org/10.1007/978-3-319-10590-1_53
38. Zhang, H., Cisse, M., Dauphin, Y.N., Lopez-Paz, D.: mixup: beyond empirical risk minimization. arXiv (2017)
39. Zhou, B., Khosla, A., Lapedriza, A., Oliva, A., Torralba, A.: Learning deep features for discriminative localization. In: Proceedings of the IEEE Conference on Computer Vision and Pattern Recognition, pp. 2921–2929 (2016)
40. Zintgraf, L.M., Cohen, T.S., Adel, T., Welling, M.: Visualizing deep neural network decisions: prediction difference analysis. arXiv (2017)

SGLP-Net: Sparse Graph Label Propagation Network for Weakly-Supervised Temporal Action Localization

Xiaoyao Wu⑩ and Yonghong Song(✉)

Faculty of Electronic and Information Engineering, Xi'an Jiaotong University, Xi'an, China
wuxiaoyao@stu.xjtu.edu.cn, songyh@xjtu.edu.cn

Abstract. The present weakly-supervised methods for Temporal Action Localization are primarily responsible for capturing the temporal context. However, these approaches have limitations in capturing semantic context, resulting in the risk of ignoring snippets that are far apart but sharing the same action categories. To address this issue, we propose an action label propagation network utilizing sparse graph networks to effectively explore both temporal and semantic information in videos. The proposed SGLP-Net comprises two key components. One is the multi-scale temporal feature embedding module, a novel method that extracts both local and global temporal features of the videos during the initial stage using CNN and self-attention and serves as a generic module. The other is an action label propagation mechanism, which uses graph networks for feature aggregation and label propagation. To avoid the issue of excessive feature completeness, we optimize training using sparse graph convolutions. Extensive experiments are conducted on THUMOS14 and ActivityNet1.3 benchmarks, among which advanced results demonstrate the superiority of the proposed method. Code can be found at https://github.com/xyao-wu/SGLP-Net.

Keywords: Weakly-Supervised Temporal Action Localization · Attention modules · Action label propagation network · Sparse graph networks

1 Introduction

Temporal Action Localization (TAL) aims to locate action instances from untrimmed long videos and provide the start time, end time, and action categories of these actions. TAL is a sub task of video content understanding and has been widely applied in various domains such as video retrieval [6], surveillance, anomaly detection, sports analysis and more.

With the rapid development of deep learning, fully-supervised methods [2, 14, 23, 27, 29, 30] have achieved significant results. However, they have high annotation costs, need a lot of data, and are subjective. Weakly-Supervised Temporal

© The Author(s), under exclusive license to Springer Nature Singapore Pte Ltd. 2024
B. Luo et al. (Eds.): ICONIP 2023, LNCS 14451, pp. 149–161, 2024.
https://doi.org/10.1007/978-981-99-8073-4_12

Action Localization (WTAL) [5,15,19,20,22,24] only requires video-level action labels, which are inexpensive and simple to get. Also, it lessens subjective errors at source, making it more appropriate to use in the real-world.

One of the WTAL methods is to adopt a pre-made action recognition network and a classifier for action localization [15,22]. These methods are simple and easy to understand, but they tend to be less accurate. The other [19,20] often formulates the problem as a multi-instance learning (MIL) [7] task. These approaches divide the video into packages and have an improvement in accuracy. However, it may overlook or make mistakes in detection because it doesn't accurately record the temporal positions of actions on the timeline. Furthermore, as it is trained just on video-level labels, it is more susceptible to noise and has lower detection accuracy.

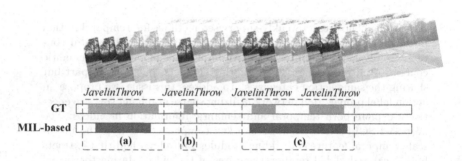

Fig. 1. Intuition behind the proposed SGLP-Net. Transparent frames represent background frames. (a) Missed detection of snippets. (b) Missed detection of actions. (c) Over-completeness.

Figure 1 demonstrates an example of a MIL-based network. Analysis has shown that MIL-based approaches consider video snippets as separate instances and ignore their underlying temporal patterns, leading to insufficient utilization of long-term information. In fact, most feature embedding layers in prior pipelines are implemented using one or more stacked conv1d, which are limited by the kernel's shape and size. While they are effective at collecting short-term dependencies, they fail to catch long-term snippet characteristics, making it difficult to cope with flexible and varied video durations. To overcome these limitations, we propose a novel weakly-supervised feature embedding method by approaching local and global temporal information in videos.

Regarding the missing action detection caused by the absence of frame-level annotations in the WTAL task, we introduce a new framework named **S**parse **G**raph **L**abel **P**ropagation **Net**work (SGLP-Net). Our approach aims to explore the inherent temporal and semantic information in the video itself. Specifically, SGLP-Net leverages correlation between snippets, propagates labels across nodes, and infers labels of unlabeled snippets to categorize actions for the entire video. Each snippet is represented as a node in the graph, and connections between them are made using context and similarity edges. Finally, adopting a sparse graph network can help with closely linked actions that can't be properly separated in the MIL-based environment due of the dense characteristics of the snippets. Layer sampling in Graph Convolutional Networks (GCN) is used in this method to reduce computational complexity. Overall, our proposed method offers encouraging results in improving the performance of action snippet detection in videos by taking a unique perspective on MIL-based video comprehension methods and addressing the identified challenges.

We summarize our main contributions as follows:

- A multi-scale temporal feature embedding module is designed to capture both local and global temporal relationships using convolution and self-attention in the initial stage. It can be easily integrated into other TAL frameworks.
- We propose a novel action label propagation mechanism for temporal action localization of untrimmed videos under weak supervision settings. It is implemented using a sparse graph network and iteratively trained using a pseudo label strategy to automatically transmit snippet feature information to adjacent snippets, in order to better utilize the inherent features and accurately locate the temporal position of each behavior in the video.
- We provide extensive experiments and the qualitative visualization results to demonstrate the efficacy of our design. And the quantitative results show that our SGLP-Net outperforms current state-of-the-art (SOTA) methods.

2 Related Work

Graph-based Temporal Action Localization. In recent years, graph-based methods have been proposed for TAL. [27] constructs a graph based on the interaction, aiming to classify and adjust each proposal's border by using contextual information. [30] proposes video self-stitching and cross-scale graph pyramid network. All of these methods follow the fully-supervised paradigm.

In WTAL, [26] supplements and enhances features by mining supplementary data across segments. In contrast, we use sparse graph convolutions for label propagation and consider various kinds of interactions between nodes while building edges. Additionally, we employ pseudo label for training to effectively exploit intrinsic information in videos.

Label Propagation Algorithm (LPA). LPA [21] is a popular method for community detection in graph data. It creates communities by spreading the labels of the nodes to their nearby nodes depending on how comparable their attributes are. It has been widely used in various applications. [8] performs label propagation on a large image dataset for few shot learning. Encouraged by these, we attempt to apply sparse graph convolutions and LPA to WTAL. To the best of our knowledge, we are the first to do this, and the experimental results support our approach.

3   Method

In this section, we first define the formulation of WTAL, then provide a detailed introduction to our SGLP-Net. Figure 2 displays the design of our model. The network comprises two essential components: (a) Multi-scale Temporal Feature Embedding, which is composed of two parts. The first part is a convolutional block with four convolutional layers for obtaining local features, while the second part, a self-attention module, is used to obtain global features. (b) Action Label Propagation, which takes the embedded features X as input and generates edges based on various associations between nodes to build a graph. The layer sampling [3] is used during training. Overall, the SGLP-Net can efficiently extract temporal features and propagate labels within videos, leading to precise temporal action localization results.

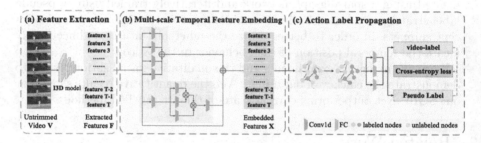

Fig. 2. The framework of our proposed SGLP-Net.

3.1   Problem Definition

Given an untrimmed video V, which contains multiple action instances $\{\psi_i = (t_i^s, t_i^e, c_i)\}_{i=1}^{N_\psi}$, where N_ψ is the number of action instances, t_i^s and t_i^e denotes the start and end time of action instance ψ_i, and $c_i \in \mathbb{R}^C$ represents the class category. WTAL aims to detect all action instances $\left\{\hat{\psi}_i = \left(\hat{t}_i^s, \hat{t}_i^e, \hat{c}_i, \hat{\phi}_i\right)\right\}_{i=1}^{M}$ only using video-level labels, where $\hat{\phi}_i$ denotes the confidence score of action instance $\hat{\psi}_i$ and M is the number of instances.

3.2   Feature Extractor

Following recent WTAL [15,25] methods, we first divide it into non-overlapping snippets based on a predefined sampling ratio σ, and then apply a Kinetics-400 [12] pre-trained I3D [1] model to extract features. The extracted snippet-level features are denoted by the $\frac{1}{2}D$-dimensional feature vectors, which can be concatenated to form the video-level representation $F \in \mathbb{R}^{T \times D}$.

3.3   Multi-scale Temporal Feature Embedding

Since the extracted video features F are not trained for the WTAL task, in order to map F to task-specific feature space, we introduce a novel feature embedding module named Multi-scale Temporal Feature Embedding (MTF). Concretely, following [15,20], the features are then fed into a temporal convolution layer and the ReLU [18] activation for feature modeling in the previous works: $X = ReLU(Conv(F))$.

However, this method is insufficient in utilizing global temporal features. Inspired by the attention [22] used in video understanding, our MTF captures the multi-resolution local temporal dependencies and the global temporal dependencies within an untrimmed video.

MTF uses four CNNs to obtain multi-scale features in video. Formally, we can denote the MTF module as following:

$$X_i = ReLU(Conv_{3 \times 3}(F, \theta_i)) \tag{1}$$

$$X_{Conv} = X_1 \otimes X_2 \otimes X_3 \otimes X_4 \tag{2}$$

where $i \in \{1, 2, 3, 4\}$; θ_i denotes the trainable convolution parameters of feature embedding layer and $ReLU$ is the non-linear activation function. The output of Conv is formed with a concatenation of X_i.

The global temporal dependencies is achieved with a self-attention module. In detail, we aim to produce an attention map $M \in \mathbb{R}^{T \times T}$ that estimates the pairwise correlation between snippets.

$$F^c = Conv_{1 \times 1}(F, \theta_5), \tag{3}$$

$$F^{ci} = Conv_{1 \times 1}(F^c, \theta_i), i \in \{1, 2, 3\} \tag{4}$$

$$M = (F^{c1})(F^{c2})^T, \tag{5}$$

$$F^{c4} = Conv_{1 \times 1}(MF^{c3}, \theta_9), \tag{6}$$

$$X_{Global} = F^c + F^{c4} \tag{7}$$

A skip connection using the original features F produces the final MTF outputs: $X = X_{Conv} + X_{Global} + F$.

3.4   Action Label Propagation

We construct SGLP-Net model using feature sequences X. Let $\varsigma(v, \varepsilon)$ be a graph of N nodes with nodes $v_i \in v$ and edge $e_{ij} = (v_i, v_j) \in \varepsilon$. Furthermore, let $A \in \mathbb{R}^{N \times N}$ be the adjacency matrix associated with ς. The weight of an edge represents the strength of the relationship between two connected nodes. A higher weight indicates a stronger association between them. In the subsequent, we introduce how to construct the graph and use it for temporal action localization.

Action Graph Construction. *Action Similarity Edges.* Due to the tendency of video patterns with the same action category to be the same, we believe that snippets with high similarity are more likely to have the same labels. Based on this prior, we establish the similarity edge between snippet s_i and s_j if $r(s_i, s_j) > \theta_{sim}$, where θ_{sim} is the similarity threshold and $r(s_i, s_j)$ is defined as follows:

$$A_{ij}^S = r(s_i, s_j) = \begin{cases} \frac{f(s_i) \cdot f(s_j)}{\|f(s_i)\| \|f(s_j)\|} & \text{if } r(s_i, s_j) \geq \theta_{sim} \\ 0 & \text{if } r(s_i, s_j) < \theta_{sim} \end{cases} \tag{8}$$

where $f(s_i)$ and $f(s_j)$ are the features of the snippet s_i and s_j. Without the threshold, the network would add all edges. Some snippets with low similarity and nothing to contribute to the classification and localization of s_i are aggregated, which might degrade accuracy and introduce noise into the network.

Action Contextual Edges. Considering the possibility of low similarity between two adjacent snippets in time due to the loss of features or action segmentation, but their contribution on the current snippet cannot be ignored, we add the action contextual edges. The contextual edge between snippets is defined as follows:

$$A_{ij}^C = d(s_i, s_j) = \begin{cases} |i - j| & \text{if } d(s_i, s_j) \leq \theta_{dis} \\ 0 & \text{if } d(s_i, s_j) > \theta_{dis} \end{cases} \tag{9}$$

where $d(s_i, s_j)$ is the distance of the snippet s_i and s_j; θ_{dis} is the distance threshold.

Sparse Graph Convolution for Action Localization Adjacency Matrix. By combining the two matrices, the adjacency matrix is defined as follows:

$$A = \frac{\alpha A^S + (1 - \alpha) A^C}{2} \tag{10}$$

where the two matrices A^S and A^C include A_{ij}^S (8) and A_{ij}^C (9) as their (i, j)-th entries respectively, and α is the hyperparameter.

Graph Convolution. Given the constructed graph, we apply the GCN to do action localization. We build L-layer graph convolutions in our implementation. For the l-th layer ($1 \leq l \leq L$), the graph convolution is implemented as follows:

$$X^{(l)} = \sigma(A X^{(l-1)} W^{(l)}) \tag{11}$$

where $X^{(l)} \in \mathbb{R}^{N \times D}$ are the hidden features at layer l; A is the adjacency matrix; $W^{(l)} \in \mathbb{R}^{D \times D}$ is the parameter matrix to be learned of the l-th layer; $X^{(0)} \in \mathbb{R}^{N \times D}$ are the input features; $\sigma(\cdot)$ is ReLU.

In addition, our experiments find it more effective by combining the output features of the GCN with the input features: $X' = X + X^{(l)}$, where X' is the obtained features, serves the subsequent localization task.

Algorithm 1: The training process of SGLP-Net.

Input: The embedded features $X = \left\{x_i^{(0)}\right\}_{i=1}^{N}$; snippet set $S = \{s_i\}_{i=1}^{N}$; graph depth L;

    Weight matrices $W^{(l)}$; the similarity threshold θ_{sim}; the distance threshold θ_{dis}.

Output: Trained SGLP-Net.

1 Initialize the nodes $v_i, v_i \in v$ by the snippets $s_i, s_i \in S$
2 establish edges $e_{ij}, e_{ij} = (v_i, v_j) \in \varepsilon$ between nodes using Eq. (8), Eq. (9)
3 obtain an action graph $\varsigma(v, \varepsilon)$
4 calculate adjacent matrix using Eq. (10)
5 **while** not converged **do**
6   **for** $l = 1...L$ **do**
7   **for** $v \in v$ **do**
8 | propagate features using Eq. (12)
9   **end**
10   **end**
11   Generate pseudo labels
12   predict and locate actions
13 **end**

Sparse GCN. If we directly use the graph convolution mentioned above, due to the large number of snippets in an untrimmed video, the network will be very large and computationally intensive. Therefore, we use the sparse graph network for optimization training. Here, we use the layer sampling. It assumes a potentially infinite graph G', whose node set V' contains countless points. For the current graph G, G is considered a subgraph of G', and all nodes V on G are independent and identically distributed (IID) samples sampled from V' based on a probability distribution P. The propagation function is as follows:

$$x_i^{(l)} = \sigma(\frac{1}{t_l} \sum_{j=1}^{t_l} A_{i,u_j} x_{u_j}^{(l-1)} W^{(l-1)}) \tag{12}$$

where t_l is the number of IID samples; $u_j, j \in \{1, ..., t_l\}$ are the samples. For better readability, Algorithm 1 depicts the algorithmic Flow of our method.

4 Experiments

4.1 Experimental Setup

Datasets. We evaluate our method on two popular TAL datasets. **THU-MOS'14** contains over 20 h of videos from 20 sports classes. The dataset [11] is challenging due to the diversified video lengths and the large number of action instances per video. The validation and test sets provide 200 and 213 untrimmed videos for action localization respectively. Following previous work, we conduct training on the validation set and perform evaluation on the test set. **ActivityNet-v1.3** is a large-scale dataset with 200 complex daily activities [10]. It has 10,024 training videos and 4,926 validation videos. We use the training set to train our model and the validation set for evaluation.

Evaluation Metrics. We adopt the standard metrics for performance evaluation of different methods, i.e., mean Average Precisions (mAPs) under different Intersection of Union (IoU) thresholds. In practice, we adopt the official evaluation code provided by ActivityNet.

Implementation Details. The Adam optimizer is used with the learning rate of 0.0001 and with the mini-batch sizes of 16, 64 for THUMOS'14 and ActivityNet-v1.3, respectively. The number of sampled snippets T is 750 for THUMOS'14 and 150 for ActivityNet-v1.3. For the layer of the graph, we use 2. For the multi-step proposal refinement, 100 and 50 epochs are set respectively. Action proposals are generated at the last epoch of each refinement step. We train the network on ten GeForce RTX 2080 Ti GPUs.

4.2 Comparison with the State of the Art

In Table 1, we compare our SGLP-Net with other methods on THUMOS'14. Selected fully-supervised methods are presented for reference. We observe that SGLP-Net outperforms all the previous WTAL methods and establishes new SOTA with 46.6% average mAP for IoU thresholds 0.1:0.7. In particular, our approach outperforms ACGNet, which also utilizes GCN to guide the model training but without explicit modeling for long-term temporal information. Even compared with the fully-supervised methods, SGLP-Net outperforms TAL-Net [2] and achieves comparable results with P-GCN when the IoU threshold is low.

Table 1. Comparison with state-of-the-art methods on THUMOS'14 dataset. The average mAPs are computed under the IoU thresholds [0.1:0.1:0.7].

Supervision	Method	Publication	mAP@IoU (%)							
			0.1	0.2	0.3	0.4	0.5	0.6	0.7	AVG
Fully	TAL-Net [2]	CVPR 2018	59.8	57.1	53.2	48.5	42.8	33.8	20.8	45.1
	P-GCN [27]	ICCV 2019	69.5	67.8	63.6	57.8	49.1	–	–	–
	VSGN [30]	ICCV 2021	–	–	66.7	60.4	52.4	41.0	30.4	–
	ActionFormer [29]	ECCV 2022	–	–	82.1	77.8	71.0	59.4	43.9	–
	TriDet [23]	CVPR 2023	–	–	83.6	80.1	72.9	62.4	47.4	–
Weakly	STPN [19]	CVPR 2018	52.0	44.7	35.5	25.8	16.9	9.9	4.3	27.0
	CMCS [15]	CVPR 2019	57.4	50.8	41.2	32.1	23.1	15.0	7.0	32.4
	DGAM [22]	CVPR 2020	60.0	54.2	46.8	38.2	28.8	19.8	11.4	37.0
	ACGNet [26]	AAAI 2021	68.1	62.6	53.1	44.6	34.7	22.6	12.0	42.5
	ACM-Net [20]	TIP 2021	68.9	62.7	55.0	44.6	34.6	21.8	10.8	42.6
	ASM-Loc [9]	CVPR 2022	71.2	65.5	**57.1**	46.8	36.6	25.2	13.4	45.1
	DELU [4]	ECCV 2022	71.5	**66.2**	56.5	47.7	**40.5**	27.2	15.3	46.4
	F3-Net [17]	TMM 2023	69.4	63.6	54.2	46.0	36.5	–	–	–
	Ours	–	**71.8**	66.0	56.8	**48.6**	40.1	**27.5**	**15.7**	**46.6**

Table 2. Comparison with state-of-the-art methods on ActivityNet-v1.3 dataset. The average mAPs are computed under the IoU thresholds [0.5:0.05:0.95].

Method	Publication	mAP@IoU (%)			
		0.5	0.75	0.95	AVG
STPN [19]	CVPR 2018	29.3	16.9	2.6	16.3
TSCN [28]	ECCV 2020	35.3	21.4	5.3	21.7
Bas-Net [13]	AAAI 2020	34.5	22.5	4.9	22.2
AUMN [16]	CVPR 2021	38.35	23.5	5.2	23.5
UGCT [25]	CVPR 2021	39.1	22.4	5.8	23.8
ACM-Net [20]	TIP 2021	37.6	24.7	6.5	24.4
ASM-Loc [9]	CVPR 2022	41.0	24.9	6.2	25.1
F3-Net [17]	TMM 2023	39.9	25.0	**6.7**	25.2
Ours	–	**41.7**	**25.3**	5.8	**26.3**

Table 3. Contribution of each component.

SGLP-Net		mAP@IoU (%)								
MTF	ALP	0.1	0.2	0.3	0.4	0.5	0.6	0.7	AVG	Δ
		68.9	62.7	55.0	44.6	34.6	21.8	10.8	42.6	0
✓		70.2	63.4	56.3	46.1	35.8	24.7	13.6	44.3	+1.7
	✓	70.8	64.9	56.4	47.0	36.7	25.8	14.1	45.1	+2.5
✓	✓	71.8	66.0	56.8	48.6	40.1	27.5	15.7	46.6	+4.0

Table 4. Numbers of layers in MTF.

Num	mAP@IoU (%)							
	0.1	0.2	0.3	0.4	0.5	0.6	0.7	AVG
1	69.3	62.9	55.7	45.0	35.1	23.6	11.3	44.3
2	70.1	65.3	56.2	47.8	38.9	26.7	14.6	45.7
3	70.6	65.6	56.5	47.9	39.2	27.0	14.8	45.9
4	71.8	66.0	56.8	48.6	40.1	27.5	15.7	46.6
5	71.4	65.7	56.6	48.2	39.8	27.3	14.9	46.3

Table 5. Validation of the generality of MTF.

Method	mAP@IoU (%)							
	0.1	0.2	0.3	0.4	0.5	0.6	0.7	AVG
DGAM [22]	60.0	54.2	46.8	38.2	28.8	19.8	11.4	37.0
DGAM [22] + MTF	62.3	56.4	47.6	40.1	30.2	20.7	11.0	38.3
ACM-Net [20]	68.9	62.7	55.0	44.6	34.6	21.8	10.8	42.6
ACM-Net [20] + MTF	71.3	64.1	55.8	45.2	35.6	22.8	11.9	43.8
ASM-Loc [9]	71.2	65.5	57.1	46.8	36.6	25.2	13.4	45.1
ASM-Loc [9] + MTF	71.6	65.8	56.7	47.5	38.2	26.0	14.3	45.7

Table 6. Efficacy of the sparse GCN.

sparse GCN	mAP@IoU (%)							
	0.1	0.2	0.3	0.4	0.5	0.6	0.7	AVG
	71.1	65.2	55.8	47.3	37.1	26.4	13.9	45.3
✓	71.8	66.0	56.8	48.6	40.1	27.5	15.7	46.6

We also conduct experiments on ActivityNet-v1.3 and the comparison results are summarized in Table 2. Our SGLP-Net obtains a new SOTA performance of 26.3% average mAP, surpassing the latest works (e.g. ASM-Loc [9], F3-Net [17]). The results on both datasets justify the effectiveness of SGLP-Net.

4.3 Ablation Studies on THUMOS'14

Contribution of Each Component. In Table 3, we conduct an ablation study to investigate the contribution of each component in SGLP-Net. It is obvious that a gain can be achieved by adding any of our proposed modules. Specially, introduce MTF in the feature embedding stage increases the performance by 1.7%, and introduce ALP significantly increases the performance by 2.5%. The two modules are complementary to each other. When incorporating these modules together, our approach boosts the final performance from 42.6% to 46.6%.

Validation of the Generality of MTF. In Table 4, we integrated MTF into networks DGAM, ACM-Net, and ASM-Loc, resulting in performance improvements of 1.3%, 1.2%, and 0.6%, respectively. These results provide empirical evidence supporting the effectiveness and generalizability of MTF.

Number of CNN Layers in MTF. Table 5 shows the results increasing the number of CNN layers in MTF. We can see that the performance improves as the number of layers increases, indicating the better localization results. We adopt 4 layers as our default setting since the performance saturates after that.

Effectiveness of the Sparse GCN in ALP. In Table 6, it shows the results increasing when adopting the saprse GCN in ALP to train the network. This may be that without utilizing sparse graph GCN, the current snippet acquired an excessive amount of irrelevant features, leading to the blurring of the most discriminative ones and consequently causing a decline in localization. While by leveraging fastGCN, we are able to alleviate this situation effectively and also accelerate network training.

4.4 Qualitative Results

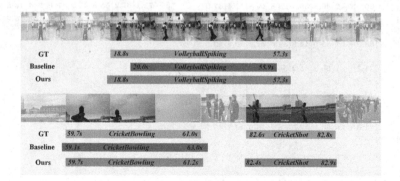

Fig. 3. Qualitative results on a video of THUMOS'14. The results of ACM-Net (Baseline), SGLP-Net (Ours), and ground truth (GT) are shown in blue, red, and green, respectively. (Color figure online)

In Fig. 3, we visualize some action detection results generated by our SGLP-Net on THUMOS'14. We can find that our model can successfully detect the action instances in these examples with precise action boundaries. For the untrimmed video with multiple action instances, the foreground actions and background scenes can be well distinguished.

5   Conclusion

In this paper, we propose a new WTAL framework, SGLP-Net. By using MTF in the feature embedding stage, local and global temporal features of the video can be obtained in the initial stage, and it can be embedded as a universal module into other networks. By using sparse graph networks to implement label propagation mechanisms, it is possible to mine more information about the video itself, including the relationships between video snippets. Extensive experiments on publicly available datasets can demonstrate the effectiveness of our method.

References

1. Carreira, J., Zisserman, A.: Quo vadis, action recognition? A new model and the kinetics dataset. In: 2017 IEEE Conference on Computer Vision and Pattern Recognition (CVPR), pp. 4724–4733 (2017). https://doi.org/10.1109/CVPR.2017.502

2. Chao, Y.W., Vijayanarasimhan, S., Seybold, B., Ross, D.A., Deng, J., Sukthankar, R.: Rethinking the faster R-CNN architecture for temporal action localization. In: 2018 IEEE/CVF Conference on Computer Vision and Pattern Recognition, pp. 1130–1139 (2018). https://doi.org/10.1109/CVPR.2018.00124

3. Chen, J., Ma, T., Xiao, C.: FastGCN: fast learning with graph convolutional networks via importance sampling. In: International Conference on Learning Representations (2018). https://openreview.net/forum?id=rytstxWAW

4. Chen, M., Gao, J., Yang, S., Xu, C.: Dual-evidential learning for weakly-supervised temporal action localization. In: Avidan, S., Brostow, G., Cissa, M., Farinella, G.M., Hassner, T. (eds.) European Conference on Computer Vision, vol. 13664, pp. 192–208. Springer, Heidelberg (2022). https://doi.org/10.1007/978-3-031-19772-7_12

5. Chen, T., Li, B., Tao, Y., Wang, Y., Zhu, Y.: Class-incremental learning with multiscale distillation for weakly supervised temporal action localization. In: Tanveer, M., Agarwal, S., Ozawa, S., Ekbal, A., Jatowt, A. (eds.) ICONIP 2022. LNCS, vol. 13623, pp. 367–378. Springer, Cham (2023). https://doi.org/10.1007/978-3-031-19772-7_12

6. Ciptadi, A., Goodwin, M.S., Rehg, J.M.: Movement pattern histogram for action recognition and retrieval. In: Fleet, D., Pajdla, T., Schiele, B., Tuytelaars, T. (eds.) ECCV 2014. LNCS, vol. 8690, pp. 695–710. Springer, Cham (2014). https://doi.org/10.1007/978-3-319-10605-2_45

7. Dietterich, T.G., Lathrop, R.H., Lozano-Pérez, T.: Solving the multiple instance problem with axis-parallel rectangles. Artif. Intell. **89**(1), 31–71 (1997). https://doi.org/10.1016/S0004-3702(96)00034-3. https://www.sciencedirect.com/science/article/pii/S0004370296000343

8. Douze, M., Szlam, A., Hariharan, B., Jégou, H.: Low-shot learning with large-scale diffusion. In: 2018 IEEE/CVF Conference on Computer Vision and Pattern Recognition, pp. 3349–3358 (2018). https://doi.org/10.1109/CVPR.2018.00353

9. He, B., Yang, X., Kang, L., Cheng, Z., Zhou, X., Shrivastava, A.: ASM-Loc: action-aware segment modeling for weakly-supervised temporal action localization. In: 2022 IEEE/CVF Conference on Computer Vision and Pattern Recognition (CVPR), pp. 13915–13925 (2022). https://doi.org/10.1109/CVPR52688.2022.01355

10. Heilbron, F.C., Escorcia, V., Ghanem, B., Niebles, J.C.: Activitynet: a large-scale video benchmark for human activity understanding. In: 2015 IEEE Conference on Computer Vision and Pattern Recognition (CVPR), pp. 961–970 (2015). https://doi.org/10.1109/CVPR.2015.7298698

11. Idrees, H., et al.: The thumos challenge on action recognition for videos "in the wild". Comput. Vision Image Underst. **155**, 1–23 (2017). https://doi.org/10.1016/j.cviu.2016.10.018. https://www.sciencedirect.com/science/article/pii/S1077314216301710

12. Kay, W., et al.: The kinetics human action video dataset (2017)

13. Lee, P., Uh, Y., Byun, H.: Background suppression network for weakly-supervised temporal action localization. In: Proceedings of the AAAI Conference on Artificial Intelligence, vol. 34, 11320–11327 (2020). https://doi.org/10.1609/aaai.v34i07.6793

14. Li, L., Kong, T., Sun, F., Liu, H.: Deep point-wise prediction for action temporal proposal. In: Gedeon, T., Wong, K.W., Lee, M. (eds.) ICONIP 2019. LNCS, vol. 11955, pp. 475–487. Springer, Cham (2019). https://doi.org/10.1007/978-3-030-36718-3_40

15. Liu, D., Jiang, T., Wang, Y.: Completeness modeling and context separation for weakly supervised temporal action localization. In: 2019 IEEE/CVF Conference on Computer Vision and Pattern Recognition (CVPR), pp. 1298–1307 (2019). https://doi.org/10.1109/CVPR.2019.00139

16. Luo, W., et al.: Action unit memory network for weakly supervised temporal action localization. In: 2021 IEEE/CVF Conference on Computer Vision and Pattern Recognition (CVPR), pp. 9964–9974 (2021). https://doi.org/10.1109/CVPR46437.2021.00984

17. Moniruzzaman, M., Yin, Z.: Feature weakening, contextualization, and discrimination for weakly supervised temporal action localization. IEEE Trans. Multimedia, 1–13 (2023). https://doi.org/10.1109/TMM.2023.3263965

18. Nair, V., Hinton, G.E.: Rectified linear units improve restricted Boltzmann machines. In: Proceedings of the 27th International Conference on International Conference on Machine Learning, ICML 2010, pp. 807–814. Omnipress, Madison (2010)

19. Nguyen, P., Liu, T., Prasad, G., Han, B.: Weakly supervised action localization by sparse temporal pooling network. In: 2017 IEEE Conference on Computer Vision and Pattern Recognition (CVPR) (2017). https://doi.org/10.1109/CVPR.2018.00706

20. Qu, S., Chen, G., Li, Z., Zhang, L., Lu, F., Knoll, A.: Acm-net: action context modeling network for weakly-supervised temporal action localization. arXiv preprint arXiv:2104.02967 (2021)

21. Raghavan, U.N., Albert, R., Kumara, S.: Near linear time algorithm to detect community structures in large-scale networks. Phys. Rev. E **76**, 036106 (2007). https://doi.org/10.1103/PhysRevE.76.036106

22. Shi, B., Dai, Q., Mu, Y., Wang, J.: Weakly-supervised action localization by generative attention modeling. In: 2020 IEEE/CVF Conference on Computer Vision and Pattern Recognition (CVPR), pp. 1006–1016 (2020). https://doi.org/10.1109/CVPR42600.2020.00109

23. Shi, D., Zhong, Y., Cao, Q., Ma, L., Li, J., Tao, D.: Tridet: temporal action detection with relative boundary modeling. arXiv:2303.07347 (2023)

24. Su, H., Zhao, X., Lin, T., Fei, H.: Weakly supervised temporal action detection with shot-based temporal pooling network. In: Cheng, L., Leung, A.C.S., Ozawa, S. (eds.) ICONIP 2018. LNCS, vol. 11304, pp. 426–436. Springer, Cham (2018). https://doi.org/10.1007/978-3-030-04212-7_37

25. Yang, W., Zhang, T., Yu, X., Qi, T., Zhang, Y., Wu, F.: Uncertainty guided collaborative training for weakly supervised temporal action detection. In: 2021 IEEE/CVF Conference on Computer Vision and Pattern Recognition (CVPR), pp. 53–63 (2021). https://doi.org/10.1109/CVPR46437.2021.00012

26. Yang, Z., Qin, J., Huang, D.: ACGNet: action complement graph network for weakly-supervised temporal action localization. In: Proceedings of the AAAI Conference on Artificial Intelligence, vol. 36, pp. 3090–3098 (2022)

27. Zeng, R., et al.: Graph convolutional networks for temporal action localization. In: 2019 IEEE/CVF International Conference on Computer Vision (ICCV), pp. 7093–7102 (2019). https://doi.org/10.1109/ICCV.2019.00719

28. Zhai, Y., Wang, L., Tang, W., Zhang, Q., Yuan, J., Hua, G.: Two-stream consensus network for weakly-supervised temporal action localization. In: Vedaldi, A., Bischof, H., Brox, T., Frahm, J.-M. (eds.) ECCV 2020. LNCS, vol. 12351, pp. 37–54. Springer, Cham (2020). https://doi.org/10.1007/978-3-030-58539-6_3

29. Zhang, C.L., Wu, J., Li, Y.: ActionFormer: localizing moments of actions with transformers. In: Avidan, S., Brostow, G., Cissé, M., Farinella, G.M., Hassner, T. (eds.) ECCV 2022. LNCS, vol. 13664, pp. 492–510. Springer Nature Switzerland, Cham (2022). https://doi.org/10.1007/978-3-031-19772-7_29

30. Zhao, C., Thabet, A., Ghanem, B.: Video self-stitching graph network for temporal action localization. In: 2021 IEEE/CVF International Conference on Computer Vision (ICCV), pp. 13638–13647 (2021). https://doi.org/10.1109/ICCV48922.2021.01340

VFIQ: A Novel Model of ViT-FSIMc Hybrid Siamese Network for Image Quality Assessment

Junrong Huang[1](✉) and Chenwei Wang[2]

[1] Department of Computer Science and Technology, Nanjing University,
Nanjing 210093, Jiangsu, China
stevehjr2003@gmail.com
[2] Google LLC, Mountain View, CA 94043, USA

Abstract. The Image Quality Assessment (IQA) is to measure how
humans perceive the quality of images. In this paper, we propose a new
model named for VFIQ – a ViT-FSIMc Hybrid Siamese Network for
Full Reference IQA – that combines signal processing and leaning-based
approaches, the two categories of IQA algorithms. Specifically, we design
a hybrid Siamese network that leverages the Vision Transformer (ViT)
and the feature similarity index measurement (FSIMc). To evaluate the
performance of the proposed VFIQ model, we first pre-train the ViT
module on the PIPAL dataset, and then evaluate our VFIQ model on
several popular benchmark datasets including TID2008, TID2013, and
LIVE. The experiment results show that our VFIQ model outperforms
the state-of-the-art IQA models in the commonly used correlation metrics of PLCC, KRCC, and SRCC. We also demonstrate the usefulness
of our VFIQ model in different vision tasks, such as image recovery and
generative model evaluation.

Keywords: FR-IQA · ViT · Siamese Network

1 Introduction

How to quantitatively assess the perceptual similarity between two images? For
example, as shown in Fig. 1, under different distortion circumstances, how can
we quantify the distortions? This question has been recently investigated in the
field of image quality assessment (IQA).

Broadly, IQA can be classified into three categories: Full-Reference, No-
Reference, and Aesthetic IQA. In particular, the goal of Full-Reference IQA
(FR-IQA) models is to design a reasonable evaluation metric that measures the
human perceptual similarity between a reference image and a distortion image.
Among these three categories, we focus on the FR-IQA models in this paper.

There are two main-stream approaches for the FR-IQA algorithms: signal
processing techniques and learning-based metrics [1]. Signal processing techniques such as PSNR, SSIM [17], and MS-SSIM [18] compare the low-level texture and stripe features, while learning-based metrics such as LPIPS [21] and

© The Author(s), under exclusive license to Springer Nature Singapore Pte Ltd. 2024
B. Luo et al. (Eds.): ICONIP 2023, LNCS 14451, pp. 162–174, 2024.
https://doi.org/10.1007/978-981-99-8073-4_13

(a) Groundtruth   (b) Noise   (c) Compression   (d) Contrast change

Fig. 1. The images under different distortion circumstances

AHIQ [9] use powerful backbones such as inception-v3, VGG, and ResNet50 to extract feature maps for comparison and evaluation.

Generative models such as Generative Adversarial Networks [6] have significantly advanced the image processing field [7,19]. Yet, they also pose new challenges to IQA, as they can produce images and videos that are difficult for the human vision system (HVS) to distinguish from real ones. In fact, it is widely observed that most generative models can create realistic-like images on global semantic features, but they often fail to accurately capture the details and textures [8]. Since the HVS is not very sensitive to subtle variations in details, current IQA algorithms that use pixel-wise comparison could be weak in reflecting the real performance of generative models on global semantic features.

We expect generative models to achieve a good balance of capturing global semantic information and detailed low-level pixel-wise information. To achieve this goal, we can use a patch-level [1] comparison based on the Vision Transformer (ViT) to extract high-level global semantic information. This is because the ViT [4] can effectively capture long-range dependencies and extract high-level semantic features with its global receptive field and multi-head attention mechanism. Moreover, instead of using Convolutional Neural Networks (CNN) as the traditional detailed spatial texture extractor for learning-based IQA models [1], we would use the Feature Similarity Index Measure with Chrominance (FSIMc) [20] as our low-level information evaluation metric. The phase congruency (PC) model of FSIMc can fine-grainedly represent local detailed spatial structure, and the gradient magnitude (GM) can depict the texture similarity in image.

To this end, in this paper we combine the two main-stream IQA algorithm ideas above and propose a ViT-FSIMc Hybrid Siamese Network for FR-IQA, abbreviated as VFIQ. In particular, we split the image information that is strongly correlated with the HVS into two domains: the abstract high-level semantic feature domain and the low-level detailed texture domain. The ViT extracts the high-level abstract information, and the FSIMc [20] measures the low-level texture. Finally, we design the VFIQ score as the weighted average of the outputs from a fully-connected layer whose inputs are the two contributions described above.

To compare the performance of different FR-IQA models, in this paper we use PyTorch to develop a FR-IQA framework as a fair benchmark. Then we retrain various FR-IQA models on the PIPAL [8] dataset and test them on cross-domain datasets including TID2008 [12], TID2013 [11] and LIVE [14].

Finally, to assess the performance of our model, we consider three widely-used metrics, including Pearson's linear correlation coefficient (PLCC), Spearman's rank-order correlation coefficient (SRCC), and Kendall's rank-order correlation coefficient (KRCC). While PLCC assesses the linear correlation between the ground truth and the IQA model's output predictions, SRCC and KRCC describe the monotonic correlation similarity. For all of them, the higher score, the better model/algorithm. To avoid the risk of overfitting bias from the training dataset, we apply VFIQ in gradient-based perceptual optimization tasks to show the generalization ability of VFIQ.

2 Related Work

2.1 FR-IQA Metrics

The FR-IQA algorithms take a pair of two images – one distortion image and one reference image – as the input to measure their perceptual similarity.

Traditionally, the error visibility methods use a direct distance metric to pixels or the transformed representations of the images, e.g., the Mean Squared Error (MSE). The MSE method has been a standard FR method for assessing signal fidelity and quality for more than half of a century, and it still plays a fundamental role in the development of signal and image processing algorithms.

Currently, the most popular FR-IQA metrics are PSNR and SSIM as they are simple and convenient in optimization. Specifically, the SSIM [17] evaluates the similarity of local image structures with correlation measures. On the other hand, recently more and more FR-IQA models appear to be learning-based, which removes the limitation of conventional IQA metrics. These models typically learn a measure from a training set of images and corresponding perceptual distances by using supervised machine learning technologies. However, due to the high dimensionality of the input space (formed by millions of pixels), learning-based models are prone to overfit the limited data.

2.2 Vision Transformer (ViT)

Based upon the self-attention mechanism, the Transformer architecture was first proposed in [16] and achieved remarkable improvements in many tasks of natural language processing. Inspired by the Transformer, the ViT proposed by Dosovitskiy [4] divides the image into patches as the input substitution of word sequences, which becomes the latest popular feature extraction backbone.

The ViT has a global receptive field and can capture global information, as it takes patches from the whole image as the input. As a comparison, the conventional CNN backbone concentrates on local features and has difficulties in extracting long-range features as the convolutional process loses some important feature information [8,15].

2.3 FSIMc

The success of SSIM [17] showed that the HVS is adapted to the structural information in images, but the visual information is often redundant for perceptual similarity. In fact, low-level features such as edges and zero crossings provide significant information for the HVS to understand and interpret the scene. As a result, an FR-IQA metric named for FSIM was proposed in [20]. This metric assumes that we can extract highly informative features at the points of high phase congruency (PC) where the Fourier waves at different frequencies have congruent phases. Thus, the PC maps are used as the main feature in computing FSIM.

Another complementary component of the FSIM metric is the Gradient Magnitude (GM) which is calculated by three different gradient operators: The Sobel operator, the Prewitt operator and the Scharr operator. Hybriding PC and GM makes the FSIM model efficient in evaluating low-level HVS perceptual similarity. We denote by $G_x = \partial f(x,y)/\partial x$ and $G_y = \partial f(x,y)/\partial y$ the partial derivatives of the image $f(x,y)$ along horizontal and vertical directions and by $G = \sqrt{G_x^2 + G_y^2}$ the GM of $f(x,y)$.

The native FSIM index is designed for gray-scale images or the luminance components of color images. To extend the FSIM algorithm to process the chrominance information, the FSIMc adopts a color conversion from the RGB space to the YIQ space. In the YIQ color space, the luminance can be separated from the chrominance. As a result, the final FSIMc can be formulated as:

$$\text{FSIMc} = \frac{\sum_{x \in \Omega} S_L(x) \cdot [S_C(x)]^\lambda \cdot PC_m(x)}{\sum_{x \in \Omega} PC_m(x)} \tag{1}$$

where $S_L(x)$ is a product of $S_{PC}(x)$ and $S_G(x)$ in the RGB space, and the $S_C(x)$ is a product of $S_I(x)$ and $S_Q(x)$ in the YIQ space.

3 Model and Methodology

In this section, we introduce the architecture of ViT-FSIMc Hybrid Siamese Network, abbreviated as VFIQ, that we proposed in this paper. Our model takes a pair of the distorted image and a reference image as the input and tells the VHIQ score at the output. As shown in Fig. 2, it consists of three important modules—a feature extraction module, an evaluation module, and a hybrid module. In the following, we introduce each module in detail, followed by the computation process.

3.1 Feature Extraction Module

The feature extraction module extracts a high-level semantic vector for comparison. To process the input, this module includes a Siamese neural network which has two branches. For the backbone, we choose the ViT owing to its efficiency

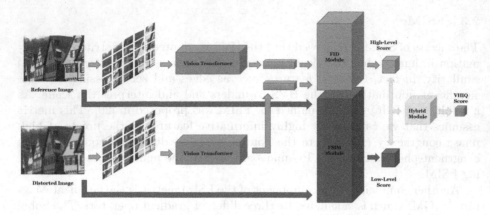

Fig. 2. VFIQ architecture

in extracting high-level semantic representations. The ViT also includes self-attention modules, which enhance its ability to capture long-distance features.

The Siamese neural network is constructed by two ViTs which share the same weights to extract the high-level semantic vector from the pair of images. During the forward process, the pair of the distortion and the reference images are fed into the Siamese neural network. To enhance the feature extraction ability, we first take out three different feature maps from intermediate layers in the transform encoder and then fuse these feature vectors in an average pooling layer to output the high-level semantic vector. See the ViT architecture in Fig. 3.

3.2   Evaluation Module

In the evaluation module, we calculate the high-level semantic and the low-level texture similarity scores, respectively.

After the ViT extracts the high-level semantic vector from the patch-wise distortion image and reference image, the similarity (and/or the difference) between the distortion semantic vector and the reference semantic vector can be characterized from their high-dimensional vector distributions. By assuming they follow a Gaussian distribution, we use the Fréchet Inception Distance (FID) to compute the high-level semantic similarity score between the two images. Specifically, let ϕ denote the feature function constructed in the feature extraction module; $\phi(P_r)$ and $\phi(P_g)$ be two Gaussian random vectors with empirical means μ_r, μ_g and covariance C_r, C_g, respectively. Then the FID between $\phi(P_r)$ and $\phi(P_g)$ is calculated as:

$$\text{FID}(\phi(P_r), \phi(P_g)) = ||\mu_r - \mu_g|| + \text{Tr}(C_r + C_g - 2(C_r C_g)^{\frac{1}{2}}). \quad (2)$$

On the other hand, for the low-level detailed texture similarity comparison, we adopt the FSIMc [20] algorithm and the equation in (1) to calculate the score. The original distortion image and reference image are fed into the FSIMc module to produce the result.

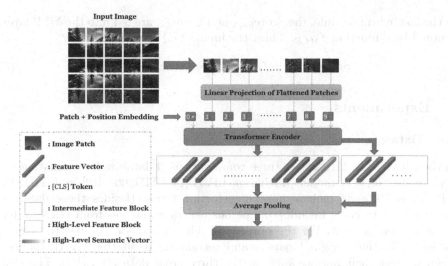

Fig. 3. Vision Transformer architecture

3.3 Hybrid Module

In the hybrid module, we apply a simple MLP ($2 \times 8 \times 1$), a fully-connected layer to predict the final score. Essentially, the MLP module essentially fuses the high-level and the low-level scores calculated in the evaluation module, and outputs the proper result score.

To train the model, we use the official Mean Opinion Score (MOS) in the dataset as the target label and apply the MSE as the loss function between the predicted score and the MOS score. Thus, it is typical supervised learning, and the hybrid weights of the two parts can be consequently learned.

3.4 Mathematical Formulation

In this section, we formulate the entire computation process of the VFIQ model. Denote by x and y the input distortion image and the reference image, respectively, where $x, y \in R^{H \times W \times C}$. In the feature extraction module, they are transformed into 224×224 and the image patches are fed into the Siamese network. We denote the extraction process as a function ϕ, and thus $\phi(x), \phi(y)$ are the high-level semantic feature maps of the distortion and the reference images. Here, $\phi(x), \phi(y) \in R^{p \times p \times 3c}$. Then in the evaluation module, the high-level semantic score is computed as

$$\text{Score}_{high} = \text{FID}(\phi(x), \phi(y)) = ||\mu_y - \mu_x|| + \text{Tr}(C_y + C_x - 2(C_y C_x)^{\frac{1}{2}}), \quad (3)$$

and the low-level texture score is computed as

$$\text{Score}_{low} = \text{FSIMc}(x, y). \quad (4)$$

In the last hybrid module, the $Score_{high}$ and $Score_{low}$ are fed into the MLP layer, denoted by a function f_{MLP}. Thus, the final VFIQ score is given by:

$$Score_{VFIQ} = f_{MLP}(Score_{high}, Score_{low}). \qquad (5)$$

4   Experiments

4.1   Datasets

In the FR-IQA field, there are three commonly used datasets for evaluating the performance of IQA models, including LIVE [14], TID2008 [12] and TID2013 [11]. These datasets include normal distortion types. Besides those datasets, the PIPAL dataset [8] including numerous generated images from GAN [6] has recently been popular. For the purpose of fair comparison, we use the official models from their original papers and retrain them on the PIPAL dataset. Then we test their performance on the three cross datasets LIVE, TID2003, and TID2008 to compare the generalization ability. Moreover, to quantify the performances of various FR-IQA models, we use the three metrics – PLCC, SRCC, and KRCC introduced in Sect. 1.

4.2   Implementation Details

The input images are normalized and the image size is randomly cropped to 224×224, as the ViT we employ is pre-trained on the ImageNet dataset. Both ViT-B/8 and ViT-B/16 are tested as the backbone. In the ViT, to enhance the feature extraction ability, we extract two more intermediate blocks to provide supplementary information. The inner blocks consist of a self-attention module and a Feed Forward Network (FFN). Hence, the feature map we extracted from the blocks belongs to $R^{p \times p \times 768}$ where $p = 14$ or 28. In the validation and testing stage, we randomly crop each pair of images twenty times and calculate the final score based on the twenty random samples. In addition, we use AdamW as the optimizer and set the initial learning rate as 10^{-4} and the weight decay rate as 10^{-5}. The final result of VFIQ applies ViT-B/8 as the backbone. We run the entire experiment on a single NVIDIA Geforce RTX 3090.

4.3   Results

To visualize the prediction scores of different IQA models and compare it against the actual MOS score, we show the scatter plots in Fig. 4. Compared to PSNR, SSIM, and AHIQ scatter plots, the proposed VFIQ scatters appear more concentrated and less scattered than the others. This implicitly indicates the VFIQ performs the best.

We also assess the performance of our model with PLCC, SRCC, and KRCC. For all of them, as we mentioned in Sect. 1, the higher score, the better. The results are shown in Table 1, Table 2, and Table 3. It can be seen that our model

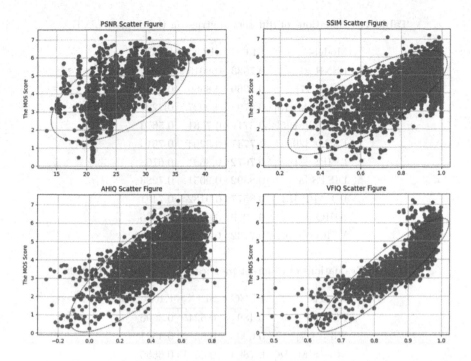

Fig. 4. The scatterplot figure of different IQA prediction scores plot with MOS score.

outperforms all the others in 7 out of the 9 comparisons (3 datasets × 3 met-rics/dataset), and the performance gains on 5 of them are significant. Specifically, on the LIVE dataset, except for ranking the 5th out of 10 models on PLCC, our model exceeds the second best by 0.64% in SRCC and 8.1% in KRCC; on the TID2008 dataset, except for ranking the 2nd out of 10 models on SRCC, our model exceeds the second best by 2.1% in PLCC and 0.04% in KRCC; on the TID2013 dataset, our model exceeds the second best by 3.5%, 2.7%, 3.87% in PLCC, SRCC, and KRCC, respectively. Thus, we can conclude that our proposed VFIQ model outperforms the state-of-the-art performance in general.

4.4 Reference Image Recovery

In this section, we investigate a native vision task by trying to avoid the risk of over-fitting bias brought by the IQA datasets. In particular, given a reference image x and an initial image y_0, our goal is to recover the image x by numerically optimizing $y^* = \arg\min_y D(x, y)$, where D denotes a FR-IQA measurement, and the higher predicted quality, the lower score. Here, we use the MSE metric as the recovering optimizer, and clearly the solution is given by $y^* = x$. Most of the current IQA algorithms are continuous and differentiable, so we can use a gradient-based optimization process to recover the image and estimate the per-formance of the IQA algorithm by perceptually observing the recovered results.

Table 1. Evaluations of different metrics on the dataset LIVE

Models	PLCC	SRCC	KRCC
PSNR	0.7633	0.8013	0.5964
SSIM [17]	0.7369	0.8509	0.6547
MS-SSIM [18]	0.679	0.9027	0.7227
CW-SSIM [5]	0.5714	0.7681	0.5673
FSIMc [20]	0.7747	0.9204	0.7515
LPIPS [21]	0.7672	0.869	0.6768
DISTS [3]	0.8392	0.9051	0.7283
PIEAPP [13]	0.8577	0.9182	0.7491
AHIQ [9]	0.8039	0.8967	0.7066
VFIQ (Ours)	0.7732	**0.9236**	**0.7873**

Table 2. Evaluations of different metrics on the dataset TID2008

Models	PLCC	SRCC	KRCC
PSNR	0.489	0.5245	0.3696
SSIM [17]	0.6003	0.6242	0.4521
MS-SSIM [18]	0.7894	0.8531	0.6555
CW-SSIM [5]	0.5965	0.6473	0.4625
FSIMc [20]	0.8341	0.884	0.6991
LPIPS [21]	0.711	0.7151	0.5221
DISTS [3]	0.7032	0.6648	0.4861
PIEAPP [13]	0.6443	0.7971	0.6089
AHIQ [9]	0.6772	0.6807	0.4842
VFIQ (Ours)	**0.8512**	0.8808	**0.6994**

Table 3. Evaluations of different metrics on the dataset TID2013

Models	PLCC	SRCC	KRCC
PSNR	0.6601	0.6869	0.4958
SSIM [17]	0.6558	0.6269	0.4550
MS-SSIM [18]	0.7814	0.7852	0.6033
CW-SSIM [5]	0.5815	0.6533	0.4715
FSIMc [20]	0.8322	0.8509	0.6665
LPIPS [21]	0.7529	0.7445	0.5477
DISTS [3]	0.7538	0.7077	0.5212
PIEAPP [13]	0.7195	0.8438	0.6571
AHIQ [9]	0.7379	0.7075	0.5127
VFIQ (Ours)	**0.8613**	**0.8739**	**0.6923**

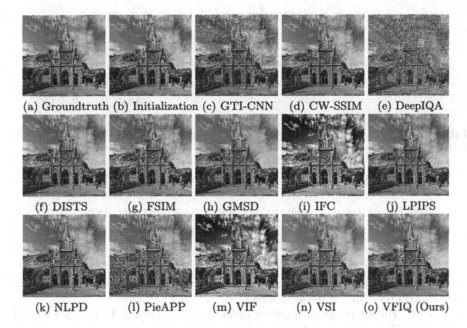

(a) Groundtruth (b) Initialization (c) GTI-CNN   (d) CW-SSIM   (e) DeepIQA

(f) DISTS   (g) FSIM   (h) GMSD   (i) IFC   (j) LPIPS

(k) NLPD   (l) PieAPP   (m) VIF   (n) VSI   (o) VFIQ (Ours)

Fig. 5. The performance of different IQA algorithms for recovering the reference image

Following the IQA algorithms used in last section for comparison, we visually compare the recovered images of PSNR, SSIM [17], MS-SSIM [18], FSIM [20], AHIQ [9], VIF [2], LPIPS [21], DISTS [3], NLPD [10], DeepIQA [1], PieAPP [13], and our VFIQ.

The images recovered from a JPEG-compressed version of a reference image are shown in Fig. 5. It can be seen that the recovered images of many IQA algorithms, such as sub-figures (c), (e), (i), (l), (m), are significantly worse than the initial image. Meanwhile, our VFIQ model appears to efficiently recover the initial image.

5   Ablation of the ViT-FSIMc Architecture

To verify the effectiveness of the ViT and FSIMc [20] hybrid architecture, we ablate them and substitute the backbone with CNN-based neural networks such as InceptionNet, VGG, and Resnet50 to compare their performances. Specifically, we first perform experiments by using a single ViT with the FID, the single FSIMc, and our proposed VFIQ, respectively. Then we replace the backbone with the CNN-FSIMc hybrid architecture where the CNN takes the Inception-3 Net, VGG-16, and ResNet50, respectively. For the fairness of evaluation, we sample three intermediate feature maps from the CNN backbone and re-train the hybrid module for different backbones, and compute the weighted-average results of 50 predictions. Again, we calculate the PLCC, SRCC, and KRCC

Table 4. The (PLCC, SRCC, KRCC) scores of ViT, FSIMc, VFIQ bankbones

Dataset	ViT	FSIMc	VFIQ (Ours)
LIVE	(0.6842, 0.7091, 0.6994)	(0.7747, 0.9204, 0.7515)	(0.7732, **0.9236**, 0.7573)
TID2008	(0.6883, 0.7762, 0.5829)	(0.8341, **0.884**, 0.6991)	(**0.8512**, 0.8808, **0.6994**)
TID2013	(0.7212, 0.7839, 0.5893)	(0.8322,0.8509,0.6665)	(**0.8613, 0.8739, 0.6993**)

Table 5. The (PLCC, SRCC, KRCC) scores of Inception-v3 Net, VGG-16, ResNet50 with FSIMc

Dataset	Inception-v3 + FSIMc	VGG-16 + FSIMc	ResNet50 + FSIMc
LIVE	(0.7245, 0.8813, 0.7541)	(0.7397, 0.8742, 0.748)	(**0.7918**, 0.8905, **0.7873**)
TID2008	(0.7692, 0.8459, 0.6256)	(0.7812, 0.8210, 0.6521)	(0.8041, 0.8491, 0.6873)
TID2013	(0.7721, 0.8213, 0.6815)	(0.7903, 0.8188, 0.6734)	(0.8012, 0.8348, 0.6948)

scores for the datasets LIVE [14], TID2008 [12] and TID2013 [11] to examine their contribution to the performance.

The experiment results are shown in Table 4 and Table 5. Compared to ViT and FSIMs, Table 4 implies that our proposed VFIQ hybrid architecture achieves the best scores in 6 out of 9 cases (3 datasets × 3 metrics/dataset). Moreover, comparing Table 5 to our proposed VFIQ column in Table 4, we can observe that the CNN-FSIMc hybrid architecture shows worse performances than the VFIQ model, except for the PLCC and KRCC scores on the LIVE dataset only. Therefore, the effectiveness of our proposed VFIQ model has been validated.

6   Conclusion

In this paper, we propose a new model called VFIQ, which is a ViT-FSIMc hybrid Siamese network for Full-Reference Image Quality Assessment. This new model leverages the latest Vision Transformer as the backbone to extract the high-level semantic feature and combines the use of a conventional FR-IQA algorithm FSIMc. The VFIQ model outperforms many state-of-the-art IQA models. We also apply our VFIQ model in the task of reference image recovery, which highlights its potential in optimizing low-level vision tasks.

Several topics are worthy of further exploration in the future work. For example, during our investigation, we found that the proposed VFIQ model can be used for inferring the potential mode collapse, which is a general and widely known stability problem in GAN. Thus, it would be of interest to develop a methodology of mode collapse inference as an application of VFIQ. In addition, besides the three commonly used datasets for evaluation, one can look for other datasets, particularly those datasets in specific domains, to further evaluate or improve the performance of the VFIQ model.

References

1. Bosse, S., Maniry, D., Müller, K.R., Wiegand, T., Samek, W.: Deep neural networks for no-reference and full-reference image quality assessment. IEEE Trans. Image Process. **27**(1), 206–219 (2017)
2. Bovik, A.C.: A visual information fidelity approach to video quality assessment (2005)
3. Ding, K., Ma, K., Wang, S., Simoncelli, E.P.: Image quality assessment: unifying structure and texture similarity. IEEE Trans. Pattern Anal. Mach. Intell. **44**(5), 2567–2581 (2022). https://doi.org/10.1109/TPAMI.2020.3045810
4. Dosovitskiy, A., et al.: An image is worth 16 × 16 words: transformers for image recognition at scale. arXiv preprint arXiv:2010.11929 (2020)
5. Gao, Y., Rehman, A., Wang, Z.: CW-SSIM based image classification. In: 2011 18th IEEE International Conference on Image Processing, pp. 1249–1252. IEEE (2011)
6. Goodfellow, I., et al.: Generative adversarial nets. In: Neural Information Processing Systems (2014)
7. Gu, J., Shen, Y., Zhou, B.: Image processing using multi-code GAN prior. In: 2020 IEEE/CVF Conference on Computer Vision and Pattern Recognition (CVPR), pp. 3009–3018 (2020). https://doi.org/10.1109/CVPR42600.2020.00308
8. Jinjin, G., Haoming, C., Haoyu, C., Xiaoxing, Y., Ren, J.S., Chao, D.: PIPAL: a large-scale image quality assessment dataset for perceptual image restoration. In: Vedaldi, A., Bischof, H., Brox, T., Frahm, J.-M. (eds.) ECCV 2020. LNCS, vol. 12356, pp. 633–651. Springer, Cham (2020). https://doi.org/10.1007/978-3-030-58621-8_37
9. Lao, S., et al.: Attentions help CNNs see better: attention-based hybrid image quality assessment network. In: Proceedings of the IEEE/CVF Conference on Computer Vision and Pattern Recognition, pp. 1140–1149 (2022)
10. Laparra, V., Ballé, J., Berardino, A., Simoncelli, E.P.: Perceptual image quality assessment using a normalized Laplacian pyramid. In: Human Vision and Electronic Imaging 2016, HVEI 2016, pp. 43–48. Society for Imaging Science and Technology (2016)
11. Ponomarenko, N., et al.: Image database TID2013: peculiarities, results and perspectives. Signal Process. Image Commun. **30**, 57–77 (2015)
12. Ponomarenko, N., Lukin, V., Zelensky, A., Egiazarian, K., Carli, M., Battisti, F.: TID2008-a database for evaluation of full-reference visual quality assessment metrics. Adv. Mod. Radioelectron. **10**(4), 30–45 (2009)
13. Prashnani, E., Cai, H., Mostofi, Y., Sen, P.: PieAPP: perceptual image-error assessment through pairwise preference. In: Proceedings of the IEEE Conference on Computer Vision and Pattern Recognition, pp. 1808–1817 (2018)
14. Sheikh, H.R., Sabir, M.F., Bovik, A.C.: A statistical evaluation of recent full reference image quality assessment algorithms. IEEE Trans. Image Process. **15**(11), 3440–3451 (2006)
15. Shi, S., et al.: Region-adaptive deformable network for image quality assessment. In: Proceedings of the IEEE/CVF Conference on Computer Vision and Pattern Recognition, pp. 324–333 (2021)
16. Vaswani, A., et al.: Attention is all you need. arXiv (2017)
17. Wang, Z., Bovik, A.C., Sheikh, H.R., Simoncelli, E.P.: Image quality assessment: from error visibility to structural similarity. IEEE Trans. Image Process. **13**(4), 600–612 (2004)

18. Wang, Z., Simoncelli, E.P., Bovik, A.C.: Multiscale structural similarity for image quality assessment. In: The Thirty-Seventh Asilomar Conference on Signals, Systems & Computers, vol. 2, pp. 1398–1402. IEEE (2003)
19. Xia, W., Yang, Y., Xue, J.H., Wu, B.: TediGAN: text-guided diverse face image generation and manipulation. In: Proceedings of the IEEE/CVF Conference on Computer Vision and Pattern Recognition, pp. 2256–2265 (2021)
20. Zhang, L., Zhang, L., Mou, X., Zhang, D.: FSIM: a feature similarity index for image quality assessment. IEEE Trans. Image Process. **20**(8), 2378–2386 (2011)
21. Zhang, R., Isola, P., Efros, A.A., Shechtman, E., Wang, O.: The unreasonable effectiveness of deep features as a perceptual metric. In: Proceedings of the IEEE Conference on Computer Vision and Pattern Recognition, pp. 586–595 (2018)

Spiking Reinforcement Learning
for Weakly-Supervised Anomaly Detection

Ao Jin[1], Zhichao Wu[1], Li Zhu[1], Qianchen Xia[2]([✉]), and Xin Yang[1]

[1] Dalian University of Technology, Dalian, China
{dllsja,wuzhch}@mail.dlut.edu.cn, {zhuli,xinyang}@dlut.edu.cn
[2] Tsinghua University, Beijing, China
qianchenxia@tsinghua.edu.cn

Abstract. Weakly-supervised Anomaly Detection (AD) has achieved significant performance improvement compared to unsupervised methods by harnessing very little additional labeling information. However, most existing methods ignore anomalies in unlabeled data by simply treating the whole unlabeled set as normal; that is, they fail to resist such noise that may considerably disturb the learning process, and more importantly, they cannot extract key anomaly features from these unlabeled anomalies, which are complementary to those labeled ones. To solve this problem, a spiking reinforcement learning framework for weakly-supervised AD is proposed, named ADSD. Compared with artificial neural networks, the spiking neural network can effectively resist input perturbations due to its unique coding methods and neuronal characteristics. From this point of view, by using spiking neurons with noise filtering and threshold adaptation, as well as a multi-weight evaluation method to discover the most suspicious anomalies in unlabeled data, ADSD achieves end-to-end optimization for the utilization of a few labeled anomaly data and rare unlabeled anomalies in complex environments. The agent in ADSD has robustness and adaptability when exploring potential anomalies in the unknown space. Extensive experiments show that our method ADSD significantly outperforms four popular baselines in various environments while maintaining good robustness and generalization performance.

Keywords: Reinforcement Learning · Robustness and Generalization · Spiking Neural Network · Weakly-supervised Anomaly Detection

1 Introduction

Anomaly Detection (AD), also known as outlier detection, is concerned with identifying rare events that significantly deviate from normal behaviors, such

This work was supported in part by National Key Research and Development Program of China (2022ZD0210500), the National Natural Science Foundation of China under Grants U21A20491/62332019/61972067, and the Distinguished Young Scholars Funding of Dalian (No. 2022RJ01).

as network attacks in cybersecurity, equipment malfunctions in industrial processes, and diseases in healthcare [2,3,15,27,30]. Timely identification of anomalies can prevent severe security breaches, reduce maintenance costs, and avoid catastrophic failures. However, since abnormal data is rare and unpredictable, obtaining all anomaly information is almost impossible in real-world scenarios [21]. One of the recent hotspots in research is how to use a small portion of labeled data in addition to large unlabeled data to learn basic patterns of normal behavior and detect anomalies. Compared to unsupervised and fully supervised methods [1], weakly-supervised methods [19,22,25,36] can significantly reduce labeling costs while maintaining high detection performance.

Currently, weakly-supervised methods focus on training model parameters using underlying information within labeled anomalies to explore latent anomaly features [23]. Although these methods show good performance in relevant fields, their performance often depends on the quality of a small subset of labeled data. When facing situations where the known abnormal features cannot cover all the data feature characteristics, their performance and generalization ability are significantly downgraded. Moreover, these methods exhibit particularly poor performance when handling noisy and contamination-polluted scene information [31]. These problems hinder the deployment of weakly-supervised methods in practical applications to some extent.

To address these issues, some researchers have shifted their attention to Reinforcement Learning (RL) [12]. RL is an emerging learning paradigm, in which an agent interacts with the environment by learning policies to maximize rewards or achieve specific goals. The recently proposed method DPLAN [20] has achieved good performance in multiple anomaly detection tasks with autonomous learning and adaptive capacity. However, DPLAN is still unable to effectively detect anomalies when applied in production scenarios for the following reasons:

1. The distance-based sampling method may result in falling into local optima;
2. The reward setting in the method ignores the positive rewards of the agent actively exploring unknown environment anomalies, which leads to poor performance of the method in complex datasets;
3. We empirically found that DPLAN has weak robustness w.r.t. unlabelled data pollution. Nevertheless, RL has great potential in anomaly detection tasks due to its unique ability to automatically and interactively fit the given anomaly data.

In this paper, A Spiking Reinforcement Learning (SRL) framework for weakly-supervised anomaly detection is proposed, which focuses on balancing exploration-exploitation in labeled and unlabeled data space to improve the robustness and cope with the practical anomaly detection problems in highly polluted environments. Given the weak robustness of unlabeled data, the Spiking Neural Network (SNN) is introduced into the field of AD by taking the advantage of spiking neurons in the network, which can mitigate the effects of noise due to their collective effects and architectural connectivity [14].

To address the sampling problem, the sampling method based on multiple angles is designed, which combines two unsupervised methods with cosine similarity metric to enhance sampling diversity to approximate the global optimal solution. In addition, the reward function is improved to encourage the agent to actively explore the unlabeled space for discovering potential anomalies, thereby improving the detection performance.

Our contributions can be summarized as follows:

1. A novel SRL framework for tabular anomaly detection is proposed to improve adaptability to unusual data tasks with complex structures, which is the first attempt for our knowledge in anomaly detection.
2. An anomaly-biased simulated environment is improved to increase the diversity and accuracy of the exploration space, which provides important support for exploring unknown anomalies in complex scenes.
3. The extensive experiments over benchmark datasets are conducted in various scenarios with different levels of pollution coverage. Our experimental results show that our proposed method outperforms previous state-of-the-art methods in terms of not only accuracy, but also adapting to complex unknown scenarios and possessing excellent generalization and stability.

2   Related Work

2.1   Weakly-Supervised Tabular Anomaly Detection

Tabular anomaly detection has been widely studied in machine learning, but most of the existing methods assume the data is fully-labeled or fully-unlabeled, which is a strong assumption and limits their practical applicability. Weakly-supervised anomaly detection has emerged as an alternative approach to address this challenge. In this setting, only a limited number of anomalies are labeled while the majority of data are unlabeled. Early attempts in this research line leverage PU-learning, self-training, and active learning as their main methodology. With the burgeoning of deep learning, deep methods (e.g., DevNet [22], Deep SAD [23], FeaWAD [36], and DPLAN [20]) are proposed by the community, showing considerable performance leaps over these techniques.

Despite these advancements, there is still much to be done to further improve the effectiveness of deep weakly supervised anomaly detection in the tabular setting. Specifically, developing new methods that can effectively handle the ambiguity and uncertainty of unlabeled data space remains an area of active research.

2.2   Reinforcement Learning

RL is a type of machine learning method in which an agent learns to interact with an environment to achieve a specific goal through trial-and-error experience. The agent learns to make decisions by receiving feedback in the form of rewards or punishments in response to its actions. This feedback enables the agent to learn from its mistakes and improve its decision-making abilities over time. RL has been successfully applied in a wide range of fields, such as gaming, finance, and robotics.

DQN [10,18,28] is a classic deep reinforcement learning method that combines deep learning and Q-learning and uses a neural network to estimate Q-values. By integrating the DQN algorithm, DPLAN [20] introduces reinforcement learning to tabular anomaly detection tasks, actively exploring unknown anomalies and resulting in joint optimization of both known and unknown anomaly detection.

2.3   Spiking Neural Network

SNN has garnered considerable attention recently due to its potential to better model biological neural networks and improve robustness compared to ANN [7]. It has been extensively employed in various fields [6,33,35]. The robustness and the low-power-consumption characteristic of SNN have found diverse applications in combination with RL, such as in robots and network architecture searches [13,26,32,34]. However, in the anomaly detection domain, the existing methods only focus on time-series AD [11] and have not incorporated SRL to develop anomaly detection methods for tabular data.

3   The Proposed Approach

3.1   Problem Statement

The proposed ADSD method focuses on tabular data and integrates a smaller set of labeled samples with a larger set of unlabeled samples to enhance the detection of anomalies in complex real-world scenarios. Our objective is to develop an anomaly score function that is fueled by a small fraction of supervised signals and a large proportion of unsupervised signals.

Specifically, the dataset $\mathcal{D}$ is defined into two types $\mathcal{D}_a$ and $\mathcal{D}_u$ (with $\mathcal{D}_a \cap \mathcal{D}_u = \varnothing$) from various scenarios in a unified manner, where $\mathcal{D}_a$ stores the small-scale anomalous datasets composed of N anomalous instances, and $\mathcal{D}_u$ comprises large-scale unlabeled datasets, primarily consisting of normal instances, with a minor fraction of unlabeled anomalous instances. Such a setup fits the actual situation and it is expected that the proposed method fully learns the exceptional features in $\mathcal{D}_a$ and explores anomalies as much as possible in the space of $\mathcal{D}_u$. Anomaly score detection function $\mathcal{L}$ is defined as $\mathcal{D} \to \mathbb{R}$, where $0 \leq \mathcal{L}(s_i) < \mathcal{L}(s_j) \leq 1, s_i, s_j \in \mathcal{D}$, s_j represents abnormal data, and s_i represents normal data.

3.2   Algorithm Framework

In this section, we introduce our method named Anomaly Detection via Spiking Deep Q-learning (ADSD) in detail. As shown in Fig. 1, our SRL-based anomaly detection approach comprises the following key modules: an SRL-based agent $\mathcal{A}$, and an anomaly-biased simulation environment E. The simulation environment E encompasses an observation sampling function $\mathcal{G}$ and a reward function r_t. The agent $\mathcal{A}$ interacts with the simulation environment E using a combined reward derived from the r_t functions, to jointly learn from $\mathcal{D}_a$ and $\mathcal{D}_u$. The proposed ADSD framework can be described as follow:

- **Observation space** consists of entire training datasets $\mathcal{D}$ and provides the current state s that the agent requires for each decision.
- **Action space** is defined to be $\{a_0, a_1\}$, a_0, and a_1 represent the binary classification labels of the agent for the current state, indicating normal and abnormal, respectively.
- **SRL-based agent** $\mathcal{A}$ is based on SNN, which aims to find the optimal execution actions under multiple environmental settings to maximize the mission-specific reward.
- **Environment** E includes the initial state, the state transition after an action is performed, the reward function, and the agent's policy. To enable agents to effectively explore unlabeled anomalies in $\mathcal{D}_u$, we design a new multi-angle anomaly-biased sampling function in this part.

3.3 SRL-Based Agent

Our target is to transform the anomaly detection problem into a reinforcement learning task. In this work, the construction of Q-network relies on SNN, with the Q-function and loss function consistent with the classical DQN algorithm, which is presented as:

$$Q^*(\mathbf{s}, a) = \max_{\pi} \mathbb{E} \left[r_t + \gamma r_{t+1} + \gamma^2 r_{t+2} + \cdots \mid s_t = \mathbf{s}, a_t = a, \pi \right] \tag{1}$$

$$L_i(\theta_i) = \mathbb{E}_{s,a \sim \rho(\cdot); s' \sim \mathcal{E}} \left[\left(r + \gamma \max_{a'} Q(s', a'; \theta_{i-1}) - Q(s, a; \theta_i) \right) \right] \tag{2}$$

where $Q^*(\mathbf{s}, a)$ can be obtained by following the optimal policy π after taking action a in state s and r_t represents the immediate reward obtained at time step t, γ is the discount factor ($0 \leq \gamma < 1$). Specifically, the loss function evaluates the performance of the current Q-network parameter θ_i. Here, $\rho(\cdot)$ is the state action distribution in the experience replay buffer, and $\mathcal{E}$ is the experience replay buffer of the environment. The goal of this equation is to gradually make the Q-network obtain the optimal Q-function during training and ultimately lead to optimal decision-making.

The integrate-and-fire (IF) [5] and leaky integrate-and-fire (LIF) spiking neurons [8] are chosen in SNN. The IF neuron is used as the encoder, it receives signal input from other neurons, and after accumulation, as long as the total input signal exceeds the threshold, the neuron will produce a single output signal, whose mathematical model can be expressed as:

$$\sum_j w_j x_j \geqslant \theta \tag{3}$$

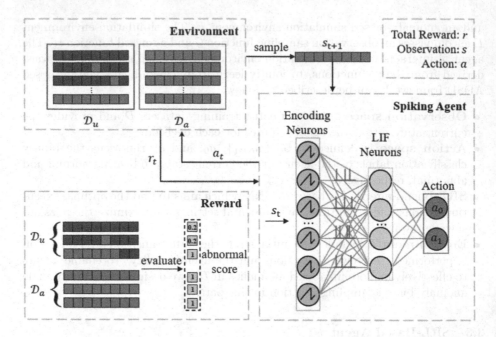

Fig. 1. The framework of the proposed ADSD method. The optimal action is determined by the combined evaluation of the sampling function and the reward function, with the former providing diverse and anomalous samples and the latter assigning appropriate rewards.

where w_j represents the weight of the input signal, x_j represents the size of the input signal, and θ represents the threshold. In addition, the LIF neuron is used as a decision-maker. When a LIF neuron receives spikes from other neurons, the spikes are scaled accordingly based on learned synaptic weights. Depolarization is achieved by summing over all the scaled spikes. A decay function over time is used to drive the potential membrane to hyperpolarization. The formal definition of LIF can be given as:

$$\tau_m \frac{dV}{dt} = -(V - E_L) + I(t)R_m \quad V_{th} = \infty \tag{4}$$

where V represents the membrane voltage, E_L is the resting potential, $I(t)$ is the input current received by the neuron, R_m is the membrane resistance, and τ_m is the membrane capacitance. Especially, V_{th} is set to infinity, so the neuron will not generate a spike output to ensure that the output of the Q-function is a continuous value.

3.4 Anomaly-Biased Environment

This environment can be divided into the following two parts:

- **Multi-perspective sampling function** is composed of two unsupervised anomaly score functions and one distance-based function. Which is used to achieve balanced exploitation and exploration of the environment.
- **Composite reward function** is defined to yield a reward based on its performance in detecting known and unknown anomalies.

Multi-perspective Sampling Function. In general, distance-based sampling methods can only find the most obvious anomalies so that they mostly fall into local optimal. To mitigate the impact of this problem, a multi-angle function representing the potential of a group of outliers is used to sample, which is of great significance for exploring the anomaly space. The sampling functions are defined as:

$$s_{t+1} = \begin{cases} \phi(s_t) & \text{with probability } p \\ \psi(s_t) & \text{with probability } 1-p \end{cases} \tag{5}$$

where $\phi(s_t)$ is a function randomly sampled from $\mathcal{D}_a$, $\psi(s_t)$ is a anomaly-biased function to find the potential anomaly state in $\mathcal{D}_u$. In order to balance the exploitation and exploration of the environment, $p = 0.5$ is set by default in this paper.

$$score_{dis}(s_t) = \begin{cases} 1 - norm_{cos\_sim}(s, s_t) & \text{if } a_t = a^1 \\ norm_{cos\_sim}(s, s_t) & \text{if } a_t = a^0 \end{cases} \tag{6}$$

$$S_{dis} = \{i_{1:5}\} \text{ where } score_{dis}(i_1) \geq \dots \geq score_{dis}(i_5) \tag{7}$$

$$S_{LOF} = \{j_{1:5}\} \text{ where } score_{LOF}(j_1) \geq \dots \geq score_{LOF}(j_5) \tag{8}$$

$$S_{iForest} = \{k_{1\cdot5}\} \text{ where } score_{iForest}(k_1) \geq \dots \geq score_{iForest}(k_5) \tag{9}$$

$$S_{INT} = S_{dis} \cap S_{LOF} \cap S_{iForest} \tag{10}$$

$$\psi(s_t) = \begin{cases} \text{argmax}(score_{dis}(n) + score_{LOF}(n) + score_{iForest}(n)) & \text{if } S_{INT} \neq \varnothing \\ \text{argmax}(score_{dis}(n)) & \text{otherwise} \end{cases} \tag{11}$$

The combination of three scores, namely $score_{dis}$ based on the normalized cosine-similarity algorithm $norm_{cos\_sim}$, $score_{LOF}$ based on the LOF [4], and $S_{iForest}$ based on the iForest algorithm, is used to comprehensively measure potential anomalies. In particular, each score is preprocessed for normalization. For each algorithm, the first five exceptional states (i, j, k) and their corresponding scores $(S_{dis}, S_{LOF}, S_{iForest})$ are computed. These exceptional states and their scores are then combined to create the final set of state space, which is subsequently intersected. If the intersection is not empty, the state whose sum of anomaly scores is the largest, is selected as s_{t+1}. Otherwise, the state with the highest distance-based anomaly score is selected as s_{t+1}. This design helps the agent to visit as many types of anomalies as possible in the state space.

Composite Reward Function. The reward function is designed to consider both external and intrinsic rewards. The external reward function is designed for the labeled dataset $\mathcal{D}_a$, while the internal reward function is designed for the unlabeled dataset $\mathcal{D}_u$. Specifically, for the intrinsic reward function, the unsupervised algorithm Copula-based Outlier Detection (COPOD) [16], inspired by the curiosity algorithm [24] is introduced, to guide the agent in exploring the environment. COPOD is used here because it is adaptable and robust to various data types and distributions. Our reward function can be defined as follows:

$$
r_t = \begin{cases} 1 & \text{if } a_t = a^1 \text{ and } s_t \in \mathcal{D}_a \\ 1 & \text{score } (s_t) > \text{ threshold} \\ 0.2 & \text{if } a_t = a^0 \text{ and } s_t \in \mathcal{D}_u \\ -1 & \text{otherwise,} \end{cases} \tag{12}
$$

where the external reward assigned to each action of the agent depends on its current state and action. If the agent chooses the action of a_1 in step t and the state is in the $\mathcal{D}_a$, it receives a large reward and indicates that the agent correctly identified anomalies in $\mathcal{D}_a$.

For the intrinsic reward, the threshold is used to adjust the effect of the unsupervised method in $\mathcal{D}_u$. Here, score (s_t) denotes the anomaly score of the current state s_t. Typically, the threshold is dynamically adjusted based on the score (s) function. If the anomaly score is higher than the set threshold, the current state s_t is likely a potential anomaly. When the agent correctly identifies an anomaly, it receives a large reward. Conversely, if the anomaly score is lower than the threshold, the current state s_t is considered normal, and the agent receives a small reward to encourage exploratory behavior.

4   Experiments

In this section, the proposed ADSD method is compared with four state-of-the-art anomaly detectors including DPLAN [20], FeaWAD [36], DevNet [22], and iForest [17]. Six real-world datasets are used in our experiments. Five datasets (*fault, InternetAds, landset, skin,* and *Waveform*) are from ADBench [9], and *DMC* is a chemical industry dataset containing 38 features including temperature, pressure, flow rate, etc. Following [20,29], we employ two evaluation metrics, the area under the ROC curve (AUC-ROC) and the area under the PR curve (AUC-PR), to impartially measure the ranking quality of anomaly scores derived from different anomaly detectors. With considering the weakly-supervised learning paradigm, we assume the number of labeled anomalies is 60. As anomalies are rare events in practical applications, the contamination rate is first fixed to 2% in Sect. 4.1, which specifically illustrates the detection performance. We further investigate the robustness w.r.t. different contamination ratio in Sect. 4.2, in which a wide range of contamination levels is considered.

4.1   Overall Performance

According to the results in Table 1 and Table 2, the performance of our proposed ADSD algorithm and the four contrastive algorithms are analyzed, and the following conclusions are drawn,

- ADSD has shown the best performance in AUC-PR on five datasets except for Waveform and achieved the best performance in AUC-ROC on five datasets except for skin where it is still the second-best performer. The results demonstrate that the SNN-based agent is capable of effectively utilizing known anomalies, while possessing the ability to explore in an unlabeled environment, thereby enabling the discovery of potential anomalies even when only a minimal amount of labeled data is available.
- For DMC dataset, as the known anomalies may not include all anomaly types, there are also unknown anomalies in the dataset, therefore the DPLAN and ADSD algorithms with active probing capabilities performed better.

Table 1. AUC-ROC performance of the proposed ADSD and four competing methods

Dataset	AUR-ROC Performance				
	iForest	DevNet	FeaWAD	DPLAN	ADSD (ours)
fault	0.6266	0.5365	0.5070	0.6092	**0.6828**
InternetAds	0.6193	0.7689	0.7596	0.7025	**0.8114**
landsat	0.5946	0.4770	0.7428	0.5941	**0.7730**
skin	0.8623	0.8393	0.7853	0.9572	0.9222
Waveform	0.6785	0.6423	0.6094	0.8345	**0.9491**
DMC	0.7406	0.7282	0.8929	0.9134	**0.9334**

Table 2. AUC-PR performance of the proposed ADSD and four competing methods

Dataset	AUR-PR Performance				
	iForest	DevNet	FeaWAD	DPLAN	ADSD (ours)
fault	0.4079	0.3343	0.3826	0.4370	**0.5887**
InternetAds	0.3817	0.4294	0.4856	0.6026	**0.6486**
landsat	0.3082	0.1897	0.4433	0.3019	**0.4783**
skin	0.4325	0.4075	0.3674	0.5977	**0.6080**
Waveform	0.0489	0.0377	0.0344	**0.5255**	0.4544
DMC	0.4660	0.6031	0.6116	0.6829	**0.8401**

4.2   Anomaly Pollution Test

We designed experiments to verify the pollution robustness and adaptability of the methods mentioned above. On the basis of the original 2% pollution, the degree of impact of six pollution rates from 2% to 12% is assessed on the algorithm. The results are shown in Fig. 2.

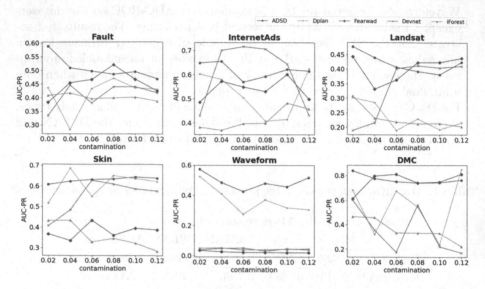

Fig. 2. Performance of methods under different pollution rate settings.

The following three remarks can be made from the results. (i) Similar to the results in Table 1 and Table 2, ADSD can maintain good performance under different pollution factor settings, which indicates that the spiking neural network used in ADSD has certain robustness for anomaly detection tasks. (ii) Compared with DPLAN, ADSD sampling mode can avoid the dilemma of local optimization, and then better explore the state space to obtain good performance. (iii) The majority of compared methods for *Fault* and *InternetAds* exhibit clear fluctuation trends. This can be attributed to the insufficient amount of data provided for both datasets, leading to overfitting and model performance instability.

4.3   Ablation Study

The effect of each part in our model is investigated in further. First, DPLAN is used as a comparison base (named INIT), and then examine the effects of different neural network types and environment improvement on the accuracy of the algorithm by setting two experiments EXP1 and EXP2. EXP1 represents the performance improvement after converting ANN to SNN, and EXP2 represents the performance improvement after improving the sampling function and reward function. Detailed results can be found in Table 3.

Table 3. AUC-ROC performance of ablation study

Dataset	INIT	EXP1	EXP2
fault	0.6092	0.6193	0.5419
InternetAds	0.7025	0.7339	0.7768
landsat	0.5941	0.6350	0.5979
Waveform	0.8345	0.8917	0.9453
DMC	0.9134	0.9150	0.9298

As can be seen from Table 3, in the vast majority of datasets, EXP1 and EXP2 can achieve higher AUC with limited labeled anomaly data, indicating the effectiveness of the contribution point of our algorithm. However, in the fault dataset, our improvement EXP2 did not perform well, which may be due to the fact that INIT's Euclidean distance-based sampling function is better at detecting anomalies in environments with limited data and low data dimensionality.

5   Conclusions

This paper proposed ADSD, a novel spiking reinforcement learning framework for weakly-supervised anomaly detection. The AD agent guided by SRL greatly enhances the robustness of the model when exploring the contaminated environment, during which the autonomous learning ability and the adaptive ability of the agent can be simultaneously guaranteed. In complex scenarios, the ability to recognize anomalies is continuously improved through the reward mechanism driven by unsupervised information and a few labeled anomalies. Furthermore, the performance of ADSD in finding potential anomalies in unlabeled datasets is significantly enhanced with the multi-angle environmental sampling method. The experiments under different settings show ADSD achieves significant performance improvement over existing methods, especially showing technical advantages when handling unknown abnormal types.

References

1. Aggarwal, C.C.: An Introduction to Outlier Analysis. In: Outlier Analysis, pp. 1–34. Springer, Cham (2017). https://doi.org/10.1007/978-3-319-47578-3_1
2. Asrori, S.S., Wang, L., Ozawa, S.: Permissioned blockchain-based XGBoost for multi banks fraud detection. In: Tanveer, M., Agarwal, S., Ozawa, S., Ekbal, A., Jatowt, A. (eds.) Neural Information Processing: 29th International Conference, ICONIP 2022, Virtual Event, 22–26 November 2022, Proceedings, Part III. LNCS, vol. 13625, pp. 683–692. Springer, Cham (2023). https://doi.org/10.1007/978-3-031-30111-7_57
3. Bhuyan, M.H., Bhattacharyya, D.K., Kalita, J.K.: Network anomaly detection: methods, systems and tools. IEEE Commun. Surv. Tutorials **16**(1), 303–336 (2013)

4. Breunig, M.M., Kriegel, H.P., Ng, R.T., Sander, J.: LOF: identifying density-based local outliers. In: Proceedings of the 2000 ACM SIGMOD International Conference on Management of Data, pp. 93–104 (2000)
5. Burkitt, A.N.: A review of the integrate-and-fire neuron model: I. homogeneous synaptic input. Biol. Cybern. **95**, 1–19 (2006)
6. Ding, J., et al.: Biologically inspired dynamic thresholds for spiking neural networks. Adv. Neural. Inf. Process. Syst. **35**, 6090–6103 (2022)
7. Friedrich, J., Urbanczik, R., Senn, W.: Spatio-temporal credit assignment in neuronal population learning. PLoS Comput. Biol. **7**(6), e1002092 (2011)
8. Gerstner, W., Kistler, W.M., Naud, R., Paninski, L.: Neuronal Dynamics: From Single Neurons to Networks and Models of Cognition. Cambridge University Press, Cambridge (2014)
9. Han, S., Hu, X., Huang, H., Jiang, M., Zhao, Y.: ADBench: anomaly detection benchmark. Adv. Neural. Inf. Process. Syst. **35**, 32142–32159 (2022)
10. Hessel, M., et al.: Rainbow: combining improvements in deep reinforcement learning. In: Proceedings of the AAAI Conference on Artificial Intelligence, vol. 32 (2018)
11. Hu, L., Liu, Y., Qiu, W.: A deep spiking neural network anomaly detection method. Comput. Intell. Neurosci. CIN **2022**, 6391750 (2022)
12. Kaelbling, L.P., Littman, M.L., Moore, A.W.: An introduction to reinforcement learning. In: The Biology and Technology of Intelligent Autonomous Agents, pp. 90–127 (1995)
13. Kim, Y., Li, Y., Park, H., Venkatesha, Y., Panda, P.: Neural architecture search for spiking neural networks. In: Avidan, S., Brostow, G., Cissé, M., Farinella, G.M., Hassner, T. (eds.) Computer Vision-ECCV 2022: 17th European Conference, Tel Aviv, Israel, 23–27 October 2022, Proceedings, Part XXIV. LNCS, vol. 13684, pp. 36–56. Springer, Cham (2022). https://doi.org/10.1007/978-3-031-20053-3_3
14. Kozma, R., Pino, R.E., Pazienza, G.E.: Advances in Neuromorphic Memristor Science and Applications, vol. 4. Springer, Cham (2012). https://doi.org/10.1007/978-94-007-4491-2
15. Lavin, A., Ahmad, S.: Evaluating real-time anomaly detection algorithms-the Numenta anomaly benchmark. In: 2015 IEEE 14th International Conference on Machine Learning and Applications (ICMLA), pp. 38–44. IEEE (2015)
16. Li, Z., Zhao, Y., Botta, N., Ionescu, C., Hu, X.: COPOD: copula-based outlier detection. In: 2020 IEEE International Conference on Data Mining (ICDM), pp. 1118–1123. IEEE (2020)
17. Liu, F.T., Ting, K.M., Zhou, Z.H.: Isolation forest. In: 2008 Eighth IEEE International Conference on Data Mining, pp. 413–422. IEEE (2008)
18. Mnih, V., et al.: Human-level control through deep reinforcement learning. Nature **518**(7540), 529–533 (2015)
19. Pang, G., Cao, L., Chen, L., Liu, H.: Learning representations of ultrahigh-dimensional data for random distance-based outlier detection. In: Proceedings of the 24th ACM SIGKDD International Conference on Knowledge Discovery & Data Mining, pp. 2041–2050 (2018)
20. Pang, G., van den Hengel, A., Shen, C., Cao, L.: Toward deep supervised anomaly detection: reinforcement learning from partially labeled anomaly data. In: Proceedings of the 27th ACM SIGKDD Conference on Knowledge Discovery & Data Mining, pp. 1298–1308 (2021)
21. Pang, G., Shen, C., Cao, L., Hengel, A.V.D.: Deep learning for anomaly detection: a review. ACM Comput. Surv. (CSUR) **54**(2), 1–38 (2021)

22. Pang, G., Shen, C., van den Hengel, A.: Deep anomaly detection with deviation networks. In: Proceedings of the 25th ACM SIGKDD International Conference on Knowledge Discovery & Data Mining, pp. 353–362 (2019)

23. Ruff, L., et al.: Deep semi-supervised anomaly detection. arXiv preprint arXiv:1906.02694 (2019)

24. Savinov, N., et al.: Episodic curiosity through reachability. arXiv preprint arXiv:1810.02274 (2018)

25. Tamersoy, A., Roundy, K., Chau, D.H.: Guilt by association: large scale malware detection by mining file-relation graphs. In: Proceedings of the 20th ACM SIGKDD International Conference on Knowledge Discovery and Data Mining, pp. 1524–1533 (2014)

26. Tang, G., Shah, A., Michmizos, K.P.: Spiking neural network on neuromorphic hardware for energy-efficient unidimensional SLAM. In: 2019 IEEE/RSJ International Conference on Intelligent Robots and Systems (IROS), pp. 4176–4181. IEEE (2019)

27. Ukil, A., Bandyoapdhyay, S., Puri, C., Pal, A.: IoT healthcare analytics: the importance of anomaly detection. In: 2016 IEEE 30th International Conference on Advanced Information Networking and Applications (AINA), pp. 994–997. IEEE (2016)

28. Van Hasselt, H., Guez, A., Silver, D.: Deep reinforcement learning with double Q-learning. In: Proceedings of the AAAI Conference on Artificial Intelligence, vol. 30 (2016)

29. Xu, H., Pang, G., Wang, Y., Wang, Y.: Deep isolation forest for anomaly detection. IEEE Trans. Knowl. Data Eng. 1–14 (2023). https://doi.org/10.1109/TKDE.2023.3270293

30. Xu, H., Wang, Y., Wei, J., Jian, S., Li, Y., Liu, N.: Fascinating supervisory signals and where to find them: deep anomaly detection with scale learning. In: Proceedings of the International Conference on Machine Learning (2023)

31. Yoon, J., Sohn, K., Li, C.L., Arik, S.O., Lee, C.Y., Pfister, T.: Self-trained one-class classification for unsupervised anomaly detection. arXiv e-prints pp. arXiv-2106 (2021)

32. Zhang, D., Zhang, T., Jia, S., Xu, B.: Multi-scale dynamic coding improved spiking actor network for reinforcement learning. In: Proceedings of the AAAI Conference on Artificial Intelligence, vol. 36, pp. 59–67 (2022)

33. Zhang, H., et al.: In the blink of an eye: event-based emotion recognition. In: ACM SIGGRAPH 2023 Conference Proceedings, pp. 1–11 (2023)

34. Zhang, J., et al.: Spiking transformers for event-based single object tracking. In: Proceedings of the IEEE/CVF conference on Computer Vision and Pattern Recognition, pp. 8801–8810 (2022)

35. Zhang, J., et al.: Frame-event alignment and fusion network for high frame rate tracking. In: Proceedings of the IEEE/CVF Conference on Computer Vision and Pattern Recognition (CVPR), pp. 9781–9790, June 2023

36. Zhou, Y., Song, X., Zhang, Y., Liu, F., Zhu, C., Liu, L.: Feature encoding with autoencoders for weakly supervised anomaly detection. IEEE Trans. Neural Networks Learn. Syst. 33(6), 2454–2465 (2021)

Resource-Aware DNN Partitioning
for Privacy-Sensitive Edge-Cloud Systems

Aolin Ding[1], Amin Hass[1], Matthew Chan[2], Nader Sehatbakhsh[3],
and Saman Zonouz[4(✉)]

[1] Security R&D, Accenture Labs, Accenture, Washington, DC, USA
[2] Rutgers University, New Brunswick, NJ, USA
[3] University of California, Los Angeles (UCLA), Los Angeles, CA, USA
[4] Georgia Institute of Technology, Atlanta, GA, USA
saman.zonouz@gatech.edu

Abstract. With recent advances in deep neural networks (DNNs), there
is a significant increase in IoT applications leveraging AI with edge-cloud
infrastructures. Nevertheless, deploying large DNN models on resource-
constrained edge devices is still challenging due to limitations in computation, power, and application-specific privacy requirements. Existing
model partitioning methods, which deploy a partial DNN on an edge
device while processing the remaining portion of the DNN on the cloud,
mainly emphasize communication and power efficiency. However, DNN
partitioning based on the privacy requirements and resource budgets of
edge devices has not been sufficiently explored in the literature. In this
paper, we propose AWARESL, a model partitioning framework that splits
DNN models based on the computational resources available on edge
devices, preserving the privacy of input samples while maintaining high
accuracy. In our evaluation of multiple DNN architectures, AWARESL
effectively identifies the split points that adapt to resource budgets of
edge devices. Meanwhile, we demonstrate the privacy-preserving capability of AWARESL against existing input reconstruction attacks without
sacrificing inference accuracy in image classification tasks.

Keywords: Edge Computing · Privacy-Preserving Machine Learning ·
Data Privacy · Split Learning

1 Introduction

Coupled with recent advances in deep learning (DL), a noticeable trend is that
machine learning (ML) models used for robotic vehicles [4] and edge computing
applications [19] have become increasingly complex with the growing size of
model parameters. As a result, it is challenging to deploy complex DNNs on
edge infrastructures, due to their restricted computational capacity and limited

---

Supported by National Science Foundation (NSF), Accenture, and Department of
Energy (DoE) Award DE-OE0000780, Cyber Resilient Energy Delivery Consortium.

B. Luo et al. (Eds.): ICONIP 2023, LNCS 14451, pp. 188–201, 2024.
https://doi.org/10.1007/978-981-99-8073-4_15

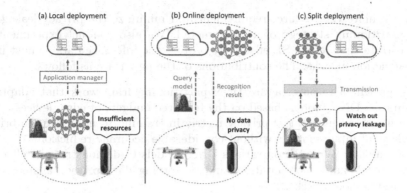

Fig. 1. The challenges of the DNN deployment for an edge-cloud system.

resources as illustrated in Fig. 1 (a). One alternative way is to deploy the entire DNN model on a cloud server, and edge devices can send queries for inference. However, as shown in Fig. 1 (b), transferring raw data to the cloud server poses privacy concerns for input samples in privacy-sensitive scenarios.

Recently, model partitioning techniques [3,9,11,12,16,17] such as split learning (SL) have been proposed to resolve the resource constraints in edge deployment. Specifically, SL approaches [2,5,14,21] split one complete DNN model into two sub-networks, uploading and training them correspondingly on the edge and cloud sides. Prior works primarily focus on reducing communication costs [11,16] and power consumption [9], thus partially addressing the resource constraints, but comparatively insufficient attention has been given to the imperative of privacy preservation. SL methods are conventionally believed to provide the private inference of a DNN in untrusted scenarios [6] since they do not require these two entities to share model structures or raw training data. However, recent works [7,13,22] have reported that existing SL frameworks are not as effective as intended. As shown in Fig. 1 (c), the attackers are still able to recover the sensitive data such as raw inputs since prior privacy metrics [1,7,10,20] only focus on the input data in the inference process. Therefore, attaining a better privacy-utility balance under computational constraints in both training and inference stages is needed to securely deploy complex DNNs on edge devices.

In this paper, we present AWARESL, a resource-aware DNN partitioning framework that offloads computation from edge devices while preserving input data privacy. AWARESL investigates the resource limitations and privacy requirements of the target edge platform to determine the split point for a DNN model, and thus fulfills the computational and privacy needs without loss of accuracy. We achieve this by designing an adaptive DNN partitioning strategy and integrating a distance correlation-based privacy metric into the model training. Compared to the conventional distance correlation work [20], our work can be easily extended to different types of neural network models, including both sequential and non-sequential architectures. We evaluate AWARESL with different DNN models on multiple image datasets. The results show that AWARESL can effectively calculate the resource requirement and minimize the privacy leakage for each partitioning strategy of a DNN model. We also conduct experiments to demonstrate that AWARESL effectively protect the inference data against input reconstruction attacks. The contributions of the paper are as follows:

- We present a resource-aware DNN partitioning framework that adaptively splits the DNN models based on the resource budgets of edge devices.
- We utilize a quantitative privacy metric in training to minimize the privacy leakage of input samples without considerable accuracy reduction.
- We evaluate AWARESL on various DNN models to demonstrate its effectiveness in offloading computations and preserving data privacy against input reconstruction attacks.

The structure of this paper is as follows: Sect. 2 provides related works. In Sect. 3, we discuss the design of our framework and the privacy metrics. We present our evaluation results in Sect. 4 and conclude in Sect. 5.

2   Related Works

Privacy Attacks on Edge-Cloud Systems. Depending on the victim that attackers exploit, privacy attacks on SL can be categorized as *model-oriented attacks* [7], which aim to extract an equivalent model and duplicate the functionality of the ML model, or *data-oriented attacks* [13,22], where the attackers attempt to recover the sensitive data such as raw inputs and membership information. Our work focuses on the *data-oriented attacks*, most specifically, data reconstruction attacks presented in [7,13,22], where an attacker can either query the model or recover the input samples from intermediate outputs. Data reconstruction attacks show that an adversary can reconstruct private input samples under a variety of threat models (e.g., attacker capabilities, query access), given knowledge of the edge model (i.e., white-box or black-box) or knowledge about the training set. This is done by minimizing the error between original input and reconstructed sample using Likelihood Estimation (rMLE) [7] or mean squared error (MSE) [13] with a shadow model.

Privacy Preservation in SL. Research efforts [1,7,8,15,20] have been made to reduce the privacy leakage of inference data in SL through homomorphic encryption and measurable privacy metrics. For instance, homomorphic encryption [8,15] allows the ML model to perform inference directly on the encrypted data without intermediate decryption or prior knowledge of the private key, which prevents the reconstruction of sensitive information. However, these cryptography-based methods introduce extra communication costs and computation overhead, which is not applicable to realistic deployment scenarios. Meanwhile, most measurable privacy metrics [1,7,10,20] only focus on input data privacy during the model inference process. However, we argue that the way of model partitioning itself and the privacy measurement during the training process also play vital roles in improving the robustness and effectiveness

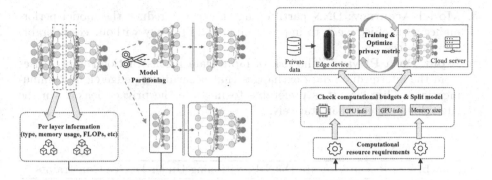

Fig. 2. The overview of AWARESL. AWARESL analyzes the memory usage and computation load for each layer as well as for two partial models, then quantifying data privacy measurements during the training process. Thus, we identify split points that achieve a balance between data privacy, resource utilization, and model accuracy.

against input reconstruction attacks. Therefore, in our paper, we discuss how to split a DNN based on the computational resources on the edge devices with a focus on privacy preservation, which is insufficiently investigated by prior works.

3 Design of AWARESL

3.1 Resource-Aware Model Partitioning

As shown in Fig. 2, we design AWARESL to examine the potential split points based on resource constraints and also minimize the privacy leakage of input samples. For an edge-cloud system, splitting a DNN topology at different layers results in various computation offloading, communication latency, and resource usage. In addition, data privacy in edge-cloud communication demands attention in model training and inference since data exchanges between DNN sub-networks might expose sensitive information of the input data. Therefore, we regard this resource-aware DNN partitioning as a search problem to identify the optimal strategy that achieves high data privacy, high model accuracy and low computational resource usage. This optimization focuses on the following four factors:

1. **Computation Complexity**: DNN partitioning should offload the computation based on the model complexity and the edge device's computational budgets. The latter partially depends on its power consumption, which should also be considered in practice.
2. **Memory Usage**: The edge device hosts the first portion of the DNN in the memory and completes the partial model executions during both training and inference processes. Meanwhile, memory usage and requirements depend on the DNN architecture and concrete layer types.

3. **Model Accuracy**: DNN partitioning should not reduce the model performance. Our objective is to improve input data privacy without considerable accuracy loss.
4. **Input Data Privacy**: The risk of input data reconstruction mainly arises from the intermediate data exchanges that occur during edge-cloud communications. Therefore, it is preferable to measure the privacy leakage in the edge-cloud system quantitatively.

Algorithm 1. Resource-aware DNN Partitioning Under Device Constraints

Input: DNN model $\mathcal{F}$ with N layers; layer information $\{L_i | i = 1 \cdots N\}$;
Hardware specification of target platform: memory $M_{platform}$ and computation power $C_{platform}$
Output: Split Point SP

```
 1: function PERLAYERANALYSIS(f)
 2:     for each i ∈ 1, 2, ··· , N do
 3:         LC_i ← GETFLOPS(L_i)                          ▷ Calculate per-layer computation cost
 4:         LM_i ← GETMEMORYSIZE(L_i)                     ▷ Calculate per-layer layer memory usage
 5:         C_edge = Σ_{j=1}^{i} LC_j                      ▷ Calculate total computation cost on edge side
 6:         M_edge = Σ_{j=i}^{N} LM_j                      ▷ Calculate total memory usage on edge side
 7:         if C_edge < C_platform and M_edge < M_platform then          ▷ Check
 8:             ValidPartitions ← APPEND(i)               ▷ Collect all valid split points
 9:     return ValidPartitions
10: function PARTITIONDECISION(f)                         ▷ Main function
11:     SplitPoints ← PERLAYERANALYSIS(f)
12:     Initialize lists PL, ACC                          ▷ Privacy leakage, Inference accuracy
13:     for each k ∈ SplitPoints do                       ▷ Split model for every valid split points
14:         F_E, F_C ← SPLITMODEL(f, k)                   ▷ Deploy split portions respectively
15:         TRAINSPLITMODELS(F_E, F_C)                    ▷ Training on tasks
16:         PL_k ← PRIVACYMEASURE(F_E)                    ▷ Examine the privacy metric after training
17:         ACC_k ← GETACCURACY(F_E, F_C)                 ▷ Measure the inference accuracy
18:     SP ← FINDMINMAX(PL, ACC)                          ▷ Determine the optimal split point
19:     return SP
```

We present our method for determining optimal split points given a set of edge constraints in Algorithm 1. Considering a DNN model $\mathcal{F}$ with N layers $\{L_i | i = 1 \cdots N\}$ that is split into two sub-networks $\mathcal{F}_E$ and $\mathcal{F}_C$, we first apply per-layer analysis to calculate the computational resource consumption for each layer by the floating-point operations (FLOPs) and memory usage, respectively (Line 3–4). Then, we accumulate the FLOPs and the memory usage for the sub-network $\mathcal{F}_E$ to be deployed on the edge device (Line 5–6). We compare the total required computation cost and memory usage of the edge device's model portion with the resource capability of the target edge platform (Line 7) and identify the split point as valid if it satisfies the resource requirements (Line 8).

After collecting all valid split points through per-layer analysis, we split the DNN $\mathcal{F}$ as $\mathcal{F}_E$ and $\mathcal{F}_C$ and deploy them onto the edge device and the cloud server for each valid split point k (Line 13–14), following by the data privacy measurement and optimization during the training process. Specifically, we integrate the privacy metric into the loss function and minimize the privacy leakage PL_k of input samples during the collaborative training (Line 16), which is discussed in detail in Sect. 3.2. Meanwhile, we also evaluate the model inference accuracy

ACC_k in comparison with conventional model training to ensure no significant accuracy drop (Line 17). Therefore, we leverage a Minmax strategy to determine the optimal split point with the highest inference accuracy and minimum input data privacy leakage (Line 18). In other words, we optimize privacy protection against reconstruction attacks on raw input samples.

3.2 Privacy Preservation via Distance Correlation

As shown in Fig. 3, a complete DNN model is split into two portions at the split point and distributed to the edge device and cloud server respectively. This scenario creates a privacy vulnerability in the form of exposed intermediate layer data. Previous works [7,22] study dynamic partitioning by calculating the privacy only during the inference process. We extend those works by investigating the impacts of incorporating privacy awareness into the training process.

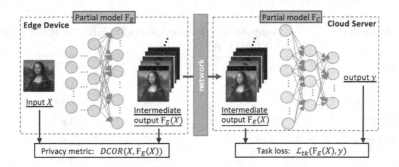

Fig. 3. Privacy metric and task loss function calculation in AWARESL.

Specifically, we apply distance correlation (DCOR) techniques introduced in [18,20] to quantitatively measure the ease of reconstructing input samples from the intermediate activation outputs at the split layer. Distance correlation represents the normalized version of distance covariance, both of which measure dependence between two vectors of arbitrary (but not necessarily equal) dimension. Unlike Pearson's correlation, distance correlation can capture both linear and non-linear dependencies. We compute DCOR between raw inputs X and intermediate activation outputs $\mathcal{F}_E(X)$ at the split layer for each batch as:

$$DCOV(X, \mathcal{F}_E(X)) = n^2 \, \text{Tr}(Cov(X) \cdot Cov(\mathcal{F}_E(X))) \tag{1}$$

$$DCOR(X, \mathcal{F}_E(X)) = \frac{DCOV^2(X, \mathcal{F}_E(X))}{\sqrt{DCOV^2(X) \, DCOV^2(\mathcal{F}_E(X))}} \tag{2}$$

During our learning protocol, the DNN model parameters are updated not only by the prediction loss values but also by the data privacy measurements, which can be mathematically represented by following loss function:

$$\mathcal{L}_{total} = \alpha_1 \cdot DCOR(X, \mathcal{F}_E(X)) + \alpha_2 \cdot \mathcal{L}_{tk}(\mathcal{F}_E(X), y) \tag{3}$$

where $DCOR$ is the distance correlation metric, $\mathcal{L}_{tk}$ is the task loss of the distributed model (e.g., cross-entropy for a classification task), and y is a suitable label for the target task (if any). In the equation, the hyper-parameters α_1 and α_2 define the relevance of distance correlation in the final loss function, creating and managing a tradeoff between data privacy (i.e., how much information an attacker can recover from the smashed data) and model's utility on the target task (e.g., inference accuracy). For attackers, this optimization increases the difficulty of reconstructing the original input samples from intermediate activation data. Note that the distance correlation term depends on just the edge's network $\mathcal{F}_E$ and the private data X. Thus, it can be computed and applied locally on the edge side without any influence from the cloud.

Table 1. Edge device examples with constrained computational resources and memory size including the computation power as GFLOPS per watt, On-chip RAM (OCRAM) and synchronous DRAM (SDRAM).

Edge Device	Computation Power (GFLOPS/W)	OCRAM (KB)	SDRAM (MB)
Google Coral SoM	24.6	160	1024
Intel Myraid 2	100	2048	1024
Intel NCS 2	667	2560	4096
GAP 8	20[a]	512	15
Raspberry Pi Zero W	0.238	288	512
Nvidia Jetson Nano	47.2	262	4096

[a] GAP 8 only consumes milliwatts of power, so we use maximum computation power.

4   Evaluation

4.1   Experimental Setup

Datasets and Models. We evaluate AWARESL on image classification tasks with multiple different neural networks VGGNet and ResNet. We have intentionally chosen these neural networks because they include both *sequential* and *non-sequential* architectures. The *non-sequential* neural networks such as ResNet, made by stacking the residual blocks, contain skip connections that limit the choices of split points since the model can only be split between residual blocks while *sequential* neural networks such as VGGNet can be split at any single layer. The datasets we evaluate include FashionMNIST, CIFAR-10/100 and ImageNet.

Edge Devices and Hardware Specifications. To evaluate AWARESL with varying computational budgets, we show the commercial edge device examples from different vendors in Table 1. We calculate the computational power of these devices per billion floating-point operations (GFLOPS) executed within a nominal 1-watt power envelope. To define the memory budgets, we consider the on-chip SRAM memory, including L1 and L2 caches. Knowing the edge hardware specifications, AWARESL optimizes the split point based on an edge device's available resources to achieve high accuracy and privacy.

4.2 Resource-Aware Computation Partitioning

To ease the deployment of a split DNN, We first investigate each layer's memory usage and computation characteristics. This information provides us with the exact resource consumption of two DNN portions to identify the valid split points under the resource budgets of the target edge device.

Per-Layer Analysis. Figure 4 shows the network architecture and the per-layer analysis results for each convolutional layer in the VGG16 model. Figure 4a presents the VGG16 network architecture as an example and shows the different split points for this model. It contains 16 weight layers including 13 convolutional layers that perform 3×3 convolutions in a homogeneous architecture and 3 fully-connected layers that contain 4096 channels for each of the first two as the third performs 1000-way ILSVRC classification with the final soft-max layer. We go through each layer and explore its parameters such as different kernel sizes, kernel numbers and activation functions.

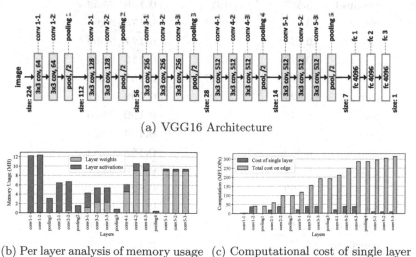

(a) VGG16 Architecture

(b) Per layer analysis of memory usage (c) Computational cost of single layer
and edge device's partial model

Fig. 4. Resource analysis of single layer and corresponding partial model on edge device

Memory Usage. We consider the memory usage of each convolutional layer in the following aspects: (1) intermediate layer outputs including the activations and their gradients. These values depend on the input image size and layer type, and they are retained for the back-propagation process. In Fig. 4b, we show the memory usage results for a single 224×224 image sampled from ImageNet as the input volume. For instance, the size of layer outputs at CONV1-1 is $224 \times 224 \times 64$, which turns out to be around 12.8 MB if we assume floating point numbers are stored in four bytes. Since this is for only one image, the total memory size for each layer's output should be multiplied by the batch size in practice; (2) neuron parameters of weights. We do not consider the biases since there are comparatively very few. The weights of each convolutional layer rely on the filter size and number of features. For instance, the total number of weights for layer CONV3-1 is $(3 \times 3 \times 128) \times 256$. Therefore, we observe that the early convolutional layers have a large number of activation calculations and the later convolutional layers have intensive weight updates. Besides, pooling layers are inserted in VGG16 to reduce the number of parameters (i.e., computation) in

Table 2. The DCOR measured for the ease of input reconstruction (lower is better) before and after the training of VGG16 on CIFAR-10 dataset when partitioning it on different split layers.

Split Layer	DCOR-before	DCOR-after	Δ^a	Accuracy[b] (%)
conv1-1	0.721	0.49	**0.231**	93.29
conv1-2	0.626	0.415	**0.211**	93.33
pooling1	0.567	0.427	**0.140**	93.53
conv2-1	0.554	0.381	**0.173**	93.9
conv2-2	0.528	0.351	**0.177**	93.99
pooling2	0.472	0.353	**0.119**	94.36
conv3-1	0.494	0.337	**0.157**	93.87
conv3-2	0.424	0.325	**0.099**	93.41
conv3-3	0.478	0.288	**0.190**	93.75
pooling3	0.449	0.315	**0.134**	92.08
conv4-1	0.43	0.277	**0.153**	93.84
conv4-2	0.427	0.251	**0.176**	93.57
conv4-3	0.369	0.249	**0.120**	94.22
pooling4	0.38	0.257	**0.123**	93.85
conv5-1	0.358	0.239	**0.119**	93.76
conv5-2	0.375	0.252	**0.123**	93.56
conv5-3	0.366	0.287	**0.079**	93.43

[a]DCOR difference before and after training.
[b]This is the final inference accuracy after model training and privacy optimizing. Note that the baseline accuracy of VGG16 on CIFAR-10 dataset is 92.95%.

the network, and thus they have zero weights (i.e., no parameter updates) and we only calculate their activation sizes as memory usage.

Computation Complexity. Figure 4c shows the computation costs of training VGG16 on CIFAR-10 dataset. We describe the computation complexity using the sum of floating-point operations (FLOPs). We calculate the FLOPs for each split layer and the total FLOPs for the edge device's model portion after splitting, which indicates the requirement of computational resources to be compared with the computation capability of edge devices (e.g., instances in Table 1). The total computation operations depend on image sizes, layer types, and back-propagation processes, and the results of Fig. 4c are calculated for the feed-forward process with the input volume as one single 224×224 image.

4.3   Balancing Model Accuracy and Privacy Preservation

Varying Split Points. Changing the split location on a DNN model affects its resource requirements, and also impacts data privacy and inference accuracy. In Table 2, we present the effectiveness of AWARESL with the training results of VGG16 on CIFAR-10 dataset when partitioning the model at different layers. We calculate the DCOR before and after training, and the final inference accuracy on the testing dataset. We observe that: (1) The different split points change the initial DCOR of a split model, with the initial DCOR decreasing as more layers are split to the edge device. This is intuitive since additional layers between the intermediate outputs and raw input samples result in reduced statistical dependence; (2) Despite each partitioning strategy having a distinct computation profile and resource consumption, inference accuracy is stable when splitting at different layers, and many split points have higher accuracy than the baseline accuracy of conventional non-split settings.

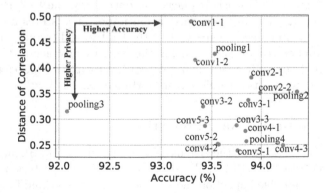

Fig. 5. Scatter plot result between inference accuracy and distance correlation.

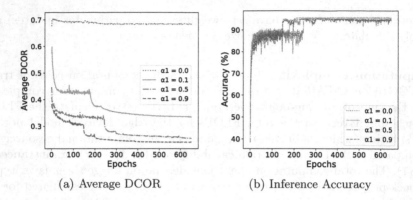

(a) Average DCOR   (b) Inference Accuracy

Fig. 6. Average DCOR among training batches and inference accuracy results when taking different importance scores α_1 at the same split point.

In Fig. 5, we present the training results of VGG16 on CIFAR-10 by different split points, which provide a visual representation that helps users determine the optimal split point by examining the trade-off between privacy protection and inference accuracy. By assessing their specific accuracy and privacy requirements, users can effectively select the optimal split point in practice.

Varying Importance Scores. In Eq. 3, the final loss function is calculated by the task loss and distance correlation with different importance scores. In Fig. 6, we present the training results obtained by splitting ResNet50 at the layer BLOCKS 3–6 and training on CIFAR-10 dataset, while varying the distance correlation's importance score α_1. We observe the following: (1) The varying distance correlation importance scores in the final loss function lead to distinct DCOR measurements. The final DCOR after model training decreases as α_1 increases as a greater emphasis is placed on optimizing the DCOR calculation during training; (2) The distance correlation importance scores do not significantly impact the inference accuracy after the split model training, but they do impact the number of epochs needed for the model convergence. Compared to the conventional non-split case (setting $\alpha_1 = 0$), AWARESL achieves comparable inference accuracy results for different importance scores.

4.4   Inference Data Privacy Protection

In this section, we explore the defense capability of AWARESL against input reconstruction attacks [7,13,22]. We show AWARESL can protect against these attacks with a negligible impact on the system's performance and functionalities.

Defense Against Reconstruction Attacks. Figure 7 shows the reconstruction attack that recovers raw input samples from intermediate outputs at the

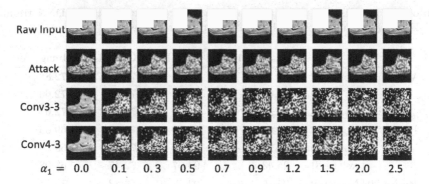

Fig. 7. Input reconstruction attack results and AWARESL defense results on FashionM-NIST dataset with different setups of split points and importance scores. The coefficient of task loss α_2 is set as 1.

split layer and AWARESL's defense results. The first row shows the original inference samples, and the second row shows the recovered images by the reconstruction attack where we observe that the adversary can accurately recover the images with high fidelity. The remaining rows show that different split points and importance scores yield different reconstruction results. Generally, we observe that the quality of recovered images decreases when the split layer becomes deeper. This is straightforward as the relationship between input and output becomes more complex and harder to revert when there are more layers. Besides, we also observe that the image quality drops significantly with the increasing importance score α_1 (i.e., coefficient of the DCOR term) in the training loss function, indicating that involving and optimizing DCOR during the training process is effective in defending input reconstruction attacks. The reason is that DNN model parameters are tuned not only based on the prediction loss values but also on the privacy metric of input samples during the training process.

Robustness Across DNN Architectures. We compare the Top-1 accuracy of AWARESL with baseline setup as conventional split learning [5,21] without privacy protection on different testing sets, as shown in Table 3. We observe that when the adversary knows either the training data or its distribution, the recovered images from Inverse-Network [7] maintain higher quality. However, our work significantly reduces the image quality reconstructed by the adversary. Meanwhile, AWARESL does not reduce the model accuracy in the inference stage compared to conventional split learning. Our experiments reveal the robustness of our privacy-preserving model partitioning approach across various DNN models and image datasets, demonstrating the feasibility of AWARESL in privacy protection and highlighting the need for further research into privacy analysis in an edge-cloud collaborative system.

Table 3. Comparison of the classification accuracy for conventional DNN training, split learning baseline [5, 21] and AWARESL on different datasets.

Model		FashionMNIST	CIFAR-10	CIFAR-100	ImageNet
VGG16	Non-split	94.71	92.95	74.20	72.88
	SL baseline	95.03	92.64	74.25	72.60
	AWARESL	94.32	94.03	74.25	72.64
ResNet18	Non-split	94.80	93.42	72.49	69.06
	SL baseline	95.29	93.19	72.69	69.30
	AWARESL	95.43	94.07	72.47	68.98
ResNet50	Non-split	95.10	94.31	73.83	75.35
	SL baseline	95.34	94.67	73.66	75.80
	AWARESL	95.50	94.55	73.81	75.44

5   Conclusion

We propose AWARESL, a DNN partitioning framework designed to address resource constraints and privacy concerns in edge-cloud systems. Compared with prior works, AWARESL optimizes privacy metrics during the training process. Our experiments show that AWARESL effectively preserves the privacy of input samples without compromising inference accuracy in practical deployments.

References

1. Abuadbba, S., et al.: Can we use split learning on 1D CNN models for privacy preserving training? In: Proceedings of the 15th ACM Asia Conference on Computer and Communications Security, pp. 305–318 (2020)
2. Banitalebi-Dehkordi, A., Vedula, N., Pei, J., Xia, F., Wang, L., Zhang, Y.: Autosplit: a general framework of collaborative edge-cloud AI. In: Proceedings of the 27th ACM SIGKDD Conference on Knowledge Discovery & Data Mining (2021)
3. Ding, A.: Trustworthy cyber-physical systems via physics-aware and AI-powered security. Ph.D. thesis, Rutgers The State University of New Jersey, School of Graduate Studies (2022)
4. Ding, A., Murthy, P., Garcia, L., Sun, P., Chan, M., Zonouz, S.: Mini-Me, you complete me! data-driven drone security via DNN-based approximate computing. In: 24th International Symposium on Research in Attacks, Intrusions and Defenses, pp. 428–441 (2021)
5. Gupta, O., Raskar, R.: Distributed learning of deep neural network over multiple agents. J. Netw. Comput. Appl. **116**, 1–8 (2018)
6. Hassanzadeh, A., Liberman, N.H., Ding, A., Salem, M.B.: Privacy-preserving collaborative machine learning training using distributed executable file packages in an untrusted environment, 29 December 2022. US Patent App. 17/356,447
7. He, Z., Zhang, T., Lee, R.B.: Model inversion attacks against collaborative inference. In: Proceedings of the 35th Annual Computer Security Applications Conference, pp. 148–162 (2019)

8. Juvekar, C., Vaikuntanathan, V., Chandrakasan, A.: {GAZELLE}: a low latency framework for secure neural network inference. In: 27th USENIX Security Symposium (USENIX Security 2018), pp. 1651–1669 (2018)
9. Kang, Y., et al.: Neurosurgeon: collaborative intelligence between the cloud and mobile edge. ACM SIGARCH Comput. Archit. News **45**(1), 615–629 (2017)
10. Khalili, H., Chien, H.J., Hass, A., Sehatbakhsh, N.: Context-aware hybrid encoding for privacy-preserving computation in IoT devices. IEEE Internet Things J. (2023)
11. Mehta, R., Shorey, R.: DeepSplit: dynamic splitting of collaborative edge-cloud convolutional neural networks. In: 2020 International Conference on COMmunication Systems & NETworkS (COMSNETS), pp. 720–725. IEEE (2020)
12. Mohammed, T., Joe-Wong, C., Babbar, R., Di Francesco, M.: Distributed inference acceleration with adaptive DNN partitioning and offloading. In: IEEE INFOCOM 2020-IEEE Conference on Computer Communications, pp. 854–863. IEEE (2020)
13. Pasquini, D., Ateniese, G., Bernaschi, M.: Unleashing the tiger: inference attacks on split learning. In: Proceedings of the 2021 ACM SIGSAC Conference on Computer and Communications Security, pp. 2113–2129 (2021)
14. Pham, N.D., Abuadbba, A., Gao, Y., Phan, T.K., Chilamkurti, N.: Binarizing split learning for data privacy enhancement and computation reduction. IEEE Trans. Inf. Forensics Secur. **18**, 3088–3100 (2023)
15. Rathee, D., et al.: CrypTFlow2: practical 2-party secure inference. In: Proceedings of the 2020 ACM SIGSAC Conference on Computer and Communications Security (2020)
16. Shao, J., Zhang, J.: Communication-computation trade-off in resource-constrained edge inference. IEEE Commun. Mag. **58**(12), 20–26 (2020)
17. Shi, C., Chen, L., Shen, C., Song, L., Xu, J.: Privacy-aware edge computing based on adaptive DNN partitioning. In: 2019 IEEE Global Communications Conference (GLOBECOM), pp. 1–6. IEEE (2019)
18. Székely, G.J., Rizzo, M.L., Bakirov, N.K.: Measuring and testing dependence by correlation of distances. Ann. Stat. **35**(6), 2769–2794 (2007)
19. Tang, M., et al.: FADE: enabling large-scale federated adversarial training on resource-constrained edge devices. arXiv preprint arXiv:2209.03839 (2022)
20. Vepakomma, P., Gupta, O., Dubey, A., Raskar, R.: Reducing leakage in distributed deep learning for sensitive health data. arXiv preprint arXiv:1812.00564 (2019)
21. Vepakomma, P., Gupta, O., Swedish, T., Raskar, R.: Split learning for health: distributed deep learning without sharing raw patient data. arXiv preprint arXiv:1812.00564 (2018)
22. Yang, Z., Zhang, J., Chang, E.C., Liang, Z.: Neural network inversion in adversarial setting via background knowledge alignment. In: Proceedings of the 2019 ACM SIGSAC Conference on Computer and Communications Security (2019)

A Frequency Reconfigurable Multi-mode Printed Antenna

Yanbo Wen[1], Huiwei Wang[1,2,3]($\boxtimes$), Menggang Chen[1], Yawei Shi[1], Huaqing Li[1], and Chuandong Li[1]

[1] College of Electronic and Information Engineering, Southwest University, Chongqing 400715, China
hwwang@swu.edu.cn
[2] Key Laboratory of Intelligent Information Processing, Chongqing Three Gorges University, Chongqing 404100, China
[3] Chongqing Innovation Center, Beijing Institute of Technology, Chongqing 401120, China

Abstract. A multi-frequency reconfigurable antenna is proposed. The designed antenna can be electronically tuned to achieve the tuning operation in the 2.4 GHz band defined by the IEEE 802.11b standard and the ultra-wideband (UWB) low frequency. An L-shaped branch and a polygon patch are used as the main radiators of the antenna. Two varactor diodes are mounted on the slots and one PIN diode is mounted on the L-shaped branch to vary the effective electrical length of the antenna. The simulation and measurement results match well. With high frequency reconfiguration stability under guaranteed miniaturization, good impedance matching (S11 > −10 dB) is obtained in several operating bands, and the overall impedance bandwidth covers 2.34–2.58 GHz and 3.11–5.14 GHz. It provides solutions for operation within WiMax, WLAN, and 5G-sub6 GHz.

Keywords: miniaturization · frequency reconfigurable · polygonal patch · pin diodes

1 Introduction

The rapid development of 5G communication technology in recent years has led to a shortage of spectrum resources, while modern wireless communication systems usually have multiple standards, and antenna modules, as an important component, undertake the task of receiving and transmitting electromagnetic waves, thus requiring antennas with the ability to be compatible with multiple communication standards. In contrast to broadband antennas [1], multi-band antennas can provide multiple frequencies while improving the isolation between

Supported in part by the China Postdoctoral Science Foundation 2022M720453, and in part by the Science and Technology Research Program of Chongqing Municipal Education Commission under Grant KJZD-M202201204.

B. Luo et al. (Eds.): ICONIP 2023, LNCS 14451, pp. 202–210, 2024.
https://doi.org/10.1007/978-981-99-8073-4_16

bands and reducing interference. Antennas with reconfigurable capabilities show good promise and research potential [2,3]. Frequency reconfigurable antennas can dynamically adjust the frequency allocation and flexibly control the electromagnetic waves in different application scenarios, which effectively solves the problem of spectrum resource constraint.

In [4], the actual electrical length of the antenna radiator was changed by using placing a PIN diode switch on the radiator of the antenna, thus enabling the operation of reconfigurable frequencies. However, this approach can only perform discrete reconfiguration of the frequency and cannot perform continuous tuning. In [5], PIN diodes loaded on two semicircular slits were implemented to switch between three modes: low-band, high-band, and dual-band. However, the number of PIN diodes loaded by the antenna reaches eight respectively, which causes the design of the bias circuit to become complicated and also increases the interference to the radiation.

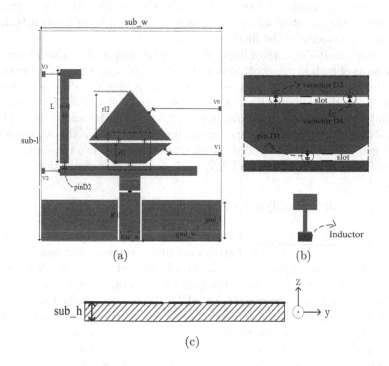

Fig. 1. Geometry of the proposed antenna. (a) (b) Top layer. (c) Sectional view.

In [6], the operation of continuous tuning of the antenna frequency is achieved by loading the varactor diodes on the antenna radiation unit. In [7], a frequency reconfigurable 1×4 patch array antenna is proposed to achieve continuous antenna tuning by loading a parasitic patch and adjusting the varactor diode DC voltage. However, the tunable range of these antennas is usually narrow in [6,7].

Table 1. THE STATE OF THE DIODE AND THE CORRESPONDING.

| State | Bias voltages (V) | | | | GroupI | GroupII | GroupIII |
	$V0$	$V1$	$V2$	$V3$	$D1$	$D2$	C
I	0	0	0	1.4	ON	OFF	
II	1.4	1.4	0	0	OFF	ON	3.2 pF
	21.4	1.4	0	0	OFF	ON	0.35 pF
III	1.4	1.4	0	1.4	ON	ON	3.2 pF
	21.4	1.4	0	1.4	ON	ON	0.35 pF

In [8], a multi-frequency reconfigurable monopole antenna was proposed using fluid channels to realize a multi-frequency reconfigurable monopole antenna, and changing the number of fluid-filled channels could accomplish switching between different operating modes in four states and having stable radiation characteristics, however, the structure of the antenna is more complicated and it is difficult to control the velocity of the fluid.

The novel multi-band reconfigurable antenna proposed in this paper solves the problems of inability to continuously tune, complex bias circuit, narrow operating band width and complex structure. By controlling the PIN diode and varactor diode on the radiator, the antenna can operate at (2.38–2.52 GHz)/(3.15–4.91 GHz) for single-band operation and at (2.34–2.58 GHz) and (3.11–5.14 GHz) for dual-band operation. The overall reconfiguration effect of this antenna is stable and can be effectively applied to wireless communication systems, such as Wireless Local Area Network (WLAN) and satellite communication.

2   Design and Analysis of Antenna

Figure 1 depicts the structure of the multi-frequency reconfigurable antenna proposed in this paper. The main radiating part of the antenna contains an L-shaped branch and a polygonal patch. The antenna is printed on an FR4 dielectric substrate of size $30 \times 26 \, mm^2$ with a thickness of 1.6 mm, a dielectric constant of 4.4 and a loss angle tangent of 0.02. Two PIN diodes are installed as switches on the L-shaped branch and polygon patches. Two varactor diodes are connected across a slot inside the polygon patch, with the positive terminal connected to the upper half of the patch. The state of PIN diode and varactor diode is controlled by bias circuit, which changes the effective electrical length of the antenna and makes the resonant frequency point of the antenna different and the impedance bandwidth varied. The antenna is fed by CPW, which can integrate the radiator and ground plate into one plane to facilitate processing and manufacturing.

Figure 2 depicts the whole design process of the multi-frequency reconfiguration antenna, Antenna I is an Lshaped branch antenna, which can be regarded as a monopole antenna. Antenna II is a hexagonal microstrip patch antenna. The impedance of the hexagonal patch changes slowly as the current passes over the edge, resulting in it resonating more easily compared to others and has

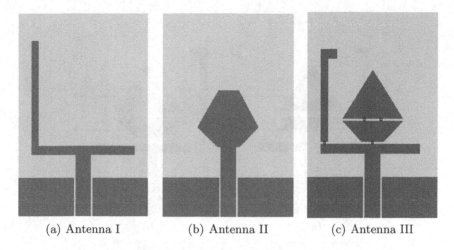

(a) Antenna I   (b) Antenna II   (c) Antenna III

Fig. 2. Frequency reconfiguration antenna design process.

advantages in return loss and operating bandwidth. Antenna III is obtained by combining the deformation on the basis of Antenna I and Antenna II. Folded operation is performed at the end of the L-shaped branch to achieve a reduction in the overall size of the antenna. A PIN diode and a varactor diode are installed on the slot. In order to extend the effective length of the current, which shifts the operating bandwidth to lower frequencies, and to improve impedance matching, a triangular patch is added to the top of the patch. The PIN diode connects the L-shaped branch to the polygon patch. The discrete reconfiguration frequency operation of the antenna is achieved by adjusting the voltages $V1$, $V2$ and $V3$ of the PIN diode to switch the PIN diode between different states. The voltages $V0$ and $V1$ across the varactor diode are controlled by the DC bias circuit. As the voltage difference changes, the capacitance of the varactor diode changes, further affecting the impedance bandwidth, which can realize the continuous tuning of the antenna frequency.

The bias voltages $V1$, $V2$ and $V3$ control the switching state of $D1$ and $D2$, and the voltage difference between $V0$ and $V1$ changes the capacitance of $D3$ and $D4$, which makes the operating mode of the antenna switch. When $D1$ is on and $D2$ is in cutoff, the current cannot pass through $D2$, the surface current of the antenna is concentrated on the L-shaped branch, which is involved in radiation by the L-shaped branch, and the antenna produces resonant point in 2.45 GHz. When $D1$ is cut off and $D2$ is on, the surface current flows along the edge of the patch through $D2$, which makes it participate in the radiation, the current is concentrated on the horizontal branch and the patch, and the antenna produces the resonant point in 4.65 GHz. When $D1$ and $D2$ are on, the current flows through the diode, making both the L-shaped branch and the patch involved in radiation, at which time the antenna produces a resonance point at 2.45 GHz and 4.56 GHz each. Figure 3 depicts the antenna surface current distribution

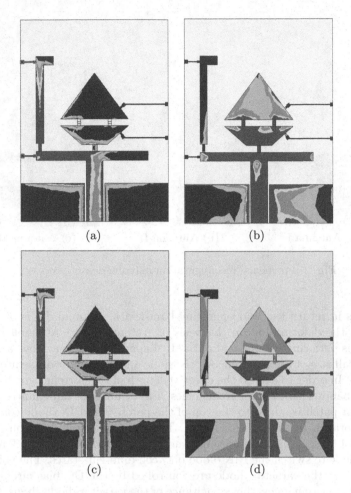

(a)   (b)

(c)   (d)

Fig. 3. Antenna surface current distribution. (a) State 1 at 2.45 GHz. (b) State 2 at 4.65 GHz. (c) State 3 at 2.45 GHz. (d) State 3 at 4.56 GHz.

in different states. The PIN diodes and varactor diodes are divided into three groups, $D1$ as the first group, $D2$ as the second group, and $D3$ and $D4$ as the third group. The states of the diodes corresponding to each bias voltage are summarized in Table 1.

Through simulation, the S11 results in different states are shown in Fig. 4. When the antenna is in state I, the antenna current is concentrated on the L-shaped branches and will not be affected by the varactor diode. When the antenna is in State II and State III, the capacitance C of the varactor diode varies from 3.2–0.35 pF by Changing the voltage difference between $V0$ and $V1$ at 0–20 V, which makes the operating bandwidth shift, thus Tuning the frequency continuously. When State II and $C = 3.2$ pF, the operating bandwidth of the antenna covers 3.14–4.72 GHz, when $C = 0.35$pF, the operating bandwidth covers

Table 2. PARAMETERS OF THE PROPOSED ANTENNA.

Parameter	sub_l	gnd_l	l	sub_w	dw	g
Value (mm)	30	5.6	13	26	1.3	0.26
Parameter	sub_h	slot	gnd_w	line_w	rl1	rl2
Value (mm)	1.6	1	11.24	3	2.8	6.9

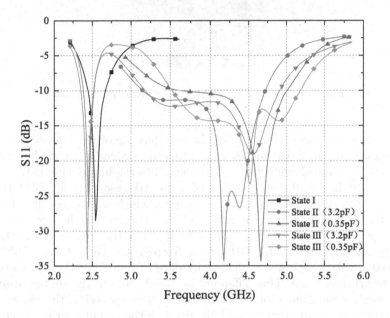

Fig. 4. Simulation $S11$ in different states.

3.61–5.01 GHz. When State III, the voltage difference of varactor diode varies from 0–20 V, the overall operating bandwidth of antenna covers 2.34–2.58 GHz and 3.12–5.14 GHz.

3 Results and Discussion

The multi-frequency reconfigured antenna proposed in this paper was fabricated and measured, and the fabricated antenna prototype is shown in Fig. 5. The impedance matching of the antenna varies considering different states, and the optimized parameters of the proposed antenna are concentrated in Table 2. The chosen PIN diode is BAR64-02V from Infineon, which according to the datasheet can be simply equivalent to a series resistor of 2.1 Ω in the on state and a parallel circuit of 0.17 pF capacitor and 3 $K\Omega$ resistor in the off state.

The varactor diode selected for frequency tuning is SMV2020-079LF from Skyworks, whose capacitance value changes from 3.2–0.35 pF as the reverse bias voltage varies from 0–20 V, accompanied by a 0.7 nH series inductor and a 2.5 Ω

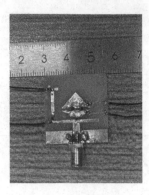

Fig. 5. Fabricated prototype of the proposed antenna.

series resistor. In Order to minimize the interference of the bias circuit to the radiation performance, the bias lines are placed perpendicular to the radiation part, while four 100 nH inductors are connected to the bias circuit as RF chokes.

As shown in Fig. 6, the antenna of state III resonates at 2.45 GHz and 4.55 GHz (C = 3.2 pF)/4.56 GHz (C = 0.35 pF), which are the resonant frequencies at the designed L-shaped branch and the designed patch. The antenna in state I, as a monopole antenna, resonates at 2.45 GHz. When it is in state II, the antenna resonates at 4.15 GHz (C = 3.2 pF)/4.65 GHz (C = 0.35 pF). The measured results match well with the simulated results, but there is a decrease in the return loss value. This difference is mainly due to the simple equivalent circuit used in the simulation to model the different states of the diode, which would have some error. This is also influenced by the manufacturing process and the device itself, which has some loss. According to the measurement results, the 10 dB bandwidth covers 2.38–2.52 GHz in state I, 3.15–4.91 GHz in state II, 2.34–2.58 GHz and 3.11–5.14 GHz in state III. Comparing the three different states, the range covered by the impedance bandwidth basically overlaps, which indicates that the reconfiguration performance of this frequency reconfigurable antenna is relatively stable.

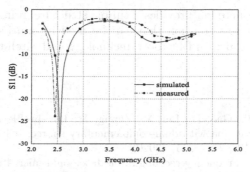

(a) Simulated and measured $S11$ in state I.

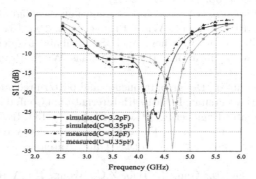

(b) Simulated and measured $S11$ in state II.

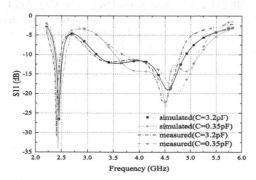

(c) Simulated and measured $S11$ in state III.

Fig. 6. Simulated and measured $S11$.

4 Conclusion

A planar printed antenna with frequency reconfigurable function was designed and fabricated. The proposed antenna combines L-shaped branch and patch, which covers 2.34–2.58 GHz and 3.11–5.14 GHz. By controlling PIN diodes and varactor diodes, both discrete and continuous frequency reconfiguration is possible. The proposed antenna provides a simple structure, miniaturization, large bandwidth and low cost solution that can be used in multi-scenario standard wireless applications such as Bluetooth technology and satellite broadcasting.

References

1. Liu, W., Wang, T., Gao, D., Liu, Y., Zhang, X.: Low-profile broadband magnetoelectric dipole antenna with dual-complementary source. IEEE Antennas Wirel. Propag. Lett. **19**(12), 2447–2451 (2020)
2. Yang, D.D., et al.: Frequency reconfigurable hexagonal microstrip antenna for 5G applications. In: 2021 International Conference on Microwave and Millimeter Wave Technology (ICMMT), pp. 1–3. IEEE (2021)
3. Mirzaei, H., Eleftheriades, G.V.: A compact frequency-reconfigurable metamaterial-inspired antenna. IEEE Antennas Wirel. Propag. Lett. **10**, 1154–1157 (2011)
4. Sun, M., Zhang, Z., Zhang, F., Chen, A.: L/S multiband frequency-reconfigurable antenna for satellite applications. IEEE Antennas Wirel. Propag. Lett. **18**(12), 2617–2621 (2019)
5. Gao, C., Lu, Z.L., Liu, M.: Design of frequency and pattern reconfigurable wideband semi-circular slot ring antenna. In: 2019 4th International Conference on Mechanical, Control and Computer Engineering (ICMCCE), pp. 13–134. IEEE (2019)
6. Nguyen-Trong, N., Piotrowski, A., Fumeaux, C.: A frequency-reconfigurable dual-band low-profile monopolar antenna. IEEE Trans. Antennas Propag. **65**(7), 3336–3343 (2017)
7. Hu, J., Yang, X., Ge, L., Guo, Z., Hao, Z.C., Wong, H.: A reconfigurable 1 × 4 circularly polarized patch array antenna with frequency, radiation pattern, and polarization agility. IEEE Trans. Antennas Propag. **69**(8), 5124–5129 (2021)
8. Singh, A., Goode, I., Saavedra, C.E.: A multistate frequency reconfigurable monopole antenna using fluidic channels. IEEE Antennas Wirel. Propag. Lett. **18**(5), 856–860 (2019)

Multi-view Contrastive Learning
for Knowledge-Aware Recommendation

Ruiguo Yu[1,2,3], Zixuan Li[2,3,4], Mankun Zhao[1,2,3], Wenbin Zhang[1,3,5], Ming Yang[6], and Jian Yu[1,2,3(✉)]

[1] College of Intelligence and Computing, Tianjin University, Tianjin, China
{rgyu,zmk,zhangwenbin,yujian}@tju.edu.cn
[2] Tianjin Key Laboratory of Cognitive Computing and Application, Tianjin, China
lzx_@tju.edu.cn
[3] Tianjin Key Laboratory of Advanced Networking, Tianjin, China
[4] Tianjin International Engineering Institute, Tianjin University, Tianjin, China
[5] Tianjin University Information and Network Center, Tianjin, China
[6] College of Computing and Software Engineering, Kennesaw State University, Kennesaw, USA
ming.yang@kennesaw.edu

Abstract. Knowledge-aware recommendation has attracted increasing attention due to its wide application in alleviating data-sparse and cold-start, but the real-world knowledge graph (KG) contains many noises from irrelevant entities. Recently, contrastive learning, a self-supervised learning (SSL) method, has shown excellent anti-noise performance in recommendation task. However, the inconsistency between the use of noisy embeddings in SSL tasks and the original embeddings in recommendation tasks limits the model's ability.

We propose a **M**ulti-view **C**ontrastive learning for **K**nowledge-aware **R**ecommendation framework (MCKR) to solve the above problems. To remove inconsistencies, MCKR unifies the input of SSL and recommendation tasks and learns more representations from the contrastive learning method. To alleviate the noises from irrelevant entities, MCKR preprocesses the KG triples according to the type and randomly perturbs of graph structure with different weights. Then, a novel distance-based graph convolutional network is proposed to learn more reliable entity information in KG. Extensive experiments on three popular benchmark datasets present that our approach achieves state-of-the-art. Further analysis shows that MCKR also performs well in reducing data noise.

Keywords: Recommender System · Knowledge Graph · Contrastive Learning · Self-Supervised Learning · Graph Convolutional Network

1 Introduction

Knowledge-aware recommendation aims to learn latent relations among items by exploiting the entity information of knowledge graph (KG) and the rich connections between entities. In recent years, many studies [10,12] have shown the influential role of KG in alleviating the data-sparse and cold-start of the recommendation system (RS). However, KG in the real-world is sparse and contains many irrelevant entities [7,14]. This problem

© The Author(s), under exclusive license to Springer Nature Singapore Pte Ltd. 2024
B. Luo et al. (Eds.): ICONIP 2023, LNCS 14451, pp. 211–223, 2024.
https://doi.org/10.1007/978-981-99-8073-4_17

brings noise that can affect the performance of RS. Specifically, when graph neural network (GNN)-based knowledge-aware recommendation aggregates neighbor representations, irrelevant entities can introduce noise to item representations.

Recently, contrastive learning, a self-supervised learning (SSL) method, has shown excellent anti-noise performance in recommendation task [15,19]. SSL tasks are employed as auxiliary tasks to enhance recommendation effectiveness. The objective is to differentiate the representations between nodes, typically achieved through two steps: (1) data augmentation, generating representations for the same node by adding noise at the data or structural level, and (2) contrastive learning, maximizing the consistency of representations for the same node. The above method forces the model to learn the more beneficial structure and reduces the dependence on a certain edge. Therefore, the SSL model adds noise first, followed by using contrastive learning to counteract the noise and can learn reliable information on the graph. However, while auxiliary task research focuses on the embeddings after adding noise, the recommendation task employs the original embeddings. This inconsistency prevents the model from fully leveraging the improvements in noise resistance offered by auxiliary tasks and limits its ability to effectively capture the information from data, ultimately impacting the quality of recommendations.[1]

Based on the aforementioned problems, we propose a general Multi-view Contrastive learning for Knowledge-aware Recommendation (MCKR) framework. Specifically, to address the inconsistency between the recommendation task and the SSL task, MCKR utilizes the more reliable user and item embeddings learned through contrastive learning methods as the embeddings for the recommendation task. To alleviate the noise from irrelevant entities on KG, we design a distance-based augmentation method to preserve more reliable structures during the data augmentation phase. Additionally, we design a Degree-based graph convolutional network (GCN) module to obtain more reliable item information. Finally, we leverage the anti-noise performance of contrastive learning to learn more robust embedding representations.

We summarize the contributions of this work as follows:

1. We propose a novel framework MCKR, which eliminates the inconsistency between the SSL task and the recommendation task, thereby enhancing the effectiveness of knowledge-aware recommendations.
2. We design a method of data augmentation and neighborhood information aggregation for the structure of KG.
3. We conduct extensive experiments on three real-world datasets from different domains. The results demonstrate the superiority of MCKR.

2   Related Work

2.1   Knowledge-Aware Recommendation

Embedding-Based Methods. Embedding-based methods [11,13,21] use embeddings of KG [5,6] to preprocess the entities. For example, CKE [21] and KGAT [13] use

---

[1] See an empirical study in Sect. 5.3.

TransR [6] to get the entity features associated with the user's historical interaction data. DKN [11] uses TransD [5] to get entity representations. The main problem with this embedding-based method is that the main optimization goal is the link prediction task rather than the recommendation task.

GNN-Based Methods. GNN-based methods [12–14] focus on modeling remote connectivity by aggregating the node representations of multi-hop neighbors through graph neural networks (GNNs) to obtain node information and graph structure information within several hops. KGAT [13] combines the user-item graph with KG as a heterogeneous graph, and GCN is then applied for aggregation. KGCN [12] uses GCN to obtain the items embedding by iteratively aggregating the neighborhood information of items. KGIN [14] models user-item interactions at the intent level and performs GNN aggregation on user-item-entity in combination with KG triples.

Path-Based Methods. Path-based methods [4,9] typically construct meta-paths of information propagation to explore the potential connections of items on the KG. RKGE [9] uses a recurrent neural network for path modeling, and MCRec [4] uses a convolutional neural network for path modeling. High-quality meta-paths can capture rich KG information, but building these paths takes much time. It's difficult to reasonably use the graph information on large-scale KG through a few meta-paths.

2.2 Contrastive Learning for Recommender Systems

Contrastive learning [2,17,18] is designed for learning representations by distinguishing between pairs of positive and negative samples. Recently, integrating the contrastive learning into RS [15,19,20,22] demonstrated the potential to enhance RS performance. SGL [15] performs random noise perturbation for data augmentation at the data level while SimGCL [20] at the representation level, respectively. MCCLK [22] considers user and item representation learning from a global, local, and semantic view. Then uses contrastive learning to explore feature and structural information. KGCL [19] uses contrastive learning to suppress noise in KG in information aggregation.

3 Preliminaries

We denote the user set and item set by $\mathcal{U} = \{u_1, u_2, ..., u_n\}$ and $\mathcal{I} = \{i_1, i_2, ..., i_m\}$, respectively. $\mathcal{O}^+ = \{y_{ui} | u \in \mathcal{U}, i \in \mathcal{I}\}$ indicates the observed user interaction with the item, where y_{ui} indicates the interaction between user u and item i. Based on the $\mathcal{O}^+$, we construct the user-item graph $\mathcal{G}_u = \{\mathcal{V}, \mathcal{E}\}$, where the node set $\mathcal{V} = \mathcal{U} \cup \mathcal{I}$ and edge $\mathcal{E} = \mathcal{O}^+$. $\mathcal{G}_k = \{(h, r, t) | h, t \in E, r \in R\}$ represents the KG, where E, R are the set of entities and relations respectively. Each KG triplet (h, r, t) characterizes the semantic correlation between the head and tail entity h and t with the relation r. Same as other knowledge-aware recommendations, we establish a set of item-entity alignments $A = \{(i, e) | i \in \mathcal{I}, e \in E\}$, where (i, e) indicates that the item i can be aligned with an entity e in the KG. To be specific, the inputs are the user-item graph $\mathcal{G}_u$ and the KG $\mathcal{G}_k$, the output is the learned function $\mathcal{F} = (u, i | \mathcal{G}_u, \mathcal{G}_k, \theta)$ to predict the item $i(i \in \mathcal{I})$ that user $u(u \in \mathcal{U})$ would like to interact with, where θ denotes the model parameters.

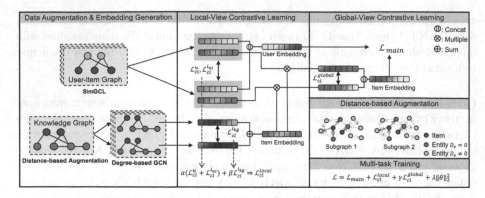

Fig. 1. The overall architecture of MCKR. $\mathcal{L}_{cl}^{local}$, $\mathcal{L}_{cl}^{global}$ SSL task share the embedding with the recommendation task and are jointly optimized by multi-task training. KG Subgraphs 1 and 2 are obtained by using two times of the distance-based Augmentation method. Better in color.

4   Methodology

We design MCKR as shown in Fig. 1. The framework consists of three parts: (1) Multi views generation, (2) Local-view contrastive learning, (3) Global-view contrastive learning. The MCKR effectively utilizes KG and user-item graph information to enhance recommendation accuracy. The recommendation task of MCKR utilizes data-augmented KG as inputs and adds noise to the embeddings learned on the user-item graph. This is similar to the impact of dropout on models, where the appropriate addition of noise can provide better generalization capabilities while preserving most of the semantics. Furthermore, we only add noise to the framework in training. During testing, the recommendation task of MCKR uses the original graph and embeddings without noise.

4.1   Multi Views Generation

To preserve the key structural and semantic information of different graph structures, we employ different data augmentation methods for the user-item graph and the KG.

User-Item View Generation. The user-item graph $\mathcal{G}_u$ is a bipartite graph, with edges indicating the user's interest in the item. If the graph structure is augmented with data using random perturbation, the user's embedding no longer represent the true interest. To avoid the above problems, we use SimGCL [20] to learn the user-item view representations $(z_u^1, z_{i_{ui}}^1)$ and $(z_u^2, z_{i_{ui}}^2)$ from the user-item graph. z_u represents the embedding of the user, and $z_{i_{ui}}$ represents the embedding of the item in the user-item graph. SimGCL [20] first utilizes LightGCN [3] to obtain the user-item graph embeddings, then applies random noise for data augmentation. More technical details can be found in the original papers.

Knowledge Graph View Generation. We explore two ways to improve the reliability of item representations from the KG: data augmentation and the design of the MCKR framework. (1) **Distance-based Augmentation**: To keep the computational pattern of each batch fixed and more efficient or handle large-scale KG, GNN-based knowledge-aware recommendation [10,12] usually uniformly random samples a fixed-size set of neighbors for each entity instead of using its full neighbors. It is similar to the data augmentation of random perturbation of the KG graph structure. However, the importance of different KG triples is different, and random edge dropout cannot guarantee the preservation of important information. Therefore, we classify the KG triples and apply different dropout probabilities to different triples. (2) **Degree-based GCN**: In real-world KG, there is a significant difference in the number of neighbors between unpopular and popular entities, which cannot be eliminated by data augmentation or fixed neighborhood size. This difference affects the representation of entities. Popular entities cannot accurately represent the item they are connected to, as they are associated with many irrelevant entities. In contrast, unpopular entities can be considered more indicative of the characteristics of their neighborhood items. Therefore, we design a Degree-based GCN, which considers the degree of the connected tail entities in weight calculation, thereby assigning higher weights to unpopular entities. We next present the implementation of these two parts in detail.

Distance-Based Augmentation. To retain more reliable KG information before structural damage, we preprocess the KG and divide the types of KG triples into item-entity and entity-entity. Specifically, we set $D_i = 0$ for any item i and use the Breadth-First-Search (BFS) algorithm to preprocess the distance of all entities D_e to the nearest item after linking. For KG triple $T = (h, r, t)$, the distance to the closest item is

$$\hat{D}_T = min(D_h, D_t) \tag{1}$$

$\hat{D}_T$ is the distance to the closest item for KG triple T. D_h, D_t is the distance to the closest item for entities h and t. The first type is item-entity connection if $\hat{D}_T = 0$. Otherwise, it is an entity-entity connection. We perform random edge dropout perturbations with different probabilities ϵ_1, ϵ_2 for the two types of triples.

According to the above method, we perform two data augmentations on the KG. Then we uniformly and randomly sample a fixed-size set of neighbors for each entity.

Degree-Based GCN. We start by introducing how degree-based GCN captures first-order neighbor information. Formally, we denote an item as $i(i \in \mathcal{I})$, and we represent the set of neighbors v and their corresponding relationships r directly connected to i as $N_i = \{(r, v)|(i, r, v) \in \mathcal{G}_k\}$. We employ the function $c(i, r)$ to compute the score of this relationship with the item:

$$c(i, r) = \frac{e_i \cdot e_r}{|\mathcal{N}_v|} \tag{2}$$

where e_i and e_r represent the original embeddings of entities with IDs i and r, respectively. The symbol $\cdot$ denotes the inner product operation between the two embeddings, and $|\mathcal{N}_v|$ represents the number of neighbors connected to entity v. In order to mitigate the impact of irrelevant entities introduced by popular entities, when calculating the score of relation r and central item i, the score $c(i, r)$ should decay for items with higher tail entity degrees.

We use the function $\tilde{c}$ to normalize the scores of item-relationship pairs.

$$\tilde{c}(i,r) = \frac{\exp(c(i,r))}{\sum_{r,v \in \mathcal{N}_i} \exp(c(i,r))} \tag{3}$$

To obtain the structural information of item i, we compute a linear combination of its neighborhood:

$$e_{\mathcal{N}_i} = \sum_{r,v \in \mathcal{N}_i} \tilde{c}(i,r) \cdot e_v \tag{4}$$

where $e_{\mathcal{N}_i}$ represents the aggregated information from the neighborhood of item i, and e_v denotes the original embedding of entity with ID v. By computing the score between item i and relationship r, we assign different weights to the neighbors v. Finally, we combine the current item embedding e_i with the neighborhood embedding $e_{\mathcal{N}_i}$ to update the current representation of the item e_i.

$$e_i = \sigma(W \cdot (e_i + e_{\mathcal{N}_i}) + b) \tag{5}$$

e_i aggregates the neighbor information $e_{\mathcal{N}_i}$ through a Multi-Layer Perceptron, where W and b are trainable parameters, and σ is an activation function similar to ReLU.

The above process only models 1-order neighbors. In order to capture complex semantic relationships and high-order connectivity on the KG, we extend it to multiple hops. Specifically, at the l-order, we expand Eq. (5) to include entities within l hops. The l-order embeddings of item i and entity v can be computed as follows:

$$
\begin{aligned}
e_i^{(l)} &= \sigma(W^{(l)} \cdot (e_i^{(l-1)} + e_{\mathcal{N}_i}^{(l-1)}) + b^{(l)}) \\
e_v^{(l)} &= \sigma(W^{(l)} \cdot (e_v^{(l-1)} + e_{\mathcal{N}_v}^{(l-1)}) + b^{(l)})
\end{aligned}
\tag{6}
$$

where $e_i^{(l-1)}$ and $e_v^{(l-1)}$ represent the $(l-1)$-order embeddings of item i and entity v, respectively. $W^{(l)}$ and $b^{(l)}$ are trainable parameters. They can be seen as a mixture of the initial embeddings of themselves and their neighbors. We consider the l-order embedding $e_i^{(l)}$ of the item as the final representation $z_{i_{kg}}$ learned by the item. Finally, through training on two augmented KGs, we get the representations of the item on the KG view $(z_{i_{kg}}^1, z_{i_{kg}}^2)$.

4.2   Multi-view Contrastive Learning

We use contrastive learning at the local-view and global-view to capture features at different granularities.

Local-View Contrastive Learning. After the above training, we get three pairs of representations: a pair of user representations and two pairs of item representations. The processing methods of representations are consistent. Here is an example of the item pair $(z_{i_{ui}}^1, z_{i_{ui}}^2)$ in the user-item graph view:

As the representations of the same item in both contrastive views are in the same space, we consider the two representations, denoted as $z_{i_{ui}}^1$ and $z_{i_{ui}}^2$, as a positive sample pair. In contrast, the representations of other items with this item representation are considered negative pairs. To optimize this contrastive learning objective, we employ the InfoNCE [1] loss function to maximize the agreement between positive pairs and minimize the agreement between negative pairs.

$$\mathcal{L}_{cl}^{i_{ui}} = -\log \frac{\exp\left(s\left(z_{i_{ui}}^1, z_{i_{ui}}^2\right)/\tau\right)}{\sum_{k \in \mathcal{I}, k \neq i} \exp\left(s\left(z_{i_{ui}}^1, z_{k_{ui}}^2\right)/\tau\right)} \tag{7}$$

s denotes the cosine similarity between two representations, while τ is a controllable hyperparameter representing the temperature coefficient. Similarly, by replacing the item pair with the user pair (z_u^1, z_u^2) and $(z_{i_{kg}}^1, z_{i_{kg}}^2)$ in Eq. (7), we can obtain $\mathcal{L}_{cl}^u$ and $\mathcal{L}_{cl}^{i_{kg}}$. Finally, we express the loss function for contrastive learning of local-view as:

$$\mathcal{L}_{cl}^{local} = \alpha\left(\mathcal{L}_{cl}^u + \mathcal{L}_{cl}^{i_{ui}}\right) + \beta\mathcal{L}_{cl}^{i_{kg}} \tag{8}$$

α and β are hyperparameters that control the weight of contrastive learning loss for different views. By tuning these hyperparameters, we can adjust the relative contribution of the user-item graph view and KG view to the overall loss function. This adjustment allows us to control the degree of regularization during training.

Global-View Contrastive Learning. Based on the Sect. 4.1, representations for two pairs of items $(z_{i_{ui}}^1, z_{i_{ui}}^2)$ and $(z_{i_{kg}}^1, z_{i_{kg}}^2)$ are obtained. Since the graph structure is modified on KG, $(z_{i_{kg}}^1, z_{i_{kg}}^2)$ represents part of the entity information. By adding the representations of the two entities, the information of the common nodes are augmented, while each independent piece of information is retained. We use a Hadamard product as an aggregation method to fuse the representations from different views.

$$\begin{aligned} z_{i_{kg}} &= z_{i_{kg}}^1 + z_{i_{kg}}^2 \\ z_{i_{global}}^1 &= z_{i_{kg}} \odot z_{i_{ui}}^1 \\ z_{i_{global}}^2 &= z_{i_{kg}} \odot z_{i_{ui}}^2 \end{aligned} \tag{9}$$

We also apply the InfoNCE [1] loss function to optimize this global-view contrastive learning objective $\mathcal{L}_{cl}^{global}$. $(z_{i_{global}}^1, z_{i_{global}}^2)$ as a positive pair, and different items serve as negative pairs. The specific formula is similar to the Eq. (7).

4.3 Model Predict

We combine the user representations (z_u^1, z_u^2) from user-item view generation and combine the two item representations $(z_{i_{global}}^1, z_{i_{global}}^2)$ to obtain the final user and item representations (z_u^*, z_i^*). We then use the inner product results as predict scores $\hat{y}(u, i)$.

$$\begin{aligned} z_u^* &= z_u^1 \| z_u^2 \\ z_i^* &= z_{i_{global}}^1 \| z_{i_{global}}^2 \\ \hat{y}(u, i) &= z_u^{*\top} z_i^* \end{aligned} \tag{10}$$

4.4 Multi-task Training

To integrate the recommendation task with the SSL task, we employ multi-task training to optimize the entire framework. For the recommendation task, we adopt the Bayesian Personalized Ranking (BPR) loss, which aims to maximize the margin between the observed user-item interaction scores and the unobserved ones:

$$\mathcal{L}_{main} = \sum_{(u,i,j) \in \mathcal{O}} -\log \sigma \left(\hat{y}_{ui} - \hat{y}_{uj} \right) \tag{11}$$

where $\mathcal{O} = \{(u,i,j)|(u,i) \in \mathcal{O}^+, (u,j) \in \mathcal{O}^-\}$ and $\mathcal{O}^- = \mathcal{U} \times \mathcal{I} \backslash \mathcal{O}^+$ is the set of unobserved interactions. The loss function for the SSL task comprises the local-view $\mathcal{L}_{cl}^{local}$ and the global-view $\mathcal{L}_{cl}^{global}$. The integrative optimization loss of our MCKR is:

$$\mathcal{L} = \mathcal{L}_{main} + \mathcal{L}_{cl}^{local} + \gamma \mathcal{L}_{cl}^{global} + \lambda \|\theta\|_2^2 \tag{12}$$

γ controls the weight of the global-view SSL task, θ represents the framework parameters for the recommendation task, and λ modulates the strength of L_2 regularization.

5 Experiment

To evaluate the performance of the MCKR framework, we conduct extensive experiments and provide detailed analyses to answer the following research questions:

1. How does MCKR compare to other state-of-the-art models?
2. How do the framework and key modules in MCKR impact performance?
3. What is the recommended effect of MCKR on KG noise problems?

5.1 Datasets and Experiment Settings

As shown in Table 1, we use the following three datasets in our experiments: Book-Crossing, MovieLens-1M, and Alibaba-iFashion [10, 14]. The datasets used in our experiments are publicly available and come from different domains with different sizes and sparsity, strengthening our experimental results' reliability and robustness. To evaluate topK recommendation, we use two commonly used metrics [13, 14], Recall@N and NDCG@N, with N set to 20 as the number of items to be sorted. The evaluation is carried out for all users in the test set, and the final result is the average metric.

To verify the superiority of our framework, We compare the proposed MCKR with the state-of-the-art baselines. As shown in Table 2, they are BPRMF [8], CKE [21], RippleNet [10], KGCN [12], KGAT [13], KGIN [14], and MCCLK [22].

We conduct our experiments using the unified benchmark experiment platform, an open source framework RecBole [16]. To ensure a fair comparison, we optimize all methods using the Adam optimizer with a batch size of 4096 and a learning rate of 0.001. All parameters are initialized using the default Xavier distribution, and the embedding size is set to 64. We use an early stopping strategy of 10 epochs to prevent overfitting and set Recall@20 as the verification index. The hyperparameters of all baselines are set through empirical studies or following the original papers.

Table 1. Summary of the recommendation datasets used in the experiments

		Book-Crossing	MovieLens-1M	Alibaba-iFashion
User-Item Interaction	# Users	17,860	6,036	114,737
	# Items	14,967	2,445	30,040
	# Interactions	139,746	753,772	1,781,093
Knowledge Graph	# Entities	77,903	182,011	59,156
	# Relations	25	12	51
	# Triples	151,500	1,241,995	279,155

5.2 Performance Comparison (RQ1)

We report the overall performance evaluation of all methods in Table 2, and our observations are summarized below.

Table 2. Overall performance comparison.

	Book-Crossing		MovieLens-1M		Alibaba-iFashion	
	Recall@20	NDCG@20	Recall@20	NDCG@20	Recall@20	NDCG@20
BPR	0.0388	0.0212	0.1456	0.1384	0.0997	0.0398
CKE	0.0441	0.0376	0.1481	0.1449	0.1058	0.0411
RippleNet	0.0613	0.0428	0.1535	0.1457	0.1124	0.0507
KGCN	0.0642	0.0463	0.1528	0.1486	0.1152	0.0513
KGAT	0.0608	0.0448	0.1521	0.1473	0.1149	0.0507
KGIN	0.0692	0.0508	0.1542	0.1488	0.1258	0.0546
MCCLK	0.0714	0.0531	0.1598	0.1513	0.1294	0.0574
MCKR	**0.0759**	**0.0545**	**0.1619**	**0.1548**	**0.1299**	**0.0580**

- Compared with the strongest baseline MCCLK, MCKR has increased by 6.30%, 1.31%, and 0.39% in the Recall indicators, respectively. These improvements demonstrate the effectiveness of the MCKR framework.
- MCKR exhibits the most remarkable improvement on the Book-Crossing dataset compared to other datasets. We attribute this to a possible factor, the sparser density of the KG in this dataset. MCKR's contrastive learning helps mitigate this sparsity and capture more reliable information.
- The method of contrastive learning, MCKR and MCCLK [22], is better than GNN-based knowledge-aware recommendation, which proves that contrastive learning helps improve the performance of RS.
- GNN-based knowledge-aware recommendation performs better than CKE [21], indicating the importance of high-order connectivity on KG. Multi-hop information on the KG greatly improves RS performance.

5.3   Ablation Studies and Sensitivity Analysis (RQ2)

In order to evaluate the effectiveness of the framework and the contributions of the main components to the final performance, we design several variants to compare with MCKR, and the specific results are shown in Table 3.

Table 3. Effect of ablation study.

Model	Book-Crossing		MovieLens-1M		Alibaba-iFashion	
	Recall@20	NDCG@20	Recall@20	NDCG@20	Recall@20	NDCG@20
$MCKR_{w/o}$ noise	0.0728	0.0516	0.1551	0.1487	0.1255	0.0563
$MCKR_{w/o}$ GC	0.0744	0.0528	0.1608	0.1539	0.1265	0.0571
$MCKR_{w/o}$ KGC	0.0703	0.0514	0.1582	0.1516	0.1238	0.0553
$MCKR_{w/o}$ UIC	0.0732	0.0524	0.1557	0.1491	0.1189	0.0532
MCKR	**0.0759**	**0.0545**	**0.1619**	**0.1548**	**0.1299**	**0.0580**

- $MCKR_{w/o}$ noise: As introduced in Sect. 1, this variant has inconsistencies between tasks. Specifically, it discards the noise in the recommendation task, and the SSL task remains unchanged. The variant replaces the input for the recommendation task with user-item view embeddings trained by LightGCN [3], and the KG view embedding trained on the KG without data augmentation.
- $MCKR_{w/o}$ GC: Discards the module of global-view contrastive learning, and the information from two different graphs are aggregated for prediction.
- $MCKR_{w/o}$ KGC: Discards the local-view contrastive learning of KG, and uses the GCN encoder to obtain neighborhood information.
- $MCKR_{w/o}$ UIC: Discards the local-view contrastive learning of user-item graph, and uses the LightGCN [3] to obtain user and item representations.

From the results in Table 3, MCKR outperforms the $MCKR_{w/o}$ noise variant, demonstrating that the inconsistency between the recommendation task and the SSL task limits the framework's ability to generalize and effectively capture the underlying information. Additionally, MCKR performs better than the variants with critical modules removed, indicating that the excellent performance of MCKR is closely related to each module, which shows that our proposed distance-based augmentation method, degree-based GCN, and the multi-view SSL task are all suitable for recommender systems.

We further investigate the impact of different hyperparameters on framework performance, as shown in Fig. 2. We study the hyperparameters α, β in Eq. (8) and γ in Eq. (12) by altering their values within $\{0.1, 0.05, 0.01, 0.005, 0.001, 0.005\}$ in MovieLens-1M dataset. Results reveal that optimal weights differ among SSL tasks and main task loss. Hyperparameter α performs best at 0.1, while the β and γ peak at $5e-3$ and $5e-2$. We attribute this to a possible factor, the user-item graph contrastive learning method simultaneously enhances both user and item representations.

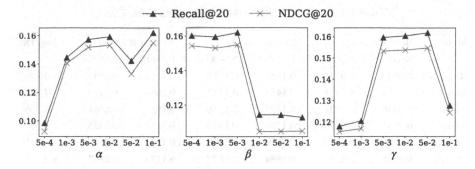

Fig. 2. Impact of hyperparameters α, β, and γ on the MovieLens-1M dataset.

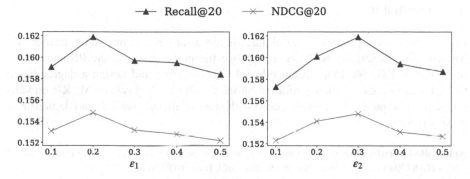

Fig. 3. Impact of different KG triples random edge dropout probabilities ϵ_1, ϵ_2 on the MovieLens-1M dataset.

We also explore random edge dropout perturbations at probabilities ϵ_1, ϵ_2 (as introduced in Sect. 4.1) of $\{0.1, 0.2, 0.3, 0.4, 0.5\}$ during the KG data augmentation phase for two distinct types of triples on the MovieLens-1M dataset. As shown in Fig. 3, the KG triples connecting items and entities are more important, with ϵ_1 achieving optimal performance at 0.2 while ϵ_2 optimal performance at 0.3.

5.4 Benefits of MCKR in Alleviating Data Noise Effect (RQ3)

To evaluate the impact of KG noise on different models, we generate a set of noise triples by randomly selecting head and tail entities and relationships, then adding them to the test set of each dataset. The number of noise triples is set to 10% of the total number of triples in each dataset. The experimental results are presented in Table 4.

MCKR outperforms all baselines in all results, demonstrating its robustness against noise and ability to utilize KG information effectively. Compared to MCCLK [22], distance-based augmentation, and degree-based GCN enables MCKR to retain valuable information, preserving high-order connectivity. In contrast, the model, which relies on attention for aggregating neighbor information, introduces more noise into the final item representation, thereby affecting RS performance.

Table 4. Benefits of MCKR in Alleviating Data Noise Effect.

Model	Book-Crossing		MovieLens-1M		Alibaba-iFashion		Avg. Dec.
	Recall@20	NDCG@20	Recall@20	NDCG@20	Recall@20	NDCG@20	
RippleNet	0.0539	0.0375	0.1384	0.1319	0.1010	0.0452	−10.79%
KGCN	0.0588	0.0423	0.1416	0.1382	0.1065	0.0470	−7.88%
KGAT	0.0546	0.0407	0.1405	0.1356	0.1054	0.0461	−8.73%
KGIN	0.0665	0.0488	0.1509	0.1445	0.1213	0.0525	−3.37%
MCCLK	0.0690	0.0512	0.1559	0.1478	0.1265	0.0558	−2.78%
MCKR	**0.0738**	**0.0529**	**0.1594**	**0.1527**	**0.1274**	**0.0567**	**−2.12%**

6   Conclusion

In this paper, we propose a novel knowledge-aware recommendation framework, MCKR, and investigate its ability to improve the performance of the RS and reduce the noise in KG. We adopt distance-based augmentation and design a degree-based GCN to capture more reliable information for the items. We evaluate MCKR on various real-world public datasets compared with state-of-the-art methods and demonstrate its superiority.

Acknowledgments. This work is jointly supported by National Natural Science Foundation of China (61877043) and National Natural Science of China (61877044).

References

1. Chen, T., Kornblith, S., Norouzi, M., Hinton, G.: A simple framework for contrastive learning of visual representations. In: International Conference on Machine Learning, pp. 1597–1607. PMLR (2020)
2. Chen, X., Xie, S., He, K.: An empirical study of training self-supervised vision transformers. In: Proceedings of the IEEE/CVF International Conference on Computer Vision, pp. 9640–9649 (2021)
3. He, X., Deng, K., Wang, X., Li, Y., Zhang, Y., Wang, M.: LightGCN: simplifying and powering graph convolution network for recommendation. In: Proceedings of the 43rd International ACM SIGIR Conference on Research and Development in Information Retrieval, pp. 639–648 (2020)
4. Hu, B., Shi, C., Zhao, W.X., Yu, P.S.: Leveraging meta-path based context for top-n recommendation with a neural co-attention model. In: Proceedings of the 24th ACM SIGKDD International Conference on Knowledge Discovery & Data Mining, pp. 1531–1540 (2018)
5. Ji, G., He, S., Xu, L., Liu, K., Zhao, J.: Knowledge graph embedding via dynamic mapping matrix. In: Proceedings of the 53rd Annual Meeting of the Association for Computational Linguistics and the 7th International Joint Conference on Natural Language Processing (Volume 1: Long papers), pp. 687–696 (2015)
6. Lin, Y., Liu, Z., Sun, M., Liu, Y., Zhu, X.: Learning entity and relation embeddings for knowledge graph completion. In: Proceedings of the AAAI Conference on Artificial Intelligence, vol. 29 (2015)

7. Pujara, J., Augustine, E., Getoor, L.: Sparsity and noise: where knowledge graph embeddings fall short. In: Proceedings of the 2017 Conference on Empirical Methods in Natural Language Processing, pp. 1751–1756 (2017)
8. Rendle, S., Freudenthaler, C., Gantner, Z., Schmidt-Thieme, L.: BPR: Bayesian personalized ranking from implicit feedback. arXiv preprint arXiv:1205.2618 (2012)
9. Sun, Z., Yang, J., Zhang, J., Bozzon, A., Huang, L.K., Xu, C.: Recurrent knowledge graph embedding for effective recommendation. In: Proceedings of the 12th ACM Conference on Recommender Systems, pp. 297–305 (2018)
10. Wang, H., et al.: RippleNet: propagating user preferences on the knowledge graph for recommender systems. In: Proceedings of the 27th ACM International Conference on Information and Knowledge Management, pp. 417–426 (2018)
11. Wang, H., Zhang, F., Xie, X., Guo, M.: DKN: deep knowledge-aware network for news recommendation. In: Proceedings of the 2018 World Wide Web Conference, pp. 1835–1844 (2018)
12. Wang, H., Zhao, M., Xie, X., Li, W., Guo, M.: Knowledge graph convolutional networks for recommender systems. In: The World Wide Web Conference, pp. 3307–3313 (2019)
13. Wang, X., He, X., Cao, Y., Liu, M., Chua, T.S.: KGAT: knowledge graph attention network for recommendation. In: Proceedings of the 25th ACM SIGKDD International Conference on Knowledge Discovery & Data Mining, pp. 950–958 (2019)
14. Wang, X., et al.: Learning intents behind interactions with knowledge graph for recommendation. In: Proceedings of the Web Conference 2021, pp. 878–887 (2021)
15. Wu, J., et al.: Self-supervised graph learning for recommendation. In: Proceedings of the 44th international ACM SIGIR Conference on Research and Development in Information Retrieval, pp. 726–735 (2021)
16. Xu, L., et al.: Recent advances in RecBole: extensions with more practical considerations (2022)
17. Yang, X., Zhang, X., Zhang, Z., Zhao, Y., Cui, R.: DTWSSE: data augmentation with a Siamese encoder for time series. In: U, L.H., Spaniol, M., Sakurai, Y., Chen, J. (eds.) Web and Big Data: 5th International Joint Conference, APWeb-WAIM 2021, Guangzhou, China, 23–25 August 2021, Proceedings, Part I 5. LNCS, vol. 12858, pp. 435–449. Springer, Cham (2021). https://doi.org/10.1007/978-3-030-85896-4_34
18. Yang, X., Zhang, Z., Cui, R.: TimeCLR: a self-supervised contrastive learning framework for univariate time series representation. Knowl.-Based Syst. **245**, 108606 (2022)
19. Yang, Y., Huang, C., Xia, L., Li, C.: Knowledge graph contrastive learning for recommendation. In: Proceedings of the 45th International ACM SIGIR Conference on Research and Development in Information Retrieval, pp. 1434–1443 (2022)
20. Yu, J., Yin, H., Xia, X., Chen, T., Cui, L., Nguyen, Q.V.H.: Are graph augmentations necessary? Simple graph contrastive learning for recommendation. In: Proceedings of the 45th International ACM SIGIR Conference on Research and Development in Information Retrieval, pp. 1294–1303 (2022)
21. Zhang, F., Yuan, N.J., Lian, D., Xie, X., Ma, W.Y.: Collaborative knowledge base embedding for recommender systems. In: Proceedings of the 22nd ACM SIGKDD International Conference on Knowledge Discovery and Data Mining, pp. 353–362 (2016)
22. Zou, D., et al.: Multi-level cross-view contrastive learning for knowledge-aware recommender system. In: Proceedings of the 45th International ACM SIGIR Conference on Research and Development in Information Retrieval, pp. 1358–1368 (2022)

PYGC: A PinYin Language Model Guided Correction Model for Chinese Spell Checking

Haoping Chen[✉] and Xukai Wang

Shanghai Jiao Tong University, Shanghai, China
{apple_chen,wangxukai}@sjtu.edu.cn

Abstract. Chinese Spell Checking (CSC) is an NLP task that detects and corrects erroneous characters in Chinese texts. Since people often use pinyin (pronunciation of Chinese characters) input methods or speech recognition to type text, most of these errors are misuse of phonetically or semantically similar characters. Previous attempts fuse pinyin information into the embedding layer of pre-trained language models. However, although they can learn from phonetic knowledge, they can not make good use of this knowledge for error correction. In this paper, we propose a **PinY**in language model **G**uided **C**orrection model (PYGC), which regards the Chinese pinyin sequence as an independent language model. Our model builds on two parallel transformer encoders to capture pinyin and semantic features respectively, with a late fusion module to fuse these two hidden representations to generate the final prediction. Besides, we perform an additional pronunciation prediction task on pinyin embeddings to ensure the reliability of the pinyin language model. Experiments on the widely used SIGHAN benchmark and a newly released CSCD-IME dataset with mainly pinyin-related errors show that our method outperforms current state-of-the-art approaches by a remarkable margin. Furthermore, isolation tests demonstrate that our model has the best generalization ability on unseen spelling errors. (Code available in https://github.com/Imposingapple/PYGC)

Keywords: Natural Language Understanding · Chinese Spell Checking · Pinyin Language Model

1 Introduction

Chinese Spell Checking (CSC) is a fundamental NLP task that aims to detect spelling errors in Chinese texts and make a correction. Spelling errors mainly come from human-generated errors, *e.g.*, handwriting errors or typing errors, and machine-generated errors, *e.g.*, errors caused by optical character recognition (OCR) systems or automatic speech recognition (ASR) systems [9,11].

In alphabetic languages such as English, where word boundaries are obvious, spelling errors are mainly due to character recognition errors, resulting in "out

B. Luo et al. (Eds.): ICONIP 2023, LNCS 14451, pp. 224–239, 2024.
https://doi.org/10.1007/978-981-99-8073-4_18

of dictionary" errors. Chinese spelling errors are in units of Chinese characters, where individual characters are miswritten as other phonologically similar or visually similar characters. Previous work has shown that 83% Chinese spelling mistakes are phonetic [12]. Thus, the error correction ability of the language model can be improved effectively by using the information of pinyin (pronunciation of Chinese characters).

The latest CSC approaches [1, 6, 9, 14, 24] consider integrating phonetic information into the large pre-trained language models [4, 16, 18]. However, although they can learn from phonetic knowledge, they can not make good use of this knowledge for error correction [33], and they lack the generalization ability in real scenarios as well [8]. Table 1 shows two examples. In the first example, the CSC model utilizes phonetic information to output "高", which is phonetically similar to the input character "诰", and "高贵" constitutes a common Chinese word. However, the output does not match the semantics of the whole sentence. In the second example, the misspelling "持刀" is a valid Chinese word and has similar pinyin to the ground truth "迟到". The model cannot utilize semantics for error correction in this case. The second example is very close to the real world of typing with the Pinyin Input Methods, where most spelling errors show up as replacing one word with another word with similar pronunciation.

Table 1. Examples of CSC results. Pinyin for each Chinese character is composed of three units: initial, final, and tone. The wrong/predict/correct characters with their pinyin are in red/blue/green. "Model Output" means the prediction results of PLOME, a strong CSC model which incorporates pinyin information.

	Sentence	Pinyin Units
Input	我们得签名诰贵工厂，请律师来帮我们处理。	g,ao,4
Model Output	我们得签名高贵工厂，请律师来帮我们处理。	g,ao,1
Gold	我们得签名告贵工厂，请律师来帮我们处理。	g,ao,4
Translation	We need to sign a contract to sue your factory. Please bring a lawyer to help us handle this.	
Input	...别再持刀！如果你又这样，我就不让你进来教室...	ch,i,2; d,ao,1
Model Output	...别再持刀！如果你又这样，我就不让你进来教室...	ch,i,2; d,ao,1
Gold	...别再迟到！如果你又这样，我就不让你进来教室...	ch,i,2; d,ao,4
Translation	...Don't be late again! If you do that again, I won't let you in the classroom...	

In this paper, we propose PYGC, a **PinY**in language model **G**uided **C**orrection model for CSC. Figure 1 provides an overview of PYGC. Our core idea is that the pinyin sequence corresponding to a Chinese character is itself a language model; that is, the pinyin of the current location can be inferred from the pinyin context. Therefore, we conduct a similar language model modeling for pinyin information in our model. Our model has two parallel encoders, one for encoding pinyin sequences and the other for encoding Chinese characters sequences. To make better use of pinyin information to improve the final CSC result, we introduce a fusion module to mix the Chinese character representation at each position with the hidden representations of three corresponding pinyin units. We decode the mixed representations to get the final output. Moreover, a confusion set based masking mechanism [14,30] adapted for our model is used in the pre-training stage, and an iterative prediction strategy is also adopted to improve CSC results [11].

We conduct experiments on the widely-used SIGHAN benchmark dataset [19] and a newly released CSCD-IME dataset [8]. Experimental results show that our model establishes new state-of-the-art performance on SIGHAN and CSCD-IME benchmarks. An Isolation Test on the SIGHAN dataset is conducted to measure the generalization ability of CSC models, our model outperforms the currently best-performing models by over 5%. Ablation studies further demonstrate the effectiveness of our fusion module and pinyin language model modeling.

The main contributions of our paper are summarized below: (1) We innovatively model pinyin information separately as a language model to enhance the results of the CSC model. (2) We propose our model PYGC, which has two parallel encoders, respectively encoding the pinyin sequence and Chinese characters, and a fusion module is used to combine two features for prediction. The performance is improved by adding pinyin prediction tasks, adopting a confusion set based masking pre-training strategy, and using an iterative decoding strategy. (3) We establish state-of-the-art performance on the widely used SIGHAN and newly released CSCD-IME datasets. The Isolation test demonstrates that our model has a strong generalization ability.

2   Related Work

The task of Chinese Spelling Checking consists of two sub-tasks: Error Detection: detecting the location of sentence errors, and Error Correction: The error is corrected based on the identification.

Traditional CSC methods can be unified under a framework, namely three steps: error detection, candidate generation, and evaluation of different candidates. Error detection can be done by Maximum entropy classification [5] or statistical methods such as bidirectional character-level N-gram language model [28]. After wrong positions are found, words in the confusion set of wrong words (Chinese characters similar to the wrong character [10,23]) are regarded as more possible options. With the help of a confusion set, people then apply rules [15] or use word dictionaries and statistic models to generate reasonable candidates [2,7,26]. Finally, language models [2,15,26,27] or classifiers [15,32] are generally used to evaluate correction candidates.

With the development of deep learning, transformer-based models have achieved significant improvements in CSC tasks compared to traditional methods. A mainstream of CSC works considers the integration of phonetic(pinyin) and glyph information into the deep neural model to improve the performance of spelling error correction, which can be divided into three types: (1) modification of masking strategy in pre-training, e.g., using a character in the masked token's confusion set to replace the original [MASK] token in pre-training task [11,14,30]; (2) Modeling at embedding level, *e.g.*, phonetic and glyph information are first extracted by feature extractor and then be added to word embedding in embedding layer [9,14,24,25]; (3) Modeling at decoding level, *e.g.*, The candidate word generated by the decoder will be limited in the confusion set of this word [22], optimal selection of Chinese characters considering phonetic and glyph information comprehensively [6], using GRU to fuse features of each character with visually or phonetically similar characters and use fused embedding as decoder [1].

To better utilize pinyin information, we propose PYGC with two parallel encoders to capture pinyin and semantic features, respectively. Multitasking and modification of masking strategy in pre-training are also used to improve CSC performance further.

3   Methods

This section presents the CSC problem formulation and then describes our PYGC methods in detail.

3.1   Problem Formulation

The Chinese spelling check (CSC) task aims to detect and correct spelling errors in Chinese texts. Given a text sequence $X = \{x_1, x_2, ..., x_n\}$ containing n characters, a CSC model is expected to detect and correct potential error characters in X and output a sequence $Y = \{y_1, y_2, \ldots, y_n\}$ of the same length. The task can be formulated as a conditional generation problem that maximizes the conditional probability $P(Y|X)$.

3.2   Architecture

In addition to the original semantic encoder, a parallel pinyin encoder is constructed to model the pinyin language model. Then the fusion module is used to combine features of two language models for the final prediction. Confusion set based pre-training and multitasking learning are utilized to improve model performance further. Figure 1 shows the overall architecture of PYGC.

Semantic Encoder. We use BERT [4] as the backbone of our semantic encoder. The BERT model is stacked with multiple layers of transformer structures, and each contains a multi-head attention module and a feed-forward layer with a residual connection. It is pre-trained on large corpora and can provide rich contextual features for input tokens. Give input tokens $X = \{x_1, x_2, ..., x_n\}$, they are first projected into H_0^s after embedding layer. Then, the computation of each Transformer layer can be formulated as follows:

$$\mathbf{H}_l^s = Transformer_l\left(\mathbf{H}_{l-1}^s\right), l \in [1, L] \tag{1}$$

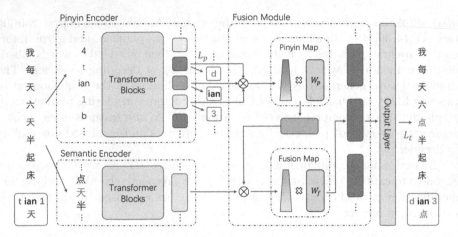

Fig. 1. Overview of the proposed PYGC. The input sentence has a spelling error "天" (trans: day, pinyin: tian1), which should be "点" (trans: o'clock, pinyin: dian3) in ground truth. The error token and its corresponding error pinyin units are marked in red, while the ground truth is marked in green. **Left:** Input characters are processed by the semantic encoder, and pinyin of the input characters are processed by the pinyin encoder. **Right:** representations obtained by the two encoders are fused by a fusion module to make final prediction.

where L is the total number of Transformer layers. The output of the final layer is used as the semantic representation of each token $H^s = H^s_L = \{h^s_1, h^s_2, \ldots, h^s_n\}$.

Pinyin Encoder. *Hanyu Pinyin* (pinyin) is an internationally recognized standard for the Latin translation of Chinese, which is used to spell Chinese characters. The pinyin of each Chinese character is composed of three units: initial, final, and tone. Both initials and finals are composed of Latin characters, while there are five kinds of tones, which are mapped as numbers $\{1, 2, 3, 4, 5\}$. Take the character "天" (pinyin: "tian1") as an example. Its initial, final, and tone are respectively "t", "ian", and "1". In our paper, we obtain the pinyin sequence corresponding to the input token sequence X through pypinyin[1]:

$$X^p = \{x_{1,initial}, x_{1,final}, x_{1,tone}, x_{2,initial}, \ldots, x_{n,initial}, x_{n,final}, x_{n,tone}\}$$

Similar to the semantic encoder, X^p first maps to H^p_0 after the embedding layer. It then comes through a stack of Transformer layers to get the pinyin representation of each pinyin unit: $H^p = \{h^p_1, h^p_2, \ldots, h^p_{3n}\}$. Since each character corresponds to three pinyin units, the length of H^p is $3n$.

Fusion Module. With semantic encoder and pinyin encoder, we get the semantic representation H and pinyin representation H^p of the input sequence respectively. We use a fusion module to mix the information of the two parts.

Specifically, the representation of the initial/final/tone pinyin unit corresponding to the i-th token in X is the $\{3i/3i + 1/3i + 2\}$-th representation

---

[1] https://pypi.org/project/pypinyin.

in H^p. We first concatenate the three representations in H^p and map them into the mixed pinyin representation of the i-th token:

$$h_i^{mp} = W^p \cdot \left[h_{3i}^p, h_{3i+1}^p, h_{3i+2}^p\right] + b^p \tag{2}$$

We then concatenate the mixed pinyin representation with the semantic representation h_i^s and project it into the fused representation for each input token:

$$h_i^f = LN\left(W^f \cdot [h_i^{mp}, h_i^s] + b^f\right) \tag{3}$$

Here $W^p \in \mathbf{R}^{D \times 3D_p}, W^f \in \mathbf{R}^{D \times 2D}, b^p \in \mathbf{R}^D, b^f \in \mathbf{R}^D$ are learnable parameters, D and D_s are the dimension of semantic and pinyin representation respectively, $[\cdot]$ means the concatenation of vectors, and $LN(\cdot)$ means layer normalization.

Output Layer. As shown in Fig. 1, our model makes two predictions.
Prediction For CSC. We get a fused representation of each position from the fusion module and make predictions for the ground truth sequence Y. Specifically, for the i-th position, we first transform the fused embedding h_i^f to the transformed feature space:

$$h_i^t = LN\left(GeLU\left(W^t \cdot h_i^f + b^t\right)\right) \tag{4}$$

Then, after linear mapping and Softmax operation, the corresponding target token is predicted:

$$p(\hat{y}_i|X) = Softmax\left(W^o \cdot h_i^t + b^o\right) \tag{5}$$

Here $W^t \in \mathbf{R}^{D \times D}, W^o \in \mathbf{R}^{V \times D}, b^t \in \mathbf{R}^D, b^o \in \mathbf{R}^V$ are learnable parameters, D is the dimension of feature space, V is the vocab size, $\hat{y}_i$ is the CSC prediction result for the i-th token, $LN(\cdot)$ is layer normalization.
Prediction for Pronunciation. To learn the erroneous input sequence information, as shown in the middle part of Fig. 1, we make predictions for each pinyin unit on top of the pinyin encoder.

Similar to the prediction for CSC, for each pinyin unit representation h_i^p, we also apply a transformation and a classification operation:

$$h_i^{pt} = LN\left(GeLU\left(W^{pt} \cdot h_i^p + b^{pt}\right)\right) \tag{6}$$

$$p(\hat{y}_i^p|X^p) = Softmax\left(W^o \cdot h_i^{pt} + b^o\right) \tag{7}$$

Here $W^{pt} \in \mathbf{R}^{D^p \times D^p}, W^o \in \mathbf{R}^{V^p \times D^p}, b^{pt} \in \mathbf{R}^{D^p}, b^o \in \mathbf{R}^{V^p}$ are learnable parameters, D^p is the dimension of pinyin representations, V^p is the total number of pinyin units (69 in our experiment), $\hat{y}_i^p$ is the pinyin prediction result for the i-th pinyin unit, $LN(\cdot)$ is layer normalization.

3.3   Multitask Learning

To train our model, we simultaneously optimize CSC prediction loss L_c and pinyin prediction loss L_p:

$$L_c = -\sum_i^n p(\hat{y}_i = y_i|X) \tag{8}$$

Table 2. Examples of different masking strategies. The chosen token to be masked is green, while the replaced token is red. "[M]" is the Mask token.

	Sentence
Original Sentence	海一般比湖泊大，盛 (sh,eng,4) 载海水。
Bert Masking	海一般比湖泊大，[M] ([M],[M],[M]) 载海水。
Complete Masking	海一般比湖泊大，成 (ch,eng,2) 载海水。
Semantic Masking	海一般比湖泊大，成 (sh,eng,4) 载海水。
Pinyin Masking	海一般比湖泊大，盛 (ch,eng,2) 载海水。
Random Masking	海一般比湖泊大，飞 (f,ei,1) 载海水。

$$L_p = -\sum_{i}^{3n} p(\hat{y_i^p} = y_i^p | X^p) \tag{9}$$

Here n is the length of the input sentence X, so the length of the input pinyin sentence X^p is $3n$. $\hat{y_i}$ and y_i^p are the ground truth CSC label and pinyin unit, respectively. We linearly combine these two functions as the objective function:

$$L = \lambda L_c + (1 - \lambda)L_p \tag{10}$$

where $\lambda \in [0,1]$ is the coefficient to balance the CSC prediction loss and pinyin prediction loss.

3.4 Confusion Set Based Pre-training Strategy

In order to obtain better initialization for downstream tasks, Bert proposed the MLM pre-training task, that is, to randomly select 15% of the tokens in the input sentences, replace them with "[MASK]" tokens, and then use a classifier to predict the original tokens of them. In the CSC task, the input sentence does not contain "[MASK]" but contains some words close to the ground truth token visually or phonetically, which is inconsistent with the pre-training task. Previous works have proved that replacing the selected tokens with Chinese characters in confusion set instead of "[MASK]" tokens in the pre-training stage could get better initialization points [14,30]. Since our PYGC model has an additional pinyin language model, we correspondingly propose a fine-grained masking strategy. As shown in Table 2, for each selected Chinese character, we perform: (1) in 40% of cases, complete masking: replace both its semantic and pinyin tokens with another character's in its confusion set; (2) in 20% of cases, semantic masking: leave the pinyin tokens unchanged, while the semantic token is replaced by another character in its confusion set; (3) in 20% of cases, pinyin masking: the semantic token remains unchanged, and pinyin tokens are replaced by pinyin units corresponding to another character in its confusion set; (4) in 10% of cases, random masking: the semantic and pinyin tokens are replaced with those of another random Chinese character; (5) the remaining 10% of cases are unchanged.

Table 3. The statistics of the SIGHAN datasets and CSCD-IME dataset, including the number of train/dev/test sentences, the average length per sentence, and the average spelling errors per sentence.

	Dataset	#Train	#Dev	#Test	Avg Len.	Avg #Err.
SIGHAN	SIGHAN13	700	/	/	41.8	0.49
	SIGHAN14	3437	/	/	49.6	1.49
	SIGHAN15	2339	/	1100	31.1	1.09
	Wang271K	271,329	/	/	42.6	1.41
CSCD-IME	CSCD-IME	30000	5000	5000	57.4	0.50

3.5 Predict Inference

According to [13,20], current state-of-the-art CSC models are non-autoregressive, assuming the predicted labels are generated independently among all positions. However, this assumption is invalid in sentences with multiple errors, especially continuum errors, leading to spelling correction failure. We adopted the iterative correction method used in [11], which, for each iteration, takes the previous iteration's output as the input and obtains better effects.

4 Experiments

4.1 Experimental Setup

Datasets. We do finetune on the SIGHAN datasets and the CSCD-IME dataset respectively: 1. *SIGHAN*: Following previous works [1,14,24], the training data contains manually annotated samples from SIGHAN13 [23], SIGH-AN14 [29], SIGHAN15 [19], as well as 271K samples automatically generated by OCR or ASR methods [21]. We employ the test sets of SIGHAN15 for evaluation. As in previous works [1,24], we use OpenCC[2] to convert SIGHAN datasets into simplified Chinese. 2. *CSCD-IME*: We adopt manually labeled splits of the newly-released CSCD-IME dataset mentioned in [8]. The CSCD-IME dataset is of higher quality, and the primary source of errors comes from the Pinyin Input Methods. Table 3 shows the statistics of the used datasets.

Evaluation Metrics. We report sentence-level precision, recall, and F1 scores to evaluate different models. The metrics are reported on both the detection and correction task. For SIGHAN, we use the widely-used scripts from [1] to calculate the metrics. For CSCD-IME, we apply the modified version proposed by the author of this dataset [8].

Baselines. We compare our methods with the following strong baselines:

- *Bert* [4] directly finetunes Bert model for CSC task.
- *Soft-Masked Bert* [31] uses a detection module to predict error probability of each token, which is used to perform soft-masking on the correction model.

---

[2] https://github.com/BYVoid/OpenCC.

- *SpellGCN* [1] fuses features of each character with visually or phonetically similar characters, and uses fused embedding as decoder.
- *REALISE* [24] uses three encoders to capture the semantic, phonetic and graphic information respectively. An additional fusion module is used to fuse these information.
- *PLOME* [14] incorporates sound and glyph information in the embedding layer, and introduces a confusion set based masking strategy.
- *SCOPE* [11] uses ChineseBert [18] as backbone, and introduces a fine-grained Chinese Pronunciation Prediction(CPP) task, paralleling to the CSC task.
- *ECGM* [17] proposes an error-guided correction model, which can attend to potential errors better with a zero-shot error detection method.

Implementation Details. The semantic encoder of our PYGC model is stacked with 12 layers of transformer modules, which is initialized using the weights of the *chinese-roberta-wwm-ext* model [3]. For the pinyin encoder, we adopt the same architecture as the *chinese-electra-180g-small-discriminator* model, with a vocabulary of 69 initials, finals, and tones. We pre-train the pinyin encoder on the wiki2019zh dataset with 512 max sequence length, 384 batch size, and an MLM task for 30 epochs.

At the pre-training stage, we use the initialization described above and further pre-train on wiki2019zh[3] for 5 epochs with a batch size of 512 and a learning rate of 5e-4. The warmup ratio is set to 0.1. The dataset contains 1 million well-structured Chinese entries, which we processed to get 9 million sentences. As Bert, we randomly mask 15% of the tokens as described above, and the masking strategy is described in Sect. 3.4. The pre-training loss is the same as described in Sect. 3.3, where we only calculate losses on the selected characters. At the finetuning stage on CSC tasks, we adopt a batch size of 32, a learning rate of 5e-5, and a warmup raion of 0.1 on both the SIGHAN and CSCD-IME datasets.

The max input sequence length is set to be 170 on both the pre-training stage and finetuning stage since the length of the pinyin unit is three times the length of the input sequence, which should be less than 512. λ is set to 0.5 in all experiments.

4.2   Main Results

Table 4 shows the sentence-level detection and correction evaluation scores on the SIGHAN15 test set. The metrics in the first group directly take the output of the transformer model as the prediction result. SpellGCN performs better than Bert by using a decoder that fuses phonetic and glyph information. REALISE, PLOME achieves better results by fusing visual and sound embedding at the embedding layer. SCOPE puts forward the fine-grained CPP task and achieves the best recall rate. The ECGM model achieves the best detection and correction precision by a zero-shot error detection method. Moreover, according to the comparison between the two groups of results, iterative decoding can improve the

---

[3] https://github.com/brightmart/nlp_chinese_corpus.

Table 4. Sentence-level performance on SIGHAN15 test set. The best results are in bold, and the second-best ones are underlined. Methods in the second group use an iterative decoding strategy on model outputs (denote with "†").

Method	Detection				Correction			
	Acc	Pre	Rec	F1	Acc	Pre	Rec	F1
Bert	82.4	74.2	78.0	76.1	81.0	71.6	75.3	73.4
Soft-Masked Bert	80.9	73.7	73.2	73.5	77.4	66.7	66.2	66.4
SpellGCN	–	74.8	80.7	77.7	–	72.1	77.7	75.9
Realise	84.7	77.3	81.3	79.3	84.0	75.9	79.9	77.8
PLOME	–	77.4	81.5	79.4	–	75.3	79.3	77.4
SCOPE	–	78.2	83.5	80.8	–	75.8	81.0	78.3
PYGC	86.2	80.3	83.0	81.6	85.3	78.5	81.1	79.8
SCOPE†	–	81.1	84.3	82.7	–	79.2	82.3	80.7
EGCM†	87.2	83.4	79.8	81.6	86.3	81.4	78.4	79.9
PYGC(ours)†	87.2	82.1	83.9	83.0	86.2	80.1	81.9	81.0

Table 5. The performance of our model and current SOTA methods on CSCD-IME. The best results are in bold.

Methods	Sentence-level						Character-level					
	Detection			Correction			Detection			Correction		
	Pre	Rec	F1	Pre	Rec	F1	Pre	Rec	F1	Pre	Rec	F1
Bert	79.16	65.83	71.88	70.55	58.66	64.06	83.00	67.01	74.15	73.59	59.41	65.75
Soft-masked Bert	80.87	64.78	71.94	74.42	59.62	66.20	84.46	65.35	73.68	77.50	59.97	67.62
PLOME	79.78	57.23	66.65	78.09	56.01	65.23	83.48	57.99	68.44	81.49	56.61	66.81
PYGC(ours)	81.72	68.34	74.43	79.23	66.26	72.17	83.88	69.78	76.18	81.22	67.56	73.76

model's performance. Compared with all previous models, our model achieves a balance between precision and recall, and establishes new state-of-the-art overall performance(0.8%/0.3% on detection, 1.5%/0.3% on correction, when without/with iterative decoding), proving our method's effectiveness.

Besides, Table 5 compares our model and the benchmark models of the CSCD-IME dataset, which is collected by some Pinyin Input Methods and manually labeled by humans [8]. We found that the PLOME model can not beat BERT even though it incorporates the visual and sound information at the embedding level and adopts confusion set based pre-training. It shows that PLOME has room for improvement in real-case spelling errors generated by Pinyin Input Methods. Our PYGC model achieves the best results in this scenario. As for detection/correction tasks, we obtained 2.55%/5.97% improvement at the sentence level and 2.03%/6.14% at the character level.

Table 6. Sentence-level correction results on SIGHAN-Isolation. Best results are in bold. **Left:** Statistics of SIGHAN-Isolation. **Right:** Sentence-level correction performance of different models on SIGHAN-Isolation.

	#Pairs	#Sents	Method	Pre	Rec	F1
Training Set	21801	277804	Bert	43.7	26.9	33.3
Test Set	1496	3162	PLOME	48.6	38.8	43.2
Train Set ∩ Test Set	1443	-	SCOPE	54.8	47.9	51.1
Isolation Train Set	20358	198050	PYGC(ours)	**64.3**	**49.0**	**56.2**
Isolation Test Set	1496	3162				

4.3   Isolation Test

It is pointed out that existing spelling error correction models achieve good results on test sets by "remembering" the errors in training sets, but their generalization ability is poor [33]. In order to measure the generalization ability of CSC models, following [33], we delete sentences containing error pairs in the SIGHAN test set from the train set to obtain a SIGHAN-Isolation dataset, the statistics for SIGHAN-Isolation are shown in Table 6(Left). We compare the sentence-level correction metrics of Bert, the current SOTA model PLOME and SCOPE, and our PYGC model on the SIGHAN-Isolation. Table 6(Right) shows that our model has the best generalization ability. Specifically, it achieves 9.1%/1.1%/5.1% improvements on Precision/Recall/F1 over the current best-performing model.

4.4   Ablation Study

This section investigates the importance of different components of our PYGC model. As shown in Table 7, when pinyin loss is removed, the performance of our model decreases by more than 3%, which indicates that the introduction of pinyin loss enables the pinyin language model to capture the ground truth pinyin information better. When the fusion module is removed, the semantic language model and the pinyin language model become independent language models, and the performance of the model also decreases by more than 1%, indicating that the fusion of semantic and pinyin information can help the final prediction. Besides, when confusion set based pre-training or iterative decoding is removed, the model performance decreases by more than 1%, proving the effectiveness of our adopted strategies.

4.5   Study of the Pinyin Language Model

The motivation of our model is that the pinyin sequence itself is a language model. To study if this hypothesis is true, we make visualizations when pre-training the pinyin language model. Figure 2(a) shows the loss and accuracy

Table 7. Ablation study on SIGHAN dataset. "-pinyin loss" means only using character correction loss at finetuning, "-fusion module" means removing fusion module, "-pretrain" means removing confusion set based pretraining, "-iterative" means removing iterative decoding.

Method	Detection			Correction		
	Pre	Rec	F1	Pre	Rec	F1
Bert	74.2	78.0	76.1	71.6	75.3	73.4
PYGC	**82.1**	**83.9**	**83.0**	**80.1**	**81.9**	**81.0**
-pinyin loss	78.2	80.0	79.1	76.9	78.7	77.8
-fusion module	81.3	82.6	81.9	79.1	80.4	79.7
-pretrain	80.0	82.3	81.1	78.2	80.4	79.3
-iterative	80.3	83.0	81.6	78.5	81.1	79.8

curves. It can be observed that the pinyin language model starts to learn valuable knowledge of pinyin when pre-training for about 10k steps. After that, the loss function continues to decline, and the accuracy rate exceeds 70% and continues to increase. This indicates that the sequence of pinyin units has the characteristics of a language model, that is, the pinyin units of the mask can be inferred from the surrounding pinyin units. Figure 2(b) shows the latent representations of pinyin units after pre-training. It can be seen that the representations of the initial consonant, final consonant, and tone are grouped into one class, respectively, indicating that the model captures the characteristics of the pinyin unit sequence.

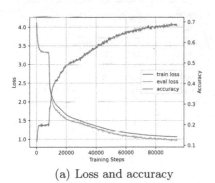

(a) Loss and accuracy

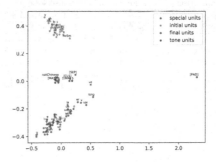

(b) Latent representations of pinyin units

Fig. 2. Visualization of the characteristics of pinyin language models. (a) The train/eval loss and eval accuracy when pre-train the pinyin language model. (b) Visualization of latent space representations after pre-training.

4.6   Case Study

Table 8 shows two real correction results of PYGC. In the first example, "很" (h,en,3), "赶" (g,an,3), "们" (m,en,2) are all valid corrections in semantic meaning. However, "很" (h,en,3) shares a common pinyin final unit(en) with the input character. The pinyin language model thus gives predictions for the corresponding initial, final, and tone units with high confidence. We can also find that the prediction probabilities for the correct pinyin units are higher for the ground truth output token, which means the pinyin language model has a "guiding" effect on the final prediction. The second example shows a sentence with continuous spelling errors. Although the pinyin language model fails to correct the tone error("3" for the input character "摆"), our PYGC model can give a correct final prediction "拜托" based on the fused pinyin and semantic information. The success of these two examples demonstrates the effectiveness of our PYGC model.

Table 8. Examples of PYGC prediction results, the incorrect input characters are marked in red and the correct predictions are marked in blue. For each incorrect character, we make predictions on its output token, pinyin initial unit, pinyin final unit, and tone. For all candidate predictions, the number in the parentheses represents its prediction probability.

Input	我回国之后，我跟(g,en,1)快去你家。		
Predict	我回国之后，我很(h,en,3)快去你家。		
	token	很(0.678),赶(0.322),们(9.32e-05)	
Top-3 Candidates	initial	h(0.999),g(8.10e-05),m(7.20e-05)	
(Probabilities)	final	en(0.999),h(3.97e-04),eng(3.20e-04)	
	tone	3(0.999),4(1.37e-04),5(1.13e-04)	
Translation	I'll come to your house soon after I return.		
Input	邻居们摆(b,ai,3)脱(t,uo,1)我当他们的代表人给李厂长写了这封信。		
Predict	邻居们拜(b,ai,4)托(t,uo,1)我当他们的代表人给李厂长写了这封信。		
	token	拜(0.995),就(1.68e-03),把(8.51e-04)	托(0.997),脱(1.83e-03),听(5.26e-05)
Top-3 Candidates	initial	b(0.999),s(1.39e-05),p(1.07e-05)	t(0.999),d(8.48e-05),sh(3.51e-05)
(Probabilities)	final	ai(0.999),o(2.30e-04),a(6.03e-05)	uo(0.999),ou(2.78e-06),i(1.40e-06)
	tone	3(0.998),4(1.66e-03),2(1.35e-04)	1(0.999),4(8.76e-06),3(4.38e-06)
Translation	The neighbors asked me to be their representative and wrote this letter to Director Li.		

5   Conclusion

we propose PYGC, a **P**in**Y**in language model **G**uided **C**orrection model for Chinese Spell Checking. PYGC captures the inherent nature of the pronunciation of Chinese sequences(Chinese pinyin) and models the pinyin sequence as an independent language model to guide the spelling error correction task. A fusion module that leverages information in textual and acoustic modality is also used to integrate semantic and pinyin representations. In addition, PYGC utilizes confusion set based pre-training and iterative decoding to improve the model performance. Experimental results demonstrate that our model establishes new state-of-art results on SIGHAN and CSCD-IME datasets. Furthermore, the metrics on the SIGHAN-Isolation dataset show that our model has the best generalization ability, which verifies the importance of pinyin in the CSC task and the effectiveness of our approach for modeling pinyin.

References

1. Cheng, X., et al.: SpellGCN: incorporating phonological and visual similarities into language models for Chinese spelling check. arXiv preprint arXiv:2004.14166 (2020)
2. Chiu, H.W., Wu, J.C., Chang, J.S.: Chinese spelling checker based on statistical machine translation. In: Proceedings of the Seventh SIGHAN Workshop on Chinese Language Processing, pp. 49–53 (2013)
3. Cui, Y., Che, W., Liu, T., Qin, B., Wang, S., Hu, G.: Revisiting pre-trained models for Chinese natural language processing. arXiv preprint arXiv:2004.13922 (2020)
4. Devlin, J., Chang, M.W., Lee, K., Toutanova, K.: Bert: pre-training of deep bidirectional transformers for language understanding. arXiv preprint arXiv:1810.04805 (2018)
5. Han, D., Chang, B.: A maximum entropy approach to Chinese spelling check. In: Proceedings of the Seventh SIGHAN Workshop on Chinese Language Processing, pp. 74–78 (2013)
6. Hong, Y., Yu, X., He, N., Liu, N., Liu, J.: Faspell: a fast, adaptable, simple, powerful Chinese spell checker based on DAE-decoder paradigm. In: Proceedings of the 5th Workshop on Noisy User-Generated Text (W-NUT 2019), pp. 160–169 (2019)
7. Hsieh, Y.M., Bai, M.H., Chen, K.J.: Introduction to CKIP Chinese spelling check system for SIGHAN bakeoff 2013 evaluation. In: Proceedings of the Seventh SIGHAN Workshop on Chinese Language Processing, pp. 59–63 (2013)
8. Hu, Y., Meng, F., Zhou, J.: CSCD-IME: correcting spelling errors generated by pinyin IME. arXiv preprint arXiv:2211.08788 (2022)
9. Huang, L., et al.: Phmospell: phonological and morphological knowledge guided Chinese spelling check. In: Proceedings of the 59th Annual Meeting of the Association for Computational Linguistics and the 11th International Joint Conference on Natural Language Processing (Volume 1: Long Papers), pp. 5958–5967 (2021)
10. Lee, L.H., Wu, W.S., Li, J.H., Lin, Y.C., Tseng, Y.H.: Building a confused character set for Chinese spell checking. In: 27th International Conference on Computers in Education (ICCE 2019), pp. 703–705. Asia-Pacific Society for Computers in Education (2019)
11. Li, J., Wang, Q., Mao, Z., Guo, J., Yang, Y., Zhang, Y.: Improving Chinese spelling check by character pronunciation prediction: the effects of adaptivity and granularity. arXiv preprint arXiv:2210.10996 (2022)
12. Liu, C.L., Lai, M.H., Tien, K.W., Chuang, Y.H., Wu, S.H., Lee, C.Y.: Visually and phonologically similar characters in incorrect Chinese words: analyses, identification, and applications. ACM Trans. Asian Lang. Inf. Process. 10(2), 1–39 (2011)
13. Liu, S., et al.: Craspell: a contextual typo robust approach to improve Chinese spelling correction. In: Findings of the Association for Computational Linguistics (ACL 2022), pp. 3008–3018 (2022)

14. Liu, S., Yang, T., Yue, T., Zhang, F., Wang, D.: Plome: pre-training with misspelled knowledge for Chinese spelling correction. In: Proceedings of the 59th Annual Meeting of the Association for Computational Linguistics and the 11th International Joint Conference on Natural Language Processing (Volume 1: Long Papers), pp. 2991–3000 (2021)

15. Liu, X., Cheng, F., Duh, K., Matsumoto, Y.: A hybrid ranking approach to Chinese spelling check. ACM Trans. Asian Low Resour. Lang. Inf. Process. **14**(4), 1–17 (2015)

16. Liu, Y., et al.: Roberta: a robustly optimized bert pretraining approach. arXiv preprint arXiv:1907.11692 (2019)

17. Sun, R., Wu, X., Wu, Y.: An error-guided correction model for Chinese spelling error correction. arXiv preprint arXiv:2301.06323 (2023)

18. Sun, Z., et al.: Chinesebert: chinese pretraining enhanced by glyph and pinyin information. arXiv preprint arXiv:2106.16038 (2021)

19. Tseng, Y.H., Lee, L.H., Chang, L.P., Chen, H.H.: Introduction to SIGHAN 2015 bake-off for Chinese spelling check. In: Proceedings of the Eighth SIGHAN Workshop on Chinese Language Processing, pp. 32–37 (2015)

20. Wang, B., Che, W., Wu, D., Wang, S., Hu, G., Liu, T.: Dynamic connected networks for Chinese spelling check. In: Findings of the Association for Computational Linguistics (ACL-IJCNLP 2021), pp. 2437–2446 (2021)

21. Wang, D., Song, Y., Li, J., Han, J., Zhang, H.: A hybrid approach to automatic corpus generation for Chinese spelling check. In: Proceedings of the 2018 Conference on Empirical Methods in Natural Language Processing, pp. 2517–2527 (2018)

22. Wang, D., Tay, Y., Zhong, L.: Confusionset-guided pointer networks for Chinese spelling check. In: Proceedings of the 57th Annual Meeting of the Association for Computational Linguistics, pp. 5780–5785 (2019)

23. Wu, S.H., Liu, C.L., Lee, L.H.: Chinese spelling check evaluation at SIGHAN bake-off 2013. In: Proceedings of the Seventh SIGHAN Workshop on Chinese Language Processing, pp. 35–42 (2013)

24. Xu, H.D., et al.: Read, listen, and see: leveraging multimodal information helps Chinese spell checking. arXiv preprint arXiv:2105.12306 (2021)

25. Yang, S., Yu, L.: COSPA: an improved masked language model with copy mechanism for Chinese spelling correction. In: Uncertainty in Artificial Intelligence, pp. 2225–2234. PMLR (2022)

26. Yang, T.H., Hsieh, Y.L., Chen, Y.H., Tsang, M., Shih, C.W., Hsu, W.L.: Sinica-IASL Chinese spelling check system at SIGHAN-7. In: Proceedings of the Seventh SIGHAN Workshop on Chinese Language Processing, pp. 93–96 (2013)

27. Yeh, J.F., Li, S.F., Wu, M.R., Chen, W.Y., Su, M.C.: Chinese word spelling correction based on n-gram ranked inverted index list. In: Proceedings of the Seventh SIGHAN Workshop on Chinese Language Processing, pp. 43–48 (2013)

28. Yu, J., Li, Z.: Chinese spelling error detection and correction based on language model, pronunciation, and shape. In: Proceedings of the Third CIPS-SIGHAN Joint Conference on Chinese Language Processing, pp. 220–223 (2014)

29. Yu, L.C., Lee, L.H., Tseng, Y.H., Chen, H.H.: Overview of SIGHAN 2014 bake-off for Chinese spelling check. In: Proceedings of the Third CIPS-SIGHAN Joint Conference on Chinese Language Processing, pp. 126–132 (2014)

30. Zhang, R., et al.: Correcting Chinese spelling errors with phonetic pre-training. In: Findings of the Association for Computational Linguistics (ACL-IJCNLP 2021), pp. 2250–2261 (2021)

31. Zhang, S., Huang, H., Liu, J., Li, H.: Spelling error correction with soft-masked bert. arXiv preprint arXiv:2005.07421 (2020)

32. Zhang, S., Xiong, J., Hou, J., Zhang, Q., Cheng, X.: Hanspeller++: a unified framework for Chinese spelling correction. In: Proceedings of the Eighth SIGHAN Workshop on Chinese Language Processing (2015)
33. Zhang, X., Zheng, Y., Yan, H., Qiu, X.: Investigating glyph phonetic information for Chinese spell checking: what works and what's next. arXiv preprint arXiv:2212.04068 (2022)

Empirical Analysis of Multi-label Classification on GitterCom Using BERT and ML Classifiers

Bathini Sai Akash[1], Lov Kumar[2], Vikram Singh[2]([✉]), Anoop Kumar Patel[2], and Aneesh Krishna[3]

[1] BITS-Pilani Hyderabad, Hyderabad, India
[2] Department of Computer Engineering, NIT, Kurukshetra, India
{lovkumar,viks,akp}@nitkkr.ac.in
[3] Curtin University Perth, Bentley, Australia
A.Krishna@curtin.edu.au

Abstract. To maintain development consciousness, simplify project coordination, and prevent misinterpretation, communication is essential for software development teams. Instant private messaging, group chats, and sharing code are just a few of the capabilities that chat rooms provide to assist and meet the communication demands of software development teams. All of this is capacitated to happen in real-time. Consequently, chat rooms have gained popularity among developers. Gitter is one of these platforms that has gained popularity, and the conversations it contains may be a treasure trove of data for academics researching open-source software systems. This research made use of the GitterCom dataset, The largest collection of Gitter developer messages that have been carefully labelled and curated and perform multi-label classification for the 'Purpose' category in the dataset. An extensive empirical analysis is performed on 6 feature selection techniques, 14 machine learning classifiers, and BERT transformer layer architecture with layer-by-layer comparison. Consequently, we achieve proficient results through our research pipeline involving Extra Trees Classifier and Random Forest classifiers with AUC (OvR) median performance of 0.94 and 0.92 respectively. Furthermore, The research proposed research pipeline could be utilized for generic multi-label text classification on software developer forum text data.

Keywords: GitterCom · BERT analysis · Data Imbalance Methods · Feature Selection · Ensemble models · Sentence Embedding

1 Introduction

The developer chatroom has two types: "public" and "private" varieties. Open source projects typically employ public chat rooms and a lot of value is placed on the creation of a welcoming community that converses and shares expertise. GitteCom, also known as Gitter, is a communication platform designed for

B. Luo et al. (Eds.): ICONIP 2023, LNCS 14451, pp. 240–252, 2024.
https://doi.org/10.1007/978-981-99-8073-4_19

developers and communities, that offers both public and private chat rooms. Public chat rooms allow developers to connect with like-minded individuals, share knowledge, and contribute to discussions. Private chat rooms provide a secure space for teams and organizations to communicate internally. Several factors contribute to Gitter's popularity among open-sourced developers. First off, Gitter allows anybody to browse user-generated content.

GitterCom [1] contains information on 10,000 messages gathered from 10 Gitter communities for open-source development projects. It provides valuable data for examining developer behaviors. However, the 'Purpose' category for each message in the dataset was manually labelled by experts. Our research primarily focuses on generating a pipeline that performs comprehensive comparative analysis on feature selection, machine learning classification performance, and BERT in-depth layer performance on the aforementioned dataset.

Extrapolating our approach could save prime resources and time for developer teams to categorize discussions into their valid roles and purpose groups. The research would consequentially contribute to the research community giving exposure and standpoint from feature selection to BERT performance for text classification in the field of Software development forum discussion. The approach in the research can be further applied in a generic sense for text classification in the domain and use cases analogous to ours. The research questions (RQs) posed and answered in our work are as follows:

- **RQ1**: Is there statistical equality between the results of the feature selection?
- **RQ2**: Does the application of the Synthetic Minority Oversampling Technique (SMOTE) significantly contribute to an increase in prediction performance?

  RQ3: What is the proficiency in the performance of Machine Learning classifiers on the text classification problem?
- **RQ4**: What are the statistical variability and impact in the performance of the 13 BERT transformer layers on multi-label classification respectively?

To the best of our knowledge, we find our work to be original and perform meticulous checks on the validity of the information provided in the research.

2   Related Work

The following section explains the related work concerned with the research. Text classification has been widely performed using supervised learning time and again. From deep learning approaches, Markov-based language models, and context-based language models to pre-trained language models, it's a widely spread field of study [2–4]. However, it is vital that an integration of these techniques is cast light upon.

Gitter has been in notable studies, particularly in text and empirical analysis. The study in [5], discovers when developers connect, comprehend the community dynamics on Gitter, pinpoint the subjects covered, and examine developer interactions. The study aims to close the awareness and comprehension gap in this field. Esteban et al. [6] in 2022 worked on contrasting the communication behaviors of Gitter with those of other platforms, developing automated techniques for message intent assessment, and, based on the results, recommended potential areas for further study.

In order to fully realize their potential as efficient communication platforms for developers, research in [7] seeks to offer insights into how Slack and Gitter's forums are used in software development teams, comprehend their influence on projects, and suggest possibilities for improvement. However, there has been little to no research on the integration of BERT embeddings with compatible feature selection techniques and classification models on the GitterCom dataset. Our research bridges this gap by generating an effective pipeline that not only performs comprehensive comparative analysis but also proposes a proficient design for the automatic labeling of The Purpose Category on the dataset.

3   Dataset Description

GitterCom [1] contains information on 10,000 messages gathered from 10 Gitter[1] communities for open-source development projects (each community comprises 1000 messages). Each communication (message) in the dataset was labeled manually to categorize the purpose of communication according to the categories distinguished by Lin et al. [8]. There are 7 features for each message in the dataset: *channel name, a unique messageID, Date-time of the message postage, message author, Purpose of the message, Purpose category, and Purpose subcategory*. The last two features were manually labeled by the authors of the research. The distribution of messages per category is given in detail in [1]. We made use of messages with only the Purpose Category for our research. The distribution of the dataset by *'Purpose' category* as: Communication with 5274 Message counts, Dev-ops with 2776 Message counts, Customer support with 1431 Message counts, and others with a 519 Message counts.

4   Research Methodology

The following section presents an exposition of the various methodologies employed in the research. This includes feature selection techniques, sampling algorithms, and machine learning classifiers.

4.1   Feature Selection

The study made use of the feature selection methods [9] include Information Gain (FR-2), Gain Ratio (FR-3), OneR (FR-4), Chi-squared (FR-5), One-way ANOVA (Fr-6), and PCA extraction (FR-7). All the preceding techniques apart from PCA are ranking feature selection techniques, PCA is a feature extraction technique. OneR is a rules-driven technique that develops rules for the dependent

---

[1] https://gitter.im/.

variables and orders the features according to classification rates. The Gain Ratio is a type of information gain strategy that aids in the elimination of bias. This method takes into consideration the split's inherent characteristics. Information gain is determined by contrasting the dataset's entropy before and after the modifications, and it represents the reduction in entropy. The relevance of the characteristic may be ranked using the chi-square statistic value.

4.2 Data Balancing

The data used for training machine learning models is prime for performance enhancement. As shown in Sect. 3, there is a class imbalance in the Gittercom dataset after multi-labeling is performed manually. If the categorization categories are not approximately evenly represented in a dataset, it is unbalanced. We made use of vanilla SMOTE (Synthetic Minority Over-sampling Technique) [10] for the removal of class imbalance in the dataset. SMOTE is an oversampling technique that balances the minority class on par with the majority.

4.3 Machine Learning Classifiers

In order to classify BERT embeddings into the multi-labels of the dataset we made use of 14 established machine-learning models. The models along with the naming convention are given in Table 1. Aiming to suitably analyze the embeddings in conjunction with SMOTE and feature selection, the models applied are diverse in approach and architecture.

Table 1. Naming convention of adapted ML classifiers

Abr	Classification Model	Abr	Classification Model
MBC	Multinomial Naive Bayes	BBC	Bernoulli Naive Bayes
GBC	Gaussian Naive Bayes	CBC	Complement Naive Bayes
DT	Decision Tree Classifier	LSV	Support Vector Classifier (Linear)
PSV	Support Vector Classifier (Poly)	KNN	K Neighbors Classifier
RSV	Support Vector Classifier (RBF)	EXT	Extra Trees Classifier
RFC	Random Forest Classifier	BAC	Bagging Classifier
GRB	Gradient Boosting Classifier	ADB	AdaBoost Classifier

5   BERT Architecture

We made use of BERT$_{base}$ architecture with 12 Layers, 12 attention heads with total parameters in the architecture of 110 M [11]. It is based on the encoder-decoder model introduced in [12]. The architecture is shown in Fig. 1. It is to be noted that the initial layer represented by the Encoder Layer in Fig. 1 is not part of the '12' transformers layers in BERT. This layer is also labelled as LR-1 in Sect. 6.4. Hence we have 13 layers including the initial encoder layer that are analyzed in the aforementioned Section.

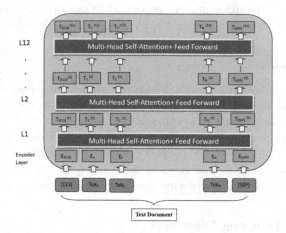

Fig. 1. Scheme of BERT architecture

6 Comparative Analysis

The following section gives comprehensive descriptive statistics and analysis of the various feature selection techniques, BERT layer performances, and Classification techniques. We made use of 6 feature ranking techniques and 14 classification techniques respectively. Our extensive comparison was performed on metrics Area Under curve (AUC) with two variants: One-vs-One and One-vs-Rest respectively. We also made use of SMOTE to overcome the class imbalance problem existing in the dataset. In total, we generated 182 datasets in our data warehouse. Each dataset corresponds to a set of techniques used for analysis. To employ statistical studies apart from AUC studies, two tests were employed, *Wilcoxon Signed Rank* [13] Test and *Friedman Test* [14]. This would evaluate the importance and dependability of the various methods used.

6.1 Feature Selection

To draw a conclusion, the 6 feature selection were compared comprehensively in this section for impact on prediction performance.

RQ1: Is There Statistical Equality Between the Results of the Feature Selection?

AUC Performance Analysis of Feature Ranking Techniques Using Descriptive Statistics and Boxplots: The AUC values for the 6 Feature selection technique used are detailed in Fig. 2. As shown in Fig. 2, it is to be noted that feature extraction and selection were performed over on unbalanced dataset and balanced data using SMOTE was applied after feature selection. The Original dataset (OD) without feature selection is referred to as FR-1 and consecutive FS techniques are labeled from 2 to 6 in numerical order. The data presented in box plots Fig. 2 leads to subsequent observations:

- All the feature selection techniques perform relatively competent with median AUC values in the range of 0.94 to 0.95.
- Chi Value Ranking (FR-5) under-performs relatively, even in comparison to the original dataset with all features. It
- Feature selection technique PCA (FR-7) seems to perform on par with the original dataset, even though it comprises of notably fewer features.

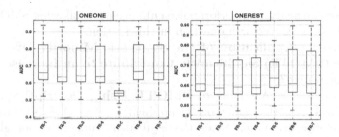

Fig. 2. Box Plots for Feature selection technique performance

Statistical Hypothesis Testing for Comparison of Feature Selection Performance: As aforementioned, we made use of two techniques for Hypothesis testing, the Wilcoxon Signed Rank Test, and the Friedman Test. The results from Hypothesis testing (p-values) are presented in Table 2. On inspection of Table 2, we see that FR-6 and FR-7 have $p \geq 0.05$ which null hypothesis is not rejected and there is no significant difference in performance over the original dataset. FR-5 is significantly different from all techniques under consideration and has minimal p-values. Further, we employ the Friedman Test to conclude the quantitative performance of the techniques. As aforementioned, the lower the rank higher the performance of the respective technique. From Table 2, we note that FR-6 has the lowest rank and performs slightly better than the original dataset without feature selection. FR-7 performs on par with the original data. Furthermore, the mean-rank values of all techniques under consideration barring FR-5 are fairly good.

Table 2. Feature selection p values AUC, for one-v-one

	FR1	FR2	FR3	FR4	FR5	FR6	FR7	Mean-Rank
FR-1	1.0	0.0	0.01	0.0	0.0	0.47	0.6	2.87
FR-2	0.0	1.0	0.71	0.86	0.0	0.0	0.0	4.31
FR-3	0.01	0.71	1.0	0.9	0.0	0.0	0.0	4.34
FR-4	0.0	0.86	0.9	1.0	0.0	0.0	0.0	4.14
FR-5	0.0	0.0	0.0	0.0	1.0	0.0	0.0	6.94
FR-6	0.47	0.0	0.0	0.0	0.0	1.0	0.95	2.55
FR-7	0.6	0.0	0.0	0.0	0.0	0.95	1.0	2.85

6.2   Effect of Oversampling

A comprehensive analysis of performance enhancement following SMOTE is performed to answer the research question in this section.

RQ2: Does the Application of the Synthetic Minority Oversampling Technique (SMOTE) Significantly Contribute to an Increase in Pediction Performance?

AUC Performance Analysis of SMOTE Using Descriptive Statistics and Boxplots: The AUC values on the application of SMOTE is shown in Fig. 3. The performance evaluation takes the equivalent cumulative for all datasets including the original dataset without feature selection. The dataset without SMOTE is referred to as OD in this section. SMD refers to SMOTE. The data presented in box plots as shown in Fig. 3 leads to subsequent observations:

- There seems to be a clear improvement in median AUC performance after the application of SMOTE.
- The median AUC value notably improved from 0.69 to 0.95 in the case of One-vs-One.

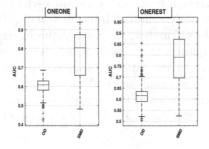

Fig. 3. Box Plot for SMOTE performance

Statistical Hypothesis Testing to Check for SMOTE Enhancement: The results from hypothesis testing (p-values) are presented in Table 3. Table

3 corresponds to hypothesis testing performed using AUC (OvO) as the metric, the research also computed p-values with AUC (OvR) which gave similar results. Further, we employ the Friedman Test to conclude on the quantitative performance enhancement generated by applying SMOTE. From Table 3, we note that there is a notable drop in Mean rank from OD to SMD. Therefore, we deduce that the application of SMOTE significantly improved the performance in classification. **Hence we conclude that the application of SMOTE over BERT embeddings of text dataset significantly enhanced the performance for multi-class prediction using Machine learning models.**

Table 3. SMOTE p values AUC one-v-one

	OD	SMD	Mean-Rank
OD	1.00	0.00	1.885
SMD	0.00	1.00	1.115

6.3 Performance of Classifiers

To draw a conclusion, 14 machine learning models used in our classification pipeline, shown in Fig. 4, were compared comprehensively in this section for impact on prediction performance.

RQ3: What is the Proficiency in the Performance of Machine Learning Classifiers on the Text Classification Problem?

AUC Performance Analysis of Classification Models Using Descriptive Statistics and Boxplots: The AUC values for the 14 machine learning models used are detailed in Fig. 4. As stated before, for descriptive statistics comparison of performance, we use median values as compared to the mean. The data presented in box plots as shown in Fig. 4 leads to subsequent observations:

- All the classification models perform relatively competent with median AUC values ranging from 0.80 to 0.95.
- Models LSV and BBC underperform relatively in comparison to others with AUC values close to 0.8.
- EXT and RFC appear to outperform all other models attaining median AUC values close to 0.95.

Statistical Hypothesis Testing for Comparison of Model Classification Performance: The results from hypothesis testing (p-values) are presented in Table 4 for AUC (OvO). On inspection of Table 4, we see that MBC, BBC, GBC, CBC, and DT largely have $p¿0.05$ for which the null hypothesis is not rejected and there is no significant difference in performance for each of these models. This could be because the first four models are variants of Naive Bayes and perhaps perform akin [15]. Further, we employ Friedman Test to conclude the quantitative performance of the technique. From Table 4, we note that

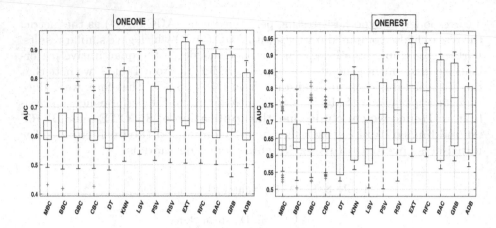

Fig. 4. Performance parameter Box-Plot for Classification Model

EXT and RFC achieve the lowest ranks and perform significantly better than all other models. **Hence, we conclude that Extra Trees Classifier (EXT) and Random Forest Classifer (RFC) outperform all other models significantly in our case of text multi-label classification using BERT sentence embeddings.**

Table 4. ML classifier p-values AUC one-v-one

	MBC	BBC	GBC	CBC	DT	KNN	LSV	PSV	RSV	EXT	RFC	BAC	GRB	ADB	Mean-Rank
MBC	1.0	0.42	0.12	1.0	0.41	0.0	0.0	0.0	0.0	0.0	0.0	0.0	0.0	0.01	10.88
BBC	0.42	1.0	0.52	0.33	0.52	0.0	0.0	0.0	0.0	0.0	0.0	0.0	0.0	0.02	10.75
GBC	0.12	0.52	1.0	0.11	0.49	0.0	0.0	0.0	0.0	0.0	0.0	0.01	0.0	0.02	9.7
CBC	1.0	0.33	0.11	1.0	0.49	0.0	0.0	0.0	0.0	0.0	0.0	0.0	0.0	0.0	11.02
DT	0.41	0.52	0.49	0.49	1.0	0.0	0.0	0.0	0.0	0.0	0.0	0.0	0.0	0.0	10.93
KNN	0.0	0.0	0.0	0.0	0.0	1.0	0.21	0.33	0.27	0.0	0.0	0.02	0.0	0.13	7.96
LSV	0.0	0.0	0.0	0.0	0.0	0.21	1.0	0.41	0.65	0.0	0.01	0.74	0.06	0.07	5.69
PSV	0.0	0.0	0.0	0.0	0.0	0.33	0.41	1.0	0.85	0.0	0.0	0.46	0.03	0.15	5.54
RSV	0.0	0.0	0.0	0.0	0.0	0.27	0.65	0.85	1.0	0.0	0.01	0.7	0.12	0.17	5.35
EXT	0.0	0.0	0.0	0.0	0.0	0.0	0.0	0.0	0.0	1.0	0.0	0.0	0.0	0.0	2.44
RFC	0.0	0.0	0.0	0.0	0.0	0.0	0.01	0.0	0.01	0.0	1.0	0.0	0.0	0.0	3.89
BAC	0.0	0.0	0.01	0.0	0.0	0.02	0.74	0.46	0.7	0.0	0.0	1.0	0.09	0.0	7.1
GRB	0.0	0.0	0.0	0.0	0.0	0.0	0.06	0.03	0.12	0.0	0.0	0.09	1.0	0.0	5.03
ADB	0.01	0.02	0.02	0.0	0.0	0.13	0.07	0.15	0.17	0.0	0.0	0.0	0.0	1.0	8.7

6.4   BERT Layer Performance Comparison (RQ4)

To draw a conclusion, the performance of 12 BERT transformer layers and the input encoding layer are compared comprehensively in this section for impact on prediction performance.

RQ4: What is the Statistical Variability and Impact in the Performance of the 13 BERT Transformer Layers Respectively? Which Layer Performs Best for Classification?

AUC Performance Analysis BERT Layers Using Descriptive Statistics and Boxplots: The AUC values for the 12 BERT layers are detailed in Fig. 5. LR-1 corresponds to the input encoding layer and each consecutive label represents each BERT layer. The data presented in box plots shown in Fig. 5 lead to subsequent observations:

– No clear insights can be derived on variability in the performance of BERT layers.
– All the layers perform with a median AUC in the range of 0.68 to 0.71

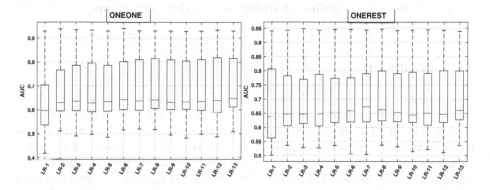

Fig. 5. Performance Parameter Box Plot for ML classifier

Table 5. BERT layer p-values AUC one-v-one

	LR1	LR2	LR3	LR4	LR5	LR6	LR7	LR8	LR9	LR10	LR11	LR12	LR13	Mean-Rank
LR-1	1.0	0.0	0.0	0.0	0.0	0.0	0.0	0.0	0.0	0.0	0.0	0.0	0.0	9.42
LR-2	0.0	1.0	0.99	0.91	0.99	0.46	0.4	0.23	0.67	0.49	0.65	0.84	0.11	7.49
LR-3	0.0	0.99	1.0	0.87	1.0	0.4	0.39	0.24	0.79	0.58	0.66	0.74	0.1	7.63
LR-4	0.0	0.91	0.87	1.0	0.93	0.34	0.32	0.18	0.63	0.42	0.56	0.71	0.08	7.12
LR-5	0.0	0.99	1.0	0.93	1.0	0.42	0.37	0.22	0.73	0.54	0.66	0.77	0.1	7.86
LR-6	0.0	0.46	0.4	0.34	0.42	1.0	0.91	0.71	0.69	0.84	0.73	0.68	0.39	6.11
LR-7	0.0	0.4	0.39	0.32	0.37	0.91	1.0	0.76	0.6	0.78	0.66	0.58	0.43	5.66
LR-8	0.0	0.23	0.24	0.18	0.22	0.71	0.76	1.0	0.36	0.56	0.45	0.43	0.59	5.51
LR-9	0.0	0.67	0.79	0.63	0.73	0.69	0.6	0.36	1.0	0.8	0.92	0.95	0.18	7.42
LR-10	0.0	0.49	0.58	0.42	0.54	0.84	0.78	0.56	0.8	1.0	0.87	0.81	0.29	7.53
LR-11	0.0	0.65	0.66	0.56	0.66	0.73	0.66	0.45	0.92	0.87	1.0	0.88	0.21	6.97
LR-12	0.0	0.84	0.74	0.71	0.77	0.68	0.58	0.43	0.95	0.81	0.88	1.0	0.22	7.04
LR-13	0.0	0.11	0.1	0.08	0.1	0.39	0.43	0.59	0.18	0.29	0.21	0.22	1.0	5.24

Statistical Hypothesis Testing for Comparison of BERT Layer Performance: The results from hypothesis testing (p-values) are presented in Table 5 for AUC (OvO). On inspection of Table 5, we see that only LR-1 (encoder layer) has a significant difference in performance with BERT layers with $p \leq 0.05$. The 12 BERT layers do not show any significant intra-performance variation according to p values. The research also computed p-values with AUC (OvR) which gave similar results. Further, we employ the Friedman Test to conclude on the quantitative performance of the technique. From Table 5, we see that LR-7, LR-8, and LR-13 perform the best with a rank close to 5.5. While we analyze the 12 layers of BERT, it is not a guarantee that the final layer would collect the most pertinent data relating to the classification objective. **Hence, we conclude that although BERT layer 12 (LR-13) outperforms all other layers, layers 5 (LR-6) and 6 (LR-7) perform similarly in our case of text multi-label classification using BERT sentence embeddings. There is no significant difference in the performance of the 12 BERT layers.**

7   Conclusion

By effectively addressing the class imbalance and feature redundancy issues, this research offers a substantial advancement in the analysis of GitterCom text classification and for generic multi-label text classification on software developers forum text data. The effectiveness of the created models utilizing 14 distinct classifiers has been evaluated by the creation of 182 datasets with varying feature selection techniques, SMOTE, and BERT layers. To address the issues of data coherence, reliability, feature irrelevance, and unbalanced data, the study uses SMOTE and 6 feature selection techniques. Extensive comparative analysis using the Wilcoxon sign rank test and Friedman test is performed to answer 4 research questions which lead us to the following conclusions:

- There is no significant difference in the performance of the 12 BERT layers. However, layer 12 outperforms all other layers, and layers 5 (LR-6) and 6 (LR-7) perform similarly in our case of text multi-label classification.
- In our scenario of text multi-label classification utilizing BERT sentence embedding, Extra Trees Classifier (EXT) and Random Forest Classifier (RFC) significantly beat all other models.
- The performance of multi-class prediction using machine learning models was significantly improved by employing SMOTE over BERT embeddings of text dataset.
- Even though there is no significant improvement in performance, one-way ANOVA (FR-6) and PCA feature extraction (FR-7) could be employed on BERT embeddings for performance enhancement.

The aforementioned conclusions could lead to intelligent decision-making for the research community over text classification using BERT and individual BERT transformer layer performance analysis. Finally, with AUC (OvR) median performance of 0.94 and 0.92, respectively, our research pipeline using Extra Trees classifier and Random Forest classifiers yield competent outcomes. Additionally, the research pipeline and design could be used for general multi-label text categorization on text data from software development forums.

References

1. Parra, E., Ellis, A., Haiduc, S.: Gittercom: a dataset of open source developer communications in gitter. In: Proceedings of the 17th International Conference on Mining Software Repositories, pp. 563–567 (2020)
2. Sun, C., Qiu, X., Xu, Y., Huang, X.: How to fine-tune BERT for text classification? In: Sun, M., Huang, X., Ji, H., Liu, Z., Liu, Y. (eds.) CCL 2019. LNCS (LNAI), vol. 11856, pp. 194–206. Springer, Cham (2019). https://doi.org/10.1007/978-3-030-32381-3_16
3. Lu, Z., Du, P., Nie, J.-Y.: VGCN-BERT: augmenting BERT with graph embedding for text classification. In: Jose, J.M., et al. (eds.) ECIR 2020. LNCS, vol. 12035, pp. 369–382. Springer, Cham (2020). https://doi.org/10.1007/978-3-030-45439-5_25
4. Li, W., Gao, S., Zhou, H., Huang, Z., Zhang, K., Li, W.: The automatic text classification method based on bert and feature union. In: 2019 IEEE 25th International Conference on Parallel and Distributed Systems (ICPADS), pp. 774–777. IEEE (2019)
5. Shi, L., et al.: A first look at developers' live chat on gitter. In: Proceedings of the 29th ACM Joint Meeting on European Software Engineering Conference and Symposium on the Foundations of Software Engineering, pp. 391–403 (2021)
6. Parra, E., Alahmadi, M., Ellis, A., Haiduc, S.: A comparative study and analysis of developer communications on slack and gitter. Empir. Softw. Eng. 27(2), 40 (2022)
7. El Mezouar, M., da Costa, D.A., German, D.M., Zou, Y.: Exploring the use of chatrooms by developers: an empirical study on slack and gitter. IEEE Trans. Softw. Eng. 48(10), 3988–4001 (2021)
8. Lin, B., Zagalsky, A., Storey, M.-A., Serebrenik, A.: Why developers are slacking off: understanding how software teams use slack. In: Proceedings of the 19th ACM Conference on Computer Supported Cooperative Work and Social Computing Companion, pp. 333–336 (2016)
9. Dash, M., Liu, H.: Feature selection for classification. Intell. Data Anal. 1(1–4), 131–156 (1997)
10. Chawla, N.V., Bowyer, K.W., Hall, L.O., Kegelmeyer, W.P.: Smote: synthetic minority over-sampling technique. J. Artif. Intell. Res. 16, 321–357 (2002)
11. Devlin, J., Chang, M.-W., Lee, K., Toutanova, K.: Bert: pre-training of deep bidirectional transformers for language understanding. arXiv preprint arXiv:1810.04805 (2018)
12. Vaswani, A., et al.: Attention is all you need. Adv. Neural Inf. Process. Syst. 30 (2017)
13. Woolson, R.F.: Wilcoxon signed-rank test. In: Wiley Encyclopedia of Clinical Trials, pp. 1–3 (2007)

14. Zimmerman, D.W., Zumbo, B.D.: Relative power of the Wilcoxon test, the Friedman test, and repeated-measures Anova on ranks. J. Exp. Educ. **62**(1), 75–86 (1993)
15. Al-Aidaroos, K.M., Bakar, A.A., Othman, Z.: Naive bayes variants in classification learning. In: 2010 International Conference on Information Retrieval and Knowledge Management (CAMP), pp. 276–281. IEEE (2010)

A Lightweight Safety Helmet Detection Network Based on Bidirectional Connection Module and Polarized Self-attention

Tianyang Li[1,2], Hanwen Xu[1(✉)], and Jinxu Bai[1]

[1] Computer Science, Northeast Electric Power University, Jilin 132012, Jilin, China
tianyangli@neepu.edu.cn , 1429969970@qq.com
[2] Jiangxi New Energy Technology Institute, Xinyu, Jiangxi, China

Abstract. Safety helmets worn by construction workers in substations can reduce the accident rate in construction operations. With the mature development of smart grid and target detection technology, automatic monitoring of helmet wearing by using the cloud-side collaborative approach is of great significance in power construction safety management. However, existing target detectors have a large number of redundant calculations in the process of multi-scale feature fusion, resulting in additional computational overhead for the detectors. To solve this problem, we propose a lightweight target detection model PFBDet. First, we design cross-stage local bottleneck module FNCSP, and propose an efficient lightweight feature extraction network PFNet based on this combined with Polarized Self-Attention to optimize the computational complexity while obtaining more feature information. Secondly, to address the redundancy overhead brought by multi-scale feature fusion, we design BCM (bidirectional connection module) based on GSConv and lightweight upsampling operator CARAFE, and propose an efficient multi-scale feature fusion structure BCM-PAN based on this combined with single aggregation cross-layer network module. To verify the effectiveness of the method, we conducted extensive experiments on helmet image datasets such as Helmeted, Ele-hat and SHWD, and the experimental results show that the proposed method has better recognition accuracy with less computational effort. And it is higher than most high-performance target detectors, which can meet the real-time detection of construction personnel wearing helmets in the construction scenarios of power systems.

Keywords: Helmet · Self-attention · Object detection · BCM-PAN · FNCSP

1 Introduction

During daily power inspection and maintenance processes, power construction personnel frequently come into contact with high-voltage power equipment. Failure to wear helmets increases the likelihood of safety accidents, posing significant

B. Luo et al. (Eds.): ICONIP 2023, LNCS 14451, pp. 253–264, 2024.
https://doi.org/10.1007/978-981-99-8073-4_20

security risks. To ensure that electric power construction personnel wear helmets and enhance on-site safety, monitoring of helmet usage among construction personnel is crucial. However, due to the expansive working areas within substations and constraints such as personnel fatigue, the effectiveness of manual supervision is often inadequate [1].

In recent years, in the field of computer vision, target detection technology has made great progress, and a series of target detection algorithms with superior performance have emerged to make it possible in helmet detection. Automatic detection of helmets worn by substation construction personnel using a monitoring system has the advantages of low cost [2] and full time work [3]. However, the high computational complexity and poor real-time performance of existing models in the task of helmet detection make it more difficult to detect helmet. The above factors make the helmet detection task one of the most challenging tasks in computer vision.

Therefore, in order to solve the above problems, a lightweight helmet wearing detection model PFBDet is proposed. The main contributions are summarized as follows:

(1) We present a novel cross-stage localized bottleneck module, FNCSP, which is integrated with Polarized Self-Attention, to design a lightweight feature extraction network, PFNet.
(2) By leveraging the benefits of GSConv convolution and lightweight upsampling operators, we introduce bidirectional connection module, which allows for the extraction of more precise image localization information without introducing unnecessary computational complexity.
(3) To achieve efficient multi-scale feature fusion, we propose BCM-PAN, a structure that combines the bidirectional connection module, with single aggregation cross-layer network module.

2   Related Work

Ross Girshick [4] et al. successfully introduced convolutional neural networks. Subsequently, relevant scholars in this field made continuous improvements on this basis and proposed target detection methods with better accuracy and speed, which made target detection algorithms more and more meet the requirements of landing applications [5]. [6] further refined the feature information of the target region of the helmet by using a non-parametric attention mechanism in the SSD model. However, the high-level features were downsampled multiple times, resulting in the loss of some detailed information.

In addition, [7] added deformable convolution with switchable null convolution, [8] added a residual network module, [9] replaced the original depth feature-based network model algorithm with a depth cascade network model, and [10] added a lightweight ACNet module. [11] introduced a compressed excitation layer and efficient channel attention. [12] used depth-separable convolution to reduce the number of parameters in the network and used the Hswish loss function to improve the helmet detection performance. However, these methods suffer from computationally redundant structures in the helmet wearing detection task and still need improvement in real-time detection.

After analyzing the aforementioned literature, it is observed that several improvement schemes have demonstrated varying degrees of enhancement in terms of accuracy and recall. However, it is worth noting that these improvements often come at the cost of compromised detection speed and real-time performance of the models, resulting in an increase in computational overhead for the detector. Therefore, in order to address this issue, this paper presents a lightweight method for detecting helmet wearing. The primary objective of this method is to enable real-time detection of helmet usage among construction personnel in power system construction scenarios while minimizing computational burden.

3   Approach

The backbone structure of our model is borrowed from the design principles of CSPDarknet [13], which is named PFNet as shown in Fig. 1, this feature extraction network follows two main modular designs: the FNCSP residual structure and the Polarized Self-Attention.

3.1   Lightweight Feature Extraction Networks

Most target detection models use residual structure, the main reason is that the target detection backbone is a deep network structure, adding residual structure can increase the gradient value of interlayer backpropagation, so as to extract finer features.

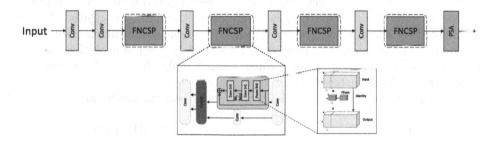

Fig. 1. PFNet adopts Polarized Self-Attention and FNCSP structure.

Therefore, leveraging the benefits of the residual structure, we have devised a novel design called FNCSP, which combines the lightweight convolution Pconv [14] and FasterNet. Illustrated in Fig. 1, the reason why Pconv is used in the residual structure is that it reduces computational redundancy and memory access. As can be seen in Fig. 1, Pconv only needs to apply regular convolution on a portion of the input channels for spatial feature extraction, and the remaining channels remain unchanged. For continuous or regular memory access, the first or last consecutive c_p^2 channel is considered representative of the entire feature map for computation, and without loss of generality, the input and output feature maps are considered to have the same number of channels, and for a typical r=1/4, Pconv is only 1/16th of the computation of regular convolution thus the FLOPs of Pconv is:

$$h \times w \times k^2 \times c_p^2$$

In addition, Pconv has fewer memory accesses:

$$h \times w \times 2c_p + k^2 \times c_p^2 \approx h \times w \times 2c_p$$

3.2   Polarized Self-attention

In comparison to other attention, The effectiveness of the polarized self-attention [15] in accomplishing fine-grained pixel-level tasks is attributed to its ability to minimize information loss through limited compression in both the spatial and channel dimensions. This characteristic enables the mechanism to sustain relatively high performance levels.

As shown in Fig. 2, the PSA is divided into two branches, the channel branch and the spatial branch, where the weights of the channel branch are calculated as follows:

$$A^{ch}(X) = F_{SG} \left[W_{z|\theta_1} \left((\sigma_1(W_v(X)) \times F_{SM}(\sigma_2(W_q(X)))) \right) \right]$$

As can be seen from the channel branch, the input feature X is first converted into q and v using 1×1 convolution, where the channel of q is completely compressed, while the channel dimension of v remains at a relatively high level. Since the channel dimension of q is compressed, the information needs to be augmented by HDR, so we choose softmax to augment the information of q, Then q and K are matrix multiplied and followed by 1×1 convolution, LN to change the $C/2$ dimension on the channel to C. Finally, the Sigmoid function is used to keep all parameters between 0 and 1.

The formula for calculating weights for spatial branching is as follows:

$$A^{sp}(X) = F_{SG} \left[\sigma_3 \left(F_{SM}(\sigma_1(F_{GP}(W_q(X)))) \times \sigma_2(W_v(X)) \right) \right]$$

Similar to the channel branch, the spatial branch first converts the input feature X into q and v using 1×1 convolution, where for the q feature, the spatial dimension is also compressed using GlobalPooling. So that it is converted to size of 1×1; while the spatial dimension of the v feature is kept at a relatively large HxW level. Since the spatial dimension of q is compressed, the q information is augmented using softmax. Then matrix multiplication is performed, and finally reshape and sigmoid are connected so that all parameters are kept between 0 and 1.

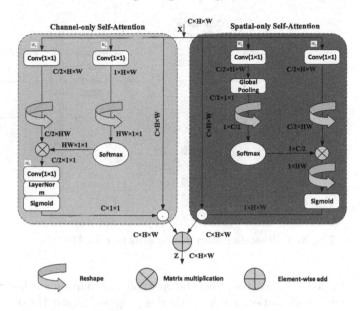

Fig. 2. Polarized Self-Attention.

3.3   Structure of BCM-PAN

In target detection tasks, the integration of multiscale features has proven to be a crucial and effective aspect of achieving accurate results.

PANet [16] introduces a bottom-up pathway operation in addition to the top-down feature fusion. This approach helps establish a more direct information flow between different feature levels, enabling improved integration of features across scales and enhancing the overall performance of target detection. Inspired by the above work, we designed a modified feature pyramid network model based on PANet as the neck of our detector. This improved PANet mainly consists of feature pyramid and bidirectional connection module. In the helmet wearing detection task, Feature information integration is an important part of the detection task, but we find that many initial features are lost in this process in different feature layers, which are beneficial to enhance the feature representation.

Therefore, to enhance the localization signal without adding too much computation, combining the lightweight convolution block GSConv with the lightweight upsampling [18–20] operator CARAFE [17] we propose BCM to integrate the feature maps of three adjacent layers.

The bidirectional connection module is shown in Fig. 3. This module fuses an additional low-level feature output from the backbone C_{i-1} into P_i and provides initial feedback information for each adjacent feature map. In this case, more accurate localization signal can be retained, which is important for small resolution target localization.

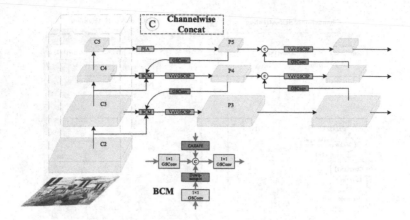

Fig. 3. Multi-scale feature fusion structure BCM-PAN.

In order to make the output of depth-separable convolution as close as possible to the original convolution, a new method is proposed to mix the original convolution, depth-separable convolution, and shuffle convolution, called GSConv. We apply this convolution to the BCM module, which can minimize the computational cost. Based on GSConv, [21] designed a single aggregation cross-layer network module VoVGSCSP based on the GSbottleneck of GSConv.

In Fig. 3 we propose an efficient multi-scale feature fusion structure BCM-PAN using bidirectional connection module and single aggregation cross-layer network module.

4   Experiment

To validate the effectiveness of PFBDet. We compare our method with several high-performance target detectors on two substation helmet datasets and the helmet detection dataset. In addition, to verify the robustness and generalizability of our proposed method, we conducted a series of comparative experiments on the safety cap public dataset and analyzed the ablation experiments of different components of the proposed method.

4.1   Experimental Details

We used an RTX3090 graphics card (PyTorch 1.11.0, CUDA11.3) for training and evaluation our method on three datasets. We use employ adaptive momentum estimation (Adam) to optimize the training models and perform online data enhancement processes such as random scaling, flipping, randomly changing contrast, and randomly changing luminance. Finally we train these models using 10^{-3} as the learning rate and experimentally train 200 epochs with batch size set to 32.

4.2 DataSets

We conducted experiments on three helmet datasets. They are briefly described below.

HelmetDet [22]: This dataset is a public dataset of helmets, 2114 helmets from the electric power industry, with a total of 9619 label samples, of which 8550 are wearing helmet labels and 1069 are not wearing helmet labels.

Elec-hat [31]: This dataset is a self-collected dataset. The total number of helmet images in it is 7000, all from construction workers wearing helmets in the power industry. The dataset contains two category labels (hat and no-hat) and overlaps and obscures each other heavily.

SHWD [23]: A publicly available dataset for helmet wearing detection. A total of 7581 images are included with 9044 human helmet wearers (positive) and 11,1514 normal head items (unworn or negative).

Evaluation metrics: In the manner of commonly evaluated metrics for target detection, we chose mAP(0.5) and mAP(0.5:0.95) to evaluate the performance of our method and baseline and other comparative high-performance target detectors. In addition, we removed the symbol % from the experimental results for a concise representation.

4.3 Results and Discussions

Ablation Study: The results of the ablation experiments for PFBDet are shown in Table 1. All experimental results were performed on the validation set of SHWD.

PFNet. We first obtained the base model similar to yolov5-lite-s. yolov5-lite-s used ShuffleNetV2-1x [32] for the backbone and PAN structure for the neck, and the obtained mAP(0.5) was 85.7. Further, we used the PFNet backbone structure to replace the original ShuffleNetV2- 1x, the mAP(0.5) increased to 83.3 and the FPS increased to 99.5. The results are shown in Table 1.

PSA. We adds Polarized Self-Attention to the end part of the PFNet backbone structure. The number of parameters was increased by less than 50 K. mAP(0.5) was further increased to 84.3. The results are shown in Table 1.

Table 1. Different configurations of ablation study on PFBDet.

Model	Params	mAP(0.5)	FPS
Base	1.64M	83.8	86.9
+PFNet	1.20M	83.3	99.5
+PSA	1.24M	84.3	93.7
+BCM-PAN	0.92M	85.7	103.9
+Replace BCM(GSconv+CARAFE) with BCM	0.92M	87.4	110.2

BCM-PAN. In the previous section, we compared the impact of using the backbone network ShuffleNetV2-1x and PFNet in the same configuration the inference speed of PFNet is slightly better than ShuffleNetV2-1x in this section, we add the multiscale feature fusion structure BCM-PAN based on the backbone network PFNet and replace the traditional PAN with BCM-PAN, mAP(0.5) improves from 84.3 to 85.7, and the FPS reaches 103.9. Further, we replace the BCM in the neck network with the BCM with GSConv with the lightweight upsampling operator CARAFE, and the accuracy and speed reach a higher level. The results are shown in Table 1.

We verified the effectiveness of the BCM and PSA through ablation experiments. As can be seen from Table 2, the BCM and PSA have made significant progress in the data set.

Table 2. Effectiveness of the BCM and PSA on PFBDet.

Model	PSA	BCM	Params	mAP(0.5)	mAP(0.5:0.95)
PFBDet	✗	✗	0.85M	85.7	49.7
PFBDet	✔	✗	0.90M	86.8	51.5
PFBDet	✗	✔	0.89M	86.4	51.2
PFBDet	✔	✔	0.92M	87.4	52.4

We verify the effectiveness of using GSConv and lightweight upsampling operator in BCM through ablation experiments. As can be seen from Table 3, CARAFE and GSConv achieved significant progress on our dataset with only a small amount of additional parameters and computational effort. This also proves that the BCM proposed in this paper can provide initial feedback information for each adjacent feature map with fewer parameters, while retaining more accurate positioning signals.

Table 3. Ablation study about CARAFE and GSConv on BCM.

Model	BCM		Params	mAP(0.5)	mAP(0.5:0.95)
	GSConv	CARAFE			
PFBDet	✗	✗	0.921M	85.7	50.6
PFBDet	✔	✗	0.91M	86.8	51.9
PFBDet	✗	✔	0.93M	86.5	51.7
PFBDet	✔	✔	0.924M	87.4	52.4

Comparison with Other Methods: We tested the speed performance of all official models of fp16 accuracy on the same RTX3090 GPU. Based on the results, we can see that the proposed model outperforms all other methods, where on

Table 4. The performance of our method on dataset.

Method	Elec-hat			HelmetDet		
	mAP(0.5)	mAP(0.5:0.95)	FPS	mAP(0.5)	mAP(0.5:0.95)	FPS
YOLOv5lite-c	89.9	68.0	86.2	82.6	65.4	85.9
YOLOv5lite-e	89.1	63.6	93.0	80.3	60.7	84.3
YOLOv5lite-s	89.7	65.2	81.9	81.7	62.4	81.3
Nanodet-plus-m	85.9	58.3	–	75.9	45.9	–
PP-PicoDet-S(320)	85.4	58.2	91.5	71.7	46.0	79.5
PP-PicoDet-S(416)	88.9	62.3	87.9	76.1	50.3	74.1
PP-PicoDet-XS(320)	83.7	54.8	91.7	63.9	38.8	69.8
PP-PicoDet-XS(416)	86.2	58.6	55.7	70.9	45.2	72.2
PP-YOLO-Tiny(320)	83.6	49.8	70.4	75.9	43.9	76.1
PP-YOLO-Tiny(416)	87.2	53.8	70.5	78.4	45.6	58.0
YOLOX-Nan	88.3	62.1	62.5	72.4	44.6	58.5
YOLOX-Tiny	89.7	61.6	80.5	80.1	45.9	54.9
PFBDet	**91.0**	62.4	**96.2**	**84.9**	69.8	**88.2**

the publicly available dataset HelmetDet, we compare PFBDet with YOLOv5-lite [24], YOLOX-Tiny [25], YOLOX-Nano [26], PPYOLO-Tiny [27], PP-Picodet [28], Nanodet-plus [29] were compared and the model achieved 84.9% map and 88.2 FPS, which are 1.1% and 2.3 higher than the state-of-the-art methods, respectively. The results are shown in Table 4. Meanwhile in Table 4, our results on the private dataset Ele-hat showed that the model achieved 91% map and 90.2 FPS, which are higher than the state-of-the-art methods, respectively higher than its 0.7% and 3.2.

As can be seen in Fig. 4, our PFBDet has a good detection effect and can effectively match the criteria for manual detection.

(a) Detection results on HelmetDet   (b) Detection results on Elec-hat

Fig. 4. The PFBDet detection results.

Comparison with the SOTA: In addition, in order to test the robustness and universality of our proposed method, we evaluated the SOTA model on the validation set of the public data set SHWD, focusing on the accuracy and speed performance of our proposed method on the public data set. The evaluation was consistent with the experimental equipment and environment used in the previous two datasets. We compared PFBDet with YOLOv5-lite, YOLOX-Tiny, YOLOX-Nano, PPYOLO-Tiny, PP-Picodet, and YOLOv6-lite [30] in Table 5.

Table 5. The performance of our method on SHWD dataset.

Model	Input Size	mAP(0.5)	mAP(0.5:0.95)	FPS	Params	FLOPs
YOLOv5-lite-c	512 × 512	84.5	52.3	87.7	4.57M	5.92G
YOLOv5-lite-e	320 × 320	83.0	51.2	93.4	0.78M	0.73G
YOLOv5-lite-g	640 × 640	85.9	53.7	78.7	5.39M	15.6G
YOLOv5-lite-s	416 × 416	83.8	52.6	86.9	1.64M	1.66G
PP-PicoDet-S	320 × 320	77.2	46.8	97.0	0.99M	0.73G
PP-PicoDet-S	416 × 416	84.9	52.3	107.9	0.99M	1.24G
PP-PicoDet-XS	320 × 320	73.1	44.0	105.4	0.70M	0.67G
PP-PicoDet-XS	416 × 416	81.4	48.9	104.7	0.70M	1.13G
PP-YOLO-Tiny	320 × 320	75.2	38.8	97.2	1.08M	0.58G
PP-YOLO-Tiny	416 × 416	82.3	43.0	101.1	1.08M	1.02G
YOLOX-Nan	640 × 640	79.3	46.4	79.1	0.91M	1.08G
YOLOX-Tiny	640 × 640	84.9	48.3	68.6	5.06M	6.45G
YOLOv6-lite-s	320 × 320	72.2	44.5	60.1	0.55M	0.56G
YOLOv6-lite-m	320 × 320	72.9	45.3	61.7	0.79M	0.67G
YOLOv6-lite-l	224 × 128	73.5	45.8	63.1	1.09M	0.24G
PFBDet	640 × 640	**87.4**	52.4	**110.2**	0.92M	1.32G

5 Conclusion

To address the issue of redundant computations in existing target detectors, this paper proposes a lightweight single-stage target detector called PFBDet. The main contributions of this work are as follows: 1) Designing a cross-stage local bottleneck module called FNCSP, which serves as the basis for the efficient lightweight feature extraction network PFNet. By incorporating Polarized Self-Attention, the computational complexity is optimized while capturing more feature information, thereby enhancing the detection performance of the model.2) Introducing a bidirectional connection module BCM based on GSConv and the lightweight upsampling operator CARAFE. Building upon this, an efficient multi-scale feature fusion structure named BCM-PAN is proposed, combined with a single aggregated cross-layer network module. This structure reduces redundancy overhead associated with multi-scale feature fusion and improves the learning capability of the network.

The effectiveness of the proposed method is validated using a safety cap dataset. Experimental results demonstrate that our detector achieves faster model inference while obtaining more feature information, outperforming other detectors. In summary, the method successfully achieves real-time and accurate identification of workers wearing helmets, playing a vital role in the safety monitoring of power systems.

It is important to note that in the task of helmet detection in power construction scenes, detection in complex haze scenes, uneven lighting environments, and irregularly shaped targets is also crucial. Therefore, in future work, we plan to carefully investigate these aspects by incorporating image datasets with data enhancement techniques tailored to different scenes.

Acknowledgement. This work was supported by the Scientific Research Funds of Northeast Electric Power University (No. BSZT07202107).

References

1. Wang, Z., Wu, Y., Yang, L., Thirunavukarasu, A., Evison, C., Zhao, Y.: Fast personal protective equipment detection for real construction sites using deep learning approaches. Sensors **21**(10), 3478 (2021). https://doi.org/10.3390/s21103478
2. Foudeh, H., Luk, P., Whidborne, J.: An advanced unmanned aerial vehicle (UAV) approach via learning-based control for overhead power line monitoring: a comprehensive review. IEEE Access **2021**(9), 130410–130433 (2021)
3. Liang, W., et al.: Research on detection algorithm of helmet wearing state in electric construction. In: 14th National Conference on Signal and Intelligent Information Processing and Application, pp. 508–512 (2021)
4. Girshick, R.: Fast r-cnn. In: Proceedings of the IEEE International Conference on Computer Vision, pp. 1440–1448 (2020)
5. Wang, H., Hu, Z., Guo, Y., Yang, Z., Zhou, F., Xu, P.: A real-time safety helmet wearing detection approach based on CSYOLOv3. Appl. Sci. **10**(19), 6732 (2020). https://doi.org/10.3390/app10196732
6. Han, G., et al.: Method based on the cross-layer attention mechanism and multiscale perception for safety helmet-wearing detection. Comput. Electric. Eng. **95**, 107–128 (2021)
7. Li, P.: Research and Implementation of Key Technology of On-site Safety Early Warning Based on Object Detection and Depth Estimation. University of Electronic Science and Technology (2021)
8. Wang, Y.S., Gu, Y.W., Feng, X.C.: Research on detection method of helmet wearing based on attitude estimation. Appl. Res. Comput. **38**(3), 937–940 (2021)
9. Yang, Z., et al.: Industrial safety helmet detection algorithm based on depth cascade model. Comput. Modern. **1**, 91–97 (2022)
10. Wang, D., et al.: MusiteDeep: a deep-learning based webserver for protein post translational modification site prediction. Nucl. Acids Res. **48**, 140–146 (2020)

11. Yue, H., et al.: Helmet- wearing detection based on improved YOLOv5. Comput. Modern. **6**, 104–108 (2022)

12. Zhang, M., Han, Y., Liu, Z.F.: Detection method of high altitude safety protective equipment for construction workers based on deep learning. China Saf. Sci. J. **32**(5), 140–146 (2022)

13. Glenn, J.: YOLOv5 release v6.1 (2022). https://github.com/ultralytics/yolov5/releases/tag/v6.1

14. Chen, J., et al.: Run, don't walk: chasing higher FLOPS for faster neural networks. In: Proceedings of the IEEE Conference on Computer Vision and Pattern Recognition (CVPR2023) (2023). https://doi.org/10.48550/arXiv.2303.03667

15. Liu, H.J., et al.: Polarized self attention: towards high-quality pixel-wise regression. arXiv preprint arXiv:2107.00782 (2021)

16. Liu, S., et al.: Path aggregation network for instance segmentation. In: Proceedings of the IEEE Conference on Computer Vision and Pattern Recognition, pp. 8759–8768 (2018)

17. Wang, J., et al.: CARAFE: content-aware reassembly of features. In: Proceedings of the IEEE/CVF International Conference on Computer Vision, pp. 3007–3016 (2019)

18. Zoph, B., et al.: Learning transferable architectures for scalable image recognition. In: Proceedings of the IEEE Conference on Computer Vision and Pattern Recognition, pp. 8697–8710. IEEE, New Jersey (2018)

19. Ji, X., et al.: Dead zone compensation for proportional directional valve based on bilinear interpolation control strategy. Chin. Hydraul. Pneumat. **45**(6), 56–62 (2021)

20. Long, J., Shelhamer, E., Darrell, T.: Fully convolutional networks for semantic segmentation. In: 2015 IEEE Conference on Computer Vision and Pattern Recognition (CVPR), pp. 3431–3440 (2015)

21. Li, H., Li, J., Wei, H., Liu, Z., Zhan, Z., Ren, Q.: Slim-neck by GSCONY: a better design paradigm of detector architectures for autonomous vehicles. arXiv preprint arXiv:2206.02424 (2022)

22. https://aistudio.baidu.com/aistudio/datasetdetail/183436

23. https://github.com/njvisionpower/Safety-Helmet-Wearing-Dataset

24. Glenn jocher et al.yolov5 (2021). https://github.com/ultralytics/yolov5

25. Tiny, Y., Ge, Z., Liu, S., Wang, F., Li, Z., Sun, J.: YOLOX: exceeding yolo series in 2021. arXiv preprint arXiv:2107.08430 (2021)

26. Ge, Z., Liu, S., Wang, F., Li, Z., Sun, J.: YOLOX: exceeding yolo series in 2021. arXiv preprint arXiv:2107.08430 (2021)

27. PaddlePaddle Authors. PaddleDetection, object detection and instance segmentation toolkit based on paddlepaddle (2021). https://github.com/PaddlePaddle/PaddleDetection

28. Yu, G., et al.: PP-PicoDet: a better real-time object detector on mobile devices. arXiv preprint arXiv:2111.00902 (2021)

29. NanoDet Authors. NanoDet (2021). https://github.com/RangiLyu/nanodet

30. Li, C., et al.: YOLOV6: a single-stage object detection framework for industrial applications. arXiv preprint arXiv:2209.02976 (2022)

31. https://aistudio.baidu.com/aistudio/datasetdetail/96283

32. Ye, L.: AugShuffleNet: Improve ShuffleNetV2 via More Information Communication (2022). https://doi.org/10.48550/arXiv.2203.06589

Direct Inter-Intra View Association
for Light Field Super-Resolution

Da Yang[1,2,3], Hao Sheng[1,2,3]([✉]), Shuai Wang[1,2,3], Rongshan Chen[1,2,3],
and Zhang Xiong[1,2,3]

[1] State Key Laboratory of Virtual Reality Technology and Systems, School of
Computer Science and Engineering, Beihang University, Beijing 100191, People's
Republic of China
{da.yang,shenghao,shuaiwang,rongshan,xiongz}@buaa.edu.cn
[2] Zhongfa Aviation Institute, Beihang University, 166 Shuanghongqiao Street,
Pingyao Town, Yuhang District, Hangzhou 311115, People's Republic of China
[3] Faculty of Applied Sciences, Macao Polytechnic University, Macao, SAR 999078,
People's Republic of China

Abstract. Light field (LF) cameras record both intensity and directions of light rays in a scene with a single exposure. However, due to the inevitable trade-off between spatial and angular dimensions, the spatial resolution of LF images is limited which makes LF super-resolution (LFSR) a research hotspot. The key of LFSR is the complementation across views and the extraction of high-frequency information inside each view. Due to the high-dimensinality of LF data, previous methods usually model these two processes separately, which results in insufficient inter-view information fusion. In this paper, LF Transformer is proposed for comprehensive perception of 4D LF data. Necessary inter-intra view correlations can be directly established inside each LF Transformer block. Therefore it can handle complex disparity variations of LF. Then based on LF Transformers, 4DTNet is designed which comprehensively performs inter-intra view high-frequency information extraction. Extensive experiments on public datasets demonstrate that 4DTNet outperforms the current state-of-the-art methods both numerically and visually.

Keywords: Light field · Super-resolution · Transformer

1 Introduction

Light field (LF) imaging captures both intensity and directions of light rays in the scene and has recently gained considerable attention as a promising technology in a wide range of applications, such as depth estimation [21,25], semantic segmentation [2,19,42], saliency detection [20,22], de-occlusion [32], rain removal [41], object tracking [23,24,29,30], virtual reality [8,45], etc. By inserting a microlens array between the sensor and the main lens, commercial hand-held LF cameras [1,14,16] capture 4D LF images in a single exposure with the size of an ordinary camera. However, the restricted sensor resolution results in a trade-off between the spatial and angular dimensions. To ensure sufficient sampling rate in angular

© The Author(s), under exclusive license to Springer Nature Singapore Pte Ltd. 2024
B. Luo et al. (Eds.): ICONIP 2023, LNCS 14451, pp. 265–278, 2024.
https://doi.org/10.1007/978-981-99-8073-4_21

domain, spatial resolution is sacrificed in LF cameras, which poses difficulties in potential applications. Therefore light field super-resolution (LFSR), a task to improve spatial resolution of LF images, has become a research hotspot.

LFSR pays great attention to mixing valid complementary information across sub-aperture images (SAIs). Due to complexity caused by high-dimensionality, previous methods usually decouple LF into lower-dimensional data for easier complementary information extraction. Some prior works achieve this by stacking SAIs along specific directions, like the horizontal/vertical SAI stacks in LFNet [33], the star-like input of resLF [46] and MEG-Net [48], etc. Without full access to all the SAIs, these methods easily reach the limit. Others try direct computations in specific dimensions, like the spatial-angular separable convolution (SAS) in LFSSR [43], the spatial-angular interaction in LF-InterNet [35], the alternate angular-spatial Transformers in LFT [12], etc. The alternate spatial-angular operations model the LF structure indirectly and hence it is hard to fully exploit valid information across views under complex disparity variations.

In this paper, we first propose LF Transformers for direct modeling of 4D LF. To be specific, spatial and angular dimensions are processed simultaneously, guided with 4D positional encoding. Any inter-intra view correlation can be constructed directly inside each LF Transformer block. Therefore valid complementary information can be extracted regardless of the complexity in disparity pattern of the LF. Then 4DTNet is designed based on LF Transformers. In 4DTNet, the details neglected by alternate angular and spatial operations are successfully recovered. As proved in the experiments, 4DTNet extracts and complements valid inter-intra view high-frequency information better than previous methods.

In summary, the main contributions of this paper are concluded as follows:

1) LF Transformer is proposed to unite information complementation across SAIs and high-frequency information extraction inside SAIs. LF Transformer is able to establish any necessary inter-view correlation inside one Transformer block thus can better handle the complex disparity variations of LF.
2) Based on LF Transformers, 4DTNet is proposed, which effectively extracts valid high-frequency information from 4D LF. With LF Transformers, the network well handles LFs with various disparity ranges.
3) Extensive experiments demonstrate the state-of-the-art performance of 4DTNet on various LF datasets and the effectiveness of LF Transformers in inter-intra view high-frequency information extraction.

2   Related Work

In this section, LFSR methods are reviewed from early traditional works to recent learning-based methods.

2.1 Traditional Methods for LFSR

Traditional methods usually perform LFSR through disparity estimation and warping. Wanner and Goldluecke [38,39] first deduced disparity maps from epipolar plane images (EPIs). Then a super-resolved novel view is synthesized with a variational model by representing the existing views with the novel view according to disparity maps. Mitra and Veeraraghavan [15] characterized pixels in a patch with a single disparity value and proposed to reconstruct patches with higher spatial resolution based on a Gaussian Mixture Model. All the SAIs are warped to the center view by Heber and Pock [6]. With the assumption that the resulting matrix was of low rank, a Robust PCA model is proposed to super-resolve the entire LF image simultaneously. Zhang et al. [47] proposed a micro-lens-based matching term for depth estimation and super-resolution. The subpixel information was recovered by matching micro-lens images with target view image. Rossi and Frossard [18] put forward a square constraint for complementing information across SAIs and a graph-based regularizer for maintaining geometric structure. In this way, they cast LFSR into a global optimization problem which super-resolved all SAIs simultaneously. Their algorithm only required an approximate disparity range, other than precise disparity estimation. However, when the input LF image exceeded the preset disparity range, it performed inferiorly.

Traditional methods are easier to implement and less data hunger. However, they fail to fully exploit the complementary information across SAIs due to disparity estimation of limited precision and the limited manually-designed priors.

2.2 Learning-Based Methods for LFSR

Recently, with superiority in feature extraction of neural networks, learning-based methods have brought LFSR to a new era.

The pioneer work LFCNN was proposed by Yoon et al. [44], which performed spatial and angular super-resolution sequentially with a structure similar to SRCNN [3]. Zhang et al. [46] proposed a residual network resLF to extract compelementary details from other SAIs. In resLF, SAIs with consistent pixel offsets in four directions were grouped into image stacks to promote the learning of inherent geometric relations. However, the star-like input structure of resLF failed in exploiting all SAIs. The authors alleviated the problem in MEG-Net [48]. For each target view, all the available SAIs in four directions are exploited for subpixel information extraction. Yeung et al. [43] adopted 4D convolution and SAS to characterize the complicated structure within and across SAIs. The authors demonstrated that simultaneous processing of spatial and angular domains was more effective than the alternate scheme. Jin et al. [9] proposed to extract complementary information for the target view through combinatorial geometry embedding. The authors put forward a structure-aware loss defined on EPI for further regularization of strucutral consistency. LF-InterNet [35] first decoupled and extracted spatial and angular information separately. Then spatial and angular features were interacted repetitively before the final fusion for super-resolution. It was further improved in DistgSSR [34] with feature extraction on EPIs. The repetitive interaction between angular and spatial domains was also adopted by LF-IINet [13]. Two special modules IntraFUM and InterFUM were designed for intra-view feature updating with the assistance of the inter-view feature and vice versa, respectively. LF-DFNet [36] was proposed by Wang *et al.* based on deformable convolution. A collect-and-distribute scheme was designed to align the center-view feature and each side-view feature bidirectionally. Recently, Vision Transformer [4] has proved its effectiveness in modeling long-range correlations which is critical in LFSR. DPT [31] treated LF as horizontal and vertical sequences. SA-LSA layers were proposed to extract global contextual information across SAIs from the original LF and its gradient maps. Then a cross-attention fusion Transformer was introduced to integrate the two features for final SR. LFT [12] proposed angular Transformer and spatial Transformer to model dependencies arcoss and inside SAIs respectively. Angular and spatial information were incorporated by alternating angular and spatial Transformers.

By decoupling LF into subspaces, the high-frequency information across views underneath can be discovered more easily. However, the artificial fragmentation of LF data also destroys part of the internal correlations and therefore limits the performance of these methods.

3   Architecture of 4DTNet

A 4D LF is denoted as $L \in \mathbb{R}^{U \times V \times X \times Y}$, where $U \times V$ represents angular resolution and $X \times Y$ represents spatial resolution. With α denoting the upscaling factor, the task of LFSR is to recover high-resolution (HR) SAIs $\{L_{(u,v)}^{HR} \in \mathbb{R}^{\alpha X \times \alpha Y}\}$ from their low-resolution (LR) counterparts $\{L_{(u,v)}^{LR} \in \mathbb{R}^{X \times Y}\}$.

The key of LFSR mainly lies in two aspects: high-frequency information complementation across SAIs and high-frequency information extraction inside each SAI. Previous methods tend to model the above processes respectively. However, the strict division of spatial and angular domain increases the difficulty in building effective inter-view correlations. Hence in this paper, LF Transformer for simultaneous modeling of the two processes is proposed. In LF Transformer, any correlation between SAIs can be constructed directly, which both reduces the loss in angular information and improves the efficiency in inter-view high-frequency information complementation. Based on LF Tranformers, 4DTNet is designed for high-quality LFSR.

In this section, we first illustrate the overall architecture of our 4DTNet. Then the detailed implementation of LF Transformer is introduced.

3.1   Overall Architecture

As shown in Fig. 1, 4DTNet is designed in a global residual structure. The network focuses on learning the high-frequency details absent in the LR LF.

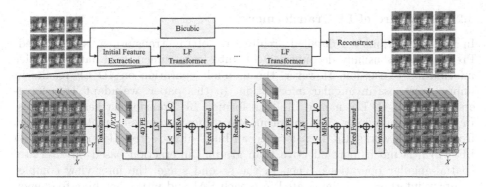

Fig. 1. The overall architecture of 4DTNet. Initial features are first extracted with cascaded 3×3 convolutions. LF Tranformers are sequentially placed for comprehensive inter-intra view high-frequency information extraction. Combining the outputs of the LF Tranformer sequence and bicubic interpolation, the final super-resolved LF is reconstructed.

Initial Feature Extraction. The network first processes LR LF L^{LR} with cascased 3×3 convolutions on each of its SAIs for local spatial embedding:

$$F_{init} = f_{init}(L^{LR}), \qquad (1)$$

where $F_{init} \in \mathbb{R}^{U \times V \times X \times Y \times d}$ and $d = 64$.

Inter-Intra View High-Frequency Information Extraction. The initial features are then fed into sequential LF Transformers for inter-intra view high-frequency information extraction:

$$F_{LF} = Trans_{LF}(\cdots (Trans_{LF}(F_{init}))), \qquad (2)$$

where $F_{LF} \in \mathbb{R}^{U \times V \times X \times Y \times d}$. With full access to all the pixels from all the SAIs, LF Transformers can model the correlations across and inside views more comprehensively. Different from previous methods which adopt an indirect inter-view information association scheme, LF Transformers can well handle complex disparity variation. The details of LF Transformer is introduced in Sect. 3.2. The number of LF Transformers is set to 8 in our experiments.

Super-Resolved LF Reconstruction. The output of the sequence of LF Transformers is first up-sampled based on PixelShuffle [26]:

$$F_{LF}^{SR} = Up(F_{LF}), \qquad (3)$$

where $F_{LF}^{SR} \in \mathbb{R}^{U \times V \times \alpha X \times \alpha Y}$. And the final super-resolved LF is obained through direct combination of the up-sampled feature and the bicubic interpolation of the LR LF:

$$L^{SR} = F_{LF}^{SR} + Bicubic(L^{LR}), \qquad (4)$$

where $L^{SR} \in \mathbb{R}^{U \times V \times \alpha X \times \alpha Y}$.

3.2   Structure of LF Transformer

In this subsection, the detailed architecture of LF Transformer is illustrated. Previous works usually decouple 4D LF into lower-dimensional data for easier information extraction. However, the manual separation on 4D LF results in unnecessary loss in angular information. In this paper, we adapt Transformers directly to 4D LF, named LF Transformer. LF Transformer has two stages. In the first stage, inter-view complementary information is extracted. Different from previous works, LF Transformer can build direct correlation inside the 4D LF which is much more effective than gradual interaction based on alternate spatial-angular operations. Then in the second stage, the inter-view complementary information is integrated into each SAI and intra-view high-frequency information is further excavated. With LF Transformers, inter-intra view correlations can be well established and hence better performance is achieved as shown in Sect. 4.

In the first stage, the input $F_{in} \in \mathbb{R}^{U \times V \times X \times Y \times d}$ of LF Transformer is first reshaped into pixel-level LF tokens $T_{LF} \in \mathbb{R}^{UVXY \times d}$. Then 4D positional encoding is performed to embed accurate positional information both angularly and spatially:

$$
\begin{aligned}
P_{LF}(p_u, p_v, p_x, p_y, 2i) &= sin\left(p_u/\alpha^{2i/d}\right) + sin\left(p_v/\alpha^{2i/d}\right) \\
&\quad + sin\left(p_x/\alpha^{2i/d}\right) + sin\left(p_y/\alpha^{2i/d}\right), \\
P_{LF}(p_u, p_v, p_x, p_y, 2i+1) &= cos\left(p_u/\alpha^{2i/d}\right) + cos\left(p_v/\alpha^{2i/d}\right) \\
&\quad + cos\left(p_x/\alpha^{2i/d}\right) + cos\left(p_y/\alpha^{2i/d}\right),
\end{aligned}
\tag{5}
$$

where $p_u = 1, \cdots, U$, $p_v = 1, \cdots, V$, $p_x = 1, \cdots, X$, $p_y = 1, \cdots, Y$, and i indicates potision in the embedding dimension. Following [28], α is set to 10000.

Query Q_{LF} and key K_{LF} are generated by layer-normalizing the summation of tokens T_{LF} and positional encoding vectors P_{LF}, and T_{LF} is directly used as value V_{LF}:

$$
\begin{aligned}
Q_{LF} = K_{LF} &= LN(T_{LF} + P_{LF}), \\
V_{LF} &= T_{LF}.
\end{aligned}
\tag{6}
$$

Multi-head self-attention (MHSA) is used to explore the correlation among LF tokens, i.e., the correlation among all the pixels recorded by the 4D LF:

$$
\begin{aligned}
T_{LF}^{MHSA} &= MHSA(Q_{LF}, K_{LF}, V_{LF}) \\
&= Concat(H_1, ..., H_{N_H})W_O,
\end{aligned}
\tag{7}
$$

where $W_O \in \mathbb{R}^{d \times d}$. Any necessary inter-view correlation can be found here for complementary information extraction. The h_{th} head H_h is computed as:

$$
H_h = Softmax\left(\frac{Q_{LF,h}W_{Q,h}\left(K_{LF,h}W_{K,h}^T\right)}{\sqrt{d/N_H}}\right) V_{LF}W_{V,h},
\tag{8}
$$

where $h = 1, ..., N_H$ denotes the index of heads. $W_{Q,h}$, $W_{K,h}$ and $W_{V,h} \in \mathbb{R}^{d/N_H \times d/N_H}$ controls the linear projection of query Q_{LF}, key K_{LF} and value V_{LF}, respectively.

Then T_{LF}^{MHSA} is added with the original LF tokens T_{LF}:

$$T'_{LF} = \left(T_{LF}^{MHSA} + T_{LF} \right). \tag{9}$$

Another layer-normalization is performed before it is fed into a multi-layer perceptron:

$$T^*_{LF} = MLP(LN(T'_{LF})) + T'_{LF}, \tag{10}$$

where T^*_{LF} is of same shape as T_{LF}.

In the second stage, to further integrate the extracted inter-view complementary information and excavate intra-view high-frequency details, LF Transformer performs information fusion inside each SAI. T^*_{LF} was reshaped into SAI-level:

$$T_{SAI} = Reshape(T^*_{LF}), \tag{11}$$

where $T_{SAI} \in \mathbb{R}^{UV \times XY \times d}$. Then the same structure as the first stage is applied on each SAI, other than that 2D positional encoding is imposed here:

$$P_{SAI}(p_x, p_y, 2i) = sin\left(p_x/\alpha^{2i/d}\right) + sin\left(p_y/\alpha^{2i/d}\right),$$
$$P_{SAI}(p_x, p_y, 2i+1) = cos\left(p_x/\alpha^{2i/d}\right) + cos\left(p_y/\alpha^{2i/d}\right), \tag{12}$$

where $p_x = 1, \cdots, X$, $p_y = 1, \cdots, Y$, and i indicates potision in the embedding dimension. α is also set to 10000. In the end, untonkenization is performed and the feature maps are reshaped back to $F_{uul} \in \mathbb{R}^{U \times V \times X \times Y \times d}$.

In the first stage of LF Transformer, inter-view correlations with complex disparity variations are established, benefiting from the full access to the 4D LF data and 4D positional encoding. Then in the second stage, the inter-view complementary information is integrated and the intra-view high-frequency details are further excavated. Different from previous indirect inter-view information association schemes, LF Transformer handles complex LF data with complicated disparity variations more adequately. The effectiveness of LF Transformer and 4DTNet is discussed in Sect. 4.

4   Experiments

In this section, we first introduce the implementation details of the experiments. Then comparison with the state-of-the-art methods is conducted. Our 4DTNet outperforms other methods on various datasets. Ablation study is performed in the end to validate the contribution of our designs.

4.1   Implementation Details

We adopt 5 public datasets [7,11,17,27,40] in our experiments. The separation of training set and test set is the same as previous works [12,34]. For fair comparison, we also use the 5×5 angular resolution. During training and testing,

$5 \times 5 \times 32 \times 32/5 \times 5 \times 64 \times 64$ patches are cropped from LF images for $2\times/4\times$ LFSR respectively and $5 \times 5 \times 16 \times 16$ input patches are generated with bicubic interpolation. The training data is augmented by 8 times with random transposition and random flipping (left-right, up-down).

Following previous works, super-resolution was performed on Y-channel only. Both peak signal-to-noise ratio (PSNR) and structural similarity (SSIM) [37] were adopted for quantitative evaluation.

The network is implemented in Pytorch (1.13.0) and trained with one NVIDIA RTX A6000 GPU. The weights of the model are initialized with He initialization [5]. Our network is trained with L1 loss and Adam optimizer [10]. The learning rate is initialized to 8e-4 and decreased by a factor of 0.5 every 20 epochs, which stops at 80 epochs. A minibatch of 8 is used to accelerate training speed and smooth training curve.

4.2 Comparison with State-of-The-Art Methods

10 state-of-the-art LFSR methods [9,12,13,31,34–36,43,46,48] are used for comparison. For all the methods, we use their official published models and weights. Both upscaling factors of ×2 and ×4 are adopted in the comparisons.

Table 1. Comparison with the state-of-the-art methods on 5 public datasets. Red texts indicate the best results and blue texts indicate the second best results. 4DTNet outperforms other methods on all datasets in both metrics.

Method	Scale	EPFL [17]	HCInew [7]	HCIold [40]	INRIA [11]	STFgantry [27]	Average
resLF [46]	×2	33.62/0.971	36.69/0.974	43.42/0.993	35.40/0.980	38.35/0.990	37.49/0.982
LFSSR [43]		33.67/0.974	36.80/0.975	43.81/0.994	35.28/0.983	37.94/0.990	37.50/0.983
LF-ATO [9]		34.27/0.976	37.24/0.977	44.21/0.994	36.17/0.984	39.64/0.993	38.31/0.985
LF-InterNet [35]		34.11/0.976	37.17/0.976	44.57/0.995	35.83/0.984	38.44/0.991	38.02/0.984
LF-DFNet [36]		34.51/0.976	37.42/0.977	44.20/0.994	36.42/0.984	39.43/0.993	38.39/0.985
MEG-Net [48]		34.31/0.977	37.42/0.978	44.10/0.994	36.10/0.985	38,77/0.992	38.14/0.985
LF-IINet [13]		34.73/0.977	37.77/0.979	44.85/0.995	36.57/0.985	39.89/0.994	38.76/0.986
DPT [31]		34.49/0.976	37.36/0.977	44.30/0.994	36.41/0.984	39.43/0.993	38.40/0.985
LFT [12]		34.80/0.978	37.84/0.979	44.52/0.995	36.59/0.986	40.51/0.994	38.85/0.986
DistgSSR [34]		34.81/0.979	37.96/0.980	44.94/0.995	36.59/0.986	40.40/0.994	38.94/0.987
4DTNet (Ours)		34.93/0.979	38.00/0.981	44.99/0.995	36.85/0.987	40.65/0.995	39.08/0.987
resLF [46]	×4	28.26/0.904	30.72/0.911	36.71/0.968	30.34/0.941	30.19/0.937	31.24/0.932
LFSSR [43]		28.60/0.912	30.93/0.915	36.91/0.970	30.59/0.947	30.57/0.943	31.52/0.937
LF-ATO [9]		28.51/0.912	30.88/0.914	37.00/0.970	30.71/0.948	30.61/0.943	31.54/0.937
LF-InterNet [35]		28.81/0.916	30.96/0.916	37.15/0.972	30.78/0.949	30.37/0.941	31.61/0.939
LF-DFNet [36]		28.77/0.917	31.23/0.920	37.32/0.972	30.83/0.950	31.15/0.949	31.86/0.942
MEG-Net [48]		28.75/0.916	31.10/0.918	37.29/0.972	30.67/0.949	30.77/0.945	31.72/0.940
LF-IINet [13]		29.04/0.919	31.33/0.921	37.62/0.973	31.03/0.952	31.26/0.950	32.06/0.943
DPT [31]		28.94/0.917	31.20/0.919	37.41/0.972	30.96/0.950	31.15/0.949	31.93/0.941
LFT [12]		29.26/0.921	31.46/0.922	37.63/0.974	31.21/0.952	31.86/0.955	32.28/0.945
DistgSSR [34]		28.99/0.920	31.38/0.922	37.56/0.973	30.99/0.952	31.65/0.954	32.12/0.944
4DTNet(Ours)		29.42/0.922	31.51/0.923	37.71/0.974	31.49/0.953	32.21/0.957	32.47/0.946

Quantitative Results. Table 1 lists quantitative comparisons with the state-of-the-art methods on 5 datasets. For both ×2 and ×4, 4DTNet achieves the highest PSNR and SSIM scores on all the 5 datasets. As shown in Table 1, the performance improvements are larger on real-world datasets (EPFL, INRIA and STFgantry) than synthetic datasets (HCInew and HCIold). For example in ×4 tasks, the average gain in PSNR on real-world datasets is 0.27dB (vs 0.06dB on synthetic datasets). In general, the real-world datasets are much more complicated with complex scenes, varying imaging quality and complex disparity ranges. However, benefiting from the direct inter-view correlation construction of LF Transformer, our method achieves better performance regardless of complexity in LF data.

Note that although 4DTNet only uses 16 × 16 patches, it still surpasses other methods which use 32 × 32 patches on STFgantry with larger disparity range. This indicates that 4DTNet can recognize and organize useful complementary information across SAIs more effectively.

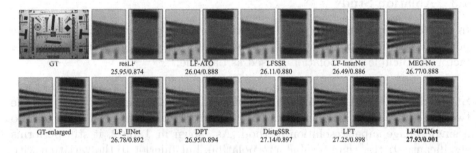

Fig. 2. The ×4 SR results of small-disparity *ISO Chart 1* in EPFL. Two typical hard areas are enlarged for better observation. 4DTNet is the only one that clearly reconstructs the thin stripes in the green box. It is also the only one that achieves partial success in reconstructing the repetitive patterns in the red box.

Qualitative Results. Visual results on small-disparity dataset EPFL and large-disparity dastaset STFgantry are shown in Fig. 2 and Fig. 3, respectively. The harder ×4 task is chosen for demonstration. In the classic *ISO Chart 1*, our method is the only one that manages to reconstruct the thin stripes in the green box. And it also approximately recovers the repetitive patterns in the red box, in which other methods totally fail. As for scene *LegoKnights* in STFgantry, the relatively large disparity hinders the other methods from extracting useful complementary details. While 4DTNet better reconstructs the fine structures like the brick joints in the green box in Fig. 3. The large disparity also makes the area in the red box invisible in most SAIs as it move out of field of view. Compared with other methods, the global perception of LF makes 4DTNet free from the influence of this invisibility. It only extracts reliable complementary information and successfully reconstructs the shadows on the toy bricks.

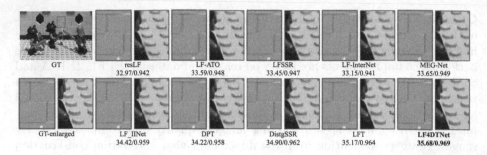

Fig. 3. The ×4 SR results of large-disparity *Lego Knights* in STFgantry. Two typical hard areas are enlarged for better observation. In spite of the large disparity of the scene, 4DTNet clearly reconstructs the thin stripes in the green box. It is also the only one that successfully reconstructs the shadows in the red box which is invisible in most SAIs.

4.3 Ablation Study

In this subsection, we compare 4DTNet with several variations to investigate the potential benefits of our designs. The experiments are performed under upscaling factor×4. Bicubic interpolation is used as baseline.

Structure of LF Transformer. We first investigate the contribution of the two stages in LF Transformer. With only the second stage, 4DTNet degrages into a single image super-resolution method. As shown in Table 2, it still performs significantly better than bicubic interpolation, but inferior to the variation with only the first stage. This shows that inter-view complementation plays an important role in LFSR. Note that without either stage, the model performs far inferior to the original design. This indicates that although the first stage can partially model intra-view information extraction, it mainly focuses on inter-view information complementation compared with the second stage. The experimnents prove the complementarity between the two stages and the effectiveness of the structure of LF Transformer.

Effectiveness of Positional Encoding. Then we investigate the effectiveness of positional encoding in LF Transformer. As shown in Table 2, the average PSNR drops by 0.08 dB and 0.13 dB when working without 2D positional encoding and 4D positional encoding, respectively. On the one hand, it proves the effectiveness of positional encoding in guiding the perception of inter-intra view information. On the other hand, compared with 2D positional encoding functioning inside SAI space, the 4D positional encoding contributes more to the overall performance for its guidance role in the comprehensive perception of the LF.

Table 2. Performance of variants of 4DTNet with factor ×4.

Method	Scale	EPFL [17]	HCInew [7]	HCIold [40]	INRIA [11]	STFgantry [27]	Average
Bicubic	×4	25.26/0.832	27.71/0.852	32.58/0.934	26.95/0.887	26.09/0.845	27.72/0.870
only 1st stage	×4	28.72/0.910	30.85/0.913	36.85/0.969	30.90/0.946	31.04/0.946	31.67/0.937
only 2nd stage	×4	27.82/0.886	29.64/0.887	35.20/0.954	29.72/0.926	28.98/0.911	30.27/0.913
w/o 2D Pos	×4	29.38/0.921	31.47/0.923	37.70/0.973	31.39/0.952	31.99/0.956	32.39/0.945
w/o 4D Pos	×4	29.35/0.921	31.45/0.922	37.68/0.973	31.33/0.952	31.87/0.955	32.34/0.945
4DTNet(Ours)	×4	29.42/0.922	31.51/0.923	37.71/ 0.974	31.49/ 0.953	32.21/ 0.957	32.47/0.946

5 Conclusion

In this paper, LF Transformer is proposed for comprehensive inter-intra view high-frequency information extraction. With full access to all pixels in 4D LF, LF Transformer is able to build any necessary inter-intra view correlation. Based on LF Transformers, 4DTNet is proposed and achieves state-of-the-art performance, which is demonstrated by extensive experiments.

As LF Transformer is proved to be effective in inter-intra view correlation extraction, in the future, we will further dig into other tasks that rely on comprehensive perception of LF such as semantic segmentation.

Acknowledgements. This study is partially supported by the National Key R&D Program of China (No. 2022YFC3803600), the National Natural Science Foundation of China (No.62372023), and the Open Fund of the State Key Laboratory of Software Development Environment (No. SKLSDE-2023ZX-11). Thank you for the support from HAWKEYE Group.

References

1. Adelson, E.H., Wang, J.Y.A.: Single lens stereo with a plenoptic camera. IEEE Trans. Pattern Anal. Mach. Intell. **14**(2), 99–106 (1992). https://doi.org/10.1109/34.121783
2. Cong, R., Yang, D., Chen, R., Wang, S., Cui, Z., Sheng, H.: Combining implicit-explicit view correlation for light field semantic segmentation. In: Proceedings of the IEEE/CVF Conference on Computer Vision and Pattern Recognition (CVPR), pp. 9172–9181 (June 2023)
3. Dong, C., Loy, C.C., He, K., Tang, X.: Learning a deep convolutional network for image super-resolution. In: Fleet, D., Pajdla, T., Schiele, B., Tuytelaars, T. (eds.) ECCV 2014. LNCS, vol. 8692, pp. 184–199. Springer, Cham (2014). https://doi.org/10.1007/978-3-319-10593-2_13
4. Dosovitskiy, A., et al.: An image is worth 16x16 words: transformers for image recognition at scale. In: International Conference on Learning Representations (ICLR) (2021). https://openreview.net/forum?id=YicbFdNTTy
5. He, K., Zhang, X., Ren, S., Sun, J.: Delving deep into rectifiers: surpassing human-level performance on imagenet classification. In: Proceedings of the 2015 IEEE International Conference on Computer Vision (ICCV), pp. 1026–1034 (2015). https://doi.org/10.1109/ICCV.2015.123

6. Heber, S., Pock, T.: Shape from light field meets robust PCA. In: 13th European Conference on Computer Vision (ECCV), pp. 751–767 (2014)
7. Honauer, K., Johannsen, O., Kondermann, D., Goldluecke, B.: A dataset and evaluation methodology for depth estimation on 4d light fields. In: 13th Asian Conference on Computer Vision (ACCV), pp. 19–34 (2016)
8. Huang, F.C., Chen, K., Wetzstein, G.: The light field stereoscope: immersive computer graphics via factored near-eye light field displays with focus cues. ACM Trans. Graph. **34**(4), 60:1–60:12 (2015). https://doi.org/10.1145/2766922
9. Jin, J., Hou, J., Chen, J., Kwong, S.: Light field spatial super-resolution via deep combinatorial geometry embedding and structural consistency regularization. In: 2020 IEEE/CVF Conference on Computer Vision and Pattern Recognition (CVPR), pp. 2257–2266 (2020). https://doi.org/10.1109/CVPR42600.2020.00233
10. Kingma, D.P., Ba, J.: Adam: a method for stochastic optimization. In: 3rd International Conference on Learning Representations (ICLR), pp. 1–15 (2015)
11. Le Pendu, M., Jiang, X., Guillemot, C.: Light field inpainting propagation via low rank matrix completion. IEEE Trans. Image Process. **27**(4), 1981–1993 (2018). https://doi.org/10.1109/TIP.2018.2791864
12. Liang, Z., Wang, Y., Wang, L., Yang, J., Zhou, S.: Light field image super-resolution with transformers. IEEE Signal Process. Lett. **29**, 563–567 (2022). https://doi.org/10.1109/LSP.2022.3146798
13. Liu, G., Yue, H., Wu, J., Yang, J.: Intra-inter view interaction network for light field image super-resolution. IEEE Trans. Multim. **25**, 256–266 (2023). https://doi.org/10.1109/TMM.2021.3124385
14. Lumsdaine, A., Georgiev, T.: The focused plenoptic camera. In: 2009 IEEE International Conference on Computational Photography (ICCP), pp. 1–8 (April 2009). https://doi.org/10.1109/ICCPHOT.2009.5559008
15. Mitra, K., Veeraraghavan, A.: Light field denoising, light field superresolution and stereo camera based refocussing using a gmm light field patch prior. In: 2012 IEEE Computer Society Conference on Computer Vision and Pattern Recognition Workshops (CVPRW), pp. 22–28 (2012). https://doi.org/10.1109/CVPRW.2012.6239346
16. Ng, R., Levoy, M., Brédif, M., Duval, G., Horowitz, M., Hanrahan, P.: Light field photography with a hand-held plenoptic camera. Comput. Sci. Tech. Rep. **2**(11), 1–11 (2005)
17. Rerabek, M., Ebrahimi, T.: New light field image dataset (2016). https://infoscience.epfl.ch/record/218363
18. Rossi, M., Frossard, P.: Geometry-consistent light field super-resolution via graph-based regularization. IEEE Trans. Image Process. **27**(9), 4207–4218 (2018). https://doi.org/10.1109/TIP.2018.2828983
19. Sheng, H., Cong, R., Yang, D., Chen, R., Wang, S., Cui, Z.: Urbanlf: a comprehensive light field dataset for semantic segmentation of urban scenes. IEEE Trans. Circuits Syst. Video Technol. **32**(11), 7880–7893 (2022). https://doi.org/10.1109/TCSVT.2022.3187664
20. Sheng, H., Liu, X., Zhang, S.: Saliency analysis based on depth contrast increased. In: 2016 IEEE International Conference on Acoustics, Speech and Signal Processing (ICASSP), pp. 1347–1351 (2016). https://doi.org/10.1109/ICASSP.2016.7471896
21. Sheng, H., Zhang, S., Cao, X., Fang, Y., Xiong, Z.: Geometric occlusion analysis in depth estimation using integral guided filter for light-field image. IEEE Trans. Image Process. **26**(12), 5758–5771 (2017). https://doi.org/10.1109/TIP.2017.2745100

22. Sheng, H., Zhang, S., Liu, X., Xiong, Z.: Relative location for light field saliency detection. In: 2016 IEEE International Conference on Acoustics, Speech and Signal Processing (ICASSP), pp. 1631–1635 (2016). https://doi.org/10.1109/ICASSP.2016.7471953

23. Sheng, H., Zhang, Y., Chen, J., Xiong, Z., Zhang, J.: Heterogeneous association graph fusion for target association in multiple object tracking. IEEE Trans. Circuits Syst. Video Technol. **29**(11), 3269–3280 (2019). https://doi.org/10.1109/TCSVT.2018.2882192

24. Sheng, H., et al.: Hypothesis testing based tracking with spatio-temporal joint interaction modeling. IEEE Trans. Circuits Syst. Video Technol. **30**(9), 2971–2983 (2020). https://doi.org/10.1109/TCSVT.2020.2988649

25. Sheng, H., Zhao, P., Zhang, S., Zhang, J., Yang, D.: Occlusion-aware depth estimation for light field using multi-orientation epis. Pattern Recogn. **74**, 587–599 (2018). https://doi.org/10.1016/j.patcog.2017.09.010

26. Shi, W., et al.: Real-time single image and video super-resolution using an efficient sub-pixel convolutional neural network. In: 2016 IEEE Conference on Computer Vision and Pattern Recognition (CVPR), pp. 1874–1883 (2016). https://doi.org/10.1109/CVPR.2016.207

27. Vaish, V., Adams, A.: The (New) Stanford Light Field Archive. Stanford University, Computer Graphics Laboratory (2008)

28. Vaswani, A., et al.: Attention is all you need. In: Proceedings of the 31st International Conference on Neural Information Processing Systems (NIPS 2017), pp. 6000–6010. Curran Associates Inc., Red Hook (2017)

29. Wang, S., Sheng, H., Yang, D., Zhang, Y., Wu, Y., Wang, S.: Extendable multiple nodes recurrent tracking framework with rtu++. IEEE Trans. Image Process. **31**, 5257–5271 (2022). https://doi.org/10.1109/TIP.2022.3192706

30. Wang, S., Sheng, H., Zhang, Y., Yang, D., Shen, J., Chen, R.: Blockchain-empowered distributed multi-camera multi target tracking in edge computing. In: IEEE Transactions on Industrial Informatics, pp. 1–10 (2023). https://doi.org/10.1109/TII.2023.3261890

31. Wang, S., Zhou, T., Lu, Y., Di, H.: Detail-preserving transformer for light field image super-resolution. Proc. AAAI Conf. Artif. Intell. **3**, 2522–2530 (2022). https://doi.org/10.1609/aaai.v36i3.20153

32. Wang, X., Liu, J., Chen, S., Wei, G.: Effective light field de occlusion network based on swin transformer. In: IEEE Transactions on Circuits and Systems for Video Technology, p. 1 (2022). https://doi.org/10.1109/TCSVT.2022.3226227

33. Wang, Y., Liu, F., Zhang, K., Hou, G., Sun, Z., Tan, T.: Lfnet: a novel bidirectional recurrent convolutional neural network for light-field image super-resolution. IEEE Trans. Image Process. **27**(9), 4274–4286 (2018). https://doi.org/10.1109/TIP.2018.2834819

34. Wang, Y., et al.: Disentangling light fields for super-resolution and disparity estimation. In: IEEE Transactions on Pattern Analysis and Machine Intelligence, p. 1 (2022). https://doi.org/10.1109/TPAMI.2022.3152488

35. Wang, Y., Wang, L., Yang, J., An, W., Yu, J., Guo, Y.: Spatial-angular interaction for light field image super-resolution. In: 16th European Conference on Computer Vision (ECCV), pp. 290–308 (2020)

36. Wang, Y., Yang, J., Wang, L., Ying, X., Wu, T., An, W., Guo, Y.: Light field image super-resolution using deformable convolution. IEEE Trans. Image Process. **30**, 1057–1071 (2021). https://doi.org/10.1109/TIP.2020.3042059

37. Wang, Z., Bovik, A., Sheikh, H., Simoncelli, E.: Image quality assessment: from error visibility to structural similarity. IEEE Trans. Image Process. **13**(4), 600–612 (2004). https://doi.org/10.1109/TIP.2003.819861
38. Wanner, S., Goldluecke, B.: Variational light field analysis for disparity estimation and super-resolution. IEEE Trans. Pattern Anal. Mach. Intell. **36**(3), 606–619 (2014). https://doi.org/10.1109/TPAMI.2013.147
39. Wanner, S., Goldluecke, B.: Spatial and angular variational super-resolution of 4d light fields. In: 12th European Conference on Computer Vision (ECCV), pp. 608–621 (2012)
40. Wanner, S., Meister, S., Goldlücke, B.: Datasets and benchmarks for densely sampled 4d light fields. In: Vision, Modeling & Visualization, pp. 225–226 (2013)
41. Yan, T., Li, M., Li, B., Yang, Y., Lau, R.W.H.: Rain removal from light field images with 4d convolution and multi-scale gaussian process. IEEE Trans. Image Process. **32**, 921–936 (2023). https://doi.org/10.1109/TIP.2023.3234692
42. Yang, D., Zhu, T., Wang, S., Wang, S., Xiong, Z.: Lfrsnet: a robust light field semantic segmentation network combining contextual and geometric features. Front. Environ. Sci. **10** (2022). https://doi.org/10.3389/fenvs.2022.996513
43. Yeung, H.W.F., Hou, J., Chen, X., Chen, J., Chen, Z., Chung, Y.Y.: Light field spatial super-resolution using deep efficient spatial-angular separable convolution. IEEE Trans. Image Process. **28**(5), 2319–2330 (2019). https://doi.org/10.1109/TIP.2018.2885236
44. Yoon, Y., Jeon, H., Yoo, D., Lee, J., Kweon, I.S.: Learning a deep convolutional network for light-field image super-resolution. In: 2015 IEEE International Conference on Computer Vision Workshop (ICCVW), pp. 57–65 (2015). https://doi.org/10.1109/ICCVW.2015.17
45. Yu, J.: A light-field journey to virtual reality. IEEE Multimedia **24**(2), 104–112 (2017). https://doi.org/10.1109/MMUL.2017.24
46. Zhang, S., Lin, Y., Sheng, H.: Residual networks for light field image super-resolution. In: 2019 IEEE/CVF Conference on Computer Vision and Pattern Recognition (CVPR), pp. 11038–11047 (2019). https://doi.org/10.1109/CVPR.2019.01130
47. Zhang, S., Sheng, H., Yang, D., Zhang, J., Xiong, Z.: Micro-lens-based matching for scene recovery in lenslet cameras. IEEE Trans. Image Process. **27**(3), 1060–1075 (2018). https://doi.org/10.1109/TIP.2017.2763823
48. Zhang, S., Chang, S., Lin, Y.: End-to-end light field spatial super-resolution network using multiple epipolar geometry. IEEE Trans. Image Process. **30**, 5956–5968 (2021). https://doi.org/10.1109/TIP.2021.3079805

Responsive CPG-Based Locomotion Control for Quadruped Robots

Yihui Zhang[1], Cong Hu[2], Binbin Qiu[3], and Ning Tan[1(✉)]

[1] School of Computer Science and Engineering, Sun Yat-sen University,
Guangzhou, China
tann5@mail.sysu.edu.cn
[2] Guangxi Key Laboratory of Automatic Detecting Technology and Instruments,
Guilin University of Electronic Technology, Guilin, China
[3] School of Intelligent Systems Engineering, Sun Yat-sen University, Shenzhen, China

Abstract. Quadruped robots with flexible movement are gradually replacing traditional mobile robots in many spots. To improve the motion stability and speed of the quadruped robot, this paper presents a responsive gradient CPG (RG-CPG) approach. Specifically, the method introduces a vestibular sensory feedback mechanism into the gradient CPG (central pattern generators) model and uses a differential evolution algorithm to optimize the vestibular sensory feedback parameters. Simulation results show that the movement stability and linear movement velocity of the quadruped robot controlled by RG-CPG are effectively improved, and the quadruped robot can cope with complex terrains. Prototype experiments demonstrate that RG-CPG works for real quadruped robots.

Keywords: Gradient CPG · Vestibular sensory feedback · Differential evolution algorithm · Quadruped robots

1 Introduction

Quadruped robots are a class of robots that imitate the movement of animals and use legs to complete the move, offering advantages such as flexible movement, active vibration isolation, and low energy consumption. They are increasingly being used to replace traditional mobile robots (such as wheeled robots and tracked robots) due to their flexibility and adaptability [16,22,23].

It is important for animals to move flexibly in complex environments, and the same is true for robots [3]. Thus, the study of motion control should be closely integrated with robotics and biology [7]. Organisms with fixed forms of motion have widespread periodic movements, such as walking, running, swimming, jumping, flying, and other physical movements, as well as physiological

This work is partially supported by the National Natural Science Foundation of China (62173352, 62006254), the Guangxi Key Laboratory of Automatic Detecting Technology and Instruments (YQ23207), and the Guangdong Basic and Applied Basic Research Foundation (2021A1515012314).

behaviors, such as chewing, breathing, heartbeat, and gastrointestinal peristalsis, which are characterized by periodic movements. Biologists typically assume that the rhythmic motion of animals is self-produced behavior of low-level nerve centers regulated by the CPG (central pattern generators) located at the thoracic-ventral ganglion of vertebrates or invertebrates [4,21]. CPGs are distributed oscillatory networks made up of intermediate neurons, which generate multiple or single periodic signals with stable phase interlocking relationships by mutual suppression among neurons to regular the rhythmic movements of the limbs or relevant parts of the body. The synaptic connections between the neurons in the CPG are adaptable and can generate different output patterns, allowing the animal to show a variety of rhythmic motor behaviors [17]. Using CPG models as motion controllers for various types of robots has become the preferred choice of many researchers [13,15]. Many CPG models have been proposed previously, such as Van der Pol oscillator model [14], Matsuoka oscillator model [18], Hopf oscillator model [5], Kuramoto oscillator model [9,10].

Using the integrated CPG model as the core of the control system also has many limitations. One is that getting the robot to the desired motion pattern requires too many parameters in a wide search space. Comprehending the connections between parameters and output results, like waveforms, phase hysteresis, and frequency, is challenging. The Trial-and-error approach is one of the solutions. In this approach, the parameters are acquired by instinctive principles and improved under the simulations and prototype experiments help. The evolutionary computation method [12,20] is also a common solution. In recent years, supervised and unsupervised learning have been increasingly applied to parameter searches in CPG networks. Hu et al. [8] suggested a numerical approach to synthesize the parameters of a CPG network to achieve the required movement modes. This approach transforms the CPG parameters into dynamic systems that are integrated into the CPG network dynamics. The CPG network with the suggested learning rules can encode the frequency, amplitude, and relations between the phase of sample signals. Farzaneh et al. [6] suggested a supervised learning approach named Fourier-based automated learning central pattern generators (FAL-CPG). For learning rhythmic signals, Fourier analysis is used to analyze the signal, and the CPG parameters are established by contrasting them with the Fourier series.

So far, gradient CPG [2] has nice performance in execution efficiency, but it is only applied to snake robots and does not introduce any feedback. To apply it to a quadruped robot and maximize its effectiveness, we propose a responsive gradient CPG (RG-CPG) control system for it. Specifically, the main contributions of this paper are summarized as follows.

- A method based on a gradient CPG model is proposed for the motion control problem of quadruped robots.
- To improve the motion stability and velocity of the quadruped robot, a vestibular sensory feedback mechanism was integrated into the gradient CPG model.

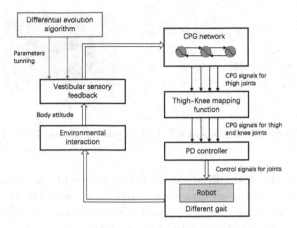

Fig. 1. Framework of RG-CPG control scheme.

- Use the DE (Differential Evolution) algorithm to adjust the feedback parameters automatically.

2 Control Framework

In this section, we present the framework of RG-CPG, as we can see in Fig. 1. First, we show in Sect. 2.1 the CPG oscillator model we use, which has a convergence behavior based on a gradient system. The thigh-knee mapping function of the quadruped robot is presented in Sect. 2.2. To make the movements of the quadruped robot more stable and faster, the vestibular sensory feedback is introduced in Sect. 2.3. Finally, to tune the vestibular sensory feedback parameters quickly, use the differential evolution algorithm in Sect. 2.4 to help us.

2.1 CPG Model Based on Gradient System

The gradient CPG model we use was proposed by Bing *et al.* [1]. The oscillator can be described by the following set of differential equations:

$$
\begin{cases}
\dot{\varphi}_i = \omega_i + M_i \cdot \Phi + N_i \cdot \tilde{\Theta} \\
\ddot{r}_i = a_i \left[\dfrac{a_i}{4}(R_i - r_i) - \dot{r}_i \right] \\
\ddot{c}_i = a_i \left[\dfrac{a_i}{4}(C_i - c_i) - \dot{c}_i \right] \\
x_i = c_i + r_i \sin(\varphi_i);
\end{cases}
\tag{1}
$$

Here, the subscript i denotes the ith oscillator. φ_i is the phase of the CPG. ω_i is the frequency of the CPG signal. M_i and N_i are convergence matrices. Φ denotes the phase vector of the CPG neuron, and the phase difference vector of the CPG neurons is $\tilde{\Theta}$. The positive coefficient a_i is employed to modify the

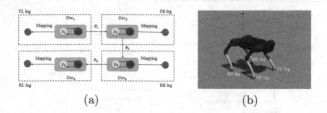

Fig. 2. (a) Structure of gradient CPG. (b) Quadruped robot and leg number.

convergence rate of the amplitude. In addition, the stable amplitude of the CPG curve is regulated by the parameter R_i, and C_i decides the bias value of the curve. r_i and c_i represent the ith oscillator's amplitude and bias, respectively. The variable x_i is the rhythmic CPG output curve.

Figure 2(a) illustrates the architecture of oscillator network of a quadruped robot (Fig. 2(b)). The network consists of four oscillators that control each of the four thigh joints of the quadruped robot. The output curve of each thigh joint can be mapped to the corresponding knee joint, and we will introduce it in Sect. 2.2.

2.2   CPG Signal Mapping

For a quadruped animal, the thigh and knee joints of the same leg have a fixed phase relationship under normal walking conditions. It can ensure that the swing leg height is always smaller than the support leg, and the foot end of the swing leg is always higher than the ground, which will not affect the body's movement.

According to the thigh-knee fixed phase relationship, the thigh position curve is flipped and translated to obtain the knee position curve to realize the thigh-knee coupling, and can write the knee joint function as:

$$
\begin{cases}
\theta_k(t) = \begin{cases} k(t)(A_h - |\theta_h(t)|), & (\dot{\theta}_h \geq 0, \text{swing phase}) \\ 0, & (\dot{\theta}_h < 0, \text{support phase}) \end{cases} \\
k(t) = k_0(1 + \dfrac{|\theta_h(t)|}{A_h}), \\
k_0 = \dfrac{A_k}{A_h}
\end{cases}
\tag{2}
$$

where θ_h and θ_k are thigh and knee CPG control signals, $k(t)$ is variable gain, A_h and A_k are thigh and knee angle amplitudes, and k_0 is knee-to-thigh amplitude ratio.

After mapping, the CPG motion curves of the thigh and knee joints are shown in Fig. 3. This function produces a knee control curve with motion in the swing phase and no motion in the support phase, forming a half-wave. Therefore it is called the half-wave function.

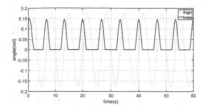

Fig. 3. Thigh-Knee mapping output signals.

2.3 Vestibular Sensory Feedback

In nature, when a quadruped animal moves on an uneven surface, it will alter the rotation range of its legs to maintain balance. Take a full-elbow quadruped as an example (as shown in Fig. 2(b)). When it goes upslope (the body is tilted back), the front legs flex more than the hind legs (As shown in Fig. 4). When it goes downslope (the body is leaned forward), the front legs flex less than the hind legs. In this work, the vestibular sensory feedback mechanism is simulated and connected to the CPG network to produce adaptive control signals.

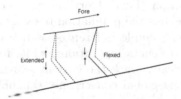

Fig. 4. Body posture adjustment when going upslope.

Combined with the sensory feedback from the interaction between the robot and the environment, the CPG network can produce adaptive walking control signals. We designed a feedback mechanism to regulate the walking posture of the quadruped robot. In this work, we added the vestibular sensory feedback term to the gradient CPG model for the first time. We rewrite (1) as follows:

$$\begin{cases} \dot{\varphi}_i = \omega_i + M\left\{i,:\right\} \cdot \varPhi + N\left\{i,:\right\} \cdot \tilde{\varTheta} \\ \ddot{r}_i = a_i \left[\dfrac{a_i}{4}(R_i - r_i) - \dot{r}_i\right] \\ \ddot{c}_i = a_i \left[\dfrac{a_i}{4}(C_i - c_i) - \dot{c}_i\right] \\ x_i = c_i + r_i \sin\left(\varphi_i\right) + feed_i; \end{cases} \tag{3}$$

where $feed_i$ is a feedback term determined by the three-axis attitude angle obtained from IMU (Inertial measurement unit) measurements. It can be expanded as:

$$feed_i = \sigma_p(leg) \cdot k_p \theta_{pitch} + \sigma_r(leg) \cdot k_r \theta_{roll} + \sigma_y(leg) \cdot k_y \theta_{yaw} \tag{4}$$

where,

$$\sigma_p(leg) = \begin{cases} 1, \ leg = \text{RL} \ or \ \text{RR} \\ -1, \ leg = \text{FL} \ or \ \text{FR} \end{cases}$$

$$\sigma_r(leg) = \begin{cases} 1, \ leg = \text{FR} \ or \ \text{RR} \\ -1, \ leg = \text{FL} \ or \ \text{RL} \end{cases}$$

$$\sigma_y(leg) = \begin{cases} 1, \ leg = \text{FL} \ or \ \text{RL} \\ -1, \ leg = \text{FR} \ or \ \text{RR} \end{cases}$$

k_p, k_r and k_y denote the gain. θ_{pitch}, θ_{roll} and θ_{yaw} indicate the pitch angle, roll angle, and yaw angle of the robot body.

2.4   Differential Evolution Algorithm

The parameters of vestibular sensory feedback are difficult to adjust manually due to the unclear relationship between the vestibular sensory feedback parameters k_p, k_r, k_y and complex environments. Therefore, we use a differential evolution algorithm to adjust the parameters.

The differential evolution (DE) algorithm was presented by R.Storn and K.Price in 1995 [11]. It retains the population-based global search strategy with real number encoding, and simple difference-based variation operation, which reduces the complexity of genetic operations, while its unique memory capability allows it to dynamically track the current search situation to adjust the search strategy with strong global convergence capability and robustness. The model we use consists of four steps, as follows:

Generate Initial Population. M individuals satisfying the constraints are randomly generated in the n-dimensional space,

$$y_{ij}(0) = \text{rand}_{ij}(0,1)(y_{ij}^U - y_{ij}^L) + y_{ij}^L \tag{5}$$

where y_{ij}^U and y_{ij}^L are the upper and lower limits of the jth chromosome, and $\text{rand}_{ij}(0,1)$ is a stochastic fractional number between $[0,1]$.

Mutation. Three individuals y_{p1}, y_{p2} and y_{p3} are randomly chosen from the population, and $i \neq p_1 \neq p_2 \neq p_3$,

$$s_{ij}(t+1) = y_{p_1j}(t) + F(y_{p_2j}(t) - y_{p_3j}(t)) \tag{6}$$

where $y_{p_2j}(t) - y_{p_3j}(t)$ is the differentiation vector, the difference operation is the key to the differential evolution algorithm, F is the zoom factor, p_1, p_2, p_3 are stochastic integers, indicating the ordinal number of individuals in the population.

Crossover. Crossover is to increase the diversity of the group,

$$u_{ij}(t+1) = \begin{cases} s_{ij}(t+1), & \text{rand } l_{ij} \leq CR \\ y_{ij}(t), & \text{otherwise} \end{cases} \tag{7}$$

where rand l_{ij} is a stochastic small number between $[0,1]$ and CR is the crossover probability, $CR \in [0,1]$.

Selection. To decide if $y_i(t)$ will be part of the following generation, the test vector $u_i(t+1)$ and the target vector $y_i(t)$ are evaluated using the fitness function,

$$y_i(t+1) = \begin{cases} u_i(t+1), & f(u_i(t+1)) > f(y_i(t)) \\ y_i(t), & \text{otherwise} \end{cases} \tag{8}$$

Repeat steps 2) to 4) until the maximum value of evolutionary generations G is reached.

We want to improve the motion stability and velocity of the quadruped robot, so we write the fitness function according to S.Gay *et al* [19] as:

$$Fitness = \bar{V} \cdot \Theta_P \cdot \Theta_R \tag{9}$$

$$\Theta_P = \left(\frac{1}{1 + \frac{1}{\tau} \sum_{t=0}^{\tau} |\theta_P(t)|} \right)^{\beta_P} \tag{10}$$

$$\Theta_R = \left(\frac{1}{1 + \frac{1}{\tau} \sum_{t=0}^{\tau} |\theta_R(t)|} \right)^{\beta_R} \tag{11}$$

where $\bar{V}$ is the average velocity, $\theta_P(t)$ and $\theta_R(t)$ are the pitch and roll angle of the body at time t, τ is the total simulation time. β_P and β_R are coefficients used to prioritize the minimization of the angles over the maximization of the velocity. In this paper, we usually set $\beta_P - \beta_R$.

3   Simulations, Comparisons and Analysis

In this section, simulations are conducted to validate the effectiveness of RG-CPG. We show comparison experiments in Sect. 3.1 for the quadruped robot with RG-CPG and the quadruped robot without RG-CPG. The comparison method compares the quadruped robot's motion stability and straight-line travel speed under the two models. Then, we present the quadruped robot up-down slope experiments in Sect. 3.2. The parameters of RG-CPG are shown in Table 1.

Table 1. Parameters setting of RG-CPG.

Gait	Parameters										
	a	ω	R	C	φ_1	φ_2	φ_3	φ_4	k_p	k_r	k_y
Walk	2.0	$\frac{\pi}{20}$	1.05	0.0	0	π	$\frac{3\pi}{2}$	$\frac{\pi}{2}$	0.6	0.3	0.1
Trot	2.0	$\frac{\pi}{22.5}$	2.55	0.0	0	π	0	π	0.8	0.3	0.1
Gallop	2.0	$\frac{\pi}{22.5}$	1.35	0.0	0	0	π	π	0.6	0.3	0.1

3.1   Simulation of the Quadruped Robot with RG-CPG

We let the quadruped robot move on a flat surface in two states, walking and galloping. The direction is along the positive direction of the X-axis. We plot the change in body tilt during the motion of the quadruped robot. As shown in Fig. 5, the body tilt angle during walking for the quadruped robot without RG-CPG is −0.028 rad to 0.012 rad for pitch angle and −0.07 rad to 0.07 rad for roll angle. The body tilt angle during walking for the quadruped robot with RG-CPG is −0.011 rad to 0.01 rad for pitch angle and −0.03 rad to 0.03 rad for roll angle. The situation is similar for the galloping gait, which indicates that the quadruped robot with RG-CPG has higher stability during motion. Readers may notice that the body roll angle gets larger during galloping for the quadruped robot with RG-CPG, it is because in galloping gait, the two front legs and the two hind legs of the quadruped robot are in the same phase respectively, so the optimization of stability is mainly reflected in the convergence of the pitch angle.

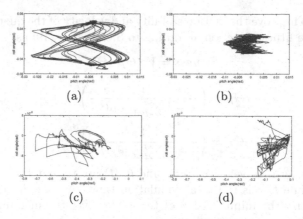

Fig. 5. Pitch and roll angles of the body of the quadruped robot in motion. (a) Walking without RG-CPG. (b) Walking with RG-CPG. (c) Galloping without RG-CPG. (d) Galloping with RG-CPG.

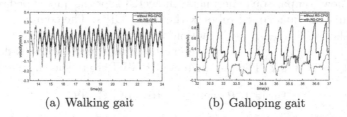

Fig. 6. The linear velocity of the quadruped robot along the positive direction of the X-axis for walking and galloping gaits. The solid red line represents the motion velocity of the quadruped robot with RG-CPG, and the blue dotted line represents the motion velocity of the quadruped robot without RG-CPG. (Color figure online)

We also separately measured the quadruped robot's linear velocity along the positive direction of the X-axis during its motion. We can observe from Fig. 6(a) that the motion velocity of the quadruped robot with RG-CPG in the positive direction along the X-axis in walking gait is higher than that of the quadruped robot without RG-CPG. This situation is even more evident in the galloping gait (as shown in Fig. 6(b)). These suggest that the quadruped robot with RG-CPG has stronger linear motion ability.

3.2 Slope Experiments

To test the performance of our proposed method in different terrains, we designed the slope experiment. In the experiment, the quadruped robot trot and gallop through a slope device that includes a 10° uphill, a platform, and a 10° downhill. Figure 7 shows Snapshots of the quadruped robot passing the slope device.

From Fig. 8(a), it can be observed that the robot is in the upslope at 14 to 19.5 s when the robot's body is tilted back (pitch angle decreases). The left foreleg (FL) and right foreleg (FR) thigh joint angles increase, and the entire leg becomes flexed, while the left hindleg (RL) and right hindleg (RR) thigh joint angles decrease, and the entire leg becomes extended, the robot body is in a low front and high rear posture, as shown in Fig. 4. The robot moves on the platform in 19.5 to 22.3 s. Then between 22.3 to 25 s, the robot is in the downslope, and body posture is the opposite of the upslope. At the trot gait (Fig. 8(b)), the robot goes upslope in 83 to 86.2 s, moves on the platform in 86.2 to 87.6 s, and goes downslope in 87.6 to 89.5 s, the state of the body and joints during this period is similar to that of the galloping gait.

4 Prototype Experiments

4.1 Technical Specifications

The quadruped robot system mainly consists of a control board, servos, brackets, power supplies, sensors, and software, Table 2 summarizes the technical specifications of the quadruped robot.

4.2 Experiments

To confirm the applicability of RG-CPG to real quadruped robots, we use the quadruped robot to execute walk and trot gait on flat ground (Fig. 9) and record the change in the angle of the thigh joints. From Fig. 10(a), we can see that the four thighs swing in the order of FR-RL-FL-RR, it is a typical walking gait. And as shown in Fig. 10(b), the four thighs are divided into two groups (FR and RL in phase, FL and RR in phase), and they swing alternately, which is a typical trot gait. The experiment demonstrates that RG-CPG allows a quadruped robot to execute specific gaits in the real world.

Fig. 7. Snapshots of the robot going upslope (left), walking on the platform (middle), and going downslope (right).

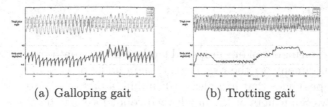

(a) Galloping gait   (b) Trotting gait

Fig. 8. Experimental data of slope in two gait states. The top curve represents the angle of change of thigh joints. The bottom curve shows the body tilt.

Table 2. Overview of the quadruped robot.

Items	Discriptions
Size	1126·467·636 mm (Standing)
Weight	55 kg
Battery voltage rating	51.8 V
Battery power rating	932.4 Wh
Hip rotation range	$-0.75 \sim 0.75$ rad
Thigh rotation range	$-1.0 \sim 3.5$ rad
Calf rotation range	$-2.6 \sim -0.6$ rad
Maximum instantaneous torque of joints	$210\,\text{N} \cdot \text{m}$

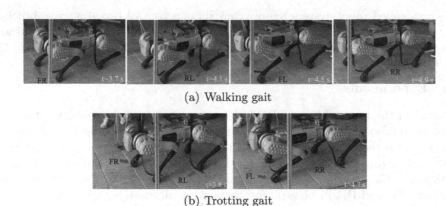

(a) Walking gait

(b) Trotting gait

Fig. 9. Snapshots of the quadruped robot performing two gaits.

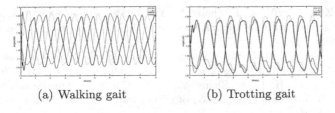

(a) Walking gait   (b) Trotting gait

Fig. 10. Variation curve of thigh joint angle of the quadruped robot.

5   Conclusion

In this paper, we present a responsive gradient CPG (RG-CPG) control method for quadruped robots. Introducing a vestibular sensory feedback term in the model and optimizing the vestibular sensory feedback parameters using a differential evolution algorithm can improve the quadruped robot's motion stability and linear motion velocity. In addition, RG-CPG enables the quadruped robot to cope with different terrains, such as slopes. Simulation results validated the effectiveness and practicality of RG-CPG. Prototype experiments demonstrate the feasibility of the method. Future work will focus on using RG-CPG for different types of legged robots and combining the method with reinforcement learning.

References

1. Bing, Z., Cheng, L., Chen, G., Röhrbein, F., Huang, K., Knoll, A.: Towards autonomous locomotion: CPG-based control of smooth 3d slithering gait transition of a snake-like robot. Bioinspir. Biomimet. **12**(3), 035001 (2017)
2. Bing, Z., Cheng, L., Huang, K., Zhou, M., Knoll, A.: CPG-based control of smooth transition for body shape and locomotion speed of a snake-like robot. In: 2017 IEEE International Conference on Robotics and Automation (ICRA), pp. 4146–4153 (2017)
3. Bruzzone, L., Quaglia, G.: Review article: locomotion systems for ground mobile robots in unstructured environments. Mech. Sci. **3**(2), 49–62 (2012)
4. Delcomyn, F.: Neural basis of rhythmic behavior in animals. Science **210**(4469), 492–498 (1980)
5. Du, S., Wu, Z., Wang, J., Qi, S., Yu, J.: Design and control of a two-motor-actuated tuna-inspired robot system. IEEE Trans. Syst. Man Cybernet. Syst. **51**(8), 4670–4680 (2021)
6. Farzaneh, Y., Akbarzadeh, A., Akbari, A.A.: New automated learning CPG for rhythmic patterns. Intel. Serv. Robot. **5**(3), 169–177 (2012)
7. Hogan, N., Sternad, D.: On rhythmic and discrete movements: reflections, definitions and implications for motor control. Exp. Brain Res. **181**(1), 13–30 (2007)
8. Hu, Y., Liang, J., Wang, T.: Parameter synthesis of coupled nonlinear oscillators for CPG-based robotic locomotion. IEEE Trans. Indust. Electron. **61**(11), 6183–6191 (2014)
9. Mao, Y., Zhang, Z.: Asymptotic frequency synchronization of Kuramoto model by step force. IEEE Trans. Syst. Man Cybernet. Syst. **50**(8), 2768–2778 (2020)

10. Pan, J., Zhou, Z., Wang, J., Zhang, P., Yu, J.: Development of a penguin-inspired swimming robot with air lubrication system. IEEE Trans. Indust. Electron. **70**(3), 2780–2789 (2023)
11. Storn, R., Price, K.: Differential evolution-a simple and efficient heuristic for global optimization over continuous spaces. J. Global Optim. **11**(4), 341–359 (1997)
12. Wang, M., Dong, H., Li, X., Zhang, Y., Yu, J.: Control and optimization of a bionic robotic fish through a combination of CPG model and pso. Neurocomputing **337**, 144–152 (2019)
13. Wang, M., Zhang, Y., Yu, J.: An SNN-CPG hybrid locomotion control for biomimetic robotic fish. J. Intell. Robot. Syst. **105**(2), 1–25 (2022)
14. Yu, H., Gao, H., Ding, L., Li, M., Deng, Z., Liu, G.: Gait generation with smooth transition using CPG-based locomotion control for hexapod walking robot. IEEE Trans. Indust. Electron. **63**(9), 5488–5500 (2016)
15. Yu, J., Wu, Z., Yang, X., Yang, Y., Zhang, P.: Underwater target tracking control of an untethered robotic fish with a camera stabilizer. IEEE Trans. Syst. Man Cybernet. Syst. **51**(10), 6523–6534 (2021)
16. Zhao, Y., Chai, X., Gao, F., Qi, C.: Obstacle avoidance and motion planning scheme for a hexapod robot octopus-III. Robot. Auton. Syst. **103**, 199–212 (2018)
17. Zheng, H., Zhang, X., Li, T., Guanghong, D.: Robot motion control method based on CPG principle. Chin. High Technol. Lett. **13**(7), 64–68 (2003)
18. Zhong, G., Chen, L., Jiao, Z., Li, J., Deng, H.: Locomotion control and gait planning of a novel hexapod robot using biomimetic neurons. IEEE Trans. Control Syst. Technol. **26**(2), 624–636 (2018)
19. Gay, S., Santos-Victor, J., Ijspeert, A.: Learning robot gait stability using neural networks as sensory feedback function for central pattern generators. In: 2013 IEEE/RSJ International Conference on Intelligent Robots and Systems, pp. 194–201 (2013)
20. Yu, J., Wu, Z., Wang, M., Tan, M.: CPG network optimization for a biomimetic robotic fish via PSO. IEEE Trans. Neural Netw. Learn. Syst. **27**(9), 1962–1968 (2016)
21. Yu, J., Tan, M., Chen, J., Zhang, J.: A survey on CPG-inspired control models and system implementation. IEEE Trans. Neural Netw. Learn. Syst. **25**(3), 441–456 (2014)
22. Wang, L., et al.: Design and dynamic locomotion control of quadruped robot with perception-less terrain adaptation. In: Cyborg and Bionic Systems (2022)
23. Huang, H.W., et al.: Mobile robotic platform for contactless vital sign monitoring. In: Cyborg and Bionic Systems (2022)

Vessel Behavior Anomaly Detection Using Graph Attention Network

Yuanzhe Zhang[1] , Qiqiang Jin[2] , Maohan Liang[2] , Ruixin Ma[3] ,
and Ryan Wen Liu[2]([✉])

[1] School of Computer and Artificial Intelligence, Wuhan University of Technology,
Wuhan 430070, China
[2] School of Navigation, Wuhan University of Technology, Wuhan 430063, China
`wenliu@whut.edu.cn`
[3] Tianjin Research Institute for Water Transport Engineering, M.O.T., Tianjin
300456, China

Abstract. Vessel behavior anomaly detection is of great significance
for ensuring navigation safety, combating maritime crimes, and maritime management. Unfortunately, most current researches ignore the
temporal dependencies and correlations between ship features. We propose a novel vessel behavior anomaly detection using graph attention
network (i.e., VBAD-GAT) framework, which characterizes these complicated relationships and dependencies through a graph attention module that consists of a time graph attention module and a feature graph
attention module. We also adopt a process of graph structure learning to
obtain the correct feature graph structure. Moreover, we propose a joint
detection strategy combining reconstruction and prediction modules to
capture the local ship features and long-term relationships between ship
features. We demonstrate the effectiveness of the graph attention module
and the joint detection strategy through the ablation study. In addition,
the comparative experiments with three baselines, including the quantitative analysis and visualization, show that VBAD-GAT outperforms all
other baselines.

Keywords: Anomaly detection · Graph attention network · AIS data
mining

1 Introduction

As the number of vessels worldwide increases, law enforcement agencies face
new challenges in ensuring navigation safety, combating maritime crimes, and
maritime management. The abnormal ship behavior often hides rich semantic
information. For example, an irregular change in a ship's course and a sudden
decrease in speed may mean that the ship is out of control due to a malfunction;
a ship's deviation from the original route may indicate that the ship is engaged
in illegal fishing or smuggling. Therefore, vessel behavior anomaly detection can

assist law enforcement agencies in promptly discovering lost, malfunctioning, and out-of-control ships and vessels engaged in illegal activities such as smuggling, overloading, and illegal fishing.

The Automatic Identification System (AIS) was developed in 1990 and originally aimed to avoid collisions of ships to ensure the safety of ship navigation [22]. Compared with radars, AIS signals have wider coverage and more stable signal quality. An AIS message usually consists of static ship data (Maritime Mobile Service Identity [MMSI], ship name, ship type, ship size, etc.) and ship motion data (position, speed, course, etc.). With the popularization of AIS equipment on various types of ships, AIS data mining [13–15], especially ship behavior anomaly detection using AIS data, has attracted widespread attention in the academic circle. We select the research object of this work as AIS data based on the above reasons.

Motivated by the definition of outliers by Hawkins et al. [6], we define an abnormal ship behavior as follows: a ship feature that differs significantly from that of other ships. The ship features studied in this work contain six types: latitude, longitude, course over ground (COG), speed over ground (SOG), turning rate, and acceleration. The latitude, longitude, COG, and SOG can be directly obtained from AIS messages. The turning rate $tr = (dir - dir_{pre})/(t - t_{pre})$ and acceleration $acc = (spd - spd_{pre})/(t - t_{pre})$ characterize the change rates of COG and SOG respectively. Here, tr, dir, t, acc, and spd denote the current turning rate, COG, timestamp, acceleration, and SOG; $*_{pre}$ represents the previous value of feature $*$.

There are complicated temporal dependencies and correlations between ship features. For instance, a large turning rate of a ship indicates that this ship is changing its course, and sailors generally reduce the ship's speed to control it better; a ship's position is usually close to where the ship was at the last few moments. In addition, the features of a ship consist of the local features generated at the latest moments and the long-term features formed since the ship set sail. The local features are more significantly associated with current ship behavior than long-term features. However, the influence of early ship features also cannot be ignored. Unfortunately, current studies do not take both factors into account.

To solve the above problems, we propose a novel Vessel Behavior Anomaly Detection using Graph Attention Network (i.e., VBAD-GAT) framework, combining graph attention and joint detection modules. The graph attention module consists of two sub-modules: a temporal graph attention module and a feature graph attention module. The temporal graph attention module captures the temporal dependencies of ship features; the feature graph attention module aims to characterize the correlations between ship features. In particular, the feature graph attention module includes a graph structure learning process to obtain a correct feature graph structure. The joint detection module consists of a reconstruction module and a prediction module. The reconstruction module is used to capture the local ship features; the prediction module aims to characterize long-term features.

In summary, the contributions of this paper are as follows:

- In our model, we introduce a graph attention module consisting of a time graph attention module and a feature graph attention module to capture temporal dependencies and correlations between ship features.
- We propose a joint detection strategy employing reconstruction and prediction modules to capture the local and long-term ship features.
- Through the ablation study on two AIS datasets, we demonstrate the effectiveness of the graph attention module and joint detection module of VBAD-GAT. In addition, the comparative experiments show that VBAD-GAT outperforms all other baselines.

2 Related Work

2.1 Deep Learning-Based Anomaly Detection

Deep learning-based anomaly detection, a popular research direction, comprises reconstruction- and forecast-based models. The two models will be reviewed in this subsection separately.

Reconstruction-based models detect anomalies based on an assumption that normal instances can be better reconstructed than anomalies. The autoencoder networks (AE) [7], one of the most common reconstruction-based models, play an essential role in anomaly detection [4,9,24]. An AE usually consists of an encoder and a decoder. The encoder extracts features from the data, while the decoder restores the features to the original input. To solve the problem of low interpretability of abnormal time series, Kieu et al. [10] proposed an autoencoder framework that divides time series into clean time series without abnormal data and abnormal time series. In addition, as an application of generative adversarial networks (GAN) [5], GAN-based anomaly detection received widespread attention [1,20,21,25]. The GAN can learn the latent space of the generative network and capture the distribution of normal data. It computes an anomaly score by the difference between generated and actual data, and anomalies usually have greater differences. A model that deals with the complicated distributions of time series is designed by Zhao et al. [27]. They employed and trained the GAN to learn the pattern of task series.

Forecast-based models work on an assumption that anomalies can generate predicted data with a greater difference between actual data than normal data. This approach works well for sequential data such as frame streams. Liu et al. [16] calculated the differences between predicted and actual frames to detect an abnormal frame. Ye et al. [23] computed anomaly scores by training a convolutional network to generate future frames given an input sequence.

However, unlike VBAD-GAT, the above methods either employ reconstruction-based or prediction-based models; therefore, the advantages of both cannot be combined. Besides, the graph attention network is introduced in our framework to capture the temporal dependencies and correlations between ship features.

2.2   Vessel Behavior Anomaly Detection

The existing studies of vessel behavior anomaly detection consist of three categories: statistical methods, clustering methods, and neural network methods. In this subsection, the three methods will be reviewed separately.

Statistical methods aim to develop models based on probabilistic statistics that reveal vessel behavior modes served as the foundations for identifying abnormal behavior. d'Afflisio et al. [3] developed an approach employing the Generalized Likelihood Ratio Test (GLRT) and the Model Order Selection (MOS) in order to judge whether a vessel is reporting false position information by AIS messages. Laxhammar et al. [12] and Ristic et al. [19] employed techniques of kernel density estimation and the Gaussian mixture model (GMM), respectively, to perform the vessel behavior anomaly detection. Most statistical methods assume AIS data follow a specific distribution, which is often incorrect.

As one of the traditional techniques of machine learning, clustering technology has been widely adopted in vessel behavior anomaly detection. Zhen et al. [28] designed a novel distance measure between tracks. They incorporated hierarchical clustering and Naive Bayes classifier algorithms to identify abnormal vessel behaviors. To reconstruct the information in original vessel trajectories affected by noises from AIS data, Liu et al. [15] utilized the density-based clustering method DBSCAN and a bidirectional long short-term memory (BLSTM)-based technique. Unfortunately, clustering-based methods have some limitations: non-incremental clustering techniques will spend too much time processing AIS data; some clustering algorithms lack robustness to noise abound in AIS data.

The deep learning-based technique is widely exploited in vessel behavior anomaly detection due to neural networks' powerful fitting capabilities. Nguyen et al. [17] proposed a novel embedding of vessel data. They adopted a variational recurrent neural network to identify abnormal vessel trajectories. He et al. [8] proposed a transfer learning-based trajectory anomaly detection strategy combing a variational self-encoder with a graph variational autoencoder. Zhao et al. [26] developed a long short-term memory network (LSTM) to forecast vessel trajectories using vessel behavior modes extracted by DBSCAN. If an actual position significantly differs from the projected positions, the vessel trajectories are regarded as abnormal trajectories.

3   Proposed Method

This section presents our problem definition and explains the proposed model VBAD-GAT. Figure 1 gives an overview of our model.

3.1   Problem Definition

A vessel trajectory is denoted by a matrix $X \in \mathbb{R}^{6 \times N} = \{x_1, x_2, ..., x_N\}$, where N is the number of timestamped points on X, and $x_i \in \mathbb{R}^6$ is the i-th timestamped point that contains six features: latitude, longitude, COG, SOG, turning

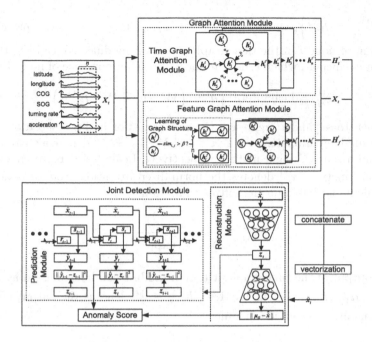

Fig. 1. An overview of VBAD-GAT.

rate and acceleration of the vessel. For a point $x_t \in X$, a sequence of points $X_t \in \mathbb{R}^{6 \times n} = \{x_{t-n+1}, x_{t-n+2}, ..., x_t\}$ can be generated through a sliding window with length n, where $x_i \in X (t - n + 1 \leq i \leq t)$. A point x_t is an anomaly for a vessel trajectory dataset $\mathcal{X}$ if and only if there exist some points in X_t whose features deviate significantly from those of other points in $\mathcal{X}$. Given a vessel trajectory dataset $\mathcal{X}$ and $X_t = \{x_{t-n+1}, x_{t-n+2}, ..., x_t\}$, the goal of vessel behavior anomaly detection is to determine whether x_t is an anomaly by outputting $\hat{y}_t \in \{0, 1\}$, where $\hat{y}_t = 1$ means x_t is an anomaly.

3.2   Graph Attention Module

Time Graph Attention Module. The time graph attention module utilizes the graph attention networks (GAT) to capture temporal dependencies of ship features. This graph contains n nodes, and each node is a representation of a timestamp; each edge represents the relationship between corresponding representations of two timestamps. Since we lack prior knowledge of temporal dependencies of vessel behaviors, the time graph is regarded as a complete graph. That is, any node is adjacent to other n-1 nodes:

$$N_i^{(t)} = \{j | 1 \leq j \leq n\},\tag{1}$$

where $N_i^{(t)}$ is a set containing node i's neighbors and itself in the time graph. We denote the input of the time graph attention module as $H^{(t)} \in \mathbb{R}^{6 \times n} =$

$\left\{ h_1^{(t)}, h_2^{(t)}, \cdots, h_n^{(t)} \right\}$, where $h_i^{(t)} \in \mathbb{R}^6 = x_{t-n+i} (1 \leq i \leq n)$ represents the embedding of node i in the time graph attention module. We call $e_{ij}^{(t)}$ the time attention coefficient, which is used to measure the importance of node j to node i:

$$e_{ij}^{(t)} = \text{LeakyReLU} \left(a^{(t)T} \left(W^{(t)} h_i^{(t)} \| W^{(t)} h_j^{(t)} \right) \right), \tag{2}$$

where operators $\|$ and $*^T$ represent the concatenation and transposition, respectively; trainable parameters $a^{(t)} \in \mathbb{R}^{12}$ and $W^{(t)} \in \mathbb{R}^{6 \times 6}$ denote a shared weight vector and a shared weight matrix, respectively; LeakyReLU represents a linear activation function. $\alpha_{ij}^{(t)}$ denotes the normalized attention score of node j to i and is calculated based on the softmax function:

$$\alpha_{ij}^{(t)} = \frac{\exp(e_{ij}^{(t)})}{\sum_{k \in N_i} \exp(e_{ik}^{(t)})}. \tag{3}$$

The output of the time graph attention module can be expressed as $H^{(t)'} \in \mathbb{R}^{6 \times n} = \left\{ h_1^{(t)'}, h_2^{(t)'}, \cdots, h_n^{(t)'} \right\}$. Here, $h_i^{(t)'} \in \mathbb{R}^6$ denotes the output of node i in the time graph attention module:

$$h_i^{(t)'} = \sigma \left(\sum_{j \in N_i^{(t)}} \alpha_{ij}^{(t)} W^{(t)} h_j^{(t)'} \right), \tag{4}$$

where σ is a non-linear activation function.

Feature Graph Attention Module. The feature graph attention module captures the correlations between vessel features. This module is a GAT that contains six nodes. Each node in the graph represents a vessel feature. A node's embedding vector is represented as $h_i^{(f)} \in \mathbb{R}^n = (x_{i,1}, x_{i,2}, \cdots, x_{i,n})^T$, where $x_{i,j}$ denotes the element at row i and column j in X_t; each edge is used to characterize the relationship between corresponding two features.

Unlike temporal dependencies, there is actually no correlation between some ship features. Therefore, we need to learn a feature graph structure and apply it to the graph attention module, that is, to determine the neighbors $N_i^{(f)} (1 \leq i \leq 6)$ of node i in the feature graph. We define $N_i^{(f)}$ as nodes with a similarity greater than β to node i:

$$N_i^{(f)} = \{j | sim_{ij} > \beta\}, \tag{5}$$

where sim_{ij} is the similarity between representations of node i and j. We calculate sim_{ij} by a Gaussian kernel function:

$$\phi(h_i^{(f)}, h_j^{(f)}) = \sqrt{(h_i^{(f)} - h_j^{(f)})^T M (h_i^{(f)} - h_j^{(f)})}, \tag{6}$$

$$sim_{ij} = \exp\left(-\frac{\phi\left(h_i^{(f)}, h_j^{(f)}\right)}{2\kappa^2}\right),\tag{7}$$

where κ denotes the size of the Gaussian kernel function; $\phi\left(h_i^{(f)}, h_j^{(f)}\right)$ represents the generalized mahalanobis distance between $h_i^{(f)}$ and $h_j^{(f)}$; $M \in \mathbb{R}^{n\times n} = W_s W_s^T$ and $W_s \in \mathbb{R}^{n\times n}$ is a weight matrix. Once the feature graph structure is learned, we can follow a similar method to time graph attention module to obtain the output $H^{(f)'} \in \mathbb{R}^{n\times 6} = \left\{h_1^{(f)'}, h_2^{(f)'}, \cdots, h_6^{(f)'}\right\}$ of the feature graph attention module, where $h_i^{(f)'} \in \mathbb{R}^n$ is the ouput of node i. Based on the Eq. (4), a masked attention mechanism is adopted when computing the $h_i^{(f)'}$, that is, only the contribution of node i's neighbors N_i to node i is considered.

Finally, the output $\widetilde{x}_t \in \mathbb{R}^{18n}$ of the graph attention layer is generated by concatenating $X_t^T, H^{(t)'}, H^{(f)'T}$ and vectorization:

$$\widetilde{x}_t = \text{vec}\left(X_t^T || H^{(t)'} || H^{(f)'T}\right).\tag{8}$$

Here, vec represents vectorization and $\text{vec}(A) = (a_{11}a_{21}\cdots a_{1m}a_{12}\cdots a_{mn})^T$, where a_{ij} denotes the element in the i-th row and j-th column of $A \in \mathbb{R}^{m\times n}$.

3.3   Joint Detection Module

Reconstruction Module. The reconstruction module utilizes a variational autoencoder network (VAE) [11] to characterize the local features of ship behaviors. We take $\widetilde{x}_t$ as the input to the VAE. The VAE consists of two components: an interference network and a generative network. The optimization objective of both networks is to maximize evidence lower bound $\text{ELBO}(q, \widetilde{x}_t; \theta, \phi)$, that is:

$$\max_{\theta,\phi} \text{ELBO}\left(q, \widetilde{x}_t; \theta, \phi\right) = \max_{\theta,\phi} \mathbb{E}_{z_t \sim q(z_t;\phi)}\left[\log\frac{p\left(\widetilde{x}_t|z_t;\theta\right)p\left(z_t;\theta\right)}{q\left(z_t;\phi\right)}\right]\tag{9}$$

$$= \max_{\theta,\phi}\mathbb{E}_{z_t \sim q(z_t|\widetilde{x}_t;\phi)}\left[\log p(\widetilde{x}_t|z_t;\theta)\right] - \text{KL}\left(q(z_t|\widetilde{x}_t;\phi), p(z_t;\theta)\right).\tag{10}$$

Here, ϕ and θ respectively represent the parameters of the interference network and the generative network, and these two networks are three-layer fully connected networks in our model; $p(z_t;\theta)$ is a prior distribution, and it is assumed to follow a standard Gaussian distribution $\mathcal{N}(0, I)$; KL represents the KL divergence; $\mathbb{E}$ denotes the expectation.

Prediction Module. The prediction module utilizes a gated recurrent unit network (GRU) [2] to characterize long-term correlations of ship behavior. The GRU is a variant of recurrent neural network, which can effectively solve the long-term dependency problem of simple recurrent neural networks. In this module, the output $\widetilde{x}_{t-1} \in \mathbb{R}^{18n}$ of graph attention network at the previous moment is considered as the current input of the GRU. The label y_t of the GRU is set to the latent variable z_t generated by the VAE. The GRU uses the reset gate $r_t \in [0,1]^{18n}$ to control the degree to which current candidate state $\tilde{h}_t$ depends on previous candidate state $\tilde{h}_{t-1}$:

$$r_t = \sigma(W_r \tilde{x}_t + U_r h_{t-1} + b_r), \tag{11}$$

$$\tilde{h}_t = \tanh(W_h \tilde{x}_t + U_h(r_t \odot h_{t-1}) + b_h), \tag{12}$$

where tanh is a non-linear activation function. Moreover, the GRU uses an update gate $s_t \in [0,1]^{18n}$ to control how much information the current state needs to retain from the historical state h_{t-1} and how much new information needs to be accepted from the candidate state $\tilde{h}_t$:

$$s_t = \sigma(W_s \tilde{x}_t + U_s h_{t-1} + b_s), \tag{13}$$

$$h_t = s_t \odot h_{t-1} + (1 - s_t) \odot \tilde{h}_t. \tag{14}$$

In Eqs. 11, 12 and 13, W_*, U_* and $b_* \in \mathbb{R}^{18n \times 18n}$ are learnable parameters, where $* \in \{r, h, s\}$. Finally, we obtain $\hat{y}_t \in \mathbb{R}^d$ from the hidden state h_t through a fully connected layer:

$$\hat{y}_t = V h_t + c, \tag{15}$$

where $V \in \mathbb{R}^{d \times 18n}, c \in \mathbb{R}^d$ are the parameters of output layers. The prediction module defines the optimization objective using the quadratic loss function:

$$\min \frac{1}{2} \|\hat{y}_t - y_t\|^2. \tag{16}$$

The joint detection module optimizes parameters by jointly training the reconstruction module and prediction module. The loss function $Loss$ of the joint detection module is as follows:

$$Loss_r = -\text{ELBO}\left(q, \tilde{x}_t; \theta, \phi\right), \tag{17}$$

$$Loss_p = \frac{1}{2} \|\hat{y}_t - y_t\|^2, \tag{18}$$

$$Loss = Loss_r + \lambda Loss_p, \tag{19}$$

where $Loss_r$ represents the loss function of the reconstruction module, $Loss_p$ denotes the loss function of the prediction module, and λ is used to adjust the weights of two loss functions.

For a given test point x_t and a trained model, the anomaly score of x_t is calculated by the following formula:

$$\text{score}(x_t) = \|\hat{y}_t - y_t\|^2 + \eta \|\hat{\tilde{x}}_t - \mu_G\|^2, \tag{20}$$

where η is used to measure the contributions of the reconstruction and prediction modules to the abnormal score, $\hat{\tilde{x}}_t$ represents samplings from the generative network in the reconstruction module. Finally, we can obtain the anomaly detection result of x_t by judging whether $\text{score}(x_t)$ is greater than a score threshold γ, that is: $\hat{y}_t = I(\text{score}(x_t) > \gamma)$, where $I(\cdot)$ outputs 1 if and only if equation · holds.

4 Experiments

4.1 Experimental Setup

Data Description. We conducted all experiments on two AIS datasets: the Yangtze River Delta dataset (YZRD) and the Florida Strait dataset (FS). Our experiments used ship trajectories instead of trajectory points for performance evaluation. This is because trajectory points are too numerous to label and analyze. We obtained 2634 and 3709 vessel trajectories through trajectory extraction from YZRD and FS, respectively. Three volunteers manually labeled all ship trajectories by MMSI.

Implementation Details. All experiments are conducted using PyTorch. Our model is trained and tested on a single NVIDIA RTX Titan/24GB. The dimension d of the latent variable z_t of the reconstruction module is set to 6, and the number n of columns of input X_t equals 8. An Adam optimization strategy with initial learning rate $\beta_1 = 0.9$ and $\beta_2 = 0.99$ is adopted for training. The batch size for the training phase is set to 64. In addition, in terms of graph structure learning, the node similarity threshold β is set to 0.25, and the size κ of the Gaussian kernel function is set to 1. A vessel trajectory is considered abnormal if its abnormal parts formed by abnormal points exceed 10% in length or 5% in time.

Table 1. Performance comparison for vessel behavior anomaly detection.

Datasets	YZRD			FS		
Evaluation Metrics	Precision	Recall	F1 Score	Precision	Recall	F1 Score
GAT-NT	0.813	0.798	0.805	0.824	0.801	0.812
GAT-NF	0.763	0.774	0.769	0.795	0.812	0.803
GAT-NL	0.904	0.893	0.898	0.915	0.898	0.906
GAT-NR	0.832	0.824	0.827	0.845	0.837	0.841
GAT-NP	0.838	0.840	0.839	0.816	0.811	0.813
GeoTracknet [17]	0.757	0.765	0.761	0.769	0.779	0.774
MADVCN [28]	0.724	0.721	0.722	0.756	0.750	0.753
TREAD [18]	0.746	0.742	0.739	0.783	0.761	0.772
VBAD-GAT	**0.973**	**0.978**	**0.975**	**0.965**	**0.966**	**0.965**

4.2 Ablation Study

Graph Attention Module. This part utilizes two degenerated versions, GAT-NT and GAT-NF, of VBAD-GAT to validate the effectiveness of the graph attention module in VBAD-GAT. The GAT-NT and GAT-NF are respectively obtained by removing the time graph attention module and the feature graph attention module from VBAD-GAT. In addition, we conducted experiments that treat the feature graph structure as a complete graph to validate the effectiveness of the graph structure learning, and its corresponding model is called GAT-NL. The precision, recall, and F1 scores obtained by four models for two datasets are shown in Table 1.

It can be seen from Table 1 that the GAT-NT that does not consider the temporal dependencies of ship features is lower than VBAD-GAT in all three indicators and two datasets. For instance, for YZRD, the GAT-NT is 16.4% and 14.6% lower in precision than VBAD-GAT and 18.4% and 17.1% lower in recall than VBAD-GAT. This is because, compared to GAT-NT, VBAD-GAT can capture the temporal dependencies between ship features in the sliding window through the time graph attention module. Also, VBAD-GAT outperforms GAT-NF that does not consider relationships between ship features significantly on both datasets. For instance, the recall of VBAD-GAT is 21.6% and 19.0% higher than that of GAT-NF on YZRD and FS, respectively. This is in line with our expectations. Unlike VBAD-GAT, GAT-NF, which removes the feature graph attention module, cannot capture the correlations between ship features. Finally, it can be seen from Table 1 that compared with GAT-NF, the performance of GAT-NL has a particular improvement on both datasets but still perform worse than VBAD-GAT. For example, the F1 Score obtained by GAT-NL is 7.9% and 6.6% lower than that of VBAD-GAT on YZRD and FS, respectively. Unlike time, there is no correlation between certain ship features. On one hand, GAT-NL simply regards the ship feature graph structure as a complete graph. This wrong graph structure incorrectly characterizes the correlations between ship features, thus degrading the detection performance. On the other hand, VBAD-GAT learns the feature graph structure by calculating the similarity between node representations through the Gaussian kernel function and the Mahalanobis distance. Therefore, VBAD-GAT can achieve better performance than GAT-NL.

Joint Detection Strategy. This part aims to validate the effectiveness of the joint detection strategy in VBAD-GAT by the degenerated versions GAT-NR and GAT-NP of VBAD-GAT. The GAT-NR and GAT-NP are obtained by removing the reconstruction and prediction modules from the VBAD-GAT. The label y_t of the GRU units in GAT-NR is the output $\tilde{x}_t$ of the graph attention module. The precision, recall, and F1 score of GAT-NR, GAT-NP, and VBAD-GAT on two datasets are shown in Table 1.

It can be seen from the experimental results that three indicators of GAT-NR are lower than those of VBAD-GAT. For example, the F1 score of GAT-NR is 15.2% lower than that of VBAD-GAT on YZRD. This is because GAT-NR does not utilize VAE to capture the local features of ship behavior. On one hand, although the GRU unit uses a gating mechanism to memorize the long-term historical information of ships, this will weaken the ability of GAT-NR to capture the local features of ships. On the other hand, VBAD-GAT, which utilizes both the prediction and reconstruction modules, can accurately characterize the local features of ships while capturing the long-term trend of ship features. As a result, the performance of VBAD-GAT is significantly improved compared with GAT-NR. Moreover, we can find that VBAD-GAT performs better than GAT-NP. Especially in terms of recall, VBAD-GAT exceeds GAT-NP by 19.1% in FS. A voyage of a ship usually lasts several hours to several days. The historical information on ship behavior is wealthy. However, GAT-NP, which only uses the prediction module to capture the local ship features, does not characterize the long-term trend of ship features. Therefore, it has a much lower recall than VBAD-GAT. In addition, it can be seen that GAT-NP performs worse on FS than on YZRD. This is because the ships in FS have a longer voyage time, so the long-term trend of ship features is more diverse and complex, and the influence of lacking the prediction module is even more significant.

4.3   Comparative Experiments

We demonstrate in this subsection the effectiveness of the VBAD-GAT by three methods: GeoTracknet [17], MADVCN [28], and TREAD [18]. The GeoTracknet uses a variational recurrent neural network to model ship trajectories and introduces a contrario algorithm to solve the problem caused by the geographical correlation of ship behavior; the MADVCN first clusters vessel trajectories and then employs a Naive Bayes Classifier to perform ship behavior anomaly detection; the TREAD detects anomalies by using DBSCAN to extract routes.

Quantitative Comparison. This part demonstrates the effectiveness of VBAD-GAT by the quantitative analysis with three baselines. The precision, recall, and F1 score of GeoTracknet, MADVCN, and TREAD on two datasets are shown in Table 1. It can be seen from Table 1 that VBAD-GAT is superior to three baselines in three indicators and two datasets. This is in line with our expectations. The current features of a ship are more dependent on the features in the recent period than the early historical features. The GeoTracknet, which uses variational recurrent neural networks to model ship behaviors, does not focus on capturing the local features of ship behavior. Therefore, compared with VBAD-GAT, the detection accuracy of GeoTracknet is lower. For instance, the precision of VBAD-GAT on the YZRD exceeds that of GeoTracknet by 28.5%. The MADVCN clusters ship trajectories by employing a novel trajectory similarity measure method. Although MADVCN considers the relationships between the features of different ships, it ignores the correlations between the features of a single ship. On the contrary, with the feature graph attention module, VBAD-GAT can effectively characterize the relationships between features of a single ship. In addition, MADVCN performs worse in YZRD than in FS. This is because the ship behavior in YZRD is more diverse and changes more frequently, so the influence of the lack of the feature graph attention module is more pronounced. Similar to MADVCN, the TREAD using the DBSCAN does not consider the dependencies between ship features of a single ship, so its performance on three indicators and two datasets is inferior to that of VBAD-GAT.

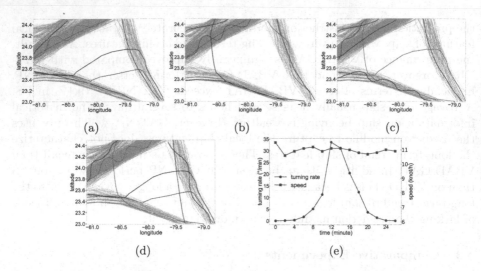

Fig. 2. Visualization of vessel behavior detection results in FS and the turning rate and speed of an abnormal vessel identified by VBAD-GAT:(a) Detection result of GeoTracknet, (b) Detection result of MADVCN, (c) Detection result of TREAD, (d) Detection result of VBAD-GAT, (e) Turning rate and speed of an abnormal vessel identified by VBAD-GAT.

Visual Comparison. This part demonstrates the effectiveness of VBAD-GAT by visualizing the detection results of VBAD-GAT and other baselines on FS. The visualization result of GeoTrackNet, MADVCN, TREAD and VBAD-GAT are shown in Figs. 2a,2b,2c and 2d, respectively. It can be seen from the visualization result that the detection results of VBAD-GAT are more accurate than those of all other baselines. For instance, a ship corresponding to an abnormal vessel trajectory marked in purple in Fig. 2d has an abnormal behavior of route switching, i.e., switching from the original route to a new route. On the contrary, from Fig. 2b, we find that MADVCN fails to recognize this vessel trajectory as an anomaly. In addition, we show a vessel's turning rate and speed corresponding to an abnormal vessel trajectory marked in red identified by VBAD-GAT in Fig. 2e. We can see that even though this ship's turning rate increases, the ship's speed does not decrease, which is clearly an abnormal behavior.

5   Conclusion and Future Work

This paper proposes a model VBAD-GAT for vessel behavior anomaly detection. The graph attention module is introduced to capture the temporal dependencies of vessel features and relationships between vessel features. It consists of two essential components: the time graph attention module and the feature graph attention module. We also employ the joint detection strategy combining the prediction and reconstruction modules. Finally, the ablation study demonstrates the effectiveness of the graph attention module and joint detection strategy, and the comparison with baselines shows VBAD-GAT outperforms other baselines. In the future, we will explore how to use the self-attention mechanism to capture ship characteristics and strive to obtain a model with better performance.

Acknowledgements. This work was supported by the National Natural Science Foundation of China (No.: 52271365).

References

1. Akcay, S., Atapour-Abarghouei, A., Breckon, T.P.: Ganomaly: semi-supervised anomaly detection via adversarial training. In: Jawahar, C.V., Li, H., Mori, G., Schindler, K. (eds.) Computer Vision - ACCV 2018, pp. 622–637. Springer International Publishing, Cham (2019)
2. Chung, J., Gulcehre, C., Cho, K., Bengio, Y.: Empirical evaluation of gated recurrent neural networks on sequence modeling (2014)
3. d'Afflisio, E., Braca, P., Willett, P.: Malicious AIS spoofing and abnormal stealth deviations: a comprehensive statistical framework for maritime anomaly detection. IEEE Trans. Aerosp. Electron. Syst. **57**(4), 2093–2108 (2021)
4. Ding, K., Li, J., Bhanushali, R., Liu, H.: Deep anomaly detection on attributed networks, pp. 594–602. https://doi.org/10.1137/1.9781611975673.67
5. Goodfellow, I.J., et al.: Generative adversarial networks (2014)
6. Hawkins, D.: Identification of Outliers. Chapman and Hall (1980)
7. Hinton, G.E., Zemel, R.: Autoencoders, minimum description length and helmholtz free energy. In: Cowan, J., Tesauro, G., Alspector, J. (eds.) Advances in Neural Information Processing Systems. vol. 6. Morgan-Kaufmann (1993)
8. Hu, J., et al.: Intelligent anomaly detection of trajectories for IoT empowered maritime transportation systems. IEEE Trans. Intell. Transp. Syst. **24**(2), 2382–2391 (2023)
9. Kieu, T., Yang, B., Guo, C., Jensen, C.S.: Outlier detection for time series with recurrent autoencoder ensembles. In: Proceedings of the 28th International Joint Conference on Artificial Intelligence, pp. 2725–2732. IJCAI'19, AAAI Press, Macao, China (2019)
10. Kieu, T., et al.: Robust and explainable autoencoders for unsupervised time series outlier detection. In: 2022 IEEE 38th International Conference on Data Engineering (ICDE), pp. 3038–3050 (2022). https://doi.org/10.1109/ICDE53745.2022.00273
11. Kingma, D.P., Welling, M.: Auto-encoding variational bayes (2022)
12. Laxhammar, R.: Anomaly detection for sea surveillance. In: 2008 11th International Conference on Information Fusion, pp. 1–8 (2008)
13. Liu, R.W., Hu, K., Liang, M., Li, Y., Liu, X., Yang, D.: QSD-LSTM: vessel trajectory prediction using long short-term memory with quaternion ship domain. Appl. Ocean Res. **136**, 103592 (2023) 10.1016/j.apor.2023.103592, https://www.sciencedirect.com/science/article/pii/S0141118723001335
14. Liu, R.W., et al.: STMGCN: mobile edge computing-empowered vessel trajectory prediction using spatio-temporal multigraph convolutional network. IEEE Trans. Industr. Inf. **18**(11), 7977–7987 (2022). https://doi.org/10.1109/TII.2022.3165886
15. Liu, R.W., Nie, J., Garg, S., Xiong, Z., Zhang, Y., Hossain, M.S.: Data-driven trajectory quality improvement for promoting intelligent vessel traffic services in 6g-enabled maritime iot systems. IEEE Internet Things J. **8**(7), 5374–5385 (2021)

16. Liu, X., van de Weijer, J., Bagdanov, A.D.: Leveraging unlabeled data for crowd counting by learning to rank. In: Proceedings of the IEEE Conference on Computer Vision and Pattern Recognition (CVPR) (June 2018)
17. Nguyen, D., Vadaine, R., Hajduch, G., Garello, R., Fablet, R.: Geotracknet-a maritime anomaly detector using probabilistic neural network representation of AIS tracks and a contrario detection. IEEE Trans. Intell. Transp. Syst. **23**(6), 5655–5667 (2022)
18. Pallotta, G., Vespe, M., Bryan, K.: Vessel pattern knowledge discovery from AIS data: a framework for anomaly detection and route prediction. Entropy **15**(6), 2218–2245 (2013)
19. Ristic, B., La Scala, B., Morelande, M., Gordon, N.: Statistical analysis of motion patterns in ais data: Anomaly detection and motion prediction. In: 2008 11th International Conference on Information Fusion, pp. 1–7 (2008)
20. Schlegl, T., Seeböck, P., Waldstein, S.M., Langs, G., Schmidt-Erfurth, U.: f-anogan: fast unsupervised anomaly detection with generative adversarial networks. Med. Image Anal. **54**, 30–44 (2019)
21. She, R., Fan, P.: From MIM-based GAN to anomaly detection: Event probability influence on generative adversarial networks. IEEE Internet Things J. **9**(19), 18589–18606 (2022). https://doi.org/10.1109/jiot.2022.3161630
22. Yang, D., Wu, L., Wang, S., Jia, H., Li, K.X.: How big data enriches maritime research - a critical review of automatic identification system (AIS) data applications. Transp. Rev. **39**(6), 755–773 (2019)
23. Ye, M., Peng, X., Gan, W., Wu, W., Qiao, Y.: Anopcn: video anomaly detection via deep predictive coding network. In: Proceedings of the 27th ACM International Conference on Multimedia, pp. 1805–1813. MM '19, Association for Computing Machinery, New York, NY, USA (2019). https://doi.org/10.1145/3343031.3350899
24. Zhang, C., et al.: A deep neural network for unsupervised anomaly detection and diagnosis in multivariate time series data. Proc. AAAI Conf. Artif. Intell. **33**(01), 1409–1416 (2019). https://doi.org/10.1609/aaai.v33i01.33011409
25. Zhang, Z., et al.: STAD-GAN: unsupervised anomaly detection on multivariate time series with self-training generative adversarial networks. ACM Trans. Knowl. Disc. Data **17**(5), 1–18 (2023). https://doi.org/10.1145/3572780
26. Zhao, L., Shi, G.: Maritime anomaly detection using density-based clustering and recurrent neural network. J. Navig. **72**(4), 894–916 (2019)
27. Zhao, Y., et al.: Outlier detection for streaming task assignment in crowdsourcing. In: Proceedings of the ACM Web Conference 2022, pp. 1933–1943. WWW '22, Association for Computing Machinery, New York, NY, USA (2022). https://doi.org/10.1145/3485447.3512067
28. Zhen, R., Jin, Y., Hu, Q., Shao, Z., Nikitakos, N.: Maritime anomaly detection within coastal waters based on vessel trajectory clustering and naïve bayes classifier. J. Navig. **70**(3), 648–670 (2017)

TASFormer: Task-Aware Image Segmentation Transformer

Dmitry Yudin[1,2(✉)] [ID], Aleksandr Khorin[2] [ID], Tatiana Zemskova[2] [ID],
and Darya Ovchinnikova[2] [ID]

[1] AIRI (Artificial Intelligence Research Institute), Moscow, Russia
yudin@airi.net
[2] Moscow Institute of Physics and Technology, Moscow, Russia
{khorin.an,zemskova.ts,ovchinnikova.da}@phystech.edu

Abstract. In image segmentation tasks for real-world applications, the number of semantic categories can be very large, and the number of objects in them can vary greatly. In this case, the multi-channel representation of the output mask for the segmentation model is inefficient. In this paper we explore approaches to overcome such a problem by using a single-channel output mask and additional input information about the desired class for segmentation. We call this information task embedding and we learn it in the process of the neural network model training. In our case, the number of tasks is equal to the number of segmentation categories. This approach allows us to build universal models that can be conveniently extended to an arbitrary number of categories without changing the architecture of the neural network. To investigate this idea we developed a transformer neural network segmentation model named TASFormer. We demonstrated that the highest quality results for task-aware segmentation are obtained using adapter technology as part of the model. To evaluate the quality of segmentation, we introduce a binary intersection over union (bIoU) metric, which is an adaptation of the standard mIoU for the models with a single-channel output. We analyze its distinguishing properties and use it to compare modern neural network methods. The experiments were carried out on the universal ADE20K dataset. The proposed TASFormer-based approach demonstrated state-of-the-art segmentation quality on it. The software implementation of the TASFormer method and the bIoU metric is publicly available at www.github.com/subake/TASFormer.

Keywords: Image segmentation · Task embedding · Segmentation quality metric · Transformer

1 Introduction

As a rule, the output of an image segmentation neural network is a multi-channel mask, where the channel count is equal to the number of semantic categories [28,34,35] (see Fig. 1a). This is a problem for models that need to recognize

B. Luo et al. (Eds.): ICONIP 2023, LNCS 14451, pp. 305–317, 2024.
https://doi.org/10.1007/978-981-99-8073-4_24

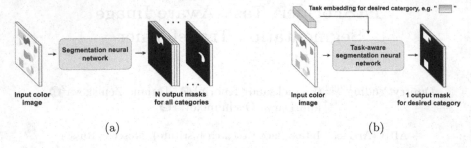

(a)   (b)

Fig. 1. A simplified scheme of the considered approaches: (a) conventional neural network for image segmentation, which generates masks for all object categories at the output at once, and (b) task-aware segmentation model, which allows, taking into account task embedding, to issue a mask of only the desired category.

real world objects, which can have a huge number of varieties. For example, the modern ADE20K dataset [41] originally contained 3,688 categories and for practical use this number was reduced to 150 to get a more realistic benchmark for analyzing scenes in images.

In this paper, we explore an approach to solving this problem based on performing segmentation taking into account the embeddings of object categories. We call them task embeddings. The selection of a separate embedding under a separate category allows us to form a single-channel mask at the output of the neural network model (see Fig. 1b). This approach is similar to the usage of text embeddings in multi-modal transformer models [1,12], but in our work we are focused on methods that work with only one modality of images.

Such task-aware models make it possible to quickly segment only the desired categories, and not all at once. This can be useful for practical applications, for example, in robotics, when a robot needs to interact with an object of only one or several types at a particular time.

The first and foremost contribution of this paper is the new transformer neural network architecture using adapter technology to solve the task-aware semantic segmentation problem. Secondly, our approach named TASFormer achieved state-of-the-art segmentation quality during experiments on popular ADE20K [41] dataset both for indoor and outdoor scenes.

Finally, we propose a new evaluation metric for semantic segmentation - binary Intersection over Union (bIoU). This metric represents a natural adaptation of standard mean Intersection over Union for the models using a task embedding during inferences and having the output in the form of binary mask. bIoU has the following distinguishing properties: 1) predicted masks contribute equally to its value regardless the area of union between ground truth and predicted masks, 2) classes with higher occurrence in a dataset have greater contribution to the final value of bIoU, 3) bIoU is computed only for ground truth categories present in a image. We discuss why these properties represent an advantage for model evaluation.

2   Related Works

The idea of using transformer architectures (VIT [8], SWIN [19], PVT [30], BEiT [31]) for image segmentation based on encoding them using universal tokens made it possible to surpass the quality of convolutional neural networks with attention modules (e.g. HRNet+OCR [39]), which dominated this area for a long time. However, convolutional models are still actively used in scene segmentation tasks in real time (e.g. PIDNet [36], DDRNet [23]).

It should be noted separately the SegFormer [35] architecture, which combines the speed of convolutional models and the accuracy of transformer methods due to the generation of more informative multi-scale feature maps.

The quality of basic neural network models of image segmentation significantly depends on the data domain and dataset on which they were trained. One of the groups of approaches that allow solving this problem is the use of adapters (HyperFormer [13], VIT-Adapter [3]). By training them with a frozen base model, it is possible to ensure that the model is adjusted to the required data domain. Another approach to training specialized encoders and decoders with a frozen central foundation model is proposed in the Frozen Pretrained Transformer (FPT) method [20].

Significant progress in increasing the universality of segmentation models was achieved when using large multimodal data sets for model training (Ref-COCO [38], PhraseCut [32], Refer-YouTube-VOS [26], A2D-Sentences, JHMDB-Sentences [9], etc.). In this case, the transformer model generates segmentation masks at the output only for the categories specified in the input text query. Similar models exist for both single image segmentation (MDETR [12], X-Decoder [42], CLIPSeg [21], VLT [6]) and video sequence segmentation (MTTR [1], ReferFormer [33], VLT [6]). Separately, it is worth highlighting a class of methods that use a language defined label of categories for open-vocabulary segmentation (ViLD [10], OvSeg [16], X-Decoder [42], SAN [37], SAM [15]). These models are trained on datasets with a fixed number of categories (e.g. COCO Stuff [2]), but the text embedding of categories allows segmentation of unseen categories. Such approaches can solve zero-shot segmentation problems, but are still inferior in quality to single-modal state-of-the-art methods that work only with images and a fixed number of categories, pre-trained on a specific dataset (CityScapes [5], ADE20K [41], SceneNet [22] and SUN RGB-D [29], etc.).

3   TASFormer Architectures

To solve the problem of task-aware semantic segmentation, we have developed several variants of a transformer model that takes as input a three-channel color image and an embedding tensor of the task as the desired category. We named our approach TASFormer. The studied options for its construction are shown in Fig. 2.

The high-speed SegFormer [35] approach was chosen as the basic neural network model, which provides the hierarchical formation of feature maps of different scales using a small number of transformer blocks.

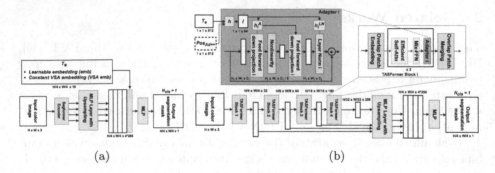

(a)   (b)

Fig. 2. TASFormer architectures: (a) with learnable task embeddings (emb) or constant embedding calculating with Vector Symbolic Analysis (VSA emb), (b) with adapter-based approach for learnable task embedding integration in transformer model.

Let's consider the proposed architectures in more detail.

TASFormer (emb). First, we considered a simple concatenation of the learnable task embedding tensor with feature maps generated in the SegFormer (MIT-B0) transformer decoder as shown in Fig. 2a. This approach is based on the assumption that the feature map before the last fully connected layer (MLP) contains a sufficient amount of various information about all possible categories of objects. Thus an additional tensor can be the condition that will help to correctly obtain the output projection of the desired category in the form of a single-channel mask.

For simplicity and convenience, the embedding spatial resolution is set to the same as the feature map of 80×80 (for input image size 320 × 320). And in general, its size is about 1% of the number of elements of the feature tensor to which concatenation is performed. The output mask of the model is also small and is only 80 × 80 pixels. This choice is made based on the desired trade-off performance and quality of the segmentation model. The use of other, larger dimensions is a possible subject for further research.

TASFormer (VSA emb). This approach of using additional task embedding is very similar to the previous one and is also shown in Fig. 2a. Its difference is the use of constant rather than learnable embeddings of semantic categories, which have the property of pairwise orthogonality of hyperdimensional vectors from the theory of Visual Symbolic Analysis (VSA) [14,25]. In practice, this property is achieved by generating high-dimensional embedding T_e, each element of which is filled with random floats sampled from a univariate Gaussian ("normal") distribution of mean 0 and variance 1:

$$T_e = \frac{E}{\|E\|}, E = RandomNormal(K),\qquad(1)$$

where K is embedding length. In our case K equals 64,000 ($80 \times 80 \times 10$). For such long vectors, the pairwise dot product is close to zero.

TASFormer (HF Adapter). This is a more complex approach to data fusion, shown in Fig. 2b. In this case, information from the learnable embedding T_e (with 512 elements) is taken into account using a neural network adapter, which is integrated into each transformer block. At the same time, unlike the original implementation of the adapter based on HyperFormer [13], we learn both the transformer blocks themselves and the adapter at the same time.

It should be noted that different adapter module weights are learned for different transformer blocks (shown in Fig. 2b with index i), except for the fully connected layer h. It acts as a task projector network for the entire model encoder to convert task embeddings into a more compact I representation.

Fully connected layers h_i^A and h_i^{LN} calculate the weight matrices respectively for the i-th adapter layers responsible for feedforward down projection and Layer Normalization.

TASFormer (HF adapter++). In order to reduce the number of model parameters while maintaining the properties of the adapter layers, we share the weights of the adapter layers, and additionally take into account the position of the adapter as it was implemented in HyperFormer [13]. In this case, we concatenate learnable task T_e and adapter position Pos_{Ad++} embeddings and feed them into a fully connected layer h as shown in Fig. 2b . Since the size of the input tensor is different for each transformer block, we cannot use one adapter for the entire network architecture, so the weight sharing occurs only inside the transformer blocks.

4 Binary Intersection over Union

4.1 Motivation

Category embedding used for inference of neural network and increasing number of categories that can be segmented inevitably lead to the fact that the use of the standard mean IoU becomes problematic. Firstly, there is a change in model output compared to conventional neural networks, and for each image the model output does not consist of a single mask with all object categories, but of a set of binary masks for the categories that were queried by embedding. This is a case of our TASFormer. In addition, it is worth noting that many of the state-of-the-art models for semantic segmentation use the paradigm of mask classification instead of pixel classification [4,11] and, therefore, have a set of masks as an output. Secondly, in a large set of language defined categories some categories can be not completely mutually exclusive. For instance, such classes are present in ADE20K dataset (e.g."skycrapper","building", "house"). Mean IoU takes into account false positive predictions of all categories. Accordingly, if the model predicts mask for some category, which doesn't correspond to the annotation, although present in reality in the image, then the value of mean IoU decreases [16].

4.2   Comparison with Similar Existing Metrics

An adaptation of the standard mean IoU for binary masks $mIoU_{Shaban}$ is described by Shaban et al. in [27] in the case of few-shot segmentation problem. Consider a set I of N images where N_{cl} classes are present. For a semantic class c in a image i the intersection I_i^c and the union U_i^c will be defined with respect to predicted and ground truth binary masks corresponding to the image i and the class c. Therefore, the definition of $mIoU_{Shaban}$ is following:

$$mIoU_{Shaban} = \frac{1}{N_{cl}} \sum_{c=1}^{N_{cl}} \frac{\sum_{i=1}^{N} I_i^c}{\sum_{i=1}^{N} U_i^c}. \tag{2}$$

Another version of IoU adaptation for the case of binary masks, used in [7,18,24], is defined as following [18]:

$$FBIoU = \frac{1}{2} \left(IoU_{fg} + IoU_{bg} \right). \tag{3}$$

$$IoU_{fg} = \frac{\sum_{i=1}^{N} \sum_{c=1}^{N_{cl}} I_i^c}{\sum_{i=1}^{N} \sum_{c=1}^{N_{cl}} U_i^c}. \tag{4}$$

IoU_{bg} is calculated in the same way but reversed the foreground and background.

The $mIoU_{Shaban}$ and $FBIoU$ have a bias towards masks having a larger union between predicted and ground-truth masks. However, the area of union between masks depends not only on the segmentation quality of a model, but also on the area of the ground truth and predicted masks. In our belief, the value of a metric should not depend on the size of a mask. Therefore, to evaluate our model we propose a modification of the standard metric of Intersection over Union. Thus, the definition of binary IoU is equivalent to the definition of overall IoU, where IoU values corresponding to different masks are averaged uniformly regardless the size of union between predicted and ground truth masks.

Binary Intersection over Union will be defined as:

$$bIoU = \frac{1}{\sum_{i=1}^{N} N_{class}(i)} \sum_{i=1}^{N} \sum_{c \in i} \frac{I_i^c}{U_i^c}. \tag{5}$$

Note that $\sum_{i=1}^{N} N_{class}(i) = \sum_{c=1}^{N_{cl}} N_{im}(c)$, where $N_{im}(c)$ - the number of images where class c occurs. Therefore the definition of $bIoU$ can be rewritten:

$$bIoU = \sum_{c=1}^{N_{cl}} \frac{N_{im}(c)}{\sum_{c'=1}^{N_{cl}} N_{im}(c')} \frac{1}{N_{im}(c)} \sum_{i=1}^{N} \frac{I_i^c \mathbf{1}_{c \in i}}{U_i^c}. \tag{6}$$

Looking at this definition one can conclude that $bIoU$ has three properties that distinguish it from $mIoU_{Shaban}$ and $FBIoU$: 1) the classes that have larger occurrence in image set have greater contribution to final value of $bIoU$, 2) all masks for one class have equal contribution to final value of $bIoU$ regardless their size, 3) $bIoU$ is computed only for categories that are present in ground

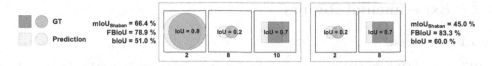

Fig. 3. An illustration of the influence of the size and frequency of occurrence of a class on the final value of various metrics for two classes: green circles and blue squares. Left: Size imbalance in the distribution of green circles leads to a higher value of $mIoU_{Shaban}$ and $FBIoU$ compared to $bIoU$. Right: Class imbalance between green circles and blue squares leads to a higher value of $bIoU$ compared to $mIoU_{Shaban}$, the small masks of green circles leads to a high value of $FBIoU$. (Color figure online)

truth image annotation. The latter fact helps to better quantify the quality of models in the case where different language defined categories refer to the same object in an image. The design of $bIoU$ allows evaluating the interaction of a model with a certain set of categories. Though, the $bIoU$ can not be used to evaluate the performance of a model in the cases where the task embedding doesn't correspond to any pixels in the image.

The differences in averaging methods influence the response of metrics to imbalances of mask sizes and class occurrences in image set. Figure 3 shows an example of how the imbalances could create cases where a marked difference between metrics can be observed.

Consider two classes: green circles and blue squares. Let there be a set of images sized 100×100 pixels, each image contains only one class from described above. Then, a case of size imbalance from the Fig. 3 (left) may occur. Let the set of images consists of: 1) two images with a large green circle that have good predicted masks (intersection area (I) equals to 6400, union area (U) equals to 8000), 2) eight images with a small green circle that is poorly predicted ($I = 160$, $U = 800$), 3) ten images with a blue square of medium size ($I = 1400$, $U = 2000$). In this case $mIoU_{Shaban}$ and $FBIoU$ have a larger value compared to $bIoU$ due to size imbalance in ground truth masks of green circles. Small segmentation masks of green circles additionally contribute to high value of IoU_{bg} in $FBIoU$.

The Fig. 3 (right) illustrates a case of class imbalance. If the set of images consists of: 1) two images with a small green circle that is poorly predicted ($I = 160$, $U = 800$), 2) eight images with a blue square of medium size ($I = 1400$, $U = 2000$). The poor performance on green circles is better captured by $mIoU_{Shaban}$ than by $bIoU$. Again $FBIoU$ has high value in the presence of small prediction and ground truth masks even if they have a little intersection. However, a large set of categories may contain some natural imbalance between categories. Then, the $bIoU$ metric would characterize the performance of a model in natural conditions for all categories at once.

5 Experimental Results

5.1 Datasets

We used ADE20K [41] as a more complex dataset to evaluate the developed approach. It is a diverse scene parsing dataset covering 150 fine-grained semantic categories consisting of 20,210 indoor and outdoor images for training and 2,000 for validation. It is currently one of the most popular benchmarks for comparing image semantic segmentation methods.

5.2 Training Setup

We train neural network models on a server with 2×Nvidia Tesla V100. We use the pre-trained MiT-B0 encoder on the Imagenet-1K dataset and randomly initialize the decoder. During training, we applied data augmentation through random brightness contrast, random horizontal flipping, and random affine transformation. We set crop size to 320×320 and a batch size of 10 on ADE20K for ablation study of our models. We trained the models using AdamW optimizer for 140K iterations on ADE20K. The learning rate was set to an initial value of 0.00005 and then used a "poly" learning rate schedule with factor 1.0 by default. For simplicity, we followed the ideas of the SegFormer [35] and did not adopt widely-used tricks such as OHEM, auxiliary losses or class balance loss. During evaluation, we scale the short side of the image to training cropping size and keep the aspect ratio for ADE20K. For comparison of state-of-the-art methods on ADE20K we additionally trained TASFormer with crop image size of 640×640 and 896×896.

We fine-tune ReferFormer [33] and VLT [6] on ADE20K in order to be able to compare them with TASFormer. We use the code provided in the corresponding repositories. We use the pre-trained ReferFormer on RefCOCO/+/g [38] with Swin-T backbone. We set a batch size of 1 and followed the fine-tuning procedure proposed by Wu et al. [33] for A2D-Sentences [9]. The fine-tuning process was extended for two additional epochs (resulting in 8 total epochs instead of 6) without drop of the learning rate. To fine-tune VLT we use the pre-trained VLT on RefCOCO with Darknet53 visual backbone and bi-GRU text encoder. We set a batch size of 8 and fine-tuned the model during 50 epochs using the same training settings as Ding et al. [6] used for original VLT training. As input text queries we used the language defined categories of objects in ADE20K. Since a category in ADE20K is described by a list of words, we chose the first words in categories definitions provided by Detectron 2 [34].

5.3 Output Adaptation

We use the following adaptation of standard output of the SegFormer model in order to maintain binary masks for each class and be able to compute the $bIoU$ metric. The output tensor of the SegFormer consists of a tensor with shape $C \times H \times W$, where C is the number of classes, H and W are height and width

Table 1. Segmentation results on different number of categories from ADE20K dataset, $bIoU$, %. All models used pre-trained MiT-B0 encoder weights and data preprocessing for mask segmentation.

Method	Params	ADE20K		
		2	12	150
SegFormer (B0)	3.8M	63.2	52.4	37.9
TASFormer (emb)	4–14M	60.8	6.1	14.1
TASFormer (VSA emb)	4.1M	48.6	0.1	0.1
TASFormer (HF adapter)	7.3M	**67.9**	**59.4**	**48.3**
TASFormer (HF adapter++)	5.7M	**67.9**	58.4	47.6

of an input image. This output tensor is normalized via min-max normalization. Next for each class i we keep all pixels that have a predicted value greater than a threshold th. We vary the threshold th in order to maximize the final value of $bIoU$. We choose this strategy since it provided the highest value of $bIoU$ compared to other adaptations. To compute the standard $mIoU$ metric we query all possible categories for each image and combine them into multichannel mask. Then we follow the standard procedure for the $mIoU$ computation.

5.4 Results on ADE20K

The TASFormer (HF adapter) approach also significantly outperformed both the SegFormer (B0) base model and the TASFormer (emb) and TASFormer (VSA emb) methods on the ADE20k dataset (see Table 1). The TASFormer (HF adapter++) demonstrates second best result and has lesser amount of parameters. It is inferior to the TASFormer (HF adapter) by 1–2%. It should be noted that the simple embedding concatenation technique (Fig. 2a) significantly underperforms on big number of classes and scores high on 2-class segmentation.

The use of learnable embeddings gives higher quality of semantic segmentation than the use of random VSA embeddings. SegFormer metric for 150 classes is significantly lower compared to other methods, since its architecture is not designed for binary mask generation. Comparison to state-of-the-art methods on ADE20K is shown in Table 2. For this purpose, we have chosen popular high-precision transformer-based models. During the experiment, we used the SegFormer pre-trained on the ADE20K dataset. We trained VLT [6] and Refer-Former [33] as described in Sect. 5.2. To calculate the bIoU metric, the approach described in Sect. 5.3 was used. Our TASFormer (HF adapter) outperforms powerful vision-language transformer models such as VLT and ReferFormer and shows an increase of 8% of the bIoU metric comparing to the baseline mobdel SegFormer (B0).

We also compared the quality of the models using the standard $mIoU$ metric. As can be seen from Table 2, models using task embedding demonstrate lower values of the $mIoU$ metric compared to the SegFormer (B0) baseline. The reason

Table 2. Comparison of state-of-the-art methods on ADE20K dataset (150 categories).

Method	Params	Crop Size	$mIoU$,%	$bIoU$,%
SegFormer (B0) [35]	3.8M	512×512	**36.8**	40.0
VLT [6]	89M	416×416	12.2	50.3
ReferFormer [33]	176M	360×640	8.7	51.08
Mobile SAM (segmentation module only) [40]	9.66M	1024×1024	4.5	48.7
SAM (segmentation module only) [15]	615M	1024×1024	5.3	**52.2**
TASFormer (HF adapter)	7.3M	320×320	14.6	48.3
TASFormer (HF adapter)	7.3M	640×640	<u>14.7</u>	51.1
TASFormer (HF adapter)	7.3M	896×896	14.4	<u>52.0</u>

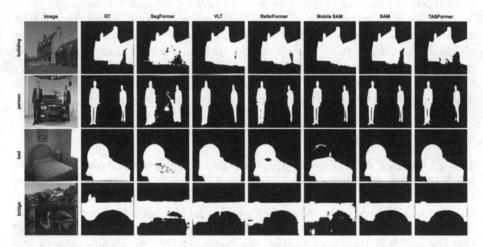

Fig. 4. Visualized segmentation results on ADE20K val set. The columns left-to-right refer to the input image, the ground truth, the outputs of various semantic segmentation models and our TASFormer (HF adapter).

behind this is that methods that use task embedding predict similar semantic categories with equal probability. Despite considering this fact, the TASFormer model outperforms VLT and ReferFormer methods.

Additionally, we assessed the segmentation quality of TASFormer compared to segmentation foundation models: SAM [15] and its real-time version Mobile SAM [40]. It should be noted that direct comparison of these models with the TASFormer is difficult, because the segmentation quality of SAM methods depends on the given box prompts. Following the pipeline of Grounded SAM we use the Grounding DINO method [17] in order to generate box prompts from text in a zero-shot manner. The TASFormer method outperforms Mobile SAM and has quality comparable to SAM despite a significantly smaller model size (see Table 2 and Fig. 4).

Figure 4 shows the visualized results of TASFormer (HF adapter) compared to other segmentation methods under different scenes. The SegFormer binary masks are noisy, however it, like TASFormer, was able to find a "plant" on the first image. The quality of masks for TASFormer generally outperforms other methods. However, it sometimes has difficulty segmenting small objects, such as the "perso"'s legs in the second image.

6 Conclusion

In this paper, we proposed and analyzed the capabilities of the transformer model for image segmentation named TASFormer with several ways to use additional task embeddings.

Based on the results of experiments on the ADE20K we can conclude that the use of additional task embeddings not only provides additional opportunities for segmenting the desired semantic categories, but also has a significant positive effect on improving the quality of segmentation of the base model.

In addition, we proposed a metric $bIoU$ that enables comparing models with output in form of a set of binary masks with the state-of-the-art approaches for semantic segmentation.

As a topic for further research can be considered a deeper study of possible sizes of task embeddings and their impact on the final quality of the model. The adapter technique has shown itself to be very promising, and experiments with other adapter architectures can also yield valuable new results. In our future work, we also plan to train TASFormer to perform segmentation in the case where a task embedding refers to a category absent in the image.

A study of the applicability of the proposed approach to other datasets, especially outdoor, is also worthwhile. A separate important promising area is also testing the possibility of integrating task-aware segmentation into robotic applications, such as visual navigation.

References

1. Botach, A., Zheltonozhskii, E., Baskin, C.: End-to-end referring video object segmentation with multimodal transformers. In: Proceedings of the IEEE/CVF CVPR (2022)
2. Caesar, H., Uijlings, J., Ferrari, V.: Coco-stuff: thing and stuff classes in context. In: Proceedings of the IEEE/CVF CVPR, pp. 1209–1218 (2018)
3. Chen, Z., et al.: Vision transformer adapter for dense predictions. In: The Eleventh International Conference on Learning Representations (2023). https://openreview. net/forum?id=plKu2GByCNW
4. Cheng, B., Misra, I., Schwing, A.G., Kirillov, A., Girdhar, R.: Masked-attention mask transformer for universal image segmentation. In: Proceedings of the IEEE/CVF CVPR. pp, 1290–1299 (2022)

5. Cordts, M., et al.: The cityscapes dataset for semantic urban scene understanding. In: Proceedings of the IEEE/CVF CVPR, pp. 3213–3223 (2016)
6. Ding, H., Liu, C., Wang, S., Jiang, X.: Vlt: vision-language transformer and query generation for referring segmentation. IEEE Trans. Patt. Anal. Mach. Intell. (2022)
7. Dong, N., Xing, E.P.: Few-shot semantic segmentation with prototype learning. In: BMVC. vol. 3 (2018)
8. Dosovitskiy, A., et al.: An image is worth 16x16 words: Transformers for image recognition at scale. In: International Conference on Learning Representations (2021), https://openreview.net/forum?id=YicbFdNTTy
9. Gavrilyuk, K., Ghodrati, A., Li, Z., Snoek, C.G.: Actor and action video segmentation from a sentence. In: Proceedings of the IEEE Conference on Computer Vision and Pattern Recognition, pp. 5958–5966 (2018)
10. Gu, X., Lin, T.Y., Kuo, W., Cui, Y.: Open-vocabulary object detection via vision and language knowledge distillation. In: International Conference on Learning Representations (2022). https://openreview.net/forum?id=lL3lnMbR4WU
11. Jain, J., Li, J., Chiu, M.T., Hassani, A., Orlov, N., Shi, H.: Oneformer: one transformer to rule universal image segmentation. In: Proceedings of the IEEE/CVF CVPR, pp. 2989–2998 (2023)
12. Kamath, A., Singh, M., LeCun, Y., Synnaeve, G., Misra, I., Carion, N.: Mdetr-modulated detection for end-to-end multi-modal understanding. In: Proceedings of the IEEE/CVF ICCV, pp. 1780–1790 (2021)
13. Karimi Mahabadi, R., Ruder, S., Dehghani, M., Henderson, J.: Parameter-efficient multi-task fine-tuning for transformers via shared hypernetworks. In: Annual Meeting of the Association for Computational Linguistics (2021)
14. Kirilenko, D., Kovalev, A.K., Solomentsev, Y., Melekhin, A., Yudin, D.A., Panov, A.I.: Vector symbolic scene representation for semantic place recognition. In: 2022 International Joint Conference on Neural Networks (IJCNN), pp. 1–8. IEEE (2022)
15. Kirillov, A., et al.: Segment anything. arXiv:2304.02643 (2023)
16. Liang, F., et al.: Open-vocabulary semantic segmentation with mask-adapted clip. In: Proceedings of the IEEE/CVF CVPR, pp. 7061–7070 (2023)
17. Liu, S., et al.: Grounding dino: Marrying dino with grounded pre-training for open-set object detection. arXiv:2303.05499 (2023)
18. Liu, W., Zhang, C., Lin, G., Liu, F.: CRCNet: few-shot segmentation with cross-reference and region-global conditional networks. Int. J. Comput. Vision 130(12), 3140–3157 (2022)
19. Liu, Z., et al..: Swin transformer: hierarchical vision transformer using shifted windows. In: Proceedings of the IEEE/CVF ICCV, pp. 10012–10022 (2021)
20. Lu, K., Grover, A., Abbeel, P., Mordatch, I.: Pretrained transformers as universal computation engines. Proc. of the AAAI Conf. Artif. Intell. 36, 7628–7636 (06 2022). https://doi.org/10.1609/aaai.v36i7.20729
21. Lüddecke, T., Ecker, A.: Image segmentation using text and image prompts. In: Proceedings of the IEEE/CVF CVPR, pp. 7086–7096 (2022)
22. McCormac, J., Handa, A., Leutenegger, S., Davison, A.J.: Scenenet RGB-D: Can 5m synthetic images beat generic imagenet pre-training on indoor segmentation? In: Proceedings. of the IEEE/CVF ICCV, pp. 2697–2706 (2017). https://doi.org/10.1109/ICCV.2017.292
23. Pan, H., Hong, Y., Sun, W., Jia, Y.: Deep dual-resolution networks for real-time and accurate semantic segmentation of traffic scenes. IEEE Trans. Intell. Transp. Syst. 24(3), 3448–3460 (2022)
24. Rakelly, K., Shelhamer, E., Darrell, T., Efros, A., Levine, S.: Conditional networks for few-shot semantic segmentation. ICLR Workshop (2018)

25. Schlegel, K., Neubert, P., Protzel, P.: A comparison of vector symbolic architectures. Artif. Intell. Rev. **55**, 4523–4555 (2021). https://doi.org/10.1007/s10462-021-10110-3
26. Seo, S., Lee, J.-Y., Han, B.: URVOS: unified referring video object segmentation network with a large-scale benchmark. In: Vedaldi, A., Bischof, H., Brox, T., Frahm, J.-M. (eds.) Computer Vision – ECCV 2020: 16th European Conference, Glasgow, UK, August 23–28, 2020, Proceedings, Part XV, pp. 208–223. Springer International Publishing, Cham (2020). https://doi.org/10.1007/978-3-030-58555-6_13
27. Shaban, A., Bansal, S., Liu, Z., Essa, I., Boots, B.: One-shot learning for semantic segmentation. In: British Machine Vision Conference (2017)
28. Shepel, I., Adeshkin, V., Belkin, I., Yudin, D.A.: Occupancy grid generation with dynamic obstacle segmentation in stereo images. IEEE Trans. Intell. Transp. Syst. **23**(9), 14779–14789 (2021)
29. Song, S., P., S., Lichtenberg, Xiao, J.: Sun RGB-D: A RGB-D scene understanding benchmark suite. Proceedings of the IEEE CVPR (2015). https://rgbd.cs.princeton.edu/
30. Wang, W., et al.: Pvt v2: improved baselines with pyramid vision transformer. Comput. Vis. Media **8**(3), 415 424 (2022)
31. Wang, W., et al.: Image as a foreign language: beit pretraining for vision and vision-language tasks. In: CVPR, pp. 19175–19186 (2023)
32. Wu, C., Lin, Z., Cohen, S., Bui, T., Maji, S.: Phrasecut: language-based image segmentation in the wild. In: Proceedings of the IEEE/CVF CVPR, pp. 10216–10225 (2020)
33. Wu, J., Jiang, Y., Sun, P., Yuan, Z., Luo, P.: Language as queries for referring video object segmentation. In: Proceedings of the IEEE/CVF CVPR, pp. 4974–4984 (2022)
34. Wu, Y., Kirillov, A., Massa, F., Lo, W.Y., Girshick, R.: Detectron2. https://github.com/facebookresearch/detectron2 (2019)
35. Xie, E., Wang, W., Yu, Z., Anandkumar, A., Alvarez, J.M., Luo, P.: Segformer: simple and efficient design for semantic segmentation with transformers. Adv. Neural. Inf. Process. Syst. **34**, 12077–12090 (2021)
36. Xu, J., Xiong, Z., Bhattacharyya, S.P.: Pidnet: a real-time semantic segmentation network inspired by PID controllers. In: CVPR, pp. 19529–19539 (2023)
37. Xu, M., Zhang, Z., Wei, F., Hu, H., Bai, X.: Side adapter network for open-vocabulary semantic segmentation. In: Proceedings of the IEEE/CVF CVP, pp. 2945–2954 (2023)
38. Yu, L., Poirson, P., Yang, S., Berg, A.C., Berg, T.L.: Modeling context in referring expressions. In: Leibe, B., Matas, J., Sebe, N., Welling, M. (eds.) Computer Vision – ECCV 2016: 14th European Conference, Amsterdam, The Netherlands, October 11-14, 2016, Proceedings, Part II, pp. 69–85. Springer International Publishing, Cham (2016). https://doi.org/10.1007/978-3-319-46475-6_5
39. Yuan, Y., Chen, X., Wang, J.: Object-contextual representations for semantic segmentation. In: ECCV (2020)
40. Zhang, C., et al.: Faster segment anything: towards lightweight SAM for mobile applications (2023)
41. Zhou, B., et al.: Semantic understanding of scenes through the ade20k dataset. Int. J. Comput. Vis. **127**(3), 302–321 (2019)
42. Zou, X., et al.: Generalized decoding for pixel, image, and language. In: Proceedings of the IEEE/CVF CVPR, pp. 15116–15127 (2023)

Unsupervised Joint-Semantics Autoencoder Hashing for Multimedia Retrieval

Yunfei Chen, Jun Long[(⊠)], Yinan Li, Yanrui Wu, and Zhan Yang[(⊠)]

Big Data Institute, School of Computer Science, Central South University,
Changsha Hunan, China
{yunfeichen,junlong,liyinan,8209200520,zyang22}@csu.edu.cn

Abstract. Cross-modal hashing has emerged as a prominent approach
for large-scale multimedia information retrieval, offering advantages in
computational speed and storage efficiency over traditional methods.
However, unsupervised cross-modal hashing methods still face challenges
in the lack of practical semantic labeling guidance and handling of
cross-modal heterogeneity. In this paper, we propose a new unsupervised cross-modal hashing method called Unsupervised Joint-Semantics
Autoencoder Hashing(UJSAH) for multimedia retrieval. First, we introduce a joint-semantics similarity matrix that effectively preserves the
semantic information in multimodal data. This matrix integrates the
original neighborhood structure information of the data, allowing it to
better capture the associations between different modalities. This ensures
that the similarity matrix can accurately mine the underlying relationships within the data. Second, we design a dual prediction network-based
autoencoder, which implements the interconversion of semantic information from different modalities and ensures that the generated binary hash
codes maintain the semantic information of different modalities. Experimental results on several classical datasets show a significant improvement in the performance of UJSAH in multimodal retrieval tasks relative to existing methods. The experimental code is published at https://
github.com/YunfeiChenMY/UJSAH.

Keywords: Cross-modal Hashing · Multimedia Retrieval ·
Joint-Semantics · Dual Prediction

1 Introduction

With the continuous advancements in science and technology, network data size
and variety are rapidly expanding. Traditional information retrieval methods
that use original data for computation suffer from high computational complexity. Therefore, achieving high retrieval efficiency while requiring minimal storage
space has become a crucial research direction. Hash learning has gained significant attention in large-scale multimodal data retrieval due to its efficient

B. Luo et al. (Eds.): ICONIP 2023, LNCS 14451, pp. 318–330, 2024.
https://doi.org/10.1007/978-981-99-8073-4_25

computation speed and low storage requirements [19]. Hash learning involves a hash function to map high-dimensional raw data into a low-dimensional binary hash code. The mainstream hashing methods primarily rely on classical hash functions based on features. At the same time, some recent works [21] have integrated semantic information extraction, hash function training, and hash code generation into the same framework.

Given the inherent heterogeneity among different modal data, directly calculating their similarity becomes difficult. Cross-modal hashing methods aim to map the original data from different modalities into a shared binary space, which enables the similarity calculations between different modalities using the Hamming distance. The data of real application scenarios are mostly unlabeled, severely hindering supervised hashing development. The unsupervised deep cross-modal hashing method utilizes deep networks' powerful feature extraction capability to fully extract deep semantic features from the raw data, enabling the generation of binary hash codes rich in semantic information. It [4] mainly integrates features of different modalities' raw data to construct similarity matrices and employs deep neural networks to construct hash functions for the generation of hash codes. Although unsupervised deep cross-modal hashing has gone well, there is still significant room for progress in constructing the similarity matrix of raw data and cross-modal feature heterogeneity. Traditional autoencoder hashing methods only consider the decoding and reconstruction of intra-modal semantic information and lack the mining of cross-modal semantic information.

We propose a new unsupervised multimedia hashing method called Unsupervised Joint-Semantics Autoencoder Hashing to address the aforementioned challenge. First, the UJSAH method design joint-semantics similarity matrices to comprehensively explore similarity relationships between multimodal data. Second, the UJSAH method uses an encoding module to generate hash codes for a given data, a decoding module to reconstruct the raw data to ensure that the resulting hash codes preserve the complete semantic information contained in the raw data, and a dual prediction module is designed in the autoencoder that explores the deep correlation relationships between different modal data. The core work of UJSAH is as follows:

1. The joint-semantic similarity matrix is constructed to explore multi-modal similarity relations and improve the hash function training guidance. The similarity matrix is constructed considering the similarity within each modality and the similarity between different modalities.
2. An autoencoder established on a dual prediction network is designed to generate hash codes. The method employs an autoencoder to ensure that the resulting hash codes preserve the complete semantic information contained in the raw data and uses a dual prediction network to achieve the exploration of semantic association relationships between multi-modal data.
3. Comprehensive experiments conducted on the MIRFlickr and NUS-WIDE datasets verify that the UJSAH method significantly exceeds the mainstream baseline methods.

2   Related Work

Current hashing can be broadly categorized into two categories: supervised hashing and unsupervised hashing.

2.1   Supervised Hashing

In supervised hashing, the semantic information present in the labels is utilized to guide the learning of semantic information in the hash codes. SASH [14] adaptively learns the similarity matrix and saves the association information in the labels to the data features to extract the label relevance for optimizing the previously mentioned matrix. [25] proposes incorporating probabilistic code balance constraints into deep supervised hashing, which enforces a discrete uniform distribution for each hash code. To guarantee that the binary hash codes generated by the model align with the semantic information classification think in the original data, DSDH [9] proposes a deeply supervised discrete hashing algorithm. Literature [17] presents an end-to-end model that effectively extracts key features and generates hash codes with precise semantic information. DPN [3] applies differentiable bit hinge-like losses to the network's output channels, ensuring their values deviate from zero. SHDCH [22] accepts hash codes by explicitly exploring hierarchical tags. DSH [11] introduces a novel approach to deep supervised hashing, aiming to retain compact hash codes that maintain similarity for large-scale image data.

Although supervised hashing methods have made significant progress in information retrieval, most data is unlabeled in real-world scenarios. In contrast, unsupervised hashing methods are more suitable for handling real-world application scenario data and reduce the expensive cost of the manual labeling process due to its property of not relying on labeled information.

2.2   Unsupervised Hashing

Unsupervised hashing fully uses the semantic information between the raw data to guarantee its maintenance in the binary hash code. To address the issue of ignoring neighboring instances and label granularity, DCH-SCR [12] digs deeper into the semantic similarity information within multimedia data. CAGAN [10] proposes an adaptive attention network model to retrieve massive multimodal data efficiently. In order to protect the privacy and security of data, [23] proposes a data-centric multimedia hash learning approach. To efficiently retrieve cross-modal remote sensing images, DACH [5] uses generative adversarial network hashing to extract fine-grained feature information in remote sensing images. To alleviate the limitations in similarity supervision and optimization strategies, DAEH [15] uses discriminative similarity matrix and adaptive self-updating optimization strategies to generate hash codes and train hash functions.

Existing cross-modal hash retrieval methods have significantly progressed around data feature extraction and cross-modal association mining. However, there is still much room for improvement in dealing with cross-modal heterogeneity and deep data feature association extraction.

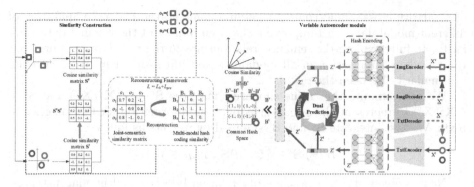

Fig. 1. The basic framework of the Unsupervised Joint-Semantics Autoencoder Hashing.

3 The Proposed Method

This section describes UJSAH method, including notations, architecture, objective function, and extensions. The framework consists of an encoding network, a dual prediction network, and a decoding network. The encoding network maintains consistency in multimedia information, the dual prediction network focuses on reconstructing the different modal data, and the decoding network combines the original data with the generated hash codes.

3.1 Notations

This paper uses bold uppercase and lowercase letters to represent matrices and vectors. Given a dataset $\mathbf{X} = \{\mathbf{X}^1, ..., \mathbf{X}^m\}|_{m=1}^{M}$ of M modalities, where $\mathbf{X}^m = \{x_1, ..., x_n\} \in \mathbb{R}^{d_m \times n}$, d_m is the dimensionality of data modality $\mathbf{X}$, n represents the size of the dataset. We use two modalities, image and text, to verify the effectiveness of the proposed method, and we set $\mathbf{X} = \{\mathbf{X}^v, \mathbf{X}^t\}$, where $\mathbf{X}^v$ and $\mathbf{X}^t$ denote the image and text feature matrices. The proposed UJSAH method is to generate compact hash code $\mathbf{B} \in \{-1, 1\}^{k \times n}$, where k represents the length of the hash code.

3.2 Architecture

As shown in Fig. 1, the proposed method UJSAH is an end-to-end framework with three main components, i.e., Encoding Network, Decoding Network, and Prediction Network, to process image and text data.

Similarity Construction

Efficient extraction of the underlying neighborhood structure and maintaining consistent hash code relationships with the original data are crucial in unsupervised cross-modal retrieval tasks. We employ cosine similarity to estimate the similarity between different data to achieve this. In this paper, we adopt the idea of similarity cross-modal computation and combine the fusion computation of different modal data with improving the deep mining of the original data similarity. To fully explore the semantic relationships in the raw data, we normalize the raw data information of all modalities and integrate them, then calculate the similarity between the raw data.

We can compute the cosine similarity matrices $\mathbf{S}^v = c(\tilde{\mathbf{X}}^v, \tilde{\mathbf{X}}^v)$ and $\mathbf{S}^t = c(\tilde{\mathbf{X}}^t, \tilde{\mathbf{X}}^t)$ to represent the semantic association information for the images and texts. We first integrate $\mathbf{S}^v$ and $\mathbf{S}^t$ summed by weights as follows:

$$\tilde{\mathbf{S}} = \mu \mathbf{S}^v + (1 - \mu)\mathbf{S}^t, s.t. \mu \in [0, 1]. \tag{1}$$

Next, we consider $\tilde{\mathbf{S}}$ as a similarity relation between multimodal instances. The unified characterization of multimodal semantic relations can be achieved by computing $\mathbf{S}^v \mathbf{S}^{t^\top}$ to dig deeply into the semantic associations between different modal data to achieve a comprehensive representation of the associative relations between $\mathbf{S}^v$ and $\mathbf{S}^t$. We propose a joint-semantics similarity matrix $\mathbf{S} = J(\mathbf{S}^v, \mathbf{S}^t) \in [-1, +1]^{n \times n}$ to construct the semantic similarity between input instances $\mathbf{X}^v$ and $\mathbf{X}^t$. To introduce the hybrid function J, we finally define the joint-semantics similarity matrix $\mathbf{S}$ as follows:

$$\mathbf{S} = J(\mathbf{S}^v, \mathbf{S}^t) = (1 - \eta)\tilde{\mathbf{S}} + \eta \frac{\mathbf{S}^v \mathbf{S}^{t^\top}}{n}, \tag{2}$$

where η is the trade-off parameter that regulates the similarity description.

Encoding Network:
Hash retrieval techniques mainly map the high-dimensional original feature into a low-dimensional information space while effectively preserving the original data's semantic information and semantic relationships. In the encoding stage, we use Image Encoder to transform image data $\mathbf{X}^v \in \mathbb{R}^{d_v \times \epsilon}$ into feature $\mathbf{Z}^v \in \mathbb{R}^{d_{ev} \times \epsilon}$ denote the d_{ev}−dimensional image feature vector with ϵ instances and further input hash layer to generate binary hash code $\mathbf{B}^v \in \{-1, 1\}^{k \times \epsilon}$. Text Encoder extracts feature $\mathbf{Z}^t \in \mathbb{R}^{d_{et} \times \epsilon}$ represents the d_{et}−dimensional text feature vector with ϵ instances from the original text data $\mathbf{X}^t \in \mathbb{R}^{d_t \times \epsilon}$ and generates binary hash code $\mathbf{B}^t \in \{-1, 1\}^{k \times \epsilon}$. The function of developing binary hash code is as follows:

$$\mathbf{Z}^m = f(\mathbf{X}^m; \theta_m), \mathbf{B}^m = sign(\mathbf{Z}^m), \tag{3}$$

where θ_v, θ_t denotes the parameter weights of the corresponding neural network.

Dual Prediction Network:
In the dual prediction stage, since different modalities of the same data have similar semantic information, we use the Image Prediction Network to convert the text information $\mathbf{Z}^t$ into image information $\bar{\mathbf{Z}}^v$. The Text Prediction Network extracts the information in image feature $\mathbf{Z}^v$ to generate the corresponding text feature $\bar{\mathbf{Z}}^t$. The dual prediction network can explore the semantic association of different modal information of the data. The process of dual prediction is defined as follows:

$$\bar{\mathbf{Z}}^v = f(\mathbf{Z}^t; \theta_{pv}), \bar{\mathbf{Z}}^t = f(\mathbf{Z}^v; \theta_{pt}), \tag{4}$$

where θ_{pv}, θ_{pt} denote the parameter weights of the corresponding network.

Decoding Network:

In the decoding stage, we use the potential features $\bar{\mathbf{Z}}^v$ which are generated by the dual prediction network as the input of the decoding network to generate the original data instances $\bar{\mathbf{X}}^v$, and input $\bar{\mathbf{Z}}^t$ to generate $\bar{\mathbf{X}}^t$ to achieve the decoding of different modalities' data, ensuring that the potential features $\bar{\mathbf{Z}}^m$ generated by the dual prediction network contain the comprehensive semantic information in the original data. Moreover, construct the similarity matrix between the original data instances to constrain the validity of the data features generated by the dual prediction network.

$$\bar{\mathbf{X}}^t = G(\bar{\mathbf{Z}}^t) = f(\bar{\mathbf{Z}}^t; \theta_{dt}), \bar{\mathbf{X}}^v = G(\bar{\mathbf{Z}}^v) = f(\bar{\mathbf{Z}}^v; \theta_{dv}), \tag{5}$$

where θ_{dv} and θ_{dt} denote the parameter weights of the corresponding networks.

3.3   Objective Function

To guarantee the quality of the resulting hash codes, we fully consider that the generated hash codes $\mathbf{B}^v$ and $\mathbf{B}^t$ maintain a similar relationship intra-modal and inter modal of the original instances to improve the performance of retrieval further. Ultimately, the hash code generation loss $\mathcal{L}_h$ is defined as follows:

$$\min_{\theta_v, \theta_t} \mathcal{L}_h = \alpha \parallel \mathbf{S} - \mathbf{B}^v \mathbf{B}^{t^\top} \parallel_F^2 + \beta \parallel \mathbf{S} - \mathbf{B}^v \mathbf{B}^{v^\top} \parallel_F^2$$
$$+ \beta \parallel \mathbf{S} - \mathbf{B}^t \mathbf{B}^{t^\top} \parallel_F^2 + \gamma \parallel \mathbf{B}^v - \mathbf{B}^t \parallel_F^2, \tag{6}$$

where θ_v, θ_t are the parameters of encoding network, and α, β, γ are the weighting factors. In order to ensure that the predicted generated data instances $\bar{\mathbf{X}}^m$ strictly maintain the similarity relationship between the raw data, the objective function is defined as follows:

$$\min_{\theta_{dv}, \theta_{dt}} \mathcal{L}_{pre} = \delta \sum_{m \in \{v,t\}} \parallel \bar{\mathbf{X}}^m - \mathbf{X}^m \parallel_F^2, \tag{7}$$

where δ is the weighting factor. The definition of the final objective function is given founded on the above several modular loss functions as follows:

$$\min_{\theta_m} \mathcal{L} = \mathcal{L}_h + \mathcal{L}_{pre}$$
$$= \alpha \parallel \mathbf{S} - \mathbf{B}^v \mathbf{B}^{t^\top} \parallel_F^2 + \beta \sum_{m \in \{v,t\}} \parallel \mathbf{S} - \mathbf{B}^m \mathbf{B}^{m^\top} \parallel_F^2$$
$$+ \gamma \parallel \mathbf{B}^v - \mathbf{B}^t \parallel_F^2 + \delta \sum_{m \in \{v,t\}} \parallel \bar{\mathbf{X}}^m - \mathbf{X}^m \parallel_F^2, \tag{8}$$

where $\theta_m \in \{\theta_v, \theta_t, \theta_{pv}, \theta_{pt}, \theta_{dv}, \theta_{dt}\}$ denote the parameter weights of network.

3.4   Extensions

More modalities: The UJSAH method can accomplish the task of multimodal scenarios, and when there are multiple modalities, a new network model can be added for each modality with appropriate modifications to the objection function Eq. 9 is shown below:

$$\min_{\theta_m} \mathcal{L} = \mathcal{L}_h + \mathcal{L}_{pre}$$

$$= \alpha \sum_{m_1, m_2 \in G} \| \mathbf{S} - \mathbf{B}^{m_1} \mathbf{B}^{m_2\top} \|_F^2 + \beta \sum_{m \in G} \| \mathbf{S} - \mathbf{B}^m \mathbf{B}^{m\top} \|_F^2$$

$$+ \gamma \sum_{m_1, m_2 \in G} \| \mathbf{B}^{m_1} - \mathbf{B}^{m_2} \|_F^2 + \delta \sum_{m \in G} \| \bar{\mathbf{X}}^m - \mathbf{X}^m \|_F^2 . \tag{9}$$

$$s.t. G = \{1, ..., M\}, m_1 \neq m_2, \theta_m \in \{\theta_1, ..., \theta_M, \theta_{p1}, ..., \theta_{pM}, \theta_{d1}, ..., \theta_{dM}\}.$$

Out-of-Sample: After the modal is fully trained, we can employ the trained model to develop binary hash codes for any sample of a new query. In detail, give a query data $x = \mathbf{X}^m \in \mathbb{R}^{d_m \times 1}$, we can obtain the hash code as follows:

$$b = sign(f(x; \theta_m)), s.t. m \in \{1, ..., M\}. \tag{10}$$

3.5   Computational Complexity Analysis

This section examines the computational complexity of the UJSAH method, as shown in Algorithm 1. During the experiments, the primary time cost lies in Eq.(8). In each iteration of the model training, we calculate the function Eq.(8), which has a time complexity of $O(n/n_b(n_b^2 d_i + n_b^2 d_t)) = O(n(n_b d_i + n_b d_t))$. Generally, the computational complexity of the algorithm for each iteration is $O((n(n_b d_i + n_b d_t))t)$, where $n_b, d_i, d_t, k, t \ll n$ and t means the number of iterations required for the model training. This time complexity can be simplified to $O(n)$, linearly correlated to the dataset size.

4   Experiments

To validate the usefulness of our UJSAH method, we have executed comprehensive experiments on MIRFlickr [7] and NUS-WIDE [1] datasets.

Algorithm 1. Unsupervised Joint-Semantics Autoencoder Hashing

---

Input: The training data: $\{\mathbf{X}^v, \mathbf{X}^t\}$, max training epoch E min-batch size: ϵ, hash code length: c, balance parameters: $\alpha, \beta, \gamma, \delta$.
Output: Parameters of the network: θ_v, θ_t.
Procedure:
Random initialization of the neural network parameters θ_m;
Extraction of image and text features from the dataset and construct similarity matrix;
Repeat:

　1: Select ϵ image-text pairs from the dataset in turn for training;
　2: Construct a similarity matrix $\mathbf{S}$ for the selected data according to Eq.2;
　3: Compute $\mathbf{Z}^v = f(\mathbf{X}^v; \theta_v), \mathbf{Z}^t = f(\mathbf{X}^t; \theta_t)$ for samples by forward-propagation;
　4: Generate binary hash codes
　5: Compute $\mathbf{B}^v, \mathbf{B}^t, \bar{\mathbf{Z}}^v, \bar{\mathbf{Z}}^t, \bar{\mathbf{X}}^v, \bar{\mathbf{X}}^t$ according to Eq.3, and Eq.4, Eq.5;
　6: Calculate the loss $\mathcal{L}$ with the Eq.8;
　7: Update the network parameter $\theta_v, \theta_t, \theta_{pv}, \theta_{pt}, \theta_{dv}, \theta_{dt}$ by using backpropagation;

Until convergent.
Return: θ_v, θ_t.

4.1 Datasets

MIRFlickr [7] contains 20,015 instances of image and text pairs and its semantic information can be classified into 24 label classes. The dataset is split into 18015 training data pairs and 2000 test data pairs, and we utilize all the available data for our experiments. **NUS-WIDE** [1] contains 270k image and text instance pairs, and this experiment selects 186,577 instance pairs from 10 of these labeled categories and 1867 image and text pairs as queries.

4.2 Baselines and Evaluation Metric

In our experiments, we experimentally analyze the proposed UJSAH method with advanced unsupervised cross-modal hash retrieval approaches, including CVH [8], IMH [16], LCMH [27], CMFH [2], LSSH [26], DBRC [6], RFDH [20], DJRH [18], AGCH [24], and DUCII [13]. All of these methods are evaluated for cross-modal retrieval, which includes retrieval of textual data by visual image information (Image-to-Text) and retrieval of visual image data by textual data (Text-to-Image). The evaluation metrics employed to estimate the retrieval accuracy of our UJSAH method and the baselines are mean average precision (mAP) and top-K accuracy. In our experiments, we set K = 50 as the value for top-K accuracy.

4.3 Implementation Detail

For the proposed UJSAH method, the parameters α, β, γ, and δ are used to balance the weights of different data items. In our experiments, when we set $\{\alpha = 0.08, \beta = 18, \gamma = 200, \delta = 0.12\}$, $\{\alpha = 1, \beta = 5, \gamma = 200, \delta = 1.2\}$ for MIRFlickr and NUS-WIDE datasets respectively. The network architecture is

Table 1. The mAP results for all methods on two datasets.

Method	I → T								T → I							
	MIRFlickr-25K				NUS-WIDE				MIRFlickr-25K				NUS-WIDE			
	16bits	32bits	64bits	128bits	16bits	32bits	64bits	128bits	16bits	32bits	64bits	128bits	16bits	32bits	64bits	128bits
CVH	0.606	0.599	0.596	0.598	0.372	0.362	0.406	0.390	0.591	0.583	0.576	0.576	0.401	0.384	0.442	0.432
IMH	0.612	0.601	0.592	0.579	0.470	0.473	0.476	0.459	0.603	0.595	0.589	0.580	0.478	0.483	0.472	0.462
LCMH	0.559	0.569	0.585	0.593	0.354	0.361	0.389	0.383	0.561	0.569	0.582	0.582	0.376	0.387	0.408	0.419
CMFH	0.621	0.624	0.625	0.627	0.455	0.459	0.465	0.467	0.642	0.662	0.676	0.685	0.529	0.577	0.614	0.645
LSSH	0.584	0.599	0.602	0.614	0.481	0.489	0.507	0.507	0.637	0.659	0.659	0.672	0.577	0.617	0.642	0.663
DBRC	0.617	0.619	0.620	0.621	0.424	0.459	0.447	0.447	0.618	0.626	0.626	0.628	0.455	0.459	0.468	0.473
RFDH	0.632	0.636	0.641	0.652	0.488	0.492	0.494	0.508	0.681	0.693	0.698	0.702	0.612	0.641	0.658	0.680
UDCMH	0.689	0.698	0.714	0.717	0.511	0.519	0.524	0.558	0.692	0.704	0.718	0.733	0.637	0.653	0.695	0.716
DJSRH	0.810	0.843	0.862	0.876	0.724	0.773	0.798	0.817	0.786	0.822	0.835	0.847	0.712	0.744	0.771	0.789
AGCH	<u>0.865</u>	<u>0.887</u>	<u>0.892</u>	<u>0.912</u>	<u>0.809</u>	<u>0.830</u>	<u>0.831</u>	<u>0.852</u>	<u>0.829</u>	0.849	0.852	<u>0.880</u>	**0.769**	<u>0.780</u>	<u>0.798</u>	<u>0.802</u>
DUCH	0.850	0.863	0.873	0.893	0.753	0.775	0.814	0.827	0.826	<u>0.855</u>	<u>0.864</u>	0.877	0.726	0.758	0.781	0.795
UJSAH	**0.884**	**0.913**	**0.927**	**0.936**	**0.812**	**0.835**	**0.858**	**0.867**	**0.853**	**0.879**	**0.881**	**0.893**	<u>0.765</u>	**0.790**	**0.803**	**0.813**

designed as follows: The image encoder ($d_v \rightarrow 4096 \rightarrow relu \rightarrow k \rightarrow tanh$), the text encoder ($d_t \rightarrow 2048 \rightarrow relu \rightarrow k \rightarrow tanh$), the dual prediction network ($k \rightarrow 1024 \rightarrow relu \rightarrow k \rightarrow tanh$), the image decoder ($k \rightarrow 4096 \rightarrow relu \rightarrow 4096 \rightarrow relu \rightarrow d_v \rightarrow relu$), the text decoder ($k \rightarrow 2048 \rightarrow relu \rightarrow 2048 \rightarrow relu \rightarrow d_t \rightarrow relu$). We conducted all experiments with the same experimental setting to ensure validity and accuracy.

4.4 Retrieval Accuracy Comparison

In this subsection, Table 1 manifests the mAP scores of our UJSAH compared to baselines in the "Image-to-Text (I→T)" and "Text-to-Image (T→I)" retrieval studies, with hash code lengths ranging from 16 to 128 bits. By analyzing Table 1, we can obtain the following conclusions:

1) Our UJSAH method reaches a satisfactory result compared to baselines with various hash code lengths and verifies the validity of the method. In particular, on the I→T task, the mean mAP scores of the proposed UJSAH are 2.9% and 1.5% higher compared to the AGCH in the MIRFlickr and NUS-WIDE datasets, respectively. On the T→I task, the mean mAP scores of the proposed UJSAH are 2.3% and 1.1% higher compared to the second highest baseline in the MIRFlickr and NUS-WIDE datasets. We propose that UJSAH outperforms baseline methods in all datasets in the cross-modal retrieval.

2) Data analysis indicates that the performance of all baseline methods shows significant improvement as the hash code length increases. Longer hash codes have the potential to capture and represent richer semantic information. However, it is essential to note that some baseline methods may experience a degradation in retrieval performance as the hash code length increases. This can be attributed to adding redundant information and introducing potential noise in more extended hash codes.

The top-K precision curves for a hash code length of 128 bits on the two datasets are depicted in Fig. 2. Experimental results show that the UJSAH

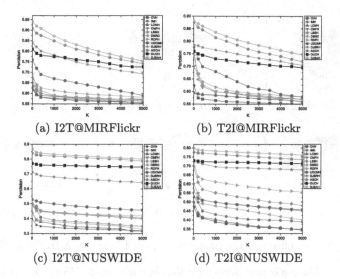

(a) I2T@MIRFlickr

(b) T2I@MIRFlickr

(c) I2T@NUSWIDE

(d) T2I@NUSWIDE

Fig. 2. The top-K precision cures of UJSAH with 128-bit on two datasets.

method outperforms the baseline hashing method at various return numbers. The experimental analysis further demonstrates the effectiveness and excellence of UJSAH method in the field of unsupervised large-scale multimedia information retrieval.

4.5 Parameter Sensitivity Analysis

Figure 3 shows the sensitivity analysis of parameters α, β, γ, and δ set from $1e^{-5}$ to $1e^4$, and parameters μ and η set from 0.1 to 1.0. 1) From Fig. 3 (a) and (b), it can be seen that parameter a is set from 0.01 to 100, and β can achieve good results regardless of the value of the model. 2) From the analysis of Fig. 3 (c) and (d), we can have the conclusion that the parameter γ is less than $1e^4$, and parameter δ is set at any value, the model effect is very stable, and the retrieval results are very satisfactory. 3) From the analysis of Fig. 3 (e) and (f), it can be concluded that the I2T of the model is greater than 0.9 and T2I is greater than 0.75 for any value of parameters μ, η. The comprehensive performance of the UJSAH method is stable and not sensitive to the changes of parameters μ, η.

4.6 Ablation Experiments

To assess the effectiveness of each component and confirm their usefulness, we conduct ablation experiments by designing various variants for each component.

UJSAH -1: We modify the similarity matrix $\mathbf{S}$ as $\mathbf{S} = \mu \mathbf{S}^v + (1 - \mu)\mathbf{S}^t$, using the traditional similarity matrix fusion as semantic constraint information, and have verified the validity of the joint-semantics similarity matrix.

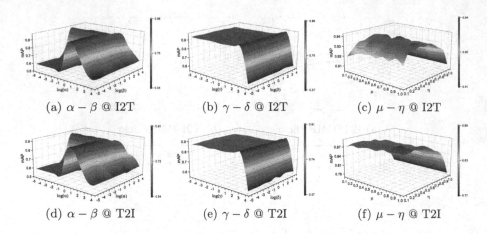

(a) $\alpha - \beta$ @ I2T   (b) $\gamma - \delta$ @ I2T   (c) $\mu - \eta$ @ I2T

(d) $\alpha - \beta$ @ T2I   (e) $\gamma - \delta$ @ T2I   (f) $\mu - \eta$ @ T2I

Fig. 3. The effects of the parameters with 128-bit on MIRFlickr-25K.

Table 2. The ablation experiments for different variants of UJSAH on MIRFlickr-25K.

Method	I → T				T → I			
	16 bits	32 bits	64 bits	128 bits	16 bits	32 bits	64 bits	128 bits
UJSAH	**0.884**	**0.913**	**0.927**	**0.936**	**0.853**	**0.879**	**0.881**	**0.893**
UJSAH-1	0.763	0.880	0.909	0.920	0.785	0.842	0.865	0.881
UJSAH-2	0.695	0.844	0.922	0.934	0.690	0.798	0.875	0.886

UJSAH -2: To verify the effectiveness of our designed dual prediction autoencoder, we modify the input ($\bar{\mathbf{Z}}^m$) of the decoding module in UJSAH framework to input ($\mathbf{Z}^m$) in the traditional way to experiment the importance of dual prediction network.

As the analysis in Table 2 can be concluded, the joint-semantics similarity matrix and dual prediction autoencoder in the proposed UJSAH model can effectively improve the retrieval accuracy.

5   Conclusion

In this paper, we present a Unsupervised Joint-Semantics Autoencoder Hashingmethod for multimedia retrieval. The generated hash codes are guaranteed to retain more information about the similarity of the raw data by autoencoder. The design of the joint-semantics similarity matrix achieves efficient mining of multimedia data similarity matrix by using a mixture of similarity matrix within each modality and the cross-modal similarity matrix construction proposed in this paper. An autoencoder established on a dual prediction network is proposed to realize the association of semantic information of cross-modal data by converting hash codes of different modal data to each other. Finally, we execute extensive experiments on the widely used datasets to prove the significance and sophistication of the UJSAH method. In future research, we plan to study fine-grained correlations, capture more complex relationships between different modalities in multimodal data, and extend the proposed architecture and similarity relation mining to shallow hash models.

Acknowledgements. This work is supported in part by the National Natural Science Foundation of China under the Grant No.62202501 and No.U2003208, in part by the National Key R&D Program of China under Grant No.2021YFB3900 902 and in part by the Science and Technology Plan of Hunan Province under Grant No.2022JJ40638.

References

1. Chua, T.S., Tang, J., Hong, R., Li, H., Luo, Z., Zheng, Y.: Nus-wide: a real-world web image database from national university of Singapore. In: Proceedings of the ACM International Conference on Image And Video Retrieval, pp. 1–9 (2009)
2. Ding, G., Guo, Y., Zhou, J.: Collective matrix factorization hashing for multimodal data. In: Proceedings of the IEEE Conference on Computer Vision and Pattern Recognition, pp. 2075–2082 (2014)
3. Fan, L., Ng, K.W., Ju, C., Zhang, T., Chan, C.S.: Deep polarized network for supervised learning of accurate binary hashing codes. In: IJCAI, pp. 825–831 (2020)
4. Fan, W., Zhang, C., Li, H., Jia, X., Wang, G.: Three-stage semisupervised cross-modal hashing with pairwise relations exploitation. IEEE Trans. Neural Netw. Learn. Syst. Early Access, 1–14 (2023)
5. Guo, J., Guan, X.: Deep adversarial cascaded hashing for cross-modal vessel image retrieval. IEEE J. Select. Topics Appl. Earth Observ. Remote Sens. **16**, 2205–2220 (2023)
6. Di, H., Nie, F., Li, X.: Deep binary reconstruction for cross-modal hashing. IEEE Trans. Multimedia **21**(4), 973–985 (2018)
7. Huiskes, M.J., Lew, M.S.: The mir flickr retrieval evaluation. In: Proceedings of the 1st ACM International Conference on Multimedia Information Retrieval, pp. 39–43 (2008)
8. Kumar, S., Udupa, R.: Learning hash functions for cross-view similarity search. In: Twenty-Second International Joint Conference on Artificial Intelligence (2011)
9. Li, Q., Sun, Z., He, R., Tan, T.: A general framework for deep supervised discrete hashing. Int. J. Comput. Vision **128**(8), 2204–2222 (2020)
10. Li, Y., Ge, M., Li, M., Li, T., Xiang, S.: Clip-based adaptive graph attention network for large-scale unsupervised multi-modal hashing retrieval. Sensors **23**(7), 3439 (2023)
11. Liu, H., Wang, R., Shan, S., Chen, X. Deep supervised hashing for fast image retrieval. In: Proceedings of the IEEE Conference on Computer Vision and Pattern Recognition, pp. 2064–2072 (2016)
12. Liu, X., Zeng, H., Shi, Y., Zhu, J., Hsia, C.H., Ma, K.K.: Deep cross-modal hashing based on semantic consistent ranking. IEEE Trans. Multimed. Early Access, 1–12 (2023)
13. Mikriukov, G., Ravanbakhsh, M., Demir, B.: Deep unsupervised contrastive hashing for large-scale cross-modal text-image retrieval in remote sensing. arXiv preprint arXiv:2201.08125 (2022)

14. Shi, Y., Nie, X., Liu, X., Zou, L., Yin, Y.: Supervised adaptive similarity matrix hashing. IEEE Trans. Image Process. **31**, 2755–2766 (2022)
15. Shi, Y., et al.: Deep adaptively-enhanced hashing with discriminative similarity guidance for unsupervised cross-modal retrieval. IEEE Trans. Circuits Syst. Video Technol. **32**(10), 7255–7268 (2022)
16. Song, J., Yang, Y., Yang, Y., Huang, Z., Shen, H.T.: Inter-media hashing for large-scale retrieval from heterogeneous data sources. In Proceedings of the 2013 ACM SIGMOD International Conference on Management of Data, pp. 785–796 (2013)
17. Su, H., Han, M., Liang, J., Liang, J., Yu, S.: Deep supervised hashing with hard example pairs optimization for image retrieval. Vis. Comput. **39**, 5405–5420 (2022)
18. Su, S., Zhong, Z., Zhang, C.: Deep joint-semantics reconstructing hashing for large-scale unsupervised cross-modal retrieval. In: Proceedings of the IEEE/CVF International Conference on Computer Vision, pp. 3027–3035 (2019)
19. Tu, R.C., et al.: Unsupervised cross-modal hashing with modality-interaction. IEEE Trans. Circ. Syst. Video Technol. (2023)
20. Wang, D., Wang, Q., Gao, X.: Robust and flexible discrete hashing for cross-modal similarity search. IEEE Trans. Circuits Syst. Video Technol. **28**(10), 2703–2715 (2017)
21. Zeng, X., Xu, K., Xie, Y.: Pseudo-label driven deep hashing for unsupervised cross-modal retrieval. Int. J. Mach. Learn. Cybern. **11**, 1–20 (2023)
22. Zhan, Y.W., Luo, X., Wang, Y., Xu, X.S.: Supervised hierarchical deep hashing for cross-modal retrieval. In: Proceedings of the 28th ACM International Conference on Multimedia, pp. 3386–3394 (2020)
23. Zhang, P.-F., Bai, G., Yin, H., Huang, Z.: Proactive privacy-preserving learning for cross-modal retrieval. ACM Trans. Inform. Syst. **41**(2), 1–23 (2023)
24. Zhang, P.-F., Li, Y., Huang, Z., Xin-Shun, X.: Aggregation-based graph convolutional hashing for unsupervised cross-modal retrieval. IEEE Trans. Multimedia **24**, 466–479 (2021)
25. Zhang, Q., Hu, L., Cao, L., Shi, C., Wang, S., Liu, D.D.: A probabilistic code balance constraint with compactness and informativeness enhancement for deep supervised hashing. In: Proceedings of the Thirty-First International Joint Conference on Artificial Intelligence. International Joint Conferences on Artificial Intelligence Organization (2022)
26. Zhou, J., Ding, G., Guo, Y.: Latent semantic sparse hashing for cross-modal similarity search. In: Proceedings of the 37th International ACM SIGIR Conference on Research Development in Information Retrieval, pp. 415–424 (2014)
27. Zhu, X., Huang, Z., Shen, H.T., Zhao, X.: Linear cross-modal hashing for efficient multimedia search. In: Proceedings of the 21st ACM International Conference on Multimedia, pp. 143–152 (2013)

TKGR-RHETNE: A New Temporal Knowledge Graph Reasoning Model via Jointly Modeling Relevant Historical Event and Temporal Neighborhood Event Context

Jinze Sun[1], Yongpan Sheng[1,2(✉)], Ling Zhan[1], and Lirong He[3]

[1] College of Computer and Information Science, Southwest University, Chongqing 400715, China
ssunjinze@outlook.com, shengyp2011@gmail.com, zl0327@email.swu.edu.cn

[2] School of Big Data & Software Engineering, Chongqing University, Chongqing 401331, China

[3] School of Information Science and Engineering, Chongqing Jiaotong University, Chongqing 400074, China
ronghe1217@gmail.com

Abstract. Temporal knowledge graph reasoning (TKGR) has been of great interest for its role in enriching the naturally incomplete temporal knowledge graph (TKG) by uncovering new events from existing ones with temporal information. At present, the majority of existing TKGR methods have attained commendable performance. Nevertheless, they still suffer from several problems, specifically their limited ability to adeptly capture intricate long-term event dependencies within the context of pertinent historical events, as well as to address the occurrence of an event with insufficient historical information or be influenced by other events. To alleviate such issues, we propose a novel TKGR method named TKGR-RHETNE, which jointly models the context of relevant historical events and temporal neighborhood events. In terms of the historical event view, we introduce an encoder based on the transformer Hawkes process and self-attention mechanism to effectively capture long-term event dependencies, thus modeling the event evolution process continuously. In terms of the neighborhood event view, we propose a neighborhood aggregator to model the potential influence between events with insufficient historical information and other events, which is implemented by integrating the random walk strategy with the TKG topological structure. Comprehensive experiments on five benchmark datasets demonstrate the superior performance of our proposed model (Code is publicly available at https://github.com/wanwano/TKGR-RHETNE).

1 Introduction

Knowledge graphs (KGs) provide significant real-world factual information using a multi-relational graph structure and have achieved remarkable promise in NLP or knowledge engineering-aware tasks. However, factual knowledge in real is continuously evolving, in the form of event knowledge, which has led to the development

B. Luo et al. (Eds.): ICONIP 2023, LNCS 14451, pp. 331–343, 2024.
https://doi.org/10.1007/978-981-99-8073-4_26

and application of temporal knowledge graphs (TKGs). TKG offers fresh perspectives and valuable insights for many time-sensitive downstream applications, including time-sensitive semantic search [1], policymaking [4], and stock prediction [5].

With the advancement of the TKG, more and more researchers are starting to pay their attention to the TKG reasoning (TKGR) task to infer new event knowledge based on existing ones for filling in natural incomplete TKG and checking these knowledge consistencies. On the benefit of their representation learning ability, various learning-based methods associate temporal information with the facts, undoubtedly, this severely limits the ability to express temporal information. While several existing embedding-based methods [9,18,23] attempt to incorporate temporal point processes such as neural Hawkes process into their models, being capable of modeling the graph-wide link formation process in a temporal graph, such as Know-Evolve [22] models the occurrence of a fact as a multi-dimensional temporal point process for calculation in a multi-relation space. Mei *et al.* [18] present extensions to the multivariate Hawkes process using continuous-time LSTM to model the asynchronous event sequence. DyRep [23] presents an inductive framework based on the temporal point process, it addresses a different problem setting of a two-time scale with both association and communication events. As an event consequence, these methods cannot fully capture the continuous event evolving dynamics and prioritize different parts of the consequence, especially for complicated long-term event dependencies in the relevant historical event context, which limits the model performance.

Meanwhile, there has been a growing number of works on graph neural networks (GNNs) to explore the intrinsic topology relevance between entities or between entities and relations in TKG, so as to obtain high-quality embeddings, such as RE-NET [11] uses a multi-relational graph aggregator to capture the graph structural information. RE-GCN [15] employs RGCN to aggregate messages from neighboring entities and utilizes an auto-regressive GRU to model the temporal dependency among events. CRNet [26] leverages concurrent events from both historical and the future for TKGR. TGAP [12] explores query-relevant substructure in the TKG for path-based inference. These methods achieved good performance to known events in historical, recurrence, or periodicity dependency of relevant historical events to forecast events happening in the future. However, some events may have insufficient historical information and even be affected by other events during the whole timeline. As a result, it is a challenge to effectively model the hidden impact between these types of events and other events on TKG embedding learning. The temporal neighborhood event context naturally offers a viable solution for enhancing the comprehension of event occurrences in temporal graph learning models addressing the aforementioned problem.

Based on the aforementioned analysis, we should make use of both relevant historical events and temporal neighborhood event contexts. To obtain information from context adequately, in this paper, we proposed a new temporal knowledge graph reasoning model via jointly modeling relevant historical event and temporal neighborhood event context, (short-handed as TKGR-RHETNE). It contains an encoder and an aggregator to capture both the event evolution process and latent factors influencing the occurrence of an event in the relevant historical and temporal neighborhood context of events in the TKG, respectively. Finally, the model uses the above two parts of information to calculate the conditional intensity and obtains the candidate entity score or time probability density of the next event.

In summary, our contributions are 3-fold as follows.

- We propose a new temporal knowledge graph reasoning model, called TKGR-RHETNE, which jointly models the relevant historical event and temporal neighborhood event context of events in the temporal knowledge graph.
- In the framework of the TKGR-RHETNE, we propose an encoder built on the transformer Hawkes process and self-attention mechanism to capture complex long-term event dependencies in relevant historical event context, thereby enabling continuous modeling of event evolution. We also present an aggregator that integrates the random walk strategy for modeling the potential influence of events with insufficient historical formation or are influenced by other events within their temporal neighborhood event context. The random walk strategy encompasses both the breadth-first random walk search, which aims to capture homophily information, and a depth-first search, designed to collect heterophily information.
- We carry out extensive experiments on five benchmark datasets to demonstrate the effectiveness of the TKGR-RHTNE in entity and timestamp forecasting tasks.

2 Related Work

2.1 Neural Hawkes Process-Based Method

Since the Hawkes process has been proven to model the exciting effects between events to capture the influence of historical events holistically, it is well suited for modeling the graph-wide link formation process in a temporal graph [22]. Neural Hawkes process [20] uses neural networks to parameterize the intensity function and exhibits high model capacity in complex real-life scenarios, e.g., Mei *et al.* [18] develop a neural Hawkes process based on continuous-time LSTM to model the asynchronous event sequence such that previous events are allowed to influence future events in complex and realistic ways. Know-Evolve [22] models the occurrence of a fact as a multidimensional temporal point process, it mainly learns non-linearly evolving entity representations by considering their interactions with other entities in multi-relational space, where the conditional intensity function of the process is adjusted based on the relationship score of the event. Zuo and Zhang *et al.* [28,30] apply the attention mechanism to the neural Hawkes process. However, these approaches fail to capture complicated dependencies well in the long-term event sequences.

2.2 Graph Neural Network-Based Method

Graph Neural Network (GNN)-based methods generally apply GNN to explore the intrinsic topology relevance between entities or between entities and relations in TKG. CyGNet [29], RE-NET [11], and RE-GCN [15] attempt to solve the entity prediction in the TKG from the perspective of each given query, which encodes the historical facts related to the subject entity in each query. These methods achieved good performance in entity prediction, but can not have the capacity of predicting timestamps in the

TKG. Some researchers employ temporal random walks to capture the continuous-time network dynamics, including CTDNE [19] based on time-respect random walks, and CAW-N [16] based on causal anonymous walks, however, they cannot collect homogeneous and heterogeneous information from the entire dynamic graph, which is implicitly subject to the homophily assumption.

3   The Proposed Model

The overall architecture of our proposed model is illustrated in Fig. 1. In the following, we provide a detailed introduction to three components and elaborate on the steps within each component.

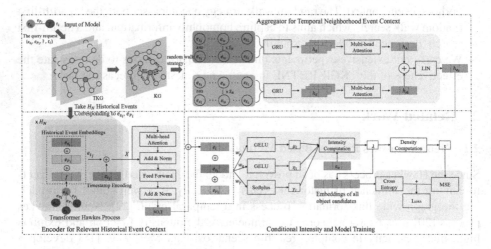

Fig. 1. The overall architecture of our proposed model.

3.1   Encoder for Relevant Historical Event Context

A TKG $\mathcal{G}$ can be formalized as a sequence of graph slices $\{\mathcal{G}_1, \mathcal{G}_2, \ldots, \mathcal{G}_T\}$, where $\mathcal{G}_t = \{(e_s, e_p, e_o, t) \in \mathcal{G}\}$ denotes a graph slice that consists of facts that occurred at the timestamp t, i.e., events. Here, e_s and e_o represent the subject and object entities, respectively, and e_p denotes the predicate as a relation type.

Taking a query $(e_{s_i}, e_{p_i}, ?, t_i)$ in the TKG for an example, we assume an object forming relationship e_{p_i} with a given subject e_{s_i} at a timestamp t_i depends in part on past events involving e_{s_i} and e_{p_i}, this historical event encoder regards the past events and timestamps as relevant historical sequence $e_i^{h,sp}$ and time sequence $t_i^{h,sp}$:

$$e_i^{h,sp} = \left\{ \bigcup_{0 \leq t_j < t_i} \mathbf{O}_{t_j}\left(e_{s_i}, e_{p_i}\right) \right\}, t_i^{h,sp} = \left\{ \bigcup_{0 \leq t_j < t_i} t_j \right\}, \quad (1)$$

where $\mathbf{O}_{t_j}(e_{s_i}, e_{p_i})$ is a set of object entities that formed a link with the subject entity e_{s_i} under the predicate e_{p_i} at a timestamp t_j $(0 \le t_j < t_i)$, $\bigcup$ is the sign for the union. To preserve the occurrence order of the events, a.k.a timestamp, in the relevant historical time sequence $t_i^{h,sp}$, the event encoder adopts a temporal encoding procedure referring to the Transformer primitive location encoding method [25]. Specifically, the positions of the timestamps in the sequence are defined as:

$$[z_{t_j}]_i = \begin{cases} \cos\left(t_j/10000^{\frac{i-1}{d}}\right), & \text{if } i \text{ is odd}, \\ \sin\left(t_j/10000^{\frac{i}{d}}\right), & \text{if } i \text{ is even}. \end{cases} \tag{2}$$

For each $t_j \in t_i^{h,sp}$, we computes $\mathbf{z}_{t_j} \in \mathbb{R}^d$, where d is the dimension of encoding. Meanwhile, to learn the correlation between each event in the relevant historical event sequence $e_i^{h,sp}$, the historical event encoder represents each event as follows:

$$\mathbf{e}_{t_j} = \mathbf{W}_\lambda\left(\mathbf{e}_{s_i} \oplus \mathbf{f}\left(e_i^{h,sp}, t_j\right) \oplus \mathbf{e}_{p_i}\right), \tag{3}$$

where $\mathbf{f}\left(e_i^{h,sp}, t_i\right) \in \mathbb{R}^r$ denotes the average embedding of the object of the current subject at timestamp t_j, and $\oplus$ represents the concatenation operator. r denotes the rank of embeddings. Because the embeddings of the events and the encodings of timestamps are defined in spaces of different natures, to ensure alignment, the historical event encoder utilizes $\mathbf{W}_\lambda$ to map the event embeddings to the same space of the temporal encodings. Then, the encoder outputs the embeddings of the historical sequence $\mathcal{H} = \{(z_{t_j}, e_{t_j})\}$ to the subsequent self-attention module by computing $\mathbf{X} = (\mathbf{E} + \mathbf{Z})$.

The self-attention module enables the model to weigh and prioritize different parts of an input historical sequence $\mathcal{H}$, allowing it to capture complex long-range event dependencies and contextual information effectively. Specifically, we compute the attention output $\mathbf{A}$ by

$$\mathbf{A} = \text{Softmax}\left(\frac{\mathbf{QK}^\top}{\sqrt{M_K}}\right)\mathbf{V}, \quad \text{where } \mathbf{Q} = \mathbf{XW}^Q, \mathbf{K} = \mathbf{XW}^K, \mathbf{V} = \mathbf{XW}^V. \tag{4}$$

Here $\mathbf{Q}, \mathbf{K}$, and $\mathbf{V}$ are the query, key, and value matrices obtained by different transformations of $\mathbf{X}$, and $\mathbf{W}^Q, \mathbf{W}^K \in \mathbb{R}^{d \times d_K}, \mathbf{W}^V \in \mathbb{R}^{d \times d_V}$ are the weights for the linear transformations, respectively. M_K is the variance of $\mathbf{QK}^\top$. The attention output $\mathbf{A}$ is then fed through a position-wise feed-forward neural network, generating hidden representations $\mathbf{h}(t)$ of the input historical event sequence:

$$\mathbf{H} = \text{ReLU}\left(\mathbf{AW}_1 + \mathbf{b}_1\right)\mathbf{W}_2 + \mathbf{b}_2, \quad \mathbf{h}(t_j) = \mathbf{H}(j,:), \tag{5}$$

where $\mathbf{W}_1 \in \mathbb{R}^{d \times d_H}, \mathbf{W}_2 \in \mathbb{R}^{d_H \times d}, \mathbf{b}_1 \in \mathbb{R}^{d_H}$, and $\mathbf{b}_2 \in \mathbb{R}^d$ are parameters of the neural network, and $\mathbf{W}_2$ has identical columns. The resulting matrix $\mathbf{H} \in \mathbb{R}^{L \times r}$ contains the hidden representation of all historical moments in the historical sequence, where each row corresponds to a specific historical moment. In practice, we employ multi-head self-attention to enhance the flexibility of the encoder, which also proves advantageous for data fitting. In addition, we incorporate masks into the attention mechanism to prevent it from accessing future information, i.e., when the attention produces an output for a specific position, we mask out all subsequent positions.

3.2 Aggregator for Temporal Neighborhood Event Context

Relying solely on the historical sequence of entities for the model reasoning can lead to diminished results, especially when confronted with insufficient historical information or disturbances from other events. The topological construction of TKG is influenced by the interactions between events. As such, integrating the graphical structure of TKG into event embeddings enables a more profound grasp of these underlying relationships.

We expect to capture the hidden semantic information associated with the temporal neighborhood events in the context of the neighborhood events of a given entity. Inspired by RAW-GNN [10], entity nodes in the neighborhood are divided according to whether they have similar behaviors with the given entity node. The hidden semantic information collected from these different entities can be divided into two categories, which we call homogeneous information and heterogeneous information.

In a slice of the temporal graph, we notice that examining the information of the broad neighborhood event sequence (BNS) of entities closely connected to or reachable with few jumps from a given entity, i.e., homogeneous information, which assists in understanding the potential impact of entity nodes with similar behaviors on a given entity, e.g., if an event is simultaneously initiated by several parallel events, extracting BNS information might be more likely to identify these latent relationships in breadth.

Besides, we can extract information from the deep neighborhood event sequence (DNS) of entities relatively distant from the designated entity, i.e., heterogeneous information. This aids in comprehending the potential impact of long or complex chains of relationships between a given entity and entities that behave differently. e.g., if an event is set off by a series of successive events, gathering DNS information might be more apt in capturing the underlying links within this extensive chain. The purpose of accumulating these two types of information is to thoroughly unearth the potential factors linked with the target event in the surrounding events of the specified target entity.

To illustrate this, we also shall consider an example consisting of a given query $(e_{s_i}, e_{p_i}, ?, t_i)$ in the TKGR, and perform the following operations:

Step 1. Obtain two types of event sequences. BNS and DNS originating from e_{s_i} can be obtained by applying breadth-first and depth-first search strategies on the graph. Consider a random walk that has traversed edge $p_{e_w e_u}$, is now on entity node e_u, and is going to visit the next entity node e_v. We set the unnormalized transition probability as follows:

$$S\left(e_{next} = e_v \mid e_{now} = e_u, e_{last} = e_w\right) = \begin{cases} 1/p, & \text{if } d_{e_w e_v} = 0 \\ 1, & \text{if } d_{e_w e_v} = 1 \\ 1/q, & \text{if } d_{e_w e_v} = 2 \\ 0, & \text{otherwise} \end{cases}, \qquad (6)$$

where $d_{e_w e_v}$ is the length of the shortest path between nodes e_u and e_w. When we set $p < 1$ and $q > 1$, the random walk prefers to go back to the previous node, generating a BNS; and when $p > 1$ and $q < 1$, the random walk tends to generate a DNS.

Step 2. Aggregate event sequence information. Unlike the existing GNN-based methods that treat all neighboring entity nodes as an unordered set, the event sequence retains the ordered connections between nodes. Therefore, when aggregating information across all entity nodes in the sequence, we also take into account the order of the entity nodes to preserve more relationship information between connected entities. The aggregation operation can be formulated as:

$$\mathbf{h}_S = GRU_\theta \left(\{ \mathbf{h}_{e_1}, \mathbf{h}_{e_2}, \ldots, \mathbf{h}_{e_K} \} \right), \tag{7}$$

where $\mathbf{h}_S \in \mathbb{R}^d$ is the embedding vector of sequence S, $\mathbf{h}_{e_k}^{(l)} \in \mathbb{R}^d$ is the feature of node e_k and θ represents all the learnable parameters in our message function GRU [3].

Step 3. Combine event sequence embeddings. Different sequences may have varying impacts on the target entity, therefore, we utilize an attention mechanism to determine the significance of each sequence embedding based on its relevance to the target entity.

$$\mathbf{h}_A = MultiHeadAttention_\theta \left(\mathbf{h}_S \right). \tag{8}$$

We use concatenation instead of sum to merge the sequential information embeddings from different strategies and feed them into a linear layer to obtain $\mathbf{h}_{e_s} \in \mathbb{R}^d$. Concatenation helps preserve unique information for each strategy.

Step 4. Obtain the final embeddings. Concatenating the representation of the historical information of the previous subsection with the representation of the neighborhood yields the final embedding used for the intensity function:

$$\mathbf{e}_i = \left(\mathbf{h} \left(t_i \right) \oplus \mathbf{h}_{e_s} \right). \tag{9}$$

3.3 Conditional Intensity

After generating final embedding of $\mathbf{e}_i$ at time t_i, we compute three parameters of the intensity function based on the $\mathbf{e}_{s_i}, \mathbf{e}_{p_i}$ and historical and neighbor hidden embedding $\mathbf{e}_i$:

$$\begin{aligned} \mu_i &= \text{GELU} \left((\mathbf{e}_i \oplus \mathbf{e}_{s_i} \oplus \mathbf{e}_{p_i}) W_\mu \right) \\ \eta_i &= \text{GELU} \left((\mathbf{e}_i \oplus \mathbf{e}_{s_i} \oplus \mathbf{e}_{p_i}) W_\eta \right), \\ \gamma_i &= \text{Softplus} \left((\mathbf{e}_i \oplus \mathbf{e}_{s_i} \oplus \mathbf{e}_{p_i}) W_\gamma \right) \end{aligned} \tag{10}$$

where η_i represents the initial intensity at time t_i, and μ_i represents the intensity when time tends to infinity. $(\eta_i - \mu_i)$ represents the rate at which the intensity changes, which can be both positive and negative. The function GELU represents the Gaussian Error Linear Unit for nonlinear activations. The Softplus is used for the decaying parameter since γ needs to be constrained to strictly positive values.

Finally, we express the intensity function of an object e_{o_i} forming relationship e_{p_i} with a given subject e_{s_i} at a timestamp t_i, as follows:

$$\lambda(e_{s_i}, e_{p_i}, t_i) = \text{Softplus} \left((\mu_i + (\eta_i - \mu_i) \exp \left(-\gamma_i \left(t - t_i \right) \right)) \cdot \mathbf{e}_o \right), \tag{11}$$

for $t \in (t_i, t_{i+1}]$, where the Softplus is employed to constrain the intensity function to be positive.

3.4   Model Training and Prediction

Entity Prediction Task. Because all object candidates share the same survival term $\lambda\left(e_{s_i}, e_{p_i}, t\right)$ and the same value of t_L, we can directly compare their intensity function $\lambda\left(e_{s_i}, e_{p_i}, t_i\right)$. We use the cross-entropy loss for learning the entity prediction:

$$\mathcal{L}_{\text{entity}} = -\sum_{i=1}^{N}\sum_{c=1}^{N_e} y_c \log\left(p\left(e_{o_i} = c \mid e_{s_i}, e_{p_i}, t_i\right)\right). \tag{12}$$

Time Prediction Task. We use the mean square error as the timestamp prediction loss:

$$\mathcal{L}_{\text{time}} = \sum_{i=1}^{N}\left(t_i - \hat{t}_i\right)^2. \tag{13}$$

Joint Learning. Different types of loss functions are linearly combined to jointly learn two tasks in an end-to-end manner:

$$\mathcal{L} = \lambda_e \mathcal{L}_e + \lambda_t \mathcal{L}_t + \lambda_l \|\Theta\|_2^2, \tag{14}$$

where λ_e, λ_t, and λ_l are weights that balance the importance of two losses. Θ contains all model parameters.

4   Experiments and Analysis

4.1   Experimental Settings and Implementation Details

Datasets. We evaluate our model and baselines on five datasets, including ICEWS14 [6], ICEWS18 [11], ICEWS05-15 [6], YAGO [17] and WIKI [14]. To ensure a fair comparison, we use the split manner provided by Sun *et al.* [21].

Baselines. For the entity prediction task, we compare our model with three categories of KG embedding models: (1) *static KG embedding models*, including TransE [2][1], DistMult [27][1], and ComplEx [24][1]. We apply these models in static KGs that ignore timestamp information. (2) *TKGR models*, including TTransE [14][2], TA-DistMult [6], DE-SimplE [7], TNTComplEx [13], RE-NET [11], CyGNet [29], RE-GCN [15] and TANGO [8]. (3) *TKGR models with Hawkes process* are most related to our model, including GHNN [9] and GHT [21]. For the timestamp prediction task, we compare our model with one temporal point process model for TKGR, i.e., GHNN [9].

Evaluation Metrics. The mean reciprocal rank (MRR) and Hits@k are standard metrics for the entity prediction task. MRR is the average reciprocal of the correct query answer rank. Hits@k indicates the proportion of correct answers among the top k candidates. For the timestamp prediction task, we use the mean absolute error (MAE) between the predicted timestamp and ground truth timestamp as the metric.

---

[1] The code for TransE, DistMult, and ComplEx is from https://github.com/thunlp/OpenKE.
[2] The code for TTransE is from https://github.com/INK-USC/RE-Net/tree/master/baselines.

Details of Baselines. We directly adopt the results of static KG embedding models, interpolated reasoning models, and CyGNet reported in earlier works [8]. For TANGO[3], RE-NET[4], RE-GCN[5], GHNN[6], and GHT[7], we use their released source codes with default hyperparameters. We rerun these models on four benchmark datasets. To ensure fairness, for RE-GCN, we don't use the module that encodes the node type information.

Details of TKGR-RHETNE. We set the dimension of all embeddings and hidden states to 200. The relevant historical sequence length for the encoder is limited to 10, and the attention head number is set to 3. For the aggregator for temporal neighborhood event context, the number and length of sequences for each strategy are 8 and 6, respectively, the GRU layer number is 1, and the attention head number is 2. In total, we train our model for 200 epochs and use the Adam optimizer with parameters, with a learning rate of 0.003 and a weight decay of 1e-5. The batch size is set to 256.

4.2 Experimental Results and Discussion

Entity Prediction. Table 1 and Table 2 report the experimental results of the entity prediction task on five TKGR datasets. We highlight the best results in bold. On the three ICEWS datasets, static KG embedding methods fell far behind our model due to their inability to capture temporal dynamics. Our model is superior to the heuristic evolutionary methods of RE-NET and RE-GCN in predicting events because their learning effect in the later time is highly dependent on the learning effect in the earlier time and will forget the previous knowledge in the process of evolution. By contrast, we introduce an encoder based on the transformer Hawkes process when encoding relevant historical event context, and it can learn the influence of the current event from the event occurring at any distance from the current time in parallel, which helps the model capture complex long-term event dependencies for modeling the event evolution continuously. Unlike existing GNN-based TKGR models, which often imply homogeneity assumptions, our model considers different aggregation strategies to collect latent factors influencing the occurrence of an event in temporal neighborhood events, which is the reason why it performs better than them. On the YAGO and WIKI datasets, it can be observed that there is a significant improvement in the performance of our model, with a boost of 13.85% and 7.47% on Hits@1. The reason for this phenomenon is that for WIKI and YAGO, a large number of quadruples contain unseen entities in the test set. This prevents many models from learning from historical information when inferencing entities, but the encoder for temporal neighborhood event context with different aggregation strategies in our model can effectively alleviate this problem, and the probability of selecting irrelevant entities is greatly reduced.

---

[3] The code for TANGO is from https://github.com/TemporalKGTeam/TANGO.

[4] The code for RE-NET is from https://github.com/INK-USC/RE-Net.

[5] The code for RE-GCN is from https://github.com/Lee-zix/RE-GCN.

[6] The code for GHNN is from https://github.com/Jeff20100601/GHNN_clean.

[7] The code for GHT is from https://github.com/JHL-HUST/GHT.

Table 1. Experimental results of entity prediction on ICEWS-style dataset. The best result in each column is boldfaced.

Models	ICEWS14				ICEWS18				ICEWS05-15			
	MRR	Hits@1	Hits@3	Hits@10	MRR	Hits@1	Hits@3	Hits@10	MRR	Hits@1	Hits@3	Hits@10
TransE	22.48	13.36	25.63	41.23	12.24	5.84	12.81	25.10	22.55	13.05	25.61	42.05
Distmult	27.67	18.16	31.15	46.96	10.17	4.52	10.33	21.25	28.73	19.33	32.19	47.54
ComplEx	30.84	21.51	34.48	49.59	21.01	11.87	23.47	39.97	31.69	21.44	35.74	52.04
TTransE	13.43	3.11	17.32	34.55	8.31	1.92	8.56	21.89	15.71	5.00	19.72	38.02
TA-DistMult	26.47	17.09	30.22	45.41	16.75	8.61	18.41	33.59	24.31	14.58	27.92	44.21
DE-SimplE	32.67	24.43	35.69	49.11	19.30	11.53	21,86	34.80	35.02	25.91	38.99	52.75
TNTComplEx	32.12	23.35	36.03	49.13	21.23	13.28	24.02	36.91	27.54	19.52	30.80	42.86
CyGNet	32.73	23.69	36.31	50.67	24.93	15.90	28.28	42.61	34.97	25.67	39.09	52.94
TANGO-TuckER	26.25	17.30	29.07	44.18	28.97	19.51	32.61	47.51	42.86	**32.73**	**48.14**	62.34
TANGO-Distmult	24.70	16.36	27.26	41.35	27.56	18.68	30.86	44.94	40.71	31.23	45.33	58.95
RE-NET	35.71	26.79	39.57	52.81	26.88	17.70	30.35	45.04	39.11	29.04	44.10	58.90
RE-GCN	36.08	27.30	39.89	53.01	27.34	18.44	30.48	44.85	39.89	30.10	44.63	56.98
GHNN	28.68	19.74	31.64	46.47	28.12	19.05	31.62	45.72	41.41	32.48	45.63	58.55
GHT	37.40	27.77	41.66	**56.19**	27.40	18.08	30.76	45.76	41.50	30.79	46.85	62.73
TKGR-RHETNE	**37.50**	**28.23**	**41.95**	55.09	**29.26**	**19.80**	**33.03**	**47.66**	**43.30**	32.48	48.08	**64.83**

Table 2. Experimental results of entity prediction on YAGO and WIKI datasets. The best result in each column is boldfaced.

Models	YAGO				WIKI			
	MRR	Hits@1	Hits@3	Hits@10	MRR	Hits@1	Hits@3	Hits@10
TransE	48.97	36.19	62.45	66.05	46.68	46.23	49.71	51.71
Distmult	54.84	47.39	59.81	68.52	49.66	46.17	52.81	54.13
ComplEx	61.29	54.88	62.28	66.82	47.84	38.15	50.08	51.39
TTransE	31.19	18.12	40.91	51.21	29.27	21.67	34.43	42.39
TA-DistMult	54.92	48.15	59.61	66.71	44.53	39.92	48.73	51.71
DE-SimplE	54.91	51.64	57.30	60.17	45.43	42.60	47.71	49.55
TNTComplEx	57.98	52.92	61.33	66.69	45.03	40.04	49.31	52.03
CyGNet	52.07	45.36	56.12	63.77	33.89	29.01	36.10	41.86
TANGO-TuckER	62.50	58.77	64.73	68.63	51.60	49.61	52.45	54.87
TANGO-Distmult	63.34	60.04	65.19	68.79	53.04	51.52	53.84	55.46
RE-NET	58.02	53.06	61.08	66.29	49.66	46.88	51.19	53.48
RE-GCN	58.27	59.98	65.62	75.94	39.84	39.82	44.43	53.88
GHNN	62.79	60.29	63.93	66.85	-	-	-	-
GHT	64.60	57.67	69.39	76.95	58.85	55.22	61.53	64.72
TKGR-RHETNE	**79.19**	**74.14**	**83.71**	**85.76**	**66.17**	**62.69**	**68.03**	**72.22**

Timestamp Prediction. In the timestamp prediction task, our model exhibits mean absolute errors of 1.76, 14.88, and 4.88 on the ICEWS14, ICEWS18, and ICEWS05-15 datasets, respectively. These results surpass those of GHNN by margins of 4.33, 6.67, and 2.86, underscoring the enhanced capabilities of our model, which leverages the transformer for constructing the Hawkes process. Conversely, GHNN employs a continuous-time LSTM to gauge the intensity function. However, this approach neglects to directly integrate timestamp information and falls short in capturing intricate long-term relationships. Consequently, our model consistently outshines GHNN in the timestamp prediction task.

4.3 Ablation Study

In this subsection, we choose YAGO to investigate the effectiveness of different aggregation strategies for entity prediction in the aggregator for temporal neighborhood event context. Specifically, we separately experiment with the impact of using only the deep aggregation strategy, and the breadth aggregation strategy, and not using the aggregator for temporal neighborhood event context on the model performance. Table 3 shows the ablation results. As can be seen from the results in the table, using only the depth strategy to aggregate neighbor information results in a decrease of 3.07% on Hits@1. This is because the model lacks the capture of concurrent events in the temporal neighborhood event context or the potential impact of entities with similar behavior on the current entity. If only the breadth strategy is used to aggregate neighbor information, the model loses its ability to capture complex information for longer event sequences, resulting in a decrease of 1.68% in the results on Hits@1. Therefore, it is necessary to use both aggregation strategies to ensure the best model performance.

Table 3. Ablation study of our model on YAGO dataset.

Models	YAGO			
	MRR	Hits@1	Hits@3	Hits@10
TKGR-RHTNE-w/o - DNS & BNS	76.44	70.75	81.20	85.03
TKGR-RHTNE-w/o - DNS	77.74	72.46	82.48	84.91
TKGR-RHTNE-w/o - BNS	76.28	71.07	81.11	84.79
TKGR-RHTNE	**79.19**	**74.14**	**83.71**	**85.76**

5   Conclusion

In this paper, we introduce a novel TKGR model, named TKGR-RHETNE, considering the modeling in the context of relevant historical events and temporal neighborhood events. Utilizing an encoder built on the transformer Hawkes process and self-attention mechanism, this model can effectively grasp complex long-term event dependencies in the relevant historical event context, enabling continuous modeling of event evolution.

Additionally, employing a neighborhood aggregator designed to integrate the random walk strategy within the TKG topological structure, the model can learn the potential influence between events with insufficient historical information and other events for modeling the temporal neighborhood event context. Our experimental results on five popular datasets demonstrate the superior performance of our model on both entity prediction and timestamp prediction tasks. Future work includes incorporating logical rules into the proposed model to enhance its interpretability in the TKGR procedure.

Acknowledgements. This work was supported by the National Natural Science Foundation of China (Grant No. 62202075, No. 62171111, No. 62376058, No. 62376043, No. 62002052). the Natural Science Foundation of Chongqing, China (2022NSCQ-MSX3749), Sichuan Science and Technology Program (Grant No. 2022YFG0189), China Postdoctoral Science Foundation (Grant No. 2022M710614), Key Laboratory of Data Science and Smart Education, Hainan Normal University, Ministry of Education (Grant No. 2022NSCQ-MSX3749), Anhui Provincial Engineering Laboratory for Beidou Precision Agriculture Information (Grant No. BDSY2023004).

References

1. Barbosa, D., Wang, H., Yu, C.: Shallow information extraction for the knowledge web. In: ICDE, pp. 1264–1267 (2013)
2. Bordes, A., Usunier, N., Garcia-Duran, A., Weston, J., Yakhnenko, O.: Translating embeddings for modeling multi-relational data. In: NIPS, pp. 2787–2795 (2013)
3. Cho, K., et al.: Learning phrase representations using rnn encoder-decoder for statistical machine translation. arXiv preprint arXiv:1406.1078 (2014)
4. Deng, S., Rangwala, H., Ning, Y.: Dynamic knowledge graph based multi-event forecasting. In: SIGKDD, pp. 1585–1595 (2020)
5. Feng, F., He, X., Wang, X., Luo, C., Liu, Y., Chua, T.S.: Temporal relational ranking for stock prediction. ACM Trans. Inform. Syst. (TOIS) **37**(2), 1–30 (2019)
6. García-Durán, A., Dumančić, S., Niepert, M.: Learning sequence encoders for temporal knowledge graph completion. arXiv preprint arXiv:1809.03202 (2018)
7. Goel, R., Kazemi, S.M., Brubaker, M., Poupart, P.: Diachronic embedding for temporal knowledge graph completion. In: AAAI, vol. 34, pp. 3988–3995 (2020)
8. Han, Z., Ding, Z., Ma, Y., Gu, Y., Tresp, V.: learning neural ordinary equations for forecasting future links on temporal knowledge graphs. In: EMNLP, pp. 8352–8364 (2021)
9. Han, Z., Ma, Y., Wang, Y., Günnemann, S., Tresp, V.: Graph hawkes neural network for forecasting on temporal knowledge graphs. arXiv preprint arXiv:2003.13432 (2020)
10. Jin, D., et al.: Raw-gnn: Random walk aggregation based graph neural network. arXiv preprint arXiv:2206.13953 (2022)
11. Jin, W., Qu, M., Jin, X., Ren, X.: Recurrent event network: Autoregressive structure inference over temporal knowledge graphs. arXiv preprint arXiv:1904.05530 (2019)
12. Jung, J., Jung, J., Kang, U.: Learning to walk across time for interpretable temporal knowledge graph completion. In: SIGKDD, pp. 786–795 (2021)
13. Lacroix, T., Obozinski, G., Usunier, N.: Tensor decompositions for temporal knowledge base completion. arXiv preprint arXiv:2004.04926 (2020)
14. Leblay, J., Chekol, M.W.: Deriving validity time in knowledge graph. In: The Web Conference, pp. 1771–1776 (2018)
15. Li, Z., et al.: Temporal knowledge graph reasoning based on evolutional representation learning. In: SIGIR, pp. 408–417 (2021)

16. Liu, M., Liu, Y.: Inductive representation learning in temporal networks via mining neighborhood and community influences. In: SIGIR, pp. 2202–2206 (2021)

17. Mahdisoltani, F., Biega, J., Suchanek, F.: Yago3: a knowledge base from multilingual wikipedias. In: 7th Biennial Conference on Innovative Data Systems Research. CIDR Conference (2014)

18. Mei, H., Eisner, J.M.: The neural hawkes process: a neurally self-modulating multivariate point process. In: NeurIPS 30 (2017)

19. Nguyen, G.H., Lee, J.B., Rossi, R.A., Ahmed, N.K., Koh, E., Kim, S.: Continuous-time dynamic network embeddings. In: The Web Conference, pp. 969–976 (2018)

20. Shchur, O., Türkmen, A.C., Januschowski, T., Günnemann, S.: Neural temporal point processes: a review. arXiv preprint arXiv:2104.03528 (2021)

21. Sun, H., Geng, S., Zhong, J., Hu, H., He, K.: Graph hawkes transformer for extrapolated reasoning on temporal knowledge graphs. In: EMNLP, pp. 7481–7493 (2022)

22. Trivedi, R., Dai, H., Wang, Y., Song, L.: Know-evolve: deep temporal reasoning for dynamic knowledge graphs. In: ICML, pp. 3462–3471. PMLR (2017)

23. Trivedi, R., Farajtabar, M., Biswal, P., Zha, H.: Dyrep: learning representations over dynamic graphs. In: ICLR (2019)

24. Trouillon, T., Welbl, J., Riedel, S., Gaussier, É., Bouchard, G.: Complex embeddings for simple link prediction. In: ICML, pp. 2071–2080 (2016)

25. Vaswani, A., et al.: Attention is all you need. In: NeurIPS, vol. 30 (2017)

26. Wang, S., Cai, X., Zhang, Y., Yuan, X.: CRNet: modeling concurrent events over temporal knowledge graph. In: Sattler, U., et al. (eds.) The Semantic Web – ISWC 2022: 21st International Semantic Web Conference, Virtual Event, October 23–27, 2022, Proceedings, pp. 516–533. Springer International Publishing, Cham (2022). https://doi.org/10.1007/978-3-031-19433-7_30

27. Yang, B., Yih, W.t., He, X., Gao, J., Deng, L.: Embedding entities and relations for learning and inference in knowledge bases. https://arxiv.org/abs/1412.6575 (2014)

28. Zhang, Q., Lipani, A., Kirnap, O., Yilmaz, E.: Self-attentive hawkes process. In: ICML, pp. 11183–11193. PMLR (2020)

29. Zhu, C., Chen, M., Fan, C., Cheng, G., Zhang, Y.: Learning from history: modeling temporal knowledge graphs with sequential copy-generation networks. In: AAAI. vol. 35, pp. 4732–4740 (2021)

30. Zuo, S., Jiang, H., Li, Z., Zhao, T., Zha, H.: Transformer hawkes process. In: ICML, pp. 11692–11702. PMLR (2020)

High-Resolution Self-attention with Fair Loss for Point Cloud Segmentation

Qiyuan Liu[1], Jinzheng Lu[2], Qiang Li[2(✉)], and Bingsen Huang[1]

[1] School of Computer Science and Technology, Southwest University of Science and Technology, Mianyang, China
[2] School of Information Engineering, Southwest University of Science and Technology, Mianyang, China
liqiangsir@swust.edu.cn

Abstract. Applying deep learning techniques to analyze point cloud data has emerged as a prominent research direction. However, the insufficient spatial and feature information integration within point cloud and unbalanced classes in real-world datasets have hindered the advancement of research. Given the success of self-attention mechanisms in numerous domains, we apply the High-Resolution Self-Attention (HRSA) module as a plug-and-play solution for point cloud segmentation. The proposed HRSA module preserve high-resolution internal representations in both spatial and feature dimensions. Additionally, by affecting the gradient of dominant and weak classes, we introduce the Fair Loss to address the problem of unbalanced class distribution on a real-world dataset to improve the network's inference capabilities. The introduced modules are seamlessly integrated into an MLP-based architecture tailored for large-scale point cloud processing, resulting in a new segmentation network called PointHR. PointHR achieves impressive performance with mIoU scores of 69.8% and 74.5% on S3DIS Area-5 and 6-fold cross-validation. With a significantly smaller number of parameters, these performances make PointHR highly competitive in point cloud semantic segmentation.

Keywords: 3D point cloud · Semantic segmentation · Self-attention · Loss function

1 Introduction

As the 3D data sensors have grown by leaps and bounds over the past few years, point cloud understanding tasks, especially semantic segmentation, have recently become a primary focus of computer vision research. Unlike the regular arrangement of pixels in images, the point cloud independently expresses each point's spatial and feature information in the metric space without a significant connection [1]. Therefore, exploring advanced 3D point cloud processing strategies is urgent.

PointNet [2], the first study directly handled the original point cloud, introduced MLPs and symmetric functions to process point clouds, while PointNeXt

---

The original version of this chapter was revised: In the PDF version of this chapter, both Fig. 4 and Fig. 5 erroneously showed the same figure. A correction to this chapter can be found at https://doi.org/10.1007/978-981-99-8073-4_46

© The Author(s), under exclusive license to Springer Nature Singapore Pte Ltd. 2024, corrected publication 2024
B. Luo et al. (Eds.): ICONIP 2023, LNCS 14451, pp. 344–356, 2024.
https://doi.org/10.1007/978-981-99-8073-4_27

[3] improved upon PointNet's [4] effectiveness and achieved top performance using diverse techniques. Moreover, plenty of networks [5–10] deliver remarkable results. However, these methods have shown limited capability in segmenting small objects, also fail to balance the concentration between global features and local features.

Additionally, our empirical investigation reveals significant differences in the distribution of the six categories, representing distinct parts of rooms, in the Stanford Large-Scale 3D Indoor Spaces (S3DIS) dataset [11]. The categories of floor and ceiling contain a large number of points, whereas the other classes have a smaller number of points. This imbalanced class distribution poses substantial challenges during model training and evaluation, necessitating the use of an appropriate loss function to account for this problem.

To handle the above problems, we introduce a novel segmentation network, PointHR. First, we design a plug-and-play module named High Resolution Self-Attention (HRSA) to optimize the spatial and feature branches extraction. To alleviate the potential loss of detailed information due to pooling and down-sampling, the HRSA preserves high-resolution internal representations in both spatial and feature dimensions during attention computation. Each branch of HRSA can maintain complete spatial and feature information while compressing the other dimension efficiently. Besides, we introduce Fair Loss with two complementary factors to help the network better dynamically balance the weights between dominant and weak classes in datasets. The experimental results and ablation study on real-world dataset S3DIS [11] demonstrate that our method achieves substantial and consistent improvements compared to several baseline approaches. Notably, PointHR's performance is almost on par with the state-of-the-art (SOTA) method PointNeXt-XL [3], which has a much larger network size (41.6M Params., 84.8G FLOPS), whereas PointHR only requires a **significantly smaller network size** (8.5M Params., 15.7G FLOPS).

In summary, our contributions are as follows:

- We introduce a plug-and-play HRSA module to boost baselines, which preserves high-resolution internal representations in both spatial and feature channels without adding more parameters and computational complexity.
- We design the improved loss function named Fair Loss by adding stabilization and enhancement factors to fine-tune sample weights for dominant and weak classes.
- We conduct extensive experiments and ablation studies to demonstrate the performance boosted by the modification.

2   Related Work

Many pioneers have made bold attempts to analyze and process raw point cloud data. Currently, three existing 3D point cloud segmentation learning-based paradigms exist: intermediate representation-based methods, point-based methods, and Self-Attention or Transfomer-based methods.

Intermediate representation-based methods [12–16] mostly convert the point cloud into images or voxels for processing, which will inevitably lose detailed information on a point cloud structure.

PointNet [2], a pioneer in learning features directly from point clouds, takes raw point data as input. Qi *et al.* [4] proposed PointNet++ with Farthest Point Sampling (FPS) module to sample the input point cloud and divide the areas. Based on PointNet++, PointNeXt [3] outperformed SOTA point-based method through data enhancement, optimization technology, and network structure improvement.

To further improve the performance of the segmentation network, Guo *et al.* [8] presented a Transformer based architecture, which sharpens the attention weights and reduces the influence of noise [17]. Point Transformer [9] achieves more comparative results using a vector self-attention operator to aggregate local features. Stratified Transformer [10] processes long and short-range relationships among points. Nevertheless, the vast number of parameters and high model complexity hinder the development of transformer-based methods.

In summary, existing networks inspire progress in point cloud segmentation but lose detailed spatial and feature information during encoder compression and decoder upsampling. Our study presents a solution by promoting the comprehensive integration of spatial and feature information while simultaneously optimizing the learning of under-segmented classes.

3   Method

3.1   Segmentation Pipeline

As shown in the Fig. 1, the general pipeline for point cloud semantic segmentation involves utilizing DNN [18] to generate feature maps by learning point clouds' position and feature information. Subsequently, the decoder conducts interpolation to increase the resolution of the feature maps, followed by assigning the predicted semantic labels to each point. The unordered point set $P = \{p_i \in \mathbb{R}^{N \times D}, i = 1, \ldots, N\}$ contains N points, each with D features describing its properties. At each local area centered at point p_i, we identify the K nearest points $\{p_{i,1}, p_{i,2}, \ldots, p_{i,k}\}$ within a specified radius. To describe the integration of the subsampling layer, grouping layer, MLPs, and reduction layer, we define a function that maps the k nearest neighbors around a center point p_i to a vector, as follows:

$$F = \Gamma\Big(\text{MLP}\big(\sum_{k=1}^{K} \sigma(s_{i,k}, s_i, f_{i,k}, f_i)\big)\Big), \tag{1}$$

where F is the resulting feature map. Γ is the reduction layer that integrates features in different local areas. The function of σ () consists of a sampling and grouping operation for efficient local feature extraction of point clouds. f_i and $f_{i,k}$ are the features of point p_i and its neighbor, while s_i and $s_{i,k}$ denote their corresponding spatial coordinates.

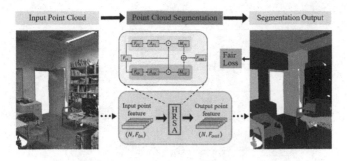

Fig. 1. The segmentation pipeline of our PointHR.

The resulting feature map F_{in} is inputted into the HRSA module, which adaptively captures the localized geometric and feature related attributes. HRSA generates a new feature map F_{out} with the exact dimensions. Section 3.2 provides comprehensive information about the HRSA module. We adopt a symmetric decoder similar to the one utilized in PointNet++ [4].

3.2 High Resolution Self-attention Module

As illustrated in Fig. 2, the HRSA aims to dynamically create a strong correlation between spatial and feature information, and squeeze out the information of the two branches as much as possible.

The HRSA module takes an input tensor F_{in} from the preceding layer to adaptively enhance or suppress features, which resembles the image capture process. Inspired by photography's compensation for underexposed regions, we introduce HRSA, a novel self-attention mechanism that includes two parallel high-resolution feature-only and spatial-only branches. In general, HRSA: **(i)** fully preserves high-resolution information from a specific channel (e.g., spatial channel in this scenario) while folding all information from the other channel (i.e., feature channel), respectively, thus keeping the high internal resolution of this specific channel (i.e., spatial channel) and reducing information loss, **(ii)** assigns the channel-specific attention distribution to the entire scene, facilitating better representation in point cloud segmentation, **(iii)** utilizes a $Softmax - Sigmoid$ combination on both channels to focus on a broader range of features and enable non-linear mapping.

Formally, given an input point features $F_{in} \in \mathbb{R}^{N \times (s+d)}$, which can be partitioned into two sub-tensors F_{ft} and F_{sp}, as illustrated in Fig. 2. Subsequently, we pass F_{ft} and F_{sp} into corresponding branch.

Feature-only branch:

$$\mathbf{M}_{ft} = f_{sg}\left[\mathbf{W}_Z\left((\mathbf{F}_{ft}\mathbf{W}_V) \times f_{sm}(\mathbf{F}_{ft}\mathbf{W}_Q)\right)\right] \odot F_{in}, \qquad (2)$$

where $\mathbf{W}_Q$, $\mathbf{W}_V$, and $\mathbf{W}_Z$ are 1*1 convolution layers. $f_{sm}()$ is a $Softmax$ operator, $f_{sg}()$ is a $Sigmoid$ function, $\times$ is the matrix multiplication operator.

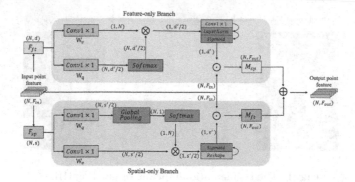

Fig. 2. Detailed structure design of HRSA module.

$\odot$ is a matrix dot-product operator. $\mathbf{M}_{ft} \in \mathbb{R}^{N \times (s' + d')}$ is the output of feature-only branch.

Within the branch, we first use two 1×1 convolutions to convert the input feature F_{ft} into Q and V, respectively. The channel of V is entirely compressed, whereas the channel dimension of Q remains a high internal resolution of $N \times d'/2$ with nearly full of feature information. Afterward, we use *Softmax* normalization to enhance the information of Q. Then Q and V are matrix multiplied, followed by 1×1 convolution and layer norm to increase the dimension of feature channel $d'/2$ to d'. Finally, we use the *Sigmoid* function to normalize the parameters and multiply them with the original input feature channel to complete the attention distribution.

Spatial-only branch:

$$\mathbf{M}_{sp} = f_{sg} \left[f_{sm} \left(f_{gp} \left(\mathbf{F}_{sp} \mathbf{W}_Q \right) \right) \times \mathbf{F}_{sp} \mathbf{W}_V \right] \odot F_{in}, \tag{3}$$

where $\mathbf{W}_Q$ and $\mathbf{W}_V$ are 1×1 convolution layers. $f_{gp}()$ is the global pooling function. $\mathbf{M}_{sp} \in \mathbb{R}^{N \times (s' + d')}$ is the output of spatial-only branch with spatial attention distribution.

The computational flow of the spatial-only attention mechanism is similar to feature-only attention. However, there is a crucial difference in which we incorporate global pooling to compress the spatial dimension. The spatial dimension of V is kept at a higher resolution to ensure that fine spatial details are preserved. This operation indicates that the similarity between any two points is determined exclusively by their spatial distance.

Integration: We exploit +, an element-wise addition operator, to integrate two attention branches. The integration is defined as such:

$$HRSA(\mathbf{F}) = \mathbf{M}_{ft} + \mathbf{M}_{sp} \tag{4}$$

3.3   Fair Loss Function

Most existing frameworks use the Cross-Entropy (CE) Loss function for semantic segmentation of 3D point clouds. However, the uneven distribution of class labels in the S3DIS [11] dataset causes these networks to perform poorly in segmentation tasks. Specifically, instances of structural elements such as ceilings, floors, walls, and doors, which we refer to as dominant classes, are prevalent in the S3DIS dataset. Instances of objects and furniture such as chairs, sofas, and bookshelves, which we refer to as weak classes, are relatively rare. The dominant class usually has high predicted probabilities, resulting in small gradients and thus delaying the learning of the minority class.

To tackle the problem, we propose an improved loss function named Fair Loss by **introducing two balancing factors applying to the gradient and loss function computation**: stabilization factor $St_{c,d}$ and enhancement factor $En_{c,d}$. The stabilization factor is defined as follows:

$$St_{c,d} = \left(\frac{\chi_d}{\chi_c}\right)^h \text{ if } \chi_c > \chi_d, \tag{5}$$

where χ_c and χ_d represents the number of points belonging to category c and d in the dataset, respectively. h is a hyper-parameter used to adjust the magnitude of stability.

The Stabilization Factor balances class distribution by adjusting the weights between classes, which are dynamically calculated by accumulating the number of points in each category. This factor reduces the penalty through hyper-parameter h for weaker classes when another category has significantly more number of samples, mitigating unfair punishment during optimization.

The enhancement factor shares a similar design idea, which is defined according to the probability ratio of each point's prediction:

$$En_{c,d} = \left(\frac{\sum_{c=1}^{C} \phi_c}{\phi_d}\right)^{\frac{2}{h}} \text{ if } \sum_{c=1}^{C} \phi_c > \phi_d, \tag{6}$$

where c and d represent two different categories, with d being the ground truth label. ϕ_d represents the probability that the prediction is true, and $\sum_{c=1}^{C} \phi_c$ represents the sum of the probabilities predicted to be other classes.

We design an enhancement factor to raise the penalties for a specific class. We do so because when the stabilization factor balances the weight between dominant and weaker classes, it may reduce the penalty for the mis-segmentation of weaker classes. Through comprehensive ablation studies, we found the optimal solution of the hyper-parameter $\frac{2}{h}$ to adjusts the degree of enhancement.

We classify points who labeled with a specific semantic category as positive samples and those not labeled with that class as negative samples. From the perspective of the weaker class, (8) demonstrates that the gradient from the dominant class is considerable.

We propose the Fair Loss, which incorporates two factors into the CE Loss function to modify the gradient computation. We define the formula as follows:

$$L_{\text{fair}} = -\frac{1}{N} \sum_{i=1}^{N} \sum_{c=1}^{C} y_{i,c} \log(\widehat{q}_{i,c})$$

$$\text{with } \widehat{q}_{i,c} = \frac{\exp(s_{i,c})}{\sum_{d \neq c}^{C} T_{i,c} \exp(s_{i,d}) + \exp(s_{i,c})},$$

(7)

where C is the semantic label. $y_{i,c}$ is the one-hot ground truth label to indicate if class label c is the correct classification for point i. $q_{i,c}$ is the predicted probability made by the classifier. $s_{i,c}$ represents the score of point i belonging to category c. Besides, $T_{i,c} = \text{St}_{c,d} \cdot \text{En}_{c,d}$.

In this way, for a point i belonging to class c, the gradient of Fair Loss value L_{fair} to the score of its negative sample $s_{i,d}$ is:

$$\frac{\partial L_{\text{fair}}}{\partial s_{i,d}} = \frac{\exp(s_{i,c})}{T_{i,c}(\exp(s_{i,c}) + \exp(s_{i,d}))}$$

$$= T_{i,c} \cdot \frac{\widehat{q}_{i,c} \cdot \exp(s_{i,d})}{\exp(s_{i,c})}.$$

(8)

$T_{i,c}$ as the product of stabilization factor and enhancement factor, not only incorporating global category distribution and local prediction rate comparison but also addressing the misclassification of underrepresented categories by modifying the gradient of negative samples. Our experiments confirm that the Fair Loss consistently improves accuracy compared to the CE Loss.

4   Experiments and Evaluation

This section presents a comprehensive evaluation of our proposed network PointHR on the S3DIS [11] dataset, demonstrating improved performance with only a slight increase in computation (Floating-Point Operations Per Second, FLOPS) and the number of parameters (Params.).

We implement our network using Pytorch [19] framework (version 1.10.1) with CUDA version 11.3. We use a single NVIDIA RTX 3090 (24GB) to complete our experiments. Our implementation is based on the work of Guo et $al.$ [3] and follows the same data preprocessing steps. We train the PointHR for 100 epochs with an initial learning rate of 0.01, using a batch size of 8 with a fixed number of points (24,000) per batch. Average results in three random runs are recorded.

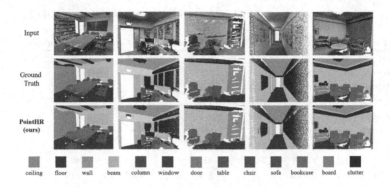

Fig. 3. Visualization of semantic segmentation results on the S3DIS ch27b16.

4.1 Semantic Segmentation in S3DIS

Table 1. Results on S3DIS, evaluated on Area-5. Evaluation metrics including OA (%), mACC (%), mIoU (%), and each individual class IoU (%)

Method	OA (%)	mAcc (%)	mIoU (%)	ceiling	floor	wall	beam	column	window	door	table	chair	sofa	bookcase	board	clutter
PointNet [2]	-	49.0	41.1	88.8	97.3	69.8	0.1	3.9	46.3	10.8	59.0	52.6	5.9	40.3	26.4	33.2
SegCloud [15]	-	57.4	48.9	90.1	96.1	69.9	0.0	18.4	38.4	23.1	70.4	75.9	40.9	58.4	13.0	41.6
PointCNN [20]	85.9	63.9	57.3	92.3	98.2	79.4	0.0	17.6	22.8	62.1	74.4	80.6	31.7	66.7	62.1	56.7
PAT [21]	-	70.8	60.1	93.0	98.5	72.3	1.0	41.5	85.1	38.2	57.7	83.6	48.1	67.0	61.3	33.6
KPConv [22]	-	72.8	67.1	92.8	97.3	82.4	0.0	23.9	58.0	69.0	81.5	91.0	75.4	75.3	66.7	58.9
PointNet++ [3]	87.5	70.1	63.0	93.0	98.0	83.1	0.0	31.0	41.8	62.1	78.0	86.1	60.0	69.8	65.8	51.7
PointNeXt-L [3]	90.0	75.1	69.0	94.4	98.4	83.9	0.0	31.7	60.3	72.1	82.2	90.1	75.8	73.9	75.9	58.2
PointHR (ours)	**90.3**	**75.6**	**69.8**	94.5	98.5	84.0	0.0	32.9	57.9	74.1	82.6	91.3	79.8	76.1	76.2	59.2

PointHR Comparision. S3DIS [11] is a vast collection of indoor 3D point clouds created by Stanford University. We employ 6-fold cross-validation on the six regions. For validation, we exclusively use Area-5. We utilize mean class-wise Intersection over Union (mIoU), mean of classwise Accuracy (mAcc), and pointwise Overall Accuracy (OA) as our evaluation metrics.

Table 1 presents the performance comparison results of various segmentation methods, where PointHR outperforms several networks. PointHR attains an OA/mAcc/mIoU of 90.3%/75.6%/69.8% on Area 5, closing to the 70% mIoU benchmark. Table 2 shows the segmentation results and the computational complexity index in the 6-fold cross-validation on the S3DIS Dataset. PointHR achieves an impressive accuracy of 74.5%, which is very close to that of PointNeXt-XL (41.6M Params., 84.8G FLOPS) while with significantly smaller network size (8.5M Params., 15.7G FLOPS).

Table 2. Results of 6-fold cross-validation experiment on S3DIS

Method	OA (%)	mIoU (%)	Params. (M)	FLOPS (G)	Throughput (ins./sec.)
PointNet [2]	78.5	47.6	3.6	35.5	162
PointNet++ [4]	81.0	54.5	1.0	7.2	186
PointCNN [20]	88.1	65.4	0.6	-	-
KPConv [22]	-	70.6	15.0	-	30
RandLA-Net [7]	88.0	70.0	1.3	5.8	159
Point Transformer [9]	90.2	73.5	7.8	5.6	34
PointNeXt-L [3]	89.8	73.9	7.1	15.2	115
PointNeXt-XL [3]	90.3	**74.9**	41.6	84.8	46
PointHR (ours)	**90.3**	74.5	8.5	15.7	116

Figure 3 depicts the predictions made by PointHR, which indicate great proximity to the ground truth. Figure 4 shows that PointHR can better segment object boundary regions than the baseline network. The network demonstrates an outstanding ability to determine intricate semantic features in complex 3D environments, such as the intersection of walls and doorways, the leg of the tables, and chairs.

Table 3. HRSA vs. baselines for semantic segmentation on the S3DIS dataset, evaluated on Area-5

Method	OA (%)	mIoU (%)	Params. (M)	FLOPS (G)
PointNet++ [4]	87.5	63.2	1.0	7.2
+ **HRSA**	89.0(+1.5)	64.7(+1.5)	1.6	7.3
PointNeXt-B [3]	89.4	67.3	3.8	8.9
+ **HRSA**	90.0(+0.6)	67.8(+0.5)	4.5	9.2
PointNeXt-L [3]	90.0	69.0	7.1	15.2
+ **HRSA**	**90.2(+0.2)**	**69.7 (+0.7)**	8.5	15.7

Plug-and-Play HRSA. We evaluate the effectiveness and robustness of the proposed HRSA module by adding it to several benchmark networks, including modernized PointNet++ [3,4], PointNeXt-B [3], and PointNeXt-L [3]. All experiments are performed on S3DIS Area-5. As shown in Table 3, HRSA consistently improves the baseline networks by 0.7% to 1.5% mIoU while maintaining roughly the same computational complexity.

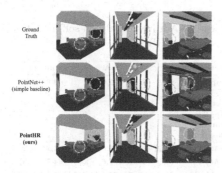

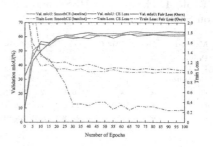

Fig. 4. Visual comparison. The yellow circles highlight where our network performs better. (Color figure online)

Fig. 5. Segmentation results for various loss functions on the S3DIS dataset during the training phase.

Loss Function Comparison. To assess the effectiveness of our proposed Fair Loss, we conduct a comparative study with two commonly employed loss functions, namely CE Loss and CE Loss with label smoothing [23]. The benchmarking network employed for our study is the modernized PointNet++ [4]– [3]. The results are illustrated in Fig. 5, where the solid line denotes the mIoU performance, and the dashed line denotes the loss value. The results indicate that Fair Loss exhibits robust and sustained learning capability. The dynamic adjustment of the gradient through the stabilization and enhancement factors accelerates the convergence rate, leading to continued improvement in mIoU even in the later stages of training. Ultimately, our proposed Fair Loss outperforms other commonly used loss functions.

4.2   Ablation Study

We conduct ablation studies on Area-5 S3DIS [11] to assess the efficacy of individual components in the PointHR architecture. The outcomes of these experiments shed light on the extent to which each component contributes towards the overall performance of PointHR.

Individual Components. The contribution of each individual module in the PointHR network is presented in Table 4. To evaluate their impact, we perform ablation studies by adding these modules incrementally to the original PointNeXt-L [3] network. Method A only adds the HRSA module, and Method B replaces CE Loss with label smoothing to Fair Loss. Method C is the PointHR proposed in this paper, which combines two modules simultaneously. The results show that both the HRSA module and Fair Loss can improve the segmentation performance of the network. Despite the competitive performance of the baseline network, which ranks among the top three networks in terms of mIoU on the

Table 4. The effect of HRSA and Fair Loss on S3DIS Area-5

Method	HRSA	Fair Loss	OA (%)	mIoU (%)
PointNeXt-L (Strong Baseline)	✗	✗	90.0	69.0
A	✔	✗	90.2(+0.2)	69.7(+0.7)
B	✗	✔	90.1(+0.1)	69.5(+0.5)
C	✔	✔	**90.3(+0.3)**	**69.8(+0.8)**

Table 5. Additive study on self-attention branches of HRSA

Method	OA (%)	mAcc (%)	mIoU (%)
Improved PointNet++ [4]– [3] (Simple Baseline)	87.5	70.6	63.2
$+M_{sp}$	88.2(+0.7)	71.6(+1.0)	64.3(+1.2)
$+M_{ft}$	89.0(+1.5)	71.6(+1.0)	64.7(+1.5)
$+(M_{sp} + M_{ft})$	**89.1(+1.6)**	**71.7(+1.1)**	**65.4(+2.2)**

S3DIS, PointHR further enhances the network to near 70% mIoU and approaches the performance of the computationally intensive PointNeXt-XL [3] network.

HRSA Ablation. We then conduct experiments to evaluate the effectiveness of spatial-only attention M_{sp} and feature-only attention M_{ft} on the baseline network, as well as the performance of the network with both HRSA and residual connection. The results are presented in Table 5. We find that both M_{sp} and M_{ft} improved the baseline network, but the feature-only attention mechanism has a more significant impact on performance. Additionally, we attempt to add a residual connection at the output of the HRSA module to preserve more original information, but the improvement is negligible.

Table 6. Ablation Study of the hyper-parameter h in Fair Loss on S3DIS Area-5

h	$\frac{2}{h}$	OA (%)	mAcc (%)	mIoU (%)
0.3	6.5	88.0(+0.5)	70.1	63.6(+0.4)
0.4	5.0	88.5(+1.0)	72.4	65.0(+1.8)
0.6	3.0	88.3(+0.8)	71.5	64.7(+1.5)
0.8	**2.5**	**88.8(+1.3)**	72.4	**65.3(+2.1)**
1.0	2.0	88.8(+1.3)	71.6	64.6(+1.4)
2.0	1.0	88.3(+0.8)	73.4	64.2(+1.0)

Hyper-parameter h in Fair Loss. As shown in Table 6, we proceed with an ablation study on the impact of the hyper-parameter h in Fair Loss on the category-imbalanced dataset S3DIS Area-5, intending to find an optimal value. In Fair Loss, h balances the category distribution by controlling the stabilization factor, thereby reducing the penalty for weaker classes. Additionally, $\frac{2}{h}$ is a complementary hyper-parameter that enhances the punishment for mis-segmentation of weak classes by controlling the enhancement factor. Thus, finding an appropriate value for h is crucial. We use improved PointNet++ proposed by Guo *et al.* [3] as the baseline network. Our experimental findings indicate that the optimal performance is attained when the hyper-parameter h is set to 0.8, resulting in a 2.1% improvement in mIoU performance compared to the baseline network.

5 Conclusion and Future Work

In this paper, we propose two novel techniques to enhance the performance of point cloud semantic segmentation. The first contribution is the introduction of a HRSA module, which preserves high-resolution internal representations in both spatial and feature dimensions during attention computation without adding more parameters and computational complexity. The second contribution is the development of an improved Fair Loss function, which leverages the idea of penalty balancing and enhancement to address the class imbalance distribution in point cloud segmentation.

Experimental results on the challenging S3DIS datasets demonstrate that our proposed PointHR architecture achieves comparative performance in terms of mIoU metric. Specifically, the accuracy of 74.5% mIoU is obtained in the 6-fold cross-validation on S3DIS. This performance is close to the state-of-the-art methods, while offering a significant computational advantage in terms of efficiency and computational complexity. In future work, we plan to extend these two modules to other areas to verify their generalization, such as shape classification or part segmentation. We also plan to optimize under-segmented regions' performance, such as object boundary regions.

Acknowledgement. This work was supported in part by the Heilongjiang Provincial Science and Technology Program under Grant 2022ZX01A16, and in part by the Sichuan Science and Technology Program under Grant 2022YFG0148.

References

1. Guo, Y., Wang, H., Hu, Q., Liu, H., Liu, L., Bennamoun, M.: Deep learning for 3D point clouds: a survey. IEEE Trans. Pattern Anal. Mach. Intell. **43**(12), 4338–4364 (2021)
2. Qi, C.R., Su, H., Mo, K., Guibas, L.J.: PointNet: deep learning on point sets for 3D classification and segmentation. In: Proceedings of the IEEE Conference Computer Vision Pattern Recognition (CVPR), vol. 1, p. 4 (2017)
3. Qian, G., et al.: PointNeXt: revisiting PointNet++ with improved training and scaling strategies(2022). arXiv:2206.04670
4. Qi, C.R., Yi, L., Su, H., Guibas, L.J.: "PointNet++: deep hierarchical feature learning on point sets in a metric space. In: Advances in neural information processing systems , pp. 5099–5108 (2017)
5. Zhao, H., Jiang, L., Fu, C.W., Jia, J.: Pointweb: enhancing local neighborhood features for point cloud processing. In: Proeedings of the IEEE Conference Computer Vision Pattern Recognition (CVPR), pp. 5565–5573 (2019)
6. Zhang, Y., Hu, Q., Xu, G., Ma, Y., Wan, J., Guo, Y.: Not all points are equal: learning highly efficient point-based detectors for 3D LiDAR point clouds. In: IEEE/CVF Conference Computer Vision Pattern Recognition (CVPR), New Orleans, LA, USA, 2022, pp. 18931–18940 (2022)

7. Hu, Q., et al.: RandLA-net: efficient semantic segmentation of large-scale point clouds. In: Proceedings of IEEE/CVF Conference Computer Vision Pattern Recognition (CVPR), pp. 11108–11117 (2020)

8. Guo, M.-H., et al.: Point cloud transformer. Comput. Vis. Media **7**(2), 187–199 (2021)

9. Zhao, H., Jiang, L., Jia, J., Torr, P.H., Koltun, V., et al.: Point transformer. In: Proceedings International Conference Computer Vision, pp. 16259–16268 (2021)

10. Lai, X., et al.: Stratified transformer for 3D point cloud segmentation. In: Proceedings of the IEEE/CVF International Conference on Computer Vision (CVPR), pp. 8500–8509 (2022)

11. Armeni, I., et al.: 3D semantic parsing of large-scale indoor spaces. In: Proceedings IEEE Conerence Computer Vision Pattern Recognition, pp. 1534–1543 (2016)

12. Boulch, A., Guerry, J., Le Saux, B., Audebert, N.: SnapNet: 3D point cloud semantic labeling with 2D deep segmentation networks. Comput. Graph. **71**, 189–198 (2018)

13. Wu, B., Wan, A., Yue, X., Keutzer, K.: SqueezeSeg: convolutional neural nets with recurrent CRF for real-time road-object segmentation from 3D LiDAR point cloud. In: Proceedings IEEE International Conference on Robotics Automation (ICRA), Brisbane, QLD, Australia, 2018, pp. 1887–1893 (2018)

14. Wu, B., Zhou, X., Zhao, S., Yue, X., Keutzer, K.: SqueezeSegV2: improved model structure and unsupervised domain adaptation for road-object segmentation from a lidar point cloud. In: Proceedings IEEE International Conference on Robotics and Automation (ICRA), Montreal, QC, Canada, 2019, pp. 4376–4382 (2019)

15. Tchapmi, L., Choy, C., Armeni, I., Gwak, J., Savarese, S.: SEGCloud: semantic segmentation of 3d point clouds. In: Proceedings Inernational Conference 3D Vision (3DV), Qingdao, China, 2017, pp. 537–547 (2017)

16. Liu, Z., Tang, H., Lin, Y., Han, S.: Point-voxel CNN for efficient 3D deep learning. In: Advances in Neural Information Processing Systems, Cambridge, MA, USA:MIT Press, 2019, pp. 965–975 (2019)

17. Vaswani, A., et al.: Attention is all you need. In: Proceedings of the 31st International Conference Neural Information Processing System, pp. 5998–6008 (2017)

18. LeCun, Y., Bengio, Y., Hinton, G.: Deep learning. Nature **521**(7553), 436–444 (2015)

19. Paszke, A., et al.: PyTorch: an imperative style high-performance deep learning library. In: Proceedings Advance Neural Information Processing System, pp. 8026–8037 (2019)

20. Lin, Y., et al.: PointCNN: Convolution on χ -transformed points. In: Proceedings of the IEEE/CVF Conference on Computer Vision and Pattern Recognition, pp. 826–836 (2018)

21. Yang, J., et al.: Modeling point clouds with self-attention and gumbel subset sampling. In: Proceedings of the IEEE/CVF Conference Computer Vision Pattern Recognition, pp. 3323–3332 (2019)

22. Thomas, H., Qi, C.R., Deschaud, J.E., Marcotegui, B., Goulette, F., Guibas, L.J.: KPConv: flexible and Deformable Convolution for Point Clouds. In: Proceedings of IEEE International Conference Computer Vision (ICCV), 2019, pp. 6411–6420 (2019)

23. Szegedy, C., Vanhoucke, V., Ioffe, S., Shlens, J., Wojna, Z.: Rethinking the inception architecture for computer vision. In: Proceedings of the IEEE Conference Computer Vision Pattern Recognition, pp. 2818–2826. (2016)

Transformer-Based Video Deinterlacing Method

Chao Song[1], Haidong Li[1], Dong Zheng[2], Jie Wang[1], Zhaoyi Jiang[1],
and Bailin Yang[1(✉)]

[1] School of Computer Science and Technology, Zhejiang Gongshang University,
Hangzhou 310018, China
ybl@zjgsu.edu.cn
[2] UNIUBI Research, Qingdao, China
zhengdong@uni-ubi.com

Abstract. Deinterlacing is a classical issue in video processing, aimed at generating progressive video from interlaced content. There are precious videos that are difficult to reshoot and still contain interlaced content. Previous methods have primarily focused on simple interlaced mechanisms and have struggled to handle the complex artifacts present in real-world early videos. Therefore, we propose a Transformer-based method for deinterlacing, which consists of a Feature Extractor, a De-Transformer, and a Residual DenseNet module. By incorporating self-attention in Transformer, our proposed method is able to better utilize the inter-frame movement correlation. Additionally, we combine a properly designed loss function and residual blocks to train an end-to-end deinterlacing model. Extensive experimental results on various video sequences demonstrate that our proposed method outperforms state-of-the-art methods in different tasks by up to 1.41~2.64dB. Furthermore, we also discuss several related issues, such as the rationality of the network structure. The code for our proposed method is available at https://github.com/Anonymous2022-cv/DeT.git.

Keywords: Deinterlacing · Transformer · Spatial information

1 Introduction

Deinterlacing is a common issue in old video processing caused by the use of interlaced scan equipment. Many valuable videos are already interlaced and cannot

The authors are grateful to Zhejiang Gongshang University for their valuable computing resources and outstanding laboratory facilities, as well as the support from the Zhejiang Provincial Natural Science Foundation of China(Grant No. LY22F020013), National Natural Science Foundation of China (Grant No. 62172366), "Pioneer" and "Leading Goose" R & D Program of Zhejiang Province (2023C01150), and "Digital+" Discipline Construction Project of Zhejiang Gongshang University (Grant No. SZJ2022B009).

B. Luo et al. (Eds.): ICONIP 2023, LNCS 14451, pp. 357–369, 2024.
https://doi.org/10.1007/978-981-99-8073-4_28

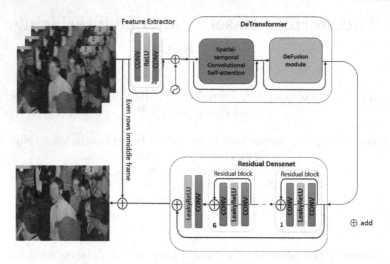

Fig. 1. De-Transformer network structure.

be reshot, and there are still interlaced scanning devices in use today, making it necessary to study deinterlacing techniques. In an interlaced video frame, the frame is divided into two fields, with one containing all odd rows and the other containing all even rows. While interlaced videos can be displayed correctly on interlaced display devices such as CRT monitors [14], they often exhibit artifacts such as horizontal stripes or water stripes when displayed on progressive devices like modern consumer LCD/LED monitors [2].

There have been various approaches proposed to address the issue of deinterlacing, such as Inter-Field Spatial Deinterlacing (IFSD) [4], Line Averaging (LA) [17], Motion Compensated Deinterlacing (MCD) [16], and others. With the success of deep learning methods in various fields, including super-resolution [7], video denoising [11], and sharpening [15], researchers have started exploring the application of deep learning to deinterlacing, and significant improvements have been achieved.

Moreover, our main contributions are:

1) We first propose a Transformer-based framework for deinterlacing, which combines feature extraction, self-attention motion compensation, and dense residual processing.
2) The designed fusion module enables better utilization of inter-frame motion correlation in interlaced videos, leading to improved deinterlacing effects.
3) The proposed method is end-to-end, and achieves superior deinterlacing results compared to previous methods through proper loss function design and dense residual block design.

2   Related Work

Traditional Methods for Video Deinterlacing. The traditional deinterlacing method is mainly used for single interlaced scanning process. However, in early video in the real world, interlacing was often mixed with many other unnatural human factors, such as blocking and noise in compression, transmission, and stored procedures, so that the performance of deinterlacing methods was often significantly reduced. Specifically, the traditional interleaved scanning mechanism can be expressed as $Y = S(X_1 \cdot X_2)$, where Y represents interlaced frames, $S(\cdot)$ represents interlaced scanning functions, and X_1 and X_2 represent odd and even fields.

Deep learning-based methods for video deinterlacing have shown promising results. However, there is still limited research in deep learning-based deinterlacing, and challenges such as interlaced artifacts caused by large motion vectors need further investigation. In recent years, deep learning technology has been developed rapidly and has shown satisfactory results in other computer vision tasks. Therefore, some researchers have tried to apply deep learning methods to solve the deinterlacing problem, and these works have achieved good results on the whole, solving most of the interlacing problems.

3   Method

3.1   Overview

This section describes the proposed Transformer-based video deinterlacing method. As shown in Fig. 1, our method consists of a feature extractor, a Transformer-based feature fusion network, and a smooth gradient network. Firstly, given a sequence of video frames, we use a stack of residual blocks to extract local features [10] from the frames. Then, the Transformer-based feature fusion network further extracts global features from spatial and temporal dimensions, which are applied to the current frame. Lastly, the smooth gradient [5] network stabilizes parameters and reconstructs the frame structure.

A progressive frame I_k consists of its odd and even rows, denoted as I_k^+ and I_k^-, respectively. Our network take three consecutive frames $\{I_{k-1}^+, I_k^-, I_{k+1}^+\}$ as input, the output is the recovered result I_k which is reconstructed from the I_k^-. Firstly the feature extraction part can be expressed as follows:

$$F_{k-1}, F_k, F_{k+1} = Fe(I_{k-1}^+, I_k^-, I_{k+1}^+) \tag{1}$$

where $Fe(\cdot)$ denotes the Feature Extractor operation which leverages the residual block to extract the local features $\{F_{k-1}, F_k, F_{k+1}\}$ from the input frames. Then the results are then fed into the De-Transformer module:

$$I_k^{de} = DeT(F_{k-1}, F_k, F_{k+1}) \tag{2}$$

where $DeT(\cdot)$ denotes the De-Transformer module, which further extracts global information from the local features $\{F_{k-1}, F_k, F_{k+1}\}$ and fuses spatial-temporal information into the F_k. Finally, we add the even rows in the middle frame and use the Residual Densenet module to reconstruct the deinterlacing result.

3.2   Feature Extractor (FE)

It consists of five residual blocks [6], each of which performs the following operations: first, a 3×3 convolutional layer is used to extract the input feature map, followed by a ReLU activation function. Then, another 3×3 convolutional layer is used to restore the channel dimension. Finally, the feature map input to the first convolutional layer is added to the output of the second convolutional layer. The specific network structure of the module is shown in Fig. 1, and the operation process of a residual block can be represented as:

$$X = I_{k-1}^+, I_k^-, I_{k+1}^+ \tag{3}$$

$$f_{k-1}, f_k, f_{k+1} = (\, Conv_{3,m} \, (\, ReLu \, (\, Conv_{3,m} \, (X) \,) \,) + X) \tag{4}$$

where $Conv_{3,m}(\cdot)$ denotes a convolution neural network with input channels of 3 and output channels of m, and $Conv_{m,m}(\cdot)$ is a convolution neural network with input channels of m and output channels of m. $\{f_{k-1}, f_k, f_{k+1}\}$ is the output of a residual block, and the module ultimately outputs three groups of features$\{F_{k-1}, F_k, F_{k+1}\}$.

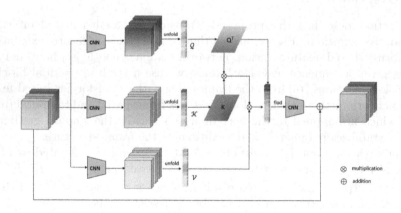

Fig. 2. Architecture of the DeT.

3.3   De-Transformer (DeT)

Self-attention [8] does not capture the position of the corresponding pixels. In our experiment, there is a poorer visual effect when the position information is missing. Therefore, overcome this, we use 3D fixed position encoding [13] and add them to the input of the self-attention layer. The position encoding contains two spatial position messages (i.e. horizontal and vertical) and one temporal position message. The spatial-temporal position encoding (PE) as $PE(pos, 2k) = sin(pos/10000^{2k/(d/3)})$ and $PE(pos, 2k + 1) = cos(pos/10000^{2k/(d/3)})$, where k is an integer in $[0, d/6)$, pos is the position in the corresponding dimension, and d is the size of the channel dimension, which needs to be divisible by 3. The De-Transformer module's architecture is shown in Fig. 2.

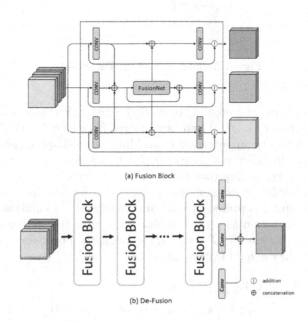

(a) Fusion Block

(b) De-Fusion

Fig. 3. Architecture of the De-Fusion module.

Spatial-Temporal Convolutional Self-attention (STCS). The Spatial-Temporal Convolutional Self-Attention (STCS) module captures the spatial and temporal location features of the interlaced frame sequences. Inspired by VSR [3], It uses three independent convolutional neural networks $\{W_q, W_k, W_v\}$ to extract the spatial information of each frame separately. Instead of taking a complete video frame as input and using a linear projection to extract patches like in ViT [8], we obtain three sets of 3D patches, each with $N = TWH/(W_pH_p)$ patches with dimension $d = C \times W_p \times H_p$, where T, W, H denote the number of frames, width and height of the frame, C, W_p, H_p are the image channel, the width of the patch, and the height of the patch. Then the operation generates the matrix of query, key, and value, respectively

$$Q = k_1 (W_q * X), K = k_1 (W_k * X), V = k_1 (W_v * X) \qquad (5)$$

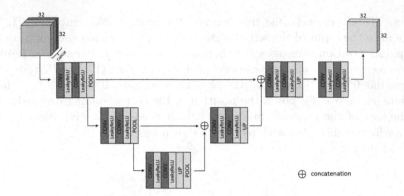

Fig. 4. Architecture of the FusionNet.

where $k_1(\cdot)$ represents an unfolding operation. Next, we reshape each patch into a new query matrix $Q = \delta(Q)$ and key matrix $K = \delta(K)$ with dimensions of $d \times N$. Here, $\delta(\cdot)$ denotes an integrated unfolding and reshaping operation. Then, we calculate the similarity matrix $\sigma_1(Q^T K)$ by aggregating it with the values matrix V. It is worth noting that the similarity matrix measures the relevance between all embedding tokens of the entire video frames, indicating that the layer has successfully captured spatial and temporal information. Finally, we use the folding operation $k_2(\cdot)$ to combine these updated sliding local patches into a feature map of size $C \times T \times W \times H$, which is then processed by the output layer W_0 to obtain the final feature map

$$
\begin{aligned}
f(X) = X + \sum_{i=1}^{h} W_0^i * k_2 \big(\\
\sigma_1 \big(\underbrace{k_1^\tau (W_q^i * X)}_{Q}{}^T \underbrace{k_1^\tau (W_k^i * X)}_{K} \big) \underbrace{k_1 (W_v^i * X)}_{V} \big)
\end{aligned}
\tag{6}
$$

where $k_1^\tau(\cdot)$ is a composition of the reshape operation τ and the unfold operation k_1.

De-Fusion Module (DeF). The feature map, which includes spatial-temporal information from the output of the STCS module, is fused by the De-Fusion module. The De-Fusion module is composed of six fusion blocks, each of which consists of a splicing fusion module and six convolutional neural networks, as shown in Fig. 3. The feature mapping $f(X)$ obtained from the spatiotemporal self-attention convolutional module, which includes the spatiotemporal features of three video frames, is input into the anti-interference fusion module to achieve the best visual effects. The output of the anti-interference fusion module is a feature mapping with complete and continuous information for odd and even rows of pixels in a frame (Fig. 4).

Specifically, the three features $\{F_{k-1}, F_k, F_{k+1}\}$ obtained from the previous step are input into the anti-interference fusion module, and the final output is obtained by combining the three features $\{De\_f_{6,k-1}, De\_f_{6,k}, De\_f_{6,k+1}\}$ through the channel-based fusion method after passing through six fusion blocks. The specific operation of the deinterlacing fusion module is as follows:

$$De\_f_{1,k-1}, De\_f_{1,k}, De\_f_{1,k+1} = FB_1\left(F_{k-1}, F_k, F_{k+1}\right) \qquad (7)$$

$$\begin{aligned} De\_f_{i,k-1}, De\_f_{i,k}, De\_f_{i,k+1} = \\ FB_i\left(De\_f_{i-1,k-1}, De\_f_{i-1,k}, De\_f_{i-1,k+1}\right) \end{aligned} \qquad (8)$$

$$f_{De} = Concat\left(De\_f_{6,k-1}, De\_f_{6,k}, De\_f_{6,k+1}\right) \qquad (9)$$

where $FB_1\left(\cdot\right)$ represents the operation of the first fusion module, $FB_i\left(\cdot\right)$ represents the operation of the i-th fusion block, $Concat\left(\cdot\right)$ represents the fusion operation, f_De represents the output of the deinterlacing fusion module, and k represents the frame number of the video frame to be repaired.

3.4   Residual Densnet(RD)

The deinterlacing self-attention network module can extract a large amount of feature information, but because the information contains many redundant and complex contents. The residual dense network (RD) module is used to further process the feature information obtained from the deinterlacing fusion module. This is done to remove redundant and complex information from the output and improve the overall generation effect of the network, resulting in deinterlaced video frames.

The RD network consists of N residual blocks, and in this paper, it is experimentally verified that the best results are achieved when N is set to 6. The operation of a single residual block can be expressed as:

$$\begin{aligned} f\left(X\right)_{RB} = Concat\left(\, Conv_{m,m}\left(\right.\right. \\ LeakyReLU\left(\, Conv_{m,m}\left(\, X\,\right)\,\right)\right), X\,) \end{aligned} \qquad (10)$$

where $Conv_{m,m}$ denotes a convolutional layer with m channels, and $LeakyReLU$ is the leaky rectified linear unit activation function. $f\left(X\right)_{RB}$ denotes the output of a single residual block.

$$f\left(X\right)_{RD} = LeakyReLU\left(Conv_{m,m}\left(N \times f\left(X\right)_{RB}\right)\right) + X \qquad (11)$$

where $f\left(X\right)_{RD}$ denotes the output of the RD module, and X is the input from the deinterlacing fusion module. The output of the RD module is obtained by passing the input through N residual blocks and then adding it back to the input, which helps to remove redundant information and retain important features for generating deinterlaced video frames.

4   Experiments and Results

4.1   Dataset and Training

The dataset used for training the deinterlacing model is created by following a similar approach as the paper [1]. Specifically, odd rows of the *(k-1)*-th frame and even rows of the *k*-th frame are combined to synthesize one frame, and then the odd rows of the *k*-th frame and the even rows of the *(k+1)*-th frame are combined to synthesize the next frame, and so on, resulting in interlaced frames. The Vimeo90K [18] dataset, which is a large-scale, high-quality video dataset covering a variety of scenes and actions, is used for training. The progressive sequences (ground truth) from the Vimeo90K dataset are used to create the interlaced input frames for training. The training/testing split provided with the dataset is used.

4.2   Loss Function

The network is trained to minimize the loss $\mathcal{L}$ on the dataset $\mathcal{D}$. For the image loss, we utilize the $\ell2$-norm [12] of pixel differences, defined as follows:

$$\theta^* = \arg\min_{\theta} \mathbb{E} Ik \in \mathcal{D} \left[\mathcal{L} \left(\mathcal{F}\theta \left(Ik - 1^+, I_k^-, I_{k+1}^+ \right), I_k \right) \right] \qquad (12)$$

where $\mathcal{L} \left(\mathcal{F}\theta \left(Ik - 1^+, I_k^-, I_{k+1}^+ \right), I_k \right)$ represents the $\ell2$-norm of the difference between the output of the method $\mathcal{F}\theta \left(Ik - 1^+, I_k^-, I_{k+1}^+ \right)$ and the ground-truth I_k.

In the above equations, $\mathcal{F}_\theta \left(\cdot \right)$ denote our method.

4.3   Results and Comparison

Through a comparison of the deep learning-based video deinterlacing methods Real-time Deep Video Deinterlacing (RDVD) [19] and Deep Deinterlacing (DD) [1], the effectiveness of the model proposed in this paper was proven in terms of both performance and experimental results. In addition, ablation experiments were conducted to demonstrate the importance of the self-attention module and residual dense module in this algorithm. The performance of the models was compared using computational PSNR and SSIM indices in the experimental section.

(1) The Experimental Results on the Vimeo90K Dataset After training using the Vimeo90K dataset, several different video sequences of different types were randomly selected from the dataset's validation set to test the performance of the testing model. Out of all the experimental results obtained, a random selection of 8 groups of videos were chosen and their PSNR and SSIM indices were measured, as well as the average PSNR and SSIM indices for the methods on the dataset. The testing PSNR and SSIM indices are shown in Table 1.

From Tables 1, it can be seen that the indicators achieved by the method designed in this paper on the validation set are higher by about 10% than those

Table 1. Comparison of PSNR (dB) between our method and existing methods on the validation set.

Methonds	Man	Reading	Board	Car	Pandas	Tiger	Train	Plane	PSNR
BICUBIC	20.23	27.40	18.88	21.23	30.65	26.95	24.91	25.39	24.46
SRCNN	22.10	31.26	21.33	24.85	30.14	26.92	24.81	25.04	25.81
RDVD	28.70	45.32	28.24	33.61	41.12	36.18	33.61	36.68	35.43
DD	30.67	48.27	28.66	35.43	41.86	36.88	35.43	37.07	36.78
Ours	**32.36**	**48.56**	**30.22**	**35.77**	**42.85**	**38.97**	**35.77**	**37.90**	**37.80**

Table 2. Comparison of PSNR (dB) between our method and existing methods on the test set.

Methonds	Women	Running	Writer	Jump	Apple	Draw	Talking	PSNR
BICUBIC	31.25	23.55	28.26	22.48	26.76	25.75	27.91	26.19
SRCNN	31.26	23.51	28.38	22.50	27.88	26.46	28.01	26.42
RDVD	45.32	32.04	36.79	41.15	36.64	35.50	40.68	37.75
DD	48.27	35.48	39.09	41.59	40.73	35.09	43.18	39.64
Ours	**48.56**	**36.38**	**39.22**	**42.82**	**42.14**	**36.40**	**45.81**	**40.75**

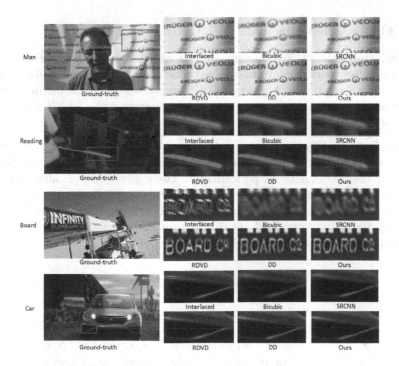

Fig. 5. Qualitative comparison on the Vimeo 90K synthetic testing set.

achieved by the RDVD [19] and DD [1] models. Similarly, to verify the effectiveness of the algorithm proposed in this paper, comparisons were made between four different algorithms and the model proposed in this paper using the testing set of the dataset. A series of randomly selected video frame images were used for verification.

From the experimental testing results, it can be seen that the model designed in this paper has achieved significant improvements in performance both on the validation set of the Vimeo90K dataset and the testing set.

(2) Visual Results Comparison

From the visualization results in Fig. 5, it can be clearly seen that the performance of the algorithm proposed in this paper in terms of video deinterlacing has been significantly improved compared with the previous methods. In addition, through partial zoom-in observation, it can be found that whether in ordinary video images or in relatively complex text images, the visualization results obtained from the model algorithm experiments in this paper are more perfect in detail processing.

To assess the performance of our method in handling different interlaced video sequences, we randomly selected a range of video frames and conducted experiments using four methods, including our proposed approach. It is evident from the figure that our method yields the best results, indicating its superior performance in handling diverse interlaced video sequences.

4.4   Ablation Experiments

We conducted ablation experiments to evaluate the contribution of the spatio-temporal self-attention convolutional module and the residual dense network module in our algorithm. The Vimeo90K dataset was used for testing, and the importance of each module was assessed by measuring the PSNR metric after retraining the model.

Spatio-Temporal Self-attention. In the ablation experiment for the spatio-temporal self-attention convolutional module, we focused on evaluating its impact on the overall network performance. Two experimental models were compared: one with the spatio-temporal self-attention convolutional module retained in the network structure, and the other with the module removed, directly inputting the results obtained by the feature extraction module to the anti-aliasing fusion module.

Table 3. The ablation experiment of spatial-temporal self-attention convolution module.

Self-attention	Man	Reading	Board	Car	Pandas	Tiger
✗	34.23	34.27	35.7	28.66	41.23	30.66
✔	**36.42**	**35.87**	**36.61**	**30.21**	**42.64**	**32.36**

Residual Blocks. During the experiments, we found that the residual dense network module can improve the PSNR metric of the experimental results. This section discusses the importance of the residual dense network module for network performance and verifies through experiments that the number of residual blocks (RBs) used in the module can balance computational cost and model performance, achieving the best experimental results.

We selected 0, 4, 6, and 8 residual blocks as experimental objects and retrained the model. The PSNR metric of the test results is shown in Table 4. Multiple sets of video frames were tested to avoid experimental errors. It can be observed from Table 4 that the experimental results with 6 RBs are better than those without using RBs or using a different number of RBs. In this work, we considered the balance between experimental effectiveness and efficiency, and setting the number of RBs to six was found to be the most suitable for our experiments.

Table 4. Effect of different residual blocks on results.

PSNR	Ours-n_RB = 6	n_RB = 0	n_RB = 4	n_RB = 8
Frame_1	**48.40**	48.32	48.28	48.32
Frame_2	36.80	36.80	**36.85**	36.75
Frame_3	**39.24**	39.24	39.10	39.18
Frame_4	42.84	42.79	**42.87**	42.85
Frame_5	**42.46**	42.12	42.33	42.38
AVERAGE	**41.95**	41.86	41.88	41.47

4.5   More Comparision

To further verify the practicability of the algorithm. Therefore, the algorithm proposed in this chapter and some existing de-interlacing methods are applied to process real-world videos with interlacing artifacts, and the results are obtained and compared. The ultimate goal is to verify the effectiveness of the proposed algorithm in practical applications.

In collaboration with the cartoon company, we compared the algorithm model proposed in this chapter with a relatively new DD model and two traditional de-interlacing filter methods (Yadif and Bwdif) in FFmpeg [9], using the *Tianyan* animation provided by them. We calculated the PSNR and SSIM between the de-interlaced video sequences generated by the above-mentioned methods and the ground truth values of the corresponding video frames provided by the cartoon company.

Table 5. Comparison four different methods in *tianyan*.

	yadif	bwdif	DD	ours
PSNR(dB)	32.03	32.09	37.71	37.96
SSIM(%)	87.27	87.88	96.87	97.27

We also de-interlaced a continuous sequence of 40 frames from the *Tianyan* video images using four different methods.

5   Conclusion

This paper proposes a video deinterlacing algorithm based on the self-attention mechanism, which addresses the challenges faced by current video deinterlacing methods. The algorithm utilizes spatio-temporal self-attention convolution and anti-interlacing fusion modules to effectively remove interlacing artifacts in videos. Overall, the proposed algorithm shows promise in practical applications for video deinterlacing and contributes to the advancement of the field. The findings of this study provide insights for further research and development of video deinterlacing algorithms with improved computational efficiency and performance.

References

1. Bernasconi, M., Djelouah, A., Hattori, S., Schroers, C.: Deep deinterlacing. In: SMPTE Annual Technical Conference and Exhibition, pp. 1–12 (2020)
2. Bhakar, V., Agur, A., Digalwar, A., Sangwan, K.S.: Life cycle assessment of crt, lcd and led monitors. Procedia CIRP **29**, 432–437 (2015)
3. Caballero, J., et al.: Real-time video super-resolution with spatio-temporal networks and motion compensation. In: Proceedings of the IEEE Conference on Computer Vision and Pattern Recognition, pp. 4778–4787 (2017)
4. Chen, M.J., Huang, C.H., Hsu, C.T.: Efficient de-interlacing technique by interfield information. IEEE Trans. Consum. Electron. **50**(4), 1202–1208 (2004)
5. d'Aspremont, A.: Smooth optimization with approximate gradient. SIAM J. Optim. **19**(3), 1171–1183 (2008)
6. De, S., Smith, S.: Batch normalization biases residual blocks towards the identity function in deep networks. Adv. Neural. Inf. Process. Syst. **33**, 19964–19975 (2020)
7. Dong, C., Loy, C.C., He, K., Tang, X.: Image super-resolution using deep convolutional networks. IEEE Trans. Pattern Anal. Mach. Intell. **38**(2), 295–307 (2015)
8. Dosovitskiy, A., et al.: An image is worth 16x16 words: transformers for image recognition at scale. arXiv preprint arXiv:2010.11929 (2020)
9. Lei, X., Jiang, X., Wang, C.: Design and implementation of a real-time video stream analysis system based on ffmpeg. In: 2013 Fourth World Congress on Software Engineering, pp. 212–216. IEEE (2013)
10. Li, J., Allinson, N.M.: A comprehensive review of current local features for computer vision. Neurocomputing **71**(10–12), 1771–1787 (2008)

11. Mahmoudi, M., Sapiro, G.: Fast image and video denoising via nonlocal means of similar neighborhoods. IEEE Signal Process. Lett. **12**(12), 839–842 (2005)
12. Mao, X., Li, Q., Xie, H., Lau, R.Y., Wang, Z.: Multi-class generative adversarial networks with the l2 loss function, vol. 5, p. 00102. arXiv preprint arXiv:1611.04076 (2016)
13. Martın, J., Jiménez, A., Seco, F., Calderón, L., Pons, J.L., Ceres, R.: Estimating the 3d-position from time delay data of us-waves: experimental analysis and a new processing algorithm. Sens. Actuators, A **101**(3), 311–321 (2002)
14. Post, D.L., Calhoun, C.S.: An evaluation of methods for producing desired colors on crt monitors. Color Res. Appli. **14**(4), 172–186 (1989)
15. Rieder, P., Scheffler, G.: New concepts on denoising and sharpening of video signals. IEEE Trans. Consum. Electron. **47**(3), 666–671 (2001)
16. Serrano, R.S.: Deinterlacing algorithms. Albalá Ingenieros SA (2016)
17. Wang, J., Jeon, G., Jeong, J.: Efficient adaptive intra-field deinterlacing algorithm using bilateral filter. In: 2012 3rd IEEE International Conference on Network Infrastructure and Digital Content, pp. 468–472. IEEE (2012)
18. Xue, T., Chen, B., Wu, J., Wei, D., Freeman, W.T.: Video enhancement with task-oriented flow. Int. J. Comput. Vision **127**(8), 1106–1125 (2019)
19. Zhu, H., Liu, X., Mao, X., Wong, T.T.: Real-time deep video deinterlacing. arXiv preprint arXiv:1708.00187 (2017)

SCME: A Self-contrastive Method for Data-Free and Query-Limited Model Extraction Attack

Renyang Liu[1], Jinhong Zhang[1], Kwok-Yan Lam[2], Jun Zhao[2], and Wei Zhou[1]($\boxtimes$)

[1] Yunnan University, Kunming, China
{ryliu,jhnova}@mail.ynu.edu.cn, zwei@ynu.edu.cn
[2] Nanyang Technological University, Singapore, Singapore
{kwokyan.lam,junzhao}@ntu.edu.sg

Abstract. Previous studies have revealed that artificial intelligence (AI) systems are vulnerable to adversarial attacks. Among them, model extraction attacks fool the target model by generating adversarial examples on a substitute model. The core of such an attack is training a substitute model as similar to the target model as possible, where the simulation process can be categorized in a data-dependent and data-free manner. Compared with the data-dependent method, the data-free one has been proven to be more practical in the real world since it trains the substitute model with synthesized data. However, the distribution of these fake data lacks diversity and cannot detect the decision boundary of the target model well, resulting in the dissatisfactory simulation effect. Besides, these data-free techniques need a vast number of queries to train the substitute model, increasing the time and computing consumption and the risk of exposure. To solve the aforementioned problems, in this paper, we propose a novel data-free model extraction method named SCME (Self-Contrastive Model Extraction), which considers both the inter- and intra-class diversity in synthesizing fake data. In addition, SCME introduces the Mixup operation to augment the fake data, which can explore the target model's decision boundary effectively and improve the simulating capacity. Extensive experiments show that the proposed method can yield diversified fake data. Moreover, our method has shown superiority in many different attack settings under the query-limited scenario, especially for untargeted attacks, the SCME outperforms SOTA methods by 11.43% on average for five baseline datasets.

Keywords: Adversarial Attacks · Model Extraction Attacks · Black-Box Attacks · Model Robustness · Information security

1 Introduction

Recently, Trusted AI, which contains fairness, trustworthiness and explainability, has received increasing attention and plays an essential role in the AI development process. The security of the AI models, however, is being doubted and has bought concerns in academia and industry. A lot of research has shown that AI models (including Machine Learning (ML) models and Deep Learning (DL) models) are vulnerable to adversarial examples [2], which are crafted by adding a virtually imperceptible perturbation to the benign input but can lead the well-trained AI model to make wrong decisions. For example, in the physical world, the attackers can maliciously alter traffic signs by sticking a small patch [12], changing the content style [4] and shooting a laser on it [5]. Although these modifications do not affect human senses, but can easily trick autonomous vehicles. Therefore, it is imperative to devise effective attack techniques to identify

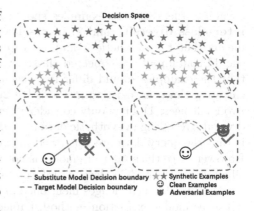

Fig. 1. Synthetic example distribution, decision boundaries and whether the attack is successful. **Top left:** bad synthetic example distribution failed to fit the target model decision boundary. **Bottom left:** unfitting of the decision boundary leads to attack failure. **Top right:** good example distribution and decision boundary fit. **Bottom right:** good decision boundary.

the deficiencies of AI models beforehand in security-sensitive applications [13].

Existing adversarial attack methods on DL models can be categorized into white-box attacks and black-box attacks. In the white-box settings, the attackers can access the whole information of the target model, including weights, inner structures and gradients. In contrast, in the black-box one, the attackers have no permission to access the models' details but the final output [3,9]. With such rules, it is clear that black-box attacks are more challenging but practical in the physical world, where the attacker lacks details of the target models. To attack DL models in black-box settings more effectively, model extraction attacks have been proposed [17], which is implemented by training a substitute model and generating adversarial examples on such model to attack the target model successfully.

Most of the previous model extraction attacks [14,15] concentrates on training the substitute model by querying the target model with real data, called data-dependent model extraction. However, it is infeasible to the physical world, where the adversary can not access the models' training data. As the counterpart, the data-free model extraction attack solved this problem by synthesizing fake data [20,23]. In this scenario, the attackers use generators to synthesize the fake data to train substitute models. For launching attacks with a high success rate, as shown in Fig. 1, the decision boundary of the substitute model should maintain a very high similarity to the target black-box model. Besides, training the substitute model with synthetic data is challenging to the problem of *How to generate valuable synthetic data for the substitute model training?* Generally speaking, the synthetic examples should have the following two properties: 1) **inter-class diversity** and 2) **intra-class diversity**. The inter-class diversity means that the synthetic examples' categories classified by the target model should contain all the expected classes, while the intra-class diversity indicates that the examples should differ from each other, even they belong to the same category. However, existing methods [18, 20, 22, 23] still suffer from the following two challenges: 1) They only consider inter-class diversity but ignore intra-class diversity, resulting in synthetic data not serving as well as real data. 2) The other is that the query data are generated in the substitute model's training process. However, once the substitute model is not well-trained, it can hardly provide the effective target model's decision information.

To solve these challenges mentioned above, in this paper, we propose a novel data-free model extraction method, named **S**elf-**C**ontrastive **M**odel **E**xtraction (SCME for short). SCME introduces the idea of contrastive learning [1] and proposes a self-contrastive mechanism to guide the training of the generator. Specifically, we design a self-contrastive loss to enlarge the distance of the substitute model's latent representation. Benefiting from this, the generator will be encouraged to synthesize more diversified fake data. Furthermore, SCME introduces the Mixup operation to interpolate two random images into a single one to build the query examples, which can improve the efficiency of the substitute model in learning the target model's decision boundaries in a model-independent manner. Extensive experiments illustrate SCME can synthesize fake data with diversity and improve the attack performance. Our contributions are summarized as follows:

- We propose a novel data-free model extraction method, called SCME, to generate efficient fake data for the substitute model training under query-limited settings.
- We use a self-contrastive mechanism to guide the generator to synthesize the fake data with inter- and intra-class diversity to help the substitute model imitate the target model efficiently.
- We introduce the Mixup into SCME, which can build fake data in a model-independent manner, to detect decision boundaries of the target model effectiveness and further help the imitating processing.
- Extensive empirical results show the SCME's superiority in the synthetic diversified fake data and the adversarial examples' attack performance in query-limited situations.

2   Related Work

Previous researches contend that the DL models are sensitive to adversarial attacks, which can be classified into white-box and black-box. In white-box settings, the attackers can generate adversarial examples with a nearly 100% attack success rate because they can access the target model. The black-box attack, however, is more threatening to the DL models in various realistic applications because they do not need the models' details. Among them, the model extraction

attack [21] has received much attention recently due to its high attack performance.

The success of model extraction attacks relies heavily on adversarial examples' transferability, which means the adversarial examples generate on model A
can also attack model B successfully. To implement such an attack, the attacker
first trains a surrogate local model by simulating the target models' output.
When the surrogate model is well-trained, it will have the same decision boundary as the target model, i.e., output the same results for the same input; this imi-
tation process is called model extraction. However, due to the data bias between
the query data used for surrogate model training and the real data used for the
target model training, creating a valid query dataset is the crucial point of model
extraction attacks. Papernot et al. [15] first used adversarial examples to query
the target model for model extraction. However, due to the surrogate model is
not well-trained, the adversarial examples generated on it cannot perform well
in the imitating process. Orekondy et al. [14] propose the Knockoff to try to find
valid query examples in a huge dataset, e.g., ImageNet [8], and adopt an adaptive strategy in the extraction process. Zhou et al. [23] proposed DAST, which is
the first work to use a generator-based data-free distillation technique in knowledge distillation for model extraction. Later, subsequent studies have improved
this approach to achieve better results [18]. However, the generator-based approach cannot obtain sufficient supervised information as in white-box knowledge
distillation, leading to a huge number of queries and low attack results.

Therefore, the block-box attack with adversarial examples' transferability
poses the request to guarantee that the local model is highly similar to the
target model. To achieve this goal, we know from previous studies that model
extraction can steal the target model from a decision boundary perspective,
even in a data-independent way. However, the previous data-free works can not
guarantee the synthetic data's diversity and need a massive number of queries
to the target model. Hence, we are well-motivated to develop a better model
extraction strategy adapted to data-free settings for carrying out attacks with
high performance. Besides, it can improve the diversity of the generated fake
data to be suitable for query-limited settings.

3 Preliminary

3.1 Adversarial Attack

Given a classifier $\mathcal{F}(\cdot)$ and an input x with its corresponding label y, we have
$\mathcal{F}(x) = y$. The adversarial attack aims to find a small perturbation δ added to
x, so the generated input x' misleads the classifier's output. The perturbation
δ is usually constrained by L_p-norm ($p = 1, 2, ..., \infty$), i.e., $\|\delta\|_p \leq \epsilon$. Then, the
definition of adversarial examples x' can be written as:

$$\mathcal{F}(x') \neq y_{true}, \quad s.t. \ \|x' - x\|_p \leq \epsilon, \tag{1}$$

where ϵ is the noise budget, y_{true} is the ground-truth label of example x.

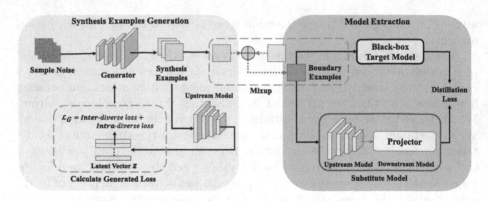

Fig. 2. Framework of SCME, where $\mathcal{L}_G$ consisting of inter- and intra-diverse loss.

3.2 Contrastive Learning

Contrastive learning models usually consist of two portions: self-supervised training in the upstream network and supervised fine-tuning in the downstream network. The upstream network $f(\cdot)$ aims to maximize the paired instance by augmenting the same data in different ways in the learned latent space while minimizing the agreement between different instances. Given a batch of examples $\{x_N\}$ without ground-truth labels, the random data transformation $\mathcal{T}$ takes each example x in $\{x_N\}$ to a paired augmented data copies x_i and x_j, resulting in $2N$ augmented examples. The trained upstream network $f(\cdot)$ encodes the paired copies to latent vectors z_i and z_j. n SimCLR [1], the contrastive loss can be formulated as:

$$\ell_{i,j} = -\log \frac{\exp\left(\text{sim}\left(z_i, z_j\right)/\tau\right)}{\sum_{k=1}^{2N} \mathbb{1}_{[k \neq i]} \exp\left(\text{sim}\left(z_i, z_k\right)/\tau\right)}, \qquad (2)$$

where the z_i and z_j are the latent vectors of positive augmented examples, and z_k indicts the latent vector of negative examples from a different class. The $sim(\cdot)$ is a similarity function, such as cosine similarity loss, $\mathbb{1}$ is the indicator function, and τ is the temperature coefficient. A well-trained upstream network $f(\cdot)$ can extract effective features and use them in the downstream network, which usually is a simple MLP network, mapping the latent vectors to different classes through supervised learning.

4 Methodology

4.1 Overview

In this part, we illustrate the framework of our proposed data-free SCME in Fig. 2, which contains the following steps: 1) Synthesised Examples Generation and 2) Model Extraction. For step 1), we use a generator $\mathcal{G}(\cdot)$ to generate the

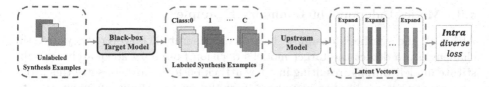

Fig. 3. The calculation process of intra-class diverse loss. C is the number of classes.

fake data $\mathcal{X}$. In step 2), we input the $\mathcal{X}$ into both substitute model $\mathcal{F}_{sub}$ and target model $\mathcal{F}_{tgt}(\cdot)$ to minimize the difference of their outputs. Notably, $\mathcal{F}_{sub}(\cdot)$ in SCME consists of upstream encoder network $\mathcal{F}_{up}(\cdot)$, a feature extraction network, projector network $F_{down}(\cdot)$, and a classifier. Mathematically, the $\mathcal{F}_{sub}(\cdot)$ can be written as:

$$\mathcal{F}_{sub}(x) = F_{down}(F_{up}(x)), \tag{3}$$

where x is an arbitrary input example.

Based on the two steps mentioned above, the F_{sub} can imitate the $\mathcal{F}_{tgt}$ in a data-free manner. Finally, we can generate adversarial examples by attacking $\mathcal{F}_{sub}$, and further attack the target model $\mathcal{F}_{tgt}$ successfully.

4.2   Intra- and Inter-class Diverse

As mentioned above, $\mathcal{X}$ should have both inter-class diversity and intra-class diversity to help the surrogate model training. Regarding this, as Fig. 3 show, we propose a self-contrastive loss to guide $\mathcal{G}(\cdot)$ in the $\mathcal{X}$ generation. Inspired by the self-supervised loss in contrastive learning, we design a self-contrastive loss in SCME. Firstly, SCME uses the generator $\mathcal{G}(\cdot)$ to sample a batch of random noise $N = \{n_1, n_2, \cdots, n_B\}$ to generate corresponding synthesize examples $\mathcal{X} = \{x_1, x_2, \cdots, x_B\}$. SCME puts the $\mathcal{X}$ into the feature extraction network $\mathcal{F}_{up}(\cdot)$ and gets the latent vectors z. Then, SCME calculates the self-contrastive intra-class diverse loss $\mathcal{L}_{intra}$ by expanding the distance of each hidden vector z_i in z. The self-contrastive loss can be formulated as:

$$\mathcal{L}_{intra} = log \sum_{i}^{B} \sum_{j}^{B} \mathbb{1}_{[i \neq j]} \cdot exp(sim(z_i, z_j)), \tag{4}$$

where B is the batch size, $\mathbb{1}$ is the indicator function and $sim(\cdot)$ is a similarity function.

In the synthesised examples generation, the loss function $\mathcal{L}_G$ of generator $\mathcal{G}(\cdot)$ contains inter-class diversity loss $\mathcal{L}_{inter}$ and intra-class diversity loss $\mathcal{L}_{intra}$. To generate inter-class diversity examples, we use the inter-class information entropy to guide the generator $\mathcal{G}(\cdot)$. That is, SCME randomly sets a batch of target label y_{tgt} and reduces the entropy between y_{tgt} and the substitute model's output of the generated examples $\mathcal{X}$. Mathematically, the inter-class loss function is:

$$\mathcal{L}_{inter} = \sum_{i=1}^{B} \mathcal{F}_{sub}(\mathcal{X}_i) log[F_{sub}(\mathcal{X}_i)], \tag{5}$$

where B is the batch size.

4.3   Model-Independent Boundary Example

Although the synthesis examples have been generated, however, they are still challenging to detect the target model's decision boundary adequately for substitute model training, resulting in a low attack performance. To solve this problem, we further modify the boundary examples by the Mixup augmentation to improve the substitute model's training efficiency. Specifically, SCME randomly selects two synthesized examples $\mathcal{X}_i$ and $\mathcal{X}_j$ first and then uses the Mixup to fuse them together to get the new boundary examples $\hat{\mathcal{X}}$, the process can be written as follows:

$$\hat{\mathcal{X}} = \lambda \mathcal{X}_i + (1 - \lambda)\mathcal{X}_j, \tag{6}$$

where the $\lambda \in [0, 1]$ is the mix weight and randomly sampled from β distribution.

4.4   Objective Function

By combining the above inter-class diversity loss $\mathcal{L}_{inter}$ and the intra-class diversity loss $\mathcal{L}_{intra}$, we obtain the generate loss $\mathcal{L}_G$ as the objective function for training the generator:

$$\mathcal{L}_G = \mathcal{L}_{intra} + \alpha \mathcal{L}_{inter}, \tag{7}$$

where the α is the hyperparameter to adjust the weight of each loss.

Once the intra-class and inter-class diverse examples are generated, we input them into both substitute model $\mathcal{F}_{sub}(\cdot)$ and the target model $\mathcal{F}_{tgt}(\cdot)$ to minimize the distance between their outputs. To craft more suitable examples for training the substitute model and make its decision boundary close to the target model in the training process, we first craft the generated examples by Mixup operation to get the boundary examples $\hat{\mathcal{X}}$. Mathematically, the objective loss function $\mathcal{L}_{train}$ of training substitute model is:

$$\mathcal{L}_{train} = \sum_{i=1}^{B} d(\mathcal{F}_{sub}(\hat{\mathcal{X}}), \mathcal{F}_{tgt}(\hat{\mathcal{X}})), \tag{8}$$

where the distance function $d(\cdot)$ is the Cross-Entropy loss in the hard label scenario and is the Mean Square Error loss in the soft label scenario.

Once the surrogate model is well-trained, we are able to generate adversarial examples on the substitute model and further attack the target black-box model.

5   Experiments

5.1   Setup

Datasets: We consider five benchmark datasets, namely MNIST [11], Fashion-MNIST [19], CIFAR-10 [7], CIFAR-100 [7], Tiny-ImageNet [10] for comprehensive experiments.

Table 1. Attack performance on MNIST and Fashion-MNIST Datasets.

Dataset	Methods	Targeted, Hard Label			Untargeted, Hard Label			Targeted, Soft Label			Untargeted, Soft Label		
		FGSM	BIM	PGD	FGSM	BIM	PGD	FGSM	BIM	PGD	FGSM	BIM	PGD
MNIST	JPBA	3.89	6.89	5.31	18.14	23.56	20.18	4.29	7.02	5.49	18.98	25.14	21.98
	Knockoff	4.18	6.03	4.66	19.55	27.32	22.18	4.67	6.86	5.26	21.35	28.56	23.34
	DaST	4.33	6.49	5.17	20.15	27.45	27.13	4.57	6.41	5.34	25.36	29.56	29.14
	Del	6.45	9.14	6.13	22.13	25.69	23.18	6.97	9.67	6.24	24.56	25.35	25.28
	EBFA	**14.45**	**28.71**	9.86	39.73	57.54	52.73	**16.99**	**36.82**	**14.55**	36.45	58.48	48.46
	SCME	9.98	9.96	**10.04**	**63.45**	**74.51**	**78.47**	10.05	10.00	10.04	**72.46**	**78.54**	**82.54**
Fashion-MNIST	JPBA	6.45	8.46	7.57	24.22	30.56	30.11	6.89	8.56	7.56	26.23	31.35	31.11
	Knockoff	6.34	8.35	7.32	28.19	36.88	35.92	6.65	8.98	8.23	30.21	36.94	36.22
	DaST	5.38	7.18	6.53	30.45	36.17	34.23	5.33	7.46	7.84	32.14	37.34	34.91
	Del	3.89	8.19	7.47	28.14	34.14	32.45	3.23	8.59	8.11	31.43	36.26	33.87
	EBFA	30.08	**76.46**	32.42	**84.85**	80.93	**89.30**	29.11	66.02	43.56	75.19	79.94	79.30
	SCME	**31.82**	70.79	**70.01**	82.26	**84.76**	85.11	**32.14**	**72.07**	**72.07**	**82.58**	**85.46**	**85.86**

Table 2. Attack performance CIFAR-10 and CIFAR-100 Datasets.

Dataset	Methods	Targeted, Hard Label			Untargeted, Hard Label			Targeted, Soft Label			Untargeted, Soft Label		
		FGSM	BIM	PGD	FGSM	BIM	PGD	FGSM	BIM	PGD	FGSM	BIM	PGD
CIFAR-10	JPBA	6.32	7.70	7.92	27.82	33.23	31.70	7.28	8.56	7.64	28.77	33.38	31.96
	Knockoff	6.26	7.02	7.04	29.61	31.86	30.68	6.46	8.27	7.35	30.02	31.98	30.35
	DaST	6.54	7.81	7.41	27.61	34.43	26.99	8.15	8.40	8.26	27.58	34.75	27.47
	Del	7.14	7.44	6.95	25.33	30.45	30.34	7.86	8.29	7.17	26.38	31.53	31.47
	EBFA	14.57	**16.95**	12.27	86.13	87.02	84.32	**31.54**	12.93	**60.14**	83.80	87.68	85.11
	SCME	**16.53**	14.76	**14.22**	**91.01**	**91.56**	**91.33**	16.22	**15.14**	15.31	**91.23**	**91.62**	**91.66**
CIFAR-100	JPBA	4.35	6.20	6.17	33.58	38.54	37.08	5.73	7.50	6.41	34.21	39.12	37.31
	Knockoff	4.40	5.86	5.25	34.84	36.92	36.34	4.88	7.05	6.18	36.01	37.61	35.47
	DaST	4.97	6.19	5.92	33.57	39.86	32.71	6.38	7.04	7.01	32.80	40.34	32.78
	Del	5.38	5.72	5.69	30.80	35.63	36.15	6.30	6.53	5.23	31.64	36.63	37.44
	EBFA	16.64	**16.88**	12.77	78.61	91.31	91.21	7.91	**16.15**	12.54	83.69	94.53	94.14
	SCME	**18.46**	14.23	**13.13**	**94.81**	**95.40**	**95.32**	**10.50**	16.02	**15.49**	**94.72**	**95.09**	**94.95**

Models: For MNIST and Fashion-MNIST datasets, we use a simple network as the target model, which has four convolution layers and pooling layers and two fully-connected layers. For CIFAR-10 and CIFAR-100, we use the ResNet-18 [6] as the target model. For Tiny-ImageNet, we use the ResNet-50 [6] as the target model. The substitute model for all the datasets is the VGG-16 [16].

Baselines: To evaluate the performance of SCME, we compare it with the data-dependent method, JPBA [15], Knockoff [14], and data-free methods, DAST [23], Del [18], EBFA [22].

Training Details: SCME and the baseline methods are trained with Adam optimizer with batch size 256. For the generator in SCME, we use an initial learning rate of 0.001 and a momentum of 0.9, and for the substitute model, we set the initial learning rate as 0.01 and momentum as 0.9. Furthermore, we set the maximal query times as 20 K, 100 K and 250 K for the MNIST dataset, CIFAR dataset and the Tiny-ImageNet dataset, respectively.

Metrics: We utilize three classical attack methods, which include FGSM, BIM and PGD, to generate adversarial examples for the surrogate model. For MNIST and Fashion-MNIST, we set perturbation budget $\epsilon = 32/255$. And for CIFAR-10, CIFAR-100 and Tiny-ImageNet, we set $\epsilon = 8/255$. In the untargeted attack

Table 3. Attack performance on Tiny-ImageNet Dataset.

Methods	Hard Label			Soft Label		
	FGSM	BIM	PGD	FGSM	BIM	PGD
JPBA	15.37	25.16	14.23	26.54	28.91	26.83
Knockoff	22.33	21.39	11.26	29.99	27.64	26.17
DaST	16.23	18.26	15.86	28.81	29.37	26.51
Del	28.31	32.54	29.73	34.28	38.49	36.72
EBFA	78.29	81.12	78.23	80.26	85.32	78.29
SCME	**90.16**	**90.25**	**89.72**	**96.44**	**96.29**	**96.32**

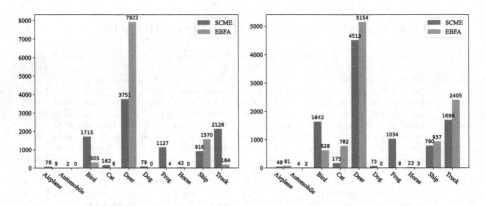

Fig. 4. Synthesised examples without (left) and with (right) data augmentation.

scenario, we only generate adversarial examples for the images which can be classified correctly by the victim model, while in targeted attacks, we only generate adversarial examples for the images which are not classified to the specific wrong labels. The **attack success rate** (ASR) is calculated by:

$$ASR = \begin{cases} \frac{1}{N} \sum_{i=1}^{N} [f(x_i^{adv}) \neq y_i], & for\ untargeted \\ \frac{1}{N} \sum_{i=1}^{N} [f(x_i^{adv}) = y_t], & for\ targeted \end{cases} \qquad (9)$$

where N is the total number of generated adversarial examples.

Besides, for given a batch of query examples $\mathcal{X}$, we input them to the target model $\mathcal{F}_{tgt}$ to get the output of each example and calculate its **Boundary Values** (BV) to verify whether the query samples are close to the decision boundary of the target model or not. The proposed BV can be calculated as follows:

$$BV = \sum_{i=1}^{B} (p(\mathcal{F}_{tgt}(\mathcal{X}_i))_{top_1} - p(\mathcal{F}_{tgt}(\mathcal{X}_i))_{top_2}), \qquad (10)$$

where p is the Soft-max function, the top_1 and top_2 are the maximum value and sub-maximal value in the output probability vector, and the B is the total example counts.

5.2   Attack Performance

Experiments on MNIST and Fashion-MNIST: We report the ASR under targeted and untargeted attacks for both label-only and probability-only scenarios. As shown in Table 1, the ASR of SCME is much higher than the SOTA baselines on MNIST and Fashion-MNIST datasets. Obviously, our method can obtain higher ASR than other baselines in most cases with a small number of queries (here is 20K). This phenomenon shows that the proposed method is more applicable to the real world than the baselines.

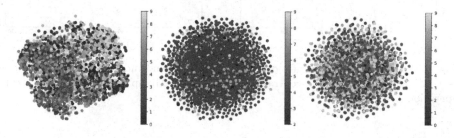

Fig. 5. The T-SNE of original CFIAR-10 data (left), synthetic data by EBFA (middle) and synthetic data by SCME (right).

Experiments on CIFAR-10, CIFAR-100 and Tiny-ImageNet: We further investigate the performance of our method on complex datasets. From the results shown in Tables 2 and 3, our method achieves the best attack performance over probability-only and label-only scenarios under all datasets. In addition, compared to the strong baselines EBFA, our method still outperforms it significantly. Although the number of categories directly affects the training of the substitute model, our method still achieves a very high ASR on the CIFAR-100 and Tiny-ImageNet datasets, which have 100 categories and 200 categories, respectively. On the Tiny-ImageNet dataset, our method even achieves the highest ASR of 96.44% in the soft label setting. These improvements effectively demonstrate the superiority of the proposed SCME.

Table 4. Boundary value of EBFA and SCME.

Methods	EBFA		SCME	
	w.o. aug	w. aug	w.o. aug	w. aug
Boundary Values	**9150.8699**	9010.6072	9469.6909	**8717.2107**

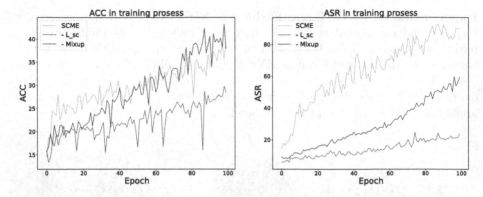

Fig. 6. The ablation results of model accuracy (left) and model ASR (right), where "- L_sc" means without self-contrastive loss, and "- Mixup" means without Mixup operation.

5.3 Evaluation on Data Diversity

To evaluate the generated data's diversity of the strong baseline EBFA and the proposed SCME, we generated 10,000 examples and fed them into the same model trained on the CIFAR-10 dataset to get the predicted labels. The results in Fig. 4 show the data with data augmentation or not. The results show most of the examples generated by EBFA were classified as "deer", while synthetic examples by SCME have preferable inter-class diversity. Further, we plot the T-SNE for real data and synthetic examples generated by EBFA and SCME, respectively, in Fig. 5. The results illustrated that our method generates examples similar to the real data, i.e., with more intra-class diversity. These phenomena strongly support that our method can generate data with high inter- and intra- diversity.

5.4 Evaluation on Boundary Value

To verify whether the query examples are closer to the decision boundary of the target model, we compared the BV of 10,000 examples generated by EBFA and SCME. The results in Table 4 show although EBFA achieves smaller BV without data augmentation, SCME can achieve substantially lower BV with data augmentation. This further demonstrates the effectiveness of the Mixup operation in SCME for generating query examples close to the decision boundary.

5.5   Ablation Study

To investigate the contribution of Self-Contrastive loss $\mathcal{L}_G$ (described in Sect. 4.2 and Mixup operation, we plot the model classification accuracy (ACC) and the model ASR in the model training process. The results in Fig. 6 shows that using both $\mathcal{L}_G$ and Mixup augmentation performs best on both ACC and ASR, besides the model training convergence faster. For instance, the standard SCME is close to convergence with 6K queries, and the ASR is also beyond 80%.

6   Conclusion

In this paper, we proposed a novel data-free model extraction attack, namely SCME, to boost the attack performance under query-limited settings. Specifically, we first design a self-contrastive loss to guide the generator to synthesize the query data with high inter- and intra-class diversity. Besides, we introduce the Mixup augmentation to combine two generated query samples as the final query input to obtain effective decision boundaries and further help the simulation process of the substitute model. Extensive empirical results show that the proposed SCME framework can achieve SOTA attack performance.

Acknowledgements. This work is supported in part by Yunnan Province Education Department Foundation under Grant No.2022j0008, in part by the National Natural Science Foundation of China under Grant 62162067 and 62101480, Research and Application of Object Detection based on Artificial Intelligence, in part by the Yunnan Province expert workstations under Grant 202205AF150145.

References

1. Chen, T., Kornblith, S., Norouzi, M., Hinton, G.E.: A simple framework for contrastive learning of visual representations. In: ICML, vol. 119, pp. 1597–1607 (2020)
2. Demuynck, K., Triefenbach, F.: Porting concepts from dnns back to gmms. In: ASRU, pp. 356–361 (2013)
3. Dong, Y., Pang, T., Su, H., Zhu, J.: Evading defenses to transferable adversarial examples by translation-invariant attacks. In: CVPR, pp. 4312–4321 (2019)
4. Duan, R., Ma, X., Wang, Y., Bailey, J., Qin, A.K., Yang, Y.: Adversarial camouflage: Hiding physical-world attacks with natural styles. In: CVPR, pp. 997–1005 (2020)
5. Duan, R., et al.: Adversarial laser beam: Effective physical-world attack to dnns in a blink. In: CVPR, pp. 16062–16071 (2021)
6. He, K., Zhang, X., Ren, S., Sun, J.: Deep residual learning for image recognition. In: CVPR, pp. 770–778 (2016)
7. Krizhevsky, A., Hinton, G., et al.: Learning multiple layers of features from tiny images (2009)
8. Krizhevsky, A., Sutskever, I., Hinton, G.E.: Imagenet classification with deep convolutional neural networks. In: NIPS, pp. 1106–1114 (2012)
9. Kurakin, A., Goodfellow, I.J., Bengio, S.: Adversarial examples in the physical world. In: ICLR (2017)

10. Le, Y., Yang, X.S.: Tiny imagenet visual recognition challenge (2015)
11. LeCun, Y., Cortes, C., Burges, C.: Mnist handwritten digit database (1998)
12. Liu, A., et al.: Perceptual-sensitive GAN for generating adversarial patches. In: AAAI, pp. 1028–1035 (2019)
13. Liu, J., Park, J.: Seeing is not always believing: detecting perception error attacks against autonomous vehicles. IEEE Trans. Dependable Sec. Comput. **18**(5), 2209–2223 (2021)
14. Orekondy, T., Schiele, B., Fritz, M.: Knockoff nets: stealing functionality of black-box models. In: CVPR, pp. 4954–4963 (2019)
15. Papernot, N., McDaniel, P.D., Goodfellow, I.J.: Practical black-box attacks against machine learning. In: Asia@CCS, pp. 506–519 (2017)
16. Simonyan, K., Zisserman, A.: Very deep convolutional networks for large-scale image recognition. In: ICLR (2015)
17. Tramèr, F., Zhang, F., Juels, A., Reiter, M.K., Ristenpart, T.: Stealing machine learning models via prediction apis. In: USENIX Security, pp. 601–618 (2016)
18. Wang, W., et al.: Delving into data: effectively substitute training for black-box attack. In: CVPR, pp. 4761–4770 (2021)
19. Xiao, H., Rasul, K., Vollgraf, R.: Fashion-mnist: a novel image dataset for benchmarking machine learning algorithms (2017)
20. Yu, M., Sun, S.: Fe-dast: fast and effective data-free substitute training for black-box adversarial attacks. Comput. Sec. **113**, 102555 (2022)
21. Yuan, X., Ding, L., Zhang, L., Li, X., Wu, D.O.: ES attack: model stealing against deep neural networks without data hurdles. IEEE Trans. Emerging Topics Comput. Intell. **6**(5), 1258–1270 (2022)
22. Zhang, J., et al.: Towards efficient data free blackbox adversarial attack. In: CVPR, pp. 15094–15104 (2022)
23. Zhou, M., Wu, J., Liu, Y., Liu, S., Zhu, C.: Dast: data free substitute training for adversarial attacks. In: CVPR, pp. 231–240 (2020)

CSEC: A Chinese Semantic Error Correction Dataset for Written Correction

Wenxin Huang[1,2], Xiao Dong[1,2], Meng-xiang Wang[3], Guangya Liu[1,2], Jianxing Yu[1,2,4(✉)], Huaijie Zhu[1,2], and Jian Yin[1,2]

[1] School of Artificial Intelligence, Sun Yat-sen University, 519082 Zhuhai, China
{huangwx67,dongx55,liugy28}@mail2.sysu.edu.cn,
{yujx26,zhuhuaijie,issjyin}@mail.sysu.edu.cn
[2] Guangdong Key Laboratory of Big Data Analysis and Processing, 510006 Guangzhou, China
[3] China National Institute of Standardization, 100088 Beijing, China
wangmx@cnis.ac.cn
[4] Pazhou Lab, Guangzhou 510330, China

Abstract. Existing research primarily focuses on spelling and grammatical errors in English, such as missing or wrongly adding characters. This kind of shallow error has been well-studied. Instead, there are many unsolved deep-level errors in real applications, especially in Chinese, among which semantic errors are one of them. Semantic errors are mainly caused by an inaccurate understanding of the meanings and usage of words. Few studies have investigated these errors. We thus focus on semantic error correction and propose a new dataset, called **CSEC**, which includes 17,116 sentences and six types of errors. Semantic errors are often found according to the dependency relations of sentences. We thus propose a novel method called **Desket** (**De**pendency **S**yntax **K**nowledge **E**nhanced **T**ransformer). Desket solves the CSEC task by (1) capturing the syntax of the sentence, including dependency relations and part-of-speech tagging, and (2) using dependency to guide the generation of the correct output. Experiments on the CSEC dataset demonstrate the superior performance of our model against existing methods.

Keywords: Writing Correction · Semantic Errors · Error Correction

1 Introduction

Textual errors are prevalent in written communication across different languages and can be broadly categorized into distinct types based on their attributes, including spelling, grammar, and semantic errors. These errors are primarily caused by human negligence or biases inherent in language models. Regardless of their specific type, these errors have significant implications for comprehension and interpretation, often resulting in miscommunication and misunderstandings of the intended message. Therefore, it is crucial to address these errors through

© The Author(s), under exclusive license to Springer Nature Singapore Pte Ltd. 2024
B. Luo et al. (Eds.): ICONIP 2023, LNCS 14451, pp. 383–398, 2024.
https://doi.org/10.1007/978-981-99-8073-4_30

Examples of Chinese Spelling and Grammar Errors			
Error Type		Wrong Sentence	Correction
Chinese Spelling Error	Near-phonetic	一平酒 (flat of wine)	一瓶酒 (bottle of wine)
	Near-type	轻过 (light through)	经过 (go through)
Chinese Grammar Error	Redundant word	忘记吃吃饭 (forget to eat eat)	忘记吃饭 (forget to eat)
	Missing words	忘记饭 (forget rice)	
	Selection errors	忘记次饭 (forget to rice)	
	Ordering errors	忘记饭吃 (forget rice eat)	

Examples of Chinese Semantic Errors

Type: 搭配不当 (Inadequate Collocatio)

Wrong Sent: 大雪纷飞，他戴着帽子和皮靴就出门了。
Correction: 大雪纷飞，他戴着帽子和穿着皮靴就出门了。
Trans: He went out in a snowstorm, putting on a hat and (wearing) leather boots.

Type: 语序错误 (Incorrect Ordering)

Wrong Sent: 老师纠正并且指出了我的错误。
Correction: 老师指出并且纠正了我的错误。
Trans: The teacher pointed out and corrected my mistakes.

Type: 成分残缺 (Missing Elements)

Wrong Sent: 通过老师的讲解，使我明白了这个道理。
Correction: 通过老师的讲解，我明白了这个道理。
Trans: Through the teacher's explanation, make I understood this truth.

Fig. 1. The left part shows examples of spelling and grammar errors. The right part shows examples of semantic errors. The words in red are errors and in blue are corresponding corrections. (Color figure online)

effective writing correction practices in various contexts, such as academic articles, publications, and business documentation [9].

Spelling errors and grammatical errors are considered shallow-level mistakes, such as missing or extra characters, incorrect word choice, or transposed word order, such as '一平酒' should be '一瓶酒' (one bottle of wine).

Instead, there are many unsolved deep-level errors in real applications, among which semantic errors are one of them. Such errors are mainly caused by an inaccurate understanding of the meanings and usage of words.

For example, in the sentence '老师纠正并且指出了我的错误' (The teacher corrected and pointed out my mistake), '纠正并且指出' (correct and point out) is incorrect in the logic sort, where '指出' (point out) should come before '纠正' (correct). The correction should be '指出并且纠正' (point out and correct).

Previous research on Chinese writing correction has predominantly focused on addressing spelling and grammar errors, often neglecting semantic errors. In order to bridge this research gap, we propose to focus on such a new task. We observe that dependency relations help to find semantic errors. As illustrated in Fig. 2, the parsing tree reveals syntactic collocational connections that hold semantic relevance. By leveraging the information provided by dependency relations and POS, we detect the erroneous components based on their contextual context. That can also facilitate the generation of potential correction options, including modifying POS, adjusting phrase order, or adding/removing keywords.

Motivated by the above observation, we introduce a novel method called **Desket** (*De*pendency *S*yntax *K*nowledge *E*nhanced *T*ransformer) to address semantic errors in Chinese sentences. We first encode the input by Bert [4], and then employ a syntactic analysis module to extract the dependency relations and POS information from the sentence. To align the character-level semantic

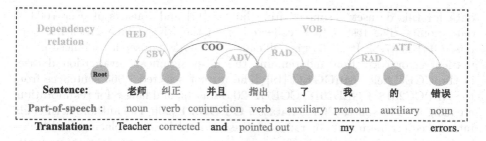

Fig. 2. The dependency syntax structure of a sentence. The words in red are errors, in blue are part-of-speech and in green are dependency relationship. For example, the label 'COO' indicates a coordination relationship, suggesting that the words 'correct' and 'point out' have the same status. (Color figure online)

information, we combine the syntactic structure and part-of-speech information to construct a character-level dependency syntax graph. The knowledge derived from this graph is captured by a Graph Transformer. We then integrate the dependency syntax knowledge with the semantic information to identify the error category of the sentence. Finally, the dependency syntax knowledge guides the generator in generating the correct sentence. To evaluate our model, we create a large-scale dataset called **CSEC**. Extensive experiments are conducted on the dataset. The results demonstrate the effectiveness of our model. In summary, our contributions in this paper include:

- We introduce a novel dataset called CSEC, which comprises 17,116 sentences and encompasses six entirely new types of Chinese semantic errors that are not covered in existing Chinese writing correction datasets;
- We propose a novel approach to address semantic errors by leveraging dependency relations, part-of-speech information, and textual semantics of words;
- Experiments demonstrate that Desket achieves the best results on the CSEC dataset.

2   Related Work

The current Chinese writing correction datasets primarily consist of **Chinese Spelling Correction (CSC)** and **Grammatical Error Correction (CGEC)**.

CSC is a fixed-length writing correction task, and there are several datasets available for CSC: **SIGHAN** [10,14,16]: SIGHAN datasets serve as the first official benchmark datasets in the field of CSC. **Wang271K** [13]: This dataset comprises 271,009 training samples and is currently the largest CSC dataset. **FA-OCR** [5]: In this dataset, the authors utilized OCR technology to convert Chinese subtitles from a video into Chinese text.

CGEC, on the other hand, is a variable-length writing correction task, and there are several datasets associated with it: **NLPCC 2018** [21]: The training data for this dataset is sourced from lang-8 [17] and consists of over 700,000 data points. The test data is collected from the PKU Chinese Learner Corpus. **CGED** [7,8]: It is the first official benchmark dataset for Chinese grammatical error correction and remains the most significant evaluation dataset in the CGEC field. **MuCGEC** [19]: The author selected 7,063 sentences from the NLPCC2018, CGED2018, CGED2020 datasets, and lang-8 for reannotation, creating a multi-source CGEC dataset. **CTC2021** [20]: The authors artificially introduced errors in texts by randomly inserting, deleting, replacing, and transposing words. **CCTC** [12]: CCTC is a cross-sentence error dataset manually annotated by the authors using 1,500 texts written by native speakers.

Unlike these datasets, our CSEC dataset focuses on identifying and correcting Chinese semantic errors.

3   Creation of Chinese Semantic Error Dataset

Data Collection and Annotation. We first manually collect and annotate this data, transforming it into Chinese semantic error correction data suitable for deep learning models. In detail, we gathered 687 publicly available documents from Baidu Wenku, Doc88, and Xueke Net. These documents include middle and high school entrance examination papers, regular examination papers, specialized training exercises, and PPT presentations. The statistical overview of the collected data is illustrated in Fig. 3(a). Subsequently, we carefully filtered and screened the documents to identify exercises containing Chinese semantic errors. The exercises are presented in the format shown in Fig. 3(b). To ensure the annotated quality, we recruited 20 undergraduate students and divided them into two groups to conduct mutual checking.

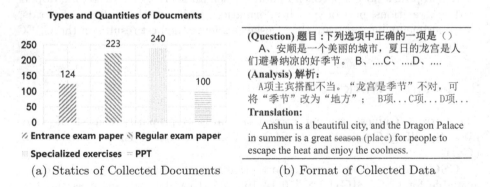

(a) Statics of Collected Documents   (b) Format of Collected Data

Fig. 3. Collected documents and data

As shown in Fig. 1, semantic errors in Chinese can be categorized into six main categories.

Table 1. Data examples. The words in red are errors and in blue are their corrections.

Type	Wrong Sentences	Correction
Inadequate Collocation	夏日的庐山是人们避暑的好季节.	夏日的庐山是人们避暑的好地方.
	Mount Lu in summer is a good season place to avoid summer heat.	
Incorrect Ordering	老师纠正并且指出了我的错误.	老师指出并且纠正了我的错误.
	He pointed out and corrected corrected and pointed out my mistakes.	
Redundant Elements	参加活动的大约有50人左右.	参加活动的大约有50人.
	There are about 50 people around in this class .	
Missing Elements	通过他的讲解，使我受到很大的启发.	通过他的讲解，我受到很大的启发.
	Through his explanation, make I was greatly inspired.	
Confused Structure	老师之所以教导我们，是因为这是她的责任的原因.	老师之所以教导我们，是因为这是她的责任
	The teacher teaches us because the reason it is her responsibility.	
Illogical Expressions	为了防止不再发生意外，我们做了很多练习.	为了防止发生意外，我们做了很多练习.
	To prevent not accidents, we have done many drills.	

(1) ***Inadequate Collocation (IC):*** IC refers to the phenomenon of combining words with conflicting semantics in a forced manner. For example, the sentence '夏日的庐山是人们避暑的好季节' demonstrates an inappropriate collocation between the subject and object, as the subject '庐山 (Mount Lu)' is erroneously combined with the object '季节 (season).' Mount Lu cannot be considered a season; instead, it should be referred to as a specific location or place.

(2) ***Incorrect Ordering (IO):*** IO occurs when the arrangement of words is disorganized, leading to a lack of coherence and logical consistency in the text.

(3) ***Redundant Elements (RE):*** RE refers to superfluous words or phrases that lead to repetition and verbosity in a sentence. For example, the sentence '参加此活动的大约有50人左右' contains both '大约' (approximately) and '左右' (around), which serve the same purpose, resulting redundancy.

(4) ***Missing Elements (ME):*** ME occurs when a sentence lacks a specific element. For example, in the sentence '王老师荣获校先进工作者。' (Teacher Wang was awarded as an advanced worker by the school). The collocation between '荣获' (awarded) and '先进工作者' (advanced worker) is incorrect as it lacks the object '称号' (title) for the verb '荣获' (awarded).

(5) ***Confused Structure (CS):*** CS happens when multiple modes of expression are arbitrarily mixed together within a single sentence. For instance, in

the sentence '这些蔬菜长得这么好，是由于社员们精心管理的结果。' (These vegetables grow so well due to the careful management of the members.), Two expressions are used: '由于...' (due to) and '...的结果' (as a result of), which suffices to retain one of these expressions.

(6) Illogical Expressions (IE): IE denotes a condition in which the structure of a sentence fails to conform to logical reasoning. For instance, in the sentence '为了防止不再发生意外，我们制定出具体的改进措施' both '防止' (prevent) and '不' (not) express negation. The presence of a double negative construction transforms the sentence into an affirmative statement, thereby deviating from its intended meaning. Hence, it is necessary to remove one of these negations.

Table 2. Comparison with natural error datasets. *CSL* means the Chinese as Second Language learners, *NCS* means Native Chinese Speakers. For NLPCC 2018, the statistics here are about their test sets.

Dataset	Task	Sent/Unit	Err.Sent.(%)	Source
SIGHAN 2013 [14]	CSC	1,700	76.1	CSL
SIGHAN 2014 [16]	CSC	4,497	88.1	CSL
SIGHAN 2015 [10]	CSC	3,439	84.0	CSL
CGED [7]	CGEC	16,055	54.2	CSL
NLPCC2018 [21]	CGEC	2,000	99.2	CSL
CCTC [12]	CGEC	1,500	9.8	NCS
MuCGEC [19]	CGEC	7,063	92.7	CSL
CSEC	**CSEC**	**17,116**	77.5	**NCS**

Data Analysis. Table 2 presents the comparison between our dataset with other ones. Our CSEC is the first large-scale dataset for Chinese semantic error correction. Figure 4 presents the statistics of the CSEC dataset. From Fig. 4(b), it can be observed that the average sentence length of the Chinese semantic error dataset is 42. According to the study by [20], the average sentence length in Chinese writing is approximately 15 characters. This indicates that Chinese semantic errors are more likely to occur in longer sentences, requiring stronger contextual awareness for error detection and correction. From Fig. 4(a) and Fig. 4(c), it can be deduced that the most prominent types of errors in Chinese semantic errors are inadequate collocations and missing elements, accounting for 19% and 17% respectively. Unlike English, Chinese does not have explicit spaces between words, and Chinese words do not have clear part-of-speech attributes. This increases the difficulty in analyzing sentence constituents. Figure 4(d) illustrates the distribution of answer counts for incorrect sentences in the CSEC dataset, showing that the majority of incorrect sentences have only one correct answer, while a smaller portion has two or even three alternative modification answers.

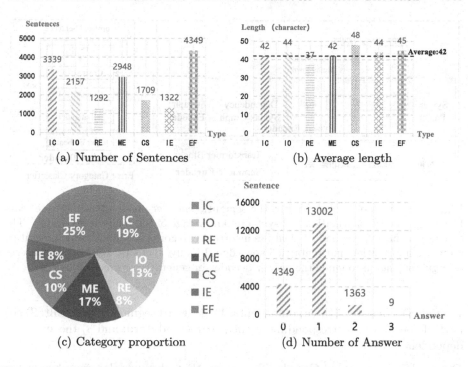

Fig. 4. Statistics of CSEC Dataset. The abbreviation 'IC' refers to Inadequate Collocation, 'IO' represents Incorrect Ordering, 'RE' signifies Redundant Elements, 'ME' denotes Missing Elements, 'CS' stands for Confused Structure, 'IE' indicates Illogical Expressions, and 'EF' means Error Free. The subsequent paper will employ these abbreviations accordingly, and further elaboration will be omitted.

4   Methodology

As illustrated in Fig. 5, we introduce the main architecture of our model.

Semantic Encoder. We use Bert [4] as the foundational architecture for our semantic encoder. Given an input sentence $Sen = (s_1, ..., s_n)$ containing n tokens, we pass it through the encoder layers of the Bert Transformer [11] to learn feature representations in the l-th layer and get the textual features $\mathbf{H}_l^{sen} = \text{Transformer}_l\left(\mathbf{H}_{l-1}^{sen}\right), l \in [1, L]$. Within the Bert transformer, each layer consists of a multi-head attention module and a feed-forward network with residual connections and layer normalization. The output of the final layer, $\mathbf{H}^{sen} = \mathbf{H}_L^{sen}$, is used as the semantic information of the input text.

Syntax Encoder. We first use the syntax parser to extract the Dependency Syntax Tree (DST) of the sentence $Sen = (s_1, ..., s_n)$. It can be represented as:

$$\mathbf{DST} = \text{SyntaxPaser}\left(Sen\right), \tag{1}$$

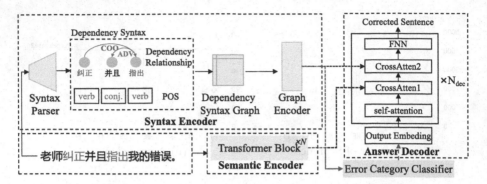

Fig. 5. Structure of Desket. Firstly, the semantic encoder and syntactic encoder capture the semantic information and syntactic knowledge of the sentence separately. The category identification module identifies incorrect categories of sentences by integrating semantic information and syntactic knowledge. Finally, syntactic knowledge will serve as a guiding signal to guide the answer decoder in generating correct sentences.

where DST consists of three components: 1) the word segmentation result; 2) the part-of-speech tags corresponding to each segmented word; and 3) the syntactic dependency relation.

Dependency Syntax Graph. To ensure that the model comprehends the meaning of each word, we address the disparity between the word-level syntax tree and character-level semantic information extracted by the syntax parser. This involves transforming the word-level syntax tree into a character-level Dependency Syntax Graph (**DSG**), aligning the semantic and syntactic knowledge of the sentence. In this process, we represent each Chinese character in the sentence as a node within the graph. The graph construction consists of two main steps. Firstly, we establish edges to incorporate part-of-speech information. When two characters can form a complete word, we introduce an edge between them to signify the corresponding part-of-speech. Secondly, we extend edges to represent dependency relations. Whenever a dependency relationship exists between two words, we create an edge connecting the last character of the first word to the first character of the second word. The weight of this edge corresponds to the syntactic relationship between the two words.

After obtaining the dependency syntax graph, we employ a graph Transformer [1] to model the DSG, allowing the model to acquire syntax knowledge of the sentence. In our **DSG**, the edges encapsulate the dependency relations and part-of-speech information of the sentence. As a result, the Graph Transformer can effectively learn the dependency relations and part-of-speech information by assigning significant importance to these edges during attention score calculation. This can be expressed as:

$$\mathbf{H}^{syn} = \text{GraphTranformer}\,(\mathbf{DSG})\,, \tag{2}$$

where $\mathbf{H}^{syn}$ is the syntactic representation of the input sentence containing the dependency relation and part-of-speech information.

Error Category Classifier. After obtaining the semantic and syntactic representations of the input sentence, we concatenate them along the feature dimension. These concatenated features are then fed into a multi-layer perceptron (MLP) for the purpose of classifying the sentence's error type. This process is formulated as follows:

$$t = \text{argmax}\left(Softmax\left(MLP(\mathbf{H}^{sen} \oplus \mathbf{H}^{syn})\right)\right), \tag{3}$$

where t represents the error type of the input sentence, $\oplus$ denotes the concatenation of vectors.

Answer Decoder. It is constructed based on a standard Transformer decoder, comprising N_{dec} identical blocks. In our proposed model, we have introduced two specialized cross-attention modules within the decoder to ensure attention to both the original sentence and its syntactic representation. One cross-attention module focuses on the interaction between semantic information and syntactic knowledge, while the other module emphasizes the relationship between semantic information and error categories. The decoder takes inputs from three sources: semantic information, syntactic knowledge, and error categories obtained from the aforementioned modules. The error category is specially encoded as a token added at the beginning of the input data. After passing the inputs through a self-attention layer, the decoder initially computes attention between the syntactic knowledge and the input using the first cross-attention mechanism. Subsequently, attention is calculated between the semantic information and the input using the second cross-attention mechanism. This design enables the model to concentrate on words that possess dependency relations and make predictions based on the combined semantic information. The learning process is shown as follows, where y represents the decoder input.

$$\mathbf{y} = \text{LN}(\mathbf{y} + \text{SelfAttn}(\mathbf{y})), \tag{4}$$

$$\mathbf{y} = \text{LN}\left(\mathbf{y} + \text{CrossAttn}\left(\mathbf{y}, \mathbf{H}^{Syn}\right)\right), \tag{5}$$

$$\mathbf{y} = \text{LN}\left(\mathbf{y} + \text{CrossAttn}\left(\mathbf{y}, \mathbf{H}^{Sen}\right)\right), \tag{6}$$

$$\mathbf{y} = \text{LN}\left(\mathbf{y} + \text{FeedForward}\left(\mathbf{y}\right)\right), \tag{7}$$

Loss Function. In our framework, we have two distinct sub-tasks: the identification task and the correction task. The identification task involves multi-classification, while the correction task entails generation. The multi-task loss functions are defined as follows:

$$\mathcal{L}_{cls} = -\sum_{i=1}^{K} y_i \log\left(p_i\right), \tag{8}$$

$$\mathcal{L}_{gen} = -\sum_{j=1}^{n} \log\left(p\left(y_j \mid y_{<j}, Sen\right)\right), \tag{9}$$

where K is the number of error types if the error type is i, then $y_i=1$, otherwise it equals 0, p_i is the probability distribution, n is the length of the corrected sentence, and Sen is the input sentence.

To optimize the model, we combine the generative loss and classification loss to obtain the overall loss: $\mathcal{L} = \mathcal{L}_{cls} + \mathcal{L}_{gen}$.

5   Experiments

Datasets: The CSEC dataset was divided into training, validation, and test sets with a ratio of 0.9:0.05:0.05. Additionally, we incorporated two pseudo datasets, namely the Lang8 data [21] and the HSK6 data [17], for pre-training purposes. Other datasets used are shown in Table 3.

Metrics: In alignment with previous studies conducted by MuCGEC [19], we use metrics such as Accuracy (A), Precision (P), Recall (R), F_1-scores or $F_{0.5}$-scores as performance indicators.

Comparing Methods: Realise [15] integrates semantic, phonological, and visual similarity information into the BERT framework. **SKBert** [18] applies a soft-mask technique to enhance the accuracy of error detection. **S2A** [6] stands as the state-of-the-art (SOTA) Chinese grammar error correction model, amalgamating seq2seq framework outputs with a token-level action sequence prediction module. **MDCSpell** [22] devises a multi-task network that leverages Transformer and BERT architectures for the execution of Chinese Spell Checking (CSC) duties. **Seq2Edit** [19] views the CGEC task as a sequence labeling problem, involving the prediction of modification labels, including insertion, deletion, and substitution. **SpellGCN** [2] combines GCN network with BERT to model the relationship between characters.

Table 3. Statistics of the used data resourcees.

Dataset	Sentence	Task	%Error	Usage
Wang271K [13]	271,329	CSC	88.8	pre-train
SIGHAN_train [10,14,16]	10,050	CSC	76.0	train
SIGHAN 2015 test [10]	1,100	CSC	55.0	test
Lang8 [17]	1,220,906	CSEC/CGEC	89.5	pre-train
HSK [3]	15,6870	CSEC/CGEC	60.8	pre-train
NLPCC_2018-test [21]	2,000	CGEC	99.2	test
CGED_2020-test [8]	1,500	CGEC	54.2	test
CSEC_train	15,116	CSEC	75.0	traing
CSEC_dev	1,000	CSEC	75.0	dev
CSEC_test	1,000	CSEC	75.0	test

5.1 Implementation Details

We utilized Bert [4] as the semantic encoder, while the graph Transformer approach [1] was employed as the syntax encoder. To extract dependency syntax
trees, we make use of the LTP tool with its official configuration. For training
our model, we employ the AdamW optimizer with a learning rate of 1e-5 and a
batch size of 32. Initially, the model was trained for 3 epochs using the Lang8
data [21] and the HSK6 data [17], followed by an additional 8 epochs using the
CSEC dataset. During the generation stage, a label smoothing loss function was
implemented with a smoothing label value of 0.1.

5.2 Experimental Results

Table 4 displays the experimental results obtained from the CSEC test sets.
We conducted two sets of experiments. In the first set, we trained our proposed
Desket model on our CSEC dataset. In the second set, we initially pre-trained the
model on pseudo data and subsequently fine-tuned it using the CSEC dataset.
In both sets of experiments, our Desket model consistently outperforms other
models in terms of both identification and correction levels. At the identification
level, it achieved F1 scores of 60.9% and 62.6%, respectively. As for the correction
level, Desket obtained $F_{0.5}$ scores of 35.3% and 38.2%, respectively.

To assess the generalization performance of the Desket model, we further
conducted experiments on NLPCC-2018, SIGHAN2015, and CGED-2020. As
shown in Table 5, Desket achieved the second-highest F1 score among all methods

Table 4. The experimental results on CSEC test sets

Training Set	Model	Identification Level				Correction Level		
		A	P	R	F1	P	R	$F_{0.5}$
CSEC	Seq2Edit [19]	60.2	62.3	56.7	57.3	**31.5**	22.6	29.2
	S2A [6]	**61.6**	**64.2**	58.4	60.1	29.5	**28.4**	**32.2**
	Realise [15]	60.6	59.0	57.1	58.4	30.2	22.5	28.2
	SKBert [18]	61.1	64.1	**59.5**	**60.4**	27.1	20.5	25.4
	SpellGCN [2]	58.3	61.5	57.6	57.2	25.4	24.4	25.2
	MDCSpell [22]	59.6	62.0	58.1	58.4	30.2	25.5	29.2
	Desket	61.9	65.4	60.3	60.9	37.1	29.8	35.3
pre-tain+CSEC	Seq2Edit [19]	61.1	63.4	57.2	59.7	33.8	29.6	32.9
	S2A [6]	62.2	**65.4**	**59.5**	**61.6**	**36.3**	**32.5**	**34.2**
	Realise [15]	61.7	63.5	58.8	60.3	32.7	24.8	30.7
	SKBert [18]	61.9	64.6	58.3	60.2	28.3	24.5	29.7
	SpellGCN [2]	60.1	63.2	57.4	59.3	32.0	27.1	31.9
	MDCSpell [22]	**62.4**	64.3	57.5	61.1	35.4	26.8	32.7
	Desket	64.2	67.4	61.9	62.6	40.7	35.7	38.2

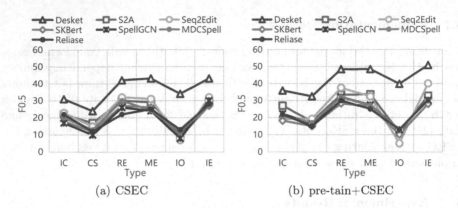

Fig. 6. The experimental results for the correction levels of various error types on the CSEC dataset

Table 5. Experiment results on SIGHAN2015 test sets.

Model	Detection Level			Correction Level		
	P	R	F1	P	R	F1
Seq2Edit [19]	76.8	73.0	69.0	71.1	76.2	68.6
S2A [6]	74.8	**80.7**	77.7	72.1	77.7	75.9
Realise [15]	76.8	65.0	63.0	70.1	72.2	70.6
SKBert [18]	73.6	75.7	66.4	72.3	66.1	57.5
SpellGCN [2]	77.1	80.9	77.9	74.5	**78.1**	76.3
MDCSpell [22]	80.8	80.6	80.7	78.4	78.2	78.3
Desket	**77.9**	78.9	**78.0**	**77.5**	75.2	**77.3**

at the Detection level, with values of 78.0% and 77.3%, respectively. Additionally, as demonstrated in Table 6, Desket attained the highest $F_{0.5}$ score and an R score of 25.24% in the NLPCC task. Moreover, in the case of CGED-2020, both the P and $F_{0.5}$ scores reach their peak values.

To further verify the advantage of Desket, we conducted an evaluation of each category. As demonstrated in Fig. 6, Desket outperformed other models in each category. That indicates Desket exhibits wide adaptability and can effectively handle various types of error data. The advantage of the Desket model may stem from its effective modeling of syntactic structure and contextual information. By leveraging syntactic parsing tools and context representation methods, Desket can accurately identify and correct errors across different categories.

Table 6. Experiment results on NLPCC-2018 test sets and CGED-2020 test sets.

Model	NLPCC-2018			CGED-2020		
	P	R	$F_{0.5}$	P	R	$F_{0.5}$
Seq2Edit [19]	**38.79**	24.03	**40.45**	33.33	**18.63**	29.17
S2A [6]	39.57	28.25	38.46	**40.72**	19.46	**32.91**
Realise [15]	38.60	24.00	27.00	30.12	15.31	26.78
SKBert [18]	35.83	22.01	29.70	30.33	14.32	27.50
SpellGCN [2]	37.60	23.70	39.41	32.1	15.02	26.73
MDCSpell [22]	38.76	23.19	34.17	36.57	14.27	27.86
Desket	36.56	**25.24**	42.11	40.97	17.05	33.90

5.3 Ablation Study

Table 7. Ablation results of the Desket model averaged on CSEC dataset. '- dep' means removing the dependency relation, ' pos' removing part-of-speech information, '- Both' means removing both of them.

Model	Identification Level				Correction Level		
	A	P	R	F1	P	R	$F_{0.5}$
Desket	**64.2**	**67.4**	**61.9**	**62.6**	**40.7**	**30.7**	**38.2**
- dep	60.1	62.5	58.4	59.7	33.8	29.6	32.9
- pos	62.7	66.9	59.1	60.2	39.3	26.9	36.9
- Both	59.1	58.1	56.8	58.5	32.7	24.8	30.7

Semantic Components: We conducted ablation study. As presented in Table 7, the Desket model outperformed all other variants in terms of identification level, achieving an F1 score of 62.6%. However, removing either the dependency structure or part-of-speech information had a detrimental effect on the model's performance, with F1 scores dropping to 59.7% and 60.2%, respectively. Notably, when both the dependency structure and part-of-speech information were removed, the model exhibited the lowest F1 score of 58.5%, underscoring the significance of these components for optimal performance.

The results showed that the full Desket model obtains the highest $F_{0.5}$ score of 38.2%. This signified the effectiveness of the complete model in performing accurate corrections. Furthermore, it was worth noting that removing the dependency structure component has a more significant negative impact on the correction level than removing the part-of-speech information. Specifically, the $F_{0.5}$ score decreased to 32.9% when the dependency structure component was eliminated and dropped to 36.9% when the part-of-speech information was removed. This emphasized the crucial role of the dependency structure component in improving the accuracy of the correction process. Moreover, when dropping both, the performance will decrease to reach its lowest point with an $F_{0.5}$ score of 30.7%. That demonstrated the importance of the dependency structure component and the part-of-speech information for our correction model.

Network Architecture Components: Table 8 provides insights into the ablation study of encoder and decoder components. The full Desket model achieved the highest identification level with an F1 score of 62.6% and the highest correction level with an $F_{0.5}$ score of 38.2%. Removing the Semantic Encoder or the Syntactic Encoder from the model resulted in lower performance across both levels. Removing the Semantic Encoder lowers the F1 score to 61.4%, whereas omitting the Syntactic Encoder yields an F1 score of 59.2%.

Table 8. Ablation study of encoder and decoder components.

Model	Identification Level				Correction Level		
	A	P	R	F1	P	R	$F_{0.5}$
Desket	**64.2**	**67.4**	**61.9**	**62.6**	**40.7**	**30.7**	**38.2**
- SemEncoder	62.0	60.4	58.3	61.4	35.6	30.2	34.2
- SynEncoder	61.7	62.9	60.1	59.2	38.4	27.8	36.5

6 Conclusion

This work focuses on Chinese semantic error correction and introduces the innovative **CSEC** dataset. By specifically targeting Chinese semantic errors, the dataset fills a gap in existing datasets that primarily concentrate on spelling and grammatical error correction. To tackle semantic errors, we propose **Desket**, a new model that leverages the graph Transformer to capture long-range dependency relationships within sentences. Experimental results demonstrate the superior performance of Desket compared to existing Chinese writing correction models.

Acknowledgements. This work is supported by the National Natural Science Foundation of China (62276279, 62002396), the Key-Area Research and Development Program of Guangdong Province (2020B0101100001), the Tencent WeChat Rhino-Bird Focused Research Program (WXG-FR-2023-06), and Zhuhai Industry-University-Research Cooperation Project(2220004002549).

References

1. Cai, E.: Graph transformer for graph-to-sequence learning. In: Proceedings of the AAAI Conference on Artificial Intelligence, vol. 34, pp. 7464–7471 (2020)
2. Cheng, X., et al.: Spellgcn: incorporating phonological and visual similarities into language models for chinese spelling check. arXiv preprint arXiv:2004.14166 (2020)
3. Cui, X., Zhang, B.I.: The principles for building the "international corpus of learner Chinese". Appli. Ling. **2**, 100–108 (2011)
4. Devlin, J., Chang, M.W., Lee, Kenton, K.: Bert: pre-training of deep bidirectional transformers for language understanding. arXiv preprint arXiv:1810.04805 (2018)
5. Hong, Y., Yu, X., He, N., Liu, N., Liu, J.: Faspell: a fast, adaptable, simple, powerful Chinese spell checker based on dae-decoder paradigm. In: Proceedings of the 5th Workshop on Noisy User-generated Text, pp. 160–169 (2019)
6. Li, e.: Sequence-to-action: grammatical error correction with action guided sequence generation. In: Proceedings of the AAAI Conference on Artificial Intelligrence, vol. 36, pp. 10974–10982 (2022)
7. Rao, G., Gong, Q., Zhang, B., Xun, E.: Overview of nlptea-2018 share task Chinese grammatical error diagnosis. In: Proceedings of the 5th Workshop on Natural Language Processing Techniques for Educational Applications, pp. 42–51 (2018)
8. Rao, G., Yang, E., Zhang, B.: Overview of nlptea-2020 shared task for Chinese grammatical error diagnosis. In: Proceedings of the 6th Workshop on Natural Language Processing Techniques for Educational Applications, pp. 25–35 (2020)
9. Sulaiman, M., Syahri, I.: Grammatical errors on descriptive academic writing. Inter. J. Educ. Res. Developm. **2**(1), 37–44 (2022)
10. Tseng, Y.H., Lee, L.H., Chang, L.P., Chen, H.H.: Introduction to sighan 2015 bake-off for Chinese spelling check. In: Proceedings of the Eighth SIGHAN Workshop on Chinese Language Processing, pp. 32–37 (2015)
11. Vaswani, A., et al.: Attention is all you need. In: Advances In Neural Information Processing Systems 30 (2017)
12. Wang, B., Duan, X., Wu, D., Che, W., Chen, Z., Hu, G.: Cctc: a cross-sentence chinese text correction dataset for native speakers. In: Proceedings of the 29th International Conference on Computational Linguistics, pp. 3331–3341 (2022)
13. Wang, D., Song, Y., Li, J., Han, J., Zhang, H.: A hybrid approach to automatic corpus generation for chinese spelling check. In: Proceedings of the 2018 Conference on Empirical Methods in Natural Language Processing, pp. 2517–2527 (2018)
14. Wu, S.H., Liu, C.L., Lee, L.H.: Chinese spelling check evaluation at sighan bake-off 2013. In: Proceedings of the Seventh SIGHAN Workshop on Chinese Language Processing, pp. 35–42 (2013)
15. Xu, H.D., et al.: Read, listen, and see: leveraging multimodal information helps chinese spell checking. arXiv preprint arXiv:2105.12306 (2021)
16. Yu, L.C., Lee, L.H., Tseng, Y.H., Chen, H.H.: Overview of sighan 2014 bake-off for Chinese spelling check. In: Proceedings of The Third CIPS-SIGHAN Joint Conference on Chinese Language Processing, pp. 126–132 (2014)
17. Zhang, B.: Features and functions of the hsk dynamic composition corpus. Inter. Chinese Lang. Educ. **4**, 71–79 (2009)
18. Zhang, S., Huang, H., Liu, J., Li, H.: Spelling error correction with soft-masked bert. arXiv preprint arXiv:2005.07421 (2020)
19. Zhang, Y., et al.: Mucgec: a multi-reference multi-source evaluation dataset for Chinese grammatical error correction. arXiv preprint arXiv:2204.10994 (2022)

20. Zhao, H., Wang, B., Wu, D., Che, W., Chen, Z., Wang, S.: Overview of ctc 2021: Chinese text correction for native speakers. arXiv preprint arXiv:2208.05681 (2022)
21. Zhao, Y., Jiang, N., Sun, W., Wan, X.: Overview of the NLPCC 2018 shared task: grammatical error correction. In: Zhang, M., Ng, V., Zhao, D., Li, S., Zan, H. (eds.) NLPCC 2018. LNCS (LNAI), vol. 11109, pp. 439–445. Springer, Cham (2018). https://doi.org/10.1007/978-3-319-99501-4_41
22. Zhu, C., Ying, Z., Zhang, B., Mao, F.: Mdcspell: a multi-task detector-corrector framework for chinese spelling correction. In: Findings of the Association for Computational Linguistics: ACL 2022, pp. 1244–1253 (2022)

Contrastive Kernel Subspace Clustering

Qian Zhang[1], Zhao Kang[1(✉)], Zenglin Xu[2,3], and Hongguang Fu[1]

[1] School of Computer Science and Engineering, University of Electronic Science and Technology of China, Chengdu, China
zkang@uestc.edu.cn
[2] Harbin Institute of Technology Shenzhen, Shenzhen, China
xuzenglin@hit.edu.cn
[3] Peng Cheng Lab, Shenzhen, Shenzhen, China

Abstract. As a class of nonlinear subspace clustering methods, kernel subspace clustering has shown promising performance in many applications. This paper focuses on the kernel selection problem in the kernel subspace clustering model. Currently, the kernel function is typically chosen by the single kernel or multiple kernel methods. The former relies on a given kernel function, which poses challenges in clustering tasks with limited prior information, making it difficult to determine a suitable kernel function beforehand. Multiple kernel methods usually assume that the optimal kernel is near a series of predefined base kernels, which limits the expressive ability of the optimal kernel. Furthermore, multiple kernel methods tend to have higher solution complexity than single kernel methods. To address these limitations, this paper utilizes contrastive learning to learn the optimal kernel adaptively and proposes the Contrastive Kernel Subspace Clustering (CKSC) method. Unlike multiple kernel approaches, CKSC is not constrained by the multiple kernel assumption. Specifically, CKSC integrates a contrastive regularization into the kernel subspace clustering model, encouraging neighboring samples in the original space to stay nearby in the reproducing kernel Hilbert space (RKHS). In this way, the resulting kernel mapping can preserve the cluster structure of the data, which will benefit downstream clustering tasks. The clustering experiments on seven benchmark data sets validate the effectiveness of the proposed CKSC method.

Keywords: Kernel subspace clustering · Kernel selection · Contrastive learning

1 Introduction

Subspace clustering is a powerful method for clustering high-dimensional data. It exploits the inherent low-dimensional subspace structure of high-dimensional

This work was partially supported by the National Key Research and Development Program of China (No. 2018AAA0100204), a key program of fundamental research from Shenzhen Science and Technology Innovation Commission (No. JCYJ20200109113403826), the Major Key Project of PCL (No. 2022ZD0115301), and an Open Research Project of Zhejiang Lab (NO.2022RC0AB04).

B. Luo et al. (Eds.): ICONIP 2023, LNCS 14451, pp. 399–410, 2024.
https://doi.org/10.1007/978-981-99-8073-4_31

data and then clusters the data into their respective subspaces. As subspace structure is ubiquitous in real-world high-dimensional data, subspace clustering has been widely used in lots of applications, including computer vision [29], heterogeneous data analysis [2], community clustering in social networks [11], hybrid system identification in control [23], etc.

Among the existing subspace clustering methods, spectral clustering-based approaches [10] have gained significant attention. These methods first construct an affinity graph based on the self-expressiveness property of data and then perform the spectral clustering algorithm [21] on the affinity graph to obtain the final data partition. The self-expressiveness property means that the feature matrix of data can be represented linearly by itself [37]. While these methods have achieved satisfactory results on many data sets, they are not effective in handling nonlinear data, which is more prevalent in real-life applications.

To deal with nonlinear data, kernelization techniques have been introduced into linear subspace clustering models. This line of work transforms the nonlinear data into a hidden high-dimensional feature space corresponding to the given kernel function, where the linearity assumption holds more often [1,22]. Linear clustering methods can then be applied. Due to its convenience, many linear subspace clustering methods have been extended to corresponding kernel-based nonlinear versions. For example, the classic sparse subspace clustering (SSC) [8] and low-rank representation (LRR) models [18] have been expanded to KSSC [24] and KLRR [32] models respectively. They both show better clustering performance than their linear counterparts. Additionally, several approaches incorporate additional regularization and constraint terms into the kernel model to enhance the quality of the affinity graph, as exemplified in references [12, 14–16]. The methods mentioned above all rely on a predefined kernel function, which introduces new challenges in modeling due to the absence of prior knowledge to guide kernel selection. Consequently, Multiple Kernel techniques [33,34] have been employed to tackle this issue. However, they often assume that the optimal kernel lies within the neighborhood of a series of predefined base kernels [35], limiting the representational capability of the optimal kernel. Moreover, multiple kernel approaches do not consistently outperform single kernel approaches [13].

In this paper, we propose a new method for kernel learning called Contrastive Kernel Subspace Clustering (CKSC). The proposed CKSC method can learn the optimal kernel adaptively from data without assuming it is close to some predefined kernel. In specific, the CKSC method incorporates a contrastive regularization term into the KSC model to jointly optimize the kernel and graph matrix. The proposed contrastive regularization encourages nearby samples from the original feature space to remain adjacent in the reproducing kernel Hilbert space (RKHS). Thus, the learned kernel can preserve the cluster structure of data. The main contributions of this paper are summarized as follows:

- This work proposes an adaptive kernel learning method, CKSC, for the subspace clustering model. The proposed CKSC method leverages contrastive learning to directly seek the optimal kernel without relying on the multiple

kernel assumption like other KSC methods. Under this assumption, the ideal kernel is forced to be located around a number of pre-specified base kernels.
- Although contrastive learning has demonstrated outstanding performance in clustering tasks, there has been limited research on its usage in KSC models. To the best of our knowledge, this is the first attempt to utilize contrastive learning to optimize the kernel matrix in the KSC model.
- Compared with several state-of-the-art KSC methods on seven benchmark data sets, the proposed CKSC method achieves the best overall performance.

Notations. In this paper, we use bold italic uppercase and lowercase letters to denote matrices and column vectors, such as matrix A and vector a. A column and a certain element of A are represented by a_i and a_{ij} respectively. A^T denotes the transpose of A, A^{-1} stands for the inverse matrix of A. $A \geq 0$ means that all elements in A are non-negative, and $A \succ 0$ means that A is positive definite. The trace of a matrix is represented as $Tr(\cdot)$. For $diag(A)$ and $Diag(a)$, they indicate the diagonal vector of matrix A and the diagonal matrix specified by the diagonal a. The squared Frobenius norm of a matrix is defined as $\|A\|_F^2 = \sum_{ij} a_{ij}^2$. $\mathbf{0}$ represents the all-zero column vector and I denotes the identity matrix.

2   Related Work

Subspace Clustering. Linear subspace clustering models usually assume that data points can be linearly represented by other data points from the same subspace and then seek the representation matrix satisfying the subspace-preserving property. This property requires that the representation coefficient between data points from different subspaces be zero [1]. These models are formulated as

$$\min_Z \ {}^1\!/_2 \|X - XZ\|_F^2 + \lambda R(Z) \quad \text{s.t. } diag(Z) = \mathbf{0}, \tag{1}$$

where $X \in \mathcal{R}^{d \times n}$ is the feature matrix, the regularization term $R(Z)$ is to promote the learned representation matrix $Z \in \mathcal{R}^{n \times n}$ to be subspace-preserving. Commonly used regularization items include sparse regularization [8], low-rank regularization [18], F-norm regularization [20,26] and block diagonal regularization [30]. Once Z is obtained, subspace clustering methods calculate the affinity graph $A = (Z + Z^T)/2$ and substitute it into the spectral clustering algorithm to obtain the final cluster partition.

The above linear subspace clustering methods cannot divide nonlinear data very well. Therefore, the kernel mapping is introduced into model (1), and the resulting kernel-based non-linear subspace clustering model is as follows:

$$\min_Z \ {}^1\!/_2 Tr(K - 2KZ + Z^T KZ) + \lambda R(Z) \quad \text{s.t. } diag(Z) = \mathbf{0}. \tag{2}$$

In model (2), the kernel matrix K, derived from the kernel mapping $\phi(\cdot)$, significantly influences the model's performance. In this paper, we adopt an adaptive strategy to select appropriate kernel matrices for different data sets. For simplicity, we choose the F-norm regularization as $R(Z)$ since it has a closed-form solution for Z.

Contrastive Learning. In recent years, contrastive learning has emerged as a highly successful approach to unsupervised representation learning. This method learns a feature representation by maximizing the similarities of positive data pairs while minimizing those of negative data pairs in latent feature space. Typically, the positive and negative pairs are constructed by data augmentation [6]. To be specific, a positive pair comprises two augmented samples generated from the same data point, while other data pairs are considered negative pairs. With the constructed positive and negative pairs, suitable contrastive loss functions can be defined. Many loss functions have been developed, such as the triplet loss [5], the noise contrastive estimation (NCE) loss [9], and the normalized temperature-scaled cross-entropy loss (NT-Xent) [4]. The excellent performance of contrastive learning in clustering has been confirmed by plenty of work [17,25,36]. In this paper, we leverage contrastive learning within the KSC framework to learn the kernel matrix. To the best of our knowledge, this is the first work that incorporates contrastive learning into a KSC model.

3   Proposed Methodology

Formulation. In the context of kernel subspace clustering models, our objective is to seek the appropriate kernel mapping $\phi(\cdot)$ in an adaptive manner, enabling the generation of the representation matrix $\boldsymbol{Z}$. To accomplish this, we treat $\phi(\cdot)$ as the target representation function to be acquired and employ contrastive learning to attain the optimal representation function. There are three main elements in contrastive learning, positive and negative pairs, and a contrastive loss. The positive and negative pairs are generally obtained by data augmentation. But for the self-expressive-based graph learning model, the computational complexity of the model increases exponentially with the number of samples. As a result, we consider selecting positive samples from the available data. In this case, we choose the k-nearest neighbors of a sample to form positive pairs with it. Based on the constructed positive and negative pairs, we define the contrastive loss as

$$
\begin{aligned}
\mathcal{J}(\boldsymbol{K}) &= \sum_{i=1}^{N} \sum_{j \in \mathbb{N}_i} -\log \frac{\exp(\phi(\boldsymbol{x}_i)^T \phi(\boldsymbol{x}_j))}{\sum_{p \neq i}^{N} \exp(\phi(\boldsymbol{x}_i)^T \phi(\boldsymbol{x}_p))} \\
&= \sum_{i=1}^{N} \sum_{j \in \mathbb{N}_i} -\log \frac{\exp(k_{ij})}{\sum_{p \neq i}^{N} \exp(k_{ip})},
\end{aligned}
\tag{3}
$$

where $\mathbb{N}_i$ is the set of k-nearest neighbors of data point $\boldsymbol{x}_i$. As we can see, this item forces the mapped features of adjacent samples in the original space to be similar and those of other non-adjacent samples to be distant. In this way, the resulting kernel mapping from (3) will benefit downstream clustering tasks because samples that are close to one another in the original space typically tend to belong to the same cluster.

Integrated (3) into the KSC framework, we get the proposed Contrastive Kernel Subspace Clustering (CKSC) model as:

$$\min_{Z,K} \frac{1}{2}Tr(K - 2KZ + Z^TKZ) + \frac{\alpha}{2}\|Z\|_F^2 + \beta \mathcal{J}(K) \qquad (4)$$

$$\text{s.t. } diag(Z) = 0, K \geq 0, K = K^T, K \succ 0.$$

Here we instantiate the basis KSC framework with the Kernel Truncated Regression Representation (KTRR) method [38] because the F-norm regularization has a closed-form solution to Z, and this method is robust to noise. In addition, to ensure that the obtained K is a valid kernel matrix, we constrain it to non-negative, symmetric, and positive definite. It is worth noting that in model (4), we did not require K to lie in the neighborhood of a series of pre-specified kernels. We only need to add a contrastive regularization term in the self-expression-based graph learning model to learn K and Z simultaneously.

Optimization. We use an alternative optimization strategy to solve variables Z and K in model (4) efficiently. With this strategy, Z and K are optimized separately, while the other variable is fixed.

(1) Fix K, Optimize Z. When K is fixed, the optimization model concerning Z is:

$$\min_Z \frac{1}{2}Tr(K - 2KZ + Z^TKZ) + \frac{\alpha}{2}\|Z\|_F^2 \quad \text{s.t. } diag(Z) = 0. \qquad (5)$$

As the objective function of (5) is separable, we can solve it by components. The sub-problem for the i-th component z_i is:

$$\min_{z_i} \frac{1}{2}\left(k_{ii} - 2z_i^T k_i + z_i^T K z_i\right) + \frac{\alpha}{2}z_i^T z_i \quad \text{s.t. } e_i^T z_i = 0, \qquad (6)$$

where e_i is a unit vector with only the i-th element being non-zero. The optimal solution for z_i can be obtained by Lagrange multiplier method as $\hat{z}_i = (K + \alpha I)^{-1}\left(k_i - \frac{[(K+\alpha I)^{-1}(k_i)]_i e_i}{[(K+\alpha I)^{-1}]_{ii}}\right)$. By formatting all the solutions $\{\hat{z}_i\}_{i=1}^n$ in matrix form, we can obtain the optimal solution to (5) as

$$\hat{Z} = Q - P Diag\left(\frac{diag(Q)}{diag(P)}\right), \qquad (7)$$

where $P = (K + \alpha I)^{-1}$, $Q = PK$.

(2) Fix Z, Optimize K. When Z is fixed, the optimization model concerning K is:

$$\min_K \frac{1}{2}Tr(K - 2KZ + Z^TKZ) + \beta \sum_{i=1}^N \sum_{j \in \mathbb{N}_i} -\log \frac{\exp(k_{ij})}{\sum_{p \neq i}^N \exp(k_{ip})} \qquad (8)$$

$$\text{s.t. } K \geq 0, K = K^T, K \succ 0.$$

Algorithm 1. The algorithm of CKSC

Input: Data matrix X, trade-off parameters α and β, number of neighbors η, truncation threshold k.
Output: The clustering results.
1: Initialize K^0 with $K^0(x_i, x_j) = \exp\left(-\frac{1}{\sigma^2} \|x_i - x_j\|_2^2\right), \sigma = n^{-2} \sum_{i,j} \|x_i - x_j\|_2$.
2: **while** $t < maxiter$ **do**
3:     Update Z^{t+1} via (7);
4:     Update K^{t+1} via AdamW;
5:     Break if $\|Z^{t+1} - Z^t\|_F \leq \varepsilon$
6: **return** the optimal Z^*;
7: Conduct the hard thresholding operation $\tau_k(\cdot)$ on each column of Z^* to only preserve the largest k elements;
8: Perform spectral clustering.

We apply the gradient descent algorithm to solve K element-wisely, the gradient for k_{ij} can be defined as $\nabla_1 + \beta \nabla_2$. The first part ∇_1 is the gradient of the self-expressive item, which is computed by $\nabla_1 = \left[I/2 - Z^T + ZZ^T/2\right]_{i,j}$. The second part ∇_2 is the gradient calculated by the contrastive loss, which is:

$$\nabla_2 = \begin{cases} \frac{\eta}{S} \exp(k_{ij}) - 1, j \in \mathbb{N}_i, \\ \frac{\eta}{S} \exp(k_{ij}), \quad j \notin \mathbb{N}_i, \end{cases} \tag{9}$$

where $S = \sum_{p \neq i}^n \exp(k_{ip})$, η is the number of neighbors. With the computed gradient, K is updated using the AdamW [19] algorithm. Additionally, after each gradient descent step, K undergoes post-processing to satisfy the constraints. The non-negative constraint is enforced by setting K to its element-wise maximum with zero ($K = \max(K, 0)$). Furthermore, the positive definite constraint is ensured through the eigenvalue decomposition operation, denoted as $[\Lambda, S_K] = eig(K)$. Specifically, we apply $\Lambda = \max(\Lambda, 10^{-14})$ to ensure non-negativity of the eigenvalues. We then reconstruct K using $K = S_K \Lambda S_K^T$.

We update Z and K alternately until the convergence condition is met. The whole procedure is outlined in Algorithm 1.

Complexity Analysis. In Algorithm 1, the primary computational cost arises from the repetitive updates of variables Z and K. The update process for Z involves inverting the matrix $(K + \alpha I)$, resulting in a computational complexity of $\mathcal{O}(n^3)$. In fact, since $(K + \alpha I)$ is a positive semi-definite matrix, Z can be solved within $\mathcal{O}(n^2)$ [31]. On the other hand, the main complexity in updating K is associated with the eigenvalue decomposition operation during the post-processing step. The computational complexity of eigenvalue decomposition for a real symmetric matrix K is approximately $\mathcal{O}(n^3)$. Consequently, the overall computational complexity of the Algorithm 1 can be expressed as $\mathcal{O}(t_1 t_2 n^3)$, where t_1 represents the number of alternating iterations, and t_2 denotes the number of gradient descents.

4   Experiments

Data Sets. We evaluate the performance of the proposed CKSC method in seven public benchmark data sets, which include three face databases (YALE, JAFFE, and ORL), two object image data sets (COIL-20 and BA), and two text corpora (TR41 and TR45). Concretely, the three face databases contain varying numbers of facial images under different environments, such as light, expression, etc. The COIL-20 data set contains 1440 toy images from 20 different subjects, pictures of the same object were taken from different poses. The BA data set contains 26 uppercase letters and 10 handwritten digits. TR41 and TR45 data sets are collected from NIST TREC Document Database.

Comparison Methods and Evaluation Metrics. To demonstrate the advantages of the proposed CKSC method, we compare it with several state-of-the-art kernel subspace clustering techniques, encompassing both single-kernel and multiple-kernel approaches. Among the single-kernel methods, we examined Kernel SSC (KSSC) [24], Kernel LRR (KLRR) [32], Similarity Learning via Kernel preserving Embedding (SLKE) [14], Structure Learning with Similarity Preserving (SLSP) [12], and Structured Graph learning framework with Single Kernel (SGSK) [16]. For the multiple-kernel methods, our evaluation included Structured Graph learning framework with multiple Kernel (SGMK) [16], Joint Robust Multiple Kernel Subspace Clustering (JMKSC) [35], Local Structural Graph and Low-Rank Consensus Multiple Kernel Learning (LLMKL) [28], and Simultaneous Learning Self-expressiveness Coefficients and Affinity Matrix Multiple Kernel Clustering (SLMKC) [27].

Two popular clustering metrics, accuracy (ACC) and normalized mutual information (NMI) [3] are applied to quantitatively evaluate the clustering quality. The higher indicator value indicates better clustering performance.

Clustering Results. We report the clustering results of all comparison methods and the proposed method in Table 1. For the single-kernel comparison methods, we adopt the strategy in [12] to assess these methods with 12 commonly used kernel functions in the literature and report the best clustering results. Furthermore, to reduce the influence of initialization, we conduct each experiment 20 times and display the results with the best objective values [7].

From Table 1, we can see that the proposed CKSC method achieves the highest performance in most cases. With respect to the classic KSSC and KLRR methods, the improvement of the proposed method is considerable. Especially on the COIL-20 dataset, the values of ACC and NMI are enhanced by 11.32% and 6.64%, respectively. Compared with other recently developed single kernel subspace clustering methods, the proposed method also demonstrates the best overall performance. Among them, SLSP is the most competitive comparison method. The ACC values on the ORL, TR41, and TR45 data sets are higher than our method, but their NMI values are much lower. On the COIL-20 data set, CKSC outperforms SLSP by 6.8% and 4.15% in terms of ACC and NMI. Since the COIL-20 data set is highly nonlinear, the excellent performance of the proposed method on this data set illustrates the advantage of adaptive kernel learning. Finally, in comparison to the four typical multiple kernel methods (i.e., SGMK, JMKSC, LLMKL, and SLMKC), the proposed method still stands out as the best, highlighting the effectiveness of contrastive kernel learning.

Table 1. Clustering results on various benchmark data sets. The best results on each data set are in bold.

(a) ACC(%)

Data	KSSC	KLRR	SLKE	SLSP	SGSK	SGMK	JMKSC	LLMKL	SLMKC	CKSC
YALE	65.45	61.21	66.24	66.60	62.75	63.62	63.00	65.50	66.70	**67.94**
JAFFE	99.53	99.53	99.85	**100**	99.53	99.53	96.70	**100**	**100**	**100**
ORL	70.50	76.50	77.00	**81.00**	70.05	70.02	72.50	80.00	73.50	80.43
COIL-20	73.54	83.19	84.03	87.71	89.31	89.31	69.60	63.60	88.40	**94.51**
BA	50.64	47.65	50.74	53.85	48.32	49.37	48.40	48.20	51.10	**54.12**
TR41	59.57	71.98	74.37	76.80	72.67	**79.38**	68.90	68.90	69.10	75.28
TR45	71.88	78.84	79.89	**83.04**	77.54	77.54	68.70	74.50	75.10	77.10

(b) NMI(%)

Data	KSSC	KLRR	SLKE	SLSP	SGSK	SGMK	JMKSC	LLMKL	SLMKC	CKSC
YALE	63.94	62.98	64.29	64.38	61.58	62.04	63.10	64.60	65.10	**67.48**
JAFFE	99.17	99.16	99.49	**100**	99.18	99.18	95.20	**100**	**100**	**100**
ORL	83.47	86.25	86.35	88.21	82.65	81.94	85.20	89.00	87.00	**90.47**
COIL-20	80.69	89.87	91.25	92.36	93.80	93.80	81.80	80.60	93.90	**96.51**
BA	62.71	61.43	63.58	64.76	61.94	62.25	62.10	61.90	64.20	**67.99**
TR41	63.36	69.63	70.89	70.50	70.59	69.85	66.00	66.60	67.90	**75.97**
TR45	69.23	77.01	**78.12**	75.27	70.70	70.92	69.00	72.60	73.00	77.61

Ablation Study. In this subsection, we investigate the importance of the proposed contrastive regularization in the CKSC model. To this end, we compared the clustering performance of the CKSC method with the CKSC method without the contrastive regularization on three representative datasets (i.e., YALE, COIL-20, and TR41). For the CKSC method without regularization, the coefficient matrix Z is obtained from the initialized RBF kernel. The experiment results are shown in Table 2. We can observe that updating the kernel matrix K using the contrastive regularization can greatly improve the clustering performance of the model. These results verify that learning a kernel mapping in which the feature representations of neighbor samples are more similar is beneficial for clustering tasks.

Parameter Analysis and Convergence Analysis. There are five parameters in our method, which include two trade-off parameters α and β, one neighbor size parameter η, one truncated threshold k, and the weight decay parameter λ for the AdamW

Table 2. Clustering performance of CKSC w/o contrastive learning.

Data	Metrics	CKSC without contrastive learning	CKSC
YALE	ACC	63.15	**67.94**
	NMI	64.67	**67.48**
COIL-20	ACC	90.63	**94.51**
	NMI	95.83	**96.51**
TR41	ACC	70.05	**75.28**
	NMI	71.23	**75.97**

optimizer. To study how these five parameters influence the model performance, we consider a wide range of candidate values for them and choose the one with the best clustering results. Taking the YALE data set as an example, we first fix $\eta = 9, k = 12, \lambda = 0.007$, and adjust $\alpha \in \{0.01, 0.1, 1, 10\}$ and $\beta \in \{0.01, 0.1, 1, 10, 100\}$. Figure 1 shows the clustering results in terms of ACC and NMI. We can see that larger $\beta(\beta \geq 10)$ performs better, illustrating the importance of the contrastive regularization. In addition, the α with better performance is around 1. Similar phenomena are also observed in other data sets. Thus we search for α from $[1, 10]$ in all experiments. Then we fix $\lambda = 0.007, \alpha = 1, \beta = 100$, and tune $\eta, k \in \{5, 7, 9, 11, 13, 15\}$. The results in Fig. 2 demonstrate that our method is insensitive to these two parameters. Finally, we fix the other four parameters and adjust the optimizer parameter λ in $\{10^{-5}, 10^{-4}, 10^{-3}, 10^{-2}, 10^{-1}\}$. From Fig. 3, we can observe that CKSC achieves reasonable results when $\lambda \in [0.01, 0.1]$. Overall, although the CKSC method has five parameters, it is not sensitive to two of them. And when an appropriate parameter range is given, CKSC is stable and easy to tune.

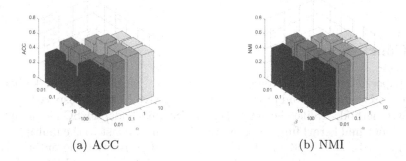

(a) ACC   (b) NMI

Fig. 1. Clustering results with respect to α and β.

For convergence analysis, we plot the logarithmic objective function value curve and the difference curve of variable Z on the ORL data set, as in Fig. 4. We can see that the proposed method tends to converge in about five steps and

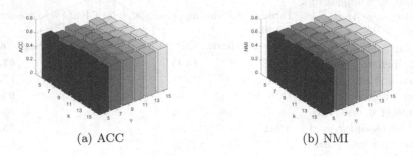

(a) ACC   (b) NMI

Fig. 2. Clustering results with respect to k and η.

remains stable as the number of iterations increases. At the same time, the difference curve also decreases rapidly and gradually tends to 0. These phenomena show that our algorithm can converge effectively.

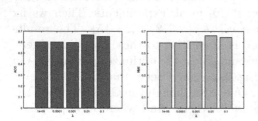

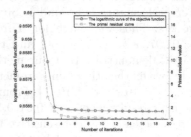

Fig. 3. Clustering results with respect to λ.

Fig. 4. Convergence curves.

5   Conclusions

This paper introduces the concept of contrastive learning into the kernel subspace clustering model and proposes the Contrastive Kernel Subspace Clustering (CKSC) method. Unlike the single kernel KSC methods, CKSC can adaptively learn an optimal kernel function from the data. In contrast to the multiple kernel KSC methods, the CKSC method does not need to restrict the optimal kernel to be around a series of predefined base kernels. Particularly, the CKSC model adopts contrastive loss as a regularization to constrain the kernel matrix so that it retains the neighborhood structure of the data in the original space. Then, combining the contrastive regularization with the self-expressive term, the CKSC model can simultaneously optimize the kernel and graph matrix. The clustering performance of the CKSC method is evaluated on multiple benchmark data sets. Compared with several state-of-the-art KSC methods, our method achieves the best overall performance. In the future, we will further explore improved strategies for positive sample selection to enhance the effectiveness of contrastive learning in kernel learning problems.

References

1. Abdolali, M., Gillis, N.: Beyond linear subspace clustering: A comparative study of nonlinear manifold clustering algorithms. Comput. Sci. Rev. **42**, 100435 (2021)
2. Bu, F.: A high-order clustering algorithm based on dropout deep learning for heterogeneous data in cyber-physical-social systems. IEEE Access **6**, 11687–11693 (2018)
3. Cai, D., He, X., Wang, X., Bao, H., Han, J.: Locality preserving nonnegative matrix factorization. In: IJCAI, pp. 1010–1015 (2009)
4. Chen, T., Kornblith, S., Norouzi, M., Hinton, G.E.: A simple framework for contrastive learning of visual representations. In: ICML. Proceedings of Machine Learning Research, vol. 119, pp. 1597–1607. PMLR (2020)
5. Chopra, S., Hadsell, R., LeCun, Y.: Learning a similarity metric discriminatively, with application to face verification. In: CVPR (1), pp. 539–546. IEEE Computer Society (2005)
6. Dosovitskiy, A., Springenberg, J.T., Riedmiller, M.A., Brox, T.: Discriminative unsupervised feature learning with convolutional neural networks. In: NIPS, pp. 766–774 (2014)
7. Du, L., et al.: Robust multiple kernel k-means using l21-norm. In: IJCAI, pp. 3476–3482. AAAI Press (2015)
8. Elhamifar, E., Vidal, R.: Sparse subspace clustering: algorithm, theory, and applications. IEEE Trans. Pattern Anal. Mach. Intell. **35**(11), 2765–2781 (2013)
9. Gutmann, M., Hyvärinen, A.: Noise-contrastive estimation: a new estimation principle for unnormalized statistical models. In: AISTATS. JMLR Proceedings, vol. 9, pp. 297–304. JMLR.org (2010)
10. Huang, K., Aviyente, S.: Sparse representation for signal classification. In: NIPS, pp. 609–616. MIT Press (2006)
11. Jalali, A., Chen, Y., Sanghavi, S., Xu, H.: Clustering partially observed graphs via convex optimization. In: ICML, pp. 1001–1008. Omnipress (2011)
12. Kang, Z., Lu, X., Lu, Y., Peng, C., Chen, W., Xu, Z.: Structure learning with similarity preserving. Neural Netw. **129**, 138–148 (2020)
13. Kang, Z., Lu, X., Yi, J., Xu, Z.: Self-weighted multiple kernel learning for graph-based clustering and semi-supervised classification. In: IJCAI, pp. 2312–2318. ijcai.org (2018)
14. Kang, Z., Lu, Y., Su, Y., Li, C., Xu, Z.: Similarity learning via kernel preserving embedding. In: AAAI, pp. 4057–4064. AAAI Press (2019)
15. Kang, Z., Peng, C., Cheng, Q.: Twin learning for similarity and clustering: A unified kernel approach. In: AAAI, pp. 2080–2086. AAAI Press (2017)
16. Kang, Z., et al.: Structured graph learning for clustering and semi-supervised classification. Pattern Recogn. **110**, 107627 (2021)
17. Li, Y., Hu, P., Liu, J.Z., Peng, D., Zhou, J.T., Peng, X.: Contrastive clustering. In: AAAI, pp. 8547–8555. AAAI Press (2021)
18. Liu, G., Lin, Z., Yan, S., Sun, J., Yu, Y., Ma, Y.: Robust recovery of subspace structures by low-rank representation. IEEE Trans. Pattern Anal. Mach. Intell. **35**(1), 171–184 (2013)

19. Loshchilov, I., Hutter, F.: Decoupled weight decay regularization. In: ICLR (Poster). OpenReview.net (2019)
20. Lu, C.-Y., Min, H., Zhao, Z.-Q., Zhu, L., Huang, D.-S., Yan, S.: Robust and efficient subspace segmentation via least squares regression. In: Fitzgibbon, A., Lazebnik, S., Perona, P., Sato, Y., Schmid, C. (eds.) ECCV 2012. LNCS, vol. 7578, pp. 347–360. Springer, Heidelberg (2012). https://doi.org/10.1007/978-3-642-33786-4_26
21. von Luxburg, U.: A tutorial on spectral clustering. Stat. Comput. **17**(4), 395–416 (2007)
22. Lv, S., Wei, L., Zhang, Q., Liu, B., Xu, Z.: Improved inference for imputation-based semisupervised learning under misspecified setting. IEEE Trans. Neural Netw. Learn. Syst. **33**(11), 6346–6359 (2022)
23. Pan, Y., Yu, H.: Biomimetic hybrid feedback feedforward neural-network learning control. IEEE Trans. Neural Netw. Learn. Syst. **28**(6), 1481–1487 (2017)
24. Patel, V.M., Vidal, R.: Kernel sparse subspace clustering. In: ICIP, pp. 2849–2853. IEEE (2014)
25. Peng, B., Zhu, W.: Deep structural contrastive subspace clustering. In: ACML. Proceedings of Machine Learning Research, vol. 157, pp. 1145–1160. PMLR (2021)
26. Peng, X., Yi, Z., Tang, H.: Robust subspace clustering via thresholding ridge regression. In: AAAI, pp. 3827–3833. AAAI Press (2015)
27. Ren, Z., Lei, H., Sun, Q., Yang, C.: Simultaneous learning coefficient matrix and affinity graph for multiple kernel clustering. Inf. Sci. **547**, 289–306 (2021)
28. Ren, Z., Li, H., Yang, C., Sun, Q.: Multiple kernel subspace clustering with local structural graph and low-rank consensus kernel learning. Knowl. Based Syst. **188** (2020)
29. Shi, X., Guo, Z., Xing, F., Cai, J., Yang, L.: Self-learning for face clustering. Pattern Recognit. **79**, 279–289 (2018)
30. Wang, L., Huang, J., Yin, M., Cai, R., Hao, Z.: Block diagonal representation learning for robust subspace clustering. Inf. Sci. **526**, 54–67 (2020)
31. Xiao, S., Tan, M., Xu, D.: weighted block-sparse low rank representation for face clustering in videos. In: Fleet, D., Pajdla, T., Schiele, B., Tuytelaars, T. (eds.) ECCV 2014. LNCS, vol. 8694, pp. 123–138. Springer, Cham (2014). https://doi.org/10.1007/978-3-319-10599-4_9
32. Xiao, S., Tan, M., Xu, D., Dong, Z.Y.: Robust kernel low-rank representation. IEEE Trans. Neural Netw. Learn. Syst. **27**(11), 2268–2281 (2016)
33. Xu, Z., Jin, R., King, I., Lyu, M.R.: An extended level method for efficient multiple kernel learning. In: Advances in Neural Information Processing Systems 21, pp. 1825–1832. Curran Associates, Inc. (2008)
34. Xu, Z., Jin, R., Yang, H., King, I., Lyu, M.R.: Simple and efficient multiple kernel learning by group lasso. In: Fürnkranz, J., Joachims, T. (eds.) Proceedings of the 27th International Conference on Machine Learning (ICML 2010), 21–24 June 2010, Haifa, Israel, pp. 1175–1182. Omnipress (2010)
35. Yang, C., Ren, Z., Sun, Q., Wu, M., Yin, M., Sun, Y.: Joint correntropy metric weighting and block diagonal regularizer for robust multiple kernel subspace clustering. Inf. Sci. **500**, 48–66 (2019)
36. Zhang, D., et al.: Supporting clustering with contrastive learning. In: NAACL-HLT, pp. 5419–5430. Association for Computational Linguistics (2021)
37. Zhang, L., Yang, M., Feng, X.: Sparse representation or collaborative representation: Which helps face recognition? In: ICCV, pp. 471–478. IEEE Computer Society (2011)
38. Zhen, L., Peng, D., Wang, W., Yao, X.: Kernel truncated regression representation for robust subspace clustering. Inf. Sci. **524**, 59–76 (2020)

UATR: An Uncertainty Aware Two-Stage Refinement Model for Targeted Sentiment Analysis

Xiaoting Guo[1], Qingsong Yin[2], Wei Yu[3(✉)], Qingbing Ji[3], Wei Xiao[4],
Tao Chang[5], and Xiaodong Wang[5]

[1] ByteDance, Zhichun Road, Haidian District, Beijing, China
[2] Unit 95835 of PLA, Urumqi, China
[3] The 30th Research Institute of China Electronics Technology Group Corporation,
Chuangye Road, Hi-tech Zone, Chengdu, China
yuwei19@nudt.edu.cn
[4] Wuhan Maritime Communication Research Institute, Canglong Avenue, WuHan,
China
[5] College of Computer, National University of Defense Technology, Deya Road,
Kaifu District, Changsha, China

Abstract. Target sentiment analysis aims to predict the fine-grained sentiment polarity of a given term. Although some achievements have been made in recent years, the accuracy of targeted sentiment multi-classification technology is still insufficient-a considerable proportion of samples are incorrectly predicted as the opposite polarity. To this end, we investigate the effectiveness of utilizing model uncertainty and propose a two-stage refinement predicting model based on uncertainty called UATR. UATR can model uncertainty by inferring the distribution of model weights and is more robust to small data learning. Experiments on standard benchmark SemEval14 show that our model can not only reduce the proportion of samples incorrectly predicted as the opposite polarity, but also improves accuracy and F1 values by more than 2% and 3% compared to the current state-of-the-art models, respectively.

Keywords: Sentiment analysis · Model uncertainty · Attention mechanism · Graph neural network · Bayesian neural network

1 Introduction

Fine-grained sentiment analysis refers to predicting the corresponding sentiment polarity to the aspect words described in the sentence. Its analysis object is more clear and more specific, which can better meet the needs of practical application. Fine-grained sentiment analysis has become a new research trend. It has a variety of different definition forms according to different application requirements,

Supported by the National Defense Science and Technology Key Laboratory Foundation under Grant 6142103032203.

Table 1. Samples with Oppsite Polarity Predictions on RGAT-BERT Model

Content	Aspect	Label	Model Prediction
Certainly not the best sushi in New York, however, it is always fresh, and the *place* is very *clean, sterile*	place	2	0
The *staff should* be a bit *more friendly*	staff	0	2
The main *draw* of this place is the *price*	price	2	0

which are specifically divided into the following three types [13]: Aspect-based sentiment analysis(ABSA) [16], also known as targeted sentiment analysis, aims to make sentiment polarity judgments for given aspect words. Aspect-oriented Opinion Words Extraction(AOWE) [6], extract sentiment words or opinion words with regard to aspect words in the sentence. End-to-End Aspect-based Sentiment Analysis(E2E-ABSA) [14], identifies the aspect terms that appear in the sentence and predict corresponding sentiment polarity for each aspect term. E2E-ABSA is closest to the requirements of the practical scene, but the task is more difficult and has not been extensively studied since it involves both sequence labeling and classification subtask. In this work, we focus on targeted sentiment multi-classification tasks.

Almost all neural network models have been explored on targeted sentiment analysis task [2, 4, 25, 26]. Although targeted sentiment analysis has been widely studied, it still faces the problem of low accuracy of multi-classification, the small size of data sets and the imbalance of category samples. The mainstream deep learning model is sensitive to data. Under certain conditions, the larger data can always have better experimental results. However, small data size and unbalanced category samples are common characteristics for datasets collected in practical applications, which will affect the performance of the model. We analyzed the prediction results of classic baseline models (including current SOTA model RGAT-BERT [1]) on benchmark dataset Restaurant [18], and found that samples incorrectly predicted as the opposite polarity accounted for 1/4 to 1/3 in misclassification prediction results of most models.

Table 1 shows some examples of experiments conducted on the current SOTA model RGAT-BERT. In realistic sentiment analysis scenarios, it is more difficult for users to tolerate that the prediction polarity is completely opposite to the label polarity, especially in applications that are sensitive to such polarity. The exact opposite prediction of sentiment polarity will bring great misleading to follow-up tasks. To this end, the effectiveness of utilizing model uncertainty has been investigated.

The introduction of uncertainty to the model is equivalent to the integration of infinite groups of neural networks on a certain weight distribution. It can effectively alleviate the problem of overfitting, have better robustness to small data [8], and reduce the impacts brought by too small data scale. The effective method of modeling model uncertainty is to modify models based on the Bayesian neural network [7,9]. Different from the standard neural network, the model based on the Bayesian neural network no longer trains fixed weight parameter values, but fits the distribution of weight parameters. In this way, the model approximates the real model as much as possible by the feedback of multiple sampling and prediction results of weights parameter distribution. Therefore, the consistent relation between model prediction results and corresponding model confidence can be better modeled by using Bayesian neural networks. Based on this, an Uncertainty-Aware Loss function (UAL) is designed to guide model training and an Uncertainty Aware Two-stage Refinement model (UATR) is proposed to further correct samples with a high possibility to be wrong. Specifically, in the first stage, the uncertainty estimation model based on Bayesian Neural Network integrates multiple groups of neural networks on the parameter distribution to obtain general predictions and corresponding uncertainty. When uncertainty is greater than a certain threshold, the sample will be selected for a secondary prediction. In the second stage, the refinement prediction model chooses the graph neural network model based on the attention mechanism. So that the dependency relationship between components in the sentence is modeled to get a better classification effect.

The main contributions of this paper include:

1. Aiming at the problem of a relatively large proportion of opposite polarity predictions that are prone to appear in small-scale datasets with unbalanced category samples, the effectiveness of model uncertainty applied to targeted sentiment analysis tasks is explored in this paper.
2. On the basis of obtaining the uncertainty of the model, an uncertainty-based loss function UAL is designed, so that the model prediction results tend to be in the interval distribution of correctly classified samples with less uncertainty, and incorrectly classified samples with greater uncertainty. In this way, model can provide a reliable explanation of uncertainty for the prediction results and reduce the proportion of samples with opposite polarity predictions.
3. On the premise of obtaining reasonable distribution of misclassified samples in the uncertainty interval, the strategy to determine the uncertainty threshold for second correction is designed and a novel targeted sentiment analysis model based on uncertainty UATR is proposed to further improve the performance of multi-classification.

Experimental results on the standard benchmark dataset SemEval14 [18] have shown that the accuracy and macro F1 value of UATR model are increased by 2.66% and 4.19% on the laptop dataset respectively, 1.7% and 2.2% on the restaurant dataset respectively, which is compared with the performance of the current SOTA model RGAT-BERT.

2   Related Work

2.1   Targeted Sentiment Analysis

In general, these variant models are either designed for a more complete mining of interdependence among components in the sentence or designed to utilize dynamic embedding representation of words. In terms of current research background, models based on graph neural networks are usually used to represent the syntax structure of sentences and have achieved good results. Hence, the related work of the word embedding and graph neural network for syntax structure encoding will be introduced.

Word Embedding. The word embedding is proposed to capture the contextual relationship and semantic similarity of words in the document. The commonly used word embedding method is distributed representation, mainly including Word2Vec [15], GloVe [17] and BERT [5]. Generally speaking, Word2Vec and GloVe are context-free distributed representation methods, which encode potential semantic information. While Bert, as an outstanding representative of pre-training models, provides context-sensitive dynamic word vectors, and people can adapt to domain words as needed vectors to fine-tune it.

Graph Neural Network. Graph Neural Network (GNN) [19] is a neural network model that operates on the graph structure to model the interdependence between nodes in the graph. In fine-grained sentiment analysis tasks, graph neural networks have been used to encode the grammatical structure of sentences in recent years, with good results. More generally, researchers use the graph convolutional neural network GCN [22] and the graph attention network GAT [11,24] to encode syntactic features so that the target word and the context can exchange information. GNNs can capture the dependency interaction between words according to the dependency tree.

The current SOTA model without extra training data on SemEval14 is RGAT, which is based on the graph attention network and integrates relationship extension. Different from the existing work, they used an independent encoder to model structured information instead of using grammatical structures to modify the contextual representation. In this paper, graph attention network is simply used to model the dependencies between the components in sentences for refinement prediction.

2.2   Model Uncertainty

Neural networks give a good result in most cases, but they are usually regarded as black boxes and cannot provide any interpretation for their results. Occasionally, a particularly bad result will be given, and the tolerance for such a particularly bad result is different in different fields. When a particularly bad result is not tolerated, the model is supposed to output this result with very high uncertainty and conduct a second intervention with the guidance of such uncertainty. A model based on the Bayesian neural network can obtain uncertainty estimation compared with a standard neural network model. Introducing uncertainty into the weights of neural networks for regularization is equivalent to integrating neural networks on the distribution of multiple groups of weights [7,9]. So that model can learn from small datasets more easily and has better robustness to overfitting of small data.

The bayesian neural network includes two processes [7], the learning process and the inference process. The learning process is to learn the distribution of parameters given the data and model, as can be seen in (1); the inference process is to predict the data on this basis, as can be seen in (2). Among these, m represents the model, D represents the given data, and θ represents the parameters.

$$p(\theta \mid D, m) = \frac{p(m \mid D, \theta)p(\theta)}{\int p(m \mid D, \theta)p(\theta)d\theta} \tag{1}$$

$$p(x \mid D, m) = \int p(x \mid \theta, D, m)p(\theta \mid D, m)d\theta \tag{2}$$

Therefore, the key to solving the problem is to deduce the posterior distribution of the parameters given data and model. However, the posterior distribution of parameters requires obtaining the prior distribution of the data. The analytical solution is very difficult to solve. Variational inference [3] and Monte Carlo sampling [20] are common approximate solutions to the posterior distribution of Bayesian neural networks. Variational inference is to select a known distribution (such as the Gaussian distribution) and continuously modify its parameters until it is close to the posterior distribution to be calculated. The goal of Variational Bayes is to find a simple distribution $q(\theta)$ with the smallest KL divergence of the posterior distribution. This distribution is parameterized. By learning the parameter distribution, the simple distribution $q(\theta)$ and the parameter posterior distribution $p(\theta \mid D)$ are as close as possible. It serves as an approximate solution of the posterior distribution. The optimization goal of variational inference is to minimize the KL divergence of the simple distribution and the parameter posterior distribution (Eq. 3).

$$\mathbb{KL}(q(\theta)\|p(\theta \mid \mathcal{D})) = \mathbb{E}_q\left(\log\left(\frac{q(\theta)}{p(\theta \mid \mathcal{D})}\right)\right) \tag{3}$$

The Monte Carlo method is a general term for a class of algorithms that solve problems through random sampling. The problem to be solved is the probability of a random event or the expectation of a random variable. Monte Carlo sampling refers to sampling many samples from a distribution, and then using these samples to replace the distribution for correlation calculations. When the analytical solution of the integral is not easy to calculate, the Monte Carlo method is used to sample a lot of samples from $p(x)$, bring the samples into the integrand, and take the average of the obtained results as the final solution. According to the Law of Large Numbers, the more samples sampled, the closer the solution obtained is to the analytical solution.

3   Approach

The standard neural network model is susceptible to the size of the dataset and the balance of categories in multi-classification task. A considerable proportion of samples are incorrectly predicted as the opposite polarity (especially for non-bert models). In response to the above problem, multiple sets of neural network models on the weight distribution are integrated by modeling uncertainty to make up for the shortcomings.

The proposed approach in this paper is as follows: the uncertainty of the model's prediction results is obtained based on the MC Dropout method, then the loss function of uncertainty perception UAL is designed to guide the model training so that the misclassified samples are distributed in a reasonable uncertainty interval. At the same time, a strategy to dynamically obtain an uncertainty threshold for secondary refinement is designed. When the uncertainty is greater than the threshold, the model based on a graph neural network is used to make refinement predictions.

3.1 Uncertainty Aware Loss

Motivation. Uncertainty estimation algorithms are usually based on Bayesian neural networks. In the related research on Bayesian neural network models, most of them simply apply Bayesian neural networks to obtain uncertainty-based models. Uncertainty is used as an observation index to express the model's confidence in the prediction results, while the feedback effect of uncertainty on the model prediction results is ignored. This section takes advantage of model uncertainty to have better robustness to small-scale data. Taking the consistency relation between model prediction results and corresponding model confidence into account, the uncertainty is fed back into the model training phase and acts on the loss function. In such cases, the model training process is guided by uncertainty and the correctness of corresponding prediction results at the same time. The special function for uncertainty perception UAL(Uncertainty-Aware Loss) is proposed. On the basis of ensuring the accuracy of multi-classification, the proportion of opposite polarity prediction is reduced.

UAL Design. In UAL, the training goal is not only to predict the sample polarity correctly but also to pair it with small uncertainty. The principle is, based on the original loss function, to reward the correct samples and impose penalties on the wrong samples. Specifically, the loss of the correctly classified sample is reduced and the reward coefficient of the sample paired with a smaller uncertainty is larger. The loss of the wrong sample is increased and the wrong sample with a smaller uncertainty will have a larger penalty coefficient. The ultimate goal is to make the model predict the sample correctly with a small uncertainty or predict the sample wrongly with a large uncertainty. In other words, the model is supposed to have a greater degree of confidence in its prediction results.

Information entropy $H(\mathbf{p})$ represents the uncertainty of random variables, as can be seen in (4), which can be used to describe the amount of information contained in random variables. The amount of information is directly related to uncertainty, and they often have a consistent trend of change. Variance focuses on the value of the random variable itself, while information entropy focuses more on the distribution of the random variable, rather than the specific value. When the number of categories that a random variable can take is certain, the probability distribution of each category is more even, the greater the entropy value. In a classification task, the information entropy of the probability vector $H(\mathbf{p})$ is used to represent the model uncertainty, as can be seen in (formula 4), where $\widehat{\mathbf{W}}_t$ is sampled from dropout weight distribution $q_\theta^*(\mathbf{W})$, marked as

$\widehat{\mathbf{W}}_t \sim q_\theta^*(\mathbf{W})$.

$$H(\mathbf{p}) = -\sum_{c=1}^{C} p_c \log p_c \tag{4}$$

The value range of uncertainty information entropy $H(p)$ can be seen in (5), where k is the number of categories in the classification task.

$$0 \leqslant H(p) \leqslant \log k \tag{5}$$

According to the previous analysis, different reward and punishment coefficients are needed to impose on the loss function in accordance with the correctness of the model prediction. This coefficient has a consistent relationship with the uncertainty of the model's prediction results for the sample. Therefore, the information entropy is normalized and used as the reward and punishment coefficient imposed on the original loss function. The samples with correct predictions are multiplied by the uncertainty, and the samples with incorrect predictions are divided by the uncertainty, expressed as $loss_{new}$, as can be seen in (7), $loss_{ori}$ is represented by the formula 6 [12], meaning the loss function without UAL guidance.

$$loss_{ori} = -\frac{1}{N}\sum_{i=1}^{N}\log p\left(\mathbf{y}_i \mid \mathbf{f}^{\widehat{\mathbf{W}}_i}(\mathbf{x}_i)\right) + \frac{1-p}{2N}\|\theta\|^2 \tag{6}$$

$$loss_{new} = \frac{loss_{ori}}{uncertainty^{(-1)^{y_{true} \Leftrightarrow y_{pred}}}} \tag{7}$$

Taking into account that the uncertainty value may be very small, when uncertainty acts on the wrong sample, the loss will be infinite. Therefore, the threshold value t needs to be set to control it. The final loss is designed as follows:

$$\text{Loss} = \begin{cases} loss_{ori} & H(p) \leqslant t \\ loss_{new} & H(p) > t \end{cases} \tag{8}$$

The sample obtains the predicted polarity and the corresponding uncertainty estimate through the MC Dropout model. According to whether the predicted polarity is consistent with the label polarity, the uncertainty estimate is guided to increase or decrease. It is different from the standard loss function, the UAL loss function has two optimization directions for the model. One is to increase the

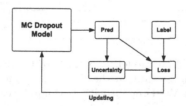

Fig. 1. UAL Guided Method. According to whether the predicted value is consistent with the label value, multiply or divide by the uncertainty value to reduce or increase the loss.

number of correctly classified samples as much as possible, the other is to make the correct predictions with a small uncertainty, and wrong predictions with large uncertainty, which is shown as Fig. 1. In the standard cross-entropy loss, the main factor influencing the loss function is the probability value corresponding to the label category, and the direction of reducing the loss is to continuously increase the probability value corresponding to the label category in the prediction result. The relation between the magnitude of the loss function value caused by different prediction results is as follows (formula 9).

$$L(predicted\ wrongly) > L(predicted\ correctly) \tag{9}$$

Under the guidance of UAL, the relation of loss function caused by different prediction results is given by the following inequality (formula 10):

$$\begin{aligned}
L(predicted\ wrongly\ with\ small\ uncertainty) > \\
L(predicted\ wrongly\ with\ large\ uncertainty) > \\
L(predicted\ correctly\ with\ large\ uncertainty) > \\
L(predicted\ correctly\ with\ small\ uncertainty)
\end{aligned} \tag{10}$$

3.2   Uncertainty Aware Two-Stage Refinement Model

Uncertainty-Aware Two-stage Refinement model UATR is introduced to further improve multi-classification performance in this section. The overall framework of the model is shown in Fig. 2. The uncertainty estimation model guided by UAL is in the first stage. The variational fully connected layer implemented by the MC Dropout method is used to obtain predictions and corresponding uncertainty. The goal of this stage is to determine a suitable uncertainty threshold

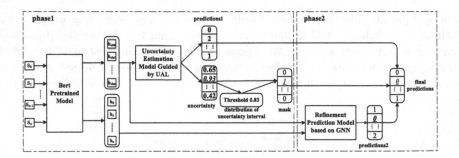

Fig. 2. An Overview of UATR Model. It consists of two phases. Bert pooled output is used to obtain prediction results in the first phase. The uncertainty is estimated through a model based on MC Dropout with UAL guidance as well. According to the distribution difference of samples predicted correctly and wrongly in the uncertainty interval, an uncertainty threshold suitable for the secondary correction is obtained. The samples with uncertainty greater than the threshold are sent to the second phase for prediction correction.

for refinement prediction in the second stage. It is determined by the distribution differences between the correct samples and the wrong samples in the uncertainty interval. So wrong predictions in the first stage have a greater probability of being sent to the second correction. While most of the samples with correct predictions are retained with an uncertainty below the threshold, effectively avoiding a large number of correct predictions being incorrectly corrected. The refinement model of the second stage is based on a graph neural network. Compared with the model of the first stage, it focuses more on modeling the dependency between components of the sentence.

Uncertainty Estimation Model Guided by UAL. The uncertainty estimation model guided by UAL is realized by a variational fully connected layer based on MC Ddropout in the first stage. The purpose of the variational fully connected layer is to transform the fully connected layer into a Bayesian-neural-network-based layer. Given the data set D, the simple distribution $q_\theta^*(\mathbf{W})$ closest to the posterior distribution $p(\mathbf{W} \mid \mathcal{D})$ of the parameters is supposed to find. Multiple rounds of sampling are performed on the parameter distribution, and the average of the sampling results is used as the final prediction result of the model. Corresponding uncertainty can be obtained by calculating information entropy, as can be seen in (4). So the uncertainty of model outputs is modeled and samples with greater uncertainty are selected for correction.

The sentence vector bert_pool_output of the input sentence is used as the input of the variational Dense layer. Keep the dropout settings of the training phase and testing phase exactly the same. The loss according to UAL, as can be seen in (7) is calculated and used to guide model training, so that model can predict correctly with small uncertainty and incorrectly with great uncertainty. When training is completed, the Bernoulli distribution closest to the posterior distribution of the parameters is obtained. In the testing phase, Monte Carlo sampling is performed on the Bernoulli distribution of parameters obtained by the training phase(take T times as an example). The final prediction result is obtained by averaging multiple sampling results on that distribution, as can be seen in (11).

$$p\left(\mathbf{y}^* \mid \mathbf{x}^*, \mathcal{D}\right) \approx \frac{1}{T} \sum_{j=1}^{T} p\left(\mathbf{y}^* \mid \mathbf{W}_j, \mathbf{x}^*\right) q_\theta^*\left(\mathbf{W}_j\right) \tag{11}$$

Strategies for Choosing Uncertainty Threshold. In a certain training batch, the corresponding uncertainty value is obtained by the prediction result. The distribution statistic of samples with correct predictions and wrong predictions in uncertainty intervals can be realized by combining sentiment classification labels. The distribution difference between wrong predictions and correct predictions is calculated by using each candidate uncertainty threshold as lower bounds. The uncertainty value corresponding to the largest difference is taken as the uncertainty threshold to guide the second refinement prediction of that batch of samples. The threshold here is not fixed but dynamically changes in each training or testing batch. It is dynamically determined by the uncertainty distribution difference between wrong and correct samples predicted by the current model. Such an adaptive adjustment strategy for determining the uncertainty threshold can make the threshold more accurate, scientific, and targeted to guide the second correction in different batches.

Figure 3 shows the difference in the distribution of correct predictions and incorrect predictions with uncertainty values between 0.7 and 0.9 as the threshold after a certain batch. Specifically, when the uncertainty threshold is selected as 0.83, samples with incorrect predictions have the largest difference with samples with correct predictions, in the uncertainty interval [0.83,1). In this case, taking the uncertainty threshold 0.83 as the lower bound, correctly predicted samples are sent to the secondary prediction with the smallest probability, while wrongly predicted samples are sent to the secondary prediction with the highest probability. So that model can achieve better results with the aid of the second refinement prediction.

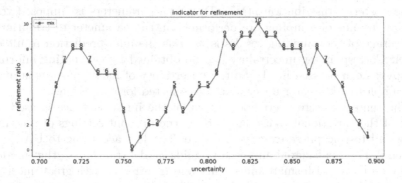

Fig. 3. Distribution difference between wrong predictions and correct predictions for uncertainty ranging from 0.7 to 0.9

A similar work on refinement based on uncertainty is the UAN model proposed in paper [10] for sequence labeling task. In addition to different models and tasks, the key difference in the use of uncertainty between the two works is that the choice of uncertainty threshold in this paper is adaptively determined by the model during the training process, not determined by historical experience. Throughout the training process, uncertainty thresholds of different training batches in different training stages are all adaptive, not fixed. This is conducive to the more targeted use of the uncertainty threshold as the criteria for sample choosing at each batch.

Modified Prediction Model Based on GNN. In the second stage of UATR, the graph neural network model based on attention mechanism (GAT) is chosen as the refinement model. GAT is a variant of the graph neural network, which uses the attention mechanism to aggregate neighbor nodes to realize the adaptive distribution of the weights of different nodes. And thereby greatly improves the expressive ability of the graph neural network model.

The GAT network constructs an unlabeled grammar graph (such as a syntactic dependency tree) $\mathcal{G} = (\mathcal{V}, \mathcal{A})$, V represents all word nodes, and A is the adjacency matrix that reflects the dependency relationship between words in a sentence.

The principle of the adjacency matrix is as follows:

1. There is an interdependence among all the words constituting the aspect word, and the label of the corresponding position adjacency matrix is set to one.
2. According to the dependency analysis, the position label of the adjacency matrix pointed to by two sentence components with the dependency relationship is set to 1, regardless of its directivity, that is, we construct an undirected graph between word nodes.
3. In self-circulation, the dependence of the word pointing to itself is supposed to exist. So the label value of the diagonal of the adjacency matrix is set to one.

The GAT layer is based on a transformer encoder as the basic framework and is mainly composed of a multi-head attention mechanism, a forward network, and a residual connection. On the basis of the transformer encoder, an adjacency matrix containing information about each word node and its semantic neighbor nodes is obtained according to the dependency relationship between the sentence components, so that it can act on the multi-head attention mechanism to explicitly represent the dependence between the target word and other words.

Inspired by Bai et al. [1], we also choose to perform feature fusion between the graph encoding output obtained by the graph attention model and the Bert_output to get richer features.

4   Experiment

This paper focuses on the benchmark dataset SemEval14, which has a relatively small data scale and three sentiment polarity categories. In this paper, research is done on targeted text sentiment analysis based on model uncertainty.

4.1   Dataset

The dataset used in the experiment is the Laptop and Restaurant three classification datasets in SemEval14[1], which are review data on computers and restaurants, respectively. The object of sentiment classification is the aspect words that appear in the review sentence. The ratio of positive and negative neutral samples is about 2:2:1 in the laptop and 3:1:1 in the restaurant. The imbalance of samples in the restaurant dataset is particularly obvious.

---

[1] https://alt.qcri.org/semeval2014/task4/.

4.2   Baseline Model

In the experiment, several classic baseline models are selected in targeted multi-classification sentiment analysis tasks, bert-based, and non-bert based, including the latest SOTA model RGAT in 2020 (without extra training data). For the fairness of results, alignment experiments on the above dataset for all baseline models are conducted. By changing the baseline model to the uncertainty model based on the Bayesian neural network and applying UAL to training, three sets of comparative experiments are performed for each model to investigate the effects of model uncertainty. The proportion of opposite polarity prediction *opp* is used as the evaluation of the effectiveness of the UAL method. Experiments are carried out on the datasets laptop and restaurant. The principle of each model is introduced as follows.:

TD-LSTM* [23]: Respectively model the former context and latter context of a given target word (including the target word), then concatenate the hidden representation of the context, and judge by the softmax output.

ATAE-LSTM* [25]: Learn the embedding representation of aspect words, use this representation and context words to do attention calculations to get the final weighted sum for sentiment classification; in the input part of the model, and combine aspect word embedding and context word embedding together.

RAM* [4]: The method of multiple attention mechanism synthesizes the important features in the difficult sentence structure so that the distant information can also be contained. The result of multiple attention is combined with the GRU network, and different behaviors are inherited from the RNN. The results of attention are combined in a non-linear method. Besides, the memory module is added between the input and attention layer to store information.

BERT-SPC [21]: It is a basic model for applying BERT to sentiment analysis problems. The process is relatively simple. First, the Bert pre-training model is used to encode the sentence and the target word, and then the vector bert_pool_output representing the sentence feature output by Bert is used for classification prediction.

RGAT-BERT [1]: On the basis of the graph attention network, the relationship label is introduced to further model the dependency relationship between sentence components. The attention of relationship perception and the attention of word perception are merged, and finally, the graph encoding vector and sentence encoding vector are feature fused to obtain the final polarity label. In the above baseline models, the non-bert model (with * indicator) is based on AlexYangLi's released code.[2]

4.3   Settings

In the experiment to verify the effectiveness of the UAL method, the setting of the comparison experiment of each baseline model variant is consistent with the baseline model. Regarding the setting of uncertainty-aware two-stage refinement model UATR, the following instructions are given:

---

[2] https://github.com/AlexYangLi/ABSA_Keras.

The Bert pre-trained model used in the experiment selects a 12-layer basic model with an embedding dimension of 768. After word embedding, the dropout rate of the variational Dense layer is set to 0.1, which is consistent with the training phase. The Monte Carlo sampling number is set to 30. Optimizer is Adam optimizer, the initial learning rate is $2 * 10^{-5}$, and the regular term is $1 * 10^{-5}$. In the GAT part of the graph attention network, two transformer encoding layers are stacked, and 4-head is set in multi-head attention. The hardware environment that the experiment relies on includes a GTX 1080Ti GPU and an Ubuntu operating system.

4.4 Main Results

Tabel 2 shows the results of all baseline models with their variant models. Specifically, *ori* represents the baseline model itself, *unc* represents the uncertainty model variant without UAL guidance, and *mix* represents the uncertainty model variant with UAL guidance. The threshold t selected by the loss function in the UAL function is set to 0.01.

Table 2. Preformance of all bascline's variants on SemEval14, *ori* represents the baseline model itself, *unc* represents the uncertainty model variant without UAL guidance, and *mix* represents the uncertainty model variant with UAL guidance, *opp* represents the proportion of samples with opposite polarity predictions in the total predicted samples.

| baseline | variant | laptop | | | variant | restaurant | | |
		accuracy	macro-F1	opp		accuracy	macro-F1	opp
TD-LSTM	ori	70.85	65.43	7.84	ori	79.11	68.51	6.70
	unc	70.85	64.52	7.68	unc	79.55	68.68	**5.80**
	mix	**72.73**	**67.47**	7.05	mix	**79.64**	**69.78**	5.89
ATAE-LSTM	ori	**71.47**	66.28	6.43	ori	77.41	66.55	7.86
	unc	68.81	62.55	7.05	unc	**78.66**	**66.82**	6.25
	mix	71.32	**66.42**	6.43	mix	**78.66**	**66.82**	6.79
RAM	ori	72.41	67.32	6.27	ori	80.71	71.18	5.27
	unc	72.57	66.59	8.31	unc	**80.98**	**71.32**	4.91
	mix	**72.73**	67.73	6.11	mix	80.54	71.18	4.91
BERT-SPC	ori	77.43	74.03	4.08	ori	84.82	77.95	3.84
	unc	77.90	73.15	4.86	unc	84.73	77.21	2.86
	mix	**78.95**	**75.02**	4.08	mix	**85.54**	**78.66**	2.68
RGAT-BERT	ori	**79.17**	**74.55**	3.45	ori	**85.89**	**79.83**	2.95
	unc	78.55	73.47	3.29	unc	85.45	79.63	2.77
	mix	78.06	74.49	2.82	mix	85.09	79.33	2.41
UATR	-	**81.83**	**78.74**	2.82	-	**87.59**	**82.03**	2.05

Table 3. Samples with Opposite Polarity Prediction Corrected by RGAT-BERT_mix Model Variant Guided by UAL

content	label	RGAT-BERT_ori	RGAT-BERT_mix
The main *draw* of this place is the *price*	2	0	2
The *decor* is what initially got me in the door	2	0	2
Save room for *deserts* - they 're to die for	2	0	2

The multi-class accuracy, the macro F1 value, and the proportion of samples with opposite polarity predictions in the total predicted samples *opp* are used as evaluation indicators. The experiment calculates the proportion of the samples with opposite polarity prediction to the total prediction samples, not the proportion of the misclassified samples. The reason for this setting is to make the results between different model variants comparable.

With the performance comparison of each model on the laptop dataset, the UATR model has achieved the best results in terms of accuracy and macro F1 value. The proportion of opposite polarity predictions *opp* is also the lowest. Compared with the SOTA model RGAT-BERT, UATR has a 2.66% improvement in accuracy and a 4.19% improvement in the macro F1 value. The macro F1 value can better reflect the multi-classification effect of the model on the imbalanced data set. The more significant increase in the macro F1 value verifies the effectiveness of the UATR model in solving the imbalance problem.

Similarly, on the restaurant dataset, our model also achieved the best results in various indicators, surpassing the performance of the SOTA model RGAT-BERT but does not introduce additional training data. Specifically, the accuracy index has increased by 1.7% points, and the macro F1 value has increased by 2.2% points. The effect improvement brought by the UATR model is more significantly manifested in the macro F1 value, which also verifies that the UATR model has an advantage of dealing with unbalanced category samples. In addition, the opposite ratio of polarity prediction *opp* on the restaurant dataset has reached a new minimum value of 2.05. It can be inferred that the positive and negative samples of the restaurant dataset have a greater degree of imbalance, so UATR has a large space for correcting the polarity prediction of the opposite sample.

On the whole, with the combination of UAL and Bayesian neural network proposed in this paper (corresponding to the mix row in the table), the variation of the ratio of opposite polarity prediction *opp* has a consistent effect on all the baseline models in the experiment. *Opp* value in the *mix* variant model is less than or equal to that of the baseline model. While the variant of *opp* value based on the Bayesian neural network without the guidance of UAL has no such properties. Figure 4 visually shows the effect of the method using uncertainty-aware loss function UAL to guide the model training. Compared with the baseline model, the uncertainty-based variant model *mix* guided by UAL achieved the smallest *opp* value on both datasets. The proportion of samples with opposite

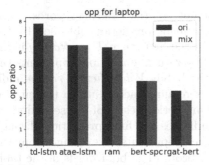
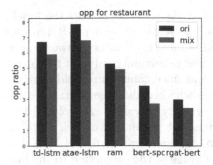

(a) *Opp* comparison chart of baseline model and mix variant model on laptop

(b) *Opp* comparison chart of baseline model and mix variant model on restaurant

Fig. 4. A comparison chart of the opposite polarity prediction between the baseline model and the UAL guided variant model *mix*.

polarity predictions to the total predicted samples reached the lowest. Take the performance of the SOTA model RGAT-BERT on the restaurant dataset as an example for analysis. Among the prediction samples of RGAT-BERT, there are 33 samples with completely opposite polarity predictions. After using uncertainty-based variant model guided by UAL to make predictions, 8 samples were predicted correctly, some of them are shown in Table 3. The error span of 5 samples between predicted polarity and label polarity becomes 1. The total number of samples with opposite polarity predictions is reduced to 27.

Table 4. UATR Model Instance Description.

sample 194: the wait *staff* was loud and inconsiderate.							
label	pred1	uncertainty	threshold	whether to refine	pred2	final pred	changes
0	0	0.229	0.4	no	2	0	\

sample 264: my favs here are the *tacos pastor* and the tostada de tinga...							
label	pred1	uncertainty	threshold	whether to refine	pred2	final pred	changes
2	2	0.433	0.5	no	1	2	\

sample 215: *Service* is known for bending over backwards to make everyone happy.							
label	pred1	uncertainty	threshold	whether to refine	pred2	final pred	changes
2	0	**0.633**	0.35	**yes**	2	2	✗ → ✓

sample 262: *food portion* was small and below average.							
label	pred1	uncertainty	threshold	whether to refine	pred2	final pred	changes
0	1	**0.845**	0.5	**yes**	0	0	✗ → ✓

Table 4 shows the refinement process of uncertainty-aware two-stage refinement prediction model UATR in the form of an example. The *label* in the header part of the table represents the label sentiment polarity of aspect words in the example sentence, and *pred1* represents sentiment prediction obtained by the UAL-guided uncertainty estimation model in the first stage. *Uncertainty* is the specific uncertainty value of the corresponding result. *Threshold* represents the current uncertainty threshold obtained in the batch. *Wheather to refine* indicates whether the second-stage correction is required according to the current uncertainty threshold. *Pred2* indicates the prediction result of the second-stage refinement prediction model. *Final pred* indicates the final prediction through two-stage refinement model UATR. *Changes* represents the change from the first stage of the prediction to the final prediction. (The bold words in the table head correspond to the aspect words in the sentence).

5   Analysis

5.1   The Distribution of Misclassified Samples

Figure 5 (a) and Fig. 5 (b) compares cumulative distribution changes of misclassification prediction samples for the RGAT-BERT model and UATR model in each uncertainty interval in the form of a line graph. Figure 5 (a) is for the laptop dataset, and Fig. 5 (b) is for the restaurant dataset. Compared with RGAT-BERT, the trend of the line chart of UATR model consistently shows a low starting point, a gentle start, and a steep increase in the later stages. In addition, the number of misclassified samples of UATR is lower than that of RGAT-BERT in the entire uncertainty interval [0,1), which provides an uncertainty explantation for model predictions and confirms the effectiveness of UATR on fine-grained multi-classification task.

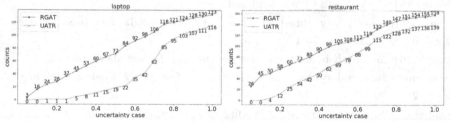

(a) Comparison of the Cumulative Distribution of Misclassified Samples of RGAT-BERT and UATR on the Laptop

(b) Comparison of the Cumulative Distribution of Misclassified Samples of RGAT-BERT and UATR on the Restaurant

Fig. 5. A Comparison of the Cumulative Distribution of Misclassified Samples of RGAT-BERT and UATR.

Table 5. Ablation Study of UAL and UATR

model	laptop		restaurant	
	accuracy	macro-F1	accuracy	macro-F1
STAGE1	78.95	75.02	85.54	78.66
STAGE2	78.75	74.45	84.91	77.74
RGAT-BERT	79.17	74.55	85.89	79.83
UATR$_{-UAL}$	81.39	78.34	86.25	79.77
UATR	**81.83**	**78.74**	**87.59**	**82.03**

5.2 Ablation Study

In this part, the effectiveness of UAL method and UATR two-stage refinement method is explored respectively. Five sets of comparative experiments are set up, which are the first stage prediction model (denoted as STAGE1) for the uncertainty estimation model, the secondary correction prediction model of UATR GAT-based (denoted as STAGE2), the current SOTA model RGAT-BERT, the UATR model without the guidance of UAL (UATR$_{-UAL}$) And the UATR model with the guidance of UAL. Table 5 records the performance of the five models on the laptop and restaurant datasets.

The results presented in Table 5 compare the performance of the STAGE1, STAGE2, RGAT-BERT, and UATR$_{-UAL}$ with the UATR model. The UATR$_{-UAL}$ model without UAL guidance achieved second only to the effect of the UATR model and also exceeded that of the current SOTA model RGAT-BERT. It is not difficult to conclude that the uncertainty-aware two-stage refinement method proposed in this paper is reasonable. It combines the uncertainty interpretation and the characteristics of the graph attention network to capture the dependency between components in the sentence. Under the premise of ensuring the overall complexity of the model, the classification accuracy and the macro F1 value are further improved.

With the experimental results of UATR$_{-UAL}$ and UATR on two datasets, the effect of UATR guided by UAL is better, which proves the effectiveness of the UAL method. And compared with the laptop dataset, the UATR model guided by UAL has a more significant improvement than UATR$_{-UAL}$ model without UAL guidance on the restaurant data set. UAL is more effective than the standard loss function for datasets with unbalanced categories.

6 Conclusion

In this paper, the utilization of model uncertainty to improve the performance of fine-grained sentiment classification tasks is explored. It aims at the common phenomenon that a considerable proportion of opposite polarities prediction samples in multi classification tasks and introduces model uncertainty considerations. It provides a new research idea for targeted sentiment analysis tasks.

Uncertainty is modeled based on a Bayesian neural network with the integration of multiple sets of neural network prediction results to weaken the impact of too small data size and imbalanced category samples. On the basis of that, reasonable and effective strategies and models are designed to deal with uncertainty. It can be used to provide a reliable interpretation of uncertainty for the model prediction results while improving the overall classification accuracy. The experiment proved that UATR has achieved better results than the current SOTA model.

References

1. Bai, X., Liu, P., Zhang, Y.: Investigating typed syntactic dependencies for targeted sentiment classification using graph attention neural network. IEEE/ACM Trans. Audio, Speech Lang. Process. **29**, 503–514 (2021)
2. Baziotis, C., Pelekis, N., Doulkeridis, C.: Datastories at semeval-2017 task 4: deep lstm with attention for message-level and topic-based sentiment analysis. In: SemEval-2017, pp. 747–754 (2017)
3. Blundell, C., Cornebise, J., Kavukcuoglu, K., Wierstra, D.: Weight uncertainty in neural network. In: International Conference on Machine Learning, pp. 1613–1622. PMLR (2015)
4. Chen, P., Sun, Z., Bing, L., Yang, W.: Recurrent attention network on memory for aspect sentiment analysis. In: Proceedings of the 2017 Conference on Empirical Methods in Natural Language Processing, pp. 452–461 (2017)
5. Devlin, J., Chang, M., Lee, K., Toutanova, K.: BERT: pre-training of deep bidirectional transformers for language understanding. In: NAACL-HLT 2019, MN, USA, 2–7 June 2019, Volume 1 (Long and Short Papers), pp. 4171–4186 (2019)
6. Fan, Z., Wu, Z., Dai, X., Huang, S., Chen, J.: Target-oriented opinion words extraction with target-fused neural sequence labeling. In: Proceedings of the 2019 Conference of the North American Chapter of the Association for Computational Linguistics: Human Language Technologies, vol. 1. pp. 2509–2518 (2019)
7. Gal, Y.: Uncertainty in Deep Learning. Ph.D. thesis, University of Cambridge (2016)
8. Gal, Y., Ghahramani, Z.: Bayesian convolutional neural networks with bernoulli approximate variational inference. arXiv preprint arXiv:1506.02158 (2015)
9. Goan, E., Fookes, C.: Bayesian neural networks: an introduction and survey. In: Mengersen, K.L., Pudlo, P., Robert, C.P. (eds.) Case Studies in Applied Bayesian Data Science. LNM, vol. 2259, pp. 45–87. Springer, Cham (2020). https://doi.org/10.1007/978-3-030-42553-1_3
10. Gui, T., et al.: Uncertainty-aware label refinement for sequence labeling. In: Webber, B., Cohn, T., He, Y., Liu, Y. (eds.) EMNLP 2020, Online, 16–20 November 2020, pp. 2316–2326 (2020)
11. Huang, B., Carley, K.M.: Syntax-aware aspect level sentiment classification with graph attention networks. In: EMNLP-IJCNLP 2019, Hong Kong, China, 3–7 November 2019, pp. 5468–5476 (2019)
12. Jordan, M.I., Ghahramani, Z., Jaakkola, T.S., Saul, L.K.: An introduction to variational methods for graphical models. Mach. Learn. **37**(2), 183–233 (1999)
13. Li, X., Bing, L., Zhang, W., Lam, W.: Exploiting BERT for end-to-end aspect-based sentiment analysis. In: W-NUT@EMNLP 2019, Hong Kong, China, 4 November 2019, pp. 34–41. Association for Computational Linguistics (2019)

14. Ma, D., Li, S., Zhang, X., Wang, H.: Interactive attention networks for aspect-level sentiment classification. In: Sierra, C. (ed.) IJCAI 2017, Melbourne, Australia, 19–25 August 2017. pp. 4068–4074 (2017)

15. Mikolov, T., Chen, K., Corrado, G., Dean, J.: Efficient estimation of word representations in vector space. In: ICLR 2013, Scottsdale, Arizona, USA, 2–4 May 2013 (2013)

16. Mitchell, M., Aguilar, J., Wilson, T., Van Durme, B.: Open domain targeted sentiment. In: EMNLP 2013, pp. 1643–1654 (2013)

17. Pennington, J., Socher, R., Manning, C.D.: Glove: global vectors for word representation. In: EMNLP 2014, pp. 1532–1543 (2014)

18. Pontiki, M., Galanis, D., Pavlopoulos, J., Papageorgiou, H., Androutsopoulos, I., Manandhar, S.: SemEval-2014 task 4: aspect based sentiment analysis. In: SemEval 2014, pp. 27–35. Association for Computational Linguistics, Dublin, Ireland (2014)

19. Scarselli, F., Gori, M., Tsoi, A.C., Hagenbuchner, M., Monfardini, G.: The graph neural network model. IEEE Trans. Neural Netw. **20**(1), 61–80 (2008)

20. Shapiro, A.: Monte carlo sampling methods. Handbooks Oper. Res. Management Sci. **10**, 353–425 (2003)

21. Song, Y., Wang, J., Jiang, T., Liu, Z., Rao, Y.: Attentional encoder network for targeted sentiment classification. arXiv preprint arXiv:1902.09314 (2019)

22. Sun, K., Zhang, R., Mensah, S., Liu, X.: Aspect-level sentiment analysis via convolution over dependency tree. In: EMNLP-IJCNLP 2019, pp. 5683–5692 (2019)

23. Tang, D., Qin, B., Feng, X., Liu, T.: Effective lstms for target-dependent sentiment classification. In: Calzolari, N., Matsumoto, Y., Prasad, R. (eds.) COLING 2016, 11–16 December 2016, Osaka, Japan, pp. 3298–3307 (2016)

24. Velickovic, P., Cucurull, G., Casanova, A., Romero, A., Liò, P., Bengio, Y.: Graph attention networks. In: ICLR 2018, Vancouver, BC, Canada (2018)

25. Wang, Y., Huang, M., Zhu, X., Zhao, L.: Attention-based lstm for aspect-level sentiment classification. In: EMNLP 2016, pp. 606–615 (2016)

26. Zhu, X., Sobihani, P., Guo, H.: Long short-term memory over recursive structures. In: International Conference on Machine Learning, pp. 1604–1612. PMLR (2015)

AttIN: Paying More Attention to Neighborhood Information for Entity Typing in Knowledge Graphs

Yingtao Wu, Weiwen Zhang$^{(\boxtimes)}$, Hongbin Zhang$^{(\boxtimes)}$, Huanlei Chen, and Lianglun Cheng

School of Computer Science and Technology, Guangdong University of Technology, Guangzhou, China
{zhangww,llcheng}@gdut.edu.cn, zhbin@mail2.gdut.edu.cn

Abstract. Entity types in knowledge graph (KG) have been employed extensively in downstream tasks of natural language processing (NLP). Currently, knowledge graph entity typing is usually inferred by embeddings, but a single embedding approach ignores interactions between neighbor entities and relations. In this paper, we propose an AttIN model that pays more attention to entity neighborhood information. More specifically, AttIN contains three independent inference modules, including a BERT module that uses the target entity neighbor to infer the entity type individually, a context transformer that aggregates information based on different contributions from the neighbor, and an interaction information aggregator (IIAgg) module that aggregates the entity neighborhood information into a long sequence. In addition, we use exponentially weighted pooling to process these predictions. Experiments on the FB15kET and YAGO43kET datasets show that AttIN outperforms existing competitive baselines while it does not need extra semantic information in the sparse knowledge graph.

Keywords: Knowledge Graphs · Entity Typing · Information Aggregation

1 Introduction

Knowledge graph (KG) is a semantic network that displays the relation between entities. Generally, a complete KG is composed of a collection of triples in the form of multiple (h, r, t) [1]. In the real world, the information in KG often faces incomplete problems [2]. There has been a lot of research on the completion of entity relation [3,4]. Nevertheless, there are relatively few studies on entity type completion.

Entity type information in the KG has been used in a variety of natural language processing (NLP) tasks, which include entity alignment [5,6], text generation [7], entity linking [8,9] and relationship extraction [10]. At present, three

B. Luo et al. (Eds.): ICONIP 2023, LNCS 14451, pp. 430–442, 2024.
https://doi.org/10.1007/978-981-99-8073-4_33

groups of approaches exist for the knowledge graph entity typing (KGET) task. First, as a representative of the embedding-based method, CET [11] designs two inference mechanisms for entity type prediction. However, it ignores the differences between neighbors. Second, GCN-based methods mainly infer possible entity types primarily through the rich semantics of entity neighbors in clustered graph convolutional networks [12]. However, they are limited by the graph structure so that the neighbors of the entity are underutilized. Third, in the category of transformer-based methods, ETE model [13] uses three different transformers to predict entity types. In more sparse knowledge graphs (such as the YAGO43kET dataset), ETE enhances the input relations with additional semantic information, which reduces the generality of the model. In addition, a major problem with these three approaches is that they do not consider more aspects of neighbor information, and do not sufficiently explore the role of known information, especially in sparse knowledge graphs.

To deal with such an issue, we propose an innovative model called AttIN. The inference module of AttIN consists of an independently encoded BERT, a context transformer and an interactive information aggregator (IIAgg). The first inference module is a separate encoding module based on BERT, which encodes each neighbor of the target entity individually through a pre-trained model. It makes efficient use of known type and relation information. The second inference module aggregates neighbors via a context transformer, which can aggregate encoded neighbors based on the contribution of them. The third inference module is IIAgg, which aggregates the encoded neighborhood information into a long sequence, so that known information can be fully considered when inferring the entity type. The above three inference modules will output multiple prediction types after a unified linear layer, which are then aggregated by exponentially weighted pooling.

We summarize the contributions of this work in the following three points:

- We propose the AttIN model consisting of three inference modules to infer the missing entity types in the knowledge graph, which pays more attention to the neighborhood information of the target entity.
- AttIN can fully consider type neighbors and relational neighbors in the sparse knowledge graph.
- Experiments on two real-world datasets show that AttIN is superior to the existing competitive baseline.

The remainder of this paper is organized as follows. In Sect. 2, we describe the work related to the existing KGET methodology. In Sect. 3 we introduce AttIN in detail. In Sect. 4, we conduct experiments to demonstrate the validity and excellence of AttIN. Finally, we conclude our work and outline our future work plans in Sect. 5.

2   Related Work

The purpose of KGC is usually to fill in the missing entities or relations in the KG. Consequently, KGET can be viewed as a KGC subtask [13]. Existing approaches used for KGET tasks can be divided into three types: embedding-based, GCN-based, and transformer-based.

2.1 Embedding-Based Methods

Knowledge graph embedding methods aim to acquire entity embeddings and type embeddings [14]. Moon et al. [2] treated entities and types in the KG as a special triple, its head is the target entity, the tail is the entity type, and the relation uniformly defined as "has type". To this end, they proposed an ETE model [15], using CONTE to obtain entity and relation embeddings in the KG, and completing the triple (entity, has type, ?) to predict missing entity types. Zhao et al. [16] improved upon ETE by embedding entities and types in separate spaces, and subsequently utilizing two distinct inference mechanisms to predict entity types. Ge et al. [17] propose a CORE model which embeds entities and types in the KG into two distinct spaces. Pan et al. [11] developed a model called CET. It utilizes two distinct inference mechanisms to address the issue of disregarding known entity types. Zhou et al. [18] invented an AttEt model employing an attention mechanism to aggregate neighborhood information of target entities. In these methods, there is only a single relation between the target entity and the known type, namely "has type". This singularity makes neighborhood information underutilized during training.

2.2 GCN-Based Methods

In the modeling of graph structures, graph convolutional networks (GCNs) have been widely utilized [19]. But directly applying GCNs to KG would produce poor results. Addressing this issue, Schlichtkrull et al. [12] extended GCNs to relational graphs and designed a model called R-GCN, which uses specific relational filters in the aggregation process. Jin et al. [20] proposed a model named HMGCN that is capable of embedding multiple semantic correlations between entities. Vashishth et al. [21] performed multi-label classification of entities obtained by embedding GCNs, and then inferred the missing type of the target entity. Zhao et al. [22] learned heterogeneous relational graphs via a multi-graph attention network, and then they used the ConnectE method to predict types. Zou et al. [23] used a method based on a relational aggregation graph attention network to predict entity types. However, some neighborhood information can negatively affect entity type prediction.

2.3 Transformer-Based Methods

ETE [13] is an entity typing model consisting of three transformer modules. The first module encodes each neighbor of an entity independently. The second module is a context transformer that aggregates the neighbors of an entity in different ways through their contribution. The third module uses the global transformer to aggregate all neighbors into a long sequence. To increase the difference between target entity types, ETE designed also clustering information. However, ETE requires additional semantic information to infer the types of entities in the sparse knowledge graph. This severely limits the versatility of ETE.

To solve the problem of noise caused by uncorrelated neighborhood information, our model uses an inference module BERT to encode each neighbor of the target entity separately. To avoid requiring additional semantics on sparse knowledge graphs, we propose a more efficient interaction information aggregator IIAgg, which can pay more attention to type neighbors and relational neighbors of the target entity.

3   Our Method

In this section, we will introduce AttIN. As shown in Fig. 1, AttIN consists of three independent type prediction modules, a linear layer and a pooling module. In the first step, BERT is used to encode each type neighbor and relational neighbor of the target entity. The encoded results will be sent to the remaining two inference modules. Context transformer module aggregates encoded neighbors based on their contribution. IIAgg module is an interaction information aggregation, which aggregates the encoded neighborhood information into a long sequence. The Gelu function is used to activate the outputs of three inference modules. These outputs are then sent to a linear layer to obtain multiple predicted type embeddings, which will be aggregated through an exponentially weighted pooling to obtain the final prediction type. Three inference modules of AttIN consider known information from three different perspectives, which allows neighbor information to be fully utilized.

In Sect. 3.1, we will introduce the notations used in AttIN, In Sect. 3.2, we will describe the three embedding modules of AttIN. Finally, we explain the pooling method and optimization strategy in Sect. 3.3 and Sect. 3.4.

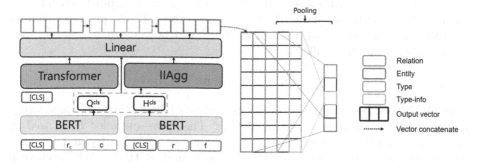

Fig. 1. The overall architecture of AttIN. The Gelu function is used as a nonlinear activation function before the linear layer to process the output of its predecessor.

3.1   Notations

In our work, a knowledge graph (KG) $\mathcal{G}$ can be regarded as a tuple $(\mathcal{E}, \mathcal{R}, \mathcal{C}, \mathcal{T})$ [24], where $\mathcal{E}$ represents the collection of entities in the KG, $\mathcal{R}$ represents the collection of relations, $\mathcal{C}$ represents the type collection of all entities, and $\mathcal{T}$ represents the collection of all triples. The triples in $\mathcal{T}$ can be divided into two categories. The first category is the relational triple (h, r, t), where $h, t \in \mathcal{E}$ stand for the head and tail entities and $r \in \mathcal{R}$ represents the edges joining h, t. The second one is the entity type triple $(e, has\ type, c)$, where $e \in \mathcal{E}$, $c \in \mathcal{C}$, has type represents the connection between the target entity and the type. For all $e \in \mathcal{E}$, neighbors of e are defined as the sum of all its relational neighbors and type neighbors, where the relational neighbor is the collection $\{(r, f) \mid (e, r, f) \in \mathcal{T}\}$, and the type neighbor is the collection $\{(has\ type, c) \mid (e, has\ type, c) \in \mathcal{T}\}$. The objective of our work is to complete the missing type c of the entity type triple in the KG.

In addition, all entities and their types are uniformly defined in KG using a unique relation *has type*. The singularity of this relation leads to the inability to capture the differences between different entity types of entities. To address this issue, this work constructs inputs for type neighbors based on the clustering information of the ETE model [13]. Specifically, it adopts the neighborhood information of the entity type to replace a single relation (*has type*), such as using *musical* to replace the *has type* relation between the target entity and the type *songwriters*.

3.2   Model Structure

BERT for Independent Encoding. Neighbors of the target entity can be useful when inferring the entity type. For example, the type triple (*Tina Arena, musical, songwriters*) can reasonably infer that *Tina Arena* has a type *musicians*. The relational triple (*Tina Arena, wasBornIn, Melbourne*) likewise infers that *TinaArena* has a type *Australian*. For the type neighbor of the target entity, We establish a sequence $H = ([CLS], r_c, c)$ as input to BERT [25], where [CLS] is a special marker that aggregates the entire sequence of characterization information. Each element $\mathbf{h}_i$ in H will be constructed as an input vector representation as follows:

$$\mathbf{h}_i = \mathbf{h}_i^{word} + \mathbf{h}_i^{pos} \tag{1}$$

where $\mathbf{h}_i^{word}$ and $\mathbf{h}_i^{pos}$ are the result of randomly initializing the word embeddings and position embeddings of type neighbors. The above obtained $\mathbf{h}_i$ is used as the input of BERT to reflect the interaction between the entity and its type. The output embedding of the BERT is represented as $\mathbf{H}^{cls} \in \mathbb{R}^{d \times 1}$, where d represents the dimension of the embedding. Each entity neighbor of the target entity corresponds to an output embedding. So a target entity with n entity types will be represented as $[\mathbf{H}_1^{cls}, \mathbf{H}_2^{cls}, \dots, \mathbf{H}_n^{cls}]$.

As with type neighbors, each relational entity neighbor of the target entity will be constructed as an input sequence $Q = ([\text{CLS}], r, f)$. The output embedding is represented as $\mathbf{Q}^{cls} \in \mathbb{R}^{d \times 1}$. For a target entity having m relational neighbors will be represented as $\left[\mathbf{Q}_1^{cls}, \mathbf{Q}_2^{cls}, \dots, \mathbf{Q}_m^{cls}\right]$.

The embedding module BERT mainly focuses on each existing neighbor of the target entity, so as to decrease the interference of unrelated neighbors on the inference results. Finally, we will use the Gelu function to nonlinearly activate $\mathbf{H}^{cls}$ and $\mathbf{Q}^{cls}$ and then feed it into the linear layer:

$$\mathbf{S}_i^{cls} = \mathbf{W} \text{ Gelu } (\mathbf{H}_i^{cls}) + b \tag{2}$$

$$\mathbf{S}_j^{cls} = \mathbf{W} \text{ Gelu } (\mathbf{Q}_j^{cls}) + b \tag{3}$$

where $\mathbf{W} \in \mathbb{R}^{L \times d}$ and $\mathbf{b} \in \mathbb{R}^L$ are the trainable parameters. L is the total number of types. $\mathbf{H}_i^{cls} \in \mathbb{R}^{d \times 1}$ and $\mathbf{Q}_j^{cls} \in \mathbb{R}^{d \times 1}$ are embeddings of the ith type neighbor and the jth relation neighbor. Their final output score is $\mathbf{S}_i^{cls}$ and $\mathbf{S}_j^{cls}$.

Context Transformer. Entities in a knowledge graph sometimes have some composite types and the contribution of neighborhood information to the type of inference will be different. To differentiate neighbors of an entity, we use a context transformer [26] to aggregate them. Specifically, context transformer receives the $\mathbf{H}^{cls}$, the $\mathbf{Q}^{cls}$ and the [CLS] embedding, then contextualizes the target entity with all neighborhood information to obtain the embedding $\mathbf{C}^{cls}$. The obtained $\mathbf{C}^{cls}$ is then subjected to nonlinear activation using the Gelu function and sent to the linear layer. The formula is as follows:

$$\mathbf{S}^{ctx} = \mathbf{W} \text{ Gelu } \left([\mathbf{C}^{cls}]\right) + b \tag{4}$$

where $\mathbf{S}^{ctx}$ is the final score for the correlation between target entity and all types.

IIAgg. It is difficult for the KG to infer a more fine-grained entity type from a single neighbor. In particular, in a sparse knowledge graph of relational neighbors like YAGO43kET, the context transformer easily ignores the influence of relational neighbors during training. Taking inspiration from Agg2T [11], we propose an interaction information aggregator IIAgg to infer the missing type. IIAgg can equally aggregate the neighborhood information of the target entity, so it can ensure that both relational neighbors and type neighbors can contribute to entity type inference. Finally, the aggregation results of IIAgg are sent to the linear layer after activation by the Gelu function. The following is the precise formula:

$$\mathbf{h}_u = \frac{1}{n+m} \sum_{n+m} \left([\mathbf{H}_1^{cls}, \dots, \mathbf{H}_n^{cls}, \mathbf{Q}_1^{cls}, \dots, \mathbf{Q}_m^{cls}]\right) \tag{5}$$

$$\mathbf{S}_u^{IIAgg} = \mathbf{W} \text{ Gelu}(\mathbf{h}_u) + b \tag{6}$$

where $\mathbf{h}_u$ is a representation of the average aggregation of all neighbors of the target entity u. $\mathbf{S}_u^{IIAgg} \in \mathbb{R}^L$ is the prediction result, which is the score of the correlation between the target entity and all types.

3.3 Pooling Method

The model above predicts multiple outcomes for the target entity. BERT module is able to infer the entity type individually from each neighbor of the target entity. For a target entity with m relation neighbors and n type neighbors, the output of the BERT module can be obtained after passing through the linear layer to obtain the prediction results of $m + n$ entity types. Combining the prediction results of the context transformer and IIAgg can result in $m+n+2$ embeddings. We define these type predictions as $\mathbf{S}_i$, where i represents the i_{th} entity type prediction result. We use exponentially weighted pooling [27] to aggregate these type results. The following is the computation method:

$$\mathbf{S}_e = \text{pool}\left([\mathbf{S}_0, \mathbf{S}_1, \ldots, \mathbf{S}_{m+n-1}, \mathbf{S}_{m+n}, \mathbf{S}_{m+n+1}]\right) = \sum_{i=0}^{m+n+1} w_i \mathbf{S}_i \tag{7}$$

$$w_i = \frac{\exp \alpha \mathbf{S}_i}{\sum_{j=0}^{m+n+1} \exp \alpha \mathbf{S}_j} \mathbf{S}_i \tag{8}$$

where α is a hyperparameter used to control temperature during pooling. After getting $\mathbf{S}_e$, we use the sigmoid function to map it to a fraction between 0 and 1:

$$\mathbf{s}_{e,k} = \sigma\left(\mathbf{S}_e\right) \tag{9}$$

where $\mathbf{s}_{e,k}$ indicates the correlation score between the target entity e and the k type. It is more likely for the target entity e to have type k the greater value of $\mathbf{s}_{e,k}$.

3.4 Optimization Approach

During the training process of AttIN, a positive sample score $\mathbf{s}_{e,k}$ indicates that the target entity e exists of type k in the KG, while the negative sample score $\mathbf{s}'_{e,k}$ indicates that the target entity e does not exist of type k in the KG. Using binary cross-entropy (BCE) as the loss function of AttIN presents a serious false negative problem [11], e.g., partial triples $(e, has\ type, k)$ are valid, but they cannot be found in existing KGs. A false negative aware loss function [11] is used here to overcome the false negative problem. It is described as:

$$\mathcal{L} = -\sum_{(e,k)\notin\mathcal{G}} \beta\left(\mathbf{s}'_{e,k} - (\mathbf{s}'_{e,k})^2\right) \log\left(1 - \mathbf{s}'_{e,k}\right) - \sum_{(e,k)\in\mathcal{G}} \log\left(\mathbf{s}_{e,k}\right) \tag{10}$$

where β is a hyper parameter, which is the total weight of the negative sample. If the negative sample correlation score is too large or too small, β will be reduced.

4   Experiments

In this section, we will perform the experiments on the knowledge graph datasets FB15kET and YAGO43kET to verify the effectiveness of AttIN.

4.1   Datasets

We evaluate AttIN on the FB15kET datasets and YAGO43kET datasets. Moon et al. [15] mapped entities in the subgraphs of Freebase [28] and YAGO [29] to corresponding entity types, thereby constructing two new datasets called FB15kET and YAGO43kET. We conduct our experiments in the same way as the dataset partitioned in the baseline. Table 1 displays the basic statistics of the datasets.

Table 1. Basic statistics for all datasets.

Dataset	FB15kET	YAGO43kET
#Entities	14951	42334
#Relations	1345	37
#Types	3584	45182
#Clusters	1081	6583
#Train.triples	483142	331686
#Train. tuples	136618	375853
#Valid	15848	3111
#Test	15847	43119

4.2   Experiment Setup

Evaluation Protocol. We calculate the target entity e and all types of relevance scores for the test sample, then rank these scores from largest to smallest. Finally, the known type of the target entity e is removed from the ranking. As evaluation metrics for model results, we use the mean reciprocal rank (MRR) and Hits at 1/3/10.

Parameter Settings. AttIN will be trained on a random minibatch of the dataset, using Adam [30] as the optimizer during the training process. We fixed the batch size at 128. Every 25 epochs are evaluated on the test set, up to a maximum of 1000 epochs. After a lot of experiments, the best hyperparameter pairings are: learning rate set to 0.001, pooling temperature set to 0.5 and embedding dim set to 100.

Baselines. AttIN will be compared with eight state-of-the-art models. They can be classified into three main types of models. The first category is embedding-based models, which are ETE [15], ConnectE [16], CET [11] and AttEt [18]. The second category of GCN-based models is CompGCN [21], RGCN [12] and RACE2T [23]. The third category of transformer-based model is TET [13].

Table 2. Evaluation of individual models on FB15kET and YAGO43kET datasets.

Datasets	FB15kET				YAGO43kET			
Model	MRR	Hit@1	Hit@3	Hit@10	MRR	Hit@1	Hit@3	Hit@10
ETE	0.500	0.385	0.553	0.719	0.230	0.137	0.263	0.422
ConnectE	0.590	0.496	0.643	0.799	0.280	0.160	0.309	0.479
CET	0.697	0.613	0.745	0.856	0.503	0.398	0.567	0.696
AttEt	0.620	0.517	0.677	0.821	0.350	0.244	0.413	0.565
CompGCN	0.657	0.568	0.704	0.833	0.357	0.274	0.384	0.520
RGCN	0.662	0.571	0.711	0.836	0.357	0.266	0.392	0.533
RACE2T	0.640	0.561	0.689	0.817	0.340	0.248	0.376	0.523
TET	0.717	0.638	**0.762**	**0.872**	0.510	0.408	0.571	0.695
AttIN	**0.721**	**0.646**	**0.762**	0.868	**0.527**	**0.423**	**0.591**	**0.716**

4.3   Performance Comparison

Table 2 shows the results of the evaluation for type prediction on FB15kET and YAGO43kET datasets. Our proposed AttIN is better than the baseline on almost all the metrics. This shows the superiority of the AttIN.

Specifically, AttIN integrates three inference modules, which together can better distinguish the contribution of different neighbors of the target entity, thereby reducing the influence of different neighbors having differential contributions to type predictions. Compared with the most competitive TET, the MRR metric of experimental results on the FB15kET and YAGO43kET has increased by 0.6% and 3.3%, and the results on the hit@1 and hit@3 indicators have also improved. The Hit@10 metric of ETE in the FB15kET dataset are slightly better than AttIN. This is due to the fact that ETE uses three transformers, which are capable of encoding neighborhood information at three different granularities.

In AttIN, we use an inference module IIAgg that focuses more on type neighbors and relational neighbors. It is able to equalize all neighbors of the target entity into a long sequence. The addition of IIAgg enables the model to better aggregate interactions between neighbors. In addition, in the sparse knowledge graph YAGO43kET, the ETE model needs additional semantic information to enhance the relational neighbor to alleviate the problem of too few relational neighbors. AttIN can get better prediction results in sparse knowledge graphs without additional semantic information.

4.4   Ablation Study

We conduct ablation study on the FB15kET and YAGO43kET datasets to investigate the impact of interaction information aggregator IIAgg in aggregating various neighborhood information. The experimental results are shown in Table 3. It can be seen that the aggregation type neighbors and relational neighbors (first row) are superior to other ablation models in all evaluation metrics.

Table 3. Results of ablation experiments with different combinations of IIAgg module.

Models			FB15kET				YAGO43kET			
Exclude-IIAgg	Type	Relation	MRR	Hit@1	Hit@3	Hit@10	MRR	Hit@1	Hit@3	Hit@10
✓	✓	✓	**0.721**	**0.646**	**0.762**	**0.868**	**0.527**	**0.423**	**0.591**	**0.716**
✓			0.713	0.633	0.760	0.867	0.512	0.405	0.577	0.706
✓	✓		0.718	0.642	0.761	0.869	0.523	0.419	0.588	0.711
✓		✓	0.716	0.638	0.760	0.869	0.509	0.410	0.570	0.684

We find that using type neighbors or relational neighbors alone for aggregation in the FB15kET datasets can also bring considerable prediction results. Analyzing the experimental ablation results on the YAGO43kET dataset, we find that aggregate neighborhood information can improve the experimental results more than the FB15kET dataset. The reason for this phenomenon is that the context transformer aggregatoo neighborhood information according to the contribution of the neighbor of the target entity. However, in the knowledge graph where this relational neighbor is much smaller than the type neighbor, the context is likely to ignore the contribution of the relational neighbor during the training process. IIAgg aggregates the neighborhood information into a sequence, ensuring that all the neighborhood information of the target entity can be taken into account in the type prediction.

Furthermore, we find that simply aggregating relational neighbors into a long sequence to predict entity types in the YAGO dataset leads to worse results. This is because there are too few relational neighbors in the YAGO dataset, making it difficult to avoid overfitting during training. Therefore, aggregating type neighbors and relational neighbors can address this problem.

5   Conclusion and Future Work

This paper proposes an entity typing inference model AttIN that pays more attention to the neighborhood information of the target entity. We use an interactive information aggregator IIAgg to aggregate neighborhood information, which ensures that all neighborhood information is not lost during training. The MRR metric of the experimental results on the FB15kET and YAGO43kET datasets was improved by 0.6% and 3.3% compared with the optimal results in the baseline, which indicates the superior performance of AttIN.

Compared to the sparse knowledge graph YAGO43kET, AttIN has less improvement in experimental results on the FB15kET dataset. In the future, we will try more knowledge graph embedding methods or use graph convolutional networks to improve our model.

Acknowledgements. Our work is supported by multiple funds in China, including Guangzhou Science and Technology Planning Project (2023B01J0001, 202201011835), the Key Program of NSFC-Guangdong Joint Funds (U2001201, U1801263), Industrial core and key technology plan of ZhuHai City (ZH22044702190034-HJL). Our work is also supported by Guangdong Provincial Key Laboratory of Cyber-Physical System (2020B1212060069).

References

1. Zhang, H., Li, A., Guo, J., Guo, Y.: Hybrid models for open set recognition. In: Vedaldi, A., Bischof, H., Brox, T., Frahm, J.-M. (eds.) ECCV 2020. LNCS, vol. 12348, pp. 102–117. Springer, Cham (2020). https://doi.org/10.1007/978-3-030-58580-8_7

2. Moon, C., Harenberg, S., Slankas, J., Samatova, N.: Learning contextual embeddings for knowledge graph completion. In: Pacific Asia Conference on Information Systems (PACIS), vol. 10 (2017)

3. Chen, K., Wang, Y., Li, Y., Li, A., Zhao, X.: Contextualise entities and relations: an interaction method for knowledge graph completion. In: Farkaš, I., Masulli, P., Otte, S., Wermter, S. (eds.) ICANN 2021. LNCS, vol. 12893, pp. 179–191. Springer, Cham (2021). https://doi.org/10.1007/978-3-030-86365-4_15

4. Yang, C., Zhang, W.: Private and shared feature extractors based on hierarchical neighbor encoder for adaptive few-shot knowledge graph completion. In: Proceedings of the 34th International Conference on Tool with Artificial Intelligence, pp. 409–416 (2022)

5. Wang, Z., Yang, J., Ye, X.: Knowledge graph alignment with entity-pair embedding. In: Proceedings of the 2020 Conference on Empirical Methods in Natural Language Processing (EMNLP), pp. 1672–1680 (2020)

6. Song, X., Zhang, H., Bai, L.: Entity alignment between knowledge graphs using entity type matching. In: Qiu, H., Zhang, C., Fei, Z., Qiu, M., Kung, S.-Y. (eds.) KSEM 2021. LNCS (LNAI), vol. 12815, pp. 578–589. Springer, Cham (2021). https://doi.org/10.1007/978-3-030-82136-4_47

7. Dong, X., Yu, W., Zhu, C., Jiang, M.: Injecting entity types into entity-guided text generation. In: EMNLP, vol. 1, pp. 734–741. Association for Computational Linguistics (2021)

8. Gupta, N., Singh, S., Roth, D.: Entity linking via joint encoding of types, descriptions, and context. In: Proceedings of the 2017 Conference on Empirical Methods in Natural Language Processing, pp. 2681–2690 (2017)

9. Le, T., Huynh, N., Le, B.: Link prediction on knowledge graph by rotation embedding on the hyperplane in the complex vector space. In: Farkaš, I., Masulli, P., Otte, S., Wermter, S. (eds.) ICANN 2021. LNCS, vol. 12893, pp. 164–175. Springer, Cham (2021). https://doi.org/10.1007/978-3-030-86365-4_14

10. Wu, C., Chen, L.: Utber: utilizing fine-grained entity types to relation extraction with distant supervision. In: 2020 IEEE International Conference on Smart Data Services (SMDS), pp. 63–71. IEEE (2020)

11. Pan, W., Wei, W., Mao, X.L.: Context-aware entity typing in knowledge graphs. In: Findings of the Association for Computational Linguistics, EMNLP 2021, pp. 2240–2250 (2021)
12. Schlichtkrull, M., Kipf, T.N., Bloem, P., van den Berg, R., Titov, I., Welling, M.: Modeling relational data with graph convolutional networks. In: Gangemi, A., et al. (eds.) ESWC 2018. LNCS, vol. 10843, pp. 593–607. Springer, Cham (2018). https://doi.org/10.1007/978-3-319-93417-4_38
13. Hu, Z., Gutiérrez-Basulto, V., Xiang, Z., Li, R., Pan, J.Z.: Transformer-based entity typing in knowledge graphs. In: EMNLP, pp. 5988–6001. Association for Computational Linguistics (2022)
14. Balažević, I., Allen, C., Hospedales, T.M.: Hypernetwork knowledge graph embeddings. In: Tetko, I.V., Kůrková, V., Karpov, P., Theis, F. (eds.) ICANN 2019. LNCS, vol. 11731, pp. 553–565. Springer, Cham (2019). https://doi.org/10.1007/978-3-030-30493-5_52
15. Moon, C., Jones, P., Samatova, N.F.: Learning entity type embeddings for knowledge graph completion. In: Proceedings of the 2017 ACM on Conference on Information and Knowledge Management, pp. 2215–2218 (2017)
16. Zhao, Y., Zhang, A., Xie, R., Liu, K., Wang, X.: Connecting embeddings for knowledge graph entity typing. In: ACL, pp. 6419–6428. Association for Computational Linguistics (2020)
17. Ge, X., Wang, Y.C., Wang, B., Kuo, C.J.: Core: a knowledge graph entity type prediction method via complex space regression and embedding. Pattern Recogn. Lett. **157**, 97–103 (2022)
18. Zhuo, J., Zhu, Q., Yue, Y., Zhao, Y., Han, W.: A neighborhood-attention fine-grained entity typing for knowledge graph completion. In: Proceedings of the Fifteenth ACM International Conference on Web Search and Data Mining, pp. 1525–1533 (2022)
19. Zhang, Z., Zhang, Y., Wang, Y., Ma, M., Xu, J.: Complex exponential graph convolutional networks. Inf. Sci. **640**, 119041 (2023)
20. Jin, H., Hou, L., Li, J., Dong, T.: Fine-grained entity typing via hierarchical multi graph convolutional networks. In: Proceedings of the 2019 Conference on Empirical Methods in Natural Language Processing and the 9th International Joint Conference on Natural Language Processing (EMNLP-IJCNLP), pp. 4969–4978 (2019)
21. Vashishth, S., Sanyal, S., Nitin, V., Talukdar, P.P.: Composition-based multi-relational graph convolutional networks. In: ICLR. OpenReview.net (2020)
22. Zhao, Y., et al.: Connecting embeddings based on multiplex relational graph attention networks for knowledge graph entity typing. IEEE Trans. Knowl. Data Eng. **35**, 4608–4620 (2022)
23. Zou, C., An, J., Li, G.: Knowledge graph entity type prediction with relational aggregation graph attention network. In: Groth, P., et al. (eds.) The Semantic Web, ESWC 2022. LNCS, vol. 13261. Springer, Cham (2022). https://doi.org/10.1007/978-3-031-06981-9_3
24. Pan, J.Z., Vetere, G., Gomez-Perez, J.M., Wu, H.: Exploiting Linked Data and Knowledge Graphs in Large Organisations. Springer, Cham (2017). https://doi.org/10.1007/978-3-319-45654-6
25. Devlin, J., Chang, M., Lee, K., Toutanova, K.: BERT: pre-training of deep bidirectional transformers for language understanding. In: NAACL-HLT, vol. 1, pp. 4171–4186. Association for Computational Linguistics (2019)
26. Chen, S., Liu, X., Gao, J., Jiao, J., Zhang, R., Ji, Y.: Hitter: hierarchical transformers for knowledge graph embeddings. In: EMNLP, vol. 1, pp. 10395–10407. Association for Computational Linguistics (2021)

27. Stergiou, A., Poppe, R., Kalliatakis, G.: Refining activation downsampling with SoftPool. In: Proceedings of the IEEE/CVF International Conference on Computer Vision, pp. 10357–10366 (2021)
28. Bollacker, K., Evans, C., Paritosh, P., Sturge, T., Taylor, J.: Freebase: a collaboratively created graph database for structuring human knowledge. In: Proceedings of the 2008 ACM SIGMOD International Conference on Management of Data, pp. 1247–1250 (2008)
29. Suchanek, F.M., Kasneci, G., Weikum, G.: YAGO: a core of semantic knowledge. In: Proceedings of the 16th International Conference on World Wide Web, pp. 697–706 (2007)
30. Kingma, D.P., Ba, J.: Adam: a method for stochastic optimization. In: ICLR (Poster) (2015)

Text-Based Person re-ID by Saliency Mask and Dynamic Label Smoothing

Yonghua Pang[1] , Canlong Zhang[1,2]([✉]) , Zhixin Li[1,2], and Liaojie Hu[3]

[1] Key Lab of Education Blockchain and Intelligent Technology, Ministry of Education, Guangxi Normal University, Guilin 541004, China
lizx@gxnu.edu.cn

[2] Guangxi Key Lab of Multi-source Information Mining & Security, Guangxi Normal University, Guilin, China
zcltyp@163.com

[3] The Experimental High School Attached to Beijing Normal University, Beijing, China

Abstract. The current text-based person re-identification (re-ID) models tend to learn salient features of image and text, which however is prone to failure in identifying persons with very similar dress, because their image contents with observable but indescribable difference may have identical textual description. To address this problem, we propose a saliency mask based re-ID model to learn non-salient but highly discriminative features, which can work together with the salient features to provide more robust pedestrian identification. To further improve the performance of the model, a dynamic label smoothing based cross-modal projection matching loss (named CMPM-DS) is proposed to train our model, and our CMPM-DS can adaptively adjust the smoothing degree of the true distribution. We conduct extensive ablation and comparison experiments on two popular re-ID benchmarks to demonstrate the efficiency of our model and loss function, and improving the existing best R@1 by 0.33% on CUHK-PEDE and 4.45% on RSTPReID.

Keywords: Person Re-identification · Cross-Modal Retrieval · Feature Extraction

1 Introduction

For text-based person re-ID, how to learn more discriminative modal features is the key. The current text-based person re-identification (re-ID) models [1,3,5,10,12] tend to learn salient features of image and text, which however is prone to failure in identifying persons with very similar dress. As shown in Fig. 1, dissimilar negative sample pair (b and c) can be well distinguished with salient features, but similar negative sample pair (b and a) is usually difficult to be distinguished by their salient features. Besides, due to the limitation of text description because, some image contents with observable but indescribable

B. Luo et al. (Eds.): ICONIP 2023, LNCS 14451, pp. 443–454, 2024.
https://doi.org/10.1007/978-981-99-8073-4_34

difference may have identical textual description. Therefore, it is very difficult to identify image-text pairs with similar content but different IDs by using their salient features. However, we notice that some non-salient but highly discriminative information can effectively distinguish such difficult samples. For example, the regions highlighted in red boxes in Fig. 1(a) and (b), such as different coloured shoes and the glasses, have more significant differences, although they are not salient overall.

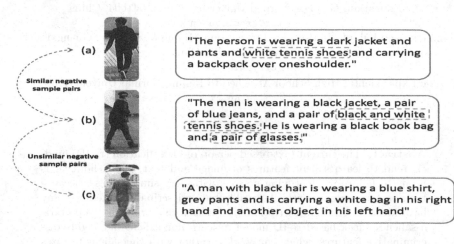

Fig. 1. (a) and (b) form similar negative sample pairs, while (b) and (c) form unsimilar negative sample pairs. We can see that for the unsimilar negative sample pairs, they can be well distinguished by the information in the salient regions, such as tops and trousers. For the similar negative sample pairs, on the one hand, they cannot be effectively distinguished from each other because the salient regions are similar. In contrast, the non-salient regions will have highly discriminative information, such as the shoes and glasses in the red box. (Color figure online)

However, the popular baseline models VIT [4] and the BERT [2] are all base on Transformer, so they are all good at capturing salient information in images and texts, but are not effective in learning non-salient but highly discriminative information that is useful for fine-grained retrieval. To let them can learn these discriminative information, we propose a new alignment paradigm, called Saliency Mask Modelling (SMM). We mask some most salient regions of image and text to force the models to attend to the non-salient regions, thus effectively learning the non-salient but highly discriminative features. Specifically, we first perform once VIT and BERT computation on original image and text, and then find the top-k tokens (i.e. salient region features) most relevant to the global tokens base on their similarity with the global token. After that, masking original image patches and words corresponding to the top-k tokens, and performing the second VIT and BERT computation on the rest image patches and words, thus forcing the model to learn non-salient but highly discriminative features.

A good loss function can often lead the network model to learn better and thus obtain good features. The traditional label smoothing method is an amendment to the cross-entropy loss function. [7] argues that encouraging the largest possible difference between positive and negative samples during the learning process of the model can lead to overconfidence in the true labels of the model, which can lead to overfitting problems. Therefore, they propose a label smoothing strategy, which aims to reduce the confidence level of the true labels and reduce the difference between positive and negative samples to solve the overfitting problem.

Cross-modal projection matching (CMPM) loss is a widely used loss for text-based person re-identification, which achieves feature projection matching for both modalities, and uses KL divergence to close the distance between the predicted distribution and the true distribution. However, the CMPM loss function suffers from a similar problem to the traditional cross-entropy loss function in that its true distribution expects the difference between positive and negative sample pairs to be as large as possible, which can easily cause overfitting problems. Inspired by [7], we propose a dynamic label smoothing based CMPM loss function. Unlike the traditional label smoothing which performs consistent smoothing of negative samples, this dynamic label smoothing performs the corresponding smoothing based on the differences between different negative samples. With this dynamic label smoothing operation, the model is able to dynamically adjust the true distribution of the data during the learning process to avoid excessive differences between positive and negative sample pairs, which we refer to CMPM-Dynamic-Smooth (CMPM-DS) for the modified CMPM. Figure 2 represents the specific difference between the two loss functions, from which it can be seen that the biggest difference between the two losses is the distance between positive and negative sample pairs.

Our contributions are as follows:

- We propose saliency mask mechanism, by which our re-ID model can focus on the non-saliency but highly discriminative information of image and text, so as to learn more discriminative features.
- To better train our the model, we propose a dynamic label smoothing based cross-modal projection matching loss function, which can perform a dynamic smoothing operation between positive and negative sample pairs to reduce their difference.
- We conduct extensive experiments on two popular re-ID benchmarks, and our R@1 on CUHK-PEDES and RSTPReid outperforms the existing best one by 0.33% and 4.45%, respectively.

2 Approach

2.1 Overview

A general overview of our model is shown in Fig. 3. Our baseline model uses the VIT pre-training model to extract visual features and the BERT pre-training

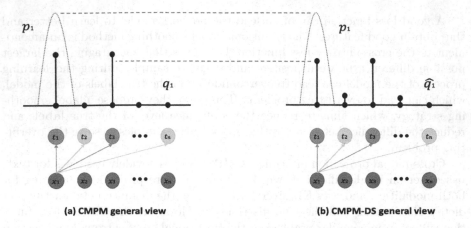

(a) CMPM general view   **(b) CMPM-DS general view**

Fig. 2. (a) represents the CMPM view; (b) represents CMPM-DS view, where green represents positive sample pairs, orange represents negative sample pairs, p is the model prediction probability, q is the true distribution probability, and KL divergence is used to reduce the difference between the two probability distributions. Compared with (a), the true distribution probability of positive sample pairs in (b) is slightly lower, and the true distribution probability value of negative sample pairs is a non-zero smaller value. Specifically, the values of different negative sample pairs in (b) are different, and their values are positively correlated with $(1 - p_{ij})$, where $p_{i}j$ represents the matching degree between x_i and t_j predicted by the model. For negative sample pairs, the setting of this value is conducive to the model to increase the punishment of difficult negative sample pairs. (Color figure online)

model to extract textual features respectively. The features extracted by the pre-training model have initial semantics, so we use these features to perform saliency mask modelling (SMM) to complete the masking of salient regions. We use the CMPM-Dynamic-Smooth (CMPM-DS) to compute loss on both the global token and the SMM global token. Each part of the model is described in detail separately below eventually.

2.2   Baseline Model

For text-based person re-identification, Our baseline model uses two pre-trained models for feature extraction of information from both modalities, where we use the pre-trained VIT model [4] for feature extraction of visual information and we select the pre-trained BERT model [2] for feature extraction of textual information. The model structure is shown in Fig. 3.

Given image x_i and text t_i as the ith image and the ith text respectively, $Encoder\_img$ denotes the visual feature extraction model and $Encoder\_txt$ denotes the text feature extraction model. The process of feature extraction is defined as follows.

$$\{X_{f1}, X_{f2}, \ldots, X_{f211}\} = Encoder\_img(x_i) \tag{1}$$

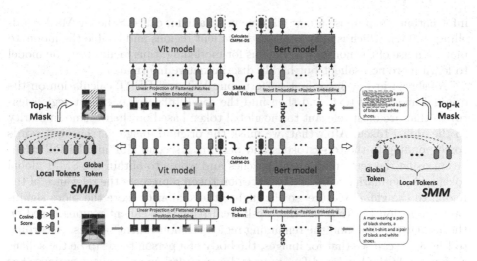

Fig. 3. Baseline model extracts visual and textual features using the VIT pre-training model and the BERT pre-training model respectively. Top-k features are selected as salient regions by the SMM module based on the similarity between the extracted global and local features. We mask the salient regions and complete global modelling using the remaining non-salient regions. Finally, we use CMPM-DS to calculate loss for both the global token and the SMM global token.

$$\{T_{f1}, T_{f2}, \ldots, T_{f64}\} = Encoder\_txt(t_i) \tag{2}$$

$$X_{global\_token} = X_{f1} \tag{3}$$

$$T_{global\_token} = T_{f1} \tag{4}$$

Where $\{X_{f1}, X_{f2}, \ldots, X_{f211}\}$ denote the 211 visual features extracted, corresponding to 210 image blocks and a learnable token respectively. Similarly, $\{T_{f1}, T_{f2}, \ldots, T_{f64}\}$ denote 64 text features extracted, since the maximum length of a sentence is defined as 64. We select the first feature of the two modal feature sequences to represent the global features, then utilize them to compute the CMPM-DS loss.

2.3 Saliency Mask Modelling

As fine-grained image-text retrieval problem, text-based person re-identification is a problem that needs to be solved in order to uncover the highly discriminative information in two similar negative sample pairs. Although the model based on the Transformer encoder architecture is good for global modelling of salient information, as shown in Fig. 3, for similar negative sample pairs, non-salient but highly discriminative information can better assist in distinguishing between them. In order to make full use of the highly discriminative non-salient information, we propose a new alignment paradigm, called Saliency Mask Modelling (SMM), which is able to mask the salient regions and enable the model to make full use of the non-salient regions for modelling, thus facilitating the model to learn more non-salient but highly discriminative information.

As shown in Fig. 3, we first perform one VIT and BERT calculation on the original image and text, and then find the top-k tokens (i.e. salient region features) that are most relevant to the global token based on their cosine similarity to the global token. After that, we mask the original image patches and words corresponding to the top-k tokens and perform a second VIT and BERT calculation using the remaining image patches and words to obtain the SMM global token. The similarity score of a local region token represents the relevance of the region to the global picture, so the higher the similarity score the more significant the region is. We select the top-k local regions as salient regions based on the similarity score, and the remaining regions as non-salient regions. According to Fig. 3, we can see that for images, the body of a person tends to be the salient region, while the head and feet tend to be regarded as non-salient regions. For text, words describing body clothing tend to be the most relevant, such as "black shorts" and "white T-shirt".

We mask the salient regions and globally model the remaining non-salient regions to obtain SMM global token. The SMM global token focuses more on the non-salient regions and can improve the model's ability to extract non-salient but highly discriminative information. Unlike Bidirectional Random Mask Modelling (BMM) proposed by [6]. which uses random mask to increase data diversity, our SMM focuses on non-salient regions of images and text, aiming to explore non-salient but highly discriminative information.

2.4   Dynamic Label Smoothing

What Is CMPM. Let us first review how the CMPM loss function implements cross-modal feature projection to enable the model extract more discriminative features. Given a mini-batch containing n image-text pairs, the image-text pairs for each image feature x_i in a mini-batch are constructed as $\{(x_i, t_i), y_{i,j}\}_{j=1}^n$. When (x_i, t_i) is a positive sample pair, the $y_{i,j}$ is 1, indicating that the true distribution of the sample pairs is true. When (x_i, t_i) is a negative sample pair, the $y_{i,j}$ is 0, it means that the true distribution of samples is false. The matching probability of image feature x_i and text feature t_j is defined as follows.

$$p_{i,j} = \frac{\exp(x_i^{\mathrm{T}} \bar{t}_j)}{\sum_{k=1}^n \exp(x_i^{\mathrm{T}} \bar{t}_k)} \quad s.t. \bar{t}_j = \frac{t_j}{\|t_j\|} \tag{5}$$

where t_j is denoted as the text feature after normalization. Geometrically, $x_i^{\mathrm{T}} \bar{t}_j$ denotes the scalar projection of image features onto the normalised text features, and the probability $p_{i,j}$ is seen as the jth scalar projection percentage of all image text pairs in the mini-batch.

For the setting of the true distribution y, [11] take into account the practical situation that a mini-batch may have multiple text features paired with

x_i, so they normalize the true distribution y. The mini-batch normalised true distribution q is defined as follows.

$$q_{i,j} = \frac{y_{i,j}}{\sum_{k=1}^{n} y_{i,k}} \tag{6}$$

[11] take advantage of KL divergence to relate the matching distribution p of sample pairs to the mini-batch normalised true distribution q to obtain the CMPM loss function, which is defined as follows.

$$L_{i\_cmpm} = \sum_{j=1}^{n} p_{i,j} \log \frac{p_{i,j}}{q_{i,j} + \epsilon} \tag{7}$$

Where ϵ is a minimum value, in order to avoid the negative infinite problem of function value. The KL divergence measures the difference between two probability distributions, and the smaller its value, the smaller the difference between the two probability distributions.

How to Achieve Dynamic Label Smoothing. Inspired by label smooth [7], we consider that label smoothing is also needed in the image-text retrieval domain to avoid the model overfitting caused by hard labels. Let us review Eq. 6, It can be found that the real distribution y of CMPM loss has the public problem of hard label in setting, that is, the absolute negation of all negative samples, which is easy to cause the model overfitting. Therefore, we propose a dynamic smoothing strategy to modify the CMPM loss to solve the overfitting caused by the hard true distribution y.

How can dynamic label smoothing be implemented? Our dynamic label smoothing differs from label smooth, which only considers how to avoid absolute negation of the true probability distribution of the negative samples, and uses a uniform distribution to describe the negative samples, without considering the discrepancies between each negative sample and the same positive samples. Our proposed dynamic label smoothing not only considers how to circumvent the absolute negation of the true probability distribution of negative samples, but also takes into account the discrepancies between different negative samples and same positive samples. Specifically, it allows for a more stringent penalty for difficult negative samples.

We consider that for the cross-modal task of image text retrieval, each pair of mismatched image-text differs to some extent. Some texts have descriptive content that is quite similar to the visual content of an image, but their true labels do not match that image; this type of text is a difficult negative sample. Therefore, smoothing operations based on how well the visual features match the text features of the negative samples are more realistic.

At the same time, we propose that dynamic label smoothing is a dynamic adaptive smoothing process, where the smoothing distribution used will be different for each step and could dynamically adjust the degree of smoothing of the true distribution as the model learns.

We refer to the smoothed modified CMPM as CMPM-Dynamic-Smooth (CMPM-DS), which is implemented as follows.

$$\hat{q}_{i,j} = \frac{\hat{y}_{i,j}}{\sum_{k=1}^{n} \hat{y}_{i,k}} \tag{8}$$

$$\hat{y}_{i,j} = \begin{cases} \alpha & y_{i,j} == True \\ 1 - p_{i,j} & y_{i,j} == False \end{cases} \tag{9}$$

$$L_{i\_cmpm-ds} = \sum_{j=1}^{n} p_{i,j} \log \frac{p_{i,j}}{\hat{q}_{i,j} + \epsilon} \tag{10}$$

It can be seen from Eq. 9 that we divide the true distribution y into two cases. To distinguish it from the original true distribution y, the real distribution after dynamic smoothing is defined $\hat{y}$, and the normalization of $\hat{y}$ is defined $\hat{q}$. We divide $\hat{y}$ into two cases.

1. Processing of the true distribution $\hat{y}$ for matched image-text pairs. For the matching true distribution $\hat{y}$, we multiply it by the hyperparameter α, a belongs to the hyperparameter that regulates the probability of the true distribution of the matched image-text pair. The larger the value of α, the closer the $\hat{q}$ of the matched image-text pair is to 1, and the closer the $\hat{q}$ of the mismatched image-text pair is to 0.
2. Processing of the true distribution $\hat{y}$ for mismatched image-text pairs. For the mismatch true distribution, we set it to $(1 - p_{i,j})$, where $p_{i,j}$ is given by Eq. 5, which reflects the degree to which the two modal features match. As value of $p_{i,j}$ ranges from 0 to 1 and the closer it is to 1, the more matching the two features are. For a mismatched image-text pair, the closer $p_{i,j}$ is to 1, the more difficult a negative matching pair is. Therefore, we set the value of the true distribution of mismatches to $(1 - p_{i,j})$, thus achieving a greater penalty for difficult negative matching pairs. It is easy to see that the magnitude of the $p_{i,j}$ value changes as the model parameters change, suggesting that the dynamic label smoothing on the true distribution $\hat{y}$ is a dynamic and adaptive process.

3   Experiments

3.1   Datasets

We conducted extensive experiments on two public datasets. CUHK-PEDES [1] dataset contains 40206 pedestrian images with 13003 pedestrian IDs and two text descriptions per image. The dataset is divided into 11,003 training IDs with 34,054 images, 1,000 validation IDs with 3,078 images and 1,000 test IDs with 3,074 images. RSTPReid [12] dataset contains 20,505 pedestrian images with 4101 IDs, each with two statements. Specifically, training set, validation set and test set is divided into 3,701, 200 and 200 IDs respectively.

Table 1. Comparison of our method with SOTA model on CUHK-PEDES (left) and RSTPReid (right)

Methods	R@1	R@5	R@10	Methods	R@1	R@5	R@10
ViTAA [5]	55.97	75.84	83.52	IMG-Net [10]	37.60	61.15	73.55
IMG-Net [10]	56.48	76.89	85.01	AMEN [8]	38.45	62.40	73.80
DSSL [12]	59.98	80.41	87.56	DSSL [12]	39.05	62.60	73.95
SSAN [3]	61.37	80.15	86.73	SSAN [3]	43.50	67.80	77.15
TIPCB [1]	63.63	82.82	89.01	IVT [6]	46.70	70.00	78.80
IVT [6]	65.59	83.11	89.21	CAIBC [9]	47.35	69.55	79.00
Ours	65.92	83.72	90.01	Ours	51.80	73.20	81.25

3.2 Evaluation Metrics and Implementation Details

Evaluation Metrics. In this paper, we use Recall@K (K = 1, 5, 10), a common evaluation metric for retrieval tasks. Recall@K (R@K) indicates rate of correct matches present in the first K images when we query images using the specified text and mAP indicates the average retrieval accuracy, which is the average of all retrieval accuracies.

Implementation Details. Models are learned on a single NVIDIA GEFORCE GTX 3090 GPU. Our baseline model is trained using the Adam optimizer for 20 epochs at a learning rate of 0.00003, followed by 60 epochs at a learning rate of 0.000003.

3.3 Comparison with State-of-the-art Methods

Table 1 presents our method compared to existing methods on CUHK-PEDES and RSTPReid. We can see that our method achieves SOTA results on both pedestrian datasets. The results of our method outperform the state-of-the-art methods by 0.33% in terms of R@1 on the CUHK-PEDES dataset and by 4.45% on the RSTPReid dataset.

3.4 Ablation Study

In Table 2, serial numbers 1 and 5 indicate the results of the baseline model without SMM and CMPM-DS. By observing serial numbers 2 and 6, we can find that after using SMM, R@1 increases by 1.29% on average in the results of the two datasets. By observing serial numbers 3 and 7, we can see that the results of the two datasets are significantly improved after the use of CMPM-DS, and the average R@1 increase is 3.39%. The serial numbers 4 and 8 show the results with the simultaneous use of SMM and CMPM-DS, indicating they can be used effectively in combination.

Table 2. Ablation experiments on two pedestrian datasets

Serial No.	Dataset	Baseline	SMM	CMPM-DS	R@1	R@5	R@10
1	CUHK-PEDES	✓	✗	✗	61.58	81.08	87.17
2	CUHK-PEDES	✓	✓	✗	62.30	82.21	88.55
3	CUHK-PEDES	✓	✗	✓	64.25	82.60	88.56
4	CUHK-PEDES	✓	✓	✓	65.92	83.72	90.01
5	RSTPReid	✓	✗	✗	45.90	71.70	80.70
6	RSTPReid	✓	✓	✗	47.75	72.85	82.30
7	RSTPReid	✓	✗	✓	50.00	74.50	81.30
8	RSTPReid	✓	✓	✓	51.80	73.20	81.25

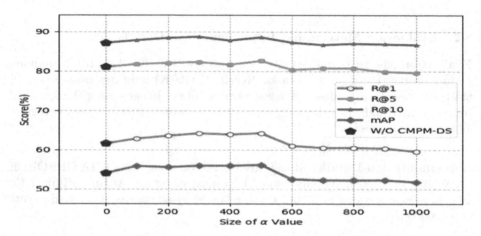

Fig. 4. Exploring the effect of different α values on performance at a batch size of 48. Different alpha values can have a significant impact on performance. When the alpha value is too large, it can lead to performance degradation.

3.5   Hyperparameter Exploration Experiments

In this section, we perform a hyperparameter comparison experiment on the CMPM-DS function to explore the effect of the hyperparameter α on performance. α is used as a hyperparameter to adjust the real probability distribution of positive matching pairs, and its value affects the degree of label smoothing on the whole. Since Eq. 8 is normalised over a mini-batch, the size of the alpha value is closely related to the size of the mini-batch.

Figure 4 reports the experimental results for baseline using different alpha values on the CUHK-PEDES dataset and for a batch-size of 48. From the results in Fig. 4, it can be seen that the model performs best when the α value is 500. Also, it is not the case that using dynamic label smoothing necessarily avoids overfitting, as the performance of the model is highly dependent on the degree of smoothing.

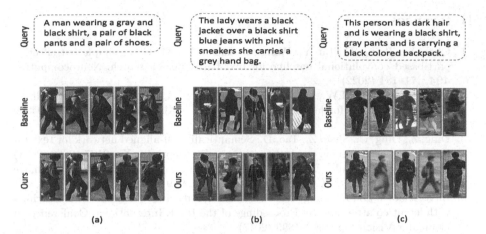

Fig. 5. Results of R@5 are shown, where the green boxes are correct results and the red boxes are incorrect results. (Color figure online)

3.6 Qualitative Results

Three examples of results of R@5 are reported, as shown in Fig. 5. In general, all retrieved images have a high correlation with the description text, even for incorrect results. Our method is effective in improving retrieval results, as shown in Fig. 5(a), and with good retrieval results in the baseline, we can still make further improvements to the retrieval results. Figure 5(b) shows the ability of our method to capture highly discriminative details, such as "a grey hand bag". For Fig. 5(c), where there is a lack of discriminative information, our method still provides better overall modelling to achieve correct retrieval.

4 Conclusion

This paper proposes a new alignment paradigm for solving the similar negative sample problem, called Saliency Mask Modelling. It masks the salient region to improve the model's ability to capture non-salient but highly discriminative information. Meanwhile, this paper draws on the idea of label smoothing and uses dynamic label smoothing for the CMPM loss function to solve the model overfitting problem. Unlike traditional label smoothing, dynamic label smoothing takes into account the differences between different negative samples and adaptively adjusts the degree of smoothing each step dynamically.

Acknowledgements. This work is supported by National Natural Science Foundation of China (Nos. 62266009, 61866004, 62276073, 61966004, 61962007), Guangxi Natural Science Foundation (Nos. 2019GXNSFDA245018), Guangxi Collaborative Innovation Center of Multi-source Information Integration and Intelligent Processing, and Guangxi "Bagui Scholar" Teams for Innovation and Research Project.

References

1. Chen, Y., Zhang, G., Lu, Y., Wang, Z., Zheng, Y.: TIPCB: a simple but effective part-based convolutional baseline for text-based person search. Neurocomputing **494**, 171–181 (2022)
2. Devlin, J., Chang, M.W., Lee, K., Toutanova, K.: Bert: pre-training of deep bidirectional transformers for language understanding. arXiv preprint arXiv:1810.04805 (2018)
3. Ding, Z., Ding, C., Shao, Z., Tao, D.: Semantically self-aligned network for text-to-image part-aware person re-identification. arXiv preprint arXiv:2107.12666 (2021)
4. Dosovitskiy, A., et al.: An image is worth 16×16 words: transformers for image recognition at scale. arXiv preprint arXiv:2010.11929 (2020)
5. Li, S., Xiao, T., Li, H., Yang, W., Wang, X.: Identity-aware textual-visual matching with latent co-attention. In: Proceedings of the IEEE International Conference on Computer Vision, pp. 1890–1899 (2017)
6. Shu, X., et al.: See finer, see more: implicit modality alignment for text-based person retrieval. arXiv preprint arXiv:2208.08608 (2022)
7. Szegedy, C., Vanhoucke, V., Ioffe, S., Shlens, J., Wojna, Z.: Rethinking the inception architecture for computer vision. In: Proceedings of the IEEE Conference on Computer Vision and Pattern Recognition, pp. 2818–2826 (2016)
8. Wang, Z., Xue, J., Zhu, A., Li, Y., Zhang, M., Zhong, C.: AMEN: adversarial multi-space embedding network for text-based person re-identification. In: Ma, H., et al. (eds.) PRCV 2021. LNCS, vol. 13020, pp. 462–473. Springer, Cham (2021). https://doi.org/10.1007/978-3-030-88007-1_38
9. Wang, Z., et al.: CAIBC: capturing all-round information beyond color for text-based person retrieval. In: Proceedings of the 30th ACM International Conference on Multimedia, pp. 5314–5322 (2022)
10. Wang, Z., Zhu, A., Zheng, Z., Jin, J., Xue, Z., Hua, G.: IMG-Net: inner-cross-modal attentional multigranular network for description-based person re-identification. J. Electron. Imaging **29**(4), 043028 (2020)
11. Zhang, Y., Lu, H.: Deep cross-modal projection learning for image-text matching. In: Ferrari, V., Hebert, M., Sminchisescu, C., Weiss, Y. (eds.) ECCV 2018. LNCS, vol. 11205, pp. 707–723. Springer, Cham (2018). https://doi.org/10.1007/978-3-030-01246-5_42
12. Zhu, A., et al.: DSSL: deep surroundings-person separation learning for text-based person retrieval. In: Proceedings of the 29th ACM International Conference on Multimedia, pp. 209–217 (2021)

Robust Multi-view Spectral Clustering with Auto-encoder for Preserving Information

Xiaojie Wang, Ye Liu[(✉)], Hongshan Pu, Yuchen Mou, and Chaoxiong Lin

The School of Future Technology, South China University of Technology,
Guangzhou 511442, China
ftwangxj@mail.scut.edu.cn, yliu03@scut.edu.cn

Abstract. Multi-view clustering is a prominent research topic in machine learning that leverages consistent and complementary information from multiple views to improve clustering performance. The graph-based multi-view clustering methods learn consistent graph with pairwise similarity between data as edges, and generate sample representation using spectral clustering. However, most existing methods seldom consider to recreate the input data using encoded representation during representation learning procedure, which result in information loss. To address this limitation, we propose a robust multi-view clustering with auto-encoder for preserving information (RMVSC-AE) that minimizes the reconstruction error between the input data and the reconstructed representation to preserve knowledge. Specifically, we discover a graph representation by jointly optimizing the graph Laplacian and auto-encoder reconstruction terms. Moreover, we introduce a sparse noisy term to further enhance the quality of the learned consistent graph. Extensive experiments on six multi-view datasets are conducted to verify the efficacy of the proposed method.

Keywords: Multi-view Clustering · Spectral Clustering · Auto-encoder · Robust Graph Learning

1 Introduction

With the rapid development of information technology nowadays, data can be collected from various sources, resulting in the generation of multi-view data. Multi-view data describes the characteristics of target object from multiple perspectives, which constitute complementary information. Therefore, multi-view data contains more comprehensive and richer information compared to single view data. For example, a picture can be described by scale-invariant feature transform [14], histograms [3], and other features. In order to mine consistent information in multi-view data, multi-view clustering has emerged.

The existing multi view clustering can be roughly divided into non-negative matrix factorization, subspace learning, and graph learning. Multi-view clustering method based on non-negative matrix decomposition decomposes the feature matrix into two matrices. One of them is a non-negative basis matrix, and

B. Luo et al. (Eds.): ICONIP 2023, LNCS 14451, pp. 455–471, 2024.
https://doi.org/10.1007/978-981-99-8073-4_35

another is a representation matrix. Multi-view subspace clustering assumes that high-dimensional data has a low-dimensional subspace structure. These methods generally establish affinity matrix through self-expression, and then apply spectral clustering on affinity matrix to obtain final clustering result. However, data in the real world does not always satisfy linear distributions, and multi-view subspace clustering cannot well capture the nonlinear structure of the data.

The graph based multi-view clustering method first constructs a similarity graph based on similarity measurement, then learns consistent graph across multiple graphs using fusion strategy, and finally outputs clustering results by adopting spectral clustering on the fused graph. The core idea of multi-view graph-based clustering is to integrate heterogeneous information from different views into a consensus embedding or graph. Existing graph-based methods only utilize similarity graph to generate graph embedding, such that manifold structure within data can be captured. However, these methods seldom consider to recreate the input data using graph embedding to measure the information loss, and hence they may not be able to fully make use of the important information in the original input data as much as possible. Furthermore, real data always contains noises, which may lead to noisy similarity graph and then degenerate final clustering results.

Motivated by the above demonstrations, in this paper, we present a novel approach called robust multi-view spectral clustering with auto-encoder for preserving information (RMVSC-AE). Our approach learn a robust consistent similarity graph and graph representation with preserved information in a unified framework. To be specific, graph Laplacian terms and auto-encoder reconstruction terms are incorporated to capture the local structure and maintain significant information in original input data. Moreover, we use L_1 regularization to capture sparse noise within consistent graph, such that a robust graph can be obtained. Mathematically, the proposed model is formulated as a minimization problem, which can be solved by alternating minimization method. The convergence and computational complexity are also shown. Extensive experiments on six datasets are conducted to illustrate the effectiveness of our proposed method. Our proposed approach represents significant improvements over state-of-the-art multi-view clustering methods.

2   Related Work

2.1   Notations

In this paper, Matrices are denoted with uppercase bold letters (e.g., $\mathbf{X}$), lowercase bold letters (e.g., $\mathbf{x}$) denote vectors, and non-bold lowercase letters (e.g., x) represent scalars. The sample size and dimension of the v-th view are denoted as n and d_v respectively. We use m to denote the number of views and c represent the number of classes. Real numbers are represented by the real space symbol $\mathbf{R}$. The trace and transpose of a square matrix $\mathbf{X}$ are represented by $Tr(\mathbf{X})$ and $\mathbf{X}^T$ respectively. The Frobenius norm and L_1 norm of a matrix $\mathbf{X}$ are represented as $\|\mathbf{X}\|_F$ and $\|\mathbf{X}\|_1$ respectively. Finally, $\mathbf{X} \geq 0$ means that each element of matrix

$\mathbf{X}$ is bigger than or equal to zero. $sgn(x)$ is the element-wise sign function. If x is greater than 0, sgn returns 1; Equal to 0, returns 0; If it is less than 0, it returns -1. $|X|$ represents taking absolute values for each element of the matrix. $max\{\mathbf{A},\mathbf{B}\}$ returns the maximum value of each corresponding element in two matrices.

2.2 Multi-view Spectral Clustering

In multi-view spectral clustering, the first step is to construct similarity matrices $\left\{\mathbf{W}^v \in \mathbf{R}^{n \times n}\right\}_{v=1}^m$ from the given multi-view data $\left\{\mathbf{X}^v \in \mathbf{R}^{n \times d_v}\right\}_{v=1}^m$ using various similarity metrics [19,29]. Next, different approaches can be employed to fuse multiple graphs. There are usually two kinds of fusion strategies, one is pre-fusion methods, they learn a consensus similarity matrix $\mathbf{W}$ and then performs spectral clustering on $\mathbf{W}$ to obtain embedding $\mathbf{F}$. Another is post-fusion, which refers to learn the consensus embedding matrix from multiple graph representation $\mathbf{F}^v$ calculated by applying the spectral clustering on $\mathbf{W}^v$ separately. In both cases, the final clustering results are obtained through k-means on graph representation.

Multi-view spectral clustering has shown great performances. The Co-reg algorithm [11], which implements a co-regularization strategy to enforce clustering consistency across multiple views. [17] proposed an automatic weighting method to assign different importance weight to different views. However, it has high computational complexity which limits its application in large-scale data. To address this limitation, [12] proposed to use bipartite graphs to approximate complete similar graphs, which can reduce computational complexity. [30] proposed a graph-learning based method to improve the quality of graphs. [18] proposed an adaptive graph learning method that fuses multiple graphs and automatically assigns reasonable weights to each graph. [1] proposed to learn a potential embedding representation from the original input data. [24] proposed to learn the data graph matrix of each view and the unified graph matrix in a mutually reinforcing manner. [21] cleverly combined adaptive graph learning and projection matrix to handle out-of-sample problems. [28] learn a consistent orthogonal embedding from multiple feature matrices and multiple similarity matrices simultaneously via the disturbed probabilistic subspace modeling and approximation. [26] proposed a multi-view binary clustering to learn a unified binary code based on auto-encoders, which optimize auto-encoder and affinity graph with low-rank constraints jointly.

Although the aforementioned methods have demonstrated remarkable performance, they encounter challenges when confronted with noise that frequently exist in real-world data. Moreover, these techniques only consider using graph to calculate representations, do not fully utilize informations embedded in the original data.

3   Methodology

In this section, we introduce our proposed RMVSC-AE method in detail. Specifically, we adopt gaussian kernel function to construct the similarity matrices of the multi-view data. Then we adaptively learn robust consensus graph from multiple views and optimize the graph embedding matrix with spectral clustering and auto-encoder. Finally, we can obtain the clustering results by simple clustering method like k-means. Besides, the computation complexity are discussed.

3.1   Graph Construction

Given a dataset $\mathbf{X}$ with m views. Denotes $\left\{\mathbf{X}^v \in \mathbf{R}^{n \times d_v}\right\}_{v=1}^m$, where d_v is the feature dimension of the v-th view and n represents the number of data samples. We use common gaussian kernel functions defined as follows to construct the similarity matrix.

$$K(x_i, x_j) = \Phi(x_i) * \Phi(x_j) = \exp(-\frac{\|x_i - x_j\|^2}{2\sigma^2}) \tag{1}$$

where σ is the bandwidth parameter. According to the literatures [12,13,15,25], K-nearest-neighbor (KNN) graph tends to perform better than the complete graph, as it builds sparse connections that helps capture the local manifold structures. Therefore, we use KNN to construct the KNN graph matrix as follows:

$$\mathbf{W}_{ij} = \begin{cases} \frac{1}{2}(\frac{K(\mathbf{x}_i, \mathbf{x}_j)}{\sum_{k \in \Phi_i} K(\mathbf{x}_i, \mathbf{x}_k)} + \frac{K(\mathbf{x}_i, \mathbf{x}_j)}{\sum_{k \in \Phi_j} K(\mathbf{x}_k, \mathbf{x}_j)}), & \mathbf{x}_i \in \Phi_j \ or \ \mathbf{x}_j \in \Phi_i \\ 0, & otherwise \end{cases} \tag{2}$$

where Φ is the KNN set of x. Then the similarity matrix $\mathbf{W}^v$ for each view is constructed independently.

3.2   Multi-view Graph Fusion

Graphs matrices of each view $\{\mathbf{W}^v\}_{v=1}^m$ can be constructed according to Eq. (2) individually. To integrate multiple graphs into a consensus graph, [21] proposed an adaptive view fusion strategy to determine the weight of each view. However, they ignored the impact of noise on the graph structure. In order to learn a more robust consistent graph, We learn the global graph $\mathbf{G}$ by

$$\min_{\mathbf{G}} \alpha^v \|\mathbf{G} + \mathbf{E}^v - \mathbf{W}^v\|_F^2 + \mu \|\mathbf{E}^v\|_1$$
$$s.t. \ \mathbf{G1} = 1, \ \mathbf{G} \geq 0, \ rank(\mathbf{L}_G) = n - c \tag{3}$$

where $\mathbf{E}^v \in \mathbf{R}^{n*n}$ is the error matrix of each view to capture the noise. The constraint $\mathbf{G} \geq 0$ limits the graph to be non-negative and $\mathbf{G1} = 1$ constrains that each row sums up to 1. μ is a trade-off parameter. $\mathbf{L}_G = \mathbf{D} - \mathbf{G}$ is the Laplacian matrix of the graph. $\mathbf{D}$ is a diagonal matrix, whose i-th diagonal element is $\sum_j \mathbf{G}_{ij}$. c is the cluster number. According to Theorem 1, the weights α are determined automatically.

Theorem 1. *If the weights α are fixed, Solving the following two problems is equivalent:*

$$\min_{\mathbf{G}} \sum_{v=1}^{m} \|\mathbf{G} - \mathbf{B}^v\|_F + \Theta(\varsigma, \mathbf{G}) \tag{4}$$

$$\min_{\mathbf{G}} \sum_{v=1}^{m} \alpha^v \|\mathbf{G} - \mathbf{B}^v\|_F^2 + \Theta(\varsigma, \mathbf{G}) \tag{5}$$

where ς is the Lagrange multiplier and $\Theta(\varsigma, \mathbf{G})$ is the formalized term derived from constraints.

Proof. Taking the derivative of Eq. (4) with respect to $\mathbf{G}$ and setting the derivative to zero, we have

$$\sum_{v=1}^{m} \frac{\partial \|\mathbf{G} - \mathbf{B}^v\|_F}{\partial \mathbf{G}} + \frac{\partial \Theta(\varsigma, \mathbf{G})}{\partial \mathbf{G}} = 0$$

$$\Leftrightarrow \sum_{v=1}^{m} \frac{2\|\mathbf{G} - \mathbf{B}^v\|_F}{2\|\mathbf{G} - \mathbf{B}^v\|_F} * \frac{\partial \|\mathbf{G} - \mathbf{B}^v\|_F}{\partial \mathbf{G}} + \frac{\partial \Theta(\varsigma, \mathbf{G})}{\partial \mathbf{G}} = 0$$

$$\Leftrightarrow \sum_{v=1}^{m} \frac{1}{2\|\mathbf{G} - \mathbf{B}^v\|_F} * \frac{\partial \|\mathbf{G} - \mathbf{B}^v\|_F^2}{\partial \mathbf{G}} + \frac{\partial \Theta(\varsigma, \mathbf{G})}{\partial \mathbf{G}} = 0 \tag{6}$$

$$\Leftrightarrow \sum_{v=1}^{m} \alpha^v * \frac{\partial \|\mathbf{G} - \mathbf{B}^v\|_F^2}{\partial \mathbf{G}} + \frac{\partial \Theta(\varsigma, \mathbf{G})}{\partial \mathbf{G}} = 0$$

where

$$\alpha^v = \frac{1}{2\|\mathbf{G} - \mathbf{B}^v\|_F} = \frac{1}{2\|\mathbf{G} + \mathbf{E}^v - \mathbf{W}^v\|_F} \tag{7}$$

If α is fixed, the derivative of the Lagrange function of Eq. (4) is equal to Eq. (5). Thus Eq. (4) is equivalent to Eq. (5). The weights α are also determined by (7)

Theorem 2. *The connected components in the graph associated with the similarity matrix $\mathbf{G}$ is equal to the multiplicity of eigenvalue zero of Laplacian matrix L_G.*

According to Theorem 2 [2,16], it is clear that if rank constraint $rank(\mathbf{L}_G) = n - c$ of the Laplacian matrix is satisfied, then the graph associated with $\mathbf{G}$ is c-connected. Directly solving the rank constraint problem is NP-hard, according to the Ky Fan's Theorem [6], we can solve the following instead.

$$\min_{\mathbf{F}} Tr(\mathbf{F}^T \mathbf{L}_G \mathbf{F}) \quad s.t. \ \mathbf{F} \in \mathbf{R}^{n*c}, \mathbf{F}^T \mathbf{F} = \mathbf{I} \tag{8}$$

3.3   Preserving Information by Auto-encoder

In general, a linear auto-encoder with a single layer is comprised of a solitary encoder and decoder. The encoder facilitates the projection of input data onto the hidden layer, while the decoder enables the projection of the data back from the hidden layer into the feature space. Specifically, Given a data matrix $\mathbf{X} \in R^{n*d}$ composed of n feature row vectors of d dimensions, the input data undergoes a projection process using a projection matrix $\mathbf{M} \in R^{d*k}$, leading to its representation in a latent space of dimensionality k. Subsequently, the latent representation is projected back to the feature space via another projection matrix $\mathbf{M}^* \in R^{k*d}$, resulting in the reconstructed data $\widehat{\mathbf{X}} \in R^{n*d}$. Moreover, we have $k < d$, it means that the latent representation reduces the dimensionality of the original data. The general form of the objective function for minimizing reconstruction errors can be expressed as follows.

$$\min_{\mathbf{M},\mathbf{M}^*} \|\mathbf{X} - \mathbf{XMM}^*\|_F^2 \tag{9}$$

Existing methods based on matrix decomposition learn a low-rank projection matrix to project data from the original feature space into a low-dimensional embedded space. In this way, the discriminant information in the original feature space is retained and used for clustering. However, the method based on one-direction projection cannot guarantee that the information in original input data are fully utilized [10]. To make the embedding space semantically meaningful, we force the embedding space to be the representation space in the linear auto-encoder. So we propose to learn the projection between the input data and the graph representation using a linear auto-encoder as follows:

$$\min_{\mathbf{M},\mathbf{M}^*} \|\mathbf{X} - \mathbf{XMM}^*\|_F^2$$
$$s.t.\ \mathbf{XM} = \mathbf{F} \tag{10}$$

To further simplify the model, we consider tied weights to be $\mathbf{M}^* = \mathbf{M}^T$ [20] and relax the constraint into a soft one. Then the objective can be rewritten as:

$$\min_{\mathbf{M}^v} \sum_{v=1}^m \|\mathbf{X}^v\mathbf{M}^v - \mathbf{F}\|_F^2 + \lambda \|\mathbf{X}^v - \mathbf{F}\mathbf{M}^{vT}\|_F^2 \tag{11}$$

where $\mathbf{M}^v \in \mathbb{R}^{d_v \times c}$ is the projection matrix of each view and λ is a weighting coefficient that controls the importance of first and second terms, which correspond to the losses of the encoder and decoder respectively. In this way, the input matrix can not only be projected into graph representation, but also be reconstructed using learned projection matrix, such that the optimized graph representation $\mathbf{F}$ can maintain both the local structure in similarity graph and main information in original data.

3.4   Overall Learning Framework

Combining Eq. (3), Eq. (8) and Eq. (11), we derive the following optimization function.

$$
\min_{\mathbf{G},\mathbf{E}^v,\mathbf{F},\mathbf{M}^v} \sum_{v=1}^{m} (\alpha^v \|\mathbf{G} + \mathbf{E}^v - \mathbf{W}^v\|_F^2 + \mu \|\mathbf{E}^v\|_1) + \beta Tr(\mathbf{F}^T \mathbf{L}_G \mathbf{F})
$$

$$
+ \sum_{v=1}^{m} \gamma(\|\mathbf{X}^v \mathbf{M}^v - \mathbf{F}\|_F^2 + \lambda \left\| \mathbf{X}^v - \mathbf{F}\mathbf{M}^{vT} \right\|_F^2) \tag{12}
$$

$$
s.t.\ \&\mathbf{G1} = 1, \mathbf{G} \geq 0, \mathbf{F}^T \mathbf{F} = \mathbf{I}
$$

where β and γ are trade-off parameters.

4   Optimization

In this section, we describe the optimization process for Eq. (12) in details. The objective function is a non-convex optimization problem, we use an alternate optimization method to solve the objective function. When one variable is updated, other variables remain fixed.

4.1   Subproblem of G

Fixing $\mathbf{E}^v$, $\mathbf{F}$ and $\mathbf{M}^v$, the optimization formula for $\mathbf{G}$ becomes:

$$
\min_{\mathbf{G}} \sum_{v=1}^{m} \alpha^v \|\mathbf{G} + \mathbf{E}^v - \mathbf{W}^v\|_F^2 + \beta Tr(\mathbf{F}^T \mathbf{L}_G \mathbf{F}) \quad s.t.\ \mathbf{G1} = 1, \mathbf{G} \geq 0 \tag{13}
$$

According to the property of Laplacian matrix [22], the trace term is further equivalent to the following formula:

$$
Tr(\mathbf{F}^T \mathbf{L}_G \mathbf{F}) = \frac{1}{2} \sum_{i,j=1}^{n} \mathbf{G}_{ij} \|\mathbf{f}_i - \mathbf{f}_j\|_2^2 = \sum_{i,j=1}^{n} \mathbf{G}_{ij} \eta_{ij} \tag{14}
$$

where $\eta_{ij} = \frac{1}{2} \|\mathbf{f}_i - \mathbf{f}_j\|_2^2$. Then, the problem (13) can be solved by each row independently. For convenience, we denote ω_i^v as the i-th row of $\mathbf{W}^v - \mathbf{E}^v$ and rewrite Eq. (13) in the vector form:

$$
\min_{\mathbf{g}_i} \left\| \mathbf{g}_i - \frac{\sum_{v=1}^{m} \alpha^v \omega_i^v - \frac{\beta}{2} \eta_i}{\sum_{v=1}^{m} \alpha^v} \right\|_2^2 \quad s.t.\ \mathbf{g}_i \mathbf{1} = 1, \mathbf{g} \geq 0 \tag{15}
$$

This problem can be solved efficiently by the algorithm proposed by [5]. Here we give the general form of Eq. (15) and its corresponding algorithm in Algorithm. 1.

$$
\min_{\mathbf{g}} \frac{1}{2} \|\mathbf{g} - \mathbf{z}\|_2^2 \quad s.t.\ \sum_i \mathbf{g} = 1, \mathbf{g} \geq 0 \tag{16}
$$

Algorithm 1. Procedure for solving Eq. (16)

Input: a vector z
Output: g
1: sort z into $\mathbf{y}$: $\mathbf{y}_1 \geq \mathbf{y}_2 \geq \cdots \geq \mathbf{y}_n$
2: find $\rho = max\left\{ j \in [n] : \mathbf{y}_j - \frac{1}{j}(\sum_{r=1}^{j}\mathbf{y}_r - 1) > 0 \right\}$
3: define $\theta = \frac{1}{\rho}(\sum_{i=1}^{\rho}\mathbf{y}_i - 1)$
4: $\mathbf{g}_i = max\left\{ \mathbf{z}_i - \theta, 0 \right\}$

4.2 Subproblem of $\mathbf{E}^v$

As for solving $\mathbf{E}^v$, it can be formulated as:

$$\min_{\mathbf{E}^v} \sum_{v=1}^{m}(\alpha^v \|\mathbf{G} + \mathbf{E}^v - \mathbf{W}^v\|_F^2 + \mu \|\mathbf{E}^v\|_1) \tag{17}$$

It can be solved separately for each view. This problem has a closed solution by soft thresholding [4].

$$\mathbf{E}^{v*} = sgn(\mathbf{W}^v - \mathbf{G})max\left\{ |\mathbf{W}^v - \mathbf{G}| - \frac{\mu}{2\alpha^v}, 0 \right\} \tag{18}$$

4.3 Subproblem of $\mathbf{M}^v$

By fixing other variables, the optimization for $\mathbf{M}^v$ can be derived as

$$\min_{\mathbf{M}^v} \sum_{v=1}^{m} \|\mathbf{X}^v\mathbf{M}^v - \mathbf{F}\|_F^2 + \lambda \|\mathbf{X}^v - \mathbf{F}\mathbf{M}^v\|_F^2 \tag{19}$$

We set the derivative of Eq. (19) with respect to $\mathbf{M}^v$ as 0, then we have the solution

$$\mathbf{M}^v = (\mathbf{X}^{vT}\mathbf{X}^v + \lambda\mathbf{I}_{d^v})^{-1}(1 + \lambda)\mathbf{X}^{vT}\mathbf{F} = \mathbf{H}^v\mathbf{F} \tag{20}$$

where $\mathbf{I}_{d^v}$ is the identify matrix of size $d^v \times d^v$.

4.4 Subproblem of $\mathbf{F}$

By substituting Eq. (20) into the objective function of $\mathbf{F}$, we have:

$$\min_{\mathbf{F}} Tr(\mathbf{F}^T(\beta\mathbf{L}_G + \mu\sum_{v=1}^{V}\mathbf{K}^v)\mathbf{F}) \quad s.t. \ \mathbf{F}^T\mathbf{F} = \mathbf{I}$$
$$\mathbf{K}^v = \mathbf{H}^{vT}\mathbf{X}^{vT}\mathbf{X}^v\mathbf{H}^v - 2(1 + \lambda)\mathbf{H}^{vT}\mathbf{X}^{vT} + \lambda\mathbf{H}^{vT}\mathbf{H}^v \tag{21}$$

As $\mathbf{K}^v$ and $\mathbf{H}^v$ depend on $\mathbf{X}^v$ and λ , they can be calculated just for once. $\mathbf{F}$ can be solved by simple eigenvalue decomposition.

In summary, the complete optimization process is listed in Algorithm 2.

Algorithm 2. Procedure for solving Eq. (12)

Input: data matrices $\{\mathbf{X}^v\}_{v=1}^m$, parameters β, μ, γ, λ
Output: Embedding matrix $\mathbf{F}$, clustering results
1: Initialization: $\mathbf{G} = \frac{1}{m}\sum_{v=1}^m \mathbf{W}^v$, $\mathbf{E}^v = \mathbf{0}$, random initialize $\mathbf{M}^v$, $\mathbf{F}$
2: **repeat**
3:    Update α^v based on Eq. (7)
4:    Update $\mathbf{G}$ based on Eq. (13)
5:    Update $\mathbf{E}^v$ based on Eq. (18)
6:    Update $\mathbf{M}^v$ based on Eq. (20)
7:    Update $\mathbf{F}$ based on Eq. (21)
8: **until** $\|\mathbf{F}^t\mathbf{F}^{t^T} - \mathbf{F}^{t-1}\mathbf{F}^{t-1^T}\|_F < 10^{-4}$
9: Run K-means on $\mathbf{F}$

4.5 Complexity Analysis

The optimization process of the proposed algorithm is achieved through alternating optimization methods. Assuming the number of iterations until convergence is T, each iteration of the algorithm involves updating five variables. The computational complexity of updating weighting coefficients α^v is $O(n^2)$. The computational complexity of updating $\mathbf{G}$ is $O(n^2)$. The computational complexity of updating $\mathbf{E}^v$ is $O(n^2V)$. The computational complexity of updating $\mathbf{M}^v$ and F is $O(cndV)$ and $O(n^3)$. The complexity of the algorithm mainly comes from the matrix eigenvalue decomposition, which is required in the Spectral clustering algorithm. Therefore, the total time complexity of the proposed method is $O(Tn^3)$, which is comparable to state-of-the-art multi-view clustering methods [1,9,17,21,28,30].

Table 1. Benchmark datasets

Dataset	Samples	Clusters	Views	Feature numbers
BBCSport	116	5	4	1991,2063,2113,2158
3Sources	169	6	3	3560,3631,3068
MSRC	210	7	5	24,576,512,256,254
NGs	500	5	3	2000,2000,2000
Caltech101-20	2386	7	6	48,40,254,1984,512,928
Hdigit	10000	10	2	784,256

5 Experiments

In this section, we evaluate our proposed method on six widely-used datasets, whose detailed information is introduced in Table 1. Moreover, convergence and parameter sensitivity analysis of our proposed method is also tested on benchmark datasets.

5.1   Comparison Algorithms and Evaluation Metrics

In our experiments, we evaluate the proposed method by comparing with eight state-of-the-art multi-view clustering methods: **Parameter-Free Auto-Weighted Multiple Graph Learning** (AMGL) [17] can automatically learn the weights of each view and achieve weighted fusion. **Graph Learning for Multi-view Clustering** (MVGL) [30] proposes an optimization framework using Laplacian rank constraints to learn graph matrices from multiple views. **Self-weighted Multi-view Clustering**(SwMC) [18] learns a graph with exactly c connected components. **Large-Scale Multi-View Spectral Clustering**(MVSC) [12] use bipartite graphs to solve large-scale problems. **Multi-View Clustering in Latent Embedding Space**(MCLES) [1] cluster the multi-view data in a learned latent embedding space. **Graph-Based Multi-View Clustering** (GMC) [24] learns a unified graph matrix and in turn improves the data graph matrix of each view. **Flexible Multi-view Spectral Clustering With Self-Adaptation** (FMSCS) [21] combines the adaptive graph learning and feature forward projection method. **Bidirectional Probabilistic Subspaces Approximation for Multi-view Clustering** (BPSA) [28] proposes to learn consistent low-dimensional embeddings from both similarity matrix and feature matrix simultaneously. To demonstrate the effectiveness of information preserving with auto-encoder, we removed auto-encoder term from the objective function (12) as another comparison method named RMVSC for comparison, the objective function of RMVSC is:

$$\min_{\mathbf{G},\mathbf{E}^v,\mathbf{F},\mathbf{M}^v} \sum_{v=1}^{m} (\alpha^v \|\mathbf{G} + \mathbf{E}^v - \mathbf{W}^v\|_F^2 + \mu \|\mathbf{E}^v\|_1) + \beta Tr(\mathbf{F}^T \mathbf{L}_G \mathbf{F})$$

$$s.t.\&\mathbf{G1} = \mathbf{1}, \mathbf{G} \geq 0, \mathbf{F}^T\mathbf{F} = \mathbf{I} \tag{22}$$

Equation (22) is also solved by alternating minimization method.

5.2   Experimental Setup

In our experiment, we set up the parameter in a grid searching manner. In our method and comparison methods, the number of KNN is fixed as 14. For a fair comparison, all the parameters in comparison methods will be tuned when optimal results are achieved. In this paper, we use accuracy (ACC), normalized mutual information (NMI), and purity [23,27] to comprehensively evaluate the performance. Higher value indicates better performance.

5.3   Experimental Results and Analysis

Table 2 shows the performance of different clustering evaluation measurements on six datasets for our proposed method and the eight SOTA methods. As indicated by Table 2, our proposed method demonstrates superior clustering performance over other methods on all datasets. Specifically, RMVSC-AE outperforms all comparison models in ACC, NMI, and Purity metrics in majority of cases. Remarkably, for bbcsport dataset, our method exhibits about 6.89%,

Table 2. Clustering Performance

	AMGL	MVGL	SwMC	MVSC	MCLES	GMC	FMSCS	BPSA	RMVSC	RMVSC-AE
NMI										
3Sources	0.5908	0.3661	0.0766	0.0780	0.5921	0.6216	0.6195	0.6960	0.6748	**0.7240**
BBCsport	0.6772	0.3811	0.0805	0.4092	0.5180	0.4771	0.6987	0.7186	0.7120	**0.8267**
MSRC	0.7386	0.5942	0.5314	0.3143	0.7915	0.7709	0.8112	0.7392	0.7981	**0.8268**
NGs	0.9165	0.3279	0.1330	0.4859	0.8802	0.9392	0.9530	0.5532	0.9218	**0.9722**
Caltech101-20	0.5445	0.4349	0.5000	0.2672	0.5950	0.4809	**0.6587**	0.6499	0.5051	0.6585
Hdigit	0.9479	0.9857	0.5416	0.9001	$-^1$	0.9939	0.9308	0.7427	0.9279	**0.9943**
ACC										
3Sources	0.6805	0.0045	0.3456	0.3846	0.6864	0.6923	0.6941	0.7746	0.7811	**0.7870**
BBCsport	0.7069	0.0323	0.3259	0.5345	0.6293	0.5603	0.7250	0.7845	0.8190	**0.8534**
MSRC	0.7362	0.5136	0.5871	0.2762	0.8810	0.7476	0.8433	0.8286	0.8762	**0.8905**
NGs	0.9740	0.0075	0.2694	0.4040	0.9600	0.9820	0.9860	0.6240	0.9760	**0.9920**
Caltech101-20	0.5337	0.2751	0.5167	0.3759	0.4520	0.4564	0.6263	**0.6388**	0.5348	0.5796
Hdigit	0.9155	0.9954	0.6695	0.9001	-	0.9981	0.9735	0.6681	0.9724	**0.9982**
Purity										
3Sources	0.7337	0.1383	0.3645	0.3965	0.7396	0.7456	0.7402	0.8112	0.8166	**0.8580**
BBCsport	0.8190	0.2261	0.3405	0.5517	0.6638	0.5862	0.7750	0.8793	0.8879	**0.9310**
MSRC	0.7648	0.7087	0.6167	0.3286	0.8810	0.7905	0.8433	0.8286	0.8762	**0.8905**
NGs	0.9740	0.1201	0.2774	0.4080	0.9600	0.9820	0.9860	0.6660	0.9760	**0.9920**
Caltech101-20	0.6539	0.5653	0.6236	0.3831	0.7310	0.5549	0.7578	0.7683	0.6513	**0.7871**
Hdigit	0.9327	0.9954	0.5908	0.9001	-	0.9981	0.9735	0.7103	0.9724	**0.9982**

The symbol '-' indicates an unobtainable result owing to a running time exceeding one day.

10.81% and 5.17% improvement in ACC, NMI, and Purity respectively compared to BPSA. Similarly, for MSRC dataset, our method achieves improvements of about 0.95%, 1.56%, and 0.95% respectively over MCLES. On the NGs dataset, we observed a similar trend yielding improvements of about 0.6%, 1.92%, and 0.6% in ACC, NMI, and Purity metrics respectively compared to MCLES. The benchmark method has achieved high performance on the Hdigit dataset, but our method has still been further improved. For Caltech 101-20 dataset. Although our method not always achieves optimal performance under the three metrics, the results is still comparable to other methods. It should be noted that the clustering results are enhanced significantly compared with RMVSC on all datasets, such results indicate that the important information can be preserved with auto-encoder.

5.4 Ablation Study

Parameter Analysis: In this section, we study the effect of parameter γ, λ, β and μ on the proposed method. β change in $\{10^{-6}, 10^{-4}, 10^{-2}, 1, 10^2, 10^4, 10^6\}$, and the range of μ is $\{10^{-2}, 10^{-1}, 1, 10^1, 10^2\}$. Both γ and λ vary from $\{10^{-4}, 10^{-2}, 1, 10^2, 10^4\}$. We fix γ and λ to 1, and analyze the impact of β and μ. Figure 1 shows the influences of the parameters β and μ in terms of NMI. For parameter β, we can see that when the value of β is too large, the

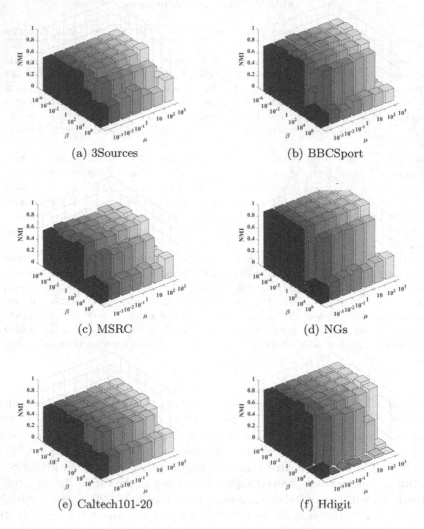

(a) 3Sources

(b) BBCSport

(c) MSRC

(d) NGs

(e) Caltech101-20

(f) Hdigit

Fig. 1. Parameter sensitivity analysis of β, μ on four datasets with γ and λ as 1, Z axis is the value of Normalized mutual information (NMI).

Laplacian matrix has too many zero eigenvalues and cannot learn the correct graph, resulting in performance degradation. Best NMI value can be achieved when β is in [0.01,1]. For the parameter μ, It performs relatively stably over a

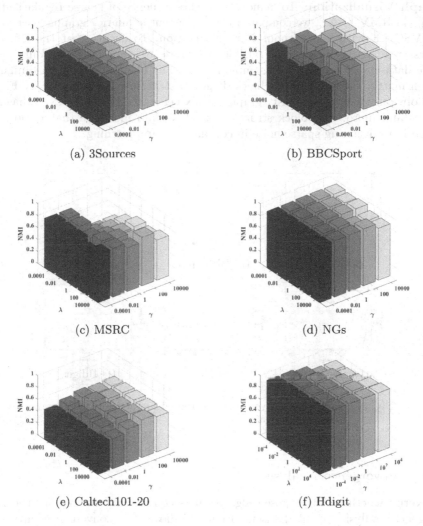

Fig. 2. Parameter sensitivity analysis of λ, γ on four datasets with β and μ as 1, Z axis is the value of Normalized mutual information (NMI).

wide range and the optimal NMI value are achieved with different μ for different datasets, because different datasets have different level of noise. Then, we fix β and μ to 1, and analyze the impact of γ and λ, the result is shown in Fig. 2. It is apparently that the impact of two parameters on different datasets is different because of the diversity of data information. On the MSRC and Hdigit datasets, the performance is better when the value of λ is large and the value of γ is small. When the value of γ is large on the Caltech101-20 dataset, the performance is better, while on the 3Sources dataset, the opposite is true.

Graph Visualization: To demonstrate the influence of sparse regularization $\| * \|_1$ in RMVSC-AE, we compared the consistent similarity graphs learned by RMVSC-AE with and without l_1 regularization. To ensure that the effect of sparse regularization is clearly visible, we chose to compare it on four small-scale datasets. When the data sample is sorted by class label, a ideal similarity graph matrix should exhibit a block-diagonal structure. As indicated by Fig. 4, it is obvious that the similarity graph matrix generated by RMVSC-AE has less noises and more clear block structure. Such results illustrate that l_1 norm is helpful in mitigating sparse noise in consensus graph learning.

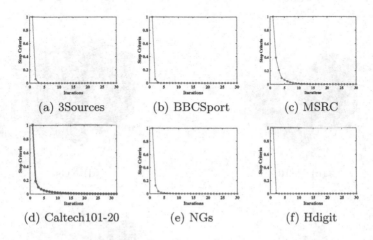

(a) 3Sources   (b) BBCSport   (c) MSRC

(d) Caltech101-20   (e) NGs   (f) Hdigit

Fig. 3. Convergence curves of the proposed method for six datasets

5.5   Convergence Analysis

To verify whether the proposed algorithm is convergent, we conduct the convergence analysis on all datasets. Figure 3 shows the convergence curves of RMVSC-AE, the x and y axes represent the number of iterations and the value of $\|\mathbf{F}^t * \mathbf{F}^{t^T} - \mathbf{F}^{t-1} * \mathbf{F}^{t-1^T}\|_F$ respectively. It is apparently that the solution of $\mathbf{F}$ decrease smoothly with the increment of iterations. Furthermore, our algorithm converge within 20 iterations for all datasets, which also guarantee the efficiency of our proposed method.

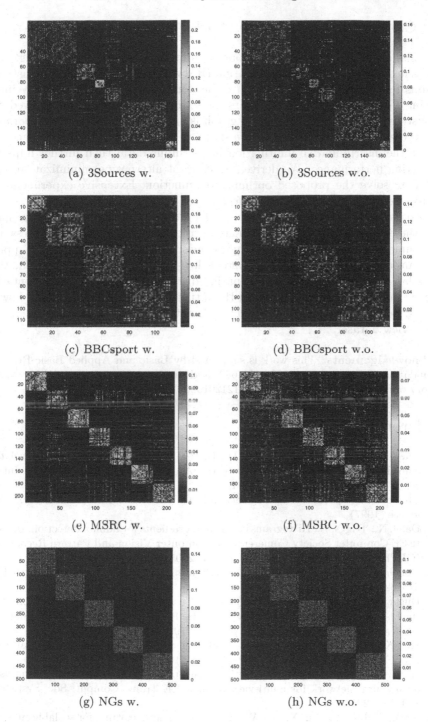

Fig. 4. Visualization of similarity graph matrix **G** learned by RMVSC-AE with(w.) and without(w.o.) l_1 regularization on benchmark datasets.

6   Conclusion and Future Work

In this article, we propose a robust multi-view spectral clustering with auto-encoder by preserving information (RMVSC-AE). RMVSC-AE constructs graphs by gaussian kernel to capture the non-linear data structure. By introducing L_1 regularization terms, sparse noise in consensus graph is removed, the robustness of the learned similarity graph is enhanced. Moreover, incorporating an auto-encoder between the feature space and the graph embedding space preserves the discriminative information in the input data, then further improves clustering performance. We derived an efficient alternating minimization algorithm to solve the proposed optimization function. Extensive experiments on real-world datasets illustrate the effectiveness of our method.

Additionally, since we have established a projection between the feature space and the embedding space, the calculated projection can be used to address out-of-sample issue [7,8]. When new sample comes, the projection can also be applied to transform new sample feature into a low-dimensional representation, then class label can be obtained by computing distances between the representation of new sample and class centers. In the future work, we will investigate how to extend our proposed method to deal with out-of-sample issue and incremental multi-view dataset.

Acknowledgements. This work is supported by Basic and Applied Basic Research Foundation of Guangzhou (2023A04J1682) and the Guangdong Provincial Key Laboratory of Human Digital Twin (2022B1212010004).

References

1. Chen, M.S., Huang, L., Wang, C.D., Huang, D.: Multi-view clustering in latent embedding space. In: Proceedings of the AAAI Conference on Artificial Intelligence, vol. 34, pp. 3513–3520 (2020)
2. Chung, F.R.: Spectral Graph Theory, vol. 92. American Mathematical Soc., Providence (1997)
3. Dalal, N., Triggs, B.: Histograms of oriented gradients for human detection. In: 2005 IEEE Computer Society Conference on Computer Vision and Pattern Recognition (CVPR 2005), vol. 1, pp. 886–893. IEEE (2005)
4. Donoho, D.L.: De-noising by soft-thresholding. IEEE Trans. Inf. Theory **41**(3), 613–627 (1995)
5. Duchi, J., Shalev-Shwartz, S., Singer, Y., Chandra, T.: Efficient projections onto the l 1-ball for learning in high dimensions. In: Proceedings of the 25th International Conference on Machine Learning, pp. 272–279 (2008)
6. Fan, K.: On a theorem of Weyl concerning eigenvalues of linear transformations I. Proc. Natl. Acad. Sci. **35**(11), 652–655 (1949)
7. Huang, S., Ota, K., Dong, M., Li, F.: MultiSpectralNet: spectral clustering using deep neural network for multi-view data. IEEE Trans. Comput. Soc. Syst. **6**(4), 749–760 (2019)
8. Kang, Z., Lin, Z., Zhu, X., Xu, W.: Structured graph learning for scalable subspace clustering: from single view to multiview. IEEE Trans. Cybernetics **52**(9), 8976–8986 (2021)

9. Kang, Z., et al.: Partition level multiview subspace clustering. Neural Netw. **122**, 279–288 (2020)
10. Kodirov, E., Xiang, T., Gong, S.: Semantic autoencoder for zero-shot learning. In: Proceedings of the IEEE Conference on Computer Vision and Pattern Recognition, pp. 3174–3183 (2017)
11. Kumar, A., Rai, P., Daume, H.: Co-regularized multi-view spectral clustering. In: Advances in Neural Information Processing Systems, vol. 24 (2011)
12. Li, Y., Nie, F., Huang, H., Huang, J.: Large-scale multi-view spectral clustering via bipartite graph. In: Proceedings of the AAAI Conference on Artificial Intelligence, vol. 29 (2015)
13. Liu, J., et al.: MPC: multi-view probabilistic clustering. In: Proceedings of the IEEE/CVF Conference on Computer Vision and Pattern Recognition, pp. 9509–9518 (2022)
14. Lowe, D.G.: Distinctive image features from scale-invariant keypoints. Int. J. Comput. Vis. **60**, 91–110 (2004)
15. Maier, M., Luxburg, U., Hein, M.: Influence of graph construction on graph-based clustering measures. In: Advances in Neural Information Processing Systems, vol. 21 (2008)
16. Mohar, B., Alavi, Y., Chartrand, G., Oellermann, O.: The Laplacian spectrum of graphs. Graph Theory Comb. Appl. **2**(871–898), 12 (1991)
17. Nie, F., Li, J., Li, X. Parameter-free auto-weighted multiple graph learning: a framework for multiview clustering and semi-supervised classification. In: IJCAI, pp. 1881–1887 (2016)
18. Nie, F., Li, J., Li, X.: Self-weighted multiview clustering with multiple graphs. In: IJCAI, pp. 2564–2570 (2017)
19. Nie, F., Wang, X., Jordan, M., Huang, H.: The constrained Laplacian rank algorithm for graph-based clustering. In: Proceedings of the AAAI Conference on Artificial Intelligence, vol. 30 (2016)
20. Ranzato, M., Boureau, Y.L., Cun, Y.: Sparse feature learning for deep belief networks. In: Advances in Neural Information Processing Systems, vol. 20 (2007)
21. Shi, D., Zhu, L., Li, J., Cheng, Z., Zhang, Z.: Flexible multiview spectral clustering with self-adaptation. IEEE Trans. Cybern. **53**, 2586–2599 (2021)
22. Von Luxburg, U.: A tutorial on spectral clustering. Stat. Comput. **17**, 395–416 (2007)
23. Wang, C.D., Lai, J.H., Philip, S.Y.: Multi-view clustering based on belief propagation. IEEE Trans. Knowl. Data Eng. **28**(4), 1007–1021 (2015)
24. Wang, H., Yang, Y., Liu, B.: GMC: graph-based multi-view clustering. IEEE Trans. Knowl. Data Eng. **32**(6), 1116–1129 (2019)
25. Wang, H., Yang, Y., Liu, B., Fujita, H.: A study of graph-based system for multi-view clustering. Knowl.-Based Syst. **163**, 1009–1019 (2019)
26. Wang, H., Yao, M., Jiang, G., Mi, Z., Fu, X.: Graph-collaborated auto-encoder hashing for multi-view binary clustering. arXiv preprint arXiv:2301.02484 (2023)
27. Wang, Q., Tao, Z., Xia, W., Gao, Q., Cao, X., Jiao, L.: Adversarial multiview clustering networks with adaptive fusion. IEEE Trans. Neural Netw. Learn. Syst. **34**, 7635–7647 (2022)
28. Wu, D., Dong, X., Cao, J., Wang, R., Nie, F., Li, X.: Bidirectional probabilistic subspaces approximation for multiview clustering. IEEE Trans. Neural Netw. Learn. Syst. (2022)
29. Zelnik-Manor, L., Perona, P.: Self-tuning spectral clustering. In: Advances in Neural Information Processing Systems, vol. 17 (2004)
30. Zhan, K., Zhang, C., Guan, J., Wang, J.: Graph learning for multiview clustering. IEEE Trans. Cybernet. **48**(10), 2887–2895 (2017)

Learnable Color Image Zero-Watermarking Based on Feature Comparison

Baowei Wang[1,2,3(✉)], Changyu Dai[4], and Yufeng Wu[2]

[1] Engineering Research Center of Digital Forensics Ministry of Education, Nanjing University of Information Science and Technology, 210044 Nanjing, China
[2] School of Computer Science, Nanjing University of Information Science and Technology, 210044 Nanjing, China
[3] Jiangsu Collaborative Innovation Center of Atmospheric Environment and Equipment Technology (CICAEET), Nanjing University of Information Science and Technology, 210044 Nanjing, China
[4] School of Software, Nanjing University of Information Science and Technology, 210044 Nanjing, China
`wbw.first@163.com,{20211221006,20211220033}@nuist.edu.cn`

Abstract. Zero-watermarking is one of the solutions to protect the copyright of color images without tampering with them. Existing zero-watermarking algorithms either rely on static classical techniques or employ pre-trained models of deep learning, which limit the adaptability of zero-watermarking to complex and dynamic environments. These algorithms are prone to fail when encountering novel or complex noise. To address this issue, we propose a self-supervised anti-noise learning color image zero-watermarking method that leverages feature matching to achieve lossless protection of images. In our method, we use a learnable feature extractor and a baseline feature extractor to compare the features extracted by both. Moreover, we introduce a combined weighted noise layer to enhance the robustness against combined noise attacks. Extensive experiments show that our method outperforms other methods in terms of effectiveness and efficiency.

Keywords: Zero-watermarking · Learnable · Feature comparison · Self-supervised

1 Introduction

In the era of web technology, digital multimedia is the main information carrier that is widely disseminated through various social media platforms. However, this also brings challenges to the protection of intellectual property rights and the prevention of unauthorized modifications. For instance, some malicious actors may tamper with the photos of political figures and spread them online [1]. Therefore, robust watermarking techniques are needed to provide a secure way of delivering information in an open network environment.

B. Luo et al. (Eds.): ICONIP 2023, LNCS 14451, pp. 472–483, 2024.
https://doi.org/10.1007/978-981-99-8073-4_36

Nevertheless, traditional robust watermarking methods are not suitable for all images. Some images, such as remote sensing images, medical images, professional photography images, etc., have high requirements for image quality and details. Embedding a large amount of watermark information into these images may cause distortion and affect their usability. For example, a distorted medical image may lead to a wrong diagnosis by the doctor and cause serious consequences.

Zero-watermarking is a special watermarking technique that creates a logical association between the image and the watermark without modifying the original image. Figure 1 illustrates the generation process of classical zero-watermarking.

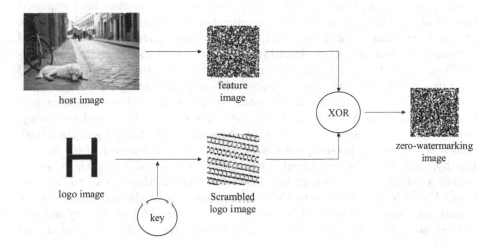

Fig. 1. The generation process of classical zero-watermarking.

Existing zero-watermarking methods have poor robustness to large distortions and cannot adapt to complex scenarios with multiple attacks. Therefore, we propose a zero-watermarking algorithm that is **"learnable"**. This means that one can simply retrain the model to achieve robustness against a new type of noise, instead of designing a new algorithm from scratch.

2   Related Work

Wen et al. [2] proposed a zero-watermarking algorithm based on the DCT transform in 2001. After that, many others have developed different traditional zero-watermarking algorithms based on the DCT transform [3–10]. However, all these conventional methods rely heavily on shallow hand-crafted image features, which suggests that they do not fully use the original image and therefore have many limitations in terms of robustness. Shao et al. [11] proposed a robust watermarking scheme based on orthogonal Fourier-Melling moments and chaotic mapping in 2016 to achieve simultaneous copyright authentication of dual images. In 2022, to address the problems of leakage of patients' personal information, theft, and tampering of cloud-transmitted medical images, Fang et al. [12] proposed a bandelet and discrete cosine transform (bandelet - dct) zero-watermarking algorithm for medical images. In 2023, Lang et al. [13] proposed a zero-watermarking algorithm based on compression-aware salient features. The algorithm segments the original image into non-overlapping blocks and then constructs a dislocated block Hadamard ensemble (SBHE) matrix as the sensing matrix for each block. These heuristics are effective in the domains for which they are designed, but they are fundamentally *static*.

Deep learning has revolutionized many fields of computer science, including image processing and watermarking. In recent years, researchers have explored the use of deep learning methods to enhance the performance and robustness of zero-watermarking algorithms [14,15]. For example, Fierro et al. [14] proposed a robust zero-watermarking algorithm based on convolutional neural networks (CNNs), where the CNN generates robust intrinsic features of an image and combines them with the owner's watermark sequence using XOR operation. Han et al. [15] extracted the deep feature maps of medical images using pre-trained VGG19 and fused them into feature images. Then, they used an average perceptual hashing algorithm to generate a set of 64-bit binary perceptual hash values based on the low-frequency part of the medical image feature matrix. Finally, they used the scrambled watermarked image and the hash values to obtain a robust zero-watermarking. However, these methods tend to use pre-trained models as feature extractors, rather than exploit the learning ability of neural networks to adapt to different types of images. Moreover, they are still vulnerable to some geometric attacks such as cropping, JPEG compression, and rotation.

Existing zero-watermarking methods only conduct robust experiments for single distortion, but the transmission of images often faces many complex scenarios, and images often suffer from multiple distortions. These methods have poor robustness in dealing with scenarios with multiple distortions.

To address the downstream task of color image zero-watermarking, we design a new model using a baseline feature extractor and a learnable feature extractor for feature comparison has been designed, which is a relatively lightweight network that accepts images from the noise layer after the attack. This model aims to maximize the robustness by making the output of this lightweight network as close as possible to the baseline feature extractor. This method is the **first** zero-watermarking algorithm that can dynamically adjust to different distortions, and also the **first** one that can handle combined noise attacks and novel noise attacks.

HiDDeN [16] first introduced the concept of noise layers and applied it to an end-to-end watermarking system with good results. Jia et al. used an improved noise layer in MBRS [17] that greatly enhanced the robustness of JPEG compression. We introduced this method to combined noise and weighted it according to the effect of the noise, obtaining good robustness to combined noise.

In summary, the contributions are listed as follows:

1. We designed a "learnable" zero-watermarking algorithm to cope with multiple complex scenarios. This method can easily adapt to new requirements, since we directly optimize for the objectives of interest. For zero-watermarking, one can simply retrain the model to obtain robustness against new types of noise when facing new complex scenarios, rather than inventing a new algorithm.
2. A weighted mini-batch combined noise layer is proposed to combat multiple attacks and improve the model's generality.
3. Various experimental results show that the scheme has excellent performance in resisting large distortion and multiple distortion.

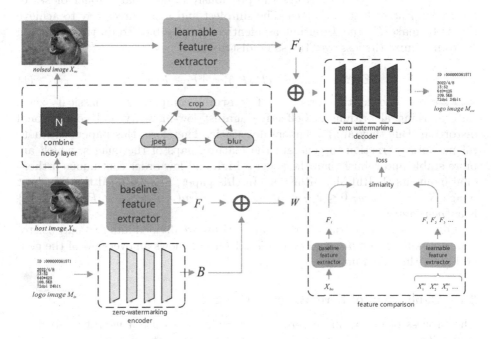

Fig. 2. An overview of the general framework diagram and the process of features comparison.

3   Methods

We aim to develop a learnable and easily portable model for color image zero-watermarking. To enhance the robustness of the image against new types of noise attacks, the only requirement is to incorporate that type of noise into the training process. To this end, our model consists of four main components. A baseline feature extractor E, a parameter-free noise layer N, a learnable feature extractor E_θ, and a feature comparator C. Additionally, we use a zero-watermarking generation was scrambling encoder E_{key} and a zero-watermarking decoder D_{key} to encrypt and decrypt the watermarking information. θ is a parameter that can be trained, and the *key* is used to encrypt and decrypt the watermarking information.

The baseline feature extractor E receives the host image X_{ho} of shape $C \times H \times W$ and extracts the host image feature F_i. The noise layer N receives the host image X_{ho}, adds noise to the host image, and outputs a noisy image X_{no}. The learnable feature extractor E_θ receives X_{no} and extracts the image feature F_i' from the noisy image. The zero-watermarking scrambler E_{key} receives F_i and a binary cipher M_{in} of length L or shape $h \times w$, to generate a zero-watermarking W. D_{key} receives W and restores the previously input binary cipher or secret image M_{out} according to the *key*. The simplest and most direct way to achieve this is to make E_θ and E output as identical as possible. To do this, we train E_θ to minimize the loss over the same distribution:

$$\mathbb{E}\left(X_{ho}, M_{in}\right)\left[\mathcal{L}_F\left(F_i, F_i'\right)\right] = \left\|F_i - F_i'\right\|^2 / L \qquad (1)$$

Figure 2 shows the process of feature comparison. Classical zero-watermarking algorithms are effective against low-intensity and conventional distortion. But they are fundamentally static. Therefore, this paper uses neural network extraction to generate the feature maps of the images, which are more stable and robust than the general high and low-order continuous orthogonal moments (FoRhFMs/loRhFMs). In this paper, it is assumed that the host image $X_{ho} = \{f(x, y), 0 \le x, y < N$ and $M_{in} = \{m(p, q), 0 \le p < P, 0 \le q < Q\}$ is a *Logo image*.

Our proposed zero-watermarking algorithms are divided into two parts: zero-watermarking *generation* and *extraction*. Figure 2 shows an overview of the general algorithm diagram.

3.1   Generation of Zero-Watermarking

The process of generating a zero-watermarking is to use a neural network to extract features of the host image to construct a zero-watermarking. The method of generating a zero-watermarking is as follows.

Generation of Chaotic Sequences. Our algorithms use the cat mapping [18] to encrypt the *Logo image*. In this paper, $M_{in} = \{m(p, q), 0 \le p < P, 0 \le q < Q\}$ is a *Logo image*, which is cat mapped to construct a chaotic sequences C_{n+1} of length $P \times Q$:

$$C_{n+1} = \begin{bmatrix} p_{n+1} \\ q_{n+1} \end{bmatrix} = \begin{bmatrix} 1 & b \\ a & ab+1 \end{bmatrix} \begin{bmatrix} p_n \\ q_n \end{bmatrix} \bmod (N) \qquad (2)$$

Where a, b, and N are positive integers, the values of a and b can be randomly generated or specified, and they are the *keys* of the zero-watermarking dislocation codec. When $p = q$, N is the width of the matrix. The mod is the function of the remainder.

Binarization of Chaotic Matrices. The chaotic sequences C are binarized to obtain the binary chaotic matrix B_n:

$$B_n = \begin{cases} 1, C_{n+1} \geq T_1 \\ 0, C_{n+1} < T_1 \end{cases} \tag{3}$$

Where T_1 refers to the threshold value of the binarization. This paper takes the average value of each element in the C_{n+1} as T_1.

Construction of Feature Matrix. The host image X_{ho} is fed into the feature extractor for feature extraction. A two-dimensional discrete Fourier transform is performed to obtain a feature matrix A_n of dimension $P \times Q$. For a detailed description of the feature extractor and the neural network, see: Sect. 3.3.

Binarization of the Feature Matrix. The obtained feature matrix A_n is binarized to get the binary feature matrix Z_n:

$$Z_n = \begin{cases} 1, A_n \geq T_2 \\ 0, A_n < T_2 \end{cases} \tag{4}$$

where T_2 refers to the binarization threshold value, this paper calculates an average value for each element of the A_n matrix as T_2.

Generation of Zero-Watermarking. XOR operation is performed on the chaotic matrix B and the feature matrix Z to obtain the final zero-watermarking $W = B \oplus Z$.

3.2   Extraction of Zero-Watermarking

The extraction process of zero-watermarking refers to retrieving the *Logo image* M_{out} from the noise-affected image $X_{no} = \{f_{no}(\dot{x}, \dot{y}), 0 \leq \dot{x}, \dot{y} < N\}$ to determine the copyright attribution of X_{no}. The detailed process of this algorithm extraction is as follows:

Resize the Image. As the noise we encounter is likely to be cropped or panned, the size and position of X_{no} may differ from X_{ho}. The image is pre-processed in three steps:

1) Equal scale scaling:

$$scale = \min\left(\frac{x}{\dot{x}}, \frac{y}{\dot{y}}\right) \tag{5}$$

2) Pan the center of the image to the coordinate origin in the upper left corner:

$$M = \begin{bmatrix} scale & 0 & -\frac{scale \times \dot{x}}{2} + \frac{x}{2} \\ 0 & scale & -\frac{scale \times \dot{y}}{2} + \frac{y}{2} \end{bmatrix} \tag{6}$$

3) Pan the picture to the center of gravity at the target location:

$$X_{no}' = M \begin{bmatrix} X_{no} \\ 1 \end{bmatrix} \tag{7}$$

The obtained X_{no}' is the equal scale scaling of X_{no}, then centered, and the excess is filled in the image.

Generation of Feature Matrix. Input X_{no}' into the feature extractor for feature extraction, and get the feature matrix A_n^* with dimension $P \times Q$.

Binarization of Feature Matrix. The feature matrix A_n^* is binarized into Z_n^*. Binarization is the same process as zero-watermarking generation.

Generation of Chaotic Matrix. XOR operation is performed on the zero-watermarking W and the binarized feature matrix Z_n^* to extract the chaotic matrix:

$$C_n^* = W \oplus Z_n^* \tag{8}$$

Disruption Reduction. Using the previously generated *keys*: a and b, construct the original matrix of dimension $P \times Q$:

$$C_n^* = \begin{bmatrix} p_n \\ q_n \end{bmatrix} = \begin{bmatrix} 1 & b \\ a & ab+1 \end{bmatrix}^{-1} \begin{bmatrix} p_{n+1} \\ q_{n+1} \end{bmatrix} \bmod (N) \tag{9}$$

Extracting the Binary *Logo* image. Binarize C_n^* and export it to binary *Logo* image M_{out}, and compare M_{in} and M_{out} to determine the ownership of X_{no}. The binarization process is the same as that of the zero-watermarking generation process.

3.3   Feature Extractor

The features of the image are extracted using a neural network. However, there are two problems in this process:

1) **Baseline.** How to measure that the features extracted by a neural network are good?
2) How to make the extracted **useful features** robust?

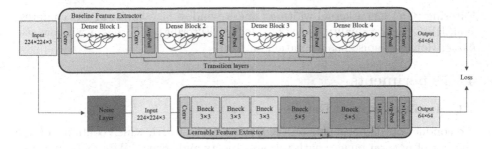

Fig. 3. Structure of baseline feature extractor and learnable feature extractor.

To address these two problems, we propose a two-step solution. First, we combine a baseline feature extractor with a learnable feature extractor. Second, we introduce a noise layer into the training process and train the model to resist noise.

As shown in the Fig. 3, we use DenseNet121 [19] as the backbone network to build our baseline feature extractor E. We use MobileNet_V3_small [20] as the backbone network to build our learnable feature extractor E_θ. We change the fully connected layer in the last layer to output features and drop the SoftMax function. The reason for choosing the MobileNet network as the learnable feature extractor is that we hope to learn the feature extraction ability of a larger model with a smaller model by this feature comparison learning method to achieve better result. The loss function is shown in Eq. 1. According to our experimental proof, this method is indeed effective.

3.4   Combined Noisy Layer

In the actual scene, X_{no} is likely to be attacked by several different noises simultaneously, making our model need to resist both single and combined noise, for which we design a weighted mini-batch combined noise layer. In each mini-batch, we give different selection weights to other noises (the higher the weight, the higher the probability of selection). The model's ability to resist different noises determines the actual consequences. For example, we give a weight of 0.1 to the noiseless layer (called the *identity layer*), 0.4 to the differentially simulated *JPEG layer*, and 0.5 to the *Crop layer* in each mini-batch. Because the impact of the Crop attack is significantly greater than that of the JPEG attack, such a design will help the model find the optimal global solution and achieve good resistance to hybrid attacks.

A differentiable JPEG compression layer is needed because the steps of DCT quantization are inseparable in the JPEG compression process. We use the differentiable JPEG compression [17] function: Jpeg-Mask, considering the portability of the model training.

Note that all non-homogeneous noise layers have a scalar hyperparameter that controls the intensity of the distortion. Crop has a scaling p; Gaussian

Filtering (GF) has a kernel width σ^2 and a window size $ksize$; JPEG compression(JPEG) has a quality factor Q; Median Filtering (MF) has a window size $ksize$; Rotation noise has a rotation angle R.

4   Experiments

4.1   Implementation Details

We conduct our experiments on the COCO dataset [21], which contains a large number of natural images with various objects and scenes. We randomly select 10,000 224×224 images from the COCO dataset for model training, 5,000 224×224 images for model validation, and another 5,000 224×224 images for model testing. The model is implemented using PyTorch and has been trained and verified on NVIDIA RTX 4000. We use the Adam optimizer throughout the experiments and set the learning rate to 10^{-3}. At the same time, a total of 100 epochs are trained.

Table 1. Algorithm robustness comparative experimental results (Average NCC from 5,000 224×224 test images).

Common attacks	Intensity of attacks	Shao [11]	Fang [12]	Lang [13]	Fierro [14]	Han [15]	Ours
JPEG (Q)	30	0.9725	0.9631	0.9455	0.9486	0.9433	**0.9758**
	50	0.9756	0.9726	0.9642	0.9696	0.9781	**0.9896**
	70	0.9766	0.9874	0.9784	0.9873	0.9862	**0.9931**
MF ($ksize$)	3	0.9849	0.9969	**0.9987**	0.9737	0.9872	0.9985
	5	0.9837	0.9731	0.9894	0.9712	0.9645	**0.9921**
	7	0.9796	0.9624	0.9732	0.9848	0.9359	**0.9899**
Rotation (R)	30	0.9565	**0.9754**	0.9636	0.9091	0.7598	0.9712
	45	0.9387	0.9436	0.9431	0.8772	0.7128	**0.9693**
GN (σ^2)	0.001	0.9736	0.9917	**0.9983**	0.9696	0.9891	0.9891
	0.01	0.9674	0.9726	0.9776	0.9242	0.9749	**0.9829**
Crop (p)	0.9	0.9768	0.9091	0.9873	0.8394	0.9425	**0.9936**
	0.7	0.9627	0.8726	0.9754	0.8242	0.9179	**0.9896**
	0.5	0.8706	0.7903	0.9621	0.8091	0.8819	**0.9841**
	Upper left corner cropping 1/4	0.8329	0.8775	0.9626	0.5644	0.7631	**0.9703**
GF ($\sigma^2 = 0.2$) ($ksize$)	1	0.9754	0.9747	0.9793	0.9631	0.9894	**0.9903**
	2	0.9660	0.9649	0.9616	0.9606	0.9721	**0.9864**
	4	0.9472	0.8027	0.9594	0.9416	0.9732	**0.9807**
RGB to Gray		0.8106	0.7175	0.7382	0.8014	0.8919	**0.9663**
Combined Noise	Crop ($p = 0.5$), Jpeg ($Q = 50$), MF ($ksize = 7$)	0.7146	0.7063	0.6661	0.7705	0.8369	**0.9156**
	GF ($ksize = 5$), MF ($ksize = 5$), Jpeg ($Q = 80$)	0.6086	0.5686	0.7011	0.7389	0.7208	**0.8578**
	GN ($\sigma^2 = 0.001$), MF ($ksize = 7$), Rotation ($R = 30$)	0.7681	0.6624	0.6452	0.7628	0.7422	**0.8968**

4.2   Metrics

We use the most commonly used **N**ormalized **C**ross-**C**orrelation (NCC) as an evaluation indicators. The higher the NCC, the higher the extraction accuracy. To make our experimental results more reliable, we compare Shao [11], Fang [12], Lang [13], Fierro [14] and Han [15] through experiments, because their goal is to explore better zero-watermarking models. All the methods use the same 5,000 224×224 images from the COCO dataset for comparison, and the size of the *Logo image* is a 64×64 binary image.

4.3   Results

Table 1 compares our method and the five methods above. From the Table 1, we can see that our model has excellent performance in robustness. Other methods did not consider these combined noise when they were designed, so their robustness is very poor. The existing zero-watermarking methods did not consider the RGB to Gray noise in their experiments, but this attack is very common in reality. As can be seen from the Table 1, these methods are not very robust because they did not take into account the change of channel number in their design. In contrast, our method can achieve good robustness by simply adding the corresponding noise layer during training.

With simple training, our method can handle complex and novel distortions and achieve remarkable robustness. By learning feature comparison and leveraging the powerful capability of neural networks, our method can effectively resist rotation, cropping, and combined noise attacks. Moreover, our model can maintain or even enhance its performance under other types of noise attacks.

Table 2. Robust performance of combined noise layer.

Metric	No noise	JPEG ($Q = 70$)	Rotation ($R = 3$)	Crop ($p = 0.7$)	GF ($\sigma^2 = 0.2\ ksize = 7$)	MF ($ksize = 5$)
NCC	1.0	0.98964	0.99804	0.99844	0.96347	0.99782

Table 2 shows the robustness of a single noise attack after training with the combined noise layer. We apply Crop ($p = 0.5$), Jpeg ($Q = 50$), MF ($ksize = 7$) as a combined noise layer.

Our method is especially robust to untrained attacks. As Table 2 shows, we did not include Rotation or GF distortion in the combined noise layer, but our method still resisted these distortions.

4.4   Ablation Study

To ensure the reliability and integrity of the experimental results, we conducted ablation experiments on the model. We added Crop ($p = 0.7$) to it for better observation. Through the experimental results in Table 3, our model achieves the best performance when combining all components. Using only the baseline feature extractor or only the learnable feature extractor is equivalent to using pre-trained models for feature extraction, which is less effective than using dual feature extractors for feature comparison. We train our model with a combined noise layer of Crop ($p = 0.5$), Jpeg ($Q = 50$), and MF ($ksize = 7$). When we use the weighted mini-batch weighted combined noise layer, we assign Crop a weight of 0.6, Jpeg a weight of 0.2, and MF a weight of 0.2. The results demonstrate that the model's robustness to Crop is significantly enhanced after weighting.

Table 3. Ablation experiment results of the proposed method.

baseline feature extractor	✓		✓	✓
learnable feature extractor		✓	✓	✓
noisy layer			✓	✓
weighted mini-batch				✓
NCC	0.93076	0.83776	0.97732	**0.99782**

5   Conclusion

We developed a deep learning zero-watermarking framework for color images. Compared with the classical zero-watermarking algorithms, this work can flexibly cope with different types of noise by changing the parameters or the noise layer during training. Instead of merely using pre-trained models to extract features, the proposed method truly exploits the powerful learning ability of neural networks. Ultimately, it enables us to deal with complex and novel noise by directly incorporating new distortions into the training process, without the need to design new specialized algorithms. In future work, we hope to see robustness against more types of image distortions (such as other lossy compression schemes) and apply the framework to other input domains (such as audio and video).

References

1. Zhou, Z., Wang, Y., Wu, Q.J., Yang, C.N., Sun, X.: Effective and efficient global context verification for image copy detection. IEEE Trans. Inf. Forensics Secur. **12**(1), 48–63 (2016)
2. Wen, Q., Sun, T., Wang, S.: Based zero-watermark digital watermarking technology. In: Proceedings of the 3rd National Conference in Information Hiding, vol. 109. Xidian University Press, Xian (2001)
3. Chen, T.H., Horng, G., Lee, W.B.: A publicly verifiable copyright-proving scheme resistant to malicious attacks. IEEE Trans. Industr. Electron. **52**(1), 327–334 (2005)
4. Chang, C.C., Lin, P.Y.: Adaptive watermark mechanism for rightful ownership protection. J. Syst. Softw. **81**(7), 1118–1129 (2008)
5. Tsai, H.H., Tseng, H.C., Lai, Y.S.: Robust lossless image watermarking based on α-trimmed mean algorithm and support vector machine. J. Syst. Softw. **83**(6), 1015–1028 (2010)
6. Tsai, H.H., Lai, Y.S., Lo, S.C.: A zero-watermark scheme with geometrical invariants using SVM and PSO against geometrical attacks for image protection. J. Syst. Softw. **86**(2), 335–348 (2013)
7. Rawat, S., Raman, B.: A blind watermarking algorithm based on fractional Fourier transform and visual cryptography. Sig. Process. **92**(6), 1480–1491 (2012)
8. Thanh, T.M., Tanaka, K.: An image zero-watermarking algorithm based on the encryption of visual map feature with watermark information. Multimedia Tools Appl. **76**(11), 13455–13471 (2017)

9. Zou, B., Du, J., Liu, X., Wang, Y.: Distinguishable zero-watermarking scheme with similarity-based retrieval for digital rights management of fundus image. Multimedia Tools Appl. **77**(21), 28685–28708 (2018)
10. Kang, X., Lin, G., Chen, Y., Zhao, F., Zhang, E., Jing, C.: Robust and secure zero-watermarking algorithm for color images based on majority voting pattern and hyper-chaotic encryption. Multimedia Tools Appl. **79**(1), 1169–1202 (2020)
11. Shao, Z., Shang, Y., Zhang, Y., Liu, X., Guo, G.: Robust watermarking using orthogonal Fourier-Mellin moments and chaotic map for double images. Sig. Process. **120**, 522–531 (2016)
12. Fang, Y., et al.: Robust zero-watermarking algorithm for medical images based on SIFT and Bandelet-DCT. Multimedia Tools Appl. **81**(12), 16863–16879 (2022). https://doi.org/10.1007/s11042-022-12592-x
13. Lang, J., Ma, C.: Novel zero-watermarking method using the compressed sensing significant feature. Multimedia Tools Appl. **82**(3), 4551–4567 (2023)
14. Fierro-Radilla, A., Nakano-Miyatake, M., Cedillo-Hernandez, M., Cleofas-Sanchez, L., Perez-Meana, H.: A robust image zero-watermarking using convolutional neural networks. In: 2019 7th International Workshop on Biometrics and Forensics (IWBF), pp. 1–5. IEEE (2019)
15. Han, B., Du, J., Jia, Y., Zhu, H.: Zero-watermarking algorithm for medical image based on VGG19 deep convolution neural network. J. Healthcare Eng. **2021**, 1–12 (2021)
16. Zhu, J., Kaplan, R., Johnson, J., Fei-Fei, L.: HiDDeN: hiding data with deep networks. In: Ferrari, V., Hebert, M., Sminchisescu, C., Weiss, Y. (eds.) ECCV 2018. LNCS, vol. 11219, pp. 682–697. Springer, Cham (2018). https://doi.org/10.1007/978-3-030-01267-0_40
17. Jia, Z., Fang, H., Zhang, W.: MBRS: enhancing robustness of DNN-based watermarking by mini-batch of real and simulated JPEG compression. In: Proceedings of the 29th ACM International Conference on Multimedia, pp. 41–49 (2021)
18. Dyson, F.J., Falk, H.: Period of a discrete cat mapping. Am. Math. Mon. **99**(7), 603–614 (1992)
19. Huang, G., Liu, Z., Van Der Maaten, L., Weinberger, K.Q.: Densely connected convolutional networks. In: Proceedings of the IEEE Conference on Computer Vision and Pattern Recognition, pp. 4700–4708 (2017)
20. Howard, A., et al.: Searching for MobileNetV3. In: Proceedings of the IEEE/CVF International Conference on Computer Vision, pp. 1314–1324 (2019)
21. Lin, T.-Y., et al.: Microsoft COCO: common objects in context. In: Fleet, D., Pajdla, T., Schiele, B., Tuytelaars, T. (eds.) ECCV 2014. LNCS, vol. 8693, pp. 740–755. Springer, Cham (2014). https://doi.org/10.1007/978-3-319-10602-1_48

P-IoU: Accurate Motion Prediction Based Data Association for Multi-object Tracking

Xinya Wu and Jinhua Xu[✉]

School of Compute Science and Technology, East China Normal University,
Shanghai, China
jhxu@cs.ecnu.edu.cn

Abstract. Multi-object tracking in complex scenarios remains a challenging task due to objects' irregular motions and indistinguishable appearances. Traditional methods often approximate the motion direction of objects solely based on their bounding box information, leading to cumulative noise and incorrect association. Furthermore, the lack of depth information in these methods can result in failed discrimination between foreground and background objects due to the perspective projection of the camera. To address these limitations, we propose a Pose Intersection over Union (P-IoU) method to predict the true motion direction of objects by incorporating body pose information, specifically the motion of the human torso. Based on P-IoU, we propose PoseTracker, a novel approach that combines bounding box IoU and P-IoU effectively during association to improve tracking performance. Exploiting the relative stability of the human torso and the confidence of keypoints, our method effectively captures the genuine motion cues, reducing identity switches caused by irregular movements. Experiments on the DanceTrack and MOT17 datasets demonstrate that the proposed PoseTracker outperforms existing methods. Our method highlights the importance of accurate motion prediction of objects for data association in MOT and provides a new perspective for addressing the challenges posed by irregular object motion.

Keywords: Multi-object tracking · Intersection over Union (IoU) · Tracking by Detection · Motion Prediction

1 Introduction

Multi-object tracking is a fundamental task in computer vision that aims to locate and track object instances across a sequence of frames in a video. This task has widespread applications including video surveillance, traffic monitoring, and autonomous driving.

Recent years have witnessed significant progress in multi-object tracking research, particularly in the tracking-by-detection paradigm (TBD) [2–4,27,32,34], which has achieved remarkable results on traditional MOT17

© The Author(s), under exclusive license to Springer Nature Singapore Pte Ltd. 2024
B. Luo et al. (Eds.): ICONIP 2023, LNCS 14451, pp. 484–496, 2024.
https://doi.org/10.1007/978-981-99-8073-4_37

dataset [17]. However, multi-object tracking in complex scenarios remains a challenging task due to occlusion, irregular object motions and indistinguishable appearances, especially in DanceTrack dataset [21].

In the presence of well-performing object detectors, data association becomes the most critical step in the tracking process. Many methods rely on cues such as bounding boxes and appearance features for data association. However, when faced with objects that exhibit similar appearances, the motion cue becomes more important. Merely relying on changes in the bounding boxes of the objects is insufficient to accurately determine their true motion. In complex scenarios, such as dance performances and sports activities, where intricate movements and limb extensions are prevalent, the bounding box of an object is susceptible to variations in direction and size. In such scenarios, the traditional motion prediction of the Kalman filter [26], which is solely updated based on the object's bounding box from an object detector, fails to accurately reflect the true motion of the object. Consequently, the predicted position obtained from this approach deviates significantly from the true position, leading to failures in data association.

Furthermore, the lack of depth information in 2D images poses challenges for general tracking algorithms which often consider objects as the same if the overlap between their bounding boxes is significant. For instance, a person standing upright in the foreground and another person in the background raising their arms may have bounding boxes of the same sizes in the 2D image. This overlap could potentially lead to identity switches.

To address the aforementioned challenges, we notice that the human body exhibits a certain level of stability, particularly in the torso region. Despite the complexities in limb movements, the central body region remains relatively stable and less affected by motion variations. This stability allows us to leverage the robustness of the human torso, where the keypoints identified by the human pose estimation provide valuable clues to identify the true motion direction of the object. Therefore we propose a Pose Intersection over Union (P-IoU) method which utilize the motion of the human body's torso as the true motion of the object. Due to the relative stability of the human body's torso, the movements and variations in the limbs do not affect the size and orientation of the body-pose box. It also addresses the front-back relationship among multiple objects by considering the body pose rather than the whole bounding box only. Based on P-IoU, we propose PoseTracker, a tracking model which combines the intersection over union (IoU) of the object's bounding boxes and P-IoU effectively during association. We effectively address the challenges posed by indistinguishable appearances and irregular motions in multi-object tracking tasks of complex visual scenarios by leveraging the proposed methods.

In summary, our work makes two key contributions. Firstly, we address the challenge of determining the true motion direction of objects by proposing the P-IoU. Secondly, we propose PoseTracker, which combines bounding box and P-IoU effectively to enhance data association accuracy, reducing identity switches caused by irregular motions. Extensive experiments on multiple datasets demonstrate that our proposed method outperforms the previous methods. These contributions significantly improve multi-object tracking performance in complex scenarios with indistinguishable appearances and irregular motions.

2   Related Work

2.1   Multi-object Tracking

There are several traditional paradigms in multi-object tracking, including the tracking-by-detection [2–4,27,32,34], tracking-by-attention [19], and tracking-by-regression [24,36]. SORT [3] employs Kalman filter [26] for motion-based multi-object tracking using observations from a given object detector. Deep-SORT [27] further integrates deep visual features [11,20] into the object association process within the SORT framework. Recently, the transformer [23] has been introduced into MOT [16,22,33,37] to learn deep representations from visual information and object trajectories, enabling end-to-end integration of both detection and tracking tasks. However, these transformer-based methods still have substantial gaps in terms of accuracy and time efficiency compared to state-of-the-art detection and tracking methods.

2.2   Motion Models

Many MOT algorithms [3,5,10,34,35] employ motion models. Some methods use motion priors, such as Kalman Filter [3,5], optical flow [29], and displacement regression [8,12], to ensure accurate distance estimations. Typically, these motion models utilize Bayesian estimation [13] to predict the next state by maximizing the posterior estimate. The Kalman filter (KF) [26] is one of the most classical motion models, which is a recursive Bayesian filter following the typical predict-update cycle. These MOT algorithms simply treat the object's bounding box as the prediction object for linear motion without considering whether the motion direction truly aligns with the bounding box representation. In this case, the predicted position can accumulate noise due to the changes in the bounding box caused by the object's limbs and irregular motion, resulting in significant errors from the actual position under the influence of the Kalman filter.

2.3   Tracking Clues

Nearly all MOT methods utilize motion cues as the primary basis for association, and many approaches achieve impressive results solely based on motion information. ByteTrack [34], built upon the traditional SORT [3] algorithm, achieves breakthrough results in crowded scenes by performing secondary associations on low-confidence detections. In addition to motion-based association methods, appearance embedding [11,20] has also been widely used in MOT methods, as exemplified by DeepSORT [27]. Appearance cues stored in memory banks facilitate the recovery of long-term lost and re-appeared trajectories. However, in cases where appearance information is similar and lacks discriminative features, these methods fail to perform well in DanceTrack [21]. As a fine-grained task in human detection, human pose estimation has been addressed by PoseTrack [1], which proposes a joint evaluation dataset for human pose recognition and tracking to enhance the synergy between body pose and tracking in videos. Nonetheless, most methods on this dataset primarily focus on human pose recognition. A representative method with outstanding performance, Simplebaseline [29], enlarges the object's bounding box using optical flow but still emphasizes the overall motion of the bounding box, without considering the object's true motion direction.

3   Method

In this section, we introduce our P-IoU and PoseTracker for multi-object tracking. Our tracking pipeline follows the tracking-by-detection paradigm [2], where object detection is performed at each frame, and the detection results are then associated to perform object tracking.

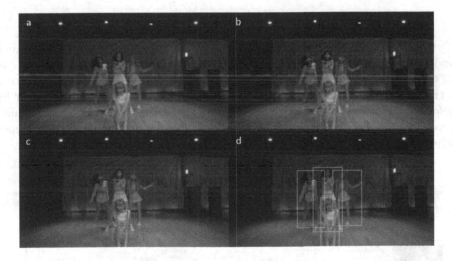

Fig. 1. The body-pose box generation process. We obtained the detection results and keypoints of the human whole body, and selected 4 points of both shoulders and both hips as the representatives of the human torso. The smallest rectangle containing the four points is used as the body-pose box of the object. (Color figure online)

3.1   P-IoU

The P-IoU means Pose Intersection over Union, which is our main contribution in this work. To obtain the true motion of the object, we use the torso motion of the object to predict the position in the next frame. It is effective to improve the accuracy of data association according to the real motion direction of the object.

We apply the well-trained object detector (i.e. YOLOX [9]) to obtain detection results at each frame, $B = \{b_1, b_2, \ldots, b_M\}$, with $b_i = \{x_b, y_b, w_b, h_b\}$ representing the bounding box in the image coordinate. These detections are then fed into the human pose recognition network Simple baseline [29] to get the N keypoints of each object, $K_i = \{k_1, \ldots, k_N\}$ (Fig. 1(b)). To represent the direction of human torso motion, we choose 4 points on both shoulders and both hips $\widetilde{K_i} = \{S_l, S_r, H_l, H_r\}$ and connect them to obtain a quadrilateral (Fig. 1(c)). Then we normalize the quadrilateral to a rectangle, by taking the smallest rectangle that includes this quadrilateral, with the edges parallel to the image borders, and get the body-pose box $p_i = \{x_p, y_p, w_p, h_p\}$ representing the object torso of each object (Fig. 1(d)red). This representation facilitates the calculation of the P-IoU and simplifies the matching process.

During the association, we calculate the IoU of the object bounding box b_i and the IoU of the body-pose box p_i separately, then fuse the two IoU distances between trajectories and detection results.

$$IoU\,(A, B) = \frac{A \cap B}{A \cup B} \tag{1}$$

Considering the case of severe occlusion, we will remove results with too much body-pose box overlap in the same frame. Specifically, we calculate the IoU of all detected body-pose boxes in the current frame, and set one of them to **None** if the IoU is above the threshold r. The choice of **None** can be made either by deleting the object with low-confidence, or by deleting the one with a smaller detection box. Following several experiments, the deletion of the low-confidence object gives better results.

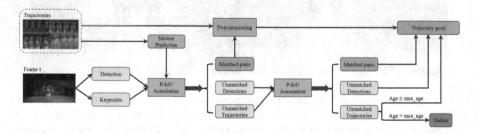

Fig. 2. Framework of our PoseTracker. The blue arrow represents the post-processing of the previous frame body-pose box of the trajectory on the match with the detection result. Green arrows indicate updating the trajectory or creating a new one. The grey box indicates the deletion of multiple unmatched trajectories.

3.2   Motion Prediction

We approximate the motion of objects between adjacent frames as linear motion and predict the position of objects in the next frame. Kalman filter [26] is a recursive Bayesian filter, providing an optimal estimation of the state of a dynamic system. The filter maintains two variables, the state $\mathbf{x}$, and the state covariance $\mathbf{P}$. In multi-object tracking, we describe the Kalman filter process with the state transition matrix $\mathbf{F}$, the observation matrix $\mathbf{H}$, the process noise covariance $\mathbf{Q}$, and the observation noise covariance $\mathbf{R}$. At each step t, given observations $\mathbf{z}_t$, Kalman filter works in an alternation of predict and update stages:

$$\text{predict:} \begin{cases} \hat{\mathbf{x}}_{t|t-1} = \mathbf{F}\hat{\mathbf{x}}_{t-1|t-1} \\ \mathbf{P}_{t|t-1} = \mathbf{F}\mathbf{P}_{t-1|t-1}\mathbf{F}^\top + \mathbf{Q}, \end{cases} \tag{2}$$

$$\text{update:} \begin{cases} \mathbf{K}_t = \mathbf{P}_{t|t-1}\mathbf{H}^\top \left(\mathbf{H}\mathbf{P}_{t|t-1}\mathbf{H}^\top + \mathbf{R}\right)^{-1} \\ \hat{\mathbf{x}}_{t|t} = \hat{\mathbf{x}}_{t|t-1} + \mathbf{K}_t \left(\mathbf{z}_t - \mathbf{H}\hat{\mathbf{x}}_{t|t-1}\right) \\ \mathbf{P}_{t|t} = (\mathbf{I} - \mathbf{K}_t\mathbf{H})\,\mathbf{P}_{t|t-1} \end{cases} \tag{3}$$

Here $\hat{\mathbf{x}}_{t|t-1}$ and $\hat{\mathbf{x}}_{t|t}$ are the prior state estimate and posteriori state estimate respectively. In our work, the state is defined as the bounding box or the body-pose box parameters $[x, y, w, h]$ and their derivatives, that is

$$\mathbf{x} = [x, y, w, h, \dot{x}, \dot{y}, \dot{w}, \dot{h}] \tag{4}$$

Similar to most tracking algorithms, we utilize Kalman filter [26] to *predict* the coordinate box location of the trajectory in the current frame and *update* Kalman filter parameters with the matched detection results, rather than directly edit the tracking results. Then we calculate the IoU of the predicted box and the detected bounding box for association.

3.3   PoseTracker

Based on P-IoU, we propose a tracking method, PoseTracker, shown in Fig. 2. PoseTracker focus on the true direction of object motion by finding the regular motion of the human torso among the irregular motion of the object.

For every frame, we acquire the bounding boxes and the body-pose boxes of all objects. Utilizing the Kalman filter, we perform motion prediction on each trajectory, enabling us to forecast its future position in next frame. Subsequently, we associate the trajectories with the objects present in the current frame by calculating their B-IoU and P-IoU distance, which means $1 - IoU$ to get cost matrix.

During the association, only B-IoU is used in SORT [3], that is,

$$C_{i,j} = d_{i,j}^{BIoU}. \tag{5}$$

The traditional methods use the minimum in each element of the matrices to combine different clues in the final cost matrix.

$$C_{i,j} = \min\left\{d_{i,j}^{BIoU}, d_{i,j}^{clue}\right\}. \tag{6}$$

Here the clue can be appearance embedding or motion such as the proposed P-IoU.

We develop a new method for combining B-IoU and P-IoU information. First, as a pre-processing, the boxes that are too far apart will be rejected and the weighting of the two IoUs will be balanced.

$$\hat{d}_{i,j}^{PIoU} = \begin{cases} \alpha_p \cdot d_{i,j}^{PIoU}, & \left(d_{i,j}^{PIoU} < \tau_p\right) \wedge \left(d_{i,j}^{BIoU} < \tau_b\right), \\ 1, & otherwise. \end{cases} \tag{7}$$

Here $d_{i,j}^{BIoU}$ is the object bounding box distance between the ith predicted trajectory bounding box and the jth detection bounding box. $d_{i,j}^{PIoU}$ is the body pose bounding box distance between the ith predicted trajectory body-pose box and the jth detection body-pose box, representing the true body motion cost. Due to the relatively small area occupied by the body-pose box in the image frame, a weight factor α_p is introduced to combine P-IoU and B-IoU.

The cost matrix is re-defined as follows:

$$C_{i,j} = \min\left\{d_{i,j}^{BIoU}, \hat{d}_{i,j}^{PIoU}\right\}. \tag{8}$$

To get the optimal match, we utilize Hungarian algorithm on the cost matrix C to match the trajectories and the detections. If the ith trajectory is matched with the jth detection, we set $M(i,j) = 1$, otherwise $M(i,j) = 0$.

In addition, in the case where the body-pose box of the previous frame of the trajectory is not **None**, we post-process the matched pairs, to make sure that the detection result intersects with the previous body-pose box of the matching trajectory.

$$\hat{M}(i,j) = \begin{cases} M(i,j), & PIoU(i,j) > 0 \\ 0, & otherwise \end{cases} \tag{9}$$

Here $\hat{M}(i,j)$ is the new matched pair after the post-processing filtering. When adjacent objects exchange positions, the post-processing $\hat{M}$ effectively reduces ID switch.

As shown in Fig. 2, associations are conducted in PoseTracker. After the first association, the unmatched trajectories and unmatched detections may not match correctly due to the large irregular motion in post-processing. So we cancel post-processing and add some lower-confidence detections to the detection set to do the second matching. After the second matching, we create new trajectories from the set of unmatched detections. The trajectory age is incremented for unmatched trajectories and reset for matched trajectories. Trajectories with an age above a maximum threshold are terminated to filter out false positives and inconsistent trajectories. The state of each trajectory is updated using the Kalman filter [26] that fuses the current state with estimated motion. This process is repeated for all frames to obtain a complete set of trajectories for all moving objects.

4 Experiments

4.1 Experimental Settings

Datasets. To evaluate the performance of our proposed object tracking method, we conducted experiments on the DanceTrack dataset [21] and the MOT17 dataset [17]. The DanceTrack dataset is a new dataset for object tracking that consists of video sequences of dancers performing various dance routines in a studio setting. The dataset contains 20 video sequences, with each sequence ranging from 30 to 60 s in length, and a total of 1,000 frames. The DanceTrack dataset is challenging due to similar appearances and irregular motions of objects. The MOT17 dataset contains 14 video sequences captured in diverse real-world settings, with challenges such as occlusions and rapid motion changes. Due to the dense flow of people, MOT17 relies more on the accuracy of detection.

Metrics. We evaluate the performance of our proposed object tracking method using the Higher Order Tracking Accuracy (HOTA) metric [15]. HOTA is a recently proposed evaluation metric that provides a comprehensive measure of object tracking performance by combining multiple performance metrics into a single score. Specifically, HOTA combines the metrics of Localization, False Positive, Missed Detection, and Identity Switch to provide a holistic measure of tracking accuracy that takes into account both the temporal and spatial accuracy of the tracked objects.

In our experiments, we report the HOTA score as the primary evaluation metric, and compare it against other state-of-the-art tracking methods. We also report MOTA and the individual component metrics that make up the HOTA score, which provide insights into the strengths and weaknesses of our method compared to other methods.

Implementation Details. Our method was implemented using PyTorch, and all the experiments ran on a desktop with 11th Gen Intel(R) Core(TM) i5-11400 @ 2.60 GHz and NVIDIA GeForce RTX 3090 GPU. To ensure a fair comparison, we used the publicly available YOLOX [9] detector, which was trained by [5,34] on the DanceTrack and MOT17 separately. For human pose estimation, we utilized the MMpose library from OpenMMLab for human pose estimation in our network and leveraged the pre-trained weights provided by the library. We selected the Simple baseline [29] model with the backbone of swin-transformer [14] for keypoints. For the elimination of the body-pose box in the same frame, we set the threshold of IoU for deletion to 0.6. Following the common practice of SORT [3], we set the detection confidence threshold to 0.6. The IoU threshold during association is 0.3. In computing the fusion matching matrix, we set the weight α_p to 0.6, and thresholds are $\tau_b = 0.5$ and $\tau_p = 0.3$ respectively.

4.2 Evaluation Results

In this section, we present the experimental results of our PoseTracker and compare its performance to mainstream MOT methods on the test sets of Dance-

Table 1. Results on DanceTrack test set [21]. Methods in the bottom blocks share the same detections.

Tracker	HOTA↑	DetA↑	AssA↑	MOTA↑	IDF1↑
CenterTrack [36]	41.8	78.1	22.6	86.8	35.7
FairMOT [35]	39.7	66.7	23.8	82.2	40.8
QDTrack [18]	45.7	72.1	29.2	83.0	44.8
TransTrack [22]	45.5	75.9	27.5	88.4	45.2
TraDes [28]	43.3	74.5	25.4	86.2	41.2
MOTR [33]	54.2	73.5	40.2	79.7	51.5
SORT [3]	47.9	72.0	31.2	91.8	50.8
DeepSORT [27]	45.6	71.0	29.7	87.8	47.9
ByteTrack [34]	47.3	71.6	31.4	89.5	52.5
OC-SORT [5]	55.1	80.3	38.3	92.0	54.6
*StrongSORT++ [7]	55.6	80.7	38.6	91.1	55.2
C-BIoU [31]	60.6	81.3	45.4	91.6	61.6
PoseTracker (Ours)	**61.3**	**81.9**	**46.7**	**92.1**	**63.5**

Table 2. Results on MOT17 test set [17]. Methods in the bottom blocks share the same detections.

Tracker	HOTA↑	MOTA↑	IDF1↑	IDs↓	AssA↑	AssR↑
FairMOT [35]	59.3	73.7	72.3	3,303	58.0	63.6
TransCenter [30]	54.5	73.2	62.2	4,614	49.7	54.2
TransTrack [22]	54.1	75.2	63.5	3,603	47.9	57.1
GRTU [25]	62.0	74.9	75.0	1,812	62.1	65.8
QDTrack [18]	53.9	68.7	66.3	3,378	52.7	57.2
MOTR [33]	57.2	71.9	68.4	2,115	55.8	59.2
TransMOT [6]	61.7	76.7	75.1	2,346	59.9	66.5
ByteTrack [34]	63.1	**80.3**	77.3	2,196	62.0	68.2
OC-SORT [5]	63.2	78.0	77.5	1,950	63.2	67.5
StrongSORT [7]	63.5	78.3	78.5	1,446	63.7	-
PoseTracker (Ours)	**64.1**	79.2	**79.3**	**1,117**	**63.9**	**68.8**

Track [21] and MOT17 (private detections) [17], as shown in Table 1 and Table 2. Each score is obtained from previous studies or by submitting the corresponding results to official evaluation servers to ensure accurate and reliable comparisons. It is worth noting that the quality of detections significantly impacts the overall tracking performance. To ensure fairness, methods in the bottom block utilize detections generated by YOLOX [9], where the YOLOX weights for the MOT17 and DanceTrack datasets are provided by ByteTrack [34] and OC-SORT [5], respectively. On the other hand, methods in the top block may employ different detection methods for their evaluations.

On the DanceTrack test set, our method shows a significant improvement in the HOTA and AssA score compared to other methods. Notably, DeepSORT [27], SORT [3], and ByteTrack [34] yield comparable results on the MOT17 test set. However, their tracking performance experiences a considerable drop on the DanceTrack test set due to the inclusion of more complex object movements and similar bounding box scales. For multi-object tracking in irregular motion and undistinguished appearance, the results on DanceTrack are strong evidence of the effectiveness of PoseTracker.

Table 3. Ablation experiments on the DanceTrack validation set [21].

Experiments	B-IoU	P-IoU	PreP	PostP	HOTA↑	MOTA↑	IDF1↑
(a)	✓				47.8	88.2	48.3
(b)		✓			51.9	86.3	50.9
(c)	✓	✓			55.4	90.9	54.5
(d)	✓	✓		✓	57.1	91.4	58.2
(e)	✓	✓	✓		59.6	91.8	60.2
PoseTracker	✓	✓	✓	✓	**60.9**	**91.9**	**63.3**

On the MOT17 test set, our method demonstrates a similar MOTA score compared to other methods. As discussed in previous research [21], the primary limitation in MOT17 lies in detection rather than tracking performance. However, our approach achieves an impressive HOTA score of 64.1, outperforming existing methods in the tracking by detection paradigm.

Our PoseTracker exhibits exceptional tracking performance, surpassing state-of-the-art methods in accurately capturing and associating objects with irregular motion patterns. The experimental results obtained on benchmark datasets validate the effectiveness of our approach.

4.3 Ablation Studies

We conducted ablation experiments on the DanceTrack [21] validation set to analyze the individual components of our proposed method. Specifically, we evaluated the impact of P-IoU, pre-processing and post-processing on tracking performance. In Table 3, P-IoU refers to associate objects by using it, PreP represents pre-processing in Eq. (7), and PostP denotes the post-processing step in Eq. (9) for reconfirming matched pairs using the previous frame's P-IoU.

As shown in Table 3, experiment (a) is the baseline method SORT [3], in which only B-IoU is used as tracking clue (Eq. (5)). In experiment (b), when we replace B-IoU with P-IoU, there is a significant improvement in the performance, especially in HOTA. In experiment (c), we combine B-IoU and P-IoU using the traditional method in Eq. (6), the HOTA and MOTA metric is improved further. Then we add post-processing in experiment (d), and identity switches are reduced significantly. In experiment (e), we apply the pre-processing to combine the B-IoU and P-IoU using Eq. (8), and the performance is better than that in (c) without pre-processing. Finally, when we use all components in our PoseTracker, the best results are achieved in all metrics.

Experimental results demonstrate the effectiveness of each component in the data association process. Notably, the use of P-IoU as a matching criterion highlights the importance of capturing the true direction of object motion, particularly in scenarios with irregular motion. Furthermore, the pre-processing which removes the boxes apart before fusion of the B-IoU and P-IoU, along with the weighting scheme, enables accurate association based on the true motion direction of the objects while also allowing the bounding box to assist with the association. Moreover, the post-processing step applied to matched trajectories and detections effectively reduces the number of identity switches, thereby further enhancing tracking accuracy.

5   Conclusion

In this paper, we have observed that many existing methods simply represent the motion of objects as the motion of their bounding boxes, which leads to the accumulation of noise over time. To address this limitation and utilize the true motion of objects as a matching criterion, we propose P-IoU to derive the true motion of humans from their torsos. Building upon this concept, we propose PoseTracker, a tracking method that combines the strengths of various elements to guide data association. PoseTracker effectively addresses the challenges associated with tracking objects exhibiting irregular motion and similar appearances. Our experimental results demonstrate the effectiveness of PoseTracker on multiple datasets, showcasing its superior performance in tracking objects with irregular motion. This research opens up new avenues for the development of tracking methods tailored to handle irregular motion, providing valuable insights for future advancements in the field.

References

1. Andriluka, M., et al.: PoseTrack: a benchmark for human pose estimation and tracking. In: CVPR (2018)
2. Andriluka, M., Roth, S., Schiele, B.: People-tracking-by-detection and people-detection-by-tracking. In: CVPR (2008)
3. Bewley, A., Ge, Z., Ott, L., Ramos, F., Upcroft, B.: Simple online and realtime tracking. In: ICIP (2016)
4. Bochinski, E., Eiselein, V., Sikora, T.: High-speed tracking-by-detection without using image information. In: AVSS (2017)
5. Cao, J., Pang, J., Weng, X., Khirodkar, R., Kitani, K.: Observation-centric sort: rethinking sort for robust multi-object tracking. In: CVPR (2023)

6. Chu, P., Wang, J., You, Q., Ling, H., Liu, Z.: TransMOT: spatial-temporal graph transformer for multiple object tracking. In: WACV (2023)
7. Du, Y., et al.: StrongSORT: make DeepSORT great again. IEEE Trans. Multimedia (2023). https://doi.org/10.1109/TMM.2023.3240881
8. Feichtenhofer, C., Pinz, A., Zisserman, A.: Detect to track and track to detect. In: ICCV (2017)
9. Ge, Z., Liu, S., Wang, F., Li, Z., Sun, J.: YOLOX: exceeding yolo series in 2021. arXiv preprint arXiv:2107.08430 (2021)
10. Han, S., Huang, P., Wang, H., Yu, E., Liu, D., Pan, X.: MAT: motion-aware multi-object tracking. Neurocomputing **473**, 75–86 (2022)
11. He, K., Zhang, X., Ren, S., Sun, J.: Deep residual learning for image recognition. In: CVPR (2016)
12. Held, D., Thrun, S., Savarese, S.: Learning to track at 100 FPS with deep regression networks. In: Leibe, B., Matas, J., Sebe, N., Welling, M. (eds.) ECCV 2016. LNCS, vol. 9905, pp. 749–765. Springer, Cham (2016). https://doi.org/10.1007/978-3-319-46448-0_45
13. Lehmann, E.L., Casella, G.: Theory of Point Estimation. Springer, New York (2006). https://doi.org/10.1007/b98854
14. Liu, Z., et al.: Swin transformer: hierarchical vision transformer using shifted windows. In: ICCV (2021)
15. Luiten, J.: HOTA: a higher order metric for evaluating multi-object tracking. Int. J. Comput. Vis. **129**, 548–578 (2020). https://doi.org/10.1007/s11263-020-01375-2
16. Meinhardt, T., Kirillov, A., Leal-Taixe, L., Feichtenhofer, C.: TrackFormer: multi-object tracking with transformers. In: CVPR (2022)
17. Milan, A., Leal-Taixé, L., Reid, I., Roth, S., Schindler, K.: MOT16: a benchmark for multi-object tracking. arXiv preprint arXiv:1603.00831 (2016)
18. Pang, J., et al.: Quasi-dense similarity learning for multiple object tracking. In: CVPR (2021)
19. Saribas, H., Cevikalp, H., Köpüklü, O., Uzun, B.: TRAT: tracking by attention using spatio-temporal features. Neurocomputing **492**, 150–161 (2022)
20. Simonyan, K., Zisserman, A.: Very deep convolutional networks for large-scale image recognition. arXiv preprint arXiv:1409.1556 (2014)
21. Sun, P., et al.: DanceTrack: multi-object tracking in uniform appearance and diverse motion. In: CVPR (2022)
22. Sun, P., et al.: TransTrack: multiple object tracking with transformer. arXiv preprint arXiv:2012.15460 (2020)
23. Vaswani, A., et al.: Attention is all you need. In: NeurIPS (2017)
24. Wan, X., Cao, J., Zhou, S., Wang, J., Zheng, N.: Tracking beyond detection: learning a global response map for end-to-end multi-object tracking. IEEE Trans. Image Process. **30**, 8222–8235 (2021)
25. Wang, S., Sheng, H., Zhang, Y., Wu, Y., Xiong, Z.: A general recurrent tracking framework without real data. In: ICCV (2021)
26. Welch, G., Bishop, G., et al.: An introduction to the Kalman filter (1995)
27. Wojke, N., Bewley, A., Paulus, D.: Simple online and realtime tracking with a deep association metric. In: ICIP (2017)
28. Wu, J., Cao, J., Song, L., Wang, Y., Yang, M., Yuan, J.: Track to detect and segment: an online multi-object tracker. In: CVPR (2021)

29. Xiao, B., Wu, H., Wei, Y.: Simple baselines for human pose estimation and tracking. In: Ferrari, V., Hebert, M., Sminchisescu, C., Weiss, Y. (eds.) ECCV 2018. LNCS, vol. 11210, pp. 472–487. Springer, Cham (2018). https://doi.org/10.1007/978-3-030-01231-1_29

30. Xu, Y., Ban, Y., Delorme, G., Gan, C., Rus, D., Alameda-Pineda, X.: TransCenter: transformers with dense queries for multiple-object tracking. arXiv e-prints, pp. arXiv-2103 (2021)

31. Yang, F., Odashima, S., Masui, S., Jiang, S.: Hard to track objects with irregular motions and similar appearances? Make it easier by buffering the matching space. In: WACV (2023)

32. Yu, F., Li, W., Li, Q., Liu, Yu., Shi, X., Yan, J.: POI: multiple object tracking with high performance detection and appearance feature. In: Hua, G., Jégou, H. (eds.) ECCV 2016. LNCS, vol. 9914, pp. 36–42. Springer, Cham (2016). https://doi.org/10.1007/978-3-319-48881-3_3

33. Zeng, F., Dong, B., Zhang, Y., Wang, T., Zhang, X., Wei, Y.: MOTR: end-to-end multiple-object tracking with transformer. In: Avidan, S., Brostow, G., Cissé, M., Farinella, G.M., Hassner, T. (eds.) Computer Vision, ECCV 2022. LNCS, vol. 13687, pp. 659–675. Springer, Cham (2022). https://doi.org/10.1007/978-3-031-19812-0_38

34. Zhang, Y., et al.: ByteTrack: multi-object tracking by associating every detection box. In: Avidan, S., Brostow, G., Cissé, M., Farinella, G.M., Hassner, T. (eds) Computer Vision, ECCV 2022. LNCS, vol. 13682, pp. 1–21. Springer, Cham (2022). https://doi.org/10.1007/978-3-031-20047-2_1

35. Zhang, Y., Wang, C., Wang, X., Zeng, W., Liu, W.: FairMOT: the fairness of detection and re-identification in multiple object tracking. IJCV **129**, 1–19 (2021)

36. Zhou, X., Koltun, V., Krähenbühl, P.: Tracking objects as points. In: Vedaldi, A., Bischof, H., Brox, T., Frahm, J.-M. (eds.) ECCV 2020. LNCS, vol. 12349, pp. 474–490. Springer, Cham (2020). https://doi.org/10.1007/978-3-030-58548-8_28

37. Zhou, X., Yin, T., Koltun, V., Krähenbühl, P.: Global tracking transformers. In: CVPR (2022)

WCA-VFnet: A Dedicated Complex Forest Smoke Fire Detector

Xingran Guo🆔, Haizheng Yu$^{(\boxtimes)}$🆔, and Xueying Liao🆔

College of Mathematics and System Sciences, Xinjiang University, Ürümqi, China
107552203480@stu.xju.edu.cn, yuhaizheng@xju.edu.cn

Abstract. Forest fires pose a significant threat to ecosystems, causing extensive damage. The use of low-resolution forest fire imagery introduces high complexity due to its multi-scene, multi-environment, multi-temporal, and multi-angle nature. This approach aims to enhance the model's generalizability across diverse and intricate fire detection scenarios. While state-of-the-art detection algorithms like YoloX, Deformable DETR, and VarifocalNet have demonstrated remarkable performance in the field of object detection, their effectiveness in detecting forest smoke fires, especially in complex scenarios with small smoke and flame targets, remains limited. To address this issue, we propose WCA-VFnet, an innovative approach that incorporates the Weld C-A component-a method featuring shared convolution and fusion attention. Furthermore, we have curated a distinctive dataset called T-SMOKE, specifically tailored for detecting small-scale, low-resolution forest smoke fires. Our experimental results show that WCA-VFnet achieves a significant improvement of approximately 35% in average precision (AP) for detecting small flame targets compared to Deformable DETR.

Keywords: Forest fire smoke detection · Small fire objects · WCA-VFnet · Low resolution images

1 Introduction

Fires are a phenomenon that extends beyond human control in terms of duration and scale, and forest fires, in particular, exhibit a highly destructive and rapidly spreading nature. Understanding the origin of forest fires forms the basis for effective fire prevention. Once the cause of a fire is clearly identified, it becomes easier to implement a range of measures aimed at controlling and preventing fires [1]. Consequently, there is an urgent need for a robust forest smoke and fire detection tool that enables efficient and rapid localization of fire-prone areas. Such a tool is crucial in minimizing the substantial ecological and societal losses caused by forest fires.

To address the issue of insufficient training data, synthetic smoke images were generated by inserting real or simulated smoke into forest backgrounds [2]. The Faster R-CNN algorithm was then employed for detection. However, this

B. Luo et al. (Eds.): ICONIP 2023, LNCS 14451, pp. 497–508, 2024.
https://doi.org/10.1007/978-981-99-8073-4_38

approach incurred an increased data processing cost. This method improved the core region proposal network (RPN) of Faster R-CNN based on the RGB color model [3]. It further utilized an ILSTM network with a color-guided anchor dropping technique to reduce the number of anchor boxes and mitigate false positive detections. However, it led to a decrease in recall rate. By combining Faster R-CNN with 3D CNN, this approach achieved higher accuracy in smoke localization and recognition [4]. This method introduced α and β parameters to adjust the dynamic foreground regions based on the optical flow algorithm [5]. The extracted dynamic regions were then passed to the YOLOv3 detector for smoke detection. It showed less sensitivity to wind speed during smoke detection, but faced challenges in extracting dynamic foreground regions when the smoke was far from the camera. ELASTIC blocks were integrated into the backbone network of YOLOv3 for fire detection [6]. BoF histograms were generated for the target fire regions and passed to a random forest classifier to determine whether the region contained fire. This approach effectively detected urban fires during nighttime, reducing false detections caused by streetlights and car lights. Instead of convolution, max pooling was utilized, and the feature maps after pooling were fused with feature maps of the same size before pooling using deconvolution [7]. This approach effectively leveraged shallow-level information for multi-scale feature fusion. However, shallow-level feature maps may lack sufficient representational power and robustness for detecting small objects. Overall, these approaches demonstrate significant progress in smoke and fire detection algorithms. However, challenges such as limited training data, varying environmental conditions, and the detection of small-scale smoke or fire instances still need to be addressed.

Based on the findings of previous studies, there are two primary challenges in the current detection of smoke and fire in forest environments. Firstly, the locations of fire detection can vary significantly, encompassing urban areas, buildings, residences, and forests, resulting in diverse smoke patterns and substantial variations in characteristics. Secondly, the detection of small fire incidents presents difficulties as their distinct features are less prominent, making early identification problematic. To address these issues, this paper proposes the integration of the Weld C-A component with the VarifocalNet backbone, presenting a novel method specifically designed for forest smoke and fire detection, namely WCA-VFnet. This approach offers exceptional detection speed and accuracy. Additionally, a custom dataset called T-SMOKE is introduced, comprising low-resolution forest fire data collected from diverse scenarios. By leveraging WCA-VFnet, it becomes possible to effectively detect and provide early warnings for small fire sources in complex forest scenes with limited resolution.

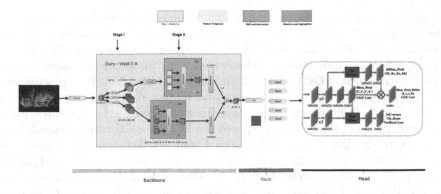

Fig. 1. Ours WCA-VFnet Model (The Weld C-A module consists of two stages. In the first stage, the input feature map is projected using three 1×1 convolutions. In the second stage, the intermediate features are employed according to two different paradigms. The features from both pathways are merged and serve as the final output. The computational complexity of each operation block is indicated in the top right corner.)

2 Related Work

Frizzi et al. [8] proposed a 9-layer convolutional neural network for classifying video frames into three categories: smoke, fire, and normal. They achieved a classification accuracy of 97.9% on a test set consisting of 1,427 fire images, 1,758 smoke images, and 2,399 normal images. However, this model was limited in detecting only red flames and exhibited suboptimal performance in detecting flames of different colors and smoke. Furthermore, it processed video frames individually without considering the motion characteristics of smoke and fire between frames. In another study, a 6-layer CNN was employed to extract static smoke texture information from original images, while the optical flow sequence was utilized to capture dynamic smoke texture information [9]. By integrating static and dynamic features of smoke texture, the detection performance in complex environments was enhanced, resulting in reduced false detection rates. This approach was specifically applied to smoke detection in complex scenes. Similarly, in [10], a motion object detection algorithm was utilized, which incorporated background dynamic updating and dark channel prior to identify suspicious smoke regions. The algorithm expanded small regions and performed feature extraction using CNN, thereby reducing the number of candidate regions and improving detection effectiveness while minimizing false positives. This method proved effective for detecting small smoke instances in multiple video scenarios. Another approach presented in [11] involved using a Gaussian mixture model to extract motion objects, followed by classification using a CNN. This method achieved higher detection accuracy compared to classical classification networks.

In reference [12], HSV channels and the Harris corner detector were employed for preprocessing flame images. The preprocessed images were then fed into an

Inceptionv3 network for feature extraction, aiming to reduce false positives and false negatives. Combined deep convolutional long short-term memory (LSTM) networks with optical flow methods for real-time fire detection [13]. The use of optical flow as input improved the detection performance compared to RGB models. Utilized the VGG16 network and made modifications by replacing three fully connected layers with two layers and adding dropout layers to prevent overfitting [14]. One fully connected layer was removed, reducing computational complexity. In [15], the authors proposed the Deep Normalization Convolutional Neural Network (DNCNN), which replaced traditional convolutional layers with batch normalization and convolutional layers. This approach addressed internal covariate shift issues, prevented gradient vanishing, and accelerated the training process. However, batch normalization was not always effective and sometimes negatively affected performance, especially with small batch sizes. Applied Deep Convolutional Generative Adversarial Networks (DCGAN) to fire detection to overcome issues such as overfitting, high false positive rates, and high false negative rates caused by limited datasets [16]. However, the generated fire images were unstable, initially appearing good but becoming blurry over time. Proposed a fire detection method based on VGG-16, where the features were extracted from video frames and fed into Integration LSTM (ILSTM) units for feature fusion [17]. ILSTM units reduced the dimensionality of the input sequence and learned different feature representations. A forest fire detection algorithm combining two streams was presented [18]. The TRPCA motion detection algorithm suppressed background interference, while the VGG16 network and BLSTM network extracted spatial and temporal features, taking into account both foreground and background factors. Additionally, introduced an attention module and an FPN feature fusion module based on VGG16 to enhance the detection of small smoke and smoke-like objects [19]. However, this approach focused solely on classification and did not provide precise smoke localization information.

3   Method

3.1   Dataset and Annotation

Quality and size of datasets play a crucial role in determining the performance of deep learning models in the field of image detection. However, there is a limited availability of publicly accessible datasets specifically focused on forest smoke fires or smoke datasets suitable for forest environments. Consequently, we addressed this issue by collecting forest smoke fire images through web scraping open data, resulting in the creation of the Forest Smoke dataset (T-SMOKE dataset). Our self-constructed dataset comprises a variety of forest smoke images captured from different perspectives and scales. We manually annotated the smoke regions in the images and transformed them into COCO format [30]. In total, the dataset consists of 2036 images, randomly divided in an 8.5:1.5 ratio, with 85% of the dataset allocated for training and 15% for validation.

3.2 WCA-VFnet

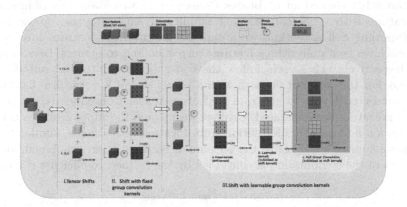

Fig. 2. Improvements in Position Shift Operations: (I) Preliminary Implementation using Tensor Shifts, (II) Efficient Implementation of Group Convolution, (III) Incorporating Learnable Kernels and Further Adaptation of Group Convolution.

Similarities Between Convolution and Self-attention. The decomposition and fusion of Convolution and Self-Attention demonstrate powerful detection performance. The core focus of this article is how to decompose and fuse the two methods. Firstly, when applying Convolution and Self-Attention to feature maps, we can observe their similarities. The first stage involves a feature learning module, where both methods share the same operation by projecting features to a deeper space using 1×1 convolutions. On the other hand, the second stage corresponds to the process of feature aggregation, although their learning paradigms differ. From a computational perspective, both the 1×1 convolution in the first stage and the self-attention module require theoretical FLOP and parameters with quadratic complexity relative to the channel size C. In contrast, both modules in the second stage are lightweight or nearly computation-free.

Based on the analysis, it is evident that Convolution and Self-Attention can share the same computational cost by using shared 1×1 convolutions to map the input feature maps. Although extracting semantic features is crucial, the aggregation stage of Self-Attention is lightweight and does not require additional learning parameters that could cause memory issues.

Fusion of Convolution and Self-attention. The fusion of Convolution and Self-Attention emerges as a compelling approach, driven by their inherent similarities. By leveraging shared 1×1 convolutions for both operations, we can effectively combine these two methods. To achieve this, we need to perform a single mapping and exploit the reuse of intermediate feature maps across different aggregation operations. In this study, we propose the Weld C-A module, as illustrated in Fig. 1. The Weld C-A module consists of two stages. In the first stage, the input features are mapped using three identical 1×1 convolutions and subsequently reshaped into N blocks. Consequently, we obtain a set of intermediate features with $3 \times N$ feature maps, significantly enriching the representation.

Regarding Self-Attention, we group the intermediate features into N sets, with each set containing three features corresponding to queries, keys, and values, respectively. These sets follow the conventional multi-head self-attention module. For the convolution path with a kernel size of k, we utilize lightweight fully connected layers to generate k^2 feature maps. Consequently, by shifting and aggregating the generated features, we process the input features through convolution and gather information from local receptive fields, similar to conventional approaches. Finally, the outputs of both paths are combined, and the intensity of the combination is controlled by the learned coefficients α and β:

$$F_{\text{out}} = \alpha F_{\text{att}} + \beta F_{\text{conv}} \tag{1}$$

3.3   Improved Offset and Summation

According to the analysis in the previous section and Fig. 1, the intermediate features in the convolution path follow the shift and summation operations performed in traditional convolution modules. Although these operations are theoretically lightweight, moving tensors in various directions can disrupt data locality and hinder vectorization. This could significantly impact the actual efficiency of our module during inference.

To address this issue, we replace the inefficient tensor shifting with a deep convolution operation using fixed kernels, as depicted in Fig. 2 (II). Taking Shift $(f, -1, -1)$ as an example, the computation of the shifted features is as follows:

$$\tilde{f}_{c,i,j} = f_{c,i-1,j-1}, \forall c, i, j, \tag{2}$$

where C represents each channel of the input.

We represent the 3×3 convolution kernel as follows:

$$K_c = \begin{bmatrix} 1 & 0 & 0 \\ 0 & 0 & 0 \\ 0 & 0 & 0 \end{bmatrix}, \forall c, \tag{3}$$

The resulting output is represented as follows:

$$\begin{aligned} f_{c,i,j}^{(\text{dwc})} &= \sum_{p,q \in \{0,1,2\}} K_{c,p,q} f_{c,i+p-\lfloor k/2 \rfloor, j+q-\lfloor k/2 \rfloor} \\ &= f_{c,i-1,j-1} = \tilde{f}_{c,i,j}, \quad \forall c, i, j. \end{aligned} \tag{4}$$

Therefore, by carefully designing kernel weights specific to the desired shift direction, the convolution output is equivalent to a simple tensor shift (Eq. (1)). To further integrate the features from different directions, we concatenate all the input features and convolution kernels and represent the shift operation as a group-wise convolution, as shown in Fig. 2 (c.a). This modification enhances the computational efficiency of our module.

Building upon this, we have introduced some modifications to enhance the flexibility of the module. As depicted in Fig. 2 (c.b), we relax the convolution kernels to be learnable weights, initialized with the shift kernels. This preserves the capability of the original shift operation while increasing the model's capacity. Additionally, we utilize multiple sets of convolution kernels to match the output channel dimensions of the convolution and self-attention paths, as shown in Fig. 2 (c.c).

3.4   The Computational Cost of the Weld C-A Module

In the first stage, the computational cost and trainable parameters are the same as self-attention and lighter compared to traditional convolutions (e.g., 3×3 convolutions). In the second stage, the Weld C-A introduces additional computational overhead with lightweight fully connected layers and group convolutions. The computational complexity is linearly dependent on the channel size C and is smaller compared to the first stage. The actual cost in the ResNet50 model exhibits a similar trend as the theoretical analysis suggests.

3.5   Weld C-A and VFnet

The Weld C-A component we propose exhibits broad applicability in two stages. We integrate this component into the feature extraction stage of VFnet [23], followed by the FPN [22] neck stage. The final classification and regression results are produced by the VFnet-heads, as illustrated in Fig. 1. One of the key advantages lies in the continuous evolution of self-attention mechanisms, with significant research efforts focused on exploring different attention operators to further enhance model performance. For instance, Patch Attention, as proposed in [31], replaces the original softmax operation by merging information from all features within a local region as attention weights. Swin-Transformer [32] introduces a window attention mechanism that maintains a consistent receptive field within the same local window, resulting in computational cost savings and faster inference speed. It is important to note that our Weld C-A component operates independently of the self-attention mechanism, as expressed by the following formula:

$$(Patchwise)\mathrm{A}\left(q_{ij}, k_{ab}\right) = \phi\left(\left[q_{ij}, [k_{ab}]_{a,b \in \mathcal{N}_k(i,j)}\right]\right), \tag{5}$$

$$(Window)\mathrm{A}\left(q_{ij}, k_{ab}\right) = \mathrm{softmax}_{a,b \in \mathcal{W}_k(i,j)}\left(q_{ij}^{\mathrm{T}} k_{ab}/\sqrt{d}\right), \tag{6}$$

$$(Global)\mathrm{A}\left(q_{ij}, k_{ab}\right) = \mathrm{softmax}_{a,b \in \mathcal{W}}\left(q_{ij}^{\mathrm{T}} k_{ab}/\sqrt{d}\right), \tag{7}$$

The formula can be simplified as follows: $[\cdot]$ denotes feature concatenation, $\phi(\cdot)$ represents two linear projection layers with intermediate non-linear activation,

$W_k(i, j)$ indicates dedicated receptive fields for each query token, and W represents the entire feature map. The resulting attention weights can be applied to the original q, k, and v calculations.

For our newly established T-SMOKE dataset, which presents challenges such as low resolution, small targets, and real-time detection, the integration of the Weld C-A module into the original VFnet network ensures improved performance. By incorporating attention operators into the base model, Weld C-A enhances both attention capabilities and lightweight performance through the two-stage shared convolution kernel operation. As a result, our model exhibits outstanding detection performance on the T-SMOKE dataset.

4   Experiment

Dataset. In the experiments, the T-SMOKE forest smoke and fire dataset mentioned in Sect. 3 was utilized. Our model underwent training on the training set and subsequent evaluation on both the validation and test sets to assess its performance.

Implementation Details. In the experiments, we use the ImageNet-pretrained ResNet-50 [20, 21]as the backbone network for ablation. The FPN layer is included to extract multi-scale feature maps [22]. The parameter initialization follows the original settings from VarifocalNet. The backbone layer is initialized as ResNet-50 with Num Stages set to 4, Out Indices as (0, 1, 2, 3), Frozen Stages as 1, and the neck layer as FPN to extract multi-scale feature maps [23]. In Channels are set to [256, 512, 1024, 2048], and Out Channels are set to 256. Start Level is set to 1, and Num Outs is set to 5. Feat Channels in the Bbox Head are set to 256. For Loss Cls, α is set to 0.75 and β is set to 2.0, with Loss Weight set to 1.0. GIoULoss is used as the Loss Bbox, with Loss Weight set to 1.5.

Additionally, we conduct fair comparison experiments with current state-of-the-art models. The default settings include Max Epochs of 100, evaluation on the validation set every 30 intervals, with the final evaluation after the 90th epoch. SGD [24] is chosen as the base optimizer with an initial learning rate of 0.01, a momentum of 0.9, and a weight decay of 0.0001. Furthermore, we compare the results with the Adam optimizer [25] in the ablation experiments. The experiments are evaluated on an NVIDIA RTX 3090 GPU.

4.1   Comparison with Advanced Models

In Table 1, we compare WCA-VFnet with several advanced models, namely CenterNet [26], YOLOv3 [27], Deformable DETR [28], VarifocalNet [23], YOLOX [29], on the T-SMOKE validation set. Our findings indicate that WCA-VFnet exhibits a slower convergence speed compared to the other models. Furthermore, it is observed that the other models generally display lower detection performance for small objects when compared to Deformable DETR. Figure 3 provides a visualization of the classification loss and bounding box loss of our proposed WCA-VFnet.

Table 1. WCA-VFnet was compared with state-of-the-art models on the T-SMOKE validation set.

Method	Epochs	bbox _mAP	AP	bbox_mAP _.75	bbox_mAP _s	AR	params	FLOPs	Training GPU hours	Inference FPS
CenterNet	90	0.222	0.432	0.214	0.303	0.343	14.21M	51.02G	3.22	12
YOLOv3	90	0.244	0.495	0.230	0.286	0.353	61.53M	193.87G	1.42	28
Deformable DETR	90	0.280	0.550	0.255	0.342	0.470	39.82M	143.69G	6.45	20
VarifocalNet	90	0.417	0.632	0.411	0.467	0.582	32.49M	189.02G	2.42	24
VarifocalNet	100	0.524	0.773	0.535	0.576	0.662	32.49M	189.02G	2.08	24
YOLOX	90	0.490	0.874	0.466	0.535	0.595	99.00M	140.76G	0.80	19
YOLOv8	90	0.601	0.887	0.532	0.617	0.628	87.35M	152.76G	0.92	23
WCA-VFnet	90	**0.646**	**0.902**	**0.696**	**0.680**	**0.728**	23.61M	165.95G	0.85	17

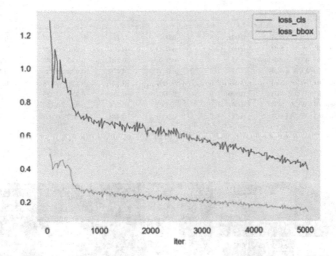

Fig. 3. WCA-VFnet's classification loss and bbox loss on the T-SMOKE validation set.

Our proposed WCA-VFnet exhibits significantly lower FLOPs compared to VarifocalNet. This is achieved through the incorporation of our specially designed Weld C-A module, which utilizes shared 1×1 convolutions to minimize computational overhead associated with traditional convolutions. It is evident that WCA-VFnet achieves nearly 1.65 times faster speeds than Deformable DETR, while also showing noticeable improvements in AP values compared to other state-of-the-art models. This performance advantage can be attributed to the presence of Transformer Attention modules in Deformable DETR, which introduces some computational overhead compared to traditional convolutions. On the other hand, other advanced models rely on traditional convolutions as their basis, leading to suboptimal detection results for small objects.

4.2 Ablation Study on WCA-VFnet

Table 2 presents a comprehensive comparison of the ablative experiments conducted in our proposed WCA-VFnet. In terms of the model, we compare the baseline VarifocalNet with our proposed WCA-VFnet. The backbones used for investigation are ResNet-50 and ResNet-101. We conduct ablative experiments to assess the advantages of the Weld C-A module and verify its superiority. By modifying the optimizer, learning rate (LR), momentum, and weights, we compare the experimental results. WCA-VFnet achieves optimal detection performance under the configuration of ResNet-101 with Weld C-A, SGD optimizer, and LR = 0.01. Figure 4 illustrates the real-time detection performance on the test set of the T-SMOKE dataset. It is evident that the small target detection performance for forest smoke fire is excellent.

Table 2. WCA-VFnet was compared with state-of-the-art models on the T-SMOKE test set.

Method	Backbone	Optimizer	LR	Momentum	weight _decay	bbox_mAP	AP	bbox_mAP_75	bbox_mAP_s	AR
VFnet	ResNet-50	SGD	0.001	0.8	0.001	0.134	0.234	0.140	0.201	0.260
VFnet	ResNet-50	SGD	0.005	0.9	0.0001	0.162	0.254	0.179	0.231	0.295
WCA-VFnet	ResNet-50+WCA	AdamW	0.01	–	0.0001	0.229	0.420	0.230	0.283	0.410
WCA-VFnet	ResNet-50+WCA	SGD	0.01	0.9	0.0001	0.678	0.922	0.738	0.707	0.746
WCA-VFnet	ResNet-101+WCA	SGD	0.05	0.9	0.0001	0.260	0.458	0.278	0.307	0.447
WCA-VFnet	ResNet-101+WCA	SGD	0.0001	0.9	0.0001	0.334	0.559	0.336	0.400	0.537
WCA-VFnet	ResNet-101+WCA	SGD	0.001	0.9	0.0001	0.724	0.951	0.796	0.749	0.779
WCA-VFnet	ResNet-101+WCA	SGD	0.01	0.9	0.0001	**0.786**	**0.977**	**0.875**	**0.803**	**0.826**

Fig. 4. WCA-VFnet Smoke and Fire Detection on test set.

5   Conclusion

WCA-VFnet is an object detector specifically developed for the detection of smoke fires, showcasing remarkable performance in terms of efficiency, real-time processing, and rapid training. It particularly excels in detecting smoke fires in forest environments, demonstrating superior capabilities in handling low-resolution fire images and detecting small smoke and flame targets. The core innovation of WCA-VFnet lies in the effective integration of Convolution and Self-Attention, enabling efficient fusion of fire-related features. Our research aims to make a valuable contribution to the prevention and mitigation of forest fires. Certainly! While we have achieved the accomplishments described in the article, further complementary research is necessary due to the dataset limited diversity and lightweight nature.

Acknowledgements. This work is supported in part by the Xinjiang Natural Science Foundation of China (2021D01C078).

References

1. Yuan, C., Zhang, Y., Liu, Z.: A survey on technologies for automatic forest fire monitoring, detection, and fighting using unmanned aerial vehicles and remote sensing techniques. Can. J. For. Res. **45**(7), 783–792 (2015)
2. Zhang, Q., Lin, G., Zhang, Y., et al.: Wildland forest fire smoke detection based on faster R-CNN using synthetic smoke images. Procedia Eng. **211**, 441–446 (2018)
3. Tingting, S.: Fire detection model based on static characteristics and dynamic behavior. J. Saf. Sci. Technol. **17**(1), 96–101 (2021)
4. Lin, G., Zhang, Y., Xu, G., et al.: Smoke detection on video sequences using 3D convolutional neural networks. Fire Technol. **55**, 1827–1847 (2019)
5. Li, P., Zhang, J., Li, W.: Smoke detection method based on optical flow improvement and YOLOv3. J. Zhejiang Univ. Technol. **49**, 9–15 (2021)
6. Park, M.J., Ko, B.C.: Two-step real-time night-time fire detection in an urban environment using Static ELASTIC-YOLOv3 and Temporal Fire-Tube. Sensors **20**(8), 2202 (2020)
7. Lijuan, L.I.U., Songnan, C.: Real-time smoke detection model based on improved SSD. J. Xinyang Normal Univ. (Nat. Sci. Ed.) **33**(2), 305–311 (2020)
8. Frizzi, S., Kaabi, R., Bouchouicha, M., et al.: Convolutional neural network for video fire and smoke detection. In: IECON 2016–42nd Annual Conference of the IEEE Industrial Electronics Society, pp. 877–882. IEEE (2016)
9. Chen, J.Z., Wang, Z.J., Chen, H.H., et al.: Dynamic smoke detection using cascaded convolutional neural network for surveillance videos. J. Univ. Electron. Sci. Technol. China **46**(6), 992–996 (2016)
10. Luo, Y., Zhao, L., Liu, P., et al.: Fire smoke detection algorithm based on motion characteristic and convolutional neural networks. Multimedia Tools Appl. **77**, 15075–15092 (2018)
11. Sun, X., Sun, L., Huang, Y.: Forest fire smoke recognition based on convolutional neural network. J. Forest. Res. **32**(5), 1921–1927 (2021)
12. Ryu, J., Kwak, D.: Flame detection using appearance-based pre-processing and Convolutional Neural Network. Appl. Sci. **11**(11), 5138 (2021)
13. Hu, C., Tang, P., Jin, W.D., et al.: Real-time fire detection based on deep convolutional long-recurrent networks and optical flow method. In: 2018 37th Chinese Control Conference (CCC), pp. 9061–9066. IEEE (2018)

14. Zhen-cun, J., Xiao-Jing, W.E.N., Zheng-xin, D., et al.: Research on fire detection of improved VGG16 image recognition based on deep learning. Fire Sci. Technol. **40**(3), 375 (2021)
15. Yin, Z., Wan, B., Yuan, F., et al.: A deep normalization and convolutional neural network for image smoke detection. IEEE Access **5**, 18429–18438 (2017)
16. Xu, Z., Guo, Y., Saleh, J.H.: Tackling small data challenges in visual fire detection: a deep convolutional generative adversarial network approach. IEEE Access **9**, 3936–3946 (2020)
17. Wei, X., Wu, S., Wang, Y.: Forest fire smoke detection model based on deep convolution long short-term memory network. J. Comput. Appl. **39**(10), 2883 (2019)
18. Qiang, X., Zhou, G., Chen, A., et al.: Forest fire smoke detection under complex backgrounds using TRPCA and TSVB. Int. J. Wildland Fire **30**(5), 329–350 (2021)
19. He, L., Gong, X., Zhang, S., et al.: Efficient attention based deep fusion CNN for smoke detection in fog environment. Neurocomputing **434**, 224–238 (2021)
20. Deng, J., Dong, W., Socher, R., et al.: ImageNet: a large-scale hierarchical image database. In: 2009 IEEE Conference on Computer Vision and Pattern Recognition, pp. 248–255. IEEE (2009)
21. He, K., Zhang, X., Ren, S., et al.: Deep residual learning for image recognition. In: Proceedings of the IEEE Conference on Computer Vision and Pattern Recognition, pp. 770–778 (2016)
22. Lin, T.Y., Dollár, P., Girshick, R., et al.: Feature pyramid networks for object detection. In: Proceedings of the IEEE Conference on Computer Vision and Pattern Recognition, pp. 2117–2125 (2017)
23. Zhang, H., Wang, Y., Dayoub, F., et al.: VarifocalNet: an IoU-aware dense object detector. In: Proceedings of the IEEE/CVF Conference on Computer Vision and Pattern Recognition, pp. 8514–8523 (2021)
24. Bottou, L., Curtis, F.E., Nocedal, J.: Optimization methods for large-scale machine learning. SIAM Rev. **60**(2), 223–311 (2018)
25. Kingma, D.P.: A method for stochastic optimization. arXiv Preprint (2014)
26. Zhou, X., Wang, D., Krähenbühl, P.: Objects as points. arXiv preprint arXiv:1904.07850 (2019)
27. Redmon, J., Farhadi, A.: YOLOv3: an incremental improvement. arXiv preprint arXiv:1804.02767 (2018)
28. Zhu, X., Su, W., Lu, L., et al.: Deformable DETR: deformable transformers for end-to-end object detection. arXiv preprint arXiv:2010.04159 (2020)
29. Ge, Z., Liu, S., Wang, F., et al.: YOLOX: exceeding YOLO series in 2021. arXiv preprint arXiv:2107.08430 (2021)
30. Lin, T.-Y., et al.: Microsoft COCO: common objects in context. In: Fleet, D., Pajdla, T., Schiele, B., Tuytelaars, T. (eds.) ECCV 2014. LNCS, vol. 8693, pp. 740–755. Springer, Cham (2014). https://doi.org/10.1007/978-3-319-10602-1_48
31. Zhao, H., Jia, J., Koltun, V.: Exploring self-attention for image recognition. In: Proceedings of the IEEE/CVF Conference on Computer Vision and Pattern Recognition, pp. 10076–10085 (2020)
32. Liu, Z., Lin, Y., Cao, Y., et al.: Swin transformer: hierarchical vision transformer using shifted windows. In: Proceedings of the IEEE/CVF International Conference on Computer Vision, pp. 10012–10022 (2021)

Label Selection Algorithm Based on Ant Colony Optimization and Reinforcement Learning for Multi-label Classification

Yuchen Pan, Yulin Xue, Jun Li, and Jianhua Xu[✉]

School of Computer and Electronic Information, School of Artificial Intelligence, Nanjing Normal University, Nanjing 210023, Jiangsu, China
{202243024,212243039}@stu.njnu.edu.cn, {lijuncst,xujianhua}@njnu.edu.cn

Abstract. Multi-label classification handles scenarios where an instance can be annotated with multiple non-exclusive but semantically related labels simultaneously. Despite significant progress, multi-label classification is still challenging due to the emergence of multiple applications leading to high-dimensional label spaces. Researchers have generalized feature dimensionality reduction techniques to label space by using label correlation information, and obtained two techniques: label embedding and label selection. There have been many successful algorithms in label embedding, but less attention has been paid to label selection. In this paper, we propose a label selection algorithm for multi-label classification: LS-AntRL, which combines ant colony optimization (ACO) and reinforcement learning (RL). This method helps ant colony algorithms search better in the search space by using temporal difference (TD) RL algorithm to learn directly from the experience of ants. For heuristic learning, we need to model the ACO problem as a RL problem, that is, to model label selection as a Markov decision process (MDP), where the label represents the state, and each ant selecting unvisited labels represents a set of actions. The state transition rules of the ACO algorithm constitute the transition function in the MDP, and the state value function is updated by TD formula to form a heuristic function in ACO. After performing label selection, we train a binary weighted neural network to recover low-dimensional label space back to the original label space. We apply the above model to five benchmark datasets with more than 100 labels. Experimental results show that our method achieves better classification performance than other advanced methods in terms of two performance evaluation metrics (Precision@n and DCG@n).

Keywords: Multi-label learning · Label selection · Ant colony algorithm · Reinforcement learning · Temporal difference

---

Supported by the Natural Science Foundation of China (NSFC) under grants 62076134 and 62173186.

B. Luo et al. (Eds.): ICONIP 2023, LNCS 14451, pp. 509–521, 2024.
https://doi.org/10.1007/978-981-99-8073-4_39

1   Introduction

Traditional supervised learning mainly deals with single label classification [19], where each instance has only one predefined label. However, many real-world classification issues involve multiple labels for many instances simultaneously [5,16], which induces multi-label classification in machine learning. Its practical applications come from image annotations with multiple labels [4] (for example, an image can be depicted by tree, sky, and grass), text classification [12] (a document involves multiple topics), and gene function prediction [20] (a gene is typically associated with multiple functions).

Nowadays, besides high-dimensional feature space and large instance size, we often encounter high-dimensional problems from label space. When processing multi-label data, the number of labels can often reach hundreds of thousands [5]. This difficulty is similar to high-dimensional feature space, which means that more computational time and internal storage are required to train classifiers, and typically results in unsatisfactory prediction performance. Currently, the main methods to solve the high-dimensional problem of labels include label embedding and label selection [22], similar to feature extraction and feature selection. However, it is not possible to simply remove labels from multi-label sets as feature extraction and feature selection methods in feature space, since all labels must exist in the predictions provided by the final classifier. Therefore, any label dimension reduction methods must have the ability to recover or reconstruct the whole label set to maintain its original structure.

Label embedding method maps the label space to a low dimensional latent space, and can get final predictions by recovering the latent label space to the original label space. Hsu et al. [10] presented the first label embedding algorithm ML-CS. This method requires a significant level of sparsity in the label space. PLST [9], similar to principal component analysis (PCA), perceives the label space of the multi-label classification problem in a geometric way, and uses the low dimensional linear subspace in the high-dimensional space to learn the key linear correlation between labels. Boolean matrix decomposition (BMD) [17] decomposes a binary matrix into two low-rank binary matrix boolean multiplication and could be applied to label embedding method. MLC-BMaD [21] applies BMD to construct a binary reduced label space. These latent labels identify and represent the dependencies between the original labels.

It is worth noting that the label space dimensionality reduction method can be improved by introducing feature information. In CPLST [6], it combines PLST and canonical correlation analysis [24], while utilizing feature information and label parts. LEML [23] models the multi-label classification problem as a empirical risk minimization problem with low-rank constraints, which not only generalizes the label and feature dimensionality reduction, but also brings the ability to support various loss function, and allows strict generalization error analysis. MoRE [15] learns low-rank projections in label space by analyzing the residual structure between features and label space. Although label embedding method can save computational costs, make the embedding space smaller and easier to handle. However, there are still some limitations. Their methods may lead to information loss and label correlation being implicitly encoded, resulting in a lack of interpretability.

Label selection is another label dimensionality reduction method. Its basic idea is to select a subset of labels rich in information from the original labels, and the complete label space can be recovered from the selected labels. MOPLMS [1] select a subset of labels based on structural sparse optimization and singular value decomposition, and then use Bayesian criteria to perform label recovery. In ML-CSSP [3], the label selection task is treated as a column subset selection problem. According to a random sampling process, k representative labels are accurately selected, and then k classifiers are learned for these selected labels. MLC-EBMD [14] obtains two low-rank matrices through precise BMD, which preserves the meaning of the lower dimensional label space. By simply using the recovery matrix, the original label space can be obtained. In LS-BaBID [11], after performing EBMD, a sequential backward selection (SBS) strategy widely used in feature selection is applied to remove some labels with less information one by one to detect fixed size column subsets. The above label selection methods are all unsupervised methods. It is worth noting that some researchers have introduced feature information to improve the performance of label selection methods.

Elham [2] proposed a submodule maximization framework with linear costs to find informative labels that are most relevant to other labels but are least redundant to each other. Its framework includes label-label and label-feature dependencies, aiming to find labels with the greatest representation and prediction capabilities. SPL-MLL [13] is different from the two-step methods. It integrates label selection, label prediction, and label recovery into a unified framework, uses a neural network to achieve feature embedding, and puts the neural network into the final unified optimization objective function.

In this paper, we propose a two-step manner method to select the most informative label subset. This method use ant colony optimization and reinforcement learning to select the most important label subset. Once the label subset is selected, we have to design a binary weighted neural network to recover the original label space from the predicted labels. The experiments illustrate that our proposed method is more effective according to two metrics (Precision@n and DCG@n, n = 1, 3 and 5), compared with four state-of-the-art techniques (LS-BaBID [11], MLC-EBMD [14], MLC-BMaD [21] and ML-CSSP [3]) on five datasets with more than 100 labels.

Summarily, there are three-fold contributions in this study:

(a) We propose a novel label selection algorithm based on ant colony optimization and reinforcement learning.
(b) To the best of our knowledge, our method is the first label selection approach to apply reinforcement learning in multi-label classification.
(c) Our proposed method is validated on five benchmark data sets with 100+ labels, compared with four state-of-the-art methods.

This paper is organized as follows. Section 2 provides some basic concepts for this paper. In Sect. 3, we detail our label selection and label recovery methods. Section 4 validates our method. Finally, the paper draws some conclusions.

2 Basic Concepts

Here we give the description of notations used through out this paper. Given training data with the form of instance-label pairs $\{\mathbf{x}_i, \mathbf{y}_i\}_{i=1}^N$ where, for the i-th instance, the $\mathbf{x}_i \in R^D$ and $\mathbf{y}_i \in \{0, 1\}^C$ denote the feature vector and label vector, where N is the number of instances, D and C are the dimensionality of feature and label space, respectively. Accordingly, the feature matrix $\mathbf{X}$ and the label matrix $\mathbf{Y}$ can be represented as

$$\begin{aligned}
\mathbf{X} &= [\mathbf{x}_1, ..., \mathbf{x}_i, ..., \mathbf{x}_N]^T \in R^{N \times D} \\
\mathbf{Y} &= [\mathbf{y}_1, ..., \mathbf{y}_i, ..., \mathbf{y}_N]^T = [\mathbf{y}^1, ..., \mathbf{y}^j, ..., \mathbf{y}^C] \in \{0, 1\}^{N \times C}
\end{aligned} \tag{1}$$

where the column vector $\mathbf{y}^j$ indicates the j-th label in $\mathbf{Y}$.

Our goal is to find the label subset $\mathbf{Y}_s$, which is depicted as a low-dimensional label subset matrix:

$$\mathbf{Y}_s = [\tilde{\mathbf{y}}_1, ..., \tilde{\mathbf{y}}_i, ..., \tilde{\mathbf{y}}_N]^T = [\tilde{\mathbf{y}}^1, ..., \tilde{\mathbf{y}}^j, ..., \tilde{\mathbf{y}}^c] \in \{0, 1\}^{N \times c} \tag{2}$$

where c is the number of selected labels $(c < C)$.

Ant colony optimization algorithm is used to distinguish the best equilibrium proof between different factors in order to limit the search space in the derivation of the final solution. It borrows from the profound strategies of ants in finding food sources, and is a tool based on meta-heuristic swarm [8].

The main objective of reinforcement learning is to maximize the reward in a complex and changeable environment. In the process of reinforcement learning, agents constantly interact with the environment to obtain the state of the environment, and then put the decision into the environment [18].

Generally speaking, reinforcement problems are expressed as Markov decision processes. It consists of four parts: a set of states (S), a set of actions (A), state transition function (T), and reward function (R).

3 The Proposed Method

In this section, based on existing research on label selection, we propose a novel ant colony optimization (ACO) with reinforcement learning (RL) for label selection method (LS-AntRL), which includes two key phases: to select a label subset which contains the most information, and to design a binary weighted neural network for recovering complete label space.

We will discuss the ACO properties and how to model the label selection search space into a Markov Decision Process (MDP). The required properties of an MDP can be satisfied as follows: the labels represent the states (S), and selecting the unvisited labels by each ant represents a set of actions (A). The state transition rule of the ant colony optimization forms the transition function (T). The reward function (R) is composed of two criteria. Next, we will provide a detailed description of the algorithm proposed in this paper.

3.1 Label Selection Algorithm Based on Ant Colony Optimization and Reinforcement Learning

In general, artificial ants in an ant colony algorithm traverse a completely connected undirected graph, where graph nodes are labels, and edges represent connections between labels. Each ant deposits pheromone values on the nodes (labels) of the graph and forms a d-dimensional vector τ called pheromone trail. We try to initialize the pheromone vector by calculating the maximum mutual information (MI) value between labels and features:

$$\tau_i = \max_i(MI(\mathbf{Y}_i, \mathbf{X})), \forall i = 1, ..., C. \tag{3}$$

Initially, ants are placed randomly on the labels, and each ant creates a solution utilizing greedy algorithms and probabilistic state transition rules while visiting different labels in the search space. At each stage, the ant visits the next label i by using the law of choice of probable or greedy action, as follows:

$$P_i^k(t) = \begin{cases} \frac{[\tau_i(t)][V_i(t)]^\beta}{\sum_{u \in N^k}[\tau_u(t)][V_u(t)]^\beta}, & \forall i \in N^k, \text{if } q > q_0 \\ 0, \text{otherwise} \end{cases} \tag{4}$$
$$i = \arg\max_{u \in N^k}\{[\tau_u][V_u]^\beta\}, \text{if } q \leq q_0$$

where τ_i is the pheromone value of label i, V_i is the state-value function associated with label i (described in detail in the next section), and N^k is the set of unseen labels accessible to ant k. The $\beta \in [0, 1]$ balances the importance between pheromone values and heuristic information. The parameter q is a random variable in the range $[0, 1]$, and q_0 is a constant numeral in the range $[0, 1]$. The two parameters q and q_0 indicate the relative importance of exploitation and exploration.

We expect ants to explore more in the initial repetitions and have more exploitation at the end of iterations. Coefficient q is set by (5), which controls the relative importance between exploration and exploitation through the number of iterations, in the following form:

$$q = \left(1 - \frac{t}{SL}\right)^{speed} \tag{5}$$

where *speed* controls the decay speed, SL is the number of labels each ant should visit in each iteration, and parameter t is the current iteration number.

The heuristic information used by the ACO algorithm usually does not change during the search process. Here, we introduce RL to try to learn the heuristic from the experience of all ants in each iteration.

As mentioned earlier, the ACO algorithm search process is modeled as an MDP. Many techniques can be used to solve MDP. In this paper, TD (0) technique is used to solve the problem, which only estimates the value of the state one step in advance:

$$V(s_t) = V(s_t) + \alpha[r_{t+1} + \gamma V(s_{t+1}) - V(s_t)] \tag{6}$$

where $\alpha \in [0, 1]$ is the learning rate, specifying how much error to accept in each step. $\gamma \in [0, 1]$ is the discount factor, which specifies the degree of influence of the next state.

We initialize the state-value vector V, whose value is the maximum cosine similarity between each label and feature, namely:

$$V_i = \max_j(\phi_{i,j}), \forall i = 1, ..., C \tag{7}$$

ϕ is the reward function described below.

The reward function consists of two metrics. The first function calculates the cosine similarity between label vector $\mathbf{y}$ and feature vector $\mathbf{x}$, and is represented as ϕ:

$$\phi(\mathbf{y}, \mathbf{x}) = \frac{|\mathbf{y}^T \mathbf{x}|}{\|\mathbf{y}\|_2 \|\mathbf{x}\|_2}. \tag{8}$$

The second function calculates the redundancy between labels using Pearson coefficient, and is indicated by φ:

$$\varphi(\bar{\mathbf{y}}_i, \bar{\mathbf{y}}_j) = \frac{|\bar{\mathbf{y}}_i^T \bar{\mathbf{y}}_j|}{\|\bar{\mathbf{y}}_i\|_2 \|\bar{\mathbf{y}}_j\|_2} \tag{9}$$

which $\bar{\mathbf{y}}_i$ and $\bar{\mathbf{y}}_j$ is two centered label vectors.

According to (6), when the agent goes from state s_t to s_{t+1} by taking action a, the formula for calculating the reward r_{t+1} is as follows:

$$r_{t+1} = \frac{\max_{s_{t+1}}(\phi_{s_{t+1}, D})}{1 + \varphi_{s_t, s_{t+1}}} \tag{10}$$

where $\phi_{s_{t+1}, D}$ is the label-feature correlation values of the next label s_{t+1} and all its corresponding features, and $\varphi_{s_t, s_{t+1}}$ is the label redundancy between the current label s_t and the next label s_{t+1}.

In every iteration, the ants update the state-value vector V locally as they build the solution. Then, at the end of repetition, the V vector of each ant k is averaged to update the state-value vector V_{s_t} globally:

$$V_{s_t} = \frac{1}{n_{ant}} \sum_{k=1}^{n_{ant}} V_{s_t}^k \tag{11}$$

where n_{ant} is the count of ants that make the solution.

We define a d-dimension vector called LC that increases the value of LC corresponding to the label by one when each ant visits a label. After ants has fully completed their solutions, the following global pheromone update rule applies to update the pheromone value for each label i:

$$\tau_i(t + 1) = (1 - \rho)\tau_i(t) + \frac{LC(i)}{\sum\limits_{j=1}^{d} LC(j)} \tag{12}$$

Algorithm 1. Label subset selection algorithm

Input: Feature matrix **X**, Label matrix **Y**.
Output: Pheromone values for each label.
Process:
1: Calculate ϕ and φ according to (8) and (9)
2: Initialize pheromones τ_i and state-value vectors V_i according to (3) and (7)
3: **for** m=1 *in nCycle* **do**
4:     Initialize LC for each label
5:     Generate ants and place them randomly on the labels
6:     **for** i=1 *in nAnt* **do**
7:         Generates an integer between $\frac{8}{C}$ and $\frac{6}{C}$ assigned to SL
8:         **for** j=1 *in SL* **do**
9:             a = Select the next label according to the state transition rule (4)
10:             Take action a; Observe the reward r and the next state s_{t+1}
11:             Each ant locally updates the state-value vector V by (6)
12:             Increment the LC for the visited label $\mathbf{y}_i$
13:         **end for**
14:     **end for**
15:     Update the state-value vector V globally by applying (11)
16:     Update the pheromone vector according to (12)
17: **end for**
18: Sort labels in descending order according to their pheromone values

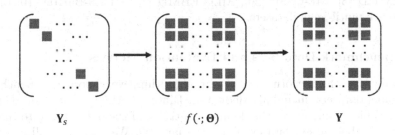

Fig. 1. The architecture of label reconstruction binary neural network.

where ρ is the pheromone decay rate. Labels with a large number of pheromones are informative labels, so c $(c < C)$ top labels are selected according to pheromone values to determine the optimal subset of labels. Please refer to **Algorithms 1** for a sketch of our approaches.

3.2 Label Recovery Operator Based on Binary Neural Network

One of the important steps of our proposed framework is to recover the original space by propagating the predicted value of the selected subset of labels to the entire label set. Therefore, our goal is to find a neural network to obtain the complete predictive label space. This paper attempts to binarize real weights used in forward propagation and backward propagation calculations in neural networks [7], so as to change these multiplications into addition and subtraction operations. Figure 1 shows the architecture of our network.

The binarization operation converts the real-valued weights to the two possible values, and a deterministic binarization operation is based on the signed function:

$$w_b = \begin{cases} +1, & \text{if } w \geq 0 \\ -1, & \text{otherwise} \end{cases} \tag{13}$$

where w_b is the binary weight, and w is the real weight.

We define a binary weighted neural network $f(\cdot; \Theta)$ used for label recovery (parameterized by Θ), and our loss function is defined as follows:

$$Q(\Theta) = \|f(\mathbf{Y}_s; \Theta) - \mathbf{Y}\|_F^2. \tag{14}$$

Please refer to [7] to see the details of neural network parameter updates. Once we obtain the trained binary weighted neural network, we can easily obtain a complete prediction label space.

4   Experiments

In this section, we evaluate the proposed method with four existing methods: MLC-BMaD [3], ML-CSSP [20], MLC-EBMD [4] and LS-BaBID [10] on five benchmark mutil-label datasets.

4.1   Benchmark Data Sets and Evaluation Metrics

For evaluating the performance of our algorithm, we choose five benchmark multi-label datasets, including Mediamill, Bibtex, CAL500, Chess and Corel5k from [1]. Table 1 summarizes the detailed statistics of the datasets. Note that the number of labels in all datasets is more than 100. We evaluated all compared algorithms using two metrics that fit the high-dimensional label applications. Precision@n, and (DisCounted Gain) DCG@n ($n = 1, 2, 3, ...$) are utilized in our experiments. Their detailed definitions are:

$$\text{Precision@}n := \frac{1}{C} \sum_{l \in rank_n(\hat{y})} y_i$$

$$\text{DCG@}n := \sum_{l \in rank_n(y)} \frac{y_i}{\log(i+1)} \tag{15}$$

where $rank_n(y)$ returns the top n label indexes of $\mathbf{y}$. Finally, their average values are calculated via averaging them over all testing instances. Additionally, the higher these two metric values are, the better the label selection techniques perform.

---

[1] https://mulan.sourceforge.net/datasets-mlc.html.

Table 1. Dataset statistics.

Dataset	Domain	#Train	#Test	#Feature	#Label	#Cardinality
Mediamill	Video	30993	12914	120	101	4.376
Bibtex	Text	4880	2515	1836	159	2.402
CAL500	Music	300	202	68	174	26.044
Chess	Text	1508	168	585	277	2.418
Corel5k	Image	4500	500	499	374	3.522

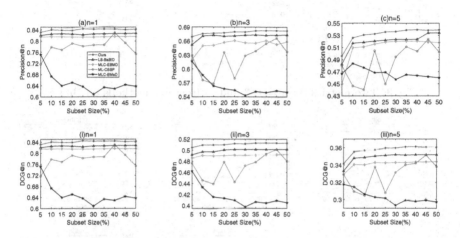

Fig. 2. Two metrics (at n = 1, 3 and 5) from five methods on Mediamill.

4.2 Experimental Settings

In our experiments, random forest with 100 trees is considered as our baseline classifier. We reduce label dimensionality with the step 5% from 5% to 50% of the number original labels (i.e., C) to verify and compare how the number of reduced labels (i.e., c) would affect classification performance after label reduction (i.e., c/C). For MLC-BMaD, its threshold for constructing label association matrix is tuned to be 0.7. For two metrics (15), we set $n = 1$, 3, and 5.

For our method, learning rate α and discount rate γ of TD (0) are set to 0.5 and 0.8. Pheromone decay rate ρ is set to 0.2. The number of iteration that the algorithm should repeat is set to 40 and the number of ants that search the features space is 10. The range of labels traversed by each ant SL is limited to $[\frac{8}{C}, \frac{6}{C}]$. For the *speed* of decay in (5), we set *speed* = 1. The trade off between pheromone and heuristic information β is tuned to be 1.

4.3 Results and Discussion

Figures 2, 3, 4, 5 and 6 show the experimental results on five datasets, Mediamill, Bibtex, CAL500, Chess and Corel5k. From these five figures, the proposed approach in most of label proportions works best compared with four other methods.

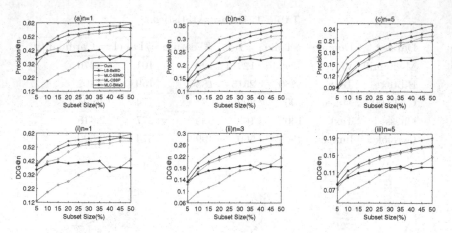

Fig. 3. Two metrics (at n = 1, 3 and 5) from five methods on Bibtex.

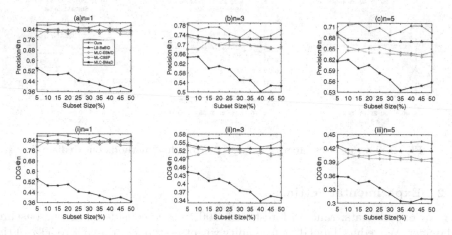

Fig. 4. Two metrics (at n = 1, 3 and 5) from five methods on CAL500.

Specifically, the performance of ML-CSSP and MLC-BMaD is unstable. Overall, MLC-EBMD perform better than MLC-CSSP and ML-BMaD. Both LS-BaBID achieved better results than MLC-EBMD, mainly due to the fact that LS-BaBID not only achieved precise approximation by removing some uninformative labels, but also applied a sequential backward selection (SBS) strategy to remove some less informative labels one by one. At most of label proportions, our method works best, compared with four existing methods, whose main reason is that we use heuristic learning to significantly improve the quality of the selected labels.

In order to compare the five methods more accurately, we use the "win" index to represent the number of times each technique achieved the best metric value. As shown in Table 2, our LS-AntRL reaches the best value of 293 times for two metrics over five datasets while the other four methods achieved a total of 7 wins.

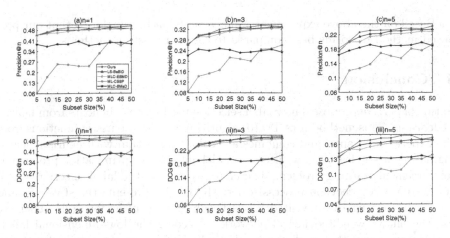

Fig. 5. Two metrics (at n = 1, 3 and 5) from five methods on Chess.

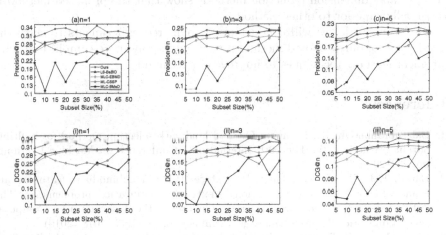

Fig. 6. Two metrics (at n = 1, 3 and 5) from five methods on Corel5k.

Table 2. The number of wins for each method and metric across five datasets.

Metric	LS-BaBID	ML-CSSP	MLC-BMaD	MLC-EBMD	LS-AntRL (ours)
Precision@1	1	0	0	0	**59**
Precision@3	2	0	0	0	**58**
Precision@5	0	0	0	0	**60**
DCG@1	1	0	0	0	**59**
DCG@3	1	1	0	0	**58**
DCG@5	0	1	0	0	**59**
Total wins	5	2	0	0	**293**

Based on the above experimental analysis, it can be concluded that our proposed method performs best compared to the four existing methods.

5   Conclusion

In this paper, we propose a new label selection method that benefits from heuristic learning. This method uses TD (0) algorithm to learn heuristic functions from the experience of ants, and combines ant colony algorithm with heuristic learning. For heuristic learning, we need to model ant colony optimization problem as reinforcement learning problem. Therefore, we modify the label selection search space to Markov decision process, where the label represents the state (S) and each ant selecting the unvisited labels becomes a set of actions (A). Then we use the binary weighted neural network to recover the low-dimensional labels back to the original high-dimensional labels. The experimental results of five different label dimension reduction methods show that the proposed method is statistically superior to other methods.

On future work, we will conduct more compared experiments to show the effectiveness of our proposed method further and apply more complicated reinforcement learning techniques to improve our technique.

References

1. Balasubramanian, K., Lebanon, G.: The landmark selection method for multiple output prediction. In: The 29th International Conference on Machine Learning, pp. 283–290 (2012)
2. Barezi, E.J., Wood, I.D., Fung, P., Rabiee, H.R.: A submodular feature-aware framework for label subset selection in extreme classification problems. In: The 2019 Conference of the North American Chapter of the Association for Computational Linguistics: Human Language Technologies, pp. 1009–1018 (2019)
3. Bi, W., Kwok, J.: Efficient multi-label classification with many labels. In: The 30th International Conference on Machine Learning, pp. 405–413 (2013)
4. Cabral, R., De la Torre, F., Costeira, J.P., Bernardino, A.: Matrix completion for weakly-supervised multi-label image classification. IEEE Trans. Pattern Anal. Mach. Intell. **37**(1), 121–135 (2014)
5. Charte, F., Rivera, A.J., Del Jesus, M.J.: Multilabel Classification: Problem Analysis, Metrics and Techniques. Springer, Cham (2016). https://doi.org/10.1007/978-3-319-41111-8
6. Chen, Y.N., Lin, H.T.: Feature-aware label space dimension reduction for multi-label classification. In: The 26th Annual Conference on Neural Information Processing Systems, vol. 25, pp. 1529–1537 (2012)
7. Courbariaux, M., Bengio, Y., David, J.P.: BinaryConnect: training deep neural networks with binary weights during propagations. In: The 28th Annual Conference on Neural Information Processing Systems, pp. 3123–3131 (2015)
8. Dorigo, M., Stützle, T.: Ant colony optimization: overview and recent advances. In: Gendreau, M., Potvin, JY. (eds.) Handbook of Metaheuristics. International Series in Operations Research & Management Science, vol. 146, pp. 227–263. Springer, Boston (2010). https://doi.org/10.1007/978-1-4419-1665-5_8

9. Tai, F., Lin, H.T.: Multilabel classification with principal label space transformation. Neural Comput. **24**(9), 2508–2542 (2012)
10. Hsu, D.J., Kakade, S.M., Langford, J., Zhang, T.: Multi-label prediction via compressed sensing. In: The 22nd Annual Conference on Neural Information Processing Systems, pp. 772–780 (2009)
11. Ji, T., Li, J., Xu, J.: Label selection algorithm based on Boolean interpolative decomposition with sequential backward selection for multi-label classification. In: The 16th International Conference on Document Analysis and Recognition, pp. 130–144 (2021)
12. Katakis, I., Tsoumakas, G., Vlahavas, I.: Multilabel text classification for automated tag suggestion. In: The 2008 European Conference on Machine Learning and Principles and Practice of Knowledge Discovery in Databases Discovery Challenge, pp. 75–83 (2008)
13. Li, J., Zhang, C., Zhu, P., Wu, B., Chen, L., Hu, Q.: SPL-MLL: selecting predictable landmarks for multi-label learning. In: Vedaldi, A., Bischof, H., Brox, T., Frahm, J.-M. (eds.) ECCV 2020. LNCS, vol. 12354, pp. 783–799. Springer, Cham (2020). https://doi.org/10.1007/978-3-030-58545-7_45
14. Liu, L., Tang, L.: Boolean matrix decomposition for label space dimension reduction: method, framework and applications. J. Phys. Conf. Ser. **1345**, 052061 (2019)
15. Liu, S., Song, X., Ma, Z., Ganaa, E.D., Shen, X.: MoRE: multi-output residual embedding for multi-label classification. Pattern Recogn. **126**, 108584 (2022)
16. Liu, W., Wang, H., Shen, X., Tsang, I.W.: The emerging trends of multi-label learning. IEEE Trans. Pattern Anal. Mach. Intell. **44**(11), 7955–7974 (2021)
17. Miettinen, P., Neumann, S.: Recent developments in Boolean matrix factorization. In: The 29th International Joint Conference on Artificial Intelligence, pp. 4922–4928 (2020)
18. Sutton, R.S., Barto, A.G.: Reinforcement Learning: An Introduction, 2nd edn. MIT Press, Cambridge (2018)
19. Theodoridis, S., Koutroumbas, K.: Pattern Recognition, 3rd edn. Elsevier, Wiley, New York (2006)
20. Wang, X., Zhang, W., Zhang, Q., Li, G.Z.: MultiP-SChlo: multi-label protein subchloroplast localization prediction with Chou's pseudo amino acid composition and a novel multi-label classifier. Bioinformatics **31**(16), 2639–2645 (2015)
21. Wicker, J., Pfahringer, B., Kramer, S.: Multi-label classification using Boolean matrix decomposition. In: The 27th Annual ACM Symposium on Applied Computing, pp. 179–186 (2012)
22. Xu, J., Mao, Z.H.: Multilabel feature extraction algorithm via maximizing approximated and symmetrized normalized cross-covariance operator. IEEE Trans. Cybern. **51**(7), 3510–3523 (2021)
23. Yu, H.F., Jain, P., Kar, P., Dhillon, I.: Large-scale multi-label learning with missing labels. In: The 31st International Conference on Machine Learning, pp. 593–601 (2014)
24. Zhang, Y., Schneider, J.: Multi-label output codes using canonical correlation analysis. In: The 14th International Conference on Artificial Intelligence and Statistics, pp. 873–882 (2011)

Reversible Data Hiding Based on Adaptive Embedding with Local Complexity

Chao Wang[ID], Yicheng Zou[ID], Yaling Zhang, Ju Zhang[ID], Jichuan Chen,
Bin Yang, and Yu Zhang[✉]

College of Computer and Information Science, Southwest University, Chongqing
400715, China
zhangyu@swu.edu.cn
http://www.swu.edu.cn/

Abstract. In recent years, most reversible data hiding (RDH) algorithms have considered the impact of texture information on embedding performance. The distortion caused by embedding secret data in the image's smooth region is much less than in the non-smooth region. It is because embedding secret data in the smooth region corresponds to fewer invalid shifting pixels (ISPs) in histogram shifting. However, though effective, the local complexity is not calculated precisely enough, which results in inaccurate texture division and does not considerably reduce distortion. Therefore, a new RDH scheme based on adaptive embedding with local complexity (AELC) is proposed to improve the embedding performance effectively. Specifically, the cover image is divided into two subsets by the checkerboard pattern. Then the local complexity of each pixel is computed by the correlation between adjacent pixels (CBAP). Finally, secret data are adaptive and preferentially embedded into the regions with lower local complexity in each subset. Experimental results show that the proposed algorithm performs best regarding invalid shifted pixels, maximum embedding capacity (EC), and peak signal-to-noise ratio (PSNR) compared to some state-of-the-art RDH methods.

Keywords: Reversible data hiding · Invalid shifting pixels · Local complexity · Histogram shifting

1 Introduction

With the rapid development of the internet, digital communication has become one of the most important ways for people to share and obtain information. At the same time, the issue of protecting private information security has garnered attention. Therefore, scholars have proposed several different information hiding techniques [13] to ensure the security of private information, such as digital watermarking [12] and steganography [14]. However, these techniques introduce distortion to the cover image, which cannot be recovered losslessly. Even minor distortions are unacceptable in the military, medical and judicial fields, where

B. Luo et al. (Eds.): ICONIP 2023, LNCS 14451, pp. 522–534, 2024.
https://doi.org/10.1007/978-981-99-8073-4_40

the accuracy of the carrier images is crucial. Nevertheless, RDH [15] can achieve error-free secret data extraction and lossless restoration of the original cover image. RDH can be classified into two categories based on whether the cover image is encrypted: RDH in the encrypted domain [5,11,25], and RDH in plaintext domain [3,4,16].

In the past two decades, RDH methods have been extensively studied in the plaintext domain. The lossless compression algorithm [24] is the earliest method of RDH. Afterwards, Tian [19] proposed the difference expansion (DE) technique, which expands the difference between a pair of pixels to embed secret data reversibly. Compared to RDH based on lossless compression, Tian's method provides higher EC and improves PSNR. Ni et al. [9] proposed the histogram shifting (HS) method in 2006, which embeds secret data into the peak point pixels of the image gray histogram and shifts others. However, the embedding capacity of this method is relatively lower. As the combination of HS and DE, prediction error expansion (PEE) was proposed by Thodi and Rodriguez in [17,18]. PEE utilizes the median edge detection (MED) predictor [21] to calculate the prediction error value of pixels. These values are then used to generate the prediction error histogram (PEH) and achieve reversible embedding through histogram shifting. PEE utilizes the local correlation between multiple adjacent pixels. Thus a better performance can be expected.

Li et al. [8] improved the conventional PEE scheme by introducing complexity measurement which sorts pixels based on the correlation between the target pixel and its adjacent pixels. Weng et al. [22] effectively distinguished different texture regions of images by two well-designed features. Fan et al. [2] proposed an average-difference-based complexity computation to cooperate with flexible patch moving to improve the embedding performance further. Peng et al. proposed an improved PVO-based method in [10], which enables more smoothing blocks to be used for embedding secret data. Wang et al. [20] designed a dynamic blocking strategy to preferentially divide the smooth region into smaller blocks to retain high EC, whereas non-smooth areas are divided into larger blocks to avoid decreasing PSNR.

With the above analysis, the impact of image texture information on embedding performance has become the focus of researchers' attention. Jia et al. [6] showed that the image distortion generated by embedding secret data based on HS could be split into two segments: the distortion caused by both valid shifting pixels (VSPs) and invalid shifting pixels (ISPs), which shown in Fig. 1. For given secret data, the distortion caused by VSPs is a fixed value. ISPs cause the real reason for image distortion. Based on this, they preferentially select smooth pixels to carry the secret data, thereby reducing the ISPs. However, Jia et al. [6] used only four adjacent pixels to calculate the target pixel's local complexity and the fluctuation value. Chu et al. [1] used the sum of the differences between four adjacent pixels to calculate the fluctuation values.

After our investigation and analysis of existing literature, We found that most scholars did not sufficiently consider the CBAP, which resulted in inaccurate measurement of image texture and insignificant reduction of ISPs, thus not

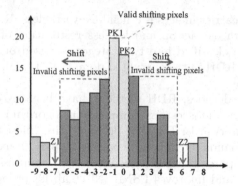

Fig. 1. An instance of division of VSPs and ISPs

substantially reducing the image distortion. This paper proposed a new RDH scheme based on adaptive embedding with local complexity to solve the above problems. Specifically, first, the cover image is divided into two subsets by the checkerboard pattern. Then we designed an effective local complexity calculation scheme to improve the local complexity's accuracy. Finally, we designed a predictor to improve the EC. Moreover, combined with the local complexity of pixels, the secret data are preferentially embedded in pixels with small local complexity to reduce distortion. The main contributions of this paper are summarized as follows.

(1) A novel variance-based local complexity calculation scheme is proposed that adopts the CBAP of cover images, which more precisely delineates pixels with lower local complexity and reduce the distortion caused by data embedding.

(2) We design a new predictor to improve the prediction performance by calculating the weighted geometric mean of neighboring pixels to reduce the effect of extreme values.

(3) To make it easier for other scholars to verify the work, we have uploaded all the code and other materials at https://github.com/WangChaoer/AELC.

The rest of this paper is organized as follows. The details of the proposed method are described in Sect. 2. Section 3 provides the experimental details and results of AELC. Finally, a conclusion is given in Sect. 4.

2   Proposed Method

In this section, we provide the implementation details of the proposed method. Figure 2 illustrates a brief flowchart of the RDH scheme proposed in this paper. In the data embedding procedure, the cover image is first processed with overflow/underflow, and then divided into two subsets by the checkerboard pattern. Next, secret data are embedded into each subset to obtain the stego-image. In the data extraction and recovery procedure, first, the stego-image is divided into two subsets by the checkerboard pattern, then secret data are extracted, and pixel values are recovered for each subset separately. Finally, the overflow/underflow pixels are recovered to obtain the original cover image.

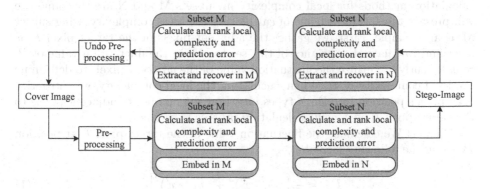

Fig. 2. The brief flowchart of the proposed AELC algorithm in this paper

2.1   Image Pre-processing

To avoid the overflow/underflow issue during embedding, we modify the pixel with a value of 0 or 255. The specific operation involves changing 0 to 1 and changing 255 to 254. Additionally, we mark these modified pixels as 1 in the location map. For the remaining pixels, we mark them as 0 in the same location map. Subsequently, the location map is losslessly compressed to reduce its size and embedded in the cover image as a part of the secret data.

As shown in Fig. 3, the cover image is partitioned into two subsets, M and N, using a checkerboard pattern. Subset M is indicated in blue, while subset N is represented in white.

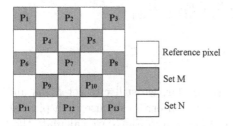

Fig. 3. An example of allocating pixels with checkerboard (Color figure online)

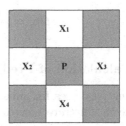

Fig. 4. Adjacent pixels of P

2.2 Local Complexity Calculation

Figure 3 and Fig. 4 depict the precise division of the target pixel P and its adjacent pixels $\{x_1, x_2, x_3, x_4\}$. In Fig. 3, we focus solely on calculating the fluctuation value of the reference pixels without calculating their local complexity. Since the calculation methods for local complexity in subsets M and N are the same, we will present the specific scheme of calculating the local complexity using subset M as an example. First, we gather the adjacent pixels of the target pixel P as a set and calculate the variance of this set to obtain the fluctuation value of P. Subsequently, we utilize the fluctuation value of the adjacent pixels to determine the local complexity of P. When calculating the local complexity, we categorize each pixel's position into three types based on the number of adjacent pixels to the target pixel. The specific calculation process is as follows:

Case 1: We calculate the fluctuation value of the target pixel P at position P_1 using the following formula:

$$F_{P_1} = \frac{1}{2}\left[(x_3 - x')^2 + (x_4 - x')^2\right] \tag{1}$$

Case 2: The fluctuation value of the target pixel P at position P_2 is computed as follows:

$$F_{P_2} = \frac{1}{3}\left[(x_2 - x')^2 + (x_3 - x')^2 + (x_4 - x')^2\right] \tag{2}$$

Case 3: The fluctuation value of the target pixel P at position P_4 is determined by the following calculation:

$$F_{P_4} = \frac{1}{4}\left[(x_1 - x')^2 + (x_2 - x')^2 + (x_3 - x')^2 + (x_4 - x')^2\right] \tag{3}$$

where x' denotes the average value of the pixels surrounding the target pixel P.

In order to accurately measure the local complexity of the pixel, we utilize the fluctuation values of the adjacent pixels of the target pixel to calculate its local complexity. Taking the pixel value in Fig. 3 as an example, the computation of the local complexity Ω for the pixel point P_4 is as following:

$$\Omega_{P_4} = \left[\frac{F_{P_4} + F_{P_1} + F_{P_2} + F_{P_6} + F_{P_7}}{5}\right] \tag{4}$$

2.3 Calculation of Prediction Errors

Since the processes for subsets M and N are similar, we still cite set M as an instance to illustrate how to calculate the prediction error. We use each pixel's four nearest neighbors to make a prediction. As shown in Fig. 3, we calculate the prediction error for all pixels in each subset, excluding the reference pixels.

Taking the target pixel P in Fig. 4 as an example, we calculate its predicted pixel value as follows:

$$P' = x_1^{f_1} x_2^{f_2} x_3^{f_3} x_4^{f_4} \tag{5}$$

where P' is the predicted value of pixel P and f_i represents the weight assigned to the corresponding pixel P_i, and $\sum_{i=1}^{4} f_i = 1$.

The prediction error is calculated by

$$Pe = P - P' \tag{6}$$

The solution for the four weights f_i in Eq. (5) is shown below:

First, we calculate the average value of the four pixel values adjacent to the target pixel P.

$$ave = \left\lfloor \frac{x_1 + x_2 + x_3 + x_4}{4} \right\rfloor \tag{7}$$

where $\lfloor \cdot \rfloor$ represents the floor function.

Then, calculate the distance between each pixel x_i and the average value ave.

$$e_i = |ave - x_i| \tag{8}$$

Finally, The corresponding weights are calculated as follows:

$$f_i' = \begin{cases} \frac{1}{4}, & \text{if } \sum_{i=1}^{4} |e_i| = 0 \\ \frac{\sum_{i=1}^{4} |e_i|}{1+|e_i|}, & otherwise \end{cases} \tag{9}$$

where $i \in [1, 4]$, normalize f_1', f_2', f_3' and f_4' to get f_1, f_2, f_3 and f_4 which represent the the weights for upper, left, right, and lower pixels, respectively.

2.4 The Data Embedding Procedure

In this procedure, the secret data is divided into two halves, with one half being embedded in subset M and the other half being embedded in subset N. The detailed embedding process is depicted as follows:

Step 1: The cover image is preprocessed according to the method described in Sect. 2.1. Subsequently, using the calculation methods described in Sects. 2.2 and 2.3, the local complexity and prediction error values of all pixels in subset M, excluding the reference pixels, are calculated.

Step 2: We sort the local complexity sequence in ascending order to obtain $\{\Omega_{M_1}, \Omega_{M_2}, \Omega_{M_3}, \cdots, \Omega_{M_i}\}$ and the sequence $\{Pe_{M_1}, Pe_{M_2}, Pe_{M_3}, \cdots, Pe_{M_i}\}$ is obtained by sorting the prediction error sequence in ascending order of the local complexity values.

Step 3: The peak points PK_{M1} and PK_{M2}, as well as their corresponding zero points Z_{M1} and Z_{M1}, are calculated by the prediction error histogram of subset M.

Step 4: We sequentially modify the prediction error values in sequence $\{Pe_{M_1}, Pe_{M_2}, Pe_{M_3}, Pe_{M_i}\}$ to embed secret data. The detailed process of modifying Pe is as follows:

$$Pe_i' = \begin{cases} Pe_i + b, & \text{if } Pe_i = \max(PK_{M1}, PK_{M2}) \\ Pe_i - b, & \text{if } Pe_i = \min(PK_{M1}, PK_{M2}) \\ Pe_i + 1, & \text{if } \max(PK_{M1}, PK_{M2}) < Pe_i < \max(Z_{M1}, Z_{M2}) \\ Pe_i - 1, & \text{if } \min(Z_{M1}, Z_{M2}) < Pe_i < \min(PK_{M1}, PK_{M2}) \\ Pe_i, & otherwise \end{cases} \tag{10}$$

where Pe_i represents the prediction error value, Pe'_i is the modified prediction error value and b is secret data, $b \in \{0, 1\}$.

Step 5: The stego-image pixel values are obtained by modifying the corresponding original pixel values in the following way:

$$P''_i = P'_i + Pe'_i \tag{11}$$

where P''_i is the modified pixel value, P'_i is the predicted value.

Step 6: According to (11), the stego-image of subset M is obtained, and on the basis of it, we implement the embedding of subset N in the same manner.

2.5   The Data Extraction and Recovery Procedure

The data extraction and recovery are the reverse process, first extract the secret data from subset N and recover subset N, then do the same on M. The detailed process is described below:

Step 1: Calculate the local complexity and modified prediction error of all pixels in subset N based on the methods described in Sects. 2.2 and 2.3.

Step 2: The sequence of local complexity is arranged in ascending order to obtain $\{\Omega_{N_1}, \Omega_{N_2}, \Omega_{N_3} \cdots, \Omega_{N_i}\}$, sorting the prediction error sequence according to the ascending order of the local complexity to obtain the modified prediction error sequence $\{Pe'_{N_1}, Pe'_{N_2}, Pe'_{N_3}, \cdots, Pe'_{N_i}\}$.

Step 3: The secret data in sequence $\{Pe'_{N_1}, Pe'_{N_2}, Pe'_{N_3}, \cdots, Pe'_{N_i}\}$ is extracted sequentially until the secret data extraction is complete. The way to extract secret data is as follows:

$$b = \begin{cases} 0, & \text{if } Pe'_i = PK_{N1} \text{ or } e'_i = PK_{N2} \\ 1, & \text{if } Pe'_i = \min(PK_{N1}, PK_{N2}) - 1 \\ & \quad \text{or } Pe'_i = \max(PK_{N1}, PK_{N2}) + 1 \end{cases} \tag{12}$$

where PK_{N1}, PK_{N2}, and Z_{N1}, Z_{N2} are the same as that in the embedding procedure, and b is the extracted secret data.

Step 4: In the process of extracting secret data and recovering the cover image, the prediction error is recovered as follows:

$$Pe_i = \begin{cases} Pe'_i - 1, & \text{if } \max(PK_{N1}, PK_{N2}) < Pe'_i \leq \max(Z_{N1}, Z_{N2}) \\ Pe'_i + 1, & \text{if } \min(Z_{N1}, Z_{N2}) \leq Pe'_i < \min(PK_{N1}, PK_{N2}) \\ Pe'_i, & \text{otherwise} \end{cases} \tag{13}$$

where Pe_i is the recovered prediction error.

Step 5: Correspondingly, the original pixel values are recovered as:

$$P_i = P'_i + Pe_i \tag{14}$$

where P_i is the recovered pixel value, P'_i is the predicted pixel value.

Step 6: After the extraction and recovery of subset N are completed, the extraction and recovery of subset M are performed similarly. In addition, the pixel values are recovered based on the decompressed location map. When the location map is marked as 1, the value 254 is changed back to 255, and the value 1 is changed back to 0. Finally, the entire image is fully recovered.

3   Experimental Results and Analyses

In this section, the efficiency of the proposed method is examined utilizing six grayscale images of the size 512×512, including Lena, Baboon, F-16, Boat, Pepper and Elaine from USC-SIPI database shown in Fig. 5. In the following sections, we present the experimental results of our proposed method and compare them with the results of some state-of-the-art schemes.

Fig. 5. Six 512 * 512 test images (Lena, Baboon, F-16, Boat, Pepper and Elaine)

3.1   The Visual Quality of Stego-Image

RDH in the plaintext domain typically uses visual quality (VQ) and EC as important metrics for evaluating the algorithm. The visual security of the stego-image is commonly assessed objectively using the PSNR. The calculation method of PSNR is as follows:

$$\begin{cases} MSE = \frac{1}{L \times W} \sum_{i=1}^{W} \sum_{j=1}^{L} \left(g_{ij} - g'_{ij} \right)^2 \\ PSNR = 10 \times \log_{10} \frac{255^2}{MSE} \end{cases} \quad (15)$$

Fig. 6. Six stego-images when the embedding capacity is maximum

Where PSNR is measured in dB, L and W represent the length and width of the image, g_{ij} represents the pixel value at position (i, j) of the cover image, and g'_{ij} represents the pixel value at position (i, j) of the stego-image.

Table 1. The maximum EC and PSNR of six test images.

Cover image	Lena	Baboon	F-16	Boat	Pepper	Elaine	Average
EC (bits)	76483	25146	96876	40001	48640	35779	53821
PSNR (dB)	48.85	48.38	49.05	48.51	48.59	48.48	48.64

In this experiment, we set the number of secret data bits to the maximum EC of each image. Table 1 shows the maximum EC and PSNR in our scheme, and its stego-image is shown in Fig. 6. The results clearly demonstrate that it is difficult for the human visual system to detect the difference between the cover images and the stego-images. Hiding data within an image to prevent detection by attackers is precisely the objective of reversible data-hiding techniques.

3.2   Comparison of the ISPs

Table 2 compares the number of ISPs between our proposed method and four other methods [1,6,7,23] on six text images when secret data consists of 10000 bits. For the smooth F-16 image, the number of ISPs is 2823 bits in our proposed method, whereas the other four methods require 18726 bits, 3181 bits, 4461 bits and 3517 bits, respectively. AS for the complex Baboon image, the number of ISPs is 41549 bits in our proposed method, while the other four methods require 98471 bits, 42914 bits, 47760 bits and 58332 bits, respectively. Table 2 shows that the number of ISPs in our proposed method is significantly lower than that in the other four methods for all six test images.

Table 2. The ISPs (bits) of different methods when 10000 bits are embedded.

Method	Lena	Baboon	F-16	Boat	Pepper	Elaine
Jung K H [7]	21377	98471	18726	53832	33282	53037
Jia et al. [6]	14088	42914	3181	32127	23359	23067
Chu et al. [1]	15604	47760	4461	33058	25286	29175
Wu H Z [23]	14932	58332	3517	30139	24346	26305
AELC (Proposed)	**13341**	**41549**	**2823**	**28658**	**22560**	**20916**

3.3   Comparison of Embedding Capacity

Table 3 provides a comparison of EC between the proposed method and three other methods [1,6,7]. The table clearly shows that the proposed method effectively improves the embedding capacity. Specifically, for the texture image Baboon, the embedding capacity is 16271 bits, 25060 bits and 24696 bits in [6,7] and [1], respectively, while the proposed method is 25146 bits. Compared to the three schemes, the proposed method exhibits an increase of 54.54%, 0.34% and 1.82%, respectively. Similarly, for the image Elaine, the proposed method surpasses the other three schemes by 49.37%, 0.41%, and 4.32%, respectively. This improvement is attributed to the proposed method utilization of a double peak selection method and a new predictor. The new predictor enhances the accuracy of predicted pixel values by calculating the weighted geometric mean of adjacent pixels, thereby reducing the impact of extreme values and improving the embedding capacity.

Table 3. Embedding capability for each image by using four schemes.

Method	Lena	Baboon	F-16	Boat	Pepper	Elaine
Jung K H [7]	38084	16271	45222	25475	32225	23953
Jia et al. [6]	76437	25060	96688	39856	48586	35635
Chu et al. [1]	74064	24696	94304	38947	47970	34296
AELC (Proposed)	**76483**	**25146**	**96876**	**40001**	**48640**	**35779**

3.4   Comparison of PSNR

To evaluate the performance of our method, we compare it with the advanced schemes proposed by Jung K H [7], Jia et al. [6], Chu et al. [1] and WU H Z [23]. The relationship between EC and PSNR is depicted in Fig. 7. Our proposed method outperforms the other methods in terms of payloads. For instance, when embedding 10000 bits into Lena, the PSNR of our proposed method is 59.68 dB, while the PSNR of the other methods is 58.16 dB, 59.48 dB, 58.84 dB and 59.16 dB. The improvement in PSNR is 1.52 dB, 0.2 dB, 0.84 dB and 0.52 dB. Similarly, When embedding 10000 bits into the texture image Baboon, our scheme outperforms [1,6,7] and [23] with an increase in PSNR by 3.51 dB, 0.17 dB, 0.61 dB and 1.41 dB. Through comparisons with four state-of-the-art strategies, we conclude that our proposed method can effectively reduce distortion compared to the other methods, whether applied to smooth or unsmooth images.

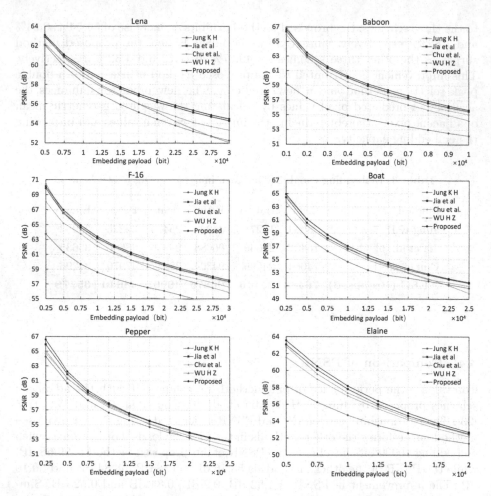

Fig. 7. Comparison of the performance with other schemes.

4   Conclusions

In this paper, we proposed a new RDH scheme called AELC, which aims to reduce the number of ISPs and increase the EC by using image texture information and CBAP. The core idea of AELC method is to partition the carrier image into two subsets using a checkerboard pattern. The local complexity of each pixel in each subset is calculated to determine the degree of smoothness, and the local complexity values are sorted in ascending order. Furthermore, the AELC method combined the prediction error with local complexity to selectively embedded the secret data in the position of the smooth pixel. It effectively reduces the number of ISPs and increases the EC of the method. A large number of experiments prove that the method proposed in this paper has very good capacity-distortion performance.

Acknowledgements. This work is supported in part by the National Natural Science Foundation of China (62032019), and the Capacity Development Grant of Southwest University (SWU116007); in part by the Chongqing Industrial Control System Information Security Technology Support Center; in part by the Chongqing Intelligent Instrument and Control Equipment Engineering Technology Research Center.

References

1. Chu, L., Wu, H., Zeng, Y., Tang, X.: Improved weighted average-based rhombus predictor in reversible data hiding using prediction error expansion. In: 2021 IEEE 6th International Conference on Computer and Communication Systems (ICCCS), pp. 6–11. IEEE (2021)
2. Fan, G., Pan, Z., Zhou, Q., Zhang, X.: Flexible patch moving modes for pixel-value-ordering based reversible data hiding methods. Expert Syst. Appl. **214**, 119154 (2023)
3. Fu, Z., Gong, M., Long, G., Gan, Z., Chai, X., Lu, Y.: Efficient capacity-distortion reversible data hiding based on combining multipeak embedding with local complexity. Appl. Intell. **52**(11), 13006–13026 (2022). https://doi.org/10.1007/s10489-022-03323-8
4. Hu, R., Xiang, S.: CNN prediction based reversible data hiding. IEEE Sig. Process. Lett. **28**, 464–468 (2021)
5. Hua, Z., Wang, Y., Yi, S., Zhou, Y., Jia, X.: Reversible data hiding in encrypted images using cipher-feedback secret sharing. IEEE Trans. Circ. Syst. Video Technol. **32**(8), 4968–4982 (2022)
6. Jia, Y., Yin, Z., Zhang, X., Luo, Y.: Reversible data hiding based on reducing invalid shifting of pixels in histogram shifting. Sig. Process. **163**, 238–246 (2019)
7. Jung, K.H.: A high-capacity reversible data hiding scheme based on sorting and prediction in digital images. Multimedia Tools Appl. **76**, 13127–13137 (2017). https://doi.org/10.1007/s11042-016-3739-x
8. Li, X., Yang, B., Zeng, T.: Efficient reversible watermarking based on adaptive prediction-error expansion and pixel selection. IEEE Trans. Image Process. **20**(12), 3524–3533 (2011)
9. Ni, Z., Shi, Y.Q., Ansari, N., Su, W.: Reversible data hiding. IEEE Trans. Circ. Syst. Video Technol. **16**(3), 354–362 (2006)
10. Peng, F., Li, X., Yang, B.: Improved PVO-based reversible data hiding. Digit. Sig. Process. **25**, 255–265 (2014)
11. Puteaux, P., Ong, S., Wong, K., Puech, W.: A survey of reversible data hiding in encrypted images-the first 12 years. J. Vis. Commun. Image Represent. **77**, 103085 (2021)
12. Quan, Y., Teng, H., Chen, Y., Ji, H.: Watermarking deep neural networks in image processing. IEEE Trans. Neural Netw. Learn. Syst. **32**(5), 1852–1865 (2020)
13. Rustad, S., Andono, P.N., Shidik, G.F., et al.: Digital image steganography survey and investigation (goal, assessment, method, development, and dataset). Sig. Process. **206**, 108908 (2022)
14. Sharafi, J., Khedmati, Y., Shabani, M.: Image steganography based on a new hybrid chaos map and discrete transforms. Optik **226**, 165492 (2021)
15. Shi, Y.Q., Li, X., Zhang, X., Wu, H.T., Ma, B.: Reversible data hiding: advances in the past two decades. IEEE Access **4**, 3210–3237 (2016)

16. Sun, J., Yang, Z., Zhang, Y., Li, T., Wang, S.: High-capacity data hiding method based on two subgroup pixels-value adjustment using encoding function. Secur. Commun. Netw. **2022** (2022)

17. Thodi, D.M., Rodriguez, J.J.: Prediction-error based reversible watermarking. In: 2004 International Conference on Image Processing, ICIP 2004, vol. 3, pp. 1549–1552. IEEE (2004)

18. Thodi, D.M., Rodríguez, J.J.: Expansion embedding techniques for reversible watermarking. IEEE Trans. Image Process. **16**(3), 721–730 (2007)

19. Tian, J.: Reversible data embedding using a difference expansion. IEEE Trans. Circ. Syst. Video Technol. **13**(8), 890–896 (2003)

20. Wang, X., Ding, J., Pei, Q.: A novel reversible image data hiding scheme based on pixel value ordering and dynamic pixel block partition. Inf. Sci. **310**, 16–35 (2015)

21. Weinberger, M.J., Seroussi, G., Sapiro, G.: The LOCO-I lossless image compression algorithm: principles and standardization into JPEG-LS. IEEE Trans. Image Process. **9**(8), 1309–1324 (2000)

22. Weng, S., Tan, W., Ou, B., Pan, J.S.: Reversible data hiding method for multi-histogram point selection based on improved crisscross optimization algorithm. Inf. Sci. **549**, 13–33 (2021)

23. Wu, H.: Efficient reversible data hiding simultaneously exploiting adjacent pixels. IEEE Access **8**, 119501–119510 (2020)

24. Xuan, G., Chen, J., Zhu, J., Shi, Y.Q., Ni, Z., Su, W.: Lossless data hiding based on integer wavelet transform. In: 2002 IEEE Workshop on Multimedia Signal Processing, pp. 312–315. IEEE (2002)

25. Yu, C., Zhang, X., Zhang, X., Li, G., Tang, Z.: Reversible data hiding with hierarchical embedding for encrypted images. IEEE Trans. Circ. Syst. Video Technol. **32**(2), 451–466 (2021)

Generalized Category Discovery with Clustering Assignment Consistency

Xiangli Yang[1], Xinglin Pan[2], Irwin King[3], and Zenglin Xu[4(✉)]

[1] School of Computer Science and Engineering, University of Electronic Science and Technology of China, Chengdu, China
xlyang@std.uestc.edu.cn
[2] Department of Computer Science, Hong Kong Baptist University, Hong Kong, China
csxlpan@comp.hkbu.edu.hk
[3] Department of Computer Science and Engineering, The Chinese University of Hong Kong, Hong Kong, China
king@cse.cuhk.edu.hk
[4] School of Computer Science and Technology, Harbin Institute of Technology, Shenzhen, China
xuzenglin@hit.edu.cn

Abstract. Generalized category discovery (GCD) is an important open-world task to automatically cluster the unlabeled samples using information transferred from the labeled dataset, given a dataset consisting of labeled and unlabeled instances. A major challenge in GCD is that unlabeled novel class samples and unlabeled known class samples are mixed together in the unlabeled dataset. To conduct GCD without knowing the number of classes in unlabeled dataset, we propose a co-training-based framework that encourages clustering consistency. Specifically, we first introduce weak and strong augmentation transformations to generate two sufficiently different views for the same sample. Then, based on the co-training assumption, we propose a consistency representation learning strategy, which encourages consistency between feature prototype similarity and clustering assignment. Finally, we use the discriminative embeddings learned from the semi-supervised representation learning process to construct an original sparse network and use a community detection method to obtain the clustering results and the number of categories simultaneously. Extensive experiments show that our method achieves state-of-the-art performance on three generic benchmarks and three fine-grained visual recognition datasets.

Keywords: Generalized category discovery · Open-world problem · Novel category discovery

B. Luo et al. (Eds.): ICONIP 2023, LNCS 14451, pp. 535–547, 2024.
https://doi.org/10.1007/978-981-99-8073-4_41

1 Introduction

Fig. 1. An illustration of the generalized category discovery task. Given a set of labeled images of some known categories, and a set of unlabeled images that contain both known and novel classes, the objective is to automatically identity known and novel classes in the test dataset.

Deep learning has proven to be remarkably successful in image classification tasks that heavily rely on vast amounts of high-quality labeled data [25,31]. However, acquiring annotations for all data is often unfeasible, which requires learning from unlabeled data. Semi-supervised learning [33,44] and few-shot learning [12,18] are two examples of such learning paradigms. Most of these learning algorithms operate in closed set settings, where known categories are predetermined. In recent years, *Novel Category Discovery* (NCD) [21,23] has received significant attention for the discovery of novel classes by transferring knowledge learned from known classes. The implementation of NCD is based on the assumption that all unlabeled samples pertain to novel categories whose number is predetermined, which is excessively idealistic and impractical in the wild. [39] relaxes these assumptions by recognizing instances of both known and novel classes in the unlabeled dataset. This new setting is named *Generalized Category Discovery* (GCD) [4,16,39]. To illustrate the setting of GCD, we consider the scenario in Fig. 1, where the training set consists of labeled instances of known classes (*e.g.*, cane and cavallo) and unlabeled instances comprising instances of known classes (*i.e.*, cane and cavallo) and novel classes (*i.e.*, gallina and elephant). An ideal model should not only be able to classify known classes (*i.e.*, cane, and cavallo), but also be able to discover novel classes (*i.e.*, gallina, and elephant).

Despite the promising aspects, GCD faces significant obstacles, namely: a) the lack of supervised prior knowledge for novel classes and b) the absence of clear distinctions between samples belonging to known and novel categories. To address these problems, [39] incorporates contrastive loss to learn discriminative representations of the unlabeled dataset. Furthermore, semi-supervised k-means [2] is applied to implement clustering. However, we find that using all unlabeled instances as negative samples in contrastive learning will treat unlabeled instances of the same category as negative pairs, which will cause the category collision problem [46]. Moreover, in real-world problems, prior information regarding the number of clustering assignments is rarely available, which makes it inappropriate to use semi-supervised k-means.

To overcome these limitations, we present a co-training consistency strategy for clustering assignments to uncover latent representations among unlabeled dataset. For the final clustering goal, we leverage *community detection* techniques to assign labels to unlabeled instances and automatically determine the optimal number of clustering categories based on the learned representations. In particular, our approach consists of two stages: semi-supervised representation learning and community detection. In semi-supervised representation learning, we adopt the supervised contrastive loss to derive the labeled information fully. Additionally, since the contrastive learning approach aligns with the co-training assumption, we introduce weak and strong augmentations of the same sample to extract two distinct views. We subsequently deploy the co-training framework to enforce the consistency of feature-prototype similarity and clustering assignment between the two views. In community detection, we construct the primary graph utilizing feature embeddings learned in semi-supervised representation learning, then apply a community detection method to obtain the outcomes. Through experimentation on six benchmarks, our approach demonstrates state-of-the-art performance and confirms the efficacy of our method. Especially in the ImageNet-100 data set, our method significantly exceeds [39] by 15.5% and 7.0% on the Novel and All classes, respectively.

2   Related Work

Novel Category Discovery. The objective of Novel Category Discovery (NCD) is to cluster unlabeled instances using transferable knowledge derived from the labeled dataset of known categories, with all unlabeled instances regarded as novel categories [38]. KCL [27] addresses the problem of transfer learning across domains, especially cross-task transfer learning, which is similar with NCD. DTC [23] utilizes a cross-entropy classifier to initialize representations on the labeled dataset. Then it maintains a list of class prototypes that represent cluster centers, and assigns instances to the closest prototype. In this way, a good clustering representation is learned. RankStats [21,22] applies RotNet [19] to perform an initialization of the encoder. Afterward, it generates pseudo-labels through dual-ranking statistics and enforces consistency between the predictions of the two branch networks. Unlike the methods that separately design loss functions for labeled and unlabeled samples, UNO [17] introduces a unified objective function to discover new classes that effectively supports the collaboration between supervised and unsupervised learning. MEDI [9] proposes a new approach using MAML [18] to solve the NCD problem and provided a solid theoretical analysis of the underlying assumptions in the NCD field. ComEx [43] modifies UNO [17] by designing two groups of experts to learn the entire dataset in a complementary way, thus mitigating the limitations of previous NCD methods. Recently, GCD [39] extends the NCD task to include the recognition of known classes in the unlabeled dataset. It leverages a well-trained visual transformer model [13] to improve visual representation.

Self-supervised Learning. Self-supervised learning [1] can simultaneously utilize supervised and unsupervised learning during the fine-tuning stage of training without the need for manual annotations. Furthermore, self-supervised learning can be divided into two types: auxiliary pretext learning and contrastive learning. When manual annotations are not available, auxiliary pretext tasks are used as supervised information to learn representations [26]. Some commonly used pretext tasks include exemplar-based methods [14], rotation [19], predicting missing pixels [35], and grayscale images [45]. However, designing a suitable auxiliary pretext task requires domain-specific knowledge in order to better serve downstream tasks. Contrastive learning minimizes the latent embedding distance between positive pairs and maximizes the distance between negative pairs [42]. SimCLR [8] learned useful representations based on contrastive loss by maximizing the similarity between the original sample and augmented views of it. MoCo [24] utilized a momentum contrast to calculate the similarity. BYOL [20] proposed a new framework for learning feature representations without the help of contrasting negative pairs. This was achieved by using a Siamese architecture, where the query branch is equipped with a predictor architecture in addition to the encoder and projector.

3   Method

3.1   Preliminaries

GCD aims to automatically classify unlabeled instances containing both known and unknown categories [39]. This is a more realistic open-world setting than the common closed-set classification, which assumes that labeled and unlabeled data belong to the same categories. Our settings follows [39]. Formally, let the train dataset be $\mathcal{D} = \mathcal{D}_L \cup \mathcal{D}_U$, where $\mathcal{D}_L = \{(x_i, y_i)|y_i \in \mathcal{Y}_{known}\}$ and $\mathcal{D}_U = \{x_i|y_i \in \mathcal{Y}_U\}$ and $\mathcal{Y}_U = \{\mathcal{Y}_{known}, \mathcal{Y}_{novel}\}$. Here, $\mathcal{Y}_U, \mathcal{Y}_{known}$ and $\mathcal{Y}_{novel}$ denote the label set of All, Known, and Novel classes, respectively. This formalization enables us to easily distinguish between the NCD setting and GCD setting. In NCD, it is assumed that $\mathcal{Y}_{known} \cap \mathcal{Y}_U = \emptyset$.

To tackle the issue of GCD, we leverage co-trained clustering assignment consistency for acquiring discriminative representations of the unlabeled data. We then employ a customized community detection technique to automatically obtain the clustering results and the number of categories. Specifically, we use supervised contrastive learning to make maximum use of the supervised information for labeled instances. As depicted in Fig. 2, we apply a co-training framework to unlabeled data to ensure the consistency between the clustering assignment and the feature-prototype similarities of the same image. Following the semi-supervised representation learning, we utilize a modified community detection strategy to allocate category labels for each unlabeled instance, regardless if it is a known one or a novel one, and to determine the number of categories.

3.2 Semi-supervised Representation Learning

Supervised Contrastive Learning. We apply supervised contrastive learning to minimize the similarity between embeddings of the same class while maximize the similarity among embeddings from different classes at the same time. Utilizing label information only in the supervised contrastive learning instead of in a cross-entropy loss results in equal treatment of labelled and unlabeled data. The supervised contrastive component nudges the network towards a semantically meaningful representation to reduce overfitting on the labeled classes. The supervised contrastive loss, as in [28,39], is defined as,

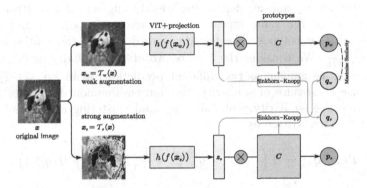

Fig. 2. Diagram of Clustering Assignment Consistency. For original image x, we apply two transformations $\mathcal{T}_w$ and $\mathcal{T}_s$ to obtain two different views x_w and x_s. Then they go through a backbone network, resulting in two embedded representations z_w and z_s. In addition, C is the matrix whose columns are the K trainable prototypes vectors, $\{c_1, c_2, \cdots, c_K\}$. $\otimes$ denotes the dot products of z_i and all prototypes in C. Then we can compute codes (clustering assignments) q_w and q_s by matching these representations to the prototypes vectors.

$$\mathcal{L}_{sup} = -\frac{1}{|\mathcal{N}(i)|} \sum_{p \in \mathcal{N}(i)} \log \frac{\exp\left(sim(z_i, z_p)/\tau_{sup}\right)}{\sum_n \mathbb{1}_{[n \neq i]} \exp\left(sim(z_i, z_n)/\tau_{sup}\right)}, \quad (1)$$

where $z_i = h(f(x_i))$, $f(\cdot)$ is the feature extractor, and $h(\cdot)$ is a multi-layer perceptron (MLP) projection head. $sim(z_i, z_p)$ denote the cosine similarity between z_i and z_p. τ is a temperature value. $\mathcal{N}(i)$ is the indices of other images having the same label as x_i. $\mathbb{1}_{[n \neq i]}$ denotes an indicator function evaluating to 1 if $n \neq i$.

Unsupervised Contrastive Learning. We found that unsupervised contrastive learning in [39] treats different unlabeled samples from the same semantic category as false negatives, resulting in a class collision problem [46]. Inspired by this, we introduce a soft approach to mitigate this problem. In contrast learning [8], two different augmentations generate two different views, and this scenario satisfies the co-training assumption, that is,

$$f_1(v_1) = f_2(v_2), \forall x = (v_1, v_2) \quad \text{(Co-Training Assumption)}, \quad (2)$$

where $f(\cdot), f_1(\cdot), f_2(\cdot)$ refer to different prediction functions. v_1 and v_2 are two different views of x. In this paper, weak augmentation and strong augmentation are utilized to generate two separate views. We rethink both the co-training assumption and the swapped prediction problem discussed in [6]. We consider the probability of feature similarity to the prototypes and promote the minimization of divergence between the soft clustering assignment and the probability in the weakly augmented case. Here, We ignore the case of strong augmentation since the induced distortions could significantly alter the image structures and make it difficult to preserve the identity of the original instances. Additionally, we measure the fit between the feature embeddings and their soft assignments.

Formally, let x_w and x_s denote the weakly augmented and strongly augmented views of the same image x, respectively. $\{c_1, c_2, \cdots, c_K\}$ is a set of K learnable prototypes. The feature is projected to the unit sphere, $i.e.$, $z_w = \frac{f(x_w)}{\|f(x_w)\|_2}$. We consider the feature-prototype similarity probability and the clustering assignment as two different predictions as in Eq. (2). Following [36], We use a measure of similarity, the Jensen-Shannon divergence between feature-prototype similarity probability p_w and clustering assignment q_w, the consistency loss is written as

$$\mathcal{L}(z_w, q_w) = H\left(\frac{1}{2}(p_w^{(k)} + q_w^{(k)})\right) - \frac{1}{2}\left(H(p_w^{(k)}) + H(q_w^{(k)})\right), \quad (3)$$

where $H(p)$ is the entropy of p, and $p_w^{(k)} = \frac{\exp(sim(z_w, c_k)/\tau_u)}{\sum_n \mathbb{1}_{[n \neq i]} \exp(sim(z_w, c_n)/\tau_u)}$. We encourage the consistency of clustering assignments between weakly and strongly augmented views of the same image, and the loss function is written as follows,

$$\mathcal{L}(z_w, z_s) = \mathcal{L}(z_w, q_s) + \mathcal{L}(z_s, q_w) \quad (4)$$
$$= H(p_w, q_s) + H(p_s, q_w) \quad (5)$$

where $H(\cdot, \cdot)$ represents the cross entropy loss, $p_s^{(k)} = \frac{\exp(sim(z_s, c_k)/\tau_u)}{\sum_n \mathbb{1}_{[n \neq i]} \exp(sim(z_s, c_n)/\tau_u)}$, and $sim(\cdot, \cdot)$ denote the cosine similarity as in Eq. (1). Similar to [5,6], we compute clustering assignment matrix $Q = [q_1, q_2, \cdots, q_B]$ with B feature embeddings $Z = [z_1, z_2, \cdots z_B]$ and the prototypes $C = [c_1, c_2, \cdots, c_K]$,

$$\max_{Q \in \mathcal{Q}} \text{Tr}(Q^T C^T Z) + \epsilon H(Q), \quad (6)$$
$$\mathcal{Q} = \left\{Q \in \mathbb{R}_+^{K \times B} | Q1_B = \frac{1}{K}1_K, Q^T1_K = \frac{1}{B}1_B\right\}, \quad (7)$$

where ϵ is a weight coefficient. $\text{Tr}(\cdot)$ denotes the trace function, and $\mathcal{Q}$ is the transportation polytope. The solution to Eq. (6) is obtaining using the Sinkhorn-Knopp algorithm [10].

Thus, the overall loss to optimize the model can be formulated as follows,

$$\mathcal{L} = \mathbb{E}_{(x,y) \in \mathcal{D}_L}\mathcal{L}_{sup} + \alpha\mathbb{E}_{x \in \mathcal{D}}\mathcal{L}(z_w, q_w) + (1 - \alpha)\mathbb{E}_{x \in \mathcal{D}}\mathcal{L}(z_w, z_s), \quad (8)$$

where α is a weight coefficient.

3.3   Label Assignment with Louvain Algorithm

Semi-supervised k-means [2] is applied in [39] for assigning cluster labels. However, this approach necessitates prior knowledge of the number of clusters, a requirement that is not realistic because such information is typically not available beforehand. The learned embeddings can be used to assign cluster labels to unlabeled instances and determine the number of clusters adaptively through a community detection method, Louvain algorithm [3,15]. The representation embeddings can be represented as an embedding graph $G = (\mathcal{V}, \mathcal{E})$, where $\mathcal{V} = \{z_i\}_{i=1}^N$, N is the number of embeddings, and $\mathcal{E}$ consists of edges that connect vertices in $\mathcal{V}$. We define a matrix W as the adjacency matrix of G, and the entry $W_{i,j}$ between vertices i and j is given by,

$$
W_{ij} = \begin{cases} 1, & \text{if } x_i, x_j \in \mathcal{D}_L, \text{and } y_i = y_j \\ sim(z_i, z_j), & \text{if } x_i \text{ or } x_j \in \mathcal{D}_U, \text{and } z_j \in \text{Neighbor}(z_i), \\ 0, & \text{otherwise} \end{cases} \tag{9}
$$

where $\text{Neighbor}(z_i) = \arg topM(\{sim(z_i, z_j)|z_j \in \mathcal{V}\})$ denotes the neighbors of z_i, i.e., the M embeddings with the greatest similarity to z_i. In this way, we construct a graph G, which represents the possible connection relationships among all instances. We can now assign category labels for all instances in the training dataset, either from the known classes or novel ones. Using the Louvain algorithm, we automatically obtain the clustering assignments and the number of categories on the constructed graph.

4   Experiments

4.1   Experimental Setup

Datasets. We evaluate our framework on three generic object recognition datasets, namely CIFAR-10 [30], CIFAR-100 [30] and ImageNet-100 [11]. These standard image recognition datasets establish the performance of different methods. We further evaluate our method on Semantic Shift Benchmark (SSB) [40], including CUB-200 [41], Stanford Cars [29] and Herbarium19 [37]. The dataset splits are described in Table 1. We follow [39] sample a subset of half the classes as **Known** categories. 50% of instances of each labeled class are drawn to form the

Table 1. Dataset splits in the experiments.

Dataset		CIFAR10	CIFAR100	ImageNet-100	CUB-200	Stanford Cars	Herbarium19
Labelled	Classes	5	80	50	100	98	341
	Images	12.5k	20k	31.9k	1498	2000	8.9k
Unlabelled	Classes	10	100	100	200	196	683
	Images	37.5k	30k	95.3k	4496	6144	25.4k

labeled set, and all the remaining data constitute the unlabeled set. For evaluation, we measure the clustering accuracy by comparing the predicted label assignment with the ground truth, following the protocol in [39]. The accuracy scores for All, Known, and Novel categories are reported.

Implementation Details. We follow the implementations and learn schedules in [39] as far as possible. Specifically, we take the ViT-B-16 pre-trained by DINO [7] on ImageNet [31] as our backbone model and we use the [CLS] token as the feature representation. We train semi-supervised contrastive learning with $\tau_{sup} = 0.07$, $\tau_u = 0.05$ and $\alpha = 0.3$. For image augmentation, we use Resized-Crop, ColorJitter, Grayscale, HorizontalFlip as the weak image augmentation. The strong transformation strategy is composed of five randomly selected from RandAugment. For model optimization, we use the AdamW provided by [32]. The initial learning rate is 0.1. The batch size are chosen based on available GPU memory. All the experiments are conducted on a single RTX-2080 and averaged over 5 different seeds.

4.2 Comparison with State-of-the-Arts

We summarize the baselines compared in our experiments, including k-means [2], RankStats+ [21], UNO+ [17] and GCD [39].

Table 2. Results on three generic datasets. Accuracy scores are reported. †denotes adapted methods.

Method	CIFAR-10			CIFAR-100			ImageNet-100		
	All	Known	Novel	All	Known	Novel	All	Known	Novel
KMeans [2]	83.6	85.7	82.5	52.0	52.2	50.8	72.7	75.5	71.3
RankStats† [21]	46.8	19.2	60.5	58.2	77.6	19.3	37.1	61.6	24.8
UNO† [17]	68.6	**98.3**	53.8	69.5	80.6	47.2	70.3	**95.0**	57.9
GCD [39]	91.5	97.9	88.2	73.0	76.2	66.5	74.1	89.8	66.3
Ours	**92.3**	91.4	**94.4**	**78.5**	81.4	**75.6**	**81.1**	80.3	**81.8**

Evaluation on Generic Datasets. The results on generic benchmarks are shown in Table 2. As we can see that our method achieves state-of-the art performance on All and Novel tested on all generic datasets, especially on ImageNet-100. Our method also achieves comparable results with other methods on Known. Specifically, for the All classes, our method beats the GCD method by 0.8%, 5.5%, and 7.0% on CIFAR-10, CIFAR-100, and ImageNet-100, respectively. For the Novel class, it is 6.2% higher on CIFAR-10, 9.1% higher on CIFAR-100, and 15.5% higher on ImageNet-100. These results experimentally show that our method learns a more compact representation on the unlabeled dataset. In addition, UNO+ uses a linear classifier, which shows strong accuracy on the Known classes, but leads to poor performance on the Novel classes.

Table 3. Results on three fine-grained datasets. Accuracy scores are reported. †denotes adapted methods.

Method	CUB-200			Stanford-Cars			Herbarium19		
	All	Known	Novel	All	Known	Novel	All	Known	Novel
k-means [2]	34.3	38.9	32.1	12.8	10.6	13.8	12.9	12.9	12.8
RankStats+ [21]	33.3	51.6	24.2	28.3	61.8	12.1	27.9	**55.8**	12.8
UNO+ [17]	35.1	49.0	28.1	35.5	70.5	18.6	28.3	53.7	14.7
GCD [39]	51.3	56.6	**48.7**	39.0	57.6	29.9	35.4	51.0	27.0
Ours	**58.0**	**65.0**	43.9	**47.6**	**70.6**	**33.8**	**36.3**	53.1	**30.7**

Evaluation on Fine-Grained Datasets. We report the results on three fine-grained datasets (as in [39] in Table 3. Our method shows optimal performance on the All classes for the three datasets tested, and achieves comparable results on the Known and Novel classes, demonstrating the effectiveness of our method for fine-grained category discovery. Specifically, on the CUB-200, Stanford-Scars, and Herbarium19 datasets, our method achieves 6.7%, 8.6%, and 0.9% improvement over the state-of-the-art method on the All classes, respectively. For the Novel classes, our method outperforms GCD by 3.9% and 3.7% on Stanford-cars and Herbarium19, respectively. Meanwhile, we find that due to the low variability between fine-grained datasets, which makes it more difficult to discover novel classes, the precision on the Novel classes is generally low in terms of results.

4.3 Ablation Study

Effectiveness of Each Component. We conduct extensive ablation experiments, and perform four experiments on the CIFAR-100 and CUB-200 datasets. Table 4 shows the contribution of introducing different components on the objective loss function, including $\mathcal{L}_{sup}$, $\mathcal{L}(z_w, q_w)$ and $\mathcal{L}(z_w, z_s)$. We can observe from the results that all components contribute significantly to our proposed approach: according to the results of experiments (1) and (2), we find that $\mathcal{L}(z_w, z_s)$ is the most important component, which proves that the co-training consistency can close the intra-class embeddings and push away the inter-class boundaries. Moreover, compared to the original DINO features, the features of our approach show more favorable clustering results on CUB-200 dataset. Comparing

Table 4. Ablation study on the components of the loss function.

Index	Component			CIFAR100			CUB-200		
	$\mathcal{L}_{sup}$	$\mathcal{L}(z_w, q_w)$	$\mathcal{L}(z_w, z_s)$	All	Known	Novel	All	Known	Novel
(1)	✗	✗	✗	34.9	36.1	33.6	15.0	13.5	21.6
(2)	✗	✗	✓	69.7	71.4	68.0	56.7	62.5	40.8
(3)	✗	✓	✓	75.7	79.2	72.5	57.5	63.9	42.8
(4)	✓	✓	✓	78.5	81.4	75.5	58.0	65.0	43.9

experiments (3) and (4), we can find that supervised contrastive learning can further improve the performance on known novel categories, demonstrating the importance of supervised information.

Table 5. Ablation study on neighborhood size.

M	CIFAR-100			CUB-200		
	All	Known	Novel	All	Known	Novel
5	78.5	81.4	75.5	58.0	65.0	43.9
10	69.6	70.4	68.7	53.5	60.8	35.1
15	64.9	65.0	64.8	46.8	54.5	26.8
20	63.5	62.3	64.8	44.5	49.8	33.7
25	59.7	60.3	59.1	39.7	44.6	30.5
30	59.8	60.4	59.2	27.3	29.3	29.7

Effectiveness of the Neighborhood Size M. Table 5 illustrates the effect of the neighborhood size for vertex $z_i \in G(\mathcal{V}, \mathcal{E})$ on the final clustering results. We select $K = 5, 10, 15, 20, 25, 30$ for the ablation experiments on both CIFAR-100 and CUB-200 datasets. We find that the final performance varies greatly depending on the number of neighbors. When $K = 5$, it can reach the optimum in all **All**, **Known** and **Novel** classes. We conjecture that too many links will negatively affect the community detection.

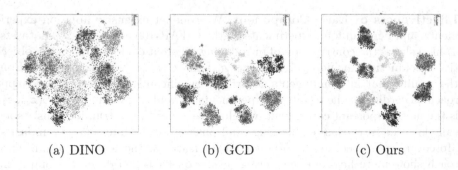

(a) DINO   (b) GCD   (c) Ours

Fig. 3. T-SNE visualization of the embeddings on CIFAR-10. The embedding clustering shows that our proposed method encourages the expansion of the distance between different clusters.

Visualization. In order to explore the clustering features on different methods more intuitively, we further visualized the features extracted by DINO, GCD and our method using T-SNE [34] on CIFAR-10. As shown in Fig. 3, compared with DINO and GCD, our method obtains clearer boundaries between different groups, and furthermore obtains more compact clusters.

5 Conclusion

In this paper, we propose a co-training strategy for GCD. In detail, we introduce a clustering assignment consistency framework that explores discriminative representations alternately. Additionally, we propose a community detection method to address the semi-supervised clustering problem in GCD. Experimental results demonstrate that our method achieves state-of-the-art performance in both generic and fine-grained tasks.

Acknowledgement. This work was partially supported by the National Key Research and Development Program of China (No. 2018AAA0100204), a key program of fundamental research from Shenzhen Science and Technology Innovation Commission (No. JCYJ20200109113403826), the Major Key Project of PCL (No. 2022ZD0115301), and an Open Research Project of Zhejiang Lab (NO.2022RC0AB04).

References

1. Albelwi, S.: Survey on self-supervised learning: auxiliary pretext tasks and contrastive learning methods in imaging. Entropy **24**(4), 551 (2022)
2. Arthur, D., Vassilvitskii, S.: k-means++: the advantages of careful seeding. In: SODA, pp. 1027–1035. SIAM (2007)
3. Blondel, V.D., Guillaume, J.L., Lambiotte, R., Lefebvre, E.: Fast unfolding of communities in large networks. J. Stat. Mech: Theory Exp. **2008**, P10008 (2008)
4. Cao, K., Brbic, M., Leskovec, J.: Open-world semi-supervised learning. In: ICLR. OpenReview.net (2022)
5. Caron, M., Bojanowski, P., Joulin, A., Douze, M.: Deep clustering for unsupervised learning of visual features. In: Ferrari, V., Hebert, M., Sminchisescu, C., Weiss, Y. (eds.) Computer Vision – ECCV 2018. LNCS, vol. 11218, pp. 139–156. Springer, Cham (2018). https://doi.org/10.1007/978-3-030-01264-9_9
6. Caron, M., Misra, I., Mairal, J., Goyal, P., Bojanowski, P., Joulin, A.: Unsupervised learning of visual features by contrasting cluster assignments. In: NeurIPS (2020)
7. Caron, M., et al.: Emerging properties in self-supervised vision transformers. In: ICCV, pp. 9630–9640. IEEE (2021)
8. Chen, T., Kornblith, S., Norouzi, M., Hinton, G.E.: A simple framework for contrastive learning of visual representations. In: ICML. Proceedings of Machine Learning Research, vol. 119, pp. 1597–1607. PMLR (2020)
9. Chi, H., et al.: Meta discovery: learning to discover novel classes given very limited data. In: ICLR. OpenReview.net (2022)
10. Cuturi, M.: Sinkhorn distances: lightspeed computation of optimal transport. In: NIPS, pp. 2292–2300 (2013)
11. Deng, J., Dong, W., Socher, R., Li, L., Li, K., Fei-Fei, L.: ImageNet: a large-scale hierarchical image database. In: CVPR, pp. 248–255. IEEE Computer Society (2009)
12. Dhillon, G.S., Chaudhari, P., Ravichandran, A., Soatto, S.: A baseline for few-shot image classification. In: ICLR. OpenReview.net (2020)
13. Dosovitskiy, A., et al.: An image is worth 16x16 words: transformers for image recognition at scale. In: ICLR. OpenReview.net (2021)

14. Dosovitskiy, A., Fischer, P., Springenberg, J.T., Riedmiller, M.A., Brox, T.: Discriminative unsupervised feature learning with exemplar convolutional neural networks. IEEE Trans. Pattern Anal. Mach. Intell. **38**(9), 1734–1747 (2016)

15. Dugué, N., Perez, A.: Directed Louvain : maximizing modularity in directed networks (2015)

16. Fei, Y., Zhao, Z., Yang, S., Zhao, B.: XCon: learning with experts for fine-grained category discovery. In: BMVC, p. 96. BMVA Press (2022)

17. Fini, E., Sangineto, E., Lathuilière, S., Zhong, Z., Nabi, M., Ricci, E.: A unified objective for novel class discovery. In: ICCV, pp. 9264–9272. IEEE (2021)

18. Finn, C., Abbeel, P., Levine, S.: Model-agnostic meta-learning for fast adaptation of deep networks. In: ICML. Proceedings of Machine Learning Research, vol. 70, pp. 1126–1135. PMLR (2017)

19. Gidaris, S., Singh, P., Komodakis, N.: Unsupervised representation learning by predicting image rotations. In: ICLR (Poster). OpenReview.net (2018)

20. Grill, J., et al.: Bootstrap your own latent - a new approach to self-supervised learning. In: NeurIPS (2020)

21. Han, K., Rebuffi, S., Ehrhardt, S., Vedaldi, A., Zisserman, A.: Automatically discovering and learning new visual categories with ranking statistics. In: ICLR. OpenReview.net (2020)

22. Han, K., Rebuffi, S., Ehrhardt, S., Vedaldi, A., Zisserman, A.: AutoNovel: automatically discovering and learning novel visual categories. IEEE Trans. Pattern Anal. Mach. Intell. **44**(10), 6767–6781 (2022)

23. Han, K., Vedaldi, A., Zisserman, A.: Learning to discover novel visual categories via deep transfer clustering. In: ICCV, pp. 8400–8408. IEEE (2019)

24. He, K., Fan, H., Wu, Y., Xie, S., Girshick, R.B.: Momentum contrast for unsupervised visual representation learning. In: CVPR, pp. 9726–9735. Computer Vision Foundation/IEEE (2020)

25. He, K., Zhang, X., Ren, S., Sun, J.: Deep residual learning for image recognition. In: CVPR, pp. 770–778. IEEE Computer Society (2016)

26. Holmberg, O.G., et al.: Self-supervised retinal thickness prediction enables deep learning from unlabelled data to boost classification of diabetic retinopathy. Nat. Mach. Intell. **2**(11), 719–726 (2020)

27. Hsu, Y., Lv, Z., Kira, Z.: Learning to cluster in order to transfer across domains and tasks. In: ICLR (Poster). OpenReview.net (2018)

28. Khosla, P., et al.: Supervised contrastive learning. In: NeurIPS (2020)

29. Krause, J., Stark, M., Deng, J., Fei-Fei, L.: 3D object representations for fine-grained categorization. In: Proceedings of the IEEE International Conference on Computer Vision Workshops, pp. 554–561 (2013)

30. Krizhevsky, A., Hinton, G.: Learning multiple layers of features from tiny images (2009)

31. Krizhevsky, A., Sutskever, I., Hinton, G.E.: ImageNet classification with deep convolutional neural networks. Commun. ACM **60**(6), 84–90 (2017)

32. Loshchilov, I., Hutter, F.: Decoupled weight decay regularization. In: ICLR (Poster). OpenReview.net (2019)

33. Lv, S., Wei, L., Zhang, Q., Liu, B., Xu, Z.: Improved inference for imputation-based semisupervised learning under misspecified setting. IEEE Trans. Neural Netw. Learn. Syst. **33**(11), 6346–6359 (2022)

34. van der Maaten, L.: Accelerating t-SNE using tree-based algorithms. J. Mach. Learn. Res. **15**(1), 3221–3245 (2014)

35. Pathak, D., Krähenbühl, P., Donahue, J., Darrell, T., Efros, A.A.: Context encoders: feature learning by inpainting. In: CVPR, pp. 2536–2544. IEEE Computer Society (2016)
36. Qiao, S., Shen, W., Zhang, Z., Wang, B., Yuille, A.: Deep co-training for semi-supervised image recognition. In: Ferrari, V., Hebert, M., Sminchisescu, C., Weiss, Y. (eds.) ECCV 2018. LNCS, vol. 11219, pp. 142–159. Springer, Cham (2018). https://doi.org/10.1007/978-3-030-01267-0_9
37. Tan, K.C., Liu, Y., Ambrose, B., Tulig, M., Belongie, S.J.: The herbarium challenge 2019 dataset. CoRR abs/1906.05372 (2019)
38. Troisemaine, C., et al.: Novel class discovery: an introduction and key concepts. CoRR abs/2302.12028 (2023)
39. Vaze, S., Han, K., Vedaldi, A., Zisserman, A.: Generalized category discovery. In: CVPR, pp. 7482–7491. IEEE (2022)
40. Vaze, S., Han, K., Vedaldi, A., Zisserman, A.: Open-set recognition: a good closed-set classifier is all you need. In: ICLR. OpenReview.net (2022)
41. Wah, C., Branson, S., Welinder, P., Perona, P., Belongie, S.: The Caltech-UCSD Birds-200-2011 dataset (2011)
42. Yang, C., An, Z., Cai, L., Xu, Y.: Mutual contrastive learning for visual representation learning. In: AAAI. pp. 3045–3053. AAAI Press (2022)
43. Yang, M., Zhu, Y., Yu, J., Wu, A., Deng, C.: Divide and conquer: compositional experts for generalized novel class discovery. In: CVPR, pp. 14248–14257. IEEE (2022)
44. Yang, X., Song, Z., King, I., Xu, Z.: A survey on deep semi-supervised learning. IEEE Trans. Knowl. Data Eng. 35(9), 8934–8954 (2023)
45. Zhang, R., Isola, P., Efros, A.A.: Colorful image colorization. In: Leibe, B., Matas, J., Sebe, N., Welling, M. (eds.) ECCV 2016. LNCS, vol. 9907, pp. 649–666. Springer, Cham (2016). https://doi.org/10.1007/978-3-319-46487-9_40
46. Zheng, M., et al.: Weakly supervised contrastive learning. In: ICCV, pp. 10022–10031. IEEE (2021)

CInvISP: Conditional Invertible Image Signal Processing Pipeline

Duanling Guo[1], Kan Chang[1(✉)], Yahui Tang[1], Mingyang Ling[1],
and Minghong Li[2]

[1] School of Computer and Electronic Information, Guangxi University, Nanning
530004, China
{duanlingguo,lingmy}@st.gxu.edu.cn, kanchang@gxu.edu.cn,
askyland@foxmail.com
[2] School of Automation, Central South University, Changsha 410083, China
minghongli233@gmail.com

Abstract. Standard RGB (sRGB) images processed by the image signal
processing (ISP) pipeline of digital cameras have a nonlinear relationship
with the scene irradiance. Therefore, the low-level vision tasks which
work best in a linear color space are not suitable to be carried out in the
sRGB color space. To address this issue, this paper proposes an approach
called CInvISP to provide a bidirectional mapping between the nonlinear sRGB and linear CIE XYZ color spaces. To ensure a fully invertible
ISP, the basic building blocks in our framework adopt the structure of
invertible neural network. As camera-style information is embedded in
sRGB images, it is necessary to completely remove it during backward
mapping, and properly incorporate it during forward mapping. To this
end, a conditional vector is extracted from the sRGB input and inserted
into each invertible building block. Experiments show that compared to
other mapping approaches, CInvISP achieves a more accurate bidirectional mapping between the two color spaces. Moreover, it is also verified that such a precise bidirectional mapping facilitates low-level vision
tasks including image denoising and retouching well.

Keywords: Image Signal Processing Pipeline · CIE XYZ Color
Space · Invertible Neural Networks · Convolutional Neural Networks

1 Introduction

In the traditional in-camera image signal processing (ISP) pipeline [1], RAW
images are first aligned to the CIE XYZ color space which is irrelevant to camera
sensors, and then transformed to the standard RGB (sRGB) color space that
are perceptually pleasant for the human visual system [2]. When converting the

---

This work was supported by the National Natural Science Foundation of China (NSFC)
[grant number 62171145], and by Guangxi Key Laboratory of Multimedia Communications and Network Technology.

B. Luo et al. (Eds.): ICONIP 2023, LNCS 14451, pp. 548–562, 2024.
https://doi.org/10.1007/978-981-99-8073-4_42

CIE XYZ color space to the sRGB color space, nonlinear operations including tone curve, color manipulation, and gamma correction are carried out in the ISP pipeline [1,2]. Therefore, the linear relationship between the captured light intensity and the realistic scene irradiance is broken.

On the other hand, as RAW images are more expensive to collect and access, most low-level vision tasks (e.g., denoising [3,4], high-dynamic-range image reconstruction [5], super-resolution [6,7]) are conducted on the nonlinear, processed sRGB images. As a result, the performance of these tasks are largely limited. To address this issue, reversing the images from the sRGB color space back to an unprocessed, linear color space becomes an attractive solution. Afterwards, we can perform low-level vision tasks in this linear color space, and then forwardly map the enhanced/restored images to the sRGB space.

To this end, Brooks et al. [8] utilized the prior knowledge of cameras to unprocess each ISP step. However, due to the fact that the specific parameters of ISP are held by camera manufactures, the realistic ISP pipeline can be considered as a black box. As a result, it is hard to precisely find the inverse function for each ISP step by applying conventional methods such as curve fitting. Several approaches have tried to model the entire ISP pipeline by using convolutional neural networks (CNN), e.g., U-net [9], DeepISP [10], PyNET [11], VisionISP [12] and Deep-FlexISP [13]. However, these approaches only model the forward mapping of the ISP pipeline, i.e., from the RAW space to the sRGB space. Recently, the *reversed ISP challenge* has been held [14], which focuses on the backward mapping of specific ISP pipelines on smartphones. Zamir et al. proposed CycleISP [15], which employs two branches of CNN to model both the backward and forward mappings, respectively. In contrast to modeling the two directions of mappings by two separate CNN networks, Xing et al. [16] introduced invertible neural network [17–19] such that completely-invertible bidirectional mapping between the two color spaces can be obtained.

Although some methods have developed the two directions of mappings between the RAW space and the sRGB space, it should be pointed out that RAW images are specific to physical color filter arrays [20]. Such a sensor-specific characteristic limits the application of CNN-based mapping approaches, as these approaches are hard to be generalized from one camera sensor to another. Similar to the RAW space, light intensity in the CIE XYZ color space is also linear to scene irradiance. More importantly, the CIE XYZ color space is device-independent [2], and thus reversing sRGB images to this space is a more suitable choice. Therefore, Afifi et al. [2] proposed a framework called CIE XYZ net to provide a pair of mappings between the CIE XYZ and sRGB spaces.

Another important issue ignored by most methods is the variety of camera styles embedded in sRGB images. For a satisfied artistic effect, different types of cameras have their own settings and profiles, thus leading to sRGB images with multiple camera styles [21,22]. These camera styles make the backward mapping a many-to-one problem, and the forward mapping an one-to-many problem. Thus to build an accurate pair of mappings, it is necessary to fully remove the camera-style information during backward mapping, and properly insert it during forward mapping. Based on the above consideration, we propose a new framework called CInvISP (conditional invertible ISP pipeline) to establish a bidirectional mapping between the sRGB and CIE XYZ spaces. Similar to InvISP [16], CInvISP also adopts the structure of invertible neural networks to deliver a faithful bidirectional mapping. Nevertheless, CInvISP additionally extracts conditional information from the input sRGB image, and incorporates it into both the backward and forward mappings, leading to obviously better performance. In summary, the main contributions of this paper are three-fold:

(1) To effectively solve both the many-to-one and one-to-many problems, we propose to extract conditional information from the input sRGB image, and then incorporate it into the invertible CNN structure. This strategy greatly helps to eliminate the camera-style information embedded in sRGB images for backward mapping, while largely promotes learning the unique camera style for forward mapping.

(2) To effectively extract conditional information, we design two parallel CNN modules to extract the features that represent sRGB image and CIE XYZ image, respectively, and then feed them both to the condition generator module (CGM) to predict the conditional vector. Besides, a paired-feature decoder is established to provide external supervision for the extraction of the two types of features.

(3) Experiments demonstrate that our CInvISP is superior to other state-of-the-art (SOTA) algorithms in providing a reliable bidirectional mapping between the sRGB and CIE XYZ spaces. Moreover, the applications of CInvISP in image denoising and image retouching have also been investigated. To facilitate further study, our source code and pre-trained model will be released on http://github.com/Duanling-Guo/CInvISP.

2   Methodology

2.1   The Framework of CInvISP

The overall structure of our network is shown in Fig. 1, which mainly includes a conditional vector extraction (CVE) sub-network, an invertible rendering (IR) sub-network, and a paired-feature decoder.

As can be seen, the IR sub-network consists of several cascaded conditional invertible blocks (CIBs). To achieve a fully invertible mapping for the ISP pipeline, similar to [16], each CIB applies an invertible CNN structure. However, as the sRGB images might be captured by different cameras, it is hard to deal with various types of camera-style information by using the normal invertible neural networks. Therefore, our CIB additionally incorporates the conditional vector $\boldsymbol{\Psi}$ extracted by the CVE sub-network, so that the camera-style information can be completely removed during backward mapping, and properly inserted during forward mapping. As a result, the backward and forward mappings of our CInvISP can be respectively represented as

$$\hat{\mathbf{I}}_{xyz} = \mathcal{G}_{\mu}^{-1}(\mathbf{I}_{srgb} \mid \boldsymbol{\Psi}) \tag{1}$$

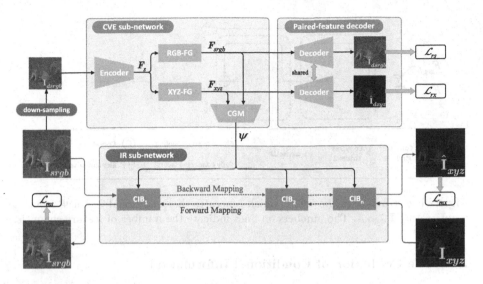

Fig. 1. The overall network structure of CInvISP. The CVE sub-network produces the conditional vector $\boldsymbol{\Psi}$ with the help of the paired-feature decoder which only exists in the training phase. $\boldsymbol{\Psi}$ is fed to the CIBs in the IR sub-network, so that the camera-style information can be completely removed during backward mapping, and properly inserted during forward mapping. To achieve a fully invertible mapping, each CIB has an invertible CNN structure.

$$\hat{\mathbf{I}}_{srgb} = \mathcal{G}_{\mu}(\mathbf{I}_{xyz} \mid \boldsymbol{\Psi}) \tag{2}$$

where $\mathcal{G}_{\mu}^{-1}(\cdot)$ and $\mathcal{G}_{\mu}(\cdot)$ denote the functions of backward and forward mappings, respectively; $\mathbf{I}_{srgb}$ and $\mathbf{I}_{xyz}$ are the input sRGB and CIE XYZ images, $\hat{\mathbf{I}}_{srgb}$ and $\hat{\mathbf{I}}_{xyz}$ are the reconstructed sRGB and CIE XYZ images, respectively. As $\mathcal{G}_{\mu}^{-1}(\cdot)$ and $\mathcal{G}_{\mu}(\cdot)$ are implemented by using invertible neural networks, both of them are parametrized by the same μ. Note that the forward mapping is applied after the backward mapping. Therefore, though the conditional vector $\boldsymbol{\Psi}$ is extracted from the sRGB image, it is available during forward mapping.

To reduce computational complexity, the $\mathbf{I}_{srgb}$ is down-sampled to a lower resolution. Since the conditional vector $\boldsymbol{\Psi}$ represents the input sRGB image globally, down-sampling the input image has little impact on the accuracy of predicting $\boldsymbol{\Psi}$. Therefore, $\boldsymbol{\Psi}$ is obtained by

$$\boldsymbol{\Psi} = \mathcal{C}_{\theta}(\mathcal{D}(\mathbf{I}_{srgb})) \tag{3}$$

where $\mathcal{D}(\cdot)$ denotes the down-sampling operation; function $\mathcal{C}_{\theta}(\cdot)$ is used to predict $\boldsymbol{\Psi}$, which is implemented by the CVE sub-network and parametrized by θ.

To help the CVE sub-network learn suitable features for predicting the conditional vector $\boldsymbol{\Psi}$, the paired-feature decoder decodes the learned features to the sRGB and CIE XYZ images, respectively. Note that such a decoder only exists in the training phase, and will be removed during testing. As a result, no additional computational burden is required.

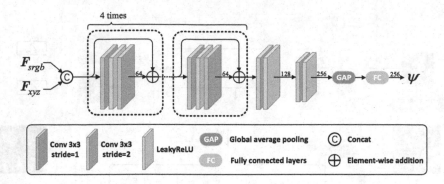

Fig. 2. The structure of the CGM, which generates the conditional vector Ψ from the input feature $\mathbf{F}_{concat}$. The numbers on lines indicate the number of feature channels.

2.2   The Prediction of Conditional Information

As a suitable guidance, the conditional vector Ψ is expected to contain camera-style information and global semantic representation of an image. To produce an optimal conditional vector Ψ, we propose to extract features $\mathbf{F}_{rgb}$ and $\mathbf{F}_{xyz}$ that represent the sRGB and CIE XYZ images, respectively, and then feed them both to the CGM. As a result, the CGM can easily recognize the differences between $\mathbf{F}_{rgb}$ and $\mathbf{F}_{xyz}$, and thus the difficulty of accurately predicting the conditional vector Ψ can be largely reduced.

To this end, as shown in Fig. 1, in the CVE sub-network, a common CNN encoder with 3 convolution layers is applied to extract shallow features $\mathbf{F}_s$ from the input, followed by an sRGB feature generator (RGB-FG) and a CIE XYZ feature generator (XYZ-FG) refining $\mathbf{F}_s$ in a parallel way. Both of the two feature generators consist of 5 cascaded residual blocks [6]. Due to the reason that the CIE XYZ image does not contain camera-style information, it is necessary to remove such information from $\mathbf{F}_{xyz}$. Therefore, inspired by the task of style transfer [23], we further incorporate the instance normalization (IN) [24] operation into each residual block in XYZ-FG, so that the embedded camera-style information can be effectively removed by the XYZ-FG.

After obtaining $\mathbf{F}_{xyz}$ and $\mathbf{F}_{rgb}$, the paired-feature decoder is applied on them to reconstruct the down-sampled CIE XYZ image $\hat{\mathbf{I}}_{dxyz}$ and the down-sampled sRGB image $\hat{\mathbf{I}}_{dsrgb}$, respectively. It should be noted that when decoding $\mathbf{F}_{xyz}$ and $\mathbf{F}_{rgb}$, the same CNN decoder which consists of 3 deconvolution layers is used. Such a setting makes $\mathbf{F}_{xyz}$ and $\mathbf{F}_{rgb}$ share similar semantic representation of an image. The differences between $\mathbf{F}_{xyz}$ and $\mathbf{F}_{rgb}$ lie in that $\mathbf{F}_{xyz}$ is trained to exclude the camera-style information, while $\mathbf{F}_{rgb}$ is required to preserve such information. These characteristics of the paired-feature decoder coincide with the objectives of the CVE sub-network, and therefore it facilitates the CVE sub-network well.

The CGM which accepts both $\mathbf{F}_{rgb}$ and $\mathbf{F}_{xyz}$ as input is responsible for predicting Ψ. The detailed structure of CGM is given in Fig. 2. As can be seen,

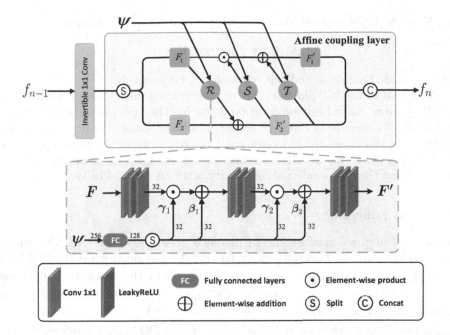

Fig. 3. The structure of the CIB. f_n denotes the features produced by the nth CIB, which contains 3 channels. $\mathbf{F}_1$ and $\mathbf{F}_2$ are features split from the output of the 1×1 invertible convolution layer. The three operators $\mathcal{R}(\cdot)$, $\mathcal{S}(\cdot)$ and $\mathcal{T}(\cdot)$ share the same structure, where the input conditional vector $\boldsymbol{\Psi}$ is used to learn parameters γ and β to modulate the features.

firstly, $\mathbf{F}_{rgb}$ and $\mathbf{F}_{xyz}$ are concatenated and sequentially refined by 4 residual blocks. Then two 3×3 convolution layers with a stride of 2 are utilized to adjust the feature dimension. Finally, $\boldsymbol{\Psi}$ is created by a global average pooling (GAP) layer and the following two fully connected layers.

2.3   The Conditional Invertible Block

Figure 3 shows the detailed structure of a CIB. As can be seen, it contains an invertible 1×1 convolution layer and an affine coupling layer [18,19]. Similar to the traditional invertible neural networks [16], there are also three operators $\mathcal{R}(\cdot)$, $\mathcal{S}(\cdot)$ and $\mathcal{T}(\cdot)$ in the affine coupling layer of CIB. However, different from [16], the conditional vector $\boldsymbol{\Psi}$ is additionally incorporated into the $\mathcal{R}(\cdot)$, $\mathcal{S}(\cdot)$ and $\mathcal{T}(\cdot)$ in CIB. By doing so, the CIB is able to notice the camera-style information and global semantic representation of an sRGB image, which greatly helps to build a more precise bidirectional mapping.

To effectively incorporate $\boldsymbol{\Psi}$ into the three operators, affine transformation is used. The conditional vector is first processed by two fully connected layers, and then split into 4 sub-vectors to form two groups of transformation parameters.

Let $\mathbf{F}$ be the input feature, the affine transformation can be represented as

$$\mathcal{M}(\mathbf{F} \mid \gamma, \beta) = \gamma \odot \mathbf{F} + \beta \tag{4}$$

where γ and β are scale and shift parameters, respectively.

Note that in the ISP pipeline, the conversion from the CIE XYZ space to the sRGB space is carried out on individual pixels. Therefore, compared with the normal 3×3 convolution which explores spatial correlation among neighbouring pixels, it is more reasonable to use 1×1 convolutions in the three operations $\mathcal{R}(\cdot)$, $\mathcal{S}(\cdot)$ and $\mathcal{T}(\cdot)$ in our CIB. Moreover, benefiting from the utilization of 1×1 convolutions, the computational complexity of a CIB can also be largely reduced.

2.4   Loss Function

During training, we measure the L_1 distance between: 1) the CIE XYZ image $\hat{\mathbf{I}}_{xyz}$ produced by the backward mapping and the ground-truth (GT) CIE XYZ image $\mathbf{I}_{xyz}^*$; 2) the sRGB image $\hat{\mathbf{I}}_{srgb}$ reconstructed by the forward mapping and the GT sRGB image $\mathbf{I}_{srgb}^*$; 3) the CIE XYZ image $\hat{\mathbf{I}}_{dxyz}$ decoded by the paired-feature decoder and its down-sampled GT, i.e., $\mathcal{D}(\mathbf{I}_{xyz}^*)$; 4) the sRGB image $\hat{\mathbf{I}}_{dsrgb}$ predicted by the paired-feature decoder and $\mathcal{D}(\mathbf{I}_{srgb}^*)$. Therefore, the loss function can be formulated by

$$\mathcal{L} = \mathcal{L}_{rs} + \lambda_1 \mathcal{L}_{rx} + \mathcal{L}_{ms} + \lambda_2 \mathcal{L}_{mx} \tag{5}$$

where λ_1 and λ_2 are trade-off parameters and

$$\mathcal{L}_{rs} = \|\hat{\mathbf{I}}_{dsrgb} - \mathcal{D}(\mathbf{I}_{srgb}^*)\|_1 \tag{6}$$

$$\mathcal{L}_{rx} = \|\hat{\mathbf{I}}_{dxyz} - \mathcal{D}(\mathbf{I}_{xyz}^*)\|_1 \tag{7}$$

$$\mathcal{L}_{ms} = \|\hat{\mathbf{I}}_{srgb} - \mathbf{I}_{srgb}^*\|_1 \tag{8}$$

$$\mathcal{L}_{mx} = \|\hat{\mathbf{I}}_{xyz} - \mathbf{I}_{xyz}^*\|_1 \tag{9}$$

Similar to [2], both λ_1 and λ_2 are set as 1.5 due to the reason that CIE XYZ images generally have intensity lower than sRGB images.

3   Experimental Results on the Accuracy of Mappings

3.1   Experimental Settings

The XYZ-sRGB dataset [2] which consists of 1265 pairs of compressed 8 bit sRGB images and uncompressed 16 bit CIE XYZ images is used for evaluation in this experiment. The images are generated from the RAW images in the MIT-Adobe FiveK dataset [25] by using the Adobe DNG Development Kit, and they are divided into 971, 244 and 50 XYZ-sRGB pairs for training, testing and validation, respectively.

Table 1. Quantitative comparison. $sRGB{\rightarrow}XYZ$ and $XYZ{\rightarrow}sRGB$ represent the backward and forward mappings, respectively. *Rec.* $XYZ{\rightarrow}sRGB$ denotes the forward mapping which takes the reconstructed CIE XYZ images as input. Note that the invertible CNN structures in InvISP and CInvISP ensure a perfect forward mapping from the reconstructed CIE XYZ images to the sRGB images. The best results are highlighted.

Mapping Type	Metric	Methods			
		U-net [9]	CIE XYZ net [2]	InvISP [16]	CInvISP
sRGB$\rightarrow$XYZ	PSNR (dB)	29.30	30.03	30.83	**31.22**
	SSIM	0.9229	0.9268	0.9222	**0.9363**
	ΔE	6.5558	7.3757	8.8974	**5.7247**
XYZ$\rightarrow$sRGB	PSNR (dB)	27.19	27.68	27.62	**28.10**
	SSIM	0.8799	0.9053	0.8949	**0.9079**
	ΔE	16.2596	6.4072	6.6371	**6.3245**
Rec. XYZ$\rightarrow$sRGB	PSNR (dB)	29.53	43.65	**100.00**[a]	**100.00**[a]
	SSIM	0.8923	0.9948	**1.0000**	**1.0000**
	ΔE	16.0512	2.7032	**0.0000**	**0.0000**

[a] When computing PSNR, the minimal MSE is set as 10^{-10} to avoid division by zero.

To demonstrate the effectiveness of CInvISP, it is compared with three representative algorithms, including CIE XYZ net [2], InvISP [16], and U-net [9]. The number of CIBs in CInvISP is set as $n = 8$ in our experiments. All the models except CIE XYZ net are retrained on the XYZ-sRGB dataset for 300 epochs by using the Adam optimizer with $\beta_1 = 0.9$, and $\beta_2 = 0.999$. During training, the batch size, the patch size and the learning rate are set as 5, 512×512, and 10^{-4}, respectively. We leverage random geometric operations (i.e., crop, flip and rotation) as data augmentation. As CIE XYZ net is trained on the XYZ-sRGB dataset, we directly use the pre-trained model released by its authors for testing.

As quality assessment metrics, peak signal-to-noise ratio (PSNR), structural similarity index (SSIM) [26], and ΔE [27] are used. Note that ΔE is a metric to evaluate the color difference between two images. All the experiments are conducted on a single NVIDIA GeForce RTX 2080Ti GPU, and all the methods are implemented with the PyTorch framework.

3.2   Comparison with SOTA Methods

The objective results of different methods are provided in Table 1. As can be seen, for both the backward and forward mappings, our CInvISP is superior to other competing methods in all quality assessment metrics. Compared with InvISP [16], CInvISP achieves more accurate mappings, which well demonstrates the effectiveness of the incorporation of conditional information. Moreover, benefiting from the utilization of invertible CNN structure, if the CIE XYZ images reconstructed by the backward mapping are used as the input of forward mapping, both CInvISP and InvISP are able to deliver exact reconstruction. Such

Table 2. Comparisons of model complexity. No additional parameters are required for the forward mapping of InvISP and CInvISP as their structures are invertible. The FLOPs are measured on images with a resolution of 512×512.

	Metric	Methods			
		U-net [9]	CIE XYZ net [2]	InvISP [16]	CInvISP
sRGB→XYZ	Parameters	2.46M	1.35M	1.41M	2.34M
	Runtime (ms)	12.72	13.42	320.18	98.02
	FLOPs	31.6G	69.0G	739.5G	19.0G
XYZ→sRGB	Parameters	2.46M	1.35M	/	/
	Runtime (ms)	12.72	13.40	320.18	80.66
	FLOPs	31.6G	69.0G	739.5G	3.5G

a characteristic makes the invertible-CNN-based methods more attractive than other approaches for the bidirectional mapping task.

Model complexity of different methods is compared in Table 2. It can be concluded that: 1) although CInvISP has the second largest model size for backward mapping, it does not consume additional parameters for forward mapping due to the invertible structure in the IR sub-network; 2) CInvISP is faster than InvISP as the IR sub-network only consists of 1×1 convolutions; 3) since the sRGB is down-sampled before being fed to the computation-intensive CVE sub-network, the floating point operations per second (FLOPs) of CInvISP are less than the other methods; 4) as the conditional vector of CInvISP has been obtained in backward mapping, it can be directly used in forward mapping, thus leading to a significant reduction in computational complexity.

3.3 Ablation Study

To verify the effectiveness of the proposed modules, ablation study is carried out, and the results of different variants of CInvISP can be found in Table 3. We have the following observations:

(1) "w/o CGM" indicates that the IR sub-network works without the guidance of the conditional vector Ψ. This result shows that removing the conditional information leads to significant performance degradation, which well demonstrates the importance of conditional information.
(2) It is obvious that the paired-feature decoder also contributes greatly to the performance, which suggests that it is hard to properly extract the conditional information without the external supervision.
(3) The results of "w/o XYZ-FG" and "w/o RGB-FG" show that combining both RGB and XYZ features is necessary. Compared to "w/o RGB-FG", "w/o XYZ-FG" results in a more obvious decrease in performance, which implies that the branch of XYZ-FG is more crucial to the extraction of camera-style information.

Table 3. Ablation Study. PF decoder is short for paired-feature decoder. The best results are highlighted.

Mapping Type	Metric	Variants of CInvISP				
		w/o CGM	w/o PF decoder	w/o XYZ-FG	w/o RGB-FG	Full Model
sRGB→XYZ	PSNR (dB)	29.37	29.46	29.74	30.96	**31.22**
	SSIM	0.8874	0.8912	0.8926	0.9147	**0.9363**
	ΔE	13.2987	10.1152	9.3912	8.5613	**5.7247**
XYZ→sRGB	PSNR (dB)	27.18	27.11	27.36	27.44	**28.10**
	SSIM	0.8768	0.8824	0.8993	0.8879	**0.9079**
	ΔE	10.9677	10.7368	9.0614	8.9332	**6.3245**

Table 4. The application of bidirectional mapping methods in the denoising task. The pre-trained SwinIR model [4] is used for blind denoising. The best results are highlighted.

Methods	PSNR (dB)	SSIM
Denoising in the sRGB space	31.30	0.6956
Denoising in the CIE XYZ space (by U-net [9])	28.81	0.8324
Denoising in the CIE XYZ space (by CIE XYZ net [2])	32.65	0.8332
Denoising in the CIE XYZ space (by InvISP [16])	28.43	0.7612
Denoising in the CIE XYZ space (by CInvISP)	**33.38**	**0.8507**

4 The Applications of CInvISP in Low-Level Vision Tasks

In this section, we evaluate the performance of CInvISP in two low-level vision tasks, including image denoising and retouching.

4.1 Image Denoising

The validation set of the SIDD dataset [28] which consists of 1280 pairs of noisy and clean images is used as the testing dataset for the denoising task. The images containing realistic noises are captured by various smartphones with different sensitivity and shutter speeds.

In this experiment, we first apply backward mapping to transfer the input sRGB images back to the CIE XYZ color space, and then utilize the pre-trained SwinIR model [4] to perform blind denoising. Finally, the denoised results are converted to the sRGB color space by using the forward mapping. The objective results of different approaches are shown in Table 4, and the visual comparisons of two examples can be found in Figs. 4 and 5, respectively.

In Table 4, we can see that denoising in the sRGB space is significantly inferior to our approach. The reason lies in that the realistic noises have a distribution much more complex than the ideal Gaussian distribution. By accurately converting the sRGB images to the CIE XYZ space, the noise distribution becomes

Fig. 4. The denoising results of image *0045_NOISY_SRGB_010* from the SIDD dataset [28]. From left to right and from top to bottom: noisy image, denoising on the sRGB image, denoising on the CIE XYZ images reconstructed by U-net [9], CIE XYZ net [2], InvISP [16] and CInvISP, respectively.

Fig. 5. The denoising results of image *0055_NOISY_SRGB_010* from the SIDD dataset [28]. From left to right and from top to bottom: noisy image, denoising on the sRGB image, denoising on the CIE XYZ images reconstructed by U-net [9], CIE XYZ net [2], InvISP [16] and CInvISP, respectively.

closer to the Gaussian distribution, and thus the SwinIR model [4] trained under the Gaussian assumption obtains better results. However, compared to denoising in the sRGB space, using U-net [9] and InvISP [16] to conduct bidirectional mapping achieves lower PSNR, which indicate that an inaccurate bidirectional mapping between the two color spaces could degrade the denoising performance.

From Figs. 4 and 5, we can see that: denoising in the sRGB cannot completely remove noises in images; the U-net-based method [9] generates obvious color distortion; both CIE XYZ net [2] and InvISP [16] produce blurred results;

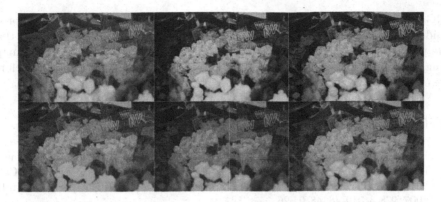

Fig. 6. The retouching results of image *a0304-dgw_137* from the MIT-Adobe FiveK dataset [25]. From left to right and from top to bottom: Expert RAW rendering, Retouching on the CIE XYZ images reconstructed by applying standard 2.2 gamma tone curve, U-net [9], CIE XYZ net [2], InvISP [16] and CInvISP, respectively.

compared with other methods, the CInvISP can effectively remove noises, while successfully preserve sharp edges and fine details.

4.2 Image Retouching

Usually, professional photographers prefer to retouch the images in the RAW space rather than the sRGB space [29]. The reason lies in that RAW photos have a wider range of tonal values and do not contain camera-style information. As the CIE XYZ space also has the same advantages as the RAW space, it is reasonable to convert an sRGB image back to the CIE XYZ space for image retouching. Therefore, we first apply different backward mapping methods to obtain the CIE XYZ version of images, and then perform the retouching method called Exposure [30] in the CIE XYZ space. As the pre-trained model of Exposure [30] provided by its authors was also trained by using the MIT-Adobe FiveK dataset [25], we directly utilize it to evaluate the effectiveness of different bidirectional mapping approaches in this experiment.

The subjective results obtained by different methods are provided in Fig. 6. Obviously, the backward mapping of our CInvISP produces retouched image much closer to human expert than the other approaches. Only unnoticeable color distortion is generated by our method.

5 Conclusion

This paper proposed a method called CInvISP to provide a bidirectional mapping between the sRGB and CIE XYZ spaces. To achieve a fully invertible mapping for the ISP pipeline, the invertible CNN structure is used in CInvISP. In addition, the feature representations of sRGB and CIE XYZ images are extracted from the sRGB input, and then combined and compressed to be a conditional vector to guide the bidirectional mapping. Experimental results have demonstrated that the proposed algorithm is able to achieve the most accurate bidirectional mapping. Moreover, the applications of CInvISP in image denoising and retouching have also been verified.

References

1. Karaimer, H.C., Brown, M.S.: A software platform for manipulating the camera imaging pipeline. In: Leibe, B., Matas, J., Sebe, N., Welling, M. (eds.) ECCV 2016. LNCS, vol. 9905, pp. 429–444. Springer, Cham (2016). https://doi.org/10.1007/978-3-319-46448-0_26
2. Afifi, M., Abdelhamed, A., Abuolaim, A., Punnappurath, A., Brown, M.S.: CIE XYZ Net: unprocessing images for low-level computer vision tasks. IEEE Trans. Pattern Anal. Mach. Intell. **44**, 4688–4700 (2022)
3. Zhang, K., Zuo, W., Chen, Y., Meng, D., Zhang, L.: Beyond a Gaussian denoiser: residual learning of deep CNN for image denoising. IEEE Trans. Image Process. **26**(7), 3142–3155 (2017)
4. Liang, J., Cao, J., Sun, G., Zhang, K., Gool, L.V., Timofte, R.: SwinIR: image restoration using swin transformer. In: Proceedings of the IEEE International Conference on Computer Vision (ICCV), Montreal, BC, Canada, pp. 1833–1844 (2021)
5. Eilertsen, G., Kronander, J., Denes, G., Mantiuk, R.K., Unger, J.: HDR image reconstruction from a single exposure using deep CNNs. ACM Trans. Graph. **36**(6), 1–15 (2017)
6. Zhang, Y., Tian, Y., Kong, Y., Zhong, B., Fu, Y.: Residual dense network for image restoration. IEEE Trans. Pattern Anal. Mach. Intell. **43**(7), 2480–2495 (2021)
7. Chang, K., Li, H., Tan, Y., Ding, P.L.K., Li, B.: A two-stage convolutional neural network for joint demosaicking and super-resolution. IEEE Trans. Circ. Syst. Video Technol. **32**(7), 4238–4254 (2022)
8. Brooks, T., Mildenhall, B., Xue, T., Chen, J., Sharlet, D., Barron, J.T.: Unprocessing images for learned raw denoising. In: Proceedings of the IEEE Conference on Computer Vision and Pattern Recognition (CVPR), Long Beach, CA, USA, pp. 11036–11045 (2019)
9. Chen, C., Chen, Q., Xu, J., Koltun, V.: Learning to see in the dark. In: Proceedings of the IEEE Conference on Computer Vision and Pattern Recognition (CVPR), Salt Lake City, UT, USA, pp. 3291–3300 (2018)
10. Schwartz, E., Giryes, R., Bronstein, A.M.: DeepISP: towards learning an end-to-end image processing pipeline. IEEE Trans. Image Process. **28**(2), 912–923 (2018)
11. Ignatov, A., Gool, L.V., Timofte, R.: Replacing mobile camera ISP with a single deep learning model. In: Proceedings of the IEEE/CVF Conference on Computer Vision and Pattern Recognition Workshops (CVPRW), Seattle, WA, USA, pp. 2275–2285 (2020)
12. Liu, S., et al.: VisionISP: repurposing the image signal processor for computer vision applications. In: Proceedings of the IEEE International Conference on Image Processing (ICIP), Taipei, Taiwan, pp. 4624–4628 (2019)
13. Liu, S., et al.: Deep-FlexISP: a three-stage framework for night photography rendering. In: Proceedings of the IEEE Conference on Computer Vision and Pattern Recognition (CVPR), New Orleans, LA, USA, pp. 1210–1219 (2022)

14. Conde, M.V., et al.: Reversed image signal processing and RAW reconstruction. AIM 2022 challenge report. In: Karlinsky, L., Michaeli, T., Nishino, K. (eds.) ECCV 2022. LNCS, vol. 13803, pp. 3–26. Springer, Cham (2023). https://doi.org/10.1007/978-3-031-25066-8_1

15. Zamir, S.W., et al.: CycleISP: real image restoration via improved data synthesis. In: Proceedings of the IEEE Conference on Computer Vision and Pattern Recognition (CVPR), Seattle, WA, USA, pp. 2696–2705 (2020)

16. Xing, Y., Qian, Z., Chen, Q.: Invertible image signal processing. In: Proceedings of the IEEE Conference on Computer Vision and Pattern Recognition (CVPR). Virtual Conference, pp. 6287–6296 (2021)

17. Kingma, D.P., Dhariwal, P.: Glow: generative flow with invertible 1×1 convolutions. In: Proceedings of the Neural Information Processing Systems (NIPS), Montreal, Canada, pp. 2722–2730 (2019)

18. Ho, J., Chen, X., Srinivas, A., Duan, Y., Abbeel, P.: Flow++: improving flow-based generative models with variational dequantization and architecture design. In: Proceedings of the International Conference on Machine Learning (ICML), Long Beach, USA, pp. 2722–2730 (2019)

19. Xiao, M., et al.: Invertible image rescaling. In: Vedaldi, A., Bischof, H., Brox, T., Frahm, J.-M. (eds.) ECCV 2020. LNCS, vol. 12346, pp. 126–144. Springer, Cham (2020). https://doi.org/10.1007/978-3-030-58452-8_8

20. Rang, N.H.M., Prasad, D.K., Brown, M.S.: Raw-to-raw: mapping between image sensor color responses. In: Proceedings of the IEEE Conference on Computer Vision and Pattern Recognition (CVPR), Columbus, OH, USA, pp. 3398–3405 (2014)

21. Liu, C., Chang, X., Shen, Y.D.: Unity style transfer for person re-identification. In: Proceedings of the IEEE Conference on Computer Vision and Pattern Recognition (CVPR). Virtual Conference, pp. 6886–6895 (2020)

22. Zhong, Z., Zheng, L., Zheng, Z., Li, S., Yang, Y.: Camera style adaptation for person re-identification. In: Proceedings of the IEEE Conference on Computer Vision and Pattern Recognition (CVPR), Salt Lake City, UT, USA, pp. 5157–5166 (2018)

23. Huang, X., Belongie, S.J.: Arbitrary style transfer in real-time with adaptive instance normalization. In: Proceedings of the IEEE International Conference on Computer Vision (ICCV), Venice, Italy, pp. 4688–4700 (2017)

24. Ulyanov, D., Vedaldi, A., Lempitsky, V.: Improved texture networks: maximizing quality and diversity in feed-forward stylization and texture synthesis. In: Proceedings of the IEEE Conference on Computer Vision and Pattern Recognition (CVPR), Honolulu, Hawaii, USA, pp. 4105–4113 (2017)

25. Bychkovsky, V., Paris, S., Chan, E., Durand, F.: Learning photographic global tonal adjustment with a database of input/output image pairs. In: Proceedings of the IEEE Conference on Computer Vision and Pattern Recognition (CVPR), Colorado Springs, CO, USA, pp. 97–104 (2011)

26. Wang, Z., Bovik, A.C., Sheikh, H.R., Simoncelli, E.P.: Image quality assessment: from error visibility to structural similarity. IEEE Trans. Image Process. **13**(4), 600–612 (2004)

27. Backhaus, W., Kliegl, R., Werner, J.S.: Color vision: perspectives from different disciplines. Optom. Vis. Sci. **76** (1999)

28. Abdelhamed, A., Lin, S., Brown, M.S.: A high-quality denoising dataset for smartphone cameras. In: Proceedings of the IEEE Conference on Computer Vision and Pattern Recognition (CVPR), Salt Lake City, UT, USA, pp. 1692–1700 (2018)

29. Zeng, H., Cai, J., Li, L., Cao, Z., Zhang, L.: Learning image-adaptive 3D lookup tables for high performance photo enhancement in real-time. IEEE Trans. Pattern Anal. Mach. Intell. **44**(4), 2058–2073 (2022)

30. Hu, Y., He, H., Xu, C., Wang, B., Lin, S.: Exposure: a white-box photo post-processing framework. ACM Trans. Graph. **37**(2), 1–17 (2018)

Ignored Details in Eyes: Exposing GAN-Generated Faces by Sclera

Tong Zhang, Anjie Peng, and Hui Zeng[✉]

School of Computer Science and Technology, Southwest University of Science and Technology, Mianyang, China
zengh5@mail2.sysu.edu.cn

Abstract. Advances in Generative adversarial networks (GAN) have significantly improved the quality of synthetic facial images, posing threats to many vital areas. Thus, identifying whether a presented facial image is synthesized is of forensic importance. Our fundamental discovery is the lack of capillaries in the sclera of the GAN-generated faces, which is caused by the lack of physical/physiological constraints in the GAN model. Because there are more or fewer capillaries in people's eyes, one can distinguish real faces from GAN-generated ones by carefully examining the sclera area. Following this idea, we first extract the sclera area from a probe image, then feed it into a residual attention network to distinguish GAN-generated faces from real ones. The proposed method is validated on the Flickr-Faces-HQ and StyleGAN2/StyleGAN3-generated face datasets. Experiments demonstrate that the capillary in the sclera is a very effective feature for identifying GAN-generated faces. Our code is available at: https://github.com/109 61020/Deepfake-detector-based-on-blood-vessels.

Keywords: Image Forensics · Generative adversarial networks · GAN-generated faces detection · Physical/physiological constraints · Capillaries

1 Introduction

With the development of Generative Adversarial Networks (GAN) [1], the quality of GAN-generated faces has been greatly improved. It is difficult for humans to distinguish these generated faces from real ones. Such easily synthetic faces can be directly leveraged for disinformation, causing profound social and ethical concerns. To avoid potential security issues caused by GAN-generated faces [2, 3], researchers have developed various novel methods to detect them [4, 5].

Existing detection methods can be roughly divided into three categories. Deep learning-based detectors [6–9] directly learn features from the raw image, alleviating the burden of constructing handcrafted features. For example, Barni et al. compute the co-occurrence of images to train a deep neural network for identifying synthesized faces [7], and Wang et al. observe that the neurons in the network 'react' differently when processing authentic and generated images [8]. The second category is based on the so-called 'GAN fingerprint' [10–13]. These methods focus on the traces left by the deconvolution

© The Author(s), under exclusive license to Springer Nature Singapore Pte Ltd. 2024
B. Luo et al. (Eds.): ICONIP 2023, LNCS 14451, pp. 563–574, 2024.
https://doi.org/10.1007/978-981-99-8073-4_43

and upsampling processes within the GAN model. Despite the high detection rate of these two types of detectors, lacking interpretability hinders their application in serious scenarios. Moreover, the artifacts in the synthesized faces on which these methods are based are prone to image post-processing operations.

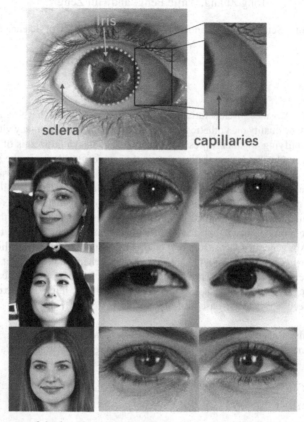

Fig. 1. The structure of the human eye, where the sclera area is highlighted (top). Below it, we show three samples of the eye area. From top to bottom, there are a real face, a face generated by ProGAN, and a face generated by StyleGAN. The details of GAN-generated faces in the sclera are not as rich as the real face.

This paper focuses on the third class of methods, exploring the differences in physical/physiological features between GAN-generated and real faces [14–18]. These methods use the defects of GAN-generated faces in the physiological characteristics, which are usually more interpretable. Specifically, we discover the common defect of lack of capillaries in the sclera of the generated faces. Capillaries in the sclera are manifestations of telangiectasia and congestion in the eyeball, which are common in humans, especially adults. However, as shown in Fig. 1, state-of-the-art GAN models often ignore such an inconspicuous yet intrinsic feature, based on which we can design a detector to recognize GAN-generated faces.

Our proposed GAN-generated face detector consists of several steps. We first segment the sclera region of the eyes. Then, a residual attention network consisting of attention blocks inspired by [19] is used to learn the distinction between real and generated faces. This residual attention network can effectively reveal artifacts of GAN-generated faces, including but not limited to capillaries. To sum up, we exploit the absence of capillaries in the GAN-generated faces as a clue for distinguishing them from real ones. According to our experiments on the Flickr-Faces-HQ dataset [20] and StyleGAN2/StyleGAN3-generated face datasets, this method is highly interpretable and effective.

2 Related Works

2.1 GAN-Generated Face Generation

GAN is one of the most popular architectures for generating realistic-looking faces [20–26]. A thorough survey of the GAN models is out of our scope, and we only focus on some classical GAN models: DCGAN [23], ProGAN [24], and StyleGAN [20, 25, 26]. As a pioneer in face generation, DCGAN is limited by the low resolution of the output images. ProGAN improves the resolution of face images and optimizes facial details by continuously adding convolutional kernels during the training process. StyleGAN decouples the input vector, feeding it into the intermediate generator layer and control the generated style. Although the faces they generate are becoming more realistic, these generators inevitably leave their own traces, such as the inconsistent color or highlights between two eyes [16, 17] and generator-specific fingerprints [10]. These artifacts are valuable clues for revealing generated faces.

2.2 Physical/Physiological-Based GAN-Generated Face Detection

Approaches based on physical/physiological features are particularly notable among various GAN-generated face detectors. These methods offer intuitive interpretability and demonstrate excellent robustness even with image post-processing. Yang et al. [14] distinguish GAN-generated faces by exploiting the distributions of facial landmarks. In [15], GAN-generated faces are detected by analyzing artifacts on the iris. Matern et al. [16] propose distinguishing GAN-generated faces by analyzing the inconsistency of the binocular color of generated faces. However, the artifact of inconsistent iris color mostly appears in ProGAN, rarely in StyleGAN or more advanced models. Hu et al. [17] spot GAN-generated faces by the inconsistency of the highlight on both eyes. Guo et al. [18] exploit the irregularity of the pupil shape in the GAN faces. However, its performance heavily depends on the accuracy of pupil contour extraction, which is challenging for dark-iris faces.

2.3 Attention Mechanism

The human visual system can quickly locate essential objects from complex scenes, which inspires the attention mechanism in various computer vision systems [27]. SENet

obtains each channel's importance by squeezing, promoting, or suppressing corresponding channels according to their importance [28]. Jaderberg et al. propose spatially transformer networks to focus on highly correlated regions [29]. Wang et al. combine attention mechanisms with residuals [19]. Specifically, a deep residual attention network is
obtained by combining channel and spatial attention.

3   Method

Despite the rapid improvement of GAN models, it is difficult for them to learn the subtle
features of human faces thoroughly. An obvious example is the absence of capillaries in
the sclera of GAN-generated faces (See Fig. 1). Inspired by this discovery, we propose a
GAN-generated face detection algorithm by examining the capillaries in the sclera. First,
we mount a face detector to locate the eyes and extract the sclera regions from them.
Then, a residual attention network [19] determines whether the extracted sclera regions
are real or generated. Figure 2 shows the overall pipeline of the proposed GAN-generated
face detector, which is detailed in the subsequent subsections.

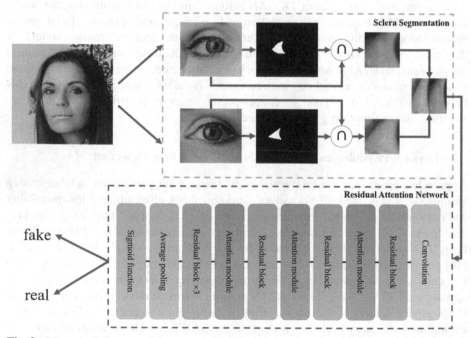

Fig. 2.   The overall diagram of the proposed GAN-generated face detector, which includes a sclera
segmentation module and a residual attention network.

3.1 Sclera Segmentation

Our detector begins with facial landmark extraction using Dlib [30]. Using the landmark points of the eye (The red dots on the eye image in Fig. 3(a)), an eye image (Fig. 3(a)) can be cropped from the probe face image. At the same time, an eye mask is obtained by calculating the convex hull image[1] based on the landmark points. Then, the Otsu algorithm [31] is adopted to binarize the eye image. The binarized image (Fig. 3(b)) is intersected with the eye mask to obtain the potential sclera mask (Fig. 3(c)). We take the maximum connected region of the potential sclera mask as the sclera mask to avoid interference from highlights. The morphological opening is applied to the sclera mask to represent the sclera region better. Finally, sclera regions of both eyes are extracted from eye images based on the sclera masks and stacked horizontally for future use.

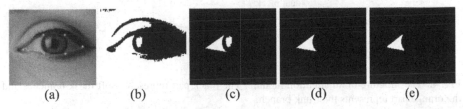

(a)	(b)	(c)	(d)	(e)

Fig. 3. Scleral segmentation process. (a) The eye image. (b) The binarized image of (a). (c) The potential sclera mask. (d) The largest connected component of (c). (e) The sclera mask.

3.2 Residual Attention Network

Our residual attention network is constructed by stacking multiple residual blocks [32] and attention modules. As shown in Fig. 4, an attention module comprises two branches: a trunk branch, which is a residual block, and a soft mask branch. The soft mask branch collects the global information of the input feature map through downsampling, and then restores the output to the size of the input through upsampling. It works as a feature selector that enhances significant features and suppresses noises of the trunk. The output of our attention module can be represented as

$$O(x) = (1 + M(x)) \times T(x). \tag{1}$$

where $T(x)$ and $M(x)$ (ranges from $[0, 1]$) represent the outputs of the trunk branch and the soft mask branch, respectively.

A Sigmoid function is used at the end of the residual attention network to complete the classification. The structure of the residual attention network is detailed in Table 1.

---

[1] https://blogs.mathworks.com/steve/2011/10/04/binary-image-convex-hull-algorithm-notes/.

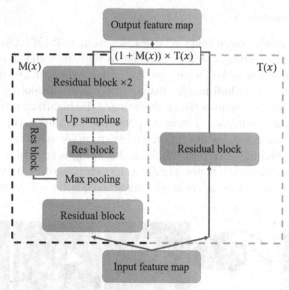

Fig. 4. The structure of an attention module. The black part represents soft mask branches and the orange part represents the trunk branch.

4 Experiments

The real faces are from the Flickr-Faces-HQ dataset, and the GAN-generated faces are created by StyleGAN2[2] and StyleGAN3. All images have a resolution of 1024 × 1024. As [16–18], images with sunglasses and closed eyes are excluded. Five hundred images from StyleGAN2 and 500 images from FFHQ, respectively, are used for training. Since a large volume of high-quality StyleGAN3-generated images is unavailable, the training dataset does not include StyleGAN3-generated images.

The size of the extracted sclera region is adjusted to 96 × 96 for training and testing. The residual attention network consists of three attention modules. The SGD optimizer with a batch size of 4 is used to train the residual attention network. We use a weight decay of 0.0002 with a momentum of 0.9 and set the initial learning rate to 0.001. The total training epochs is 100. In evaluating the robustness of the image post-processing operations, we select six common transformations, namely median filtering, resizing, noise, blur, rotation, and jpeg compression.

4.1 Comparison with the State of the Arts

The proposed method is compared with three physical/physiological-based detectors [16–18] and a data-driven-based detector [7]. In evaluating the performance of [7], we trained two detectors with different volumes of data. One is trained with the same dataset as the proposed method, 500 StyleGAN2-generated images and 500 real ones. The other one is trained with a much larger dataset, 5000 StyleGAN2-generated images and 5000

---

[2] http://thispersondoesnotexist.com.

Table 1. Details of the residual attention network.

Layer	Output Size	Network
Conv1	96 × 96	5 × 5, 64
Max pooling	48 × 48	2 × 2, stride 2
Residual block	48 × 48	$1 \times \begin{Bmatrix} 1 \times 1, 16 \\ 3 \times 3, 16 \\ 1 \times 1, 64 \end{Bmatrix}$
Attention module	48 × 48	1 × Attention
Residual block	24 × 24	$1 \times \begin{Bmatrix} 1 \times 1, 32 \\ 3 \times 3, 32, \text{stride } 2 \\ 1 \times 1, 128 \end{Bmatrix}$
Attention module	24 × 24	1 × Attention
Residual block	12 × 12	$1 \times \begin{Bmatrix} 1 \times 1, 64 \\ 3 \times 3, 64, \text{stride } 2 \\ 1 \times 1, 256 \end{Bmatrix}$
Attention module	12 × 12	1 × Attention
Residual block	12 × 12	$3 \times \begin{Bmatrix} 1 \times 1, 128 \\ 3 \times 3, 128 \\ 1 \times 1, 512 \end{Bmatrix}$
Average pooling	1 × 1	12 × 12, stride 1
FC, Sigmoid	1	

real ones, similar to that used in the original paper. We denote the first detector as [7]a and the second detector as [7]b. Although comparing Physical/physiological-based detectors with a data-driven-based one may be unfair, we are interested in how much the performance margin between them is. Figure 5(a) compares the ROCs of the compared methods on StyleGAN2. Here, 500 real images and 500 StyleGAN2-generated images are evaluated. Both [7]a and [7]b achieve nearly perfect performance when the training set and the test set are from the same source (StyleGAN2). Among the physical/physiological-based detectors, the proposed method outperforms the competitors by a large margin. The area under the curve (AUC) of our method is 0.94, whereas that of the second-best physical/physiological-based detector [16] is only 0.63. The defect of inconsistent color and highlights in both eyes has been largely eliminated in recent StyleGAN2. As a result, detectors based on corresponding artifacts [16, 17] perform unsatisfactorily in the used test set. The poor performance of [18] is because we do not exclude faces with dark irises, as done in [18]. As analyzed in Sect. 2.2, [18] only performs well for faces with light irises.

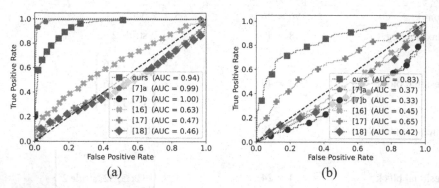

Fig. 5. ROC curves of different detectors when the systhezied images are generated with StyleGAN2 (a) and StyleGAN3 (b).

We use the StyleGAN3-generated images provided by [33] to evaluate the detectors' generalizability. Here, StyleGAN3-generated images and real images each have 250 images. As shown in Fig. 6(b), although both [7] and the proposed method suffer from the mismatch between training (StyleGAN2) and testing set (StyleGAN3), the degree of performance degradation is distinct. The proposed method achieves acceptable performance thanks to its better interpretability, whereas [18] reduces to a random guess.

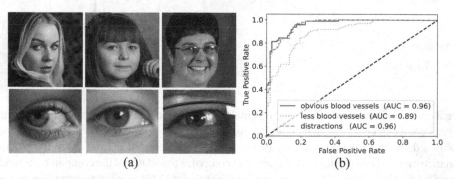

Fig. 6. The influence of eye region feature on the proposed method. (a) Faces of different eye region features. From left to right, eyes with significant capillaries, without significant capillaries, and with disturbances. (b) ROC curves on images with different eye region features.

4.2 How Do Eye Region Features Affect the Proposed Method?

To study the influence of eye region features on the proposed method, we manually divide the test set into three subsets according to the features of the eye region. The faces in the first subset have noticeable capillaries in the sclera, whereas the second has fewer capillaries. The faces in the third subset have significant distractions in the eye area, e.g., wearing glasses. Figure 6(a) gives a sample of each subset. Figure 6(b) plots the ROC curves of the proposed method under the above three scenarios. When there are apparent capillaries in the sclera, the distinguishability of the proposed method is greatly improved over the case without capillaries, AUC = 0.96 vs. AUC = 0.89, which verifies that capillaries are a critical clue for our proposed detector. Figure 6(b) also shows that the proposed method is robust to the eye area disturbance.

4.3 Robustness Analysis

Last, we evaluate the robustness of the detectors against a wide variety of image post-processing operations. For the competitors, we only report the result of [7] on StyleGAN2 since other detectors do not perform well even on non-processed images, as reported in Sect. 4.1. The AUC values after six conventional image post-processings are reported in Table 2. Overall, our method is reasonably robust against various image post-processings. The most severe performance degradation, from AUC = 0.94 to AUC = 0.88, is observed in the case of JPEG compression (QF = 85). In contrast, [7] is only robust to Gaussian blur, no matter how many images are involved in training. Such results confirm that image post-processing does not easily affect physiological characteristics [5].

4.4 Qualitative Result

Figure 7 shows heatmap (extracted from the output of the last residue block) visualizations of both real and GAN-generated sclera examples. Notably, there is a discernible contrast between the heatmap of the GAN-generated sclera and that of the real sclera. Specifically, the residual attention network predominantly focuses on the capillary regions for real sclera and emphasizes edge regions for GAN-generated sclera. This observation represents the residual attention network efficacy in discerning GAN-generated faces.

Table 2. Robustness (in terms of AUC score) of the GAN-face detectors in the presence of post-processing.

Processing	Parameter	StyleGAN2 (proposed/[7]a/[7]b)	StyleGAN3 (proposed)
Median filtering	Window size: 3×3	**0.92**/0.50/0.52	0.83
	Window size: 5×5	**0.89**/0.41/0.41	0.81
Gaussian noise	$\sigma = 0.05$	**0.94**/0.75/0.90	0.83
	$\sigma = 0.1$	**0.94**/0.58/0.65	0.82
	$\sigma = 0.2$	**0.94**/0.49/0.45	0.82
Gaussian blur	$\sigma = 0.05$	0.94/0.99/**1.00**	0.83
	$\sigma = 0.1$	0.94/0.99/**1.00**	0.83
	$\sigma = 0.2$	0.94/0.99/**1.00**	0.83
Rotation	$5°$	**0.93**/0.50/0.63	0.84
	$10°$	**0.92**/0.53/0.68	0.83
	$15°$	**0.92**/0.54/0.69	0.82
Resize	scale factor: $r = 0.8$	**0.92**/0.44/0.43	0.86
	$r = 0.9$	**0.93**/0.44/0.43	0.84
	$r = 1.1$	**0.93**/0.44/0.44	0.82
	$r = 1.2$	**0.92**/0.44/0.44	0.84
JPEG compression	quality factors: 95	**0.92**/0.67/0.73	0.79
	90	**0.90**/0.65/0.70	0.76
	85	**0.88**/0.64/0.68	0.79

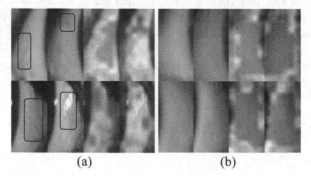

(a)   (b)

Fig. 7. Visualization of the heatmaps derived from an intermediate layer of our residual attention network. (a) and (b) represent real and StyleGAN2-generated images, respectively. This visualization offers an intuitive method for human observers to identify GAN-generated faces by examining their sclera regions.

5 Conclusion

In this work, we point out that the sclera of GAN-generated faces often lacks capillaries common in real faces. Based on this observation, a novel GAN-generated face detector is proposed. We first extract the sclera region from a probe face and then feed it into a residual attention network to determine whether it is generated or authentic. Experiments on the popular FFHQ dataset and synthesized images generated with two recent GANs demonstrate that our proposed method has achieved state-of-the-art performance in identifying GAN-generated faces. We hope the proposed method could enrich the arsenal of GAN-generated face detection.

Acknowledgements. The work is supported by the network emergency management research special topic (no. WLYJGL2023ZD003), the Opening Project of Guangdong Province Key Laboratory of Information Security Technology (Grant No. 2020B1212060078).

References

1. Goodfellow, I., et al.: Generative adversarial nets. In: Neural Information Processing Systems (2014)
2. O'Sullivan, D.: A high school student created a fake 2020 us candidate. Twitter verified it. In: CNN Business (2020). https://cnn.it/3HpHfzz
3. Sganga, N.: Is that Facebook account real? Meta reports "rapid rise" in AI-generated profile pictures. CBS News (2022)
4. Wang, X., Guo, H., Hu, S., Chang, M.C., Lyu, S.: GAN-generated faces detection a survey and new perspectives. arXiv:2202.07145 (2022)
5. Verdoliva, L.: Media forensics and DeepFakes: an overview. arXiv:2001.06564, (2020)
6. Chen, B., Tan, W., Wang, Y., Zhao, G.: Distinguishing between natural and GAN-generated face images by combining global and local features. Chin. J. Electron. **31**, 59–67 (2022)
7. Barni, M., Kallas, K., Nowroozi, E., Tondi, B.: CNN detection of GAN-generated face images based on cross-band co-occurrences analysis. In: 2020 IEEE International Workshop on Information Forensics and Security, pp. 1–6 (2020)
8. Wang, R., et al.: FakeSpotter: a simple yet robust baseline for spotting AI-synthesized fake faces. In: Twenty-Ninth International Joint Conference on Artificial Intelligence and Seventeenth Pacific Rim International Conference on Artificial Intelligence (2019)
9. Wang, S.Y., Wang, O., Zhang, R., Owens, A., Efros, A.A.: CNN-generated images are surprisingly easy to spot... for now. In: Proceedings of the IEEE/CVF Conference on Computer Vision and Pattern Recognition, pp. 8695–8704 (2020)
10. Marra, F., Gragnaniello, D., Verdoliva, L., Poggi, G.: Do GANs leave artificial fingerprints? In: IEEE Conference on Multimedia Information Processing and Retrieval, pp. 506–511 (2019)
11. Yu, N., Davis, L.S., Fritz, M.: Attributing fake images to GANs: learning and analyzing GAN fingerprints. In: Proceedings of the IEEE/CVF International Conference on Computer Vision, pp. 7556–7566 (2019)
12. Pu, J., Mangaokar, N., Wang, B., Reddy, C.K., Viswanath, B.: Noisescope: Detecting deepfake images in a blind setting. In: Annual Computer Security Applications Conference, pp. 913–927 (2020)
13. Frank, J., Eisenhofer, T., Schönherr, L., Fischer, A., Kolossa, D., Holz, T.: Leveraging frequency analysis for deep fake image recognition. In: International Conference on Machine Learning, pp. 3247–3258 (2020)

14. Yang, X., Li, Y., Qi, H., Lyu, S.: Exposing GAN-synthesized faces using landmark locations. In: Proceedings of the ACM Workshop on Information Hiding and Multimedia Security, pp. 113–118 (2019)
15. Guo, H., Hu, S., Wang, X., Chang, M.C., Lyu, S.: Robust attentive deep neural network for exposing GAN-generated faces. In: IEEE Access**10**, 32574–32583 (2022)
16. Matern, F., Riess, C., Stamminger, M.: Exploiting visual artifacts to expose deepfakes and face manipulations. In: IEEE Winter Applications of Computer Vision Workshops, pp. 83–92 (2019)
17. Hu, S., Li, Y., Lyu, S.: Exposing GAN-generated faces using inconsistent corneal specular highlights. In: IEEE International Conference on Acoustics, Speech and Signal Processing, pp. 2500–2504 (2021)
18. Guo, H., Hu, S., Wang, X., Chang, M.C., Lyu, S.: Eyes tell all: irregular pupil shapes reveal GAN-generated faces. In: IEEE International Conference on Acoustics, Speech and Signal Processing, pp. 2904–2908 (2022)
19. Wang, F., et al.: Residual attention network for image classification. In: Proceedings of the IEEE Conference on Computer Vision and Pattern Recognition, pp. 3156–3164 (2017)
20. Karras, T., Laine, S., Aila, T.: A style-based generator architecture for generative adversarial networks. In: Proceedings of the IEEE/CVF Conference on Computer Vision and Pattern Recognition, pp. 9243–9252 (2019)
21. Brock, A., Donahue, J., Simonyan, K.: Large scale GAN training for high fidelity natural image synthesis. arXiv:1809.11096 (2018)
22. Goetschalckx, L., Andonian, A., Oliva, A., Isola, P.: Ganalyze: toward visual definitions of cognitive image properties. In: Proceedings of the IEEE/CVF International Conference on Computer Vision, pp. 5744–5753 (2019)
23. Radford, A., Metz, L., Chintala, S.: Unsupervised representation learning with deep convolutional generative adversarial networks. In: International Conference on Learning Representations (2015)
24. Karras, T., Aila, Y., Laine, S., Lehtinen, J.: Progressive growing of GANs for improved quality, stability, and variation. In: International Conference on Learning Representations (2017)
25. Karras, T., Laine, S., Aittala, M., Hellsten, J., Lehtinen, J., Aila, T.: Analyzing and improving the image quality of StyleGAN. In: Proceedings of the IEEE/CVF Conference on Computer Vision and Pattern Recognition, pp. 8110–8119 (2020)
26. Karras, T., et al.: Alias-free generative adversarial networks. In: Advances in Neural Information Processing Systems, pp. 852–863 (2021)
27. Guo, M.H., et al.: Attention mechanisms in computer vision: a survey. Comput. Vis. Media **8**, 331–368 (2022)
28. Hu, J., Shen, L., Sun, G.: Squeeze-and-excitation networks. In: Proceedings of the IEEE Conference on Computer Vision and Pattern Recognition, pp. 7132–7141 (2018)
29. Jaderberg, M., Simonyan, K., Zisserman, A.: Spatial transformer networks. In: Advances in Neural Information Processing Systems (2015)
30. King, D.E.: Dlib-ml: a machine learning toolkit. J. Mach. Learn. Res. **10**, 1755–1758 (2009)
31. Pe Otsu, N.: A threshold selection method from gray-level histograms. In: IEEE Transactions on Systems, Man, and Cybernetics, pp. 62–66 (1979)
32. He, K., Zhang, X., Ren, S., Sun, J.: Identity mappings in deep residual networks. In: Leibe, B., Matas, J., Sebe, N., Welling, M. (eds.) ECCV 2016. LNCS, vol. 9908, pp. 630–645. Springer, Cham (2016). https://doi.org/10.1007/978-3-319-46493-0_38
33. Corvi, R., Cozzolino, D., Zingarini, G., Poggi, G., Nagano, K., Verdoliva, L.: On the detection of synthetic images generated by diffusion models. In: IEEE International Conference on Acoustics, Speech and Signal Processing, pp. 1–5 (2023)

A Developer Recommendation Method Based on Disentangled Graph Convolutional Network

Yan Lu[1], Junwei Du[1], Lijun Sun[1], Jinhuan Liu[1], Lei Guo[2], Xu Yu[1,3,4(✉)] (ID),
Daobo Sun[5], and Haohao Yu[6]

[1] School of Information Science and Technology, Qingdao University of Science
and Technology, Qingdao 266061, China
luyan@mails.qust.edu.cn, lijunsun@qust.edu.cn
[2] School of Information Science and Engineering, Shandong Normal University,
Jinan 250014, China
[3] Qingdao Institute of Software, China University of Petroleum,
Qingdao 266580, China
yuxu0532@upc.edu.cn
[4] Key Laboratory of Symbol Computation and Knowledge Engineering,
Jilin University, Changchun 130012, China
[5] College of Computer Science and Technology, Harbin Engineering University,
Harbin 150001, China
[6] QingDao Innovation And Development Base, Harbin Engineering University,
Harbin, China

Abstract. Crowdsourcing Software Development (CSD) solves software
development tasks by integrating resources from global developers. With
more and more companies and developers moving onto CSD platforms,
the information overload problem of the platform makes it difficult to
recommend suitable developers for the software development task. The
interaction behavior between developers and tasks is often the result of
complex latent factors. Existing developer recommendation methods are
mostly based on deep learning, where the feature representations ignores
the influence of latent factors on interactive behavior, leading to learned
feature representations that lack robustness and interpretability. To solve
the above problems, we present a Developer Recommendation Method
Based on Disentangled Graph Convolutional (DRDGC). Specifically, we
use a disentangled graph convolutional network to separate the latent
factors within the original features. Each latent factor contains specific
information and is independent from each other, which makes the features constructed by the latent factors exhibit stronger robustness and
interpretability. Extensive experiments results show that DRDGC can
effectively recommend the right developer for the task and outperforms
the baseline methods.

Keywords: Crowdsourcing Software Development · Developer
Recommendation · Disentangle Representation Learning · Graph
Representation Learning

© The Author(s), under exclusive license to Springer Nature Singapore Pte Ltd. 2024
B. Luo et al. (Eds.): ICONIP 2023, LNCS 14451, pp. 575–585, 2024.
https://doi.org/10.1007/978-981-99-8073-4_44

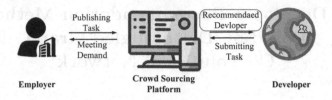

Fig. 1. The crowdsourcing model usually involves three different types of role.

1 Introduction

Crowdsourcing Software Development is a new software development mode that outsources the software development tasks of enterprises to individuals with development capabilities. Compared with traditional software development, CSD can make greater use of scattered developer resources around the world to complete complex tasks through group collaboration. Up to now, a variety of crowdsourcing platforms supporting software tasks already exist on the Internet, such as Topcoder, Upwork, GitHub, etc. In the crowdsourcing scenario, the relationship between employers, platforms, and developers as shown in Fig. 1.

For example, the Topcoder platform has more than 10 million developers who have collaborated on close to 4 million software development tasks. While bringing huge benefits, the huge number of developers and tasks inevitably creates the problem of information overload, making it difficult to find the right developers for tasks. In this background, developer recommendation methods for CSD platforms have important theoretical and applied research value and have attracted the attention of many researchers in recent years.

Recommendation systems are effective means to alleviate information overload. Refs. [9] recommend developers for tasks through similarity relations of explicit features such as interaction records and descriptive information. Refs. [10] enhance developer recommendation systems by introducing additional information, such as social network. Developer-task interaction behavior is typically driven by various latent factors. For instance, in crowdsourcing platforms, a developer may accept a task due to reasons such as technical ability, compensation, and timeliness. However, existing graph convolution-based developer recommendation methods overlook the latent factors that determine the graph structure when learning node representations by treating neighborhoods as a perceptual whole. As a result, the learned node representations lack robustness to noisy data and interpretability.

The explicit features of CSD platforms contain many programming languages and skills with high dimensionality and sparsity. Considering that disentangled representation learning can map high-dimensional sparse explicit features to a smaller number of latent factors (each latent factor consists of high-dimensional implicit features). The novel features consisting of latent factors have better robustness and interpretability [6].

Specifically, we model the task-developer interaction as a bipartite graph and propose a Developer Recommendation Method Based on Disentangled Graph Convolutional (DRDGC). First, we dynamically mine latent factors affecting task-developer interactions by the disentangled graph convolutional module combined with information from graph structures and neighboring nodes. The novel features re-modeled from latent factors have better robustness and interpretability. Further, a solution to the cold start problem is proposed for newly released tasks and newly registered developers on the bipartite graph.

2 Related Work

2.1 Crowdsourcing Software Development

In the last decade, although the research and application of recommendation systems in e-commerce, social, music, video, books, and service categories have achieved great results. However, the research on recommender systems for resource allocation and collaborative relationships in the field of software engineering has not progressed very much. It was found that due to the sufficiently large number of tasks and developers, the achievment matrix between developers and tasks is very sparse, and only a few developers and a few tasks have achievement relationships with each other, which leads to poor performance of traditional recommendation algorithms. To solve the problem of sparsity, researchers have conducted a lot of research [5].

2.2 Graph Neural Network

Recently, graph neural networks (GNN) [12] have received increasing attention in the field of recommender systems because of their powerful characterization capabilities based on node features and graph structure. Wang et al. [8] explored GNNs to capture higher-order connectivity information in bipartite graphs by propagating node features over them, which led to better performance of graph-based recommender systems. However, existing graph recommendation methods focus on and rely heavily on the descriptive features of the nodes. Hu et al. [4] modeled user news interactions as a graph in news recommendation and proposed a graph convolution model combining long-term and short-term interests to demonstrate the effectiveness of using the graph structure of user news interactions.

2.3 Disentangle Representation Learning

Disentangle representation learning aims to isolate the latent factors hidden in the interaction behavior [1]. For example, Higgins et al. [3] proposed a constrained variational auto-encoder based decentered representation for machine vision. Recently, the learning to disentangle representations of interaction behaviors has attracted increasing attention. Ma et al. [6] proposed a disentangled

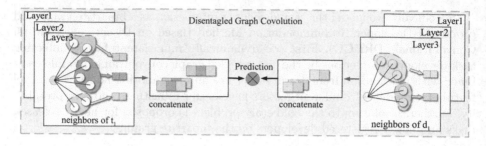

Fig. 2. DRDGC Framework Diagram.

graph convolutional network to learn node representations on graphs that use a neighbor routing mechanism to identify latent factors and aggregate features specific to that latent factor by convolution.

3   Method

In this paper, we propose a Joint Algorithm Based on Disentangled Graph Convolutional and Fuzzy C-Means for the Developer Recommendation. Figure 2 shows a high-level overview of this method.

3.1   Problem Description

In software crowdsourcing platforms, three kinds of objects exist task, developer, and crowdsourcing platform. The interaction between them is that the developer takes the task through the crowdsourcing platform, and after completing the task, this interaction will have a corresponding grade. Tasks include existing tasks and new release tasks(cold start tasks). Developers include existing developers and newly registered developers (cold start developers).

Figure 3 shows the task-developer interaction in the crowdsourcing platform visualized by the achievement matrix, where we use the size of the number to indicate the achievement of the developer in the task, and the part of the number 0 indicates that the developer is not involved in the task. The entire row and column of cold start tasks and developers are non-existent grades since there is no record of any interaction. Our goal is to fill the gap of cold start lines in the grades matrix by assigning the most suitable developer to the task with the information already available in the crowdsourcing platform.

3.2   Feature Representation

The software crowdsourcing platform provides a variety of types of features, from text to numeric. Text features of the developer include self-description, skill, and the type of developer. Numeric features of the developer include the date of registration, etc. The text feature of the task includes the title, the

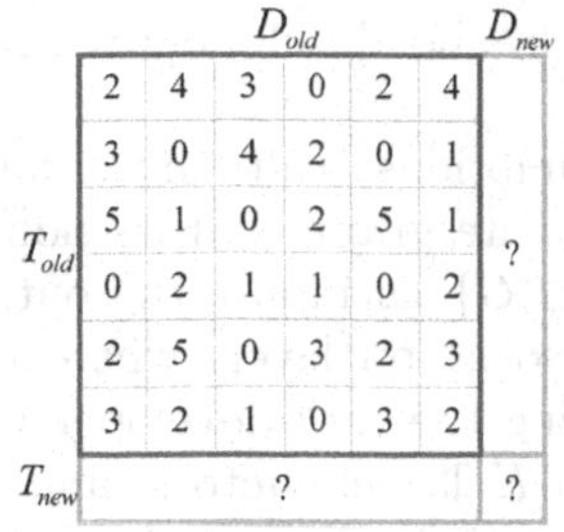

	D_{old}					D_{new}	
	2	4	3	0	2	4	
	3	0	4	2	0	1	
T_{old}	5	1	0	2	5	1	?
	0	2	1	1	0	2	
	2	5	0	3	2	3	
	3	2	1	0	3	2	
T_{new}			?				?

Fig. 3. Task-developer interaction grade matric.

description of the task, and the required skill. The numerical feature of the task includes the date of posting, date of submission, task rewards, etc.

Although the title and description contain a large amount of text feature, most of the redundant feature is not positive for model training. The set of keywords includes available information such as title, skill, type, etc. In addition, to represent unstructured data such as keywords, we use the one-hot model to encode the keywords set.

3.3 Graph Disentangle Module

Data Balancing Process. In developer recommendation systems, historical task-developer interaction is usually used as input. In this paper, we represent the historical interaction data as a bipartite graph $G = \{(d,t)|d \in D, t \in T\}$, where $T = T_{old} \cup T_{new}$ and $D = D_{old} \cup D_{new}$ represent the tasks set and developers set, respectively.

Disentangled Graph Convolutional. For most graph convolutional networks, when given the features of a node and its neighbors, the output of the node at the layer is represented as $y_t = f(x_t, \{x_d : (t,d) \in G\})$. where $f()$ denotes the convolution operation, and the output y_t is the output feature representation of the task node in the convolution layer aggregating the features of neighboring nodes. The idea is that the non-identical neighbors of a node contain rich information that can be used to describe the node more comprehensively.

In addition, y_t is a disentangled representation. In detail, the task features y_t consist of K components that represents the existence of K latent factors to be disentangled from the task-developer interaction, i.e., $y_t = [r_1, r_2, ...r_K]$, where $r_k \in R^{\frac{d_{out}}{k}} (1 \leq k \leq K)$, the kth component r_k is used to characterize the node in the kth latent aspect. The key challenge of this process is to divide the non-identical nodes with edges in the graph into K subsets, which are used to describe each of the latent factors.

We assume that in a bipartite graph, multiple latent factors influence the interaction between developers and tasks, and these latent factors may be the type of task, the skills acquired by developers, etc. There are usually many developers involved in a task, and developers with the same latent factors are clustered in the graph.

Based on the above assumptions, we take the initialized features of the task node $x_t \in R^{d_{in}}$ in the bipartite graph and its non-similar developer neighbor nodes $\{x_d \in R^{d_{in}} : (t,d) \in G\}$ as input, the output of $y_t = [r_1, r_2, ...r_K] \in R^{d_{out}}$ the convolutional dissociation layer. where r_k represents the feature of the kth latent factor affecting the task-developer interaction. We assume that the deconvolution layer has K latent factors, and map the task node and its developer neighbor nodes $i \in \{\{t\} \cup \{d : (t,d) \in G\}\}$ to K different subspaces to obtain a representation $z_{i,k}$ of each node in the subspace.

$$z_{i,k} = \frac{\sigma(\mathrm{W}_k{}^{\mathrm{T}} x_i + b_k)}{||\sigma(\mathrm{W}_k{}^{\mathrm{T}} x_i + b_k)||_2} \tag{1}$$

where $\mathrm{W}_k \in R^{d_{in} \times \frac{d_{out}}{K}}$ and $b_k \in R^{\frac{d_{out}}{K}}$ is the parameter of each channel, σ is the nonlinear activation function. We use $L2$ normalization to ensure numerical stability and to prevent neighbors with overly rich features (e.g., multiple keywords) from influencing our predictions.

$$p_{d,k}^{(t)} = \frac{\exp(z_{d,k}^T r_k^{(t)}/\tau)}{\sum\limits_{k'=1}^{K} \exp(z_{d,k'}^T r_{k'}^{(t)}/\tau)} \tag{2}$$

We let P denote the probability that neighbor node d is partitioned into the kth subset because of the latent factor k, where $p > 0$, $\sum\limits_{k=1}^{K} p_{d,k} = 1$ and τ is the hyperparameter controlling the hardness of the assignment. It is worth noting that each neighbor node can be divided into only one subset, to find the most influential latent factors in the task-developer interaction process. By continuously and iteratively dividing the neighbors into subsets and reconstructing r_k, under the assumption that neighboring nodes with the same latent factor are clustered into clusters, we can then extract the features of that latent factor in the clusters, which r_k can be considered as the center of each subspace cluster.

$$r_k^{(t)} = \frac{z_{t,k} + \sum\limits_{d:(t,d)\in G} p_{d,k}^{(t-1)} z_{d,k}}{||z_{t,k} + \sum\limits_{d:(t,d)\in G} p_{d,k}^{(t-1)} z_{d,k}||_2} \tag{3}$$

Recommendations Based on Reconstructed Individual Characteristics. The disentangled graph convolutional module requires L convolutional layers, $f^l()$ denoting the lth convolutional layer. For the task node, the output of the lth convolutional layer $y_t^l \in R^{K^l \times \Delta d}$. where K^l is the number of channels in the lth convolutional layer and Δd is the dimension of the output of each channel of fixed size. We also add the constraint $K^1 \geq K^2 \geq ... \geq K^L$.

$$y_t^{(l)} = dropout(f^l(y_t^{(l-1)}, \{y_d^{(l-1)} : (t,d) \in G\})) \tag{4}$$

where $1 \leq l \leq L$, and $y_t^{(0)} = x_t$.

The last layer of the deconvolutional modular network is the fully connected layer, denoted as $y^{(L+1)} = W^{(L+1)^T} y^{(L)} + b^{(L+1)}$, where $W^{(L+1)} \in R^{K^{(L)} \times \Delta d \times C}$, and $b^{(L+1)} \in R^C$.

$y_t^{(L+1)}$ and $y_d^{(L+1)}$ are the output of the fully connected layer, which represent the individual characteristics of the task and developer of the fused latent factors, respectively, will be used as the input of the FCM module, and their inner product will be used as the result of individual matching recommendations.

$$\hat{R}_1 = y_t y_d \tag{5}$$

Cold Start Problem. Although we have achieved comprehensive modeling of individuals and groups for the developer recommendation problem by the disentangled graph convolutional module and FCM module, the model is not yet able to solve the cold start problem at this stage. This is because cold-start nodes do not exist edges in the bipartite graph, and reconstructing the bipartite graph for training would consume a lot of computational resources.

Since only explicit features exist for the cold start nodes, we chain the cold start nodes into the bipartite graph based on the explicit features of the cold start nodes, as shown in Fig. 4. First, we initialize the features of the cold start node in the same way as in Subsect. 3.2, so that the features of the cold start node and the history node are in the same feature space. Next, the similarity between the cold start nodes and similar nodes is calculated, and we choose the commonly used cosine similarity as the similarity measure.

$$s = \frac{a \times b}{||a|| \times ||b||} \tag{6}$$

where $S_T = \{s_1, s_2, ..., s_T\}$ to denote the similarity of the cold start task nodes to the task nodes in the graph, further processed by *softmax*, $S_T = soft \max(S_T)$ as the similarity set, and the *Top k* nodes with similarity are taken as the similar node-set $N = Topk(soft \max(S_T))$. Finally, the edges of similar nodes are randomly wandered and the *Top k* with the largest weight among the nodes connected to them (non-similar) is selected as the link object of the cold start nodes.

$$\hat{R}_2^{cs} = D_{rw} R I \tag{7}$$

where D_{rw} is the matrix of indications for random wandering, R is the matrix of indications for raw scores, and I is the matrix of indications for raw scores.

This is a relatively conservative strategy but can be effective in improving recommendation results. In addition, it is relatively easier to deploy since it does not require additional training data.

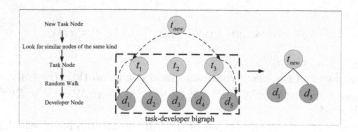

Fig. 4. Cold start node linking strategy.

4 Experiment and Results

4.1 Data Preparation

Most of the tasks on the Topcoder platform are published as competitions, and more than 10 million developers have registered with Topcoder and completed 4 million tasks. In this experiment, we collect data on tasks of the competitions, including design, development programming, module testing, and data science. Since the Topcoder platform has adopted an anti-crawler strategy in recent years, we collected data from 2015 to 2020, and although the number of tasks and developers is lower than we expected, the dataset can still support our experiment.

We randomly select 80%, 60%, 40%, and 20% of the tasks and the developers corresponding to each task as the training dataset, denoted as TR80, TR60, TR40, and TR20. Next, the remaining 20%, 40%, 60%, and 80% of tasks are filtered to ensure that the developers in the test dataset are the ones already present in the training dataset. The filtered test datasets were noted as TE20, TE40, TE60, and TE80. The statistics of the processed dataset are shown in Table 1.

Table 1. Statistical information of the dataset.

Dataset	Task	Developer	Interaction	Sparsity
TR80	10890	2586	273507	98.028%
TR70	7174	1951	124296	98.433%
TR60	6116	1534	62505	98.471%
TR50	5058	1017	37155	98.327%

4.2 Parameter Setting

For modeling the disentangled graph convolution module from explicit to implicit features, we use a uniform feature dimension, which is finally evaluated on the

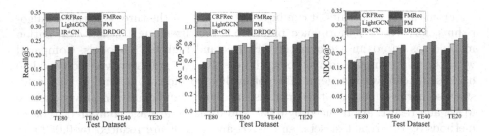

Fig. 5. Results of DRDGC compared with other methods.

validation dataset by fine-tuning the hyperparameters on the test dataset. We give the optimal parameters for the method in this paper. The task feature dimension of 100. The developer feature dimension of 150. Learning rate: 2e-05. Train batch size: 16. Eval batch size: 16. Optimizer: Adam with betas (0.9, 0.999). Number of epochs: 8* 1000.

4.3 Comparison Method

To fully validate the effectiveness of the DRDGC, **FMRec**, **LightGCN**, **PM**, **CRFRec** and **IR+CN** [2,7,10,11,13] five baseline methods were used to demonstrate the validity of our approach.

4.4 Evaluation Indicators

MAE and RMSE are often used to evaluate the accuracy of score prediction. The score prediction is derived from feature operations, and they can be used as evaluation metrics to measure the feature extraction ability of the disentangled graph convolutional module. However, developer recommendation as a *Top k* recommendation scenario, we cannot measure the performance of the model in terms of overall rating prediction accuracy, but use Recall and NDCG as evaluation metrics for the *Top k* recommendation scenario.

4.5 Results Comparison

We compared DRDGC and the comparison between DRDGC and the state-of-the-art algorithm on the TE20 test dataset is shown in Fig. 5. From the experimental results, it can be seen that DRDGC largely outperforms the compared baseline method in three evaluation indicators. NDCG, ACC-Top, and Recall, and has a 2.64%, 1.37%, and 2.3% improvement in TE20 over the current best-performing method, respectively.

In this paper, the MAE and RMSE comparison results of the DRDGC with CRFRec and LightGCN algorithms on the test dataset are shown in Table 2, where Table 3 show the MAE and RMSE comparison of the three algorithms on the test dataset in the cold start case.

From the above results, it can be seen that among the three comparison algorithms, for overall performance prediction, the disentangle graph convolution module has an advantage over the CRFRec and LightGCN models. Compared with the two comparison methods, the disentangle graph convolution module has a significant advantage in different evaluation indicators. And in both cold-start scenarios, all evaluation metrics are more significantly improved, where MAE and RMSE are reduced by 0.038 and 0.102 on average compared to other methods on different test sets, and MAE and RMSE are reduced by 0.052 and 0.113 on average in the tests under cold-start scenarios. The disentangled graph convolution can achieve more significant results than the other two methods in the cold start scenarios, further indicating that the features extracted by the disentangled graph convolution have stronger representation ability and robustness.

Table 2. The MAE and RMSE comparison results of the three algorithms.

Methods	MAE				RMSE			
	TE80	TE60	TE40	TE20	TE80	TE60	TE40	TE20
CRFRec	0.9525	0.8927	0.8494	0.8154	1.2636	1.1873	1.1272	1.0763
LightGCN	0.9158	0.8736	0.8315	0.8068	1.2051	1.1342	1.0934	1.0990
DRDGC	0.8933	0.8648	0.8205	0.7603	1.1598	1.0827	1.0497	0.9716

Table 3. The MAE and RMSE comparison results of the three methods in the cold start case.

Methods	MAE				RMSE			
	TE80	TE60	TE40	TE20	TE80	TE60	TE40	TE20
CRFRec	1.0023	0.9088	0.8603	0.8356	1.4149	1.2178	1.1352	1.1030
LightGCN	0.9548	0.8807	0.8412	0.8162	1.2367	1.1442	1.1265	1.0597
DRDGC	0.9437	0.8725	0.8359	0.7866	1.1854	1.1280	1.0716	0.9937

5   Conclusion

Aiming at the developer recommendation problem of the software crowdsourcing platform, this paper proposes a developer recommendation method based on disentangled graph convolutional. This method uses latent factors that influence developer-task interactions as features by disentangled graph convolution, which has stronger robustness and interpretability. Further, a node-linking strategy is proposed for newly released tasks or newly registered developers, which can effectively solve the cold start problem. Extensive experiments on the Topcoder platform dataset fully demonstrate the effectiveness of the developer recommendation method. In the future, we will extend this model to further explore the potential value of latent factors.

Acknowledgments. This work is jointly sponsored by National Natural Science Foundation of China (Nos. 62172249, 62202253), Natural Science Foundation of Shandong Province (Nos. ZR2021MF092, ZR2021QF074), the Fundamental Research Funds for the Central Universities, JLU (No. 93K172022K01).

References

1. Cao, J., Lin, X., Cong, X., Ya, J., Liu, T., Wang, B.: DisenCDR: learning disentangled representations for cross-domain recommendation. In: International ACM SIGIR Conference on Research and Development in Information Retrieval, pp. 267–277 (2022)
2. He, X., Deng, K., Wang, X., Li, Y., Zhang, Y., Wang, M.: LightGCN: simplifying and powering graph convolution network for recommendation. In: International Conference on Research and Development in Information Retrieval, pp. 639–648 (2020)
3. Higgins, I., et al.: Towards a definition of disentangled representations. arXiv preprint arXiv:1812.02230 (2018)
4. Hu, L., Li, C., Shi, C., Yang, C., Shao, C.: Graph neural news recommendation with long-term and short-term interest modeling. Inf. Process. Manage. **57**(2), 102142 (2020)
5. Jiang, H., et al.: DupHunter: detecting duplicate pull requests in fork-based development. IEEE Trans. Software Eng. **49**(4), 2920–2940 (2023)
6. Ma, J., Cui, P., Kuang, K., Wang, X., Zhu, W.: Disentangled graph convolutional networks. In: International Conference on Machine Learning, pp. 4212–4221 (2019)
7. Rendle, S.: Factorization machines. In: International Conference on Data Mining, pp. 995–1000 (2010)
8. Wang, X., He, X., Wang, M., Feng, F., Chua, T.S.: Neural graph collaborative filtering. In: International Conference on Research and Development in Information Retrieval, pp. 165–174 (2019)
9. Xia, X., Lo, D., Wang, X., Zhou, B.: Accurate developer recommendation for bug resolution. In: Working Conference on Reverse Engineering, pp. 72–81 (2013)
10. Yu, Y., Wang, H., Yin, G., Wang, T.: Reviewer recommendation for pull-requests in GitHub: what can we learn from code review and bug assignment? Inf. Softw. Technol. **74**, 204–218 (2016)
11. Zhang, Z., Sun, H., Zhang, H.: Developer recommendation for Topcoder through a meta-learning based policy model. Empir. Softw. Eng. **25**, 859–889 (2020)
12. Zhou, J., et al.: Graph neural networks: a review of methods and applications. AI Open **1**, 57–81 (2020)
13. Zhu, J., Shen, B., Hu, F.: A learning to rank framework for developer recommendation in software crowdsourcing. In: International Conference on Asia-Pacific Software Engineering Conference, pp. 285–292 (2015)

Novel Method for Radar Echo Target Detection

Zhiwei Chen$^{(\boxtimes)}$, Dechang Pi, Junlong Wang, and Mingtian Ping

Nanjing University of Aeronautics and Astronautics, Nanjing 211106, China
czw_czw_czw@nuaa.edu.cn

Abstract. Radar target detection, as one of the pivotal techniques in radar systems, aims to extract valuable information such as target distance and velocity from the received energy echo signals. However, with the advancements in aviation and electronic information technology, there have been profound transformations in the radar detection targets, scenarios, and environments. The majority of conventional radar target detection methods are primarily based on Constant False Alarm Rate (CFAR) techniques, which rely on certain distribution assumptions. However, when the detection scenarios become intricate or dynamic, the performance of these detectors is significantly influenced. Therefore, ensuring the robust performance of radar target detection models in complex task scenarios has emerged as a crucial concern. In this paper, we propose a radar target detection method based on a hybrid architecture of convolutional neural networks and autoencoder networks. This approach comprises clutter suppression and target detection modules. We conducted ablation experiments and comparative experiments using publicly available radar echo datasets and simulated radar echo datasets. The ablation experiments validated the effectiveness of the clutter suppression module, while the comparative experiments demonstrated the superior performance of our proposed method compared to the alternative approaches in complex background scenarios.

Keywords: Radar Target Detection · Convolutional Neural Network · AutoEncoder

1 Introduction

Radar target detection is a technology that utilizes radar signals to detect, identify, and track targets. In radar target detection, the radar emits signals that are reflected back by the targets, generating echo signals. By processing and analyzing these echo signals, valuable target information can be extracted from radar echoes that are contaminated with noise, clutter, and interference signals, enabling target detection and identification [1]. Furthermore, radar systems offer advantages over image-based data acquisition methods as they are not affected by weather conditions and lighting, providing higher accuracy and reliability. As a result, radar target detection techniques have wide-ranging applications in military, civilian, and remote sensing domains [2–4].

Currently, radar target detection methods can be categorized into two main categories: traditional radar signal processing techniques and artificial intelligence methods.

B. Luo et al. (Eds.): ICONIP 2023, LNCS 14451, pp. 586–597, 2024.
https://doi.org/10.1007/978-981-99-8073-4_45

Traditional radar target detection methods based on signal processing employ techniques such as matched filtering, Doppler processing, and clutter suppression to enhance the signal-to-noise ratio. After preprocessing, radar target detection is performed using Constant False Alarm Rate (CFAR) detectors. Generally, traditional radar target detection methods based on signal processing perform well in distinguishing targets in medium-to-high signal-to-noise ratio radar data. However, the performance of these models depends not only on the signal-to-noise ratio but also on the pre-defined parameters of the models.

In recent years, advancements in electronic information technology and aviation have led to increasingly complex radar target detection scenarios. Notably, the detection of small, slow-moving targets such as unmanned aerial vehicles (UAVs) poses significant challenges due to the complexity of the scenes and the accompanying clutter and multipath interference. These factors result in a sharp decrease in the signal-to-noise ratio of target echoes, making radar target detection in complex scenarios highly challenging. Traditional CFAR-based radar target detection methods, which are based on statistical theories and treat the environment or targets as random processes [5], face difficulties in predefining model parameters for complex scenes. This leads to significant performance variations of radar target detection models in different environmental backgrounds, indicating a lack of generality and robustness in the models.

On the other hand, with the rapid development of artificial intelligence technology, there has been an increasing exploration of related techniques in the field of radar target detection [6–8]. Machine learning and deep learning models possess strong capabilities to abstract and extract features from large datasets, and they can accomplish complex learning tasks in an end-to-end manner. Currently, machine learning and deep learning models are mainly applied in the downstream classification and detection stages of radar target detection tasks [9]. However, the suppression of clutter still relies on traditional radar signal processing methods, leading to a significant dependency of model accuracy on upstream modules, particularly in complex detection tasks.

In this paper, we propose a deep learning-based radar target detection neural network model to address the challenges of complex environmental parameter predefinition and the performance coupling issue when combining traditional signal processing with artificial intelligence models. The proposed model directly utilizes target radar echo data for target distance and velocity calculations. Comparative experiments are conducted using publicly available radar echo datasets and simulated radar echo datasets. The proposed model demonstrates superior detection performance compared to the state-of-the-art models.

2   Related Work

Research on radar target detection methods can be divided into two categories from a theoretical perspective: traditional signal processing-based methods and machine learning/deep learning-based methods. Additionally, from the perspective of input data format, it can be categorized into radar echo data, range-Doppler spectra, pulse range profiles, SAR images, PPI images, and time-frequency images.

In reference [10], Sciotti et al. validated that detectors with a constant false alarm rate (CFAR) based on the average cell under a uniform background exhibit the highest detectability. Gandhi et al. in reference [11] demonstrated that CFAR detectors with a constant false alarm rate can suffer severe performance degradation in the presence of interfering targets or clutter background transitions. As a result, they proposed the Greatest-of (GO)-CFAR and Smallest-of -CFAR(SO-CFAR) detectors based on the maximum and minimum selection of CFAR thresholds, respectively. S. Blake in [12] proposed Ordered Statistics of CFAR (OS-CFAR) for multi-target scenarios.

As for artificial intelligence based techniques, artificial intelligence approaches have also validated their potential in the field of radar target detection tasks. S. M. D. Rizvi et al. in the literature [18] used ANN for radar signal detection in a K-distribution environment. Two training algorithms were tested, back propagation (BP) and genetic algorithm (AG) for the MLP architecture. Simulation results show that the MLP architecture outperforms the classical CA-CFAR detector.Callaghan et al. explored new uses of machine learning techniques to address clutter suppression in the literature [35].

Akhtar et al. in reference [5] combined artificial neural network detectors with the GO-CFAR detector, developing and extending the use of neural networks. Through training, the neural network could identify targets in specified environments and satisfy the conditions proposed by traditional CFAR detectors. The trained neural network provided better detection performance.

Callaghan et al. explored new applications of machine learning techniques for clutter suppression in radar in reference [19]. They compared the results obtained from using two machine learning techniques for clutter suppression in a maritime clutter environment. The experimental data was collected using an experimental S-band radar system called NetRAD. The radar system was observing coastal scenes, and the collected data was classified as targets or clutter using the K-Nearest Neighbors (KNN) and Support Vector Machine (SVM) algorithms.

In reference [20] Wang et al. analyzed the potential application of Convolutional Neural Networks (CNNs) in target detection in radar. They designed a CNN-based detector and demonstrated its performance through comparisons with traditional signal processing techniques applied to the data.

Yavuz et al. proposed a deep learning-based radar target detection technique to replace conventional radar signal processing techniques in reference [24]. The proposed technique involved a Convolutional Neural Network (CNN) with the input being the range-Doppler ambiguity function. The method was compared with the classical Constant False Alarm Rate (CA-CFAR) detector, and the results showed that the proposed method outperformed CFAR by approximately a factor of 2 in detection probability (PD) and by approximately a factor of 10 in false alarm probability (PFA) with minimal computational cost.

References [26, 27], and [28] focus on preprocessing radar data or converting SAR images and PPI images to different data formats for target detection using deep learning algorithms, specifically the Faster R-CNN object detection algorithm, which relies on deep learning for noise separation and target recognition in images.

In reference [32], Jiang et al. proposed a novel CNN-based radar target detector that directly locates targets in multi-dimensional space, including range, velocity, azimuth, and elevation, using radar echo data. Compared to classical radar signal processing methods, the proposed approach does not require preprocessing time for radar signal processing and provides improved detection performance.

3 Proposed Method

In this paper, we propose a model architecture based on deep learning methods to address the limitations of traditional radar target detection methods and the significant degradation of detection performance in complex scenarios. Our approach involves directly separating the real and imaginary parts of the 2D target radar echo sequence, treating them as input channels in a multi-channel data model. Through two sub-network modules for feature extraction and target detection, the model is capable of accurately estimating the target's velocity and distance. The feature extraction network, constructed with noise-reducing self-encoders and convolutional neural networks, aims to effectively suppress clutter noise in radar echo data and greatly improve the performance of the detection module. The model architecture diagram is illustrated in Fig. 1.

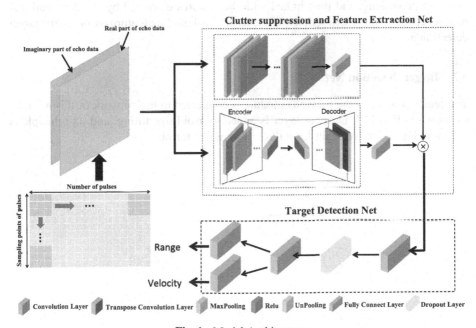

Fig. 1. Model Architecture

3.1 Clutter Suppression and Feature Extraction Net

Radar targets are often immersed in complex clutter, so the effective suppression of clutter is key to improving the detection model performance. We segment the input two-dimensional radar echo data into real and imaginary parts and treat it as multi-channel two-dimensional data similar to image data, so we use a convolutional neural network as the main feature extraction module. The clutter suppression and feature extraction network consists of two parallel sub-modules, one of which is a feature extraction structure based on a convolutional neural network, which uses a five-layer convolutional structure, chooses max pooling as the pooling method, and uses the ReLU activation function.

$$\text{Relu}(x) = \begin{cases} x, & x \geq 0 \\ 0, & x < 0 \end{cases} \tag{1}$$

The other parallel submodule consists of a convolutional autoencoder [29] where both the encoder and decoder are constructed with three layers of convolution and pooling. In the convolutional autoencoder, the encoder and decoder are both convolutional neural networks. The encoder's convolutional network learns to encode the input into a set of signals, and then the decoder convolutional neural network attempts to reconstruct the input from these signals. This can capture general features from the input and is used in our model for removing noise and suppressing clutter in the radar echo sequence.

The output features of the convolutional autoencoder are sent to the fully connected layer for processing, and then linked with the features extracted by the convolutional neural network feature extraction module as the final module feature output to the target detection network.

3.2 Target Detection Net

The feature output of the previous module is connected to the network of this module, as shown in Fig. 1. A Dropout layer is added to avoid overfitting, and its principle is shown in Fig. 2. The relevant formulas are shown in formula 2–5.

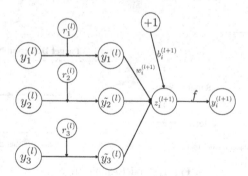

Fig. 2. Dropout Network

$$r_j^{(l)} \sim Bernoulli(p) \tag{2}$$

$$\tilde{y}^l = r^{(l)} \times y^{(l)} \tag{3}$$

$$z_i^{(l+1)} = w_i^{(l+1)} \tilde{y}^l + b_i^{(l+1)} \tag{4}$$

$$y_i^{(l+1)} = f\left(z_i^{(l+1)}\right) \tag{5}$$

3.3 Loss Function Design

The model tasks are two specific tasks, target velocity detection and target range detection, and are therefore identified as multi-task regression models. In multi-task learning it is desired that multiple related tasks are trained together, and it is desired that different tasks can promote each other to obtain better results on a single task. For the regression task, the loss functions L1-Loss and L2-Loss with Smooth L1-Loss [34] are usually used, as shown in formulas 6–8.

$$L_1(x, y) = \frac{1}{n} \sum_{i=1}^{n} |y_i - f(x_i)| \tag{6}$$

$$L_2(x, y) = \frac{1}{n} \sum_{i=1}^{n} (y_i - f(x_1))^2 \tag{7}$$

$$Smooth\, L_1(x, y) = \frac{1}{n} \sum_{i=1}^{n} \begin{cases} 0.5 \times (y_i - f(x))^2, & |y_i - f(x)| < 1 \\ |y_i - f(x)| - 0.5, & |y_i - f(x)| \geq 1 \end{cases} \tag{8}$$

L1-Loss does not suffer from gradient explosion, but it cannot be differentiated at the center point and is often used for simple models. L2-Loss is continuously smooth at each point, making it easy to differentiate and converge faster than L1-Loss, but it may cause gradient explosion and is often used for problems with low-dimensional numerical features. Smooth L1-Loss combines the advantages of L1-Loss and L2-Loss while cleverly avoiding the problems caused by both, so it is chosen as the single-task loss function. On top of that, the corresponding loss function weights were dynamically set using the Cov-Weighting method [33]. Cov-Weighting calculates the weight of each task's loss function L by the mean and standard deviation of the single loss, μ_L, σ_L, assuming that the optimization goal of a task is achieved when the variance of its loss tends to zero. The specific formulas for calculating the corresponding task loss function weights are shown in formula 9–11.

$$c_L = \frac{\sigma_L}{\mu_L} \tag{9}$$

$$r_t = \frac{L_t}{\mu_{L_{t-1}}} \tag{10}$$

$$w_{i,t} = \frac{\sigma_{r_{i,t}}}{\sum_{i=1}^{n} c_{r_{i,t}} \mu_{r_{i,t}}} \tag{11}$$

where C_L represents the relative standard deviation, r_t represents the rate of change of loss, and t is the number of training rounds to obtain the final loss function, which is given in formula 12.

$$\begin{cases} Loss_{total} = w_{1,t}Loss_{range} + w_{2,t}Loss_{velocity} \\ Loss_{range} = Loss_{velocity} = \frac{1}{n}\sum_{i=1}^{n} \begin{cases} 0.5 \times (y_i - f(x_i))^2, |y_i - f(x_i)| < 1 \\ |y_i - f(x_i)| - 0.5, |y_i - f(x_i)| \geq 1 \end{cases} \end{cases} \quad (12)$$

4 Experiments

4.1 Datasets Description

Due to the special nature of radar data, currently available radar echo data for radar target detection is scarce. Therefore, in this paper, software simulation was used to construct a radar echo dataset, along with publicly available radar echo sequence data for dim target detection and tracking datasets [13], for model validation and evaluation with multiple types of data and evaluation metrics.

Among them, the parameters of the public dataset are shown in Table 1 of relevant parameters of the dataset, and the object of data collection is fixed-wing unmanned aerial vehicles in the air. The turntable and the target unmanned aerial vehicle are equipped with GPS communication stations to guide the computer to receive and calculate the GPS information of the two, and real-time sending of the calculated results such as angles and distances is used to guide the radar servo rotation to track the unmanned aerial vehicle target to maintain it in the field of view of the radar. At the same time, the data recording equipment synchronously collects and saves the echoes. This paper will use the background sub-dataset of the data set for experiments, with a data signal-to-noise ratio of 3 dB, 7 dB, 15 dB, and 20 dB.

For the simulated data set, a linear frequency modulated pulse signal is used as the transmitted waveform. The data are simulated by setting the operating parameters, waveform parameters and target characteristics of the radar, with the clutter type as background terrain, to simulate targets with different signal-to-noise ratios.

Table 1. Relevant Parameters of the Public Dataset

Information	Parameters
Radar carrier frequency	35 GHz
Pulse repetition rate	32 kHz
Clutter background	Land
Modulation	Linear Frequency Modulation

4.2 Performance Metrics

For the radar target detection problem, the DR ndis generally used for the performance evaluation of the model, whose if the target to be detected in the data set is N, the number of model output detection targets is N_d, R_d denotes the model output distance, R_{gt} denotes the true distance of the target, R_{error} denotes the distance error, V_d denotes the model output velocity, V_{gt} denotes the true velocity of the target, V_{error} denotes the velocity error, $V_{threshold}$ denotes the acceptable velocity error threshold, $R_{threshold}$ denotes the acceptable distance error threshold. Considering that the difficulty of the model's target detection task is naturally affected when the input signal-to-noise ratio changes, we set the acceptable distance and speed error values as variables related to the signal-to-noise ratio, and the distance and speed measurement errors can be considered as valid detection targets only when they are within the acceptable error range N_{error}, and the relevant metrics are defined as shown in formulas 13–16.

$$R_{error} = |R_d - R_{gt}| \tag{13}$$

$$V_{error} = |V_d - V_{gt}| \tag{14}$$

$$N_{corr} = \left(R_{error} \leq \frac{R_{threshold}}{\sqrt{SNR}} \right) \& \left(V_{error} \leq \frac{V_{threshold}}{\sqrt{SNR}} \right) \tag{15}$$

$$DR = \frac{N_{corr}}{N} \tag{16}$$

4.3 Experimental Details

The input data of the model is two-dimensional radar echo data partitioned into time windows, where the input data size is the number of sampling points per pulse multiplied by the time interval divided by the pulse repetition frequency. The optimizer used for the model is Adam optimizer.

4.4 Ablation Experiments

To validate the effectiveness of the clutter suppression and feature extraction network modules in our proposed model, we conducted ablation experiments. The experiments involved three test models: the target detection network only (TDN), the integration of traditional signal processing methods with the target detection network (TDN with TSP), and our proposed complete model. The experiments were conducted using simulated data for verification. The experimental results are presented in Fig. 3.

It can be concluded from the experimental results that the clutter suppression and feature extraction module we designed is effective in our proposed method model and has an advantage over traditional signal processing methods when spliced with the target detection network.

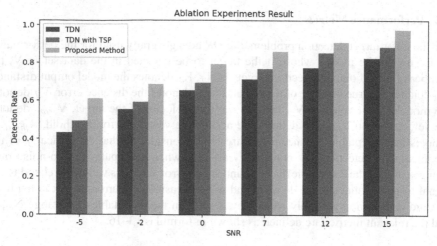

Fig. 3. Ablation Experiments Result

4.5 Comparative Experiment

The comparison methods chosen in this paper are as follows, the classical traditional radar target detection methods CA-CFAR [10], GO-CFAR [11], which are the more popular and common CFAR-based radar target detection methods at this stage, and we also chose the convolutional neural network-based radar target detection model [32], but due to the different labels of the public data set, we made some modifications on the basis of [32] and excluded the detection modules of azimuth and elevation angles from the original paper for comparison experiments.

First, we conducted experiments on a real radar echo dataset, which consisted of radar echo data with high, medium, and low signal-to-noise ratios. Due to the specific nature of deep learning models, we needed to split the dataset into training and testing data. Therefore, we used the same types of data from the complete dataset as the training and testing experiment groups. The experimental results are shown in Fig. 4.

The simulated dataset provides a more comprehensive distribution of signal-to-noise ratios. In our experiments conducted on the simulated dataset, we followed a methodology similar to that used for the real dataset. As a result, we obtained performance results for the models under different signal-to-noise ratios, as shown in Fig. 5. From the graph, we can observe that the experimental results on the simulated dataset exhibit similar trends to those obtained from the real dataset.

Based on the results obtained from the two experimental datasets, we can now analyze the findings.

In the real dataset, the target signal-to-noise ratios were around 3 dB, 7 dB, 15 dB, and 20 dB. From the results graph, it is evident that our proposed model outperformed other comparative methods, especially under lower signal-to-noise ratios, where the advantage was more pronounced. The maximum difference in detection rate reached 17.5%, with an average performance difference of 6.84%.

In the simulated dataset, the target signal-to-noise ratios were −5 dB, −2 dB, 0 dB, 7 dB, 12 dB, and 15 dB. Similar conclusions can be drawn from the results graph as in the

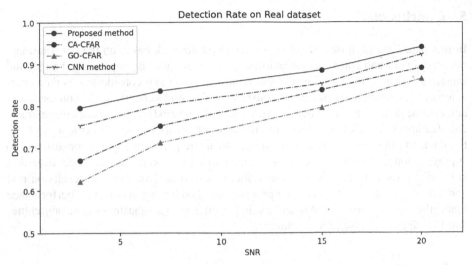

Fig. 4. Detection Rate on Real dataset

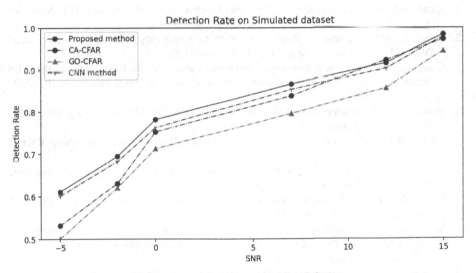

Fig. 5. Detection Rate on Simulated dataset

real dataset. Although the performance of various models approached each other under high signal-to-noise ratios, our method still exhibited the best average performance. The maximum difference in detection rate reached 10.1%, with an average performance difference of 3.89%.

We can see from the comparison results that our proposed model outperforms the comparison model in the radar target detection task, especially in the case of high clutter and low signal-to-noise ratio in the real dataset, which is more obvious than the traditional model, and our proposed model also obtains a certain performance lead in the simulated dataset.

5 Conclusions

In this paper, we propose a radar target detection network based on the convolutional neural network and convolutional autoencoder architecture. By separating the real and imaginary parts of the two-dimensional radar echo data and considering its data characteristics as image data, we use a convolutional neural network and a convolutional autoencoder to construct a clutter suppression and feature extraction network that extracts data features in parallel. The convolutional neural network extracts direct features from the data, and the convolutional autoencoder structure performs input reconstruction to suppress clutter. The output features are then fused and output to the target detection network for radar target distance and velocity detection. Experimental results on real and simulated datasets show that our proposed method has better detection performance than other comparison methods and exhibits better model robustness in strong clutter and low signal-to-noise ratio conditions.

References

1. Richards, M.A.: Fundamentals of Radar Signal Processing, 1st edn. McGraw-Hill, New York (2005)
2. Pieraccini, M., Miccinesi, L., Rojhani, N.: RCS measurements and ISAR images of small UAVs. IEEE Aero. Electron. Syst. Mag. **32**(9), 28–32 (2017)
3. Liang, C., et al.: UAV detection using continuous wave radar. In: Proceedings of the IEEE International Conference on Information Communication and Signal Processing (ICICSP), pp. 1–5. IEEE (2018)
4. Miao, Y., et al.: Efficient multipath clutter cancellation for UAV monitoring using DAB satellite-based PBR. Remote Sens. **13**(17), 3429 (2021)
5. Melvin, W.L., Scheer, J.A. (eds.): Principles of Modern Radar. SciTech Publishing, Raleigh (2013)
6. Short, R., Fukunaga, K.: The optimal distance measure for nearest neighbor classification. IEEE Trans. Inf. Theory **5**, 622–627 (1981)
7. Mountrakis, G., et al.: Support vector machines in remote sensing: a review. ISPRS J. Photogramm. Remote Sens. **66**, 247–259 (2011)
8. Youngwook, K.: Application of machine learning to antenna design and radar signal processing: a review. In: Proceedings of the 2018 International Symposium on Antennas and Propagation (ISAP), Busan, Korea (2018)
9. del Rey-Maestre, N., Jarabo-Amores, M.P., Mata-Moya, D., Humanes, J., Hoyo, P.: Machine learning techniques for coherent CFAR detection based on statistical modeling of UHF passive ground clutter. IEEE J. Sel. Top. Signal Proc. **12**, 104–118 (2018)
10. Sciotti, M., Lombardo, P.: Performance evaluation of radar detection schemes based on CA-CFAR against K-distributed Clutter. In: Proceedings of the 2001 CIE International Conference on Radar, Beijing, China, pp. 345–349 (2001)
11. Gandhi, P., Kassam, S: Optimality of the cell averaging CFAR detector. IEEE Trans. Inf. Theory **40**, 1226–1228 (1994)
12. Blake, S.: OS-CFAR theory for multiple targets and nonuniform clutter. IEEE Trans. Aerosp. Electron. Syst. **24**(6), 785–790 (1988)
13. Song, Z.Y., Hui, B.W., Fan, H.Q., et al.: A dataset for detection and tracking of dim aircraft targets through radar echo sequences, pp. 277–290. Chinese scientific data (2020)

14. Bao, Z., Jiang, Q., Liu, F., et al.: Optimization of track before detect algorithm based on particle filter. Modern Radar, 21–27 (2020)
15. Ito, N., Godsill, S.: A multi-target track-before-detect particle filter using superpositional data in non-gaussian nois. IEEE Signal Process. Lett. **27**, 1075–1079 (2020)
16. Wang, L.L., Zhou, G.J.: A unified M pseudo-spectrum for track-before-detect of targets with motion model uncertain. Digital Signal Process. **114**, 103078 (2021)
17. Yin, L., Zhang, Y., Wang, S., et al.: A review of histogram probabilistic multi-hypothesis tracking method techniques. In: System Engineering and Electronic Technology, pp. 3118–3125 (2021)
18. Rohling, H.: New CFAR-processor based on an ordered statistic. In: Proceedings of the IEEE 1985 International Radar Conference, Arlington, VA, USA (1985)
19. Callaghan, D., Burger, J., Mishra, A.K.: A machine learning approach to radar sea clutter suppression. In: Proceedings of the IEEE Radar Conference, Seattle, WA, USA 2017
20. Wang, L., Tang, J., Liao, Q.: A study on radar target detection based on deep neural networks. IEEE Sens. Lett. **3**, 1–4 (2019)
21. Gouri, A., Mezache, A., Oudira, H.: Radar cfar detection in weibull clutter based on zlog(z) estimator. Remote Sens. Lett. **11**, 581–589 (2020)
22. Zhang, B.Q., Zhou, J., Xie, J.H., et al.: Weighted likelihood CFAR detection for weibull background. Digital Signal Process. **115**, 103079 (2021)
23. Zebiri, K., Mezache, A.: Radar CFAR detection for multiple-targets situations for weibull and log-normal distributed clutter. In: Signal Image and Video Processing, p. 1671 (2021)
24. Yavuz, F., Kalfa, M.: Radar target detection via deep learning. In: 2020 28th Signal Processing and Communications Applications Conference (SIU), pp. 1–4 (2020)
25. Pan, M., Chen, J., Wang, S., Dong, Z.: A novel approach for marine small target detection based on deep learning. In: Proceedings of the IEEE 4th International Conference on Signal and Image Processing, Wuxi, China, pp. 395–399 (2019)
26. Kang, M., Leng, X., Lin, Z., Ji, K.: A modified faster R-CNN based on CFAR algorithm for SAR ship detection. In: Proceedings of the 2017 International Workshop on Remote Sensing with Intelligent Processing (RSIP), Shanghai, China (2017)
27. Mou, X., Chen, X., Guan, J.: Marine target detection based on improved faster R-CNN for navigation radar PPI images. In: Proceedings of the 2019 International Conference on Control, Automation and Information Sciences, Chengdu, China (2019)
28. Su, N., Chen, X., Guan, J., Huang, Y.: Maritime target detection based on radar graph data and graph convolutional network. IEEE Geosci. Remote Sens. Lett. **19**, 1–5 (2022)
29. Chen, L., Guan, Q., Feng, B., Yue, H., Wang, J., Zhang, F.: A multi-convolutional autoencoder approach to multivariate geochemical anomaly recognition. Minerals **9**, 270 (2019)
30. Hinton, G.E., et al.: Improving neural networks by preventing co-adaptation of feature detectors. arXiv preprint arXiv:1207.0580 (2012)
31. Liu, B.Y., Song, C.Y., Fan, H.Q.: Focus, detect, track, refocus: real-time small target detection technology based on classical pre-tracking detection framework. In: Air Weapon, pp. 10–16 (2019)
32. Jiang, W., Ren, Y., Liu, Y., et al.: A method of radar target detection based on convolutional neural network. Neural Comput. Appl. (2021)
33. Groenendijk, R., Karaoglu, S., Gevers, T., et al.: Multi-loss weighting with coefficient of variations. In: Proceedings of the IEEE/CVF Winter Conference on Applications of Computer Vision, pp. 1469–1478 (2021)
34. Girshick, R.: Fast R-CNN. In: Proceedings of the IEEE International Conference on Computer Vision, pp. 1440–1448 (2015)
35. Cheikh, K., Faozi, S.: Application of neural networks to radar signal detection in K-distributed clutter. In: First International Symposium on Control, Communications and Signal Processing, pp. 295–298 (2004)

Correction to: High-Resolution Self-attention with Fair Loss for Point Cloud Segmentation

Qiyuan Liu, Jinzheng Lu, Qiang Li$^{(\boxtimes)}$, and Bingsen Huang

Correction to:
Chapter 27 in: B. Luo et al. (Eds.): *Neural Information Processing*, LNCS 14451,
https://doi.org/10.1007/978-981-99-8073-4_27

In the originally published printed version of chapter 27, Fig. 4 and Fig. 5 both showed the same figure. This has been corrected.

The updated version of this chapter can be found at
https://doi.org/10.1007/978-981-99-8073-4_27

Author Index

Printed in the United States
by Baker & Taylor Publisher Services